2nd All-Russian Conference Thermophysics and Physical Hydrodynamics with the School for Young Scientists (TPH 2017)

Journal of Physics: Conference Series Volume 899

Yalta, Crimea
11 – 17 September 2017

ISBN: 978-1-5108-4897-9
ISSN: 1742-6588

Printed from e-media with permission by:

Curran Associates, Inc.
57 Morehouse Lane
Red Hook, NY 12571

proceedings.com

Some format issues inherent in the e-media version may also appear in this print version.

Printed by Curran Associates, Inc. (2017)

For permission requests, please contact the Institute of Physics
at the address below.

Institute of Physics
Dirac House, Temple Back
Bristol BS1 6BE UK

Phone: 44 1 17 929 7481
Fax: 44 1 17 920 0979

techtracking@iop.org

Additional copies of this publication are available from:

Curran Associates, Inc.
57 Morehouse Lane
Red Hook, NY 12571 USA
Phone: 845-758-0400
Fax: 845-758-2633
Email: curran@proceedings.com
Web: www.proceedings.com

Table of contents

Volume 899

2nd All-Russian Scientific Conference Thermophysics and Physical Hydrodynamics with the School for Young Scientists

11–17 September 2017, Yalta, Crimea
(previous conference: vol 754)

Accepted papers received: 11 September 2017
Published online: 27 September 2017

Preface

Heat transfer and hydrodynamics in single phase flows

Hydrodynamics and heat and mass transfer in multiphase flows

Reacting flow dynamics, detonation processes

Numerical methods in thermophysics and physical hydrodynamics

Techniques of thermophysics and hydrodynamics experiment

Thermophysical properties of substances and new materials

Electrophysical phenomena in gaseous and liquid media

Heat transfer and hydrodynamics in industrial processes and environment protection

10 September 2017

Preface for the TPH 2017 – Thermophysics and physical hydrodynamics conference

Proceedings of the TPH 2017 – Thermophysics and physical hydrodynamics conference

D. M. Markovich, S.V. Golovin

The origins of the Conference start from 1970 in the Soviet Union, Novosibirsk. It was organized by Kutateladze Institute of Thermophysics SB RAS. The name of the conference was "Actual problems of thermophysics and physical hydrodynamics". The conference has been organized under this name up to 2015. The conference chairs were Academician V.E. Nakoryakov, Prof. D.M. Marckovich and Prof. S.V. Alekseenko. Peer reviewed proceedings of the conference have been published in the format of printed books. In 2016 the conference is reorganized in a new format with a shorter name: "Thermophysics and physical hydrodynamics".

The conference takes place in Yalta, a beautiful city in Crimea on the bank of the Black Sea. Lavrentev Institute of Hydrodynamics and the National committee on Heat and Mass Transfer are among other conference organizers besides Kutateladze Institute of Thermophysics. The present Conference covers the following topics: heat transfer and hydrodynamics in single phase and multiphase flows, , reacting flows, detonation processes, experimental and numerical techniques in thermophysics and physical hydrodynamics, heat transfer and hydrodynamics in industrial processes. There are more than 150 participants.The proceedings contain 98 papers grouped by topic.

The scientific committee appreciates the enormous work of the editorial board and reviewers in the preparation of this volume. We would like to express our sincere thanks to all authors for their research contributions, and also to organizers of the conference for their valuable spadework.

IOP Conf. Series: Journal of Physics: Conf. Series **899** (2017) 011002 doi:10.1088/1742-6596/899/1/011002

Peer review statement

All papers published in this volume of *Journal of Physics: Conference Series* have been peer reviewed through processes administered by the proceedings Editors. Reviews were conducted by expert referees to the professional and scientific standards expected of a proceedings journal published by IOP Publishing.

Nonisothermal filling of a planar channel with a power-law fluid

E Borzenko[1], I Ryltsev[2], O Frolov[1] and G Shrager[3]

[1] Associate professor, Tomsk State University, Tomsk, Russia
[2] PhD Student, Tomsk State University, Tomsk, Russia
[3] Professor, Tomsk State University, Tomsk, Russia

E-mail: frolovoy@mail.tsu.ru

Abstract. In this paper, the fountain flow of a power-law fluid during the filling of a planar vertical channel is investigated taking into account a dissipative heating. The rheological properties of the medium are defined by Ostwald de Waele power-law with an exponential temperature dependence of effective viscosity. The problem is solved numerically by means of a finite-difference method based on the SIMPLE algorithm and method of invariants for compliance with the natural boundary conditions on a free surface. The parametrical investigations of the shape of free surface as a function of the basic dimensionless criteria and nonlinearity degree of the fluid are implemented. The effect of viscous dissipation on the formation of the kinematic characteristics of the flow is shown.

1. Introduction

In processing of polymeric materials, the injection molding based on the mold filling with a liquid medium is a widespread technology. The processed composition is characterized by a complex rheological behavior and it generally flows under nonisothermal conditions. Forecasting of the quantity of products requires the prior investigation of the technological process that can be implemented using the physical and mathematical simulation methods. The mathematical simulation method is preferable as it is more efficient and informative in comparison with the physical simulation, which is used for verification of the theoretical results on the last research stage. The process of filling is characterized by both the presence of free surface, nonisothermality connected with dissipative heating, chemical conversion, heat-exchange boundary conditions, and temperature dependence of the rheological parameters of the medium. The results of a research on nonisothermal flow of the media with a complex rheology in the channels without considering the free surface are introduced in details in [1, 2]. In the articles [3–7], there is a description of computational technologies and the results of investigation of the filling process taking into account the free surface.

The aim of this work is to study the influence of the viscous dissipation on both the shape of free surface and the kinematic of a fountain flow during the filling of a planar channel with a power-law fluid.

2. Formulation of the problem

Nonisothermal filling of the planar vertical channel with the power-law fluid in a gravity field is considered. The direction of the fluid flow is opposite to those of gravity force. The flow area is shown in figure 1. Mathematical basis of the flow includes the equations of motion, heat-transfer, and continuity written in dimensionless variables as follows:

$$\mathrm{Re}\frac{d\mathbf{V}}{dt} = -\nabla p + \nabla\bullet(2B\mathbf{E}) + \mathbf{W},\ \mathrm{Pe}\frac{d\theta}{dt} = \Delta\theta + \mathrm{Br}BA^2,\ \nabla\bullet\mathbf{V} = 0.$$

Here, $\mathbf{V}$ is the dimensionless velocity vector; p is the dimensionless pressure; t is the dimensionless time; A is the dimensionless intensity of the strain-rate tensor $\mathbf{E}$; $\mathbf{W}=\{-\mathrm{W},\ 0\}$ is the dimensionless vector; $\theta=(T-T_0)/T_0$ is the dimensionless temperature; T and T_0 are the dimensional temperature of the fluid in the flow and on a solid wall, respectively. The system of equations is completed by a rheological power-law, which includes the effective viscosity defined by the following equation [8]:

$$B = \exp[-\mathrm{C}\theta]A^{m-1}.$$

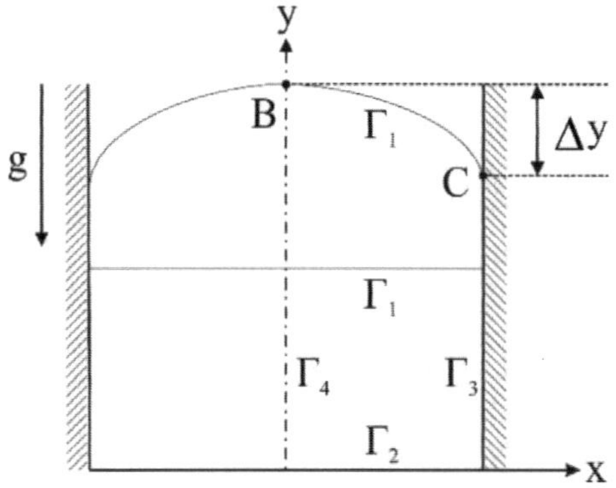

Figure 1. Calculation domain

The dimensionless similarity criteria are as follows:

$$\mathrm{Re} = \frac{\rho L^m U_0^{2-m}}{\mu_0},\ \mathrm{Pe} = \frac{c\rho U_0 L}{\lambda},\ \mathrm{W} = \frac{\rho L^{m+1} g}{\mu_0 U_0^m},\ \mathrm{Br} = \frac{\mu_0 U_0^{m+1} L^{1-m}}{\lambda T_0},\ \mathrm{C} = kT_0.$$

Here, ρ is the fluid density; L is the half-width of the channel; U_0 is the average velocity at the inlet section; m, μ_0 are the power-law rheological parameters; k is the parameter of nonisothermality; c – is the heat capacity; λ is the heat conductivity coefficient; g is the gravity acceleration.

At the inlet section Γ_2, the velocity and temperature profiles corresponding to a fully developed one-dimensional nonisothermal fluid flow with a specified constant flow rate in an infinite channel are assigned. On the solid walls Γ_3, the no-slip conditions are imposed and the dimensionless temperature is set to zero. On the free surface Γ_1, the boundary conditions are defined by both zero shear stress, the equality of normal stress to the external pressure, which can be assumed to be zero without loss of generality, and the absence of the heat flux. At the initial time moment, the free surface is plane and it is located in a distance from the inlet boundary, and the temperature of fluid is equal to those of the wall in the region of initial filling. Moreover, the free boundary moves in compliance with the kinematic condition. The surface tension forces are not considered. The problem formulation is dimensionless. The values of dimensionless criteria are assigned so that the stationary solution for nonisothermal flow in the planar infinite channel could exist [8].

Consequently, the boundary conditions are written as

$$\Gamma_1: \frac{\partial u_n}{\partial s} + \frac{\partial u_s}{\partial n} = 0,\quad p = 2B\frac{\partial u_n}{\partial n},\quad \frac{\partial\theta}{\partial n} = 0; \qquad (1)$$

$$\Gamma_2: u = 0,\quad v = f_1(x),\quad \theta = f_2(x);$$

$$\Gamma_3 : u = 0, \quad v = 0, \quad \theta = 0;$$

$$\Gamma_4 : u = 0, \quad \frac{\partial v}{\partial x} = 0, \quad \frac{\partial \theta}{\partial x} = 0.$$

The conditions (1) are written in a local Cartesian coordinate system (n, s), which is normally related to the free surface. The free boundary Γ_1 moves in compliance with the kinematic condition written in the Lagrangian representation as follows:

$$\frac{\mathrm{d}x}{\mathrm{d}t} = u, \quad \frac{\mathrm{d}y}{\mathrm{d}t} = v. \tag{2}$$

The analysis of the classical mathematical model of fluid dynamics including the motion equations, natural boundary conditions on a free surface, no-slip conditions on a moving contact line, and the values of dynamic contact angle, which is not equal to 0 and π, indicates the singularities in determination of dynamic characteristics of the flow. These singularities lead to an infinite increase in the values of dynamic characteristics as a three-phase contact line is approached [9, 10]. Therefore, the problem formulation is performed assuming that there is no shear stress (1) and the normal velocity is equal to zero at the contact point C. In the small vicinity of the contact line, the tangential velocity on the wall decreases asymptotically from the value at the point C to zero. The flow regime and the interaction behavior of the phases on the contact line are supposed to provide the edge angle equal to π with insignificant capillary effects in comparison with the inertial, viscous, and gravitational forces in the flow. Using such a boundary conditions eliminates the particularities of traditional formulation of the problem and at the same time has a little effect on the kinematic characteristics of the flow out of a small vicinity of contact point [11].

3. Method of solution

The stated problem is solved using a finite-difference method. A staggered difference grid covers the solution domain. The SIMPLE algorithm [12] is applied to calculate the velocity, temperature, and pressure fields at the internal nods. Near the free boundary, the irregular nodes appear. In these nodes, the values of unknown variables are defined using linear interpolation of data from the free surface and regular nodes. The free boundary is represented as a set of particles-markers distributed uniformly along this boundary: the first marker is located on the symmetry line and the last is on the three-phase contact line. The velocity components are calculated using the method of invariants [13] based on the combination of assumptions of no shear stresses and continuity equations for each point marked on the free surface at the given time moment. The particles-markers of the free surface, except for the last one, move in accordance with the discrete analogs of the kinematic condition (2). The motion of the contact point is implemented as provided by a slip condition and dynamical edge angle, which is equal to π.

The calculation method was tested on the problem of power-law fluid flow in a planar channel at the specified flow rate with consideration for dissipative heating and exponential temperature dependence of the viscosity at Re=0.1, Pe=100, Br=1, C=1, W=0. The parabolic profile of longitudinal velocity and zero temperature were set at the inlet of a channel, and the soft boundary conditions were specified at the outlet. On the solid walls, the no-slip boundary condition was realized and the temperature was set to zero. The length of the channel was selected to be sufficient for a steady state flow developed at the outlet.

For verification of the computational technique, the results of calculations were compared with the numerical solution of equivalent one-dimensional problem of nonisothermal power-law fluid flow in an infinite channel at the specified flow rate. As a result of testing the approximation convergence, the grid step for the following calculations was set to 1/40, which yielded the relative errors 0.08% and 0.03% for the calculations of velocity and temperature on the symmetry line of the channel, respectively [14].

4. Results and discussion

The results of nonisothermal filling of the channel at low Reynolds number and high Peclet number that are typical for the flow of polymeric fluids during the processing by a casting method are considered. The free boundary of the studied power-law fluid flow is characterized by the parameter $\chi=\Delta y/L$, which indicates the position of the point B on the symmetry line relative to the point C on three-phase contact line (figure 1). The effect of nonlinearity degree m on the parameter χ at various Br, C, and W is shown in figure 2. In the case of dilatant fluid ($m>1$) at W=0 (figure 2a), the parameter χ decreases insignificantly with an increase in m. In the case of pseudoplastic fluid ($m<1$), the convexity of the free surface increases as the m decreases. In figure 2a, b the solid lines correspond to the isothermal filling of the planar channel with the power-law fluid [15]. As the W increases the behavior of functional connection $\chi(m)$ changes.

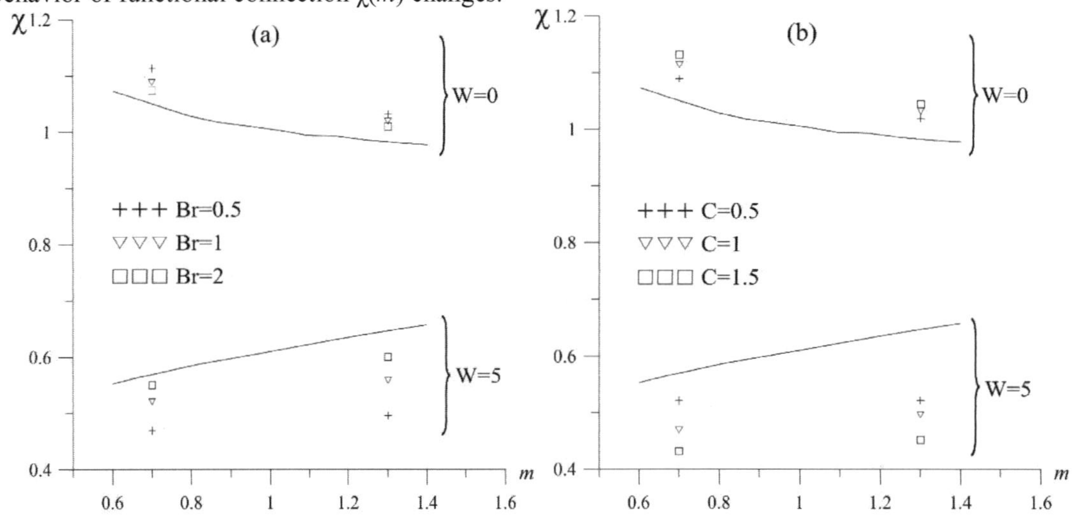

Figure 2. Parameter χ as a function of nonlinearity degree m at Re=0.1, Pe=100: (a) C=1; (b) Br=2

In considered flow, the effect of parameter Br on parameter χ is shown in figure 3. At the upper half of the figure, the curves 1, 2 represent the functional connection χ(Br) at W=0, and at the lower half, at W=5 (curves 3, 4). In the case of W=0, the value of convexity property χ of the free surface for pseudoplastic fluid is higher than for dilatant fluid, but when W=5, the χ is higher for dilatant fluid in comparison with that for pseudoplastic fluid.

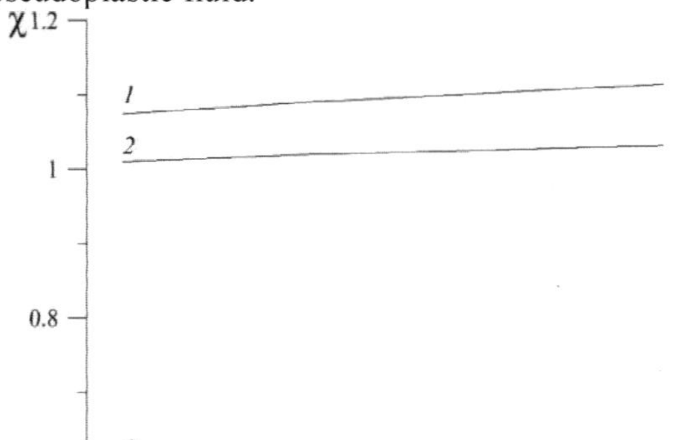

Figure 3. Parameter χ as a function of Br at Re=0.1, Pe=100: curves 1, 4 – m=0.7; curves 2, 3 – m=1.3

The fields of transverse and longitudinal velocity, and effective viscosity for both isothermal and nonisothermal filling of the planar channel with the fluid are performed in figure 4. The behavior of velocity distribution confirms the flow separation into a two-dimensional fountain flow zone in the vicinity of a free boundary and a one-dimensional flow located far away from the free boundary [6]. Heating of the fluid due to the viscous dissipation yields the viscosity decrease over the flow region (figure 4e, f). As viscosity decreases, the intensity of the fluid spreading towards the solid walls in the vicinity of free boundary increases. It in turn leads to reduction of the distance between three-phase contact point and top of the flow on the symmetry line of the channel. Dissipative heating provides the increase in dimensionless maximum of transverse and longitudinal velocity from 0.28 to 0.48 and from 1.4 to 1.6, respectively (figure 4a, b, c, d). The length of two-dimensional flow zone becomes longer than that in the case of isothermal filling of the channel. The effect of a viscous dissipation and rheology on the temperature distribution was demonstrated in [14].

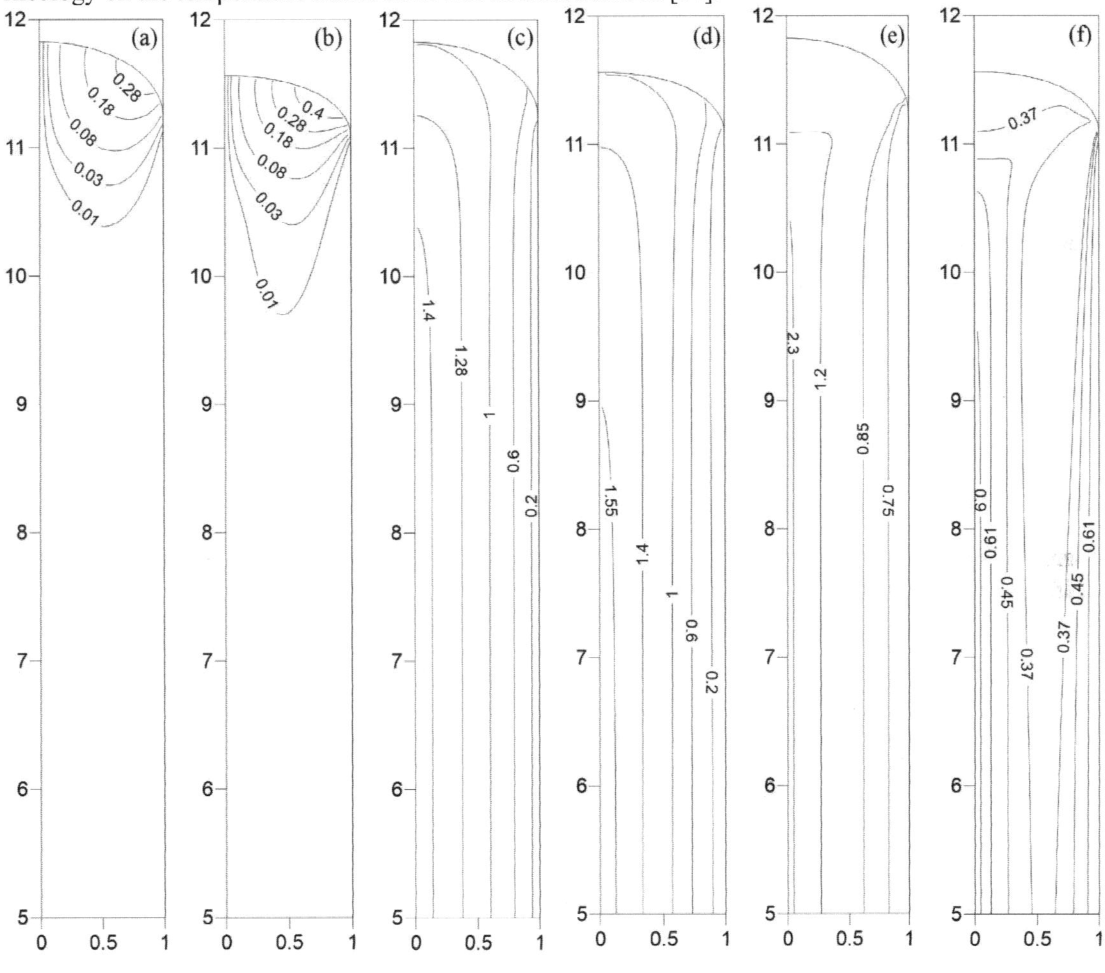

Figure 4. Distributions of transverse velocity, longitudinal velocity, and effective viscosity during the channel filling with the power-law fluid at Re=0.1, Pe=100, W=5, m=0.7:
a, c, e – isothermal filling [15]; b, d, f – nonisothermal filling at Br=2, C=1

5. Conclusions

As a result of investigation implemented, the effect of the Brinkman number Br, the W criterion, the parameters of the rheological model C and m on the kinematics of the power-law fluid flow in the planar infinite channel has been shown. Increase in the value of W from 0 to 5 leads to the changes in behavior of the functional dependence between the characteristic χ of a free boundary and nonlinearity degree *m* and the level of intensity of mechanical energy dissipation in the flow have been performed. The comparison of the results with that in the case of isothermal filling of the planar channel with non-Newtonian fluid demonstrates the increase in the length of a two-dimensional flow zone as a consequence of the decrease in the viscosity due to the fluid heating.

Acknowledgments
This work was supported by the grant of the President of the Russian Federation (MK-710.2017.1) and the Russian Foundation for Basic Research (project no. 15-08-02256).

References

[1] Baranov A V 2010 *Compos. Mech. Des.* **16** 384–399
[2] Hassan H, Regnier N, Pujos C and Defaye G 2008 *Polym. Eng. Sci.* **48** 1199–1206
[3] Nguyen-Chung T, Mennig G 2001 *Rheol. Acta* **40** 67–73
[4] Otmani R E, Zinet M, Boutaous M and Benhadid H 2011 *J. Appl. Polym. Sci.* **121** 1579–1592
[5] Wang W, Li X and Han X 2012 *Polym. Eng. Sci.* **52** 42–51
[6] Borzenko E I, Frolov O Y and Shrager G R 2014 *Fluid Dyn+* **49** 37–45.
[7] Borzenko E I, Shrager G R 2014 *Thermophys Aeromech+* **21** 211–221
[8] Kutepov A M, Polyanin A D, Zapryanov Z D et al 1996 *Chemical Hydrodynamics: Handbook* (Moscow: Byuro Kvantum)
[9] Huh C and Scriven L E 1971 *J. Colloid Interface Sci.* **35** 85–101
[10] Pukhnachev V V and Solonnikov V A 1982 *J. Appl. Math. Mech.* **46** 771–779
[11] Borzenko E I, Shrager G R 2015 *J. Appl. Mech. Tech. Ph+* **56** 167–176
[12] Patankar S *Numerical heat transfer and fluid flow. Series on computational methods in mechanics and thermal science* (New York: Taylor & Francis)
[13] Vasenin I M, Sidonsky O B and Shrager G R 1974 *Proc. USSR Acad. Sci.* **217** 295–298
[14] Borzenko E, Ryltseva K, Frolov O and Shrager G 2016 *J. Phys. Conf. Ser.* **754** 022002
[15] Borzenko E I, Shrager G R 2016 *J. Eng. Phys. Thermophys.* **89** 902–910

Behavior of heat transfer in pulsating flow in the channel entrance region

I A Davletshin, A A Paereliy

Kazan Scientific Center of RAS, Kazan, 420111, Russia

davlet60@mail.ru

Abstract. Convective heat transfer in the entrance region of a plane channel has been studied experimentally. Steady and pulsating air flow regimes have been considered. Heat transfer distributions over the channel wall have been obtained. Non-monotonous behavior of heat transfer coefficient has been revealed at some frequencies and amplitudes of forced flow pulsations. It was accompanied by heat transfer enhancement that was almost double for the local heat transfer coefficients. This may result from generation of vortices in the inlet duct in conditions of flow pulsations.

1. Introduction

Forced pulsations of flow can significantly alter the flow pattern, hydrodynamic and thermal characteristics of flow. Undoubtedly, this has an impact on the operation of different engineering devices [1] and living organisms [2]. Theoretical and experimental research of pulsating flows is motivated by the relevance of problems in this area. When setting up an experimental investigation, a certain flow velocity (pressure) law should be provided. The latter is be responsible for the pattern of pulsating flow. This problem is rather challenging on its own. First, the velocity law is to be provided by a device (pulsator) that generates forced pulsations of flow [3]. Second, acoustic properties of the working medium and channel geometry promote pulsations (resonance) or their attenuation. Third, the pulsator location within the channel can also influence the flow pattern [4]. In general, the study of processes in pulsating flows requires taking into account at least two other factors: frequency and amplitude of pulsations. There is some information on pulsating flows available nowadays that shows the effect of pulsations on velocity profiles, shear stress, laminar-turbulent transition, etc. [5-7]. But due to multi-factor nature of problems in this area, the obtained results are often partial and limited to the considered case. This can result in the situation when some results seem contradictory. For example, both heat transfer enhancement [1, 8] and deterioration [9] have been revealed at pulsating flow regimes.

One of the factors influencing the flow pattern (in particular, the one of pulsating flows) is vortex shedding. Vortices can exist even in a steady flow in its separation region. Then this region somehow interacts with forced pulsations [10] making the flow pattern even more complex. In general, vortices in pulsating flows can originate under the emerging variable moments of forces.

Flow parameters depend both on the object geometry (wall, diverging channel, orifice, etc.) and its location in the channel due to the wave structure of pulsating flow. In particular, the response of flow parameters to the forced pulsations has certain peculiarities in the entrance region of the channel and in the developed flow. Thus, investigation of the effect of flow pulsations on the distribution of heat transfer coefficients in the entrance region will contribute to the establishment of main regularities in such flows. The present work is mainly focused on interaction of flow pulsations with the processes in the inlet duct.

2. Experimental setup and procedure

An experimental setup with a 115×150 mm^2 1.2-meter long plane test section (figure 1) was used for the experiments. The inlet duct *1* in one plane was shaped according to the Bernoulli lemniscates curve.

Air flow pulsations were generated by a pulsator *4*, whose outlet cross section was periodically blocked by a rotating flap. Adjusting the blockage of the windows (with or without the flap) provided the required air flow rate and the amplitude of velocity pulsations. A receiver tank with the volume of 1.3 m^3 was installed between the test section and the compressor. It damped the velocity (pressure) pulsations further downstream. The air flow rate was measured using an ultrasonic flowmeter downstream of the receiver tank. The pulsation amplitude was estimated using the flow velocity values measured by an optical method.

The temperature of ambient air entering the channel was measured by a resistance thermometer *5* mounted in the inlet duct.

Heat transfer in the entrance section was studied on a wide wall (150 mm) of the 450-mm long section of the channel. The latter was a plate (printed circuit board) made of composite epoxy material with the width of 1.6 mm. To eliminate the heat loss, the outer surface of the plate was covered with polystyrene layer. Copper layer on the plate surface was etched to form a single strip. The strip was functionally divided into 33 sections with the length of 13 mm and width of 100 mm. Thus, the copper layer acted as 33 resistance thermometers that measured local temperatures of the wall with the streamwise spacing of 13.6 mm.

To obtain more detailed distribution of the heat transfer coefficient, the heat transfer section was installed alternately downstream of the channel inlet and in the middle part of the channel.

Heat transfer experiments were arranged as follows. Channel walls were heated by a hot air flow prior to the measurements. Then, while the wall was being cooled by the air flow at the room temperature, the temperatures of heat transfer wall were measured during 100 seconds with the frequency of 200 measurements per second. Heat balance equation together with these data yielded the distribution of heat transfer coefficient along the channel which was considered time constant. Hence time-averaged heat transfer was estimated at pulsating flow regimes.

The experiments were carried out in the range of forced pulsation frequency f=0–140 Hz and two levels of relative velocity amplitudes $\beta = A_U/U \sim 0.3$ and $\beta \sim 0.8$.

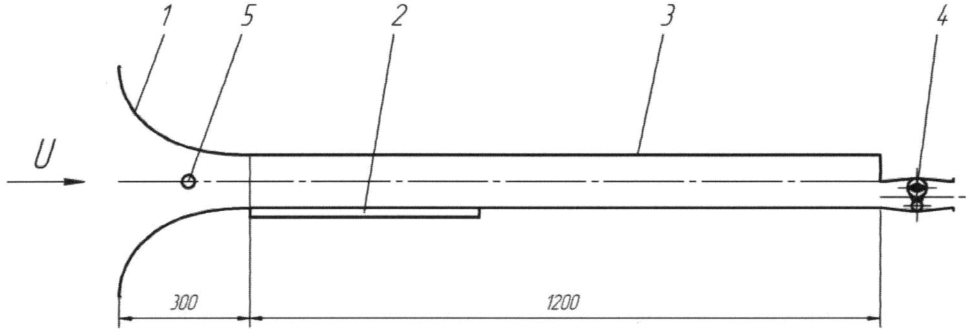

Figure 1. Experimental setup: 1 – inlet duct;
2 – heat transfer section; 3 – channel; 4 – pulsator; 5 – thermometer.

3. Results and discussion

Primary experimental data included the voltage, *V*, and the current, *I*, in the stripes of heat transfer section. Their behavior in time, τ, is shown in figure 2 for the steady flow case. When the strips were connected in series, the same current passed through them (figure 2, *b*).

This primary information yielded time variations of strip resistance $R_i=V_i/I$. Then the temperature of strips (heat transfer wall sections), $T_i=T_i(R_i)$, was estimated using the known relation between the electrical resistance of copper conductor and the temperature. Resistance for the cold condition had been measured in advance. A heat balance equation was constructed for the measured wall and air temperatures and known thermophysical properties of fiberglass laminate wall of appropriate shape. The equation yielded heat transfer coefficient on the wall $\alpha_i=\alpha_i(T_i, T_f, \tau, \rho, c, \delta)$, here T_f is the air flow temperature; ρ, *c*, δ are the density, heat capacity and thickness of the heat transfer plate. Heat transfer coefficient distribution was estimated for each value of forcing frequency.

At the first stage of experiments, heat transfer was measured in the steady air flow. The obtained distributions of heat transfer coefficient at these regimes agree well with canonical data on heat transfer in the entrance region of the channel. Peak heat transfer coefficients, α, were observed at the channel inlet with monotonous decrease further downstream.

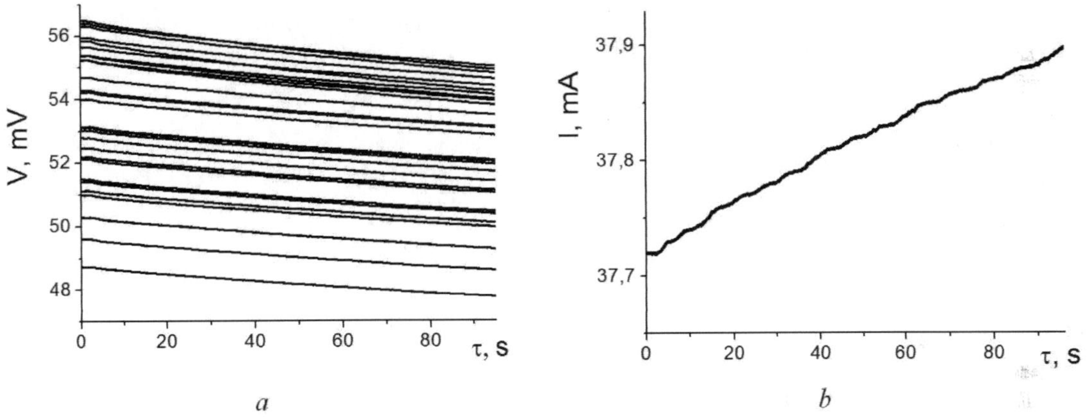

Figure 2. Voltage (a) and current (b) in the strips of heat transfer section in steady flow.

For the case of smooth-inlet channel, heat transfer coefficients were studied for the average flow velocity *U*=1.8 m/s (Reynolds number based on equivalent diameter of the channel $Re=1.5\cdot10^4$). Experimental data are demonstrated for two frequency ranges (figure 3, *a, b*) for clarity. The inlet section of the channel (where the inlet duct was attached to the channel) was taken as *x*=0. All figures contain the distribution of heat transfer coefficient in the steady flow (solid line *8*) for comparison. Two data groups along *x*-coordinate refer to two measurement cycles with different location of heat transfer wall: in the entrance region and further downstream.

In general, the experiments demonstrated heat transfer enhancement at pulsating flow regimes in the channel entrance region, in particular immediately downstream of the inlet duct. At low frequencies of up to *f*=12 Hz (figure 3, *a*) and/or low amplitudes β~0.3 (figure 3, *b*) the behavior of heat transfer coefficient distributions were similar to the ones in steady flow with slight deviations in values. The flow pattern can obviously be considered as a quasisteady one. The increase in forced pulsation frequency up to *f*=50 Hz (figure 3, *b*) significantly changes the behavior of heat transfer coefficient: it becomes non-monotonous with several maxima and minima, and double heat transfer enhancement is observed. This fact can testify to generation of large-scale vortices in the inlet duct. Their behavior in the entrance region obviously causes such non-monotonous distributions of the heat transfer coefficient. Further monotonous reduction of the heat transfer coefficient down to the steady-flow values downstream along the channel may result from breakup of these vortices.

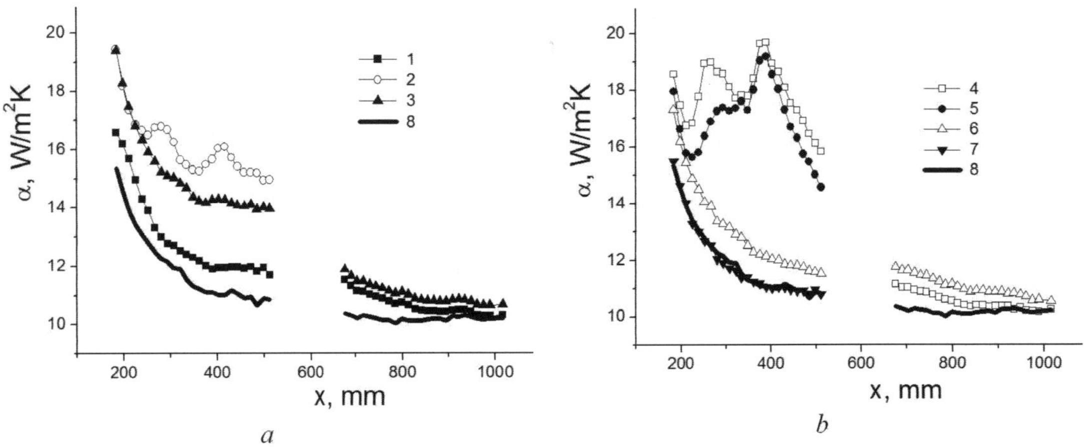

Figure 3. Heat transfer coefficient at low-frequency (a) and high-frequency (b) regimes with frequencies 12 (1), 16 (2), 20 (3), 40 (4), 50 Hz (5) at β~0.8, 70 (6), 140 Hz (7) at β~0.3 and in steady flow (8).

In general, decrease in heat transfer coefficients down to the steady-flow values was observed along the channel at all the considered frequencies. At the distance of x~1 m (~8 equivalent diameters) the difference did not exceed 8%.

4. Conclusions

Thus, when studying hydrodynamics and heat transfer from objects located in channels with forced flow pulsations, not only the wave structure of the pulsating flow should be taken into account but also the flow history. The latter largely depends on the interaction between the pulsating flow and the inlet duct at least in the entrance region. It seems likely that vortices will form here depending on the frequency and amplitude of forced flow pulsations.

Acknowledgments
This study was supported by the Russian Science Foundation (Project no.16-19-10336).

References

[1] Galitseiskii B M, Ryzhov Yu A and Yakush E V 1977 *Thermal and Hydrodynamic Processes in Oscillating Flows* (Moscow: Mashinostroenie), in Russian
[2] Helgeland A, Mardal K A, Haughton V and Reif B A 2014 *J Biomech* **47** 1082–90
[3] Chaugule V, Narayanaswamy R, Narayanan V, Jewkes J and Lucey A 2014 *Proc. 19th Australasian Fluid Mechanics Conf.* (Australia)
[4] Patel J T and Attal M H 2016 *Int. J. Current Engineering and Technology* **6** 1515–21
[5] Gündogdu M Y and Carpinlioglu M O 1999 *JSME Int. J. Series B* **42** 384–97
[6] Gundogdu M Y and Carpinlioglu M O 1999 *JSME Int. J. Series B* **42** 398–410
[7] Miau J J, Wang R H, Jian T W and Hsu Y T 2017 *Proc. R. Soc. A* vol 473 (London: Royal Society) p 20160590
[8] Wang X and Zhang 2005 *Int. J. Heat Mass Transfer* **48** 3957–70
[9] Yuan H, Tan S, Zhuang N and Tang L 2014 *Int. Communications in Heat and Mass Transfer* **53** 14–7
[10] Davletshin I A and Mikheev N I 2012 *High Temperature* **50** 412–19

IOP Conf. Series: Journal of Physics: Conf. Series **899** (2017) 022003 doi:10.1088/1742-6596/899/2/022003

An experimental study of dependence of hydro turbine vibration parameters on pressure pulsations in the flow path

D Dekterev[1,2], A Maslennikova[1] and A Abramov[1]

[1] Siberian Federal University, 79 Svobodny Ave., 660041 Krasnoyarsk, Russia
[2] Institute of Thermophysics, SB RAS, 1 Lavrentyev Ave., 630090, Novosibirsk, Russia
E-mail: dekterev_da@mail.ru

Abstract. The operation modes of the hydraulic power plant water turbine with the formation of a precessing vortex core were studied on the hydrodynamic set-up with the model of hydraulic unit. The dependence of low-frequency vibrations on flow pressure pulsations in the hydraulic unit was established. The results of the air injection effect on the vibrational parameters of the hydrodynamic set-up were presented.

1. Introduction

The operation of hydro turbine equipment, especially in high head hydroelectric power plants, can be accompanied by various problems. One of them is the operation of equipment under adverse transient modes (partial or high load). In such regimes a stable vortex structure with a pronounced precession frequency formed downstream the water turbine runner. The precessing vortex core (PVC) leads to generation of high amplitude low-frequency pressure pulsations [1 - 4], which are the most dangerous in the case of resonance with the elements of the hydraulic turbine. In this case arising vibrations leads to a faster wear of the equipment and destruction of some unit elements. This work was carried out to determine the effect of pressure pulsations in the water turbine flow path on the vibration of its construction.

2. Experimental results

This work was performed on a hydrodynamic set-up with 1/14.7 scale Francis turbine model. The pumps supply water from storage tank to the pressure tank (in this work a constant water level of 3.5 m was fixed). From the pressure tank water through the flow path enters the spiral case, then passes through the stator columns, guide vanes and then rotates the runner $D = 0.3$ m (thus the so-called "free runner" mode is realized). Then the flow leaves the water turbine model through draft tube to the drain tank.

The set-up is equipped with piezoelectric pulsation sensors, which mounted in the wall of the conical diffuser of the draft tube. The vibration/acceleration sensor was vertically mounted on the flange of the conical diffuser. Measured signals were recorded to a PC via an ADC. The frequency of the runner rotation is measured using an optical tachometer. For optical observation, the draft tube conical diffuser is made of a transparent material. In Figure 1, one can see a conical diffuser, flow with a precessing vortex and the mounted sensors (1 - vibration acceleration sensor, 2 - piezoelectric pressure sensor).

Figure 1 – Flow with a precessing vortex in the conical diffuser

Changing operation modes was performed through variation of vane's opening. The entire range of opening is divided into 15 steps. According to the integral characteristics the optimal opening is 9/15. The minimum level of flow pressure pulsations downstream the runner is also observed in this mode. Modes in range from 3/15 to 7/15 are characterized by the presence of a visible precessing vortex core in the draft tube.

2.1. One-phase flow

In this paper, the results for the guide vanes opening of 3/15 are presented. For this mode, the maximum level of vibration and most noticeable positive effect of the air injection were observed. Figure 2 shows the signal recorded by the vibration sensor.

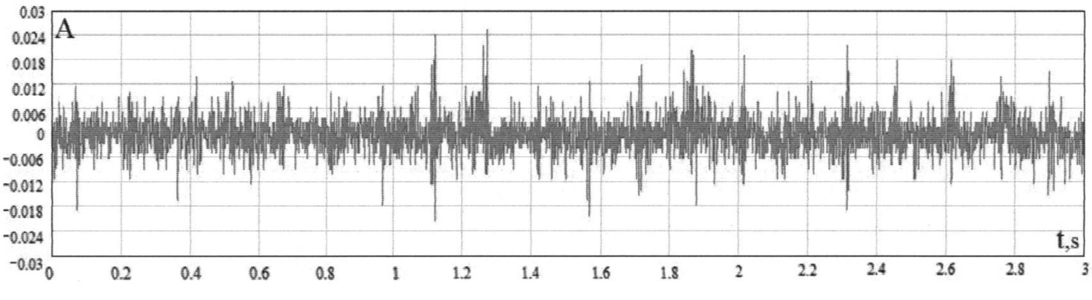

Figure 2 – Vibration sensor signal in arbitrary units

A fast Fourier transform of this signal was carried out. For a more vivid comparison of the vibration acceleration and pulsation spectrum they are presented on the same plot (Fig. 3). On the ordinate axis signal amplitude in a dimensionless form is presented. Only the low-frequency range is considered.

IOP Conf. Series: Journal of Physics: Conf. Series **899** (2017) 022003 doi:10.1088/1742-6596/899/2/022003

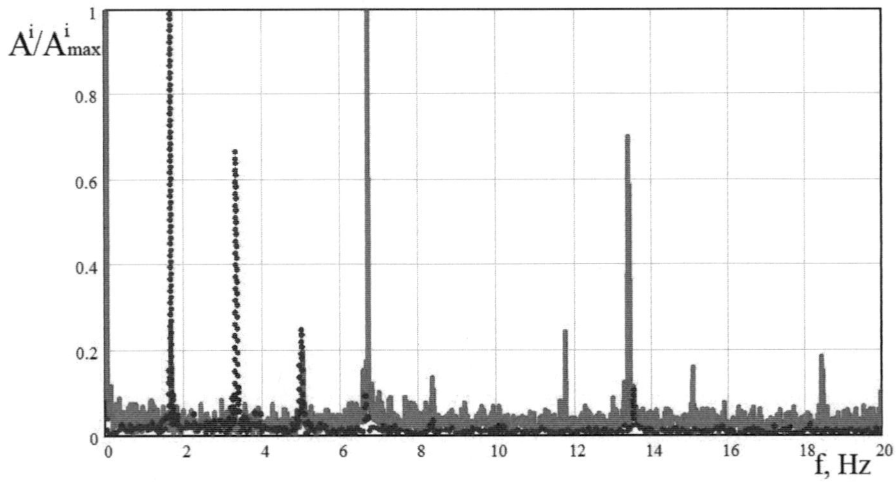

Figure 3 – Frequency spectrum
(Blue dots – pressure pulsations, red line – vibration)

Two main frequencies and their harmonics can be observed on the spectrum:
1.66 Hz – Frequency of precessing vortex core (f_{pvc})
6.68 Hz – Frequency of runner rotation (f_{rot})
It is also possible to notice combinations of these values, for example:

$$2f_{rot} - f_{pvc} = 11.7 \text{ Hz} \qquad \text{and} \qquad 2f_{rot} + f_{pvc} = 15.02 \text{ Hz}$$

In this case f_{pvc}/f_{rot}=0.25, this dependence is observed for openings 1/15 – 3/15. For other openings this parameter changes from 0.25 to 0.35.

2.2. Air injection

As is known from [3, 4], the air injection significantly affects the characteristics of a precessing vortex core. Hence, changing of the flow pressure pulsation parameters should lead to a change in the level of vibration.

Because of the technical impossibility of air injection through the shaft of the runner it was made into the spiral case close to the runner. The air flow rate was 6.3 m^3/h (2.4% of the water flow rate). The effect of air injection on the signal of vibration sensor can be seen in Figure 4.):

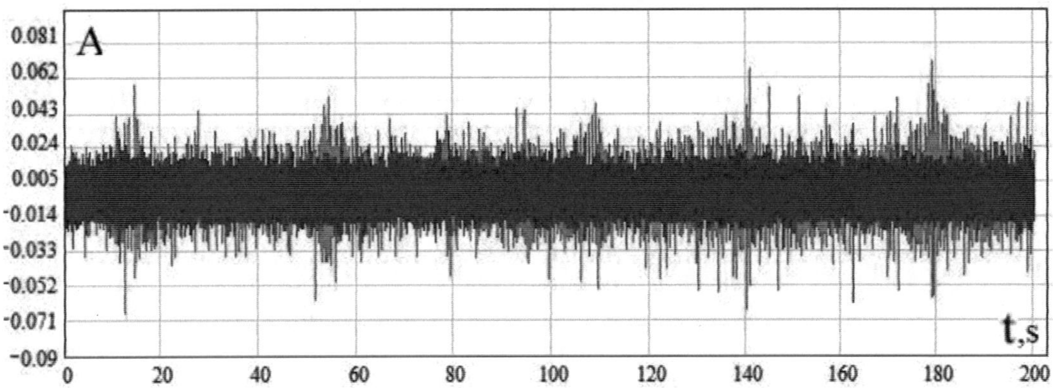

Figure 4 – Vibration sensor signal in arbitrary units
(Red line – without air, blue line – with air)

Based on a fast Fourier transformation, it can be seen that the air injection leads to the disappearance of the frequencies associated with f_{pvc}. In addition, the amplitude f_{rot} decreases.

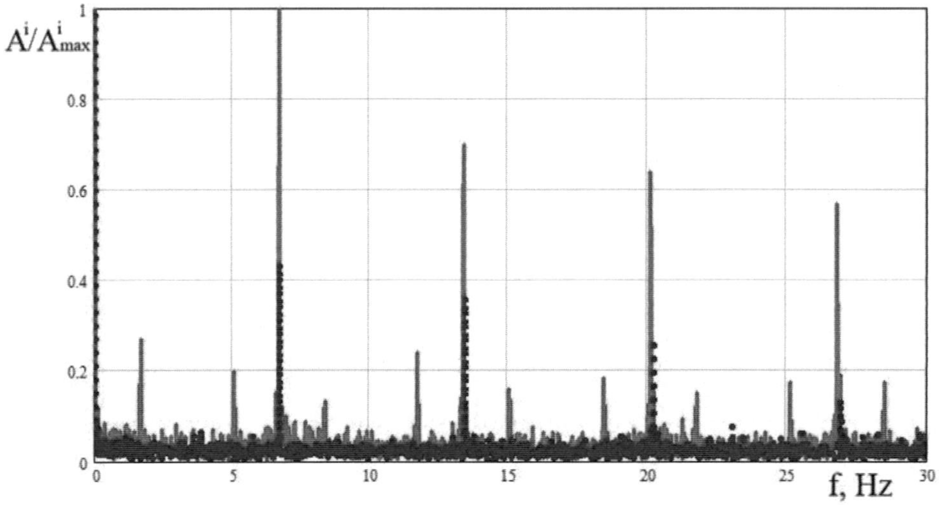

Figure 5 – Vibration frequency spectrum
(Red line – without air, blue line – with air)

It should be noted that in 6/15 and wider openings, the vibration level of the single-phase flow decreases so much that the injection of air leads to the opposite effect - an increase in the overall level of vibration. Therefore, we can talk about the advisability of air injecting only in the cases of suboptimal operating modes of the hydraulic unit.

3. Conclusions

In this paper it is shown that the air supply to the flow path leads to a decrease in the level of low-frequency vibrations arising from the pressure pulsations caused by the precession of the vortex core. However, high-frequency vibrations can increase. Thus, the air supply is expedient only in the so-called transient modes of operation of the hydraulic unit (partial or high load), when contribution of hydrodynamic processes in vibration parameters is the most significant.

Acknowledgments

The work was performed at partial support of the projects funded by the Russian Foundation for Basic Research and Krasnoyarsk Regional Fund for Support of Scientific and Scientific-Technical Activities (16-48-243076).

References

[1] Dorfler P, Sick M, Coutu A 2013 Springer-Verlag, London. 242
[2] Alekseenko S, Kuibin P, Okulov V, Shtork S 1999 JFM Vol 382 p 195–243.
[3] Platonov D, Minakov A, Dekterev D, Sentyabov A, Dekterev A 2016 JPCS V. 754
[4] Vinokurov A, Shtork S, Alekseenko S 2015 Techn Phys Lett vol 41(17) p 61-67

Numerical analysis of the Black Sea energy budget in 2011

S G Demyshev[1], **O A Dymova**[1]

[1]Marine Hydrophysical Institute RAS, 2 Kapitanskaya st., Sevastopol, 299011, Russia

E-mail: olgadym@yahoo.com

Abstract. The paper analyzes the results of the annual-averaged and seasonal-averaged variability of the Black Sea energetics in 2011. Energetic features were computed by equations describing the change rate of the kinetic and potential energy. They corresponded precisely to the finite-difference equations of the ocean model developed in the Marine Hydrophysical Institute of the Russian Academy of Sciences. A numerical experiment was run with a horizontal resolution of 1.6 km and took into account the real atmospheric forcing SKIRON for 2011. It was discovered that on average over the year the most significant components of the integral energy budget were the wind work, the dissipation due to friction, and the change of potential energy due to vertical diffusion. Seasonal variability of energy fluxes was determined by the contribution from wind and dissipation due to friction in the autumn-winter period. As a result, the vertical mixing processes enhanced and the RIM Current got stronger. In the spring and summer seasons the main energy processes were the buoyancy work and vertical turbulent diffusion due to increase in the vertical density gradient.

1. Introduction

The work of the main forces forming the circulation features in the oceans and seas was described by the budget equations of kinetic (KE) and potential energy (PE). Model and field estimations of the currents mean kinetic energy, eddy energy, available potential energy (APE) allow us to make conclusions about the generation mechanisms and dynamics of jet currents, fronts and eddies of various scales. The energetic cycle of the World Ocean was simulated in [1], where the transformations between the mean KE and APE were calculated. The authors showed that APE sank due to eddy-induced overturning circulation and the mean KE was reduced due to the vertical redistribution of geostrophic currents momentum, herewith both processes were caused by baroclinic mesoscale eddies. An analysis of energy conversions was compiled in [2] during the investigation of the generation mechanisms of the hydrophysical parameters anomalies in the Arctic Ocean. The assessment of the kinetic energy balance is used in numerical ocean models select stopping criteria for the adaptive calculations of the hydrodynamical characteristics [3]. The numerical calculations of the energetics of the Black Sea climatic circulation was first carried out in [4] and as a result the most energy-active sea zones were identified. Examination of the energy budget in the semienclosed basins with vertically separated exchange flows at the strait was presented in [5]. In this paper it was shown that the change rate of the volume-integrated mechanical energy of semienclosed seas (including the Black Sea) is directly related to the wind work, surface heat fluxes, and buoyancy fluxes. Authors concluded that the buoyancy work is one of the main factors forming the mesoscale variability in such seas.

The aim of our research was to calculate of the Black Sea energetics considering the real atmospheric forcing and to perform qualitative and quantitative estimations of the annual-averaged and seasonal-averaged contributions of the physical processes in KE and PE variations in 2011. Note that, unlike in [4], in the presented work the horizontal resolution was increased, account the water exchange on the open boundaries was taken into (in the river mouths and straits), and real forcing was used. In future it will allow us to compare the results for various years [6] and to investigate the inter-annual variability of the energy characteristics in the Black Sea.

Increase of the model resolution let to simulate mesoscale and submesoscale eddies in the velocity fields which have not been obtained in the experiments with resolution of 5 km (as in [7]). Also the variability of the sea surface height and surface currents were reconstructed more realistic especially under extreme storms. The resolution of small scales lets to more accurately calculate the energy redistribution over the movements spectrum. This is confirmed by the fact that the contribution of the turbulent dissipation term in the total energy balance has not decreased despite more quantity of eddies which were reproduced. Therefore, lower resolution had a reason for incorrect description of the energy sink in the subgrid scales.

2. Energetic model equations

The problem was solved by the eddy-resolving z-model of the Marine Hydrophysical Institute RAS (MHI-model) which is based on the motion equations in quasi-static approximation and the budget equations of temperature and salinity. Formulation of the problem, differential equations and their finite-difference equivalents, the stages of model development are described in detail in [7]. The discrete energetic equations are a consequence of the finite-difference model equations, which provides exact implementation of conservation laws at the discrete level.

To compute the circulation energetic characteristics, the MHI-model was supplemented with a subroutine for calculation of the components of KE and PE budgets [4]. The symbolic notation of the KE change equations has the following form:

$$E_t + Adv(P+E) = \Pi \leftrightarrow E + F_{Bfr}^{\tau}(E) - Diss_{Ver}(E) - Diss_{Hor}(E), \quad (1)$$

where $Adv(P)$ – advection of pressure, $Adv(E)$ – advection of KE, $\Pi \leftrightarrow E$ – buoyancy work, $F_{Bfr}^{\tau}(E)$ – KE change due to wind action and bottom friction, $Diss_{Ver}(E)$ – KE change due to vertical internal friction, and $Diss_{Hor}(E)$ – KE change due to horizontal friction, including internal horizontal friction and lateral friction. Integration of eq. (1) over the basin volume, denoted by angle brackets, gives:

$$< E_t >_V + < Adv(E) >_V + < Adv(P) >_V = < \Pi \leftrightarrow E >_V + < \tau \rightarrow E >_S -$$
$$- < Diss_{Ver}(E) >_V - < Diss_{Hor}(E) >_V . \quad (2)$$

The contributions of the advective terms is not equal to zero after integration, because the river inflows and the water exchange through the straits were taken into account in the model. The term $F_{Bfr}^{\tau}(E)$ gives the contributions on the sea surface and on the bottom. For analysis simplicity the term describing the energy dissipation due to bottom friction was introduced in term of $< Diss_{Ver}(E) >_V$. And wind contribution was allocated in the separate term $<\tau \rightarrow E>_S$ too.

According to [4] the PE change is determined by this equation:

$$\Pi_t + Adv(\Pi) = -\Pi \leftrightarrow E + Diff_{Hor}(\Pi) + Diff_{Ver}^{Fluxes}(\Pi) + Diff_{Ver}^{\kappa_V}(\Pi) +$$
$$+ Diff_{Ver}^{Bot-Sur}(\Pi) + Diff_{Ver}^{Add}(\Pi), \quad (3)$$

where $Adv(\Pi)$ – advection of PE, $Diff_{Hor}(\Pi)$ – PE change due to horizontal turbulent diffusion, $Diff_{Ver}^{Fluxes}(\Pi)$ – PE change due to buoyancy fluxes and internal vertical turbulent diffusion, $Diff_{Ver}^{Bot-Sur}(\Pi)$ – PE change due to the difference between bottom and surface density, $Diff_{Ver}^{\kappa_V}(\Pi)$ –

PE change due to the depth heterogeneity of the vertical turbulent diffusion coefficient, and $Diff_{Ver}^{Add}(\Pi)$ – additional term due to nonlinear dependence of density from the temperature and salinity. For further analysis the PE change due to heat and salinity fluxes from the atmosphere was allocated in a separate component denoted as $<Fluxes>_S$. The vertical diffusion in the system was determined by the sum of the fourth, fifth and sixth terms in eq. (3). Eq. (3) takes the next form after integration over the volume:

$$<\Pi_t>_V + <Adv(\Pi)>_V = -<\Pi \leftrightarrow E>_V + <Fluxes>_S +$$
$$+ <Diff_{Hor}(\Pi)>_V + <Diff_{Ver}(\Pi)>_V . \tag{4}$$

3. The numerical experiment setting

To calculate the hydrophysical and energetic characteristics of the Black Sea circulation, the numerical experiment was carried out under the next parameters. The boundary conditions included the wind stress fields, latent, sensible, long-wave heat fluxes, solar radiation, evaporation and precipitation obtained from the SKIRON atmospheric reanalysis with a spatial resolution of 0.1° [8]. The fields of sea surface temperature were assimilated every day. The water exchange through the straits and the inflows of the main rivers was taken into account in accordance with the mean climatic data. The climatic fields of level, temperature, salinity and current velocities obtained in [9] with a horizontal resolution of 5 km were set in the initial time moment corresponding to December 31, 2010. All boundary and initial fields were linearly interpolated on the model grid. At the initial moment the agreement procedure of hydrophysical and atmospheric fields was applied. The boundary conditions were fixated and the calculation of the model equations was carried out. As agreement criteria, we used the volume-integrated divergence of the horizontal velocity characterizing the similarity degree of the numerical velocity to geostrophic velocity. The criteria were decreased by an order after 4 days of model time. Therefore we assume that the velocity field was adapted to the density field, i.e. there was a quasi-geostrophic adjustment.

The experiment was carried out on a regular spatial grid with a horizontal step of 1.6 km, 27 vertical z-levels were used, and the time step was equal to 96 sec. Note that horizontal resolution of the model is less than the baroclinic Rossby's deformation radius [10], which is about 8 km in the Black Sea coastal zones. So we can reproduce the mesoscale and submesoscale eddies that are often detected on satellite images and field observations [11], [12]. The vertical turbulent coefficients were calculated in accordance with the Mellor-Yamada parameterization 2.5 [13]. The coefficients near the biharmonic operators in the terms that describe the horizontal turbulent viscosity in the motion equations and the horizontal turbulent diffusion in the transport equations of temperature and salinity are 10^{16} cm^4sec^{-1}, 5×10^{14} cm^4sec^{-1}, respectively. The experiment was run for the period from January 1 to December 31, 2011. Complete problem statement, boundary and initial conditions, description of the subroutine for calculation of the energy budget components are presented in detail in [6].

4. Results of the numerical experiment

As a result we obtained three-dimensional arrays of hydrophysical fields, one-dimensional arrays of all terms in eq. (2) and (4), and three-dimensional arrays of the spatial distribution of terms in eq. (1) and (3) for every day of the investigative period. The paper presents the analysis results of the Black Sea volume-integrated energetics in 2011. The angle brackets are skipped in the text and figures below.

4.1. Intra-annual variability

We considered the volume-integrated energetic characteristics (terms of eq. (2) and (4)) averaged over the year. Figure 1 presents a scheme of the directions and magnitudes of the energetic fluxes composing the KE and PE budgets. On average over 2011 there was an accumulation of KE and PE, alongside this the rate growth of PE was five times higher than of KE. The exchange between KE and PE was implemented through the buoyancy work $\Pi \leftrightarrow E$. At the mean over the year its value was positive and the share of the total PE having converted in KE was about 50 percent. The wind work $\tau \rightarrow E$ gave the maximum contribution to KE. Input from the wind was compensated by 72 percent by the vertical friction $Diss_{ver}$. The advective terms $Adv(P)$ and $Adv(E)$ are small, however, but not equal to zero because the river inflow and exchange through the Kerch Strait and the Bosphorus Strait were taken into account in the MHI-model. KE decreased due to the horizontal friction ($Diss_{hor}$) and the absolute value of this term was three times higher than the contribution from the buoyancy work.

The vertical diffusion $Diff_{ver}$ gave the main contribution to the integral PE balance and the magnitude of this term exceeded other components more than four times. $Diff_{ver}$ was mainly determined by the stratification and variability of vertical turbulent diffusion coefficient with depth. Horizontal diffusion $Diff_{hor}$ was almost five times less than vertical. This result indicates that PE increased mainly due to vertical diffusion processes and their enhancement was a consequence of the significantly heterogeneous vertical structure of the density fields in 2011. It will be illustrated below. Figure 1 demonstrates that the contribution to the PE budget due to fluxes of heat and salt from the atmosphere composed about 20 percent of the PE balance. Its value is negative so the fifth part of the PE was lost to heating of near-water layer of atmosphere. Thus, the integral KE balance was determined by the contribution from wind and dissipation due to friction, while the integral PE balance was determined the vertical diffusion on average over 2011.

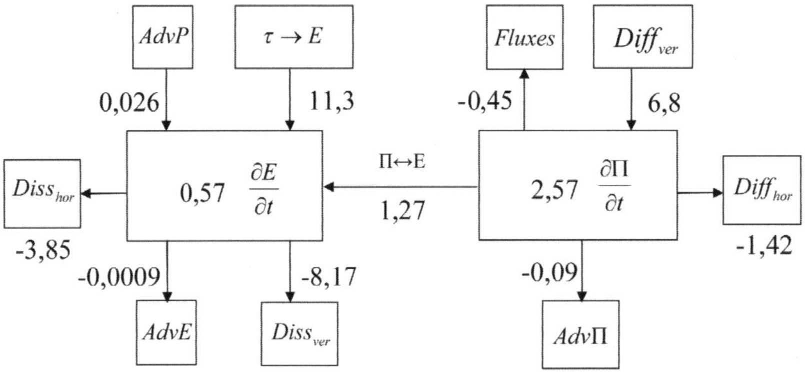

Figure 1. Annual-averaged components of the energy budget (10^{-6} erg·cm^{-3}·sec^{-1}).

4.2. Seasonal variability

The terms in eq. (2) and (4) were averaged over the seasons to investigate seasonal variability. The analysis of obtained values allowed us to identify the most energetically significant contributions depending on the time of a year. The results are presented in Table 1.

Table 1. Some seasonal average components of the energy budget (10^{-6} erg·cm^{-3}·s^{-1}).

2011	$\tau \rightarrow E$	$\Pi \leftrightarrow E$	$Diss_{ver}$	$Diff_{ver}$	*Fluxes*
winter	18.9	0.3	-14.5	-2.3	-1.9
spring	6	1.8	-3.7	13.8	1
summer	3	2.6	-2.5	15.6	1.4
autumn	17.6	0.7	-12.1	-0.2	-2.1

It can be seen that the maximum contribution to KE from the wind was detected in winter and autumn, its partial compensation occurred due to vertical friction. The increase in wind work resulted in an intensification of the modeling velocities in the RIM Current zone in the cold period, which agrees

IOP Conf. Series: Journal of Physics: Conf. Series **899** (2017) 022004 doi:10.1088/1742-6596/899/2/022004

well with the published data and field observations [14]. The buoyancy work and the vertical turbulent diffusion effects grew in spring and summer because of the seasonal variability of the density field (intensification of the river inflows and warming of the upper sea layer). These processes took place along with the relaxing wind field. The PE change due to fluxes of heat and salt from the atmosphere indicates that in the autumn and winter of 2011 the sea gave up more heat to the atmosphere than it had received in the spring and summer.

The integral profiles of temperature, salinity and conventional density averaged over the seasons were analyzed. Averaged profiles over the winter and summer of 2011 are presented in Figure 2. It shows that the profiles of salinity and conventional density are qualitatively identical in winter and the vertical temperature gradient is less than 2° C in the upper 300-m layer (Figure 2, a). Such a structure of the density field was a result of intensive winter convection and cooling. At this time the components $\Pi \leftrightarrow E$ and $Diff_{ver}$ were minimal. The significant increase in the temperature and conventional density gradients was observed in the summer compared to the winter: in the upper 50-m layer the temperature variation was equal to 17° C and the density variation was equal to 3.5 conventional units (Figure 2, b). So, during the warm year period the density of the upper quasi-homogeneous layer (and therefore its PE) was determined more by temperature than salinity. The rising of the stratification led to the situation when the terms $\Pi \leftrightarrow E$ and $Diff_{ver}$ reached their maximum values. As is clear from Table 1, the magnitude of these terms increased almost an order compared to the winter period. Thus, in the summer of 2011 the change rate of PE increased, despite the fact that the direct PE decreased.

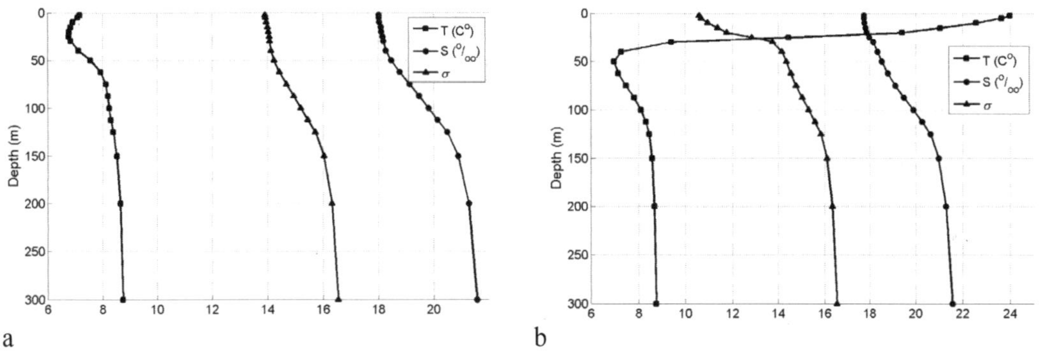

Figure 2. Averaged temperature, salinity, and conventional density profiles: a – over the winter of 2011; b – over the summer of 2011.

5. Conclusion

In the result of the present research we learned that the total energy in the Black Sea increased in 2011. The growth of KE occurred due to the wind work in the autumn-winter period. Our analysis of hydrodynamic characteristics showed that enhancement of the vertical mixing processes and the intensification of the RIM Current were observed inside the marine system at this time. The heating of the upper sea layers led to the increase in the vertical density gradients in the warm months and it caused the growth of contributions due to the buoyancy work and the vertical diffusion. Thus, in 2011 the Black Sea integral energetics was determined by influences of the wind work, the buoyancy work, the friction processes, and the vertical turbulent diffusion.

Acknowledgements
This work was supported by the Russian Foundation for Basic Research (project no. 15-05-05423).

References

[1] Aiki H et al 2016 Energetics of the global ocean: The role of mesoscale eddies *The Indo-Pacific Climate Variability and Predictability. World Scientific Series on Asia-Pacific Weather and Climate* **7** 109-134.
[2] Zalesny V B and Ivchenko V O 2010 Modeling the global circulation response and the regional response of the Arctic Ocean to the external forcing anomalies *Oceanology* **50**(6) 829-840.
[3] Sarkisyan A S 1996 On some problems and results of ocean modeling *Oceanology* **36**(5) 607-617.
[4] Demyshev S G 2004 Energy of the Black Sea climatic circulation. 1. Discrete equations of the time rate of change of kinetic and potential energy *Meteorologiya i Gidrologiya* **9** 65-80.
[5] Cessi P et al 2014 Energetics of Semienclosed Basins with Two-Layer Flows at the Strait *J. Phys. Oceanogr.* **44** 967-979.
[6] Demyshev S G and Dymova O A 2016 Analyzing intraannual variations in the energy characteristics of circulation in the Black Sea *Izv. Atmos. Ocean. Phys.* **52**(4) 386-393.
[7] Demyshev S G 2012 A numerical model of online forecasting Black Sea currents *Izv. Atmos. Ocean. Phys.* **48**(1) 120-132.
[8] Kallos G et al 1997 The regional weather forecasting system SKIRON: An overview *Proc. of the Int. Symp. on Regional Weather Prediction on Parallel Computer Environments* (Athens, Greece) 109-122.
[9] Demyshev S G, Ivanov V A and Markova N V 2009 Analysis of the Black Sea climatic fields below the main pycnocline obtained on the basis of assimilation of the archival data on temperature and salinity in the numerical hydrodynamic model *Phys. Oceanogr.* **19**(1) 1-12.
[10] Gill A 1982 Atmosphere-Ocean Dynamics (England: University of Cambridge) p 662.
[11] Karimova S 2012 Spiral eddies in the Baltic, Black and Caspian seas as seen by satellite radar data *Advances in Space Research* **50**(8) 1107-24.
[12] Zatsepin A G et al 2011 Submesoscale eddies at the Cauasus Black Sea shelf and the mechanisms of their generation *Oceanology* **51**(4) 554-567.
[13] Mellor G L and Yamada T 1982 Development of a turbulence closure model for geophysical fluid problems *Rev. Geophys. Space Phys.* **20**(4) 851-875.
[14] Ivanov V A and Belokopytov V N 2011 Oceanography of the Black Sea (Sevastopol: MHI NASU, in Russian) p 209.

IOP Conf. Series: Journal of Physics: Conf. Series **899** (2017) 022005 doi:10.1088/1742-6596/899/2/022005

Influence of secondary flows on heat transfer from a localized heat source

A. Evgrafova[1,2], **A. Sukhanovskii**[1], **M. Kuchinskii**[2]
[1]Institute of Continuous Media Mechanics, Academ. Korolyov, 1, Perm, 614013, Russia
[2]Perm State University, Bukireva St. 15, Perm, 614990, Russia

E-mail: eav@icmm.ru

Abstract. Heat transfer from a localized heat source in a cylindrical fluid layer was studied experimentally. Experiments were performed for fluids with different values of the Prandtl number and for the Rayleigh number from 10^5 to 10^7. Particular attention has been paid to investigating the influence of small-scale convective structures in the boundary layer of a basic flow on heat transfer. The dependence of the Nusselt number on the Rayleigh number was obtained by means of temperature measurements. In addition, the mean velocity of convergent flow was estimated. It has been found that changes in the structure and intensity of secondary flows have a weak effect on heat transfer. At the same time, when the Rayleigh number increases, the basic flow velocity exhibits a linear growth.

1. Introduction

Natural convection is the primary mechanism of heat transfer in many geophysical and technological processes. In this paper we focus on the case when the heated area is localized. Unlike the Rayleigh-Benard convection in the enclosed volumes heated from below and cooled from above, which was extensively studied experimentally and numerically (see references in [1]), convection from localized heat sources obtained less attention. Convective flows from a localized heat source in two-dimensional formulation for the moderate values of the Rayleigh number (up to 10^6) were studied in [2] and for turbulent regimes in [3], where Ra varied from 10^8 to 10^{12}. Three-dimensional simulation of convection due to a discrete heat source was carried out in [4] but only for Ra less than 10^6. However the appearance of secondary flows has not been investigated in these works.

When the size of the heater exceeds the fluid depth in the temperature boundary layer, the secondary convective structures in the form of horizontal rolls are generated [5]. Horizontal rolls formation over the hot plate in the case of forced convection in a rectangular channel was studied in [6,7,8] and in the case of thermal convection in a rectangular vessel in [9]. These studies showed that convective rolls substantially enhance heat transfer due to intensive mixing in the boundary layer.

Important problem is the influence of the secondary flow structure on heat transfer. In [9], it was found that a transition from longitudinal to transverse rolls leads to relatively weak changes in the local Nusselt number (less than 20%) for the same values of the Rayleigh number. Increasing of Ra leads to the remarkable change in the secondary flow structure [5]. The aim of this study is to realize how heat transfer depends on the structure of secondary flows.

IOP Conf. Series: Journal of Physics: Conf. Series **899** (2017) 022005 doi:10.1088/1742-6596/899/2/022005

2. Experimental setup

Cylindrical vessel of diameter D = 300 mm, and height L = 40 mm was used as an experimental setup. The schematic diagram of the setup is given in figure 1 a. The sides and bottom of the vessel are made of plexiglas with a thickness 3 mm and 20 mm, respectively. There was no cover or additional heat insulation on the sidewalls. The heater is a brass cylindrical plate mounted flush with the bottom. The diameter of the plate is d = 100 mm, and its thickness is 10 mm. The brass plate is heated by an electrical coil placed on the lower side of the disc. There are two ways of controlling heating which leads to different boundary conditions. It can be a constant temperature or a constant heat flux. In the experiments we changed the heating intensity by varying the heat flux. For each experiment the heat flux was fixed and controlled. The heater temperature was measured by an embedded thermocouple. Silicon oils with different values of Prandtl number, PMS-20, PMS-10 and PMS-5 (20, 10 and 5 cSt at T = 25 ^{0}C), were used as working fluids. In all experiments, the depth of the fluid layer was h = 30 mm, and the surface of the fluid was always open. Cooling of the fluid is provided mainly by the heat exchange with the surrounding air on the free surface and some heat losses through sidewalls. The temperature of the fluids was measured by a horizontal row of 12 copperconstantan thermocouples.The location of the thermocouples is also shown in figure 1. The coordinate origin is at the center of the bottom. The thermocouples can be moved through the whole layer depth with a step of 1 mm by a motorized translation stage. The data from the thermocouples were obtained by an Agilent 34970A data acquisition switch unit with a 16-channel multiplexer module 34902A. Temperature was measured in the central vertical cross-section above the heating area. To estimate the mean temperature of the fluid, one thermocouple measured temperature inside the fluid layer at mid-height (z = 15 mm), near the periphery (about 3 cm from the sidewall).

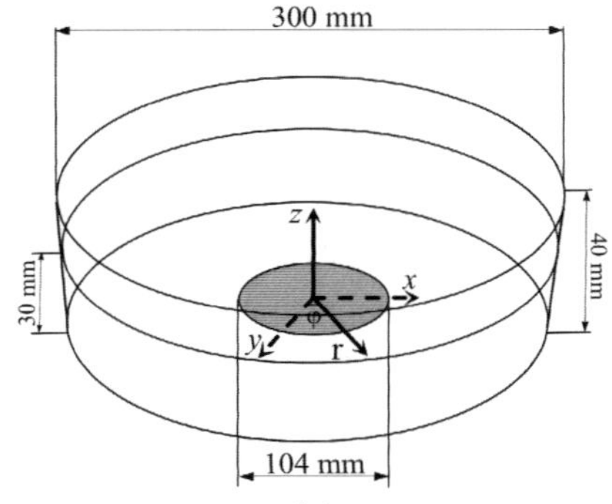

(a)

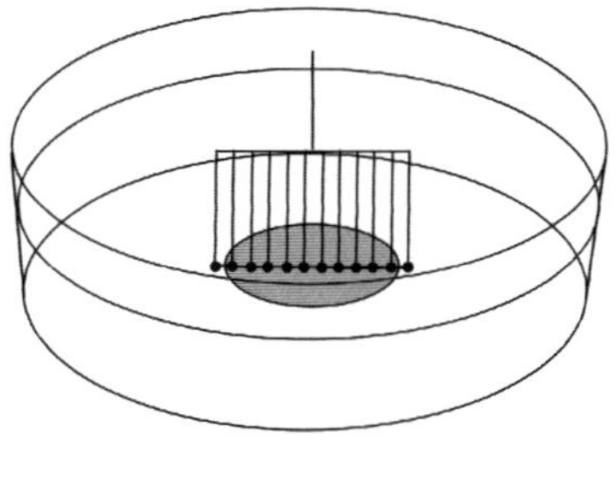
(b)

Figure 1. Experimental setup, dimensions and location of the coordinate system.

As non-dimensional parameters, we used Prandtl number Pr and Rayleigh number Ra:

$$Pr = \frac{\nu}{\chi} \tag{1}$$

$$Ra = \frac{g\beta h^3 \Delta T}{\nu\chi} \tag{2}$$

where g is the gravitational acceleration, h is the layer depth, β is the coefficient of thermal expansion, ΔT is the temperature difference between the temperature of the heater and the environment temperature, ν is the coefficient of kinematic viscosity, and χ is the thermal diffusivity.

Heat transfer was characterized by the Nusselt number defined as

IOP Conf. Series: Journal of Physics: Conf. Series **899** (2017) 022005 doi:10.1088/1742-6596/899/2/022005

$$Nu = \frac{q}{q_\lambda} \tag{3}$$

where q_λ is heat flux due to conduction, and q is overall heat flux. The heat fluxes q and q_λ were calculated as

$$q = \frac{P}{S}; q_\lambda = \lambda \frac{T_h - T_s}{h} \tag{4}$$

where P is the power of the heater, S is the heater surface area, T_h is the temperature of the heater, and T_s is the temperature of the fluid surface. The temperature of the fluid surface T_s was received by the following way. The row of thermocouples measured the temperature of fluid above the heater consistently at z = 28, 29 and 30 mm. The time of measurements was about 1000 seconds for each height. Then, the temperature was averaged over time and space in this thin near-surface layer, and the resulting temperature (T_s) was used to calculate the heat flux q_λ. Here, we neglect heat losses through the bottom and side walls.

The temperature T_s is the mean value of temperature along the thermocouple row, which is placed in the central part of the vessel above the heater. Parallel with the experiment, temperature fields of a fluid surface were obtained by a thermal imager Fluke Ti32 system for different regimes. Temperature sensitivity of the Fluke Ti32 is high (about 0.05^0C), but temperature measurement accuracy is $\pm 2\ ^0C$ and it needs to be calibrated for every regime. Therefore it was used only for getting information about temperature distribution over the entire fluid surface. Thermal imaging showed that temperature fields are not homogeneous in both radial and azimutal directions. According to our estimations, the inaccuracy of thermocouple data does not exceed 5 %.

3. Results

The heat flux in the central part of the bottom initiates the intensive upward motion above the heater. The warm fluid cools at the free surface and moves toward the periphery, where the cooled fluid moves downward along the side wall. A large-scale advective flow occupies the whole vessel. The scheme of the basic flow is presented in figure 2. The mean advective flow consists of a convergent flow in the lower part of the vessel (1) and a divergent flow in the upper layer (2).

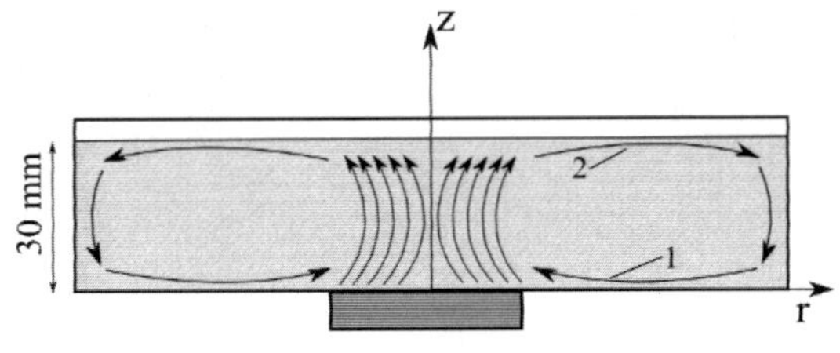

Figure 2. Large-scale circulation: 1 - convergent flow, 2 - divergent flow.

The convergent flow in the lower part of the layer leads to the formation of a boundary layer with potentially unstable temperature stratification above the heater and makes possible the generation of secondary convective flows. Examples of the experimental temperature profiles for different values of Ra are presented in figure 3.

The formation of secondary structures was described in detail in [5], where the structure and dynamics of secondary flows were found to be strongly dependent on heating intensity. Visualization of the secondary structures over the heater by the shadowgraph method is shown

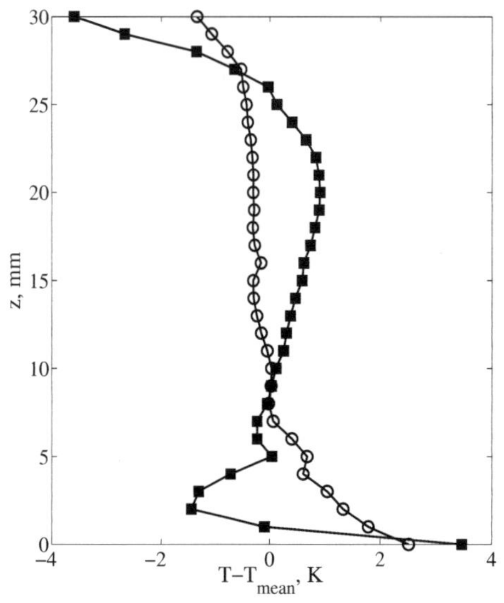

Figure 3. Temperature profiles at r = 3 cm for Ra = 0.9·10^6 (open circles) and Ra = 4·10^6(black squares).

in figure 4. Weak heating regimes (figure 4 a) are characterized by the appearance of ring-like rolls. An increase of the heat flux leads to more complex convective patterns - the superposition of a spiral and a system of radial rolls on the periphery (figure 4 b).

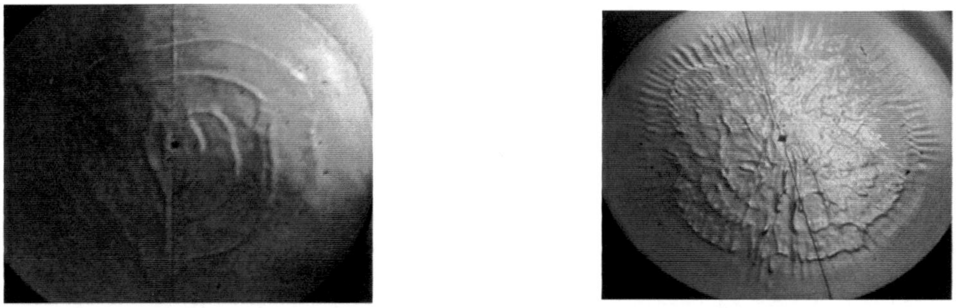

(a) (b)

Figure 4. Visualization of secondary flows with shadowgraph method: (a) $Ra = 3.3 \cdot 10^5$, (b) $Ra = 3.1 \cdot 10^6$

Velocity measurements given in [5] showed that radial rolls are robust and stationary. Ring-like rolls and spirals are oriented in transverse direction to the main flow and appear periodically on the heater periphery. Dependence of frequency F for transverse roll formation on the Rayleigh number for different fluids is presented in figure 5. The plot is done in a log-log scale and shows that F varies as $F \propto Ra^{0.60}$. This fit is shown by a solid line.

The secondary flows are responsible for intensive mixing in the boundary layer, and the variation of the secondary flow structure can lead to the change of heat transfer intensity. The dependence of Nu on Ra is shown in figure 5 b. The heat transfer grows as $Nu \propto Ra^{0.28}$ (solid line).

IOP Conf. Series: Journal of Physics: Conf. Series **899** (2017) 022005 doi:10.1088/1742-6596/899/2/022005

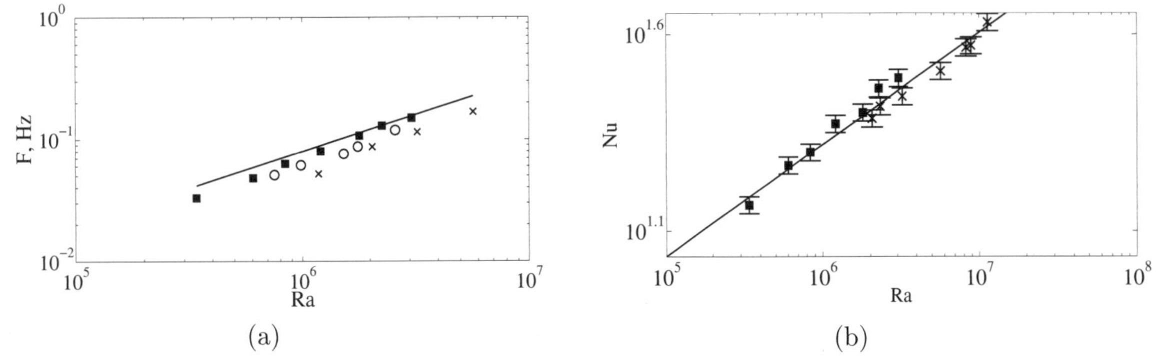

Figure 5. Characteristic frequency f (a) and Nusselt number (b) versus Rayleigh number for $Pr = 209$ (black squares), $Pr = 110$ (open circles), $Pr = 66$ (crosses).

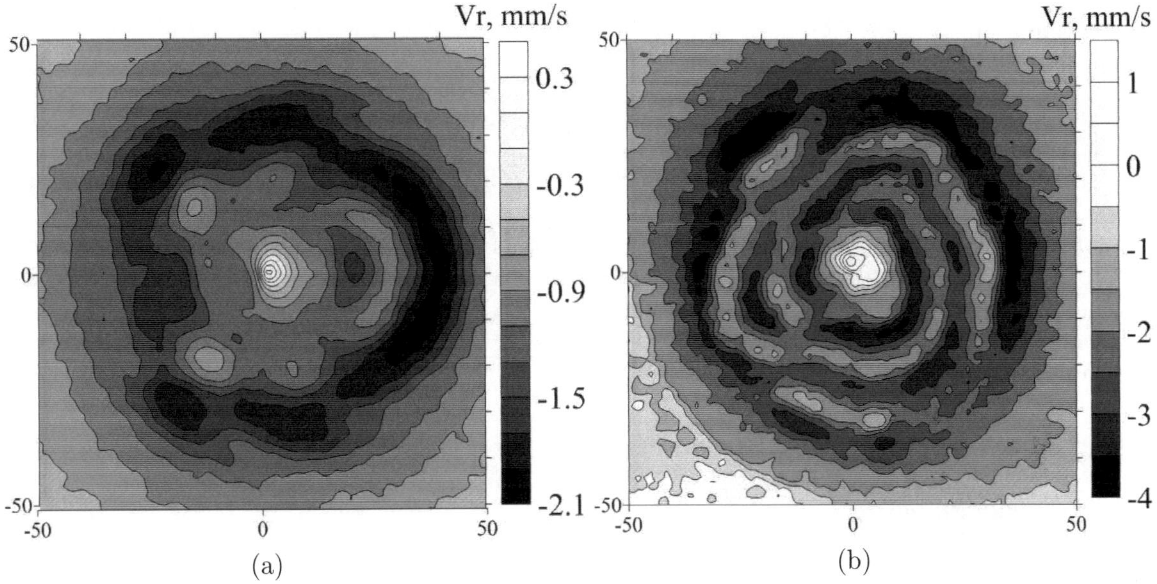

Figure 6. Instantaneous radial velocity fields over the heater at z = 3 mm for $Ra = 1.2 \cdot 10^6$ (a) and $Ra = 2.9 \cdot 10^6$ (b)

We expected a more pronounced dependence of Nu on the intensity and structure of rolls . In figure 6, the instantaneous fields of radial velocity over the heater at $z = 3$ mm are presented. The fields were obtained experimentally for $Ra = 1.2 \cdot 10^6$ (figure 6 a) and $Ra = 2.9 \cdot 10^6$ (figure 6 b). One can see that the form of transverse rolls changed noticeably and the value of velocity increased. However the heat transfer expressed by the Nusselt number weakly responded to it. The Nusselt numbers for these regimes are approximately equal to 22.2 ($Ra = 1.2 \cdot 10^6$) and 29.4 ($Ra = 2.9 \cdot 10^6$). Using the temperature measurement from the thermocouple array, the mean radial flow velocity was estimated. Figure 7 shows the dependence of the mean velocity V of convergent flow on Ra. There is almost linear dependence of V on Ra so that heat transfer for the described regimes depends mainly on the basic flow velocity.

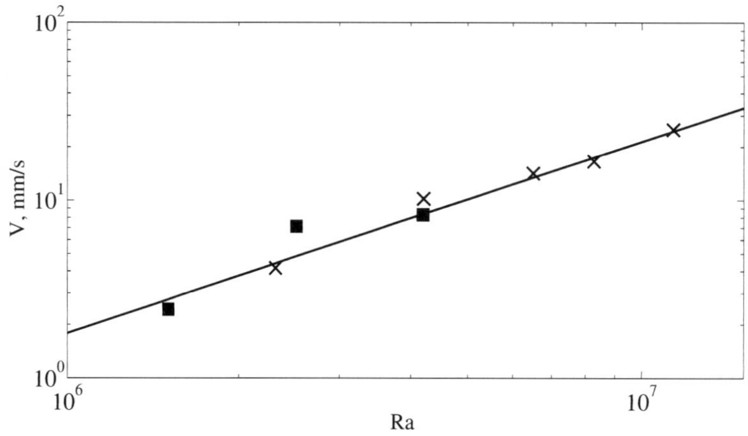

Figure 7. Dependence of mean radial flow velocity V on Ra for $Pr = 209$ (black squares), $Pr = 66$ (crosses).

4. Conclusions

Heat transfer from the localized heat source in a cylindrical layer was studied experimentally for fluids with different values of kinematic viscosity and heat flux. The flow in such a system consists of large-scale advective circulation and secondary structures appeared in the thermal boundary layer above the heated area. Our temperature measurements showed that the form and intensity of secondary flows do not cause any substantial changes in the dependence of heat transfer on the Rayleigh number. It has been also shown that the mean velocity of the convergent flow has almost linear dependence on the Rayleigh number. Hence it can be concluded that for the regimes under consideration the heat transfer is predominantly dependent on the large-scale flow intensity.

Acknowledgments

The financial support of the grant RFBR 16-31-00150 is gratefully acknowledged.

References

[1] Grossmann S and Lohse D 2008 *Phys. Rev. Lett.* **86** 3316
[2] Aydin O and Wen-Jei Y 2000 *Int. J. Numer. Methods Heat Fluid Flow* **10** 518-529
[3] Sharma A et al 2007 *Int. J. Therm. Sci.* **46(12)** 1232-1241
[4] Sezai I and Mohamad A 2000 *Int. J. Heat Mass Transfer* **43(13)** 2257-2266
[5] Sukhanovskii et al. 2016 *Physica D* **316** 23-33
[6] Maughan JR and Incropera FP 1990 *Int. J. Heat Mass Transf.* **33** 555
[7] Chiu WKS. et al 2000 *Phys. Fluids* **12** 2128
[8] Benderradji A et al 2008 *Heat Mass Transfer* **44** 1465
[9] Sukhanovsky et al. 2012 *Eur. Phys. J. B* **85** 12

IOP Conf. Series: Journal of Physics: Conf. Series **899** (2017) 022006 doi:10.1088/1742-6596/899/2/022006

Estimating local heat transfer coefficients from thin wall temperature measurements

I M Gazizov[1], **I A Davletshin**[2], **A A Paereliy**[2]

[1]Kazan National Research Technical University named after A.N.Tupolev, Kazan, 420111, Russia
[2]Kazan Scientific Center of RAS, Kazan, 420111, Russia

tapaereliy@yandex.ru

Abstract. An approach to experimental estimation of local heat transfer coefficient on a plane wall has been described. The approach is based on measurements of heat-transfer fluid and wall temperatures during some certain time of wall cooling. The wall was a thin plate, a printed circuit board, made of composite epoxy material covered with a copper layer. The temperature field can be considered uniform across the plate thickness when heat transfer is moderate and thermal resistance of the plate in transversal direction is low. This significantly simplifies the heat balance written for the wall sections that is used to estimate the heat transfer coefficient. The copper layer on the plate etched to form a single strip acted as resistance thermometers that measured the local temperature of the wall.

1. Introduction

Estimation of convective heat transfer parameters requires reliable information on the temperatures of heat-transfer fluid and the wall. It is quite challenging to obtain these temperatures as the temperature fields in general are non-uniform in space and vary in time. In practice, these measurements can be performed using the intrusive sensors that are inserted into the flow and/or the wall and can have negative effect on the hydrodynamics and heat transfer.

Different approaches to heat transfer research are employed nowadays. They include the solution to inverse heat conduction problem [1-3]. A temperature field of a solid body (a wall) is studied here. It develops depending, among others, on the boundary conditions of the solid body. Under the boundary conditions of the third kind (heat transfer coefficient h=const), there is a unique relation between the heat transfer coefficient distribution over the body surface and the temperature field within the body of given geometry and thermal properties. Under these conditions, the distribution of heat transfer coefficient along the interface between the solid body and fluid can be estimated using the measurements of their temperatures.

If thin plates (films) are employed for heat transfer research, a two-dimensional field can be considered instead of a three-dimensional one [4-6]. When the temperature is constant across the plate thickness, heat balance is considerably simplified, and so is the estimation of heat transfer coefficient. Thermocouples of different types are traditionally used for wall temperature measurements [6-8]. To obtain more detailed temperature distributions over the wall surfaces, optical methods are employed [5, 9]. However, optical methods on their own only give a qualitative pattern of a body's thermal state.

IOP Conf. Series: Journal of Physics: Conf. Series **899** (2017) 022006 doi:10.1088/1742-6596/899/2/022006

To obtain the numerical values of parameters, these methods should be calibrated e.g. using thermocouples.

The integral characteristics of heat transfer are easier to estimate (at least, in terms of the procedure). When the distributions of thermal characteristics are irrelevant or constant across the heat-transfer surface, the total heat flux and surface-average heat transfer coefficients can be estimated from the heat balance written for the relevant channel sections [10].

In general, the research of complex flows requires detailed measurements of their hydrodynamic and thermal parameters [11-13]. In this respect, acquisition of large amount of data is associated with certain technical difficulties. The development of an efficient, reliable, and accurate approach to heat transfer coefficient measurements is therefore a relevant problem of experimental research.

2. The device for measurements of thermal parameters

A printed circuit board was proposed as a main element employed in the study of convective heat transfer and measurements of local wall temperature. The circuit board was made of composite epoxy material. It had the length l_h=450 mm, the width of 200 mm and the thickness δ=1.6 mm (figure 1). Small thickness of the heat-transfer wall and its moderate thermal conductivity λ=0.3 W/(m·K) made it possible to assume that the heat flux due to thermal conductivity along the wall was many times lower than removal of heat by the heat-transfer fluid. Then, allowing for proper external thermal insulation, the heat balance for a wall section contained only a single value of heat flux that was the heat flux from the wall to the heat-transfer fluid. Under these conditions, the measured values (temperature, heat flux) can be considered local and related only to the given section of the channel.

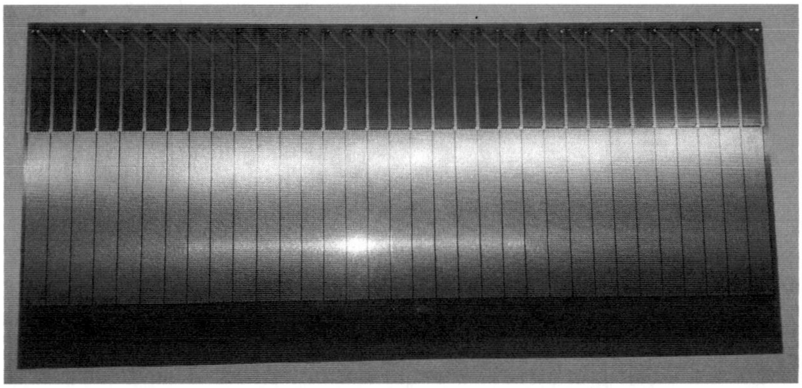

Figure 1. Printed circuit board (heat-transfer wall) employed in measurements.

Copper layer on the plate was etched to form a strip whose sections acted as 33 resistance thermometers. The area occupied by each strip (thermometer) had a length of 13 mm and a width of 100 mm. So, it was possible to measure the wall temperature local along the length and average across the channel test section with the spacing of 13.5 mm.

Thus, the circuit board took on two roles in the course of experiments. It was involved in heat transfer to the flow and provided the measurements of the wall temperature.

At low Biot numbers ($Bi = h\delta/\lambda \ll 1$), the heat transfer coefficient, h, on the wall with the area F and volume V was estimated from the wall temperature variation rate, m, in the process of cooling or heating [14]:

$m=\ln(\Delta T_1/\Delta T_2)/(\tau_2-\tau_1)$,

which was derived from the dimensions and thermal properties: $m=\psi hF/(\rho cV)$.

Here, ΔT_1 and ΔT_2 are differences between the wall and flow temperatures at the time τ_1 and τ_2; ρ, c are the density and heat capacity of the wall; the coefficient of wall temperature field non-uniformity $\psi=(B^2+1.44B+1) - 0.5$. Here, the modified Biot number is $B=hFK/(\lambda V)$, where K is the body form factor. For a thin plate:

IOP Conf. Series: Journal of Physics: Conf. Series **899** (2017) 022006 doi:10.1088/1742-6596/899/2/022006

$K=(\delta/\pi)^2$; $B=h\delta/(\lambda\,\pi^2)$; $m=\psi h/(\rho c\delta)$.

Parameters for the considered wall:

- at h=20 W/(m^2 K): B=0.011; Bi =0.1; ψ=0.99;
- at h=100 W/(m^2 K): B=0.054; Bi =0.53; ψ=0.96.

Thus, for heat transfer coefficients typical of gas flows, even if the temperature field across the wall thickness is assumed to be uniform, the error of the method lies within the range of fractions of a percent to several percent (if deviation of ψ from 1 is considered).

A circuit diagram was constructed for practical measurements (figure 2). The circuit diagram consisted of two electrical circuits:

- wall temperature measurement circuit;
- heat-transfer fluid temperature measurement circuit.

Each circuit was equipped with a DC voltage source E_w = 10 V and E_f = 1.5 V, respectively. Signals from each resistance thermometer on the board, R_1, R_2, … R_n, and a fixed resistor, R_0, were recorded in the form of voltage, U_1, U_2, … U_n and U_0, respectively, by two analog-to-digital converter units L-CARD E-14-140-M with 16 differential channels each. To measure the fluid temperature, the resistance thermometer, R_{TF}, connected in a single circuit with the resistor, R_{t0}, was used. Voltages across the thermometer, U_{TF}, and the resistor, U_{t0}, were also recorded by the analog-to-digital converter. Fixed resistors $R_0 = R_{t0}$ = 226 Ω were used to measure the electric current in the circuits of the wall temperature and fluid temperature measurement, respectively.

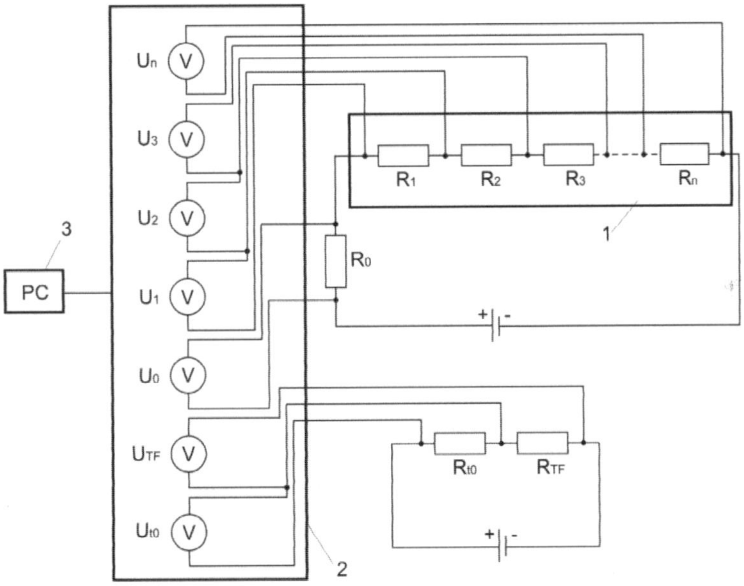

Figure 2. Measurement diagram: 1 – measurement board; 2 – analog-to-digital converter; 3 – PC (laptop).

3. Estimation of heat transfer coefficient and test measurements

The measured voltage yielded the resistance of strips of the measurement board:

$R_i(\tau)=R_0{\cdot}U_i(\tau)/U_0(\tau)$, where i is the strip number.

The current wall temperature was determined from the known relation for the electrical resistance of copper:

$T_i(\tau)=(R_i(\tau)/R_0^*(\tau)-1)/\beta+T^*$,

where β=0.004 1/K is the temperature dependence coefficient of the electrical resistance of copper; $R_i^*(\tau)$ is the resistance of strips in "cold" condition at the temperature T^*.

The heat-transfer fluid temperature, T_f, was estimated in the same way. The differences between the wall and fluid temperatures were calculated from these values:
$\Delta T_i(\tau) = T_i(\tau) - T_f$.

In the end, cooling rates of the strips, m_i, and corresponding local heat transfer coefficients, h_i, were determined from $\Delta T_i(\tau)$.

Test experiments were conducted to check the performance of the submitted approach and dedicated device. The heat-transfer wall in the experimental setup was a part of the wall of the plane channel with the cross section of 115×150 mm^2 and the length of 1.2 m (figure 3). This wall had the length of l_h=450 mm and occupied the whole width of the channel bottom wall (150 mm). It was mounted in the middle part of the channel at the distance of x=600 mm from the channel entrance. Here, the channel inlet section (where the inlet duct was attached) was taken as x=0. To guarantee the turbulent flow in the channel, a wire mesh was mounted at the channel inlet. Its cell size was 5 mm; the diameter of the steel wire was 1.2 mm. The perimeter of the 50-mm long channel section immediately downstream of the mesh was covered with an abrasive with a grit size of 0.6-1.0 mm. To minimize the heat loss to the environment, a 30-mm thick polystyrene was glued to the external surface of the heat-transfer wall.

Test heat transfer experiments were arranged as follows. The suction fan installed downstream of the channel outlet was turned on and sucked the ambient air through the channel. The required flow rate was set by a controlling gate.

At the preliminary stage of experiment, an electric blow drier was installed in front of the channel inlet. The channel walls including the inlet duct were heated by the hot air flow reaching the temperature that exceeded the ambient air temperature by 50-60°C. The heating time was ~10 min, i.e. until the temperature field of the channel walls became approximately uniform.

Then the blow dryer was turned off and moved away from the experimental setup to prevent its further impact on the air sucked by the fan. Since this moment the wall temperature and ambient air temperature were recorded. Then the heat transfer coefficient distributions along the channel wall were estimated from the temperature dynamics using the technique described above.

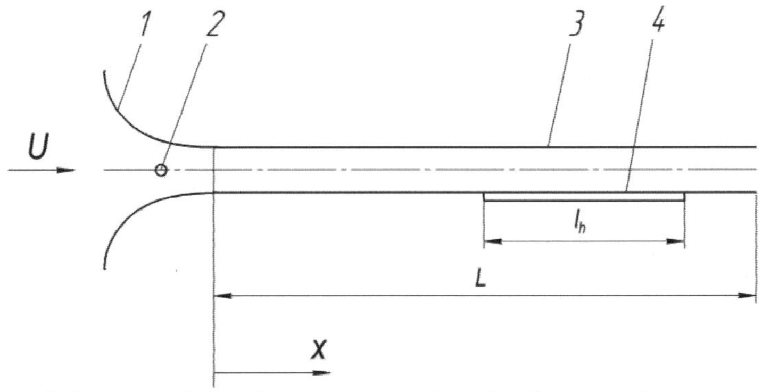

Figure 3. Schematic of experimental setup: 1 – inlet duct; 2 – thermometer; 3 – channel test section; 4 – heat-transfer section.

The studied channel was relatively short. Its length was L/d^*=9.2, where the equivalent diameter d^*=130 mm. Therefore the obtained experimental data should be considered as heat transfer coefficients of the channel entrance section. Test experiments were conducted for two mean freestream velocities $U=Q/F$=1.9 and 3.1, where Q is the volumetric flow rate through the channel; F is the channel cross-section area (figure 4). The considered Reynolds numbers were $\mathrm{Re}=U\,d^*/\nu=1.65\times10^4$ and 2.7×10^4. For reference, the dash lines show the distributions calculated from the following formula for the turbulent channel flow:

$$\mathrm{Nu}=0.021\mathrm{Re}^{0.8}\mathrm{Pr}^{0.43}\varepsilon_l,$$

IOP Conf. Series: Journal of Physics: Conf. Series **899** (2017) 022006 doi:10.1088/1742-6596/899/2/022006

where the Reynolds and Nusselt numbers were calculated using the equivalent diameter d^*, the correction for the entrance region ε_l was determined as a function of x/d^* and Re [15]. The corresponding calculated and experimental data descend streamwise and have close values of h. The calculated values are somewhat higher. This is due to the fact that the calculated heat transfer coefficients were averaged over the x length, while their local values were measured in the experiments. This being said, the calculated and experimental values of h tend to converge and become constant further downstream. Thus, the obtained test data agree with the classical concepts of heat transfer in the channel entrance region.

Figure 4. Heat transfer coefficient at mean freestream velocity of 1.9 m/s (1) and 3.1 m/s (2): symbols – experiment; lines – calculation.

4. Conclusions

The results show that the submitted method of heat transfer coefficient measurement is quite efficient. It can be used to study heat transfer in different channels with a plane wall. Making different (in terms of location and size) strips on the measurement plate, one can obtain the corresponding distributions of heat transfer coefficient on the wall.

References

[1] Zverev V G, Nazarenko V A, Pan'ko S V E and Teploukhov A V 2010 *High Temperature* **48** 741–46
[2] Yang Y C, Chen W L, Chou H M and Salazar J L 2014 *Numerical Heat Transfer, Part A: Applications* **66** 252–70
[3] Fernandes A P, dos Santos M B and Guimarães G 2015 *Applied Mathematical Modelling* **39** 6897–914
[4] Stepka F S 1983 *J. Heat Transfer* **105** 184–9
[5] Ekkad S V, Ou S and Rivir R B 2004 *ASME Turbo Expo 2004: Power for Land, Sea, and Air* (American Society of Mechanical Engineers) p 999-1005
[6] Trelewicz J R, Longtin J P, Hubble D O and Greenlaw R J 2015 *IEEE Sensors Journal* **15** 358–64
[7] Massonnet N, Carella A, Jaudouin O, Rannou P, Laval G, Celle C and Simonato J P 2014 *J. of Materials Chemistry C* **2** 1278–83
[8] Kotcioglu I, Cansiz A and Khalaji M N 2013 *Appl Therm Eng* **50** 604–13
[9] Jain D S, Rao S S and Srivastava A 2016 *Int. J. Heat Mass Transfer* **98** 697–711

IOP Conf. Series: Journal of Physics: Conf. Series **899** (2017) 022006 doi:10.1088/1742-6596/899/2/022006

[10] Grauso S, Mastrullo R, Mauro A W, Thome J R and Vanoli G P 2013 *Int. J. of Refrigeration* **36** 478–91
[11] Zumbrunnen D A 1990 *Experimental Thermal and Fluid Science* **3** 202–13
[12] Davletshin I A and Mikheev N I 2012 *High Temperature* **50** 412–19
[13] Lomascolo M, Colangelo G, Milanese M and de Risi A 2015 *Renewable and Sustainable Energy Reviews* **43** 1182–98
[14] Lykov A V 1967 *Theory of Heat Conduction* (Moscow: Vysshaya Shkola), in Russian
[15] Mikheev M A 1949 *Fundamentals of Heat Transfer* (Moscow: Gosenergoizdat), in Russian

Evaluation of the efficiency of using the swirl flow with the formation of helical vortices

Elizaveta S Gesheva[1], Sergey I Shtork[1,2] and Sergey V Alekseenko[1,2]

[1]Kutateladze Institute of Thermophysics, Siberian Branch of Russian Academy of Sciences, Lavrentyev Ave. 1, Novosibirsk, 630090 Russia

2 Novosibirsk State University, Pirogova Str. 2, Novosibirsk, 630090 Russia

gesheva_es@itp.nsc.ru

Abstract. The paper reports on the results of numerical simulation of large-scale stationary vortices forming in hydrodynamical model of a vortex combustor. In particular, the effect of boundary conditions on the residence time and path length of the fluid particles has been explored. Validity of the numerical procedure has been verified by comparison of the simulation results with the experimental data obtained by measuring the velocity fields using a PIV method. On the basis of the data obtained, it is shown that the average residence time of the fluid particles in the vortex chamber in the case of formation of a single-helix and a double-helical vortex is substantially higher than the residence time of the particles in the flow regime with a rectilinear vortex. The average distance traveled by the particles also increases noticeably in case of the flow regime with the spiral structure's formation. Thus, based on the work results, it is possible to draw a conclusion about the increased effectiveness of the use of operating modes with the formation of spiral vortex structures.

1. Introduction and experimental method

Using a swirling flow can significantly increase the efficiency of the combustion devices, accelerate the processes that occur in the combustion chambers and reduce the dimensions of the installations. As an initial stage of design of combustion chambers approximate isothermal modeling is used on scaled down laboratory models [1, 2]. In this paper the particle trajectories calculated using a numerical code were analyzed. The numerical procedure was verified by experimental data obtained with the aid of the PIV experiment and visualization of the vortices in a hydraulic model chamber with a square cross-section and a tangential inlet of the flow [3]. The scheme of the hydrodynamic bench is shown in figure 1. Water was used as a working fluid, and visualization was carried out with small air bubbles. The swirling flow was generated by means of twelve tangentially directed nozzles arranged in three layers and combined in four corner blocks. Figure 2 shows a diagram of the flow swirl, which intensity was determined using a design swirl parameter, calculated as $S = 12 \sin \alpha$. The angle α was equal to 45^0 for the flow with base and spiral vortex; and in the case of double structure, the turning angle of the nozzles α was equal to 30^0, that is in line with the values of $S = 8.5$ and $S = 6$, respectively. The height of the chamber from the bottom to the output orifice was 420 and 600 mm, depending on the position of the output diaphragm, the chamber width b is 188 mm and the diameter of the diaphragm orifice d_t equals to 60 mm. For the chamber geometry with a two-slope bottom two insert plates mounted at an angle of 50^0 to the horizon were used.

IOP Conf. Series: Journal of Physics: Conf. Series **899** (2017) 022007 doi:10.1088/1742-6596/899/2/022007

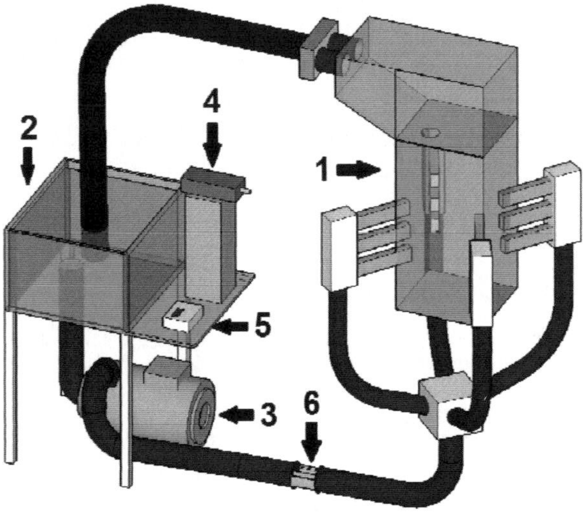

Figure 1. Scheme of the hydrodynamic stand: 1 – work area, 2 – water tank, 3 – pump, 4 – laser illumination, 5 – flowmeter, 6 – measuring section of the flowmeter.

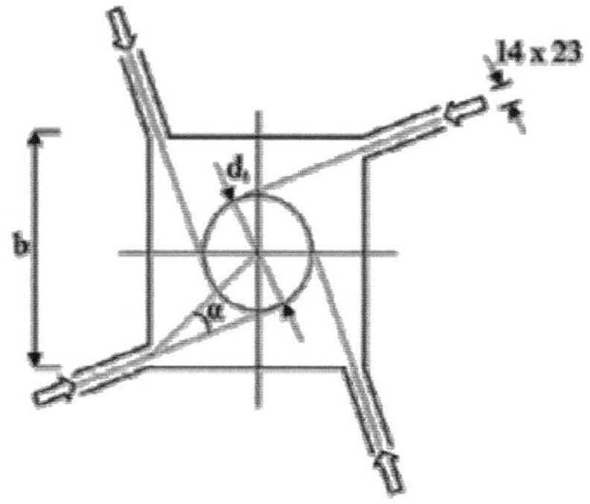

Figure 2. Diagram of the flow swirl in the chamber cross-section.

2. Visualization of the vortices and modeling results

In this model chamber various large-scale vortex structures were observed (figure 3) [4]. The shape of the vortex depends on the geometrical parameters of the chamber. At diaphragm with a symmetrical output orifice and a plane lower end the base rectilinear vortex is formed extending from the bottom of the chamber to the output orifice (figure 3, a).

When introducing asymmetry at the output, namely by shifting the output orifice, the base vortex is bended into a helical coil, thereby forming a spiral vortex (figure 3, b). For the double vortex to form,

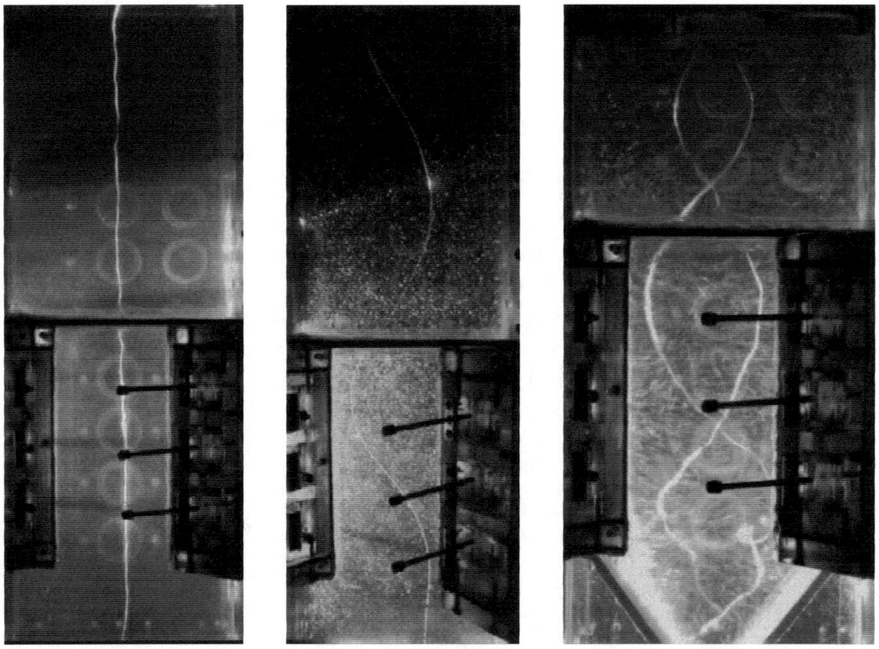

(a) (b) (c)

Figure 3. Visualization of vortex structures: a – base vortex, b – spiral vortex, c – double vortex.

IOP Conf. Series: Journal of Physics: Conf. Series **899** (2017) 022007 doi:10.1088/1742-6596/899/2/022007

there is a need in a two-slope bottom and a central diaphragm (figure 3, c). Thus, by varying the geometric parameters of the chamber, a family of different stationary vortex structures may be realized.

Numerical computations of the swirling flow were carried out using three basic modeling techniques: RANS, DES, and LES [5]. A mesh with polyhedral cells and a prismatic layer was built for the working chamber. The mesh was further refined in the vortex axis region. Thus, it consisted of 1.6 million cells, and the characteristic linear size of its cell was 3.4 mm. It was used the simulation grid with maximum possible resolution for given computer. Perhaps with a more detailed grid, the simulation results would be improved. Boundary conditions were set based on the fluid flow at the inlet and incompressibility of the fluid. In this study we used the "segregated flow" algorithm for the non-stationary implicit problem. The time step was set at 0.05, and about 15,000 iterations were calculated.

Verification of applied modeling techniques required comparison of the obtained fields and profiles of velocity with experimental data of the PIV experiment. This comparison of the fields and velocity profiles in different sections showed that the LES method the most accurately describes the pattern of the swirling flow in the working chamber figure 4). While the RANS and DES methods do not give an adequate representation of the vortex flow. Further we will consider the simulation results, obtained by LES.

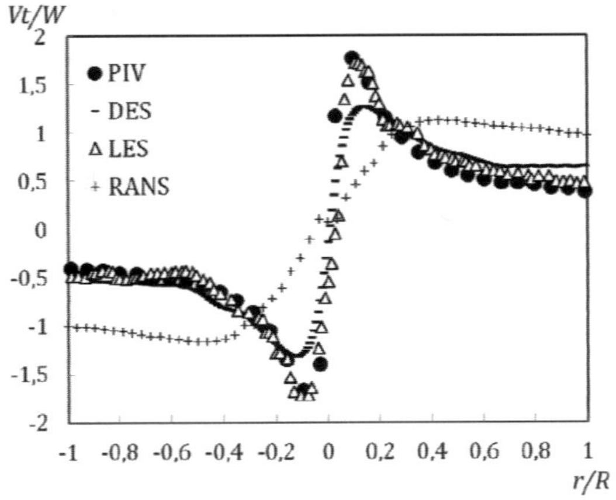

Figure 4. Profile of tangential velocity. Comparison of experimental data with results of numerical simulations.

To assess the effectiveness of the use of the flow swirling the residence time of particles released from the centers of all nozzles was calculated for each geometry. The numbering of the nozzle blocks and layers is shown in figure 5 (in the two-digit number of the nozzle *NM*, the first digit indicates the number of the nozzle block, and the second is for the layer number).

Examples of particle trajectories for base, spiral and double modes are shown in figure 6. It is seen that the lengths of trajectories for the modes with helical vortices are much larger than for the base vortex. This is proved by the analysis of the quantitative data from the table 1, which shows that in the case of the base vortex, the average residence time of particles in the chamber amounted to $T = 2.8$ s, whereas in the chamber with the shifted output orifice $T = 5.5$ s. The average path length also increased from $L = 2.63$ m to $L = 4.25$ m. Thus, the output orifice shift gives a significant increase in the residence time of particles in the chamber $T_{sp} / T_b = 5.485/2.809 = 1.953$ and in the travelled distance $L_{sp}/L_b = 4.251/2.631 = 1.513$ while the chamber height does not change ($H_{sp}/H_b = 1$). For the double mode, the ratio of the average residence time of particles in the chamber amounted to $T_{db} / T_b = 3.723/2.809 = 1.325$ (the ratio of the chamber heights at that was $H_{db}/H_b = 420/600 = 0.7$), and the ratio of the travelled distance $L_{db}/L_b = 2.86/2.631 = 1.09$. Even at a much lower chamber height in the case of a two-slope bottom, there is a noticeable increase in the average residence time of particles and in the travelled distance. In the case of the spiral vortex, the average residence time increases almost 2 times.

IOP Conf. Series: Journal of Physics: Conf. Series **899** (2017) 022007 doi:10.1088/1742-6596/899/2/022007

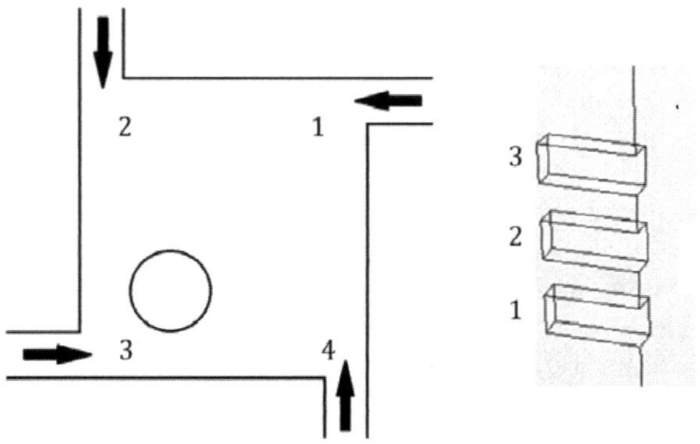

(a) (b)

Figure 5. Numbering of nozzle blocks in the layer N (a) (top view) and levels M (b). The circle indicates the position of the output orifice for the mode with a helical vortex.

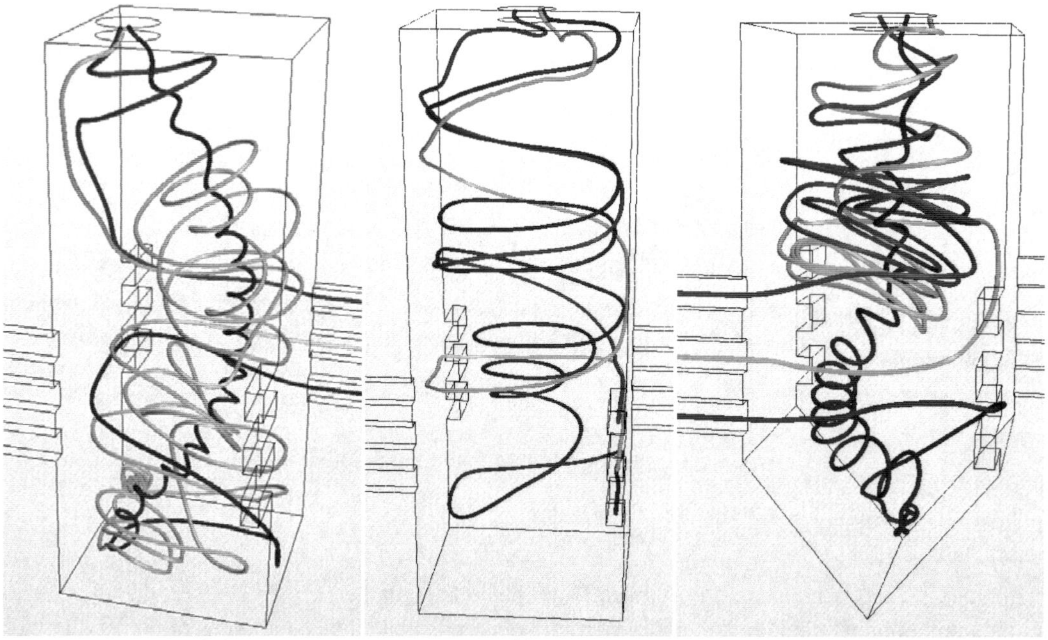

(a) *(b)* *(c)*

Figure 6. The trajectory of particles for different flow regimes: base vortex, b – spiral vortex, c – double vortex.

Table 1 below shows the values of travelled distance *L* and the residence time *T* of the particles released from the centers of the nozzles for the modes with base, spiral and double vortexes.

Thus, the study of the influence of output orifice shift and two-slope bottom on the residence time of particles in the combustion chamber and on the travelled distance confirms that the modes with helical vortex structures may be more optimal in terms of ensuring more complete combustion of the fuel.

Acknowledgments
The study was funded by the grant of the Russian Science Foundation (project No.14-29-00093).

IOP Conf. Series: Journal of Physics: Conf. Series **899** (2017) 022007 doi:10.1088/1742-6596/899/2/022007

Table 1. Values of travelled distance (path) *L* and residence time *T* in the chamber for the particles, released from the nozzle center for different flow regimes.

	base		spiral		double	
NM	*L*, м	*T*, c	*L*, м	*T*, c	*L*, м	*T*, c
11	4.62	4.6	2	1.63	4.5	5.89
12	4.75	5.3	6.2	8.84	3.85	5
13	1.1	1.36	0.84	1.02	3.27	4.47
21	3.4	4.43	7.35	10.63	2.35	2.92
22	1.57	1.91	5.13	6.31	2.05	2.55
23	1.4	1.65	2.65	4.56	2.31	3.29
31	2.8	1.53	1.7	0.72	2.2	2.36
32	3.4	3.3	5.1	5.67	2.8	3.86
33	0.84	1	7.1	10.08	2.8	3.75
41	4	4.19	7.9	10.98	2.5	3.02
42	2.5	3.39	3.8	4.35	2.9	3.82
43	1.2	1.03	1.25	1.03	2.8	3.73
mean	2.63	2.8	4.25	5.48	2.86	3.72

References

[1] Anufriev I S, Sharypov O V and Shadrin E Yu 2013 *Technical Physics Letters* **39** 466–8
[2] Alekseenko S V and Shtork S I 1994 *JETP Letters* **59** 746-50
[3] Gesheva E S, Shtork S I anf Alekseenko S V 2014 *NSU Bulletin. Series: Physics* **9**, 39–48
[4] Alekseenko S V, Kuibin P A, Okulov V L and Shtork S I 1999 *Journal of Fluid Mechanics* **382** 195-243
[5] Garbaruk A V, Strelets M Kh and Shur M L 2012 *SPb: Polytechnical University Press* 88

IOP Conf. Series: Journal of Physics: Conf. Series **899** (2017) 022008 doi:10.1088/1742-6596/899/2/022008

An investigation of a self-similarity for local vorticity and velocity components in tip vortex cores of a rotor wake

I.K. Kabardin[1], I.V. Naumov[1] and V.L. Okulov[1]

[1]Kutateladze Institute of Thermophysics, SB RAS, Novosibirsk 630090, Russia

E-mail: kabardin@itp.nsc.ru

Abstract. This experimental work is aimed at investigating the self-similarity of the local axial vorticity and azimuthal velocity in the tip vortices core in the wind turbine rotor wake. In the current work, the averaged local axial vorticity and azimuthal velocity profiles of the tip rotor vortex were measured. These profiles were tested to a self-similarity and compared with the Lamb-Oseen vortex.

1. Introduction

In recent years, the wind turbines application has grown to industrial use in wind farms. The interest to wind power tasks has increased significantly. The aerodynamics of the vortex wake behind a rotor of the wind turbines is essentially determined by dynamics of concentrated helical vortices which generated by tip of the rotor blades [1]. These helical vortices were the subject of many theoretical, experimental and numerical studies [2-9]. For example, in [10-11] the problem of the stability of a helical vortex system was considered analytically. In [12] it is shown that any two-dimensional vorticity distribution in the core tends to the Lamb-Oseen vortex. In numerical simulations of the rotor wake [13], an evolution of the tip helical vortex generated by a single-bladed rotor was tested to discover a self-similarity of the velocity and vorticity in the core of the helical vortex. The self-similar behavior of the vortex core was revealed in variables associated to the local azimuth angle changing along the center of the vortex core with an elliptical cross-section shape.

The current experimental work is aimed at investigating the self-similarity of the vorticity and axial velocity component in the core of the helical tip vortices in the rotor wake. For the first time, the azimuthally averaged vorticity profiles and the velocity component will be processed from our experimental velocity field measured behind rotor. The results will be compared with the Lamb-Oseen vortex.

2. Modelling

The profiles of velocity and vorticity components are self-similar if their profiles can be scaled and represented in the form:

$$U/U_s = f\left(r/\eta\right), \tag{1}$$

$$\omega/\omega_s = g\left(r/\eta\right). \tag{2}$$

Here U is axial velocity, ω is vorticity, r is radius, η is radial variable of similarity, f and g are some functions.

IOP Conf. Series: Journal of Physics: Conf. Series **899** (2017) 022008 doi:10.1088/1742-6596/899/2/022008

These profiles, although they are different in their amplitude and expansion, have a similar shape after the scaling. The radius, velocity, and vorticity, scaled by the suitable functions, reduce to the family of profiles describing by a single curve. The scaling profiles become independent on time and space [14].

In fluid dynamics, the Lamb–Oseen vortex models a vortex that decays due to viscosity. This vortex is named after Horace Lamb and Carl Wilhelm Oseen [15].The mathematical model for the flow velocity in the circumferential θ–direction in the Lamb–Oseen vortex is:

$$V_\theta(r,t) = \frac{\Gamma}{2\pi R}\left(1-\exp\left(-\frac{r^2}{r_c^2(t)}\right)\right), \tag{3}$$

where r is radius, $r_c(t) = \sqrt{4\nu t}$ is core radius of vortex, ν is viscosity, and Γ is circulation contained in the vortex. The associated vorticity distribution [16] of the vortex-filament is:

$$\omega_z(r,t) = \frac{\Gamma}{\pi r_c(t)^2}\exp\left(-\frac{r^2}{r_c^2(t)}\right) \tag{4}$$

An alternative definition is to use the peak azimuthal velocity in the vortex core rather than their total circulation:

$$V_\theta(r,t) = V_{\theta\max}\left(1+\frac{1}{2\alpha}\right)\frac{r_{\max}}{r}\left(1-\exp\left(-\alpha\frac{r^2}{r_{\max}^2}\right)\right), \tag{5}$$

where $r_{max} = \sqrt{\alpha}\, r_c(t)$ is the radius at which $V_{\theta max}$ is attained, and the number $\alpha = 1.25643$, see [16].

In the paper the terminology and designations like it was specified in the article [13]. The authors [13] defined a local coordinate system (fig. 1.) placed at the center of the single tip vortex (in the axial cross-section). Authors followed its evolution in a frame rotating with the blade (i.e., using the transformation $\theta = \Omega \cdot t$; $z = h \cdot \Omega \cdot t / 2\pi$, h being the helix pitch). In [13] u_ρ, u_Θ and u_ζ denote, respectively, the radial, azimuthal and axial velocities in the local coordinate system and w_ζ denotes the local vorticity normal to the axial plane (it corresponds to ω_θ in the global cylindrical coordinate system. With the transformation mentioned above the (locally) azimuthally (along Θ) averaged profiles of these different quantities are functions of ρ (the radial distance from the center of the considered vortex in the axial plane) and the time t.

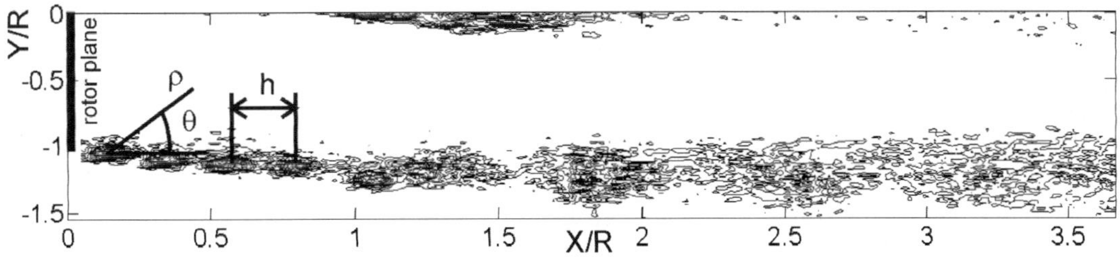

Figure. 1. The coordinate system with the center attached to a center of the tip vortex [13].

Let the averaged azimuthal velocity in the core and the azimuthal speed scale are correspondingly defined as:

$$U_\Theta = \frac{1}{2\pi}\int_0^{2\pi} u_\Theta d\Theta, \tag{6}$$

$$U_\Theta^s(t) = \max_{\rho \le h/2} U_\Theta(\rho,t). \tag{7}$$

The radial variable of similarity is defined as:

$$\eta = \rho / \rho_c \,, \tag{8}$$

where ρ_c is local radius at which U_Θ takes maximum.

$$U_\Theta(\eta,t) / U_\Theta^s(t) = \tilde{U}_\Theta(\eta) \,. \tag{9}$$

Therefore, the scaled value of $\tilde{U}$ as a function of η does not depend on t. Function $\tilde{U}_\Theta(\eta)$ is defined as:

$$\tilde{U}_\Theta(\eta) = \alpha(1 - e^{-\eta^2}) / \eta \,. \tag{10}$$

Azimuthally averaged vorticity (normal to the axial plane) in the vortex core is noted:

$$w = \frac{1}{2\pi} \int_0^{2\pi} w_\varsigma d\Theta \,. \tag{11}$$

The vorticity scale was defined as:

$$\omega_s(t) = \omega(\rho = 0,\, t). \tag{12}$$

The variable η given by equation (8) remains a similarity variable for the vorticity, so:

$$\omega(\eta,t) / \omega_s = \tilde{\omega}(\eta) \,. \tag{13}$$

Function $\tilde{\omega}(\eta)$ was defined as:

$$\tilde{\omega}(\eta) = \exp(-\eta^2) \,. \tag{14}$$

3. Experimental Methods and Results

A model of three-bladed rotor was specially manufactured for qualitative and quantitative flow visualizations downstream of the rotor [17]. The rotor diameter was $2R$ = 0.376 m, and the blades length was 0.159 m with the CD7003 blade profile [18]. The blade chord and angle of attack along span were calculated by the Glauert's theory for an optimal wind turbine [19] with the tip speed ratio (TSR) $\lambda = 5$, where $\lambda = \Omega R/V$, and Ω is angular speed of the rotor. The Reynolds number was calculated as $Re = \rho\Omega Rb/\mu$. Here ρ and μ are the density and dynamic viscosity of the working fluid (tap water), b is the length of the blade chord (1 cm). The Reynolds number for all experiments was about 20 000 at the working temperature of 20° C in the water flume. The water flume had a length of 35 m, a width of 3 m and the operative height was 0.9 m. A 3-m long test section with transparent walls of optical resolution was installed at a distance of 20 m from the channel inlet; the test section walls and bottom were made of glass.

The flow was studied by the Dantec Stereoscopic 2D-3C PIV system which gives all three velocity components throughout widow of the light sheet [20-21]. An Nd:YAG laser was used as a light source with the following characteristics: 120 mJ of energy in a single pulse, the wavelength is 532 nm, operation frequency is 15 Hz. The 2 mm thick light sheet was sent in vertically into the channel from the bottom and directly aiming at the rotor axis (see Fig. 2a). The images were recorded by two Dantec HiSense II cameras with 1344 ÷ 1024 pixels resolution. The 3D velocity field was calculated using Dantec Dynamic Studio 2.21. The cameras were placed perpendicularly to each other on the different sides of the flume and at angle of 45° to the walls (Fig. 2b). Water-filled optical prisms were installed between the cameras and the test section to reduce the distortions having from the camera inclination to the wall. Since the cameras were placed at an angle to the light sheet, the focus plane was adjusted using Scheimpflug adapters. The experimental errors has been minimized during calibration and test experiments by finding a trade-off between width of the light sheet, and the time between pulses, and the interrogation area. The stereoscopic PIV system was calibrated using a target with a well-defined dot-pattern which was translated and registered by the cameras in a number of well-defined positions

IOP Conf. Series: Journal of Physics: Conf. Series **899** (2017) 022008 doi:10.1088/1742-6596/899/2/022008

at the light sheet. The measuring error of stereoscopic PIV was at the level of 3÷5 %. The final size of the total 3D velocity field was 1.03·0.29 m.

For each measuring window the ultimate velocity field was obtained by phase averaging of 100 realizations, which were recorded in the moment of a triggered signal by a light pulse per the one revolution of the rotor. The angular encoder ROTACAM ASR58 with angular resolution 0.1° installed on the rotor hub has formed the triggered pulse when one of the blades passed through the light sheet. The stochastic errors vanish by this phase averaging. Moreover, this approach eliminates the drifting error due to non-stationary flow regimes too.

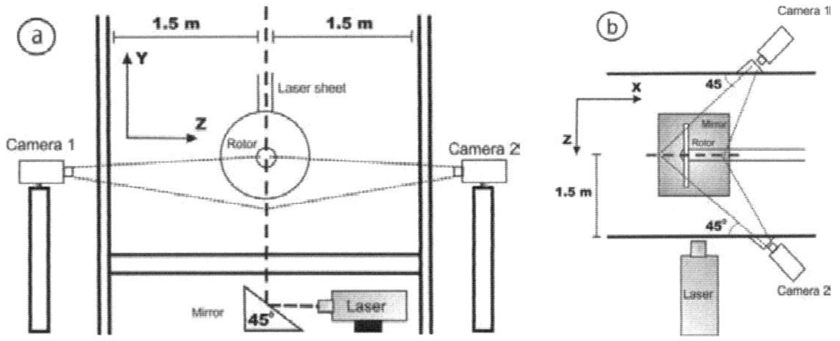

Figure 2. Sketch of stereoscopic PIV measurements in the transverse cross section and vortices visualization.

The processing of the images has been resulted in three uniform zones to yield the local velocity field. The angles from which the synchronization was done were α = 0,15,30,45,60,75,90,105 degrees of the rotor blade rotation. The more description of the experimental setup and the experimental method can be found in[22-25].Figure 3 shows the phase averaged distributions of three velocity components for two regimes with a velocity of $\lambda = 5$.

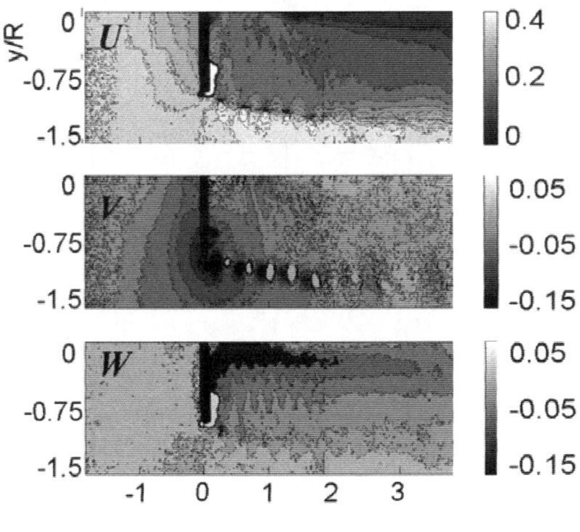

Figure 3. The iso-contours of axial (*U*), radial (*V*), tangential (*W*), velocity components for TSR $\lambda = 5$.

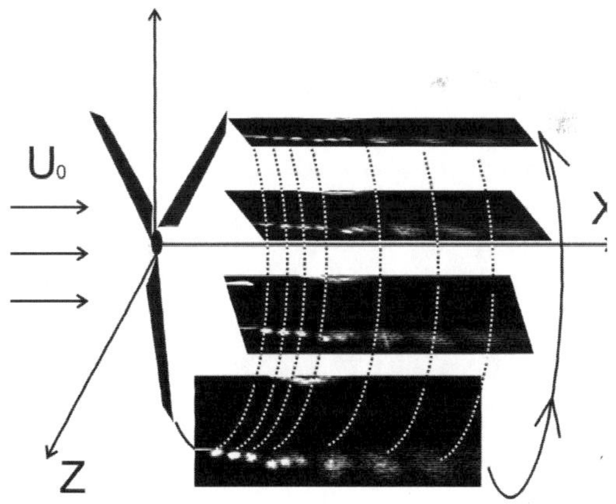

Figure 4. The evolution of tip vortices in different wake images.

The distributions of the components of the axial and radial velocity were measured in the different planes of the laser knife (Fig. 4) and also the vorticity field was calculated from the measured velocity fields. The cross sections of the tip vortex cores are well fixed(Figs. 3-4) which make it possible to determine the centers of the tip vortices in the cross-sections. The regular vortex structure destroyed for x between 1.8 R to 3.6 R and pairing with nearest turns takes place behind 3 R. If each cross-section of tip vortex cores is numbered from 1 to 9, then the vortex cores *No.* 1, 4, 7 correspond to the first blade, cores *No.* 2, 5, 8 correspond to the second blade, and cores *No.* 3, 6, 7 correspond to the third blade. It can be seen that diffusion of tip vortices is take place. First 6 core images have strong concentration of vorticity and located approximately at equal distances, and the vorticity concentration in the cores *No.* 7, 8, 9 became smaller and placed at much more distances than the first 6 ones.

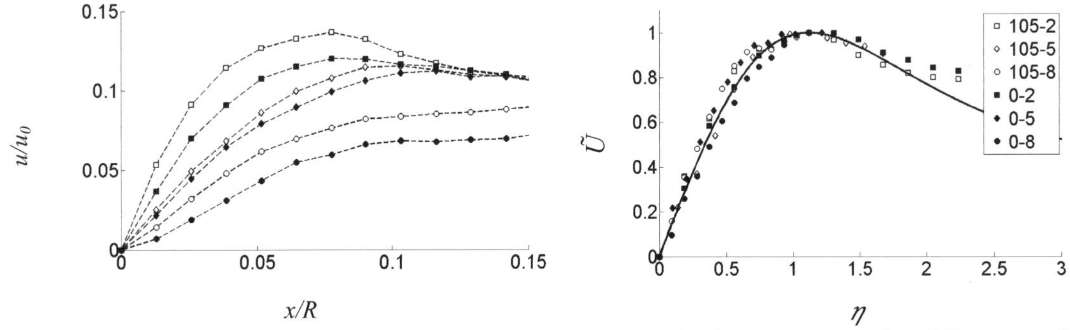

Figure 5. *In left*: Original profiles of the azimuthal velocity in the vortex core for different angles along the helical vortex. *In right*: The same profiles are scaled by equation (9). Equation (10) is plotted by solid line.

At first step the self-similarity of the local azimuthal velocity profiles were investigated. The angles 0 ° and 105 ° were taken as an interval of the changing the azimuth angle along the helical vortex. The numbers of the cross-sections of the vortex core were taken 2, 5, and 8. The velocity profiles are presented on fig. 5. The original profiles of the local azimuthal velocity in the vortex core cross-sections of different rotational angles are presented on the left plot. The profiles scaled using equation (9) is presented in the right plot. The comparison of the profiles shows a good agreement of the self-similarity for the azimuthal velocity in the core.

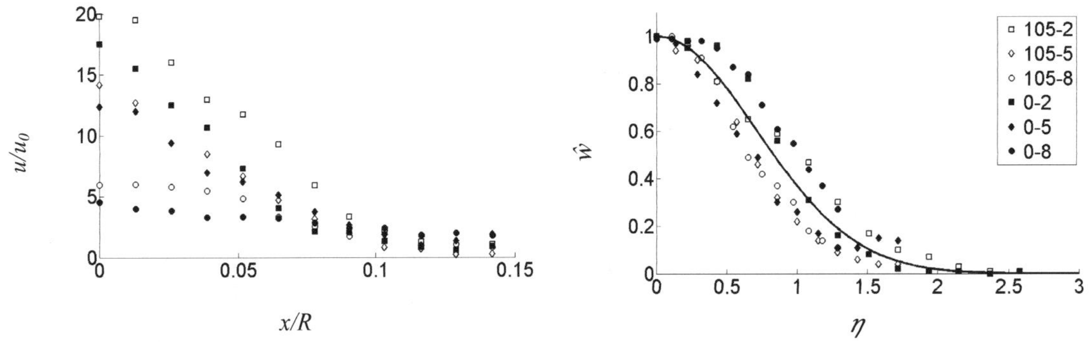

Figure 6. *In left*: Profiles of the original vorticity in the vortex core for different for different angles along the helical vortex. *In right*: The same profiles are scaled by equation (13). Vorticity of a Lamb-Oseen vortex (equation (14)) is plotted by solid line.

The original vorticity profiles in the tip vortex of the rotor operating with tip speed ratio $\lambda = 5$ is presented on figure 6 (left). The same profiles scaled using equation (13) are presented on the right side. The comparison of the profiles indicates much more inclination from the self-similarity as it was found for the velocity which can be explained by the additional error of the processing the vorticity from the measured velocity. It must be also mentioned that for the local velocity profiles the comparison with the Lamb-Oseen vortex is rather good, though, there is a little discrepancy in region $\eta = 1.8$-2.4. For the local vorticity we could not see the same perfect agreement with the Lamb-Oseen vortex, and the discrepancy of experimental points and the solution is not so small. This fact can be also explained by the additional errors of the vortex field reconstructions. So in further experiments it is necessary to measure velocity fields with less error and better spatial resolution.

4. Conclusions

The near wake flow with the strong tip helical vortices behind the three-bladed rotor has been studied by PIV measurements. The local azimuthal vorticity profiles and the local axial velocity component were investigated to discover the self-similarity in the tip vortex cores and also the good correlation with the Lamb-Oseen vortex was found.

Acknowledgments
Research was supported by the Russian Science Foundation (Project № 14-19-00487).

References

[1] Okulov V L, Sørensen J N and Wood D H 2015 *Prog. Aerospace Sci.* **73** 19-46.
[2] Fukumuto Y, Okulov V L and Wood D H 2015 *Appl. Mech. Rev.* **67** (060801) 1-7
[3] Alekseenko S V, Kuibin P A and Okulov V L 2007 *Theory of Concentrated Vortices: An Introduction* (Springer, Berlin) 400
[4] Wood D H, Okulov V L and Bhattacharjee D 2016 *Renew. Energy* **95** 269-276
[5] Fukumoto Y, Okulov V L and Wood D H 2015 *ASME. Appl. Mech. Rev.* **67** (6) 060801
[6] Okulov V L , Sørensen J N and Shen W Z 2016 *J. Phys.: Conf. Ser.* **753** 022018
[7] Kuik G A M van, Sørensen J N and Okulov V L 2015 *Prog. Aerospace Sci.* 2015. Vol.73. p.1-18.
[8] Sørensen J N, Okulov V L, Mikkelsen R F, Litvinov I V and Naumov I V 2016 *J. Phys.: Conf.* Ser. **753** 022020
[9] Okulov V L 2014 *J. Phys.: Conf. Ser.* **524** 012129
[10] Okulov V L 2004 *J. Fluid Mech.* **521** 319-342
[11] Okulov V L and Sørensen J N 2007 *J. Fluid Mech.* **576** 1-25
[12] Wayne E 2011 *Notice of the AMS* **58(**1) 10-19
[13] Ali M and Abid M 2014 *J. Fluid Mech.* **740** (R1) 1 - 11
[14] Reynolds A J 1974 *Turbulent flows in engineering* (John Wiley and Sons, London) 408
[15] Wu J Z, Ma H Y, Zhou M D 2006 *Vorticity and Vortex Dynamics*(Berlin: Springer-Verlag) 119
[16] Devenport W J, Rife M C, Liapis S I, Follin G J, 1996 *J. Fluid Mech.* **312** 67–106
[17] Okulov V L, Naumov I V, Mikkelsen R F, Kabardin I K and Sørensen J N 2014 *J. Fluid Mech.* **747** 369-380
[18] Okulov V L, Naumov I V, Sørensen J N 2007 *Technical Physics***77** 47-57.
[19] Naumov I V, Mikkelsen R F, Okulov V L and Sorensen J N 2014 *J.Phys.:Conf. Ser.* **524** 0121686
[20] Okulov V L, Naumov I V, Kabardin I K, Mikkelsen R F, Sørensen J N 2014 *J. Phys.: Conf.* Ser. **555**(1) 1-12.
[21] Naumov I V, Kabardin I K, Mikkelsen R F, Okulov V L and Sørensen J N 2016 *J. Phys.: Conf.* Ser. **753** 032051
[22] Okulov V L, Mikkelsen R F, Naumov I V, Litvinov I V, Gesheva E. and Sørensen J N 2016 *J. Phys.: Conf.* Ser. **753** 032060
[23] Litvinov I V , Naumov I V , Okulov V L , Mikkelsen R F 2015 *J. Flow Visualiz. and Img. Proc.* **22**(4) 175-183
[24] Okulov V L, Naumov I V, Mikkelsen R F and Sørensen J N 2015 *J. Phys.:Conf. Ser.***625** 012011
[25] Okulov V L, Mikkelsen R F, Sørensen J N, Naumov I V and Tsoy M A 2017 *J. Energy. Res. Tech.* **139**(5) 051210

Vortex Breakdown under Laminar Flow of Pseudoplastic Fluid

A I Kadyirov[1] and B R Abaydullin[1]

[1]KazanSC of RAS, Kazan 420111, Lovachevsky 2, Russia

Email: aidarik@rambler.ru

Abstract. The numerical investigation was carried out to study vortex breakdown for pseudoplastic fluid flow in circular pipe with twisted tape inserts. 0.67%, 1.5% and 3% aqueous solutions of Na-CMC are chosen as a pseudoplastic fluid. The numerical results are compared with available data in literature.

1. Introduction

It is well known that using the swirl inserts is an effective method to enhance the heat and mass transfer [1-2]. Both experimental and theoretical studies conducted by Yu.G. Nasmeev's revealed their effectiveness to laminar flow of nonlinear viscous media too [3]. The periodically located twisted tape inserts have been recently intensively studied [4-5]. In this case it is important to investigate the effective distance between the inserts, including the mechanism of vortex breakdown. Lucca-Negro O. observed some existing results devoted to this problem [6]. Most of represented reports are devoted to investigation of vortex breakdown under laminar flow of Newtonian fluid.

2. Problem definition

The stationary laminar flow of pseudoplastic fluid in circular pipe with inserts is conducted (Fig. 1). It is assumed that thermophysical properties of fluid such as specific heat capacity, density and thermal conductivity are independent on temperature; the mass forces are negligible.

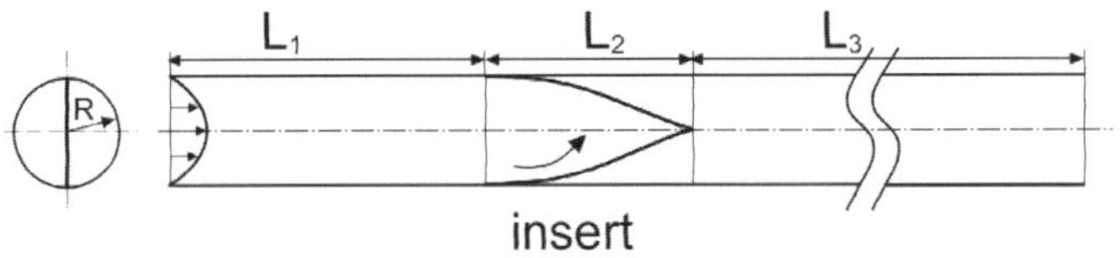

Figure 1. Problem definition-geometrical configuration

Three kinds of twisted tapes are used as screw inserts (fig. 2) with various twist ratio s/d=3…5, $L_2 = 0.5 \cdot s/d$; thickness 0.01 (m).

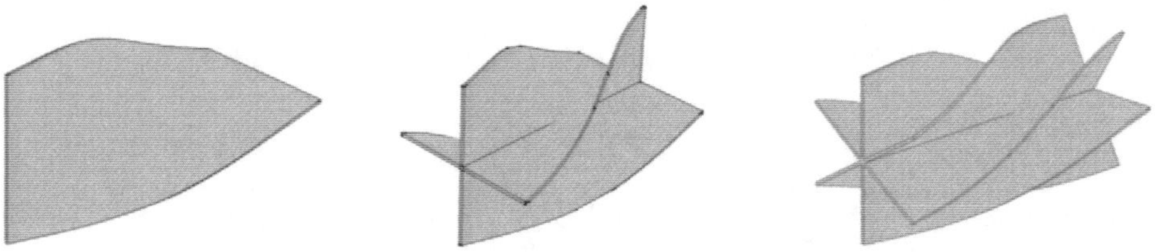

(a) (b) (c)

Figure 2. Geometry of twisted tapes

Numerical results are provided by COMSOL Multiphysics® with following governing equations:

$$\rho\left(\vec{V}\cdot\nabla\right)\vec{V}=\nabla\cdot\left(-p\mathbf{I}+\mu\left(\nabla\vec{V}+\left(\nabla\vec{V}\right)^{T}\right)\right) \quad (1)$$

$$\nabla\cdot\vec{\mathrm{V}}=0 \quad (2)$$

p – pressure, $\mathbf{I}$ - unit tensor, μ - dynamic viscosity, ρ - fluid density, $\vec{V}=\{u,v,w\}$ - vector of velocity.

Boundary conditions are follows:

1) inlet velocity

$$w=V_a\left(\frac{3n+1}{n+1}\right)\left[1-\left(\frac{r}{R}\right)^{(n+1)/n}\right] \quad (3)$$

here V_a - average velocity, r - pipe radius, n – rheological constant.

2) It is assumed a free flow of fluid at the outlet of pipe;
3) no slip flow conditions are taken at other boundaries of pipe.

The Power low model is used to predict the rheological properties of pseudoplastic fluid [7]

$$\mu=k\left|\dot{\gamma}\right|^{n-1}, \quad (4)$$

k, n – rheological constant; $\dot{\gamma}$ - shear rate.

Based on literature review the rheological and thermophysical properties of aqueous solutions of Na-CMC are listed in tables 1, 2. Fig.2 illustrates dependence of dynamic viscosity on shear rate for some fluids. Verification of numerical results is approved by comparison with experimental results for following problem: a laminar flow of water in a circular pipe with full length tape insert. Geometric parameters are follows: $D=2R$=0.025 (m); tape insert thickness 0.001 (m), twist ratio s/d=4 (s – a length with a twist angle of 180 degrees), the length of pipe with full length tape insert is 1 (m). Thermophysical properties are taken at 293 K [9]. Uniform velocity profile is used as inlet velocity. Fig. 4 shows good agreement, relative tolerance is less than 4%.

Table 1. Rheological properties of fluid at 293 K

Fluid	Power low
0.67% Na-Cmc	k=0.2041 ($\mathrm{Pa}\cdot\mathrm{s}^{\mathrm{n}}$), n=0.716 [7]
1.5% Na-Cmc	k=2.1011 ($\mathrm{Pa}\cdot\mathrm{s}^{\mathrm{n}}$), n=0.554 [7]
3% Na-Cmc	k=6.2515 ($\mathrm{Pa}\cdot\mathrm{s}^{\mathrm{n}}$), n=0.566 [7]

IOP Conf. Series: Journal of Physics: Conf. Series **899** (2017) 022009 doi:10.1088/1742-6596/899/2/022009

Table 2. Thermophysical characteristics of fluid at 293 K

Fluid	Thermophysical characteristics based on [8]
0.67% Na-Cmc	$\rho = 1014\ (kg/m^3);\ c_p = 3591\ (J/kg\cdot K);\ \lambda = 0.5618\ (W/m\cdot K)$
1.5% Na-Cmc	$\rho = 1025\ (kg/m^3);\ c_p = 3560\ (J/kg\cdot K);\ \lambda = 0.56\ (W/m\cdot K)$
3% Na-Cmc	$\rho = 1034\ (kg/m^3);\ c_p = 3315\ (J/kg\cdot K);\ \lambda = 0.5535\ (W/m\cdot K)$

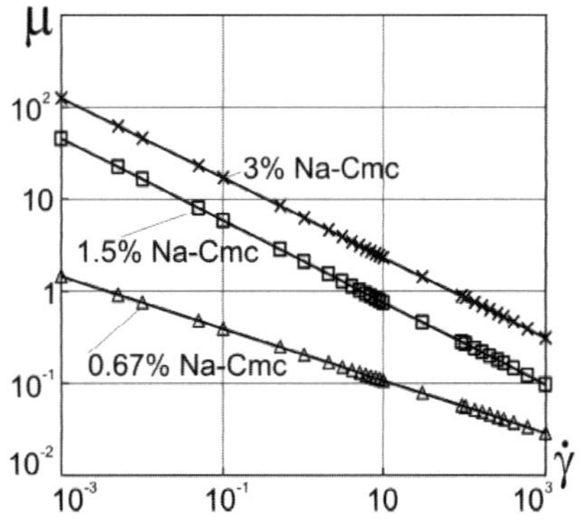

Figure 3. Dynamic viscosity versus shear rate (date from [7])

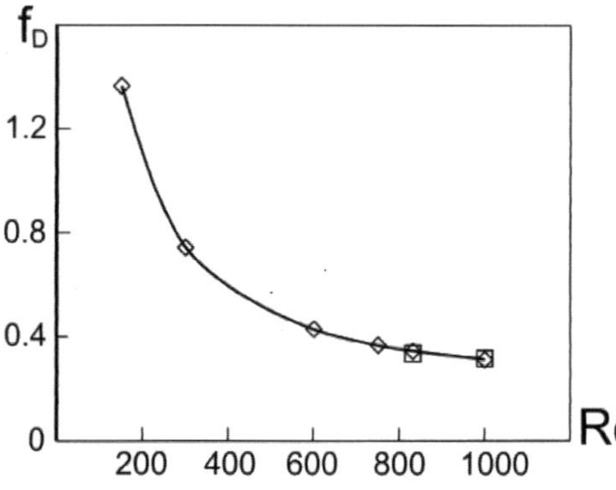

Figure 4. Darcy friction factor versus Reynolds number:
◇ – COMSOL Multiphysics®,
◇ □ – experimental data [10]

3. Numerical results

It is well known that Dean vortices appears in swirl flow. Fig.5 shows relative velocity profiles V_φ / V_z in the cross section of the pipe, where Reynolds number is calculated by [7]

$$\mathrm{Re}_n = \frac{D^n V_a^{2-n} \rho}{k' 8^{n-1}}, \qquad (5)$$

$k' = k\left(\frac{3n+1}{4n}\right)^n$, k and n are rheological constants (see table 1).

The analysis of the numerical results showed that the vortices of a swirling flow are most pronounced at distance $D^+ \le D^+_{кр}$, where $D^+_{кр}$ is a relative length of pipe (amount of diameters) calculated in downstream direction after insert while $\sqrt{v_x^2 + v_y^2} \ge 0.2 V_a$. Furthermore dependences of $D^+_{кр}$ on Re for water and various aqueous solutions of Na-Cmc are the same if Re number is calculated based on formula (5), fig. 6.

Fig. 7 shows that increasing parameter of twist ratio (*s*/*d*) leads to longer distance $D^+_{кр}$ comparing with the case when insert consists double or triple tapes.

IOP Conf. Series: Journal of Physics: Conf. Series **899** (2017) 022009 doi:10.1088/1742-6596/899/2/022009

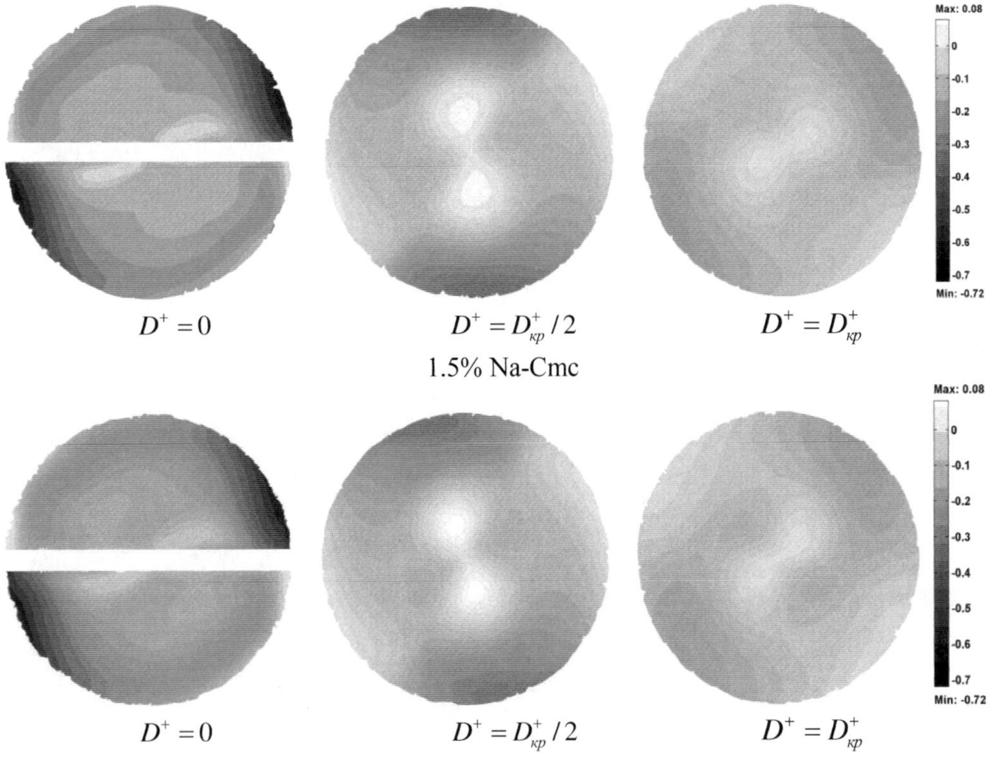

Figure 5. Relative velocity profiles (V_φ / V_z) in cross-section area of pipe: $\mathrm{Re_n} = 701.5$, $D^+ = 7.24$ (geometry of insert fig.2a with *s*/*d*=4)

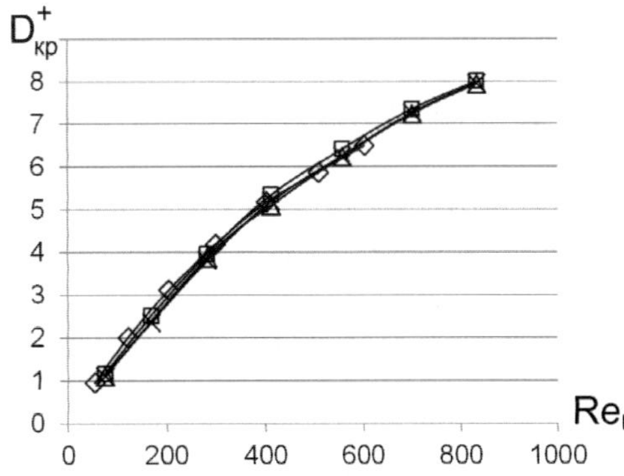

Figure 6. Relative distance $D^+_{кр}$ versus $\mathrm{Re_n}$ number:
water; 0.67% Na-Cmc;
1.5% Na-Cmc; 3% Na-Cmc;
(geometry of insert fig. 2a with s/d=4)

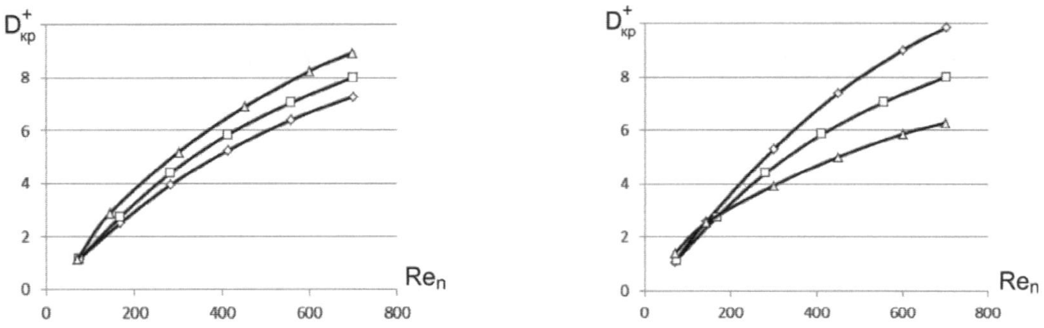

(a) (b)

Figure 7. Relative distance $D^+_{кр}$ versus $\mathrm{Re_n}$ number: (a)-◇- geometry of fig. 2a, -□- geometry of fig. 2b; -△- geometry of fig. 2c; *s*/*d*=4; (b) -◇-s/d=3, -□- s/d=4, -△- s/d=5 for geometry fig. 2b

4. Summary

The paper represents the numerical results of effective distance while $\sqrt{v_x^2 + v_y^2} \geq 0.2V_a$. It is obtained that dependences of $D^+_{кр}$ on $\mathrm{Re_n}$ for various aqueous solutions of Na-Cmc can be described with one relationship.

Acknowledgments
The reported study was funded by RFBR № 16-31-00425 mol_a

References

[1] Mitrofanova OV 2003 Hydrodynamics and heat exchange under swirl flows in channels with twisted tapes (Review) *High temperature* **41**(4) 587-633.

[2] Mohsen S, Mofid GB, Davood DG 2015 Review of heat transfer enhancement methods: Focus on passive methods using swirl flow devices *Renewable and Sustainable Energy Reviews* **49** 444-469.

[3] Nazmeev YuG, Nikolaev NA 1979 Estimation of efficiency of using twisted tapes to enhance heat transfer *Engineering and physical journal* **36**(4) 145-152.

[4] Saha SK, Dutta A, Dhal SK 2001 Friction and heat transfer characteristics of laminar swirl flow through a circular tube fitted with regularly spaced twisted-tape elements *International Journal of Heat and Mass Transfer* **44** 4211-4223.

[5] Eiamsa-ard S, Thianpong Ch, Promvonge P 2006 Experimental investigation of heat transfer and flow friction in a circular tube fitted with regularly spaced twisted tape elements *International Communications in Heat and Mass Transfer* **33** 1225–1233.

[6] Lucca-Negro O, O'Doherty T 2001 Vortex breakdown: a review *Progress in Energy and Combustion Science* **27** 431481.

[7] Wilkinson WL 1964 *Non-Newtonian fluids* (Mir, Moscow)

[8] Kadyirov AI, Vachagina EK 2013 The influence of swirl flow on flow characteristics in curved tube *Transactions of Academenergo* **2** 7–16.

[9] http://www.highexpert.ru/content/liquids/water.html

[10] Wongcharee K, Eiamsa-ard S 2011 Friction and heat transfer characteristics of laminar swirl flow through the round tubes inserted with alternate clockwise and counter-clockwise twisted-tapes *International Communications in Heat and Mass Transfer* **38** 348–352.

IOP Conf. Series: Journal of Physics: Conf. Series **899** (2017) 022010 doi:10.1088/1742-6596/899/2/022010

Initial temperatures effect on the mixing efficiency and flow modes in T-shaped micromixer

A S Lobasov[1,2,3] and A A Shebeleva[1]

[1]Siberian Federal University, 660074, Krasnoyarsk, Russia
[2]Institute of Thermophysics SB RAS, 630090, Novosibirsk, Russia
[3]Novosibirsk State University of Architecture and Civil Engineering (Sibstrin), 630008, Russia

perpetuityrs@mail.ru

Abstract. Flow patterns and mixing of liquids with different initial temperatures in T-shaped micromixers are numerically investigated on the Reynolds number range from 1 to 250. The temperature of the one of mixing media was set equal to 20°C, while the temperature of the another mixing media was varied from 10°C to 50°C; its effect on the flow structure and the mixing was studied. The dependences of the mixing efficiency and the pressure difference in this mixer on the difference in initial temperatures of miscible fluids and the Reynolds number were obtained. It was shown that the presence of a difference in initial temperatures of miscible fluids leads to a shift of flow regimes and the flow and mixing of two fluids with different initial temperatures can be considered as self-similar pattern with regard to the reduced Reynolds number.

1. Introduction

The miniaturization of technological processes has been actively promoted in recent years in the physical and chemical industry, and thus micro-mechanics has become a rapidly developing and promising research area. Microchannel devices are widely used in various fields of science and technology as microreactors, micro-scale heat exchangers, micro-mixers, etc. Many studies have noted that the use of microdevices allows significantly enhancing the physicochemical processes in comparison to classical space consuming reactors [1-3]. Control over pressure, temperature, reaction time and flow velocities in reactors with small volumes is now realized much easier and more efficiently. This implies the main undeniable advantages of microreactor microsystems: safety of highly exothermic reactions and work with toxic or explosive reagents, possible reactions in supercritical conditions, significantly reduction of research costs, as well as implementation and scaling of chemical processes. In this paper a T-shaped micromixer, which is one of the simplest in the manufacture, but at the same time, quite effective form of a microfluidic mixers was considered. Thus, flow regimes depending on the Reynolds number in such micromixers were investigated in many works [4,5]. At that, quite an interesting hydrodynamic phenomenon, namely the flow reversal or engulfment flow regime [5-8] upon reaching the critical Reynolds number equal to 130-160 was observed in the T-shaped mixer. The aim of the present work is a systematic study of the effects of difference in the initial temperatures of miscible fluids on the flow and mixing regimes in the T-shaped micromixer.

2. Mathematical model and numerical algorithm

We consider incompressible flows of multi-component Newtonian fluids, which are described using a hydrodynamic approach based on the solution of the Navier-Stokes equations. Currently, numerous experiments show that such description for fluids works well up to the channel size of 1 micron.

In general, the Navier-Stokes equations system has the following form:

$$\frac{\partial \rho}{\partial t} + \nabla(\rho \mathbf{v}) = 0$$

$$\frac{\partial \rho \mathbf{v}}{\partial t} + \nabla(\rho \mathbf{v}\mathbf{v}) = -\nabla p + \nabla \mathbf{T} \quad (1)$$

Here ρ is the fluid density, p is the pressure, $\mathbf{v}$ is the velocity, and $\mathbf{T}$ is the tensor of viscous stresses. The evolution of mass concentration is determined by the equation:

$$\frac{\partial \rho f}{\partial t} + \nabla(\rho f \mathbf{v}) = \nabla(\rho D \nabla f) \quad (2)$$

where D is the diffusion coefficient and f is the mass fraction of the passive component.

The energy conservation equation is considered as follows:

$$\frac{\partial}{\partial t}(\rho E) + \nabla(\mathbf{v}(\rho E + p)) = \nabla(\lambda \nabla T) \quad (3)$$

where $E = h - p\rho^{-1} + 0.5v^2$, λ is the thermal conductivity coefficient and h is the enthalpy, which determined by equation:

$$h = \int_{T_0}^{T} C_P dT \quad (4)$$

The mixture temperature T is calculated in each point from equation (4) by using the mixture enthalpy h found by equation (3). The viscosity coefficient and the density of the mixture as well as the thermal conductivity coefficient and the specific heat capacity are dependent on the temperature and are determined by using the well-known thermophysical properties [9-11] of pure water. The specific heat capacity of the pure water is defined in the form of a fourth-degree polynomial of temperature.

In our calculations we used a three-block grid consisting of 9.5 million grid nodes. Preliminary calculations have shown that this level of grid particularization is acceptable in terms of calculation accuracy. The fluid flow rate determined at the inlet of mixer channels was constant with steady-state velocity profile. At the mixing channel outlet, Neumann conditions were set, meaning that the derivative normal to the outlet surface, taken of all scalar quantities, is equal to zero. The walls of the mixer were considered as insulated. For velocity vector components, the no-slip condition was taken as boundary condition on the channels walls. The applicability of this type of boundary conditions for channels with a size of 50 μm had been demonstrated in [1,4]. The channel dimensions are shown in figure 1. The channel thickness is 200 μm, while the width of its narrow and wide parts is 200 μm and 400 μm, respectively. In our calculations we determined the pressure drop between each of the mixer inlets and the outlet, as well as mixing efficiency. In the literature, mixing efficiency is usually quantified using the parameter $M = 1 - (\sigma/\sigma_0)^{1/2}$, where $\sigma = V^{-1}\int_V (f - \langle f \rangle)^2 dV$ is the root-mean-square deviation of the mass fraction of mixture component f from its average value $\langle f \rangle$, $\sigma_0 = \langle f \rangle \cdot (1 - \langle f \rangle)$ is the maximum root-mean-square deviation, and V is the volume of the computational domain.

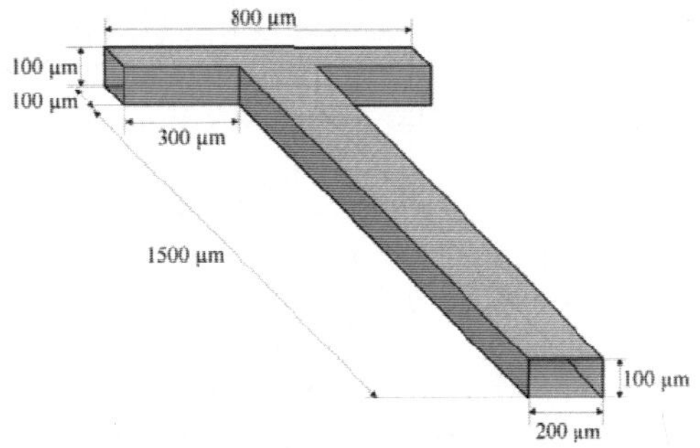

Figure 1. Geometric configuration of the task.

3. Results and discussion

The thermal mixing of Newtonian fluids in T-shaped micromixer was studied. The effect of different initial temperatures of the fluids at the inlets on the flow regimes was investigated. The temperature of one of the fluids was set equal to 20°C, while the temperature of other fluid tinted with rhodamine varied and was equal to: 10°C, 30°C, 40°C, and 50°C. Pure water with the well-known temperature-dependent thermophysical properties [9-11] was used.

The change in the flow regimes in the microchannel is characterized by the Reynolds number, determined as $\mathrm{Re} = \rho U d_h/\mu$, where $U = Q\,(2\rho H^2)^{-1} = Q_{in}\,(\rho H^2)^{-1}$ is the superficial velocity in the mixing channel, $H = 200$ μm is the channel height, $d_h = 267$ μm is the hydraulic diameter.

Fluids have different thermal properties at various fluid temperatures at the channel inlet. Mainly the mixing mode of two fluids in micromixer will be influenced by the fluid viscosity and density. As shown in [5], increasing the viscosity of one of the fluids shifts the region of transition from steady symmetrical vortex flow to the engulfment regime and increasing the density of one of the fluids causes the appearance of asymmetry in the flow of this fluid relative to the other. Both of these effects manifest themselves in the case of mixing of two fluids with different temperatures at the inlet to micromixer. The fluid with a greater viscosity starts gradually "flowing around" the fluid with a lower viscosity, as it was shown for the case of different viscosities. However, since densities of fluids are also different, therefore, with the increase in Reynolds number, the fluid with a higher density starts in turn "flowing around" the fluid with a lower density, as was shown above for the case of different densities. In addition, it was shown that the beginning of the engulfment regime depends only on the viscosity ratio and does not depend on the density ratio. Further, with the increase in Reynolds number the engulfment flow regime begin and continue to develop. This is accompanied by increase of the vortices size, the interface area and, accordingly, the mixing efficiency.

Furthermore, the difference in the fluids densities may cause some changes in flow and mixing patterns due to the influence the natural convection in non-zero gravity case. There is the dimensionless number that expresses the ratio of the buoyancy term to the flow shear term and called the Richardson number (Ri). In thermal convection problems, Richardson number represents the importance of natural convection relative to the forced convection. The Richardson number in this context is defined as $\mathrm{Ri} = g\beta(T_h - T_c)LU^{-2}$, where g is the gravitational acceleration, β is the thermal expansion coefficient, T_h is the hot fluid temperature, T_c is the cold fluid temperature, L is the characteristic length, and U is the characteristic velocity. The Richardson number can also be expressed by using a combination of the Grashof number and Reynolds number, $\mathrm{Ri} = \mathrm{Gr}\,\mathrm{Re}^{-2}$. The Grashof number (Gr) is a dimensionless number in fluid dynamics and heat transfer which approximates the ratio of the buoyancy to viscous force acting on a fluid. For pipes and channels the Grashof number is defined as $\mathrm{Gr} = g\beta(T_h - T_c)L^3\nu^{-2}$, where ν is the kinematic viscosity. Typically, the natural convection is negligible when $\mathrm{Ri} < 0.1$, forced convection is negligible when $\mathrm{Ri} > 10$, and neither is negligible when $0.1 < \mathrm{Ri} < 10$. In our case the maximum temperature difference was 30°C ($T_h = 50$°C, $T_c = 20$°C). In these conditions the average thermal expansion coefficient $\beta_{av} = 3.155$ K^{-1}, the average kinematic viscosity $\nu_{av} = 0.781 \cdot mm^2/s$ and $L = h = 200$ μm. Eventually, Gr = 1.218 and the Reynolds number in the calculations was ranged from 10 to 300, so the Richardson number was ranged from $1.353 \cdot 10^{-5}$ (for Re = 300) to 0.01218 (for Re = 10). Based on this, the natural convection can be neglected.

The distribution of temperature and concentration of mixing components in various sections of the mixer for $\Delta t = 10$°C and different Reynolds numbers is shown in figure 2. As is obvious, the distributions of concentration and temperature are not absolutely identical. The thickness of the mixing layer of temperature field is considerably higher than that of the concentration field. This is due to the different scales of the heat conduction and diffusion processes. Thus, the case of variable-temperature fluid cannot be considered absolutely identical to the cases of variable viscosity and density of mixing media.

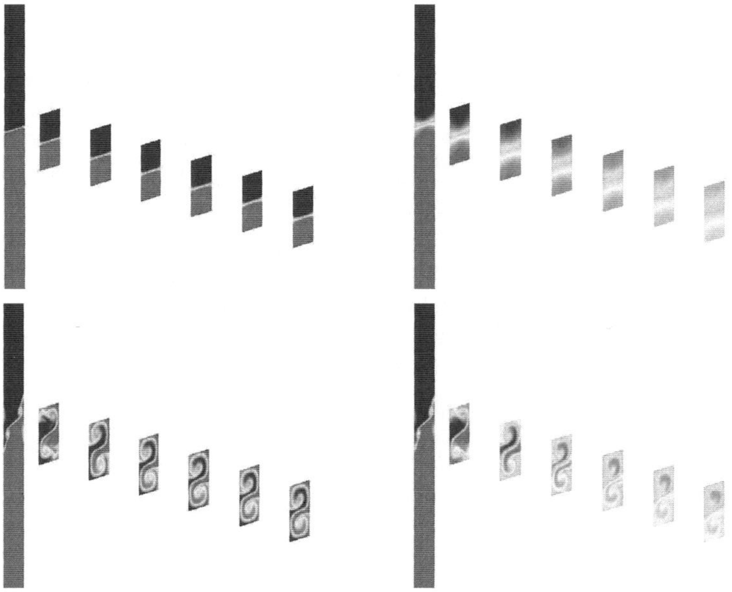

Figure 2. Concentration (left) and temperature (right) distribution at Re = 10 (upper) and Re = 150 (lower)

Figure 3a represented mixing efficiencies depending on the Reynolds number for different initial temperatures of the fluids. Analysis of the calculation results shows that change of flow regime at the Reynolds number equal to 162 occurs at a temperature difference equal to -10°C, while at Re = 126 for $\Delta t = 10$°C, at Re = 118 for $\Delta t = 20$°C, and at Re = 114 for $\Delta t = 30$°C. Certainly, this is largely due to the viscosity effect. An interesting effect can be obtained if the reduced Reynolds number is calculated using the arithmetic mean value of $\mu_m = 0.5 \cdot (\mu_0 + \mu_1)$, where μ_0 is the viscosity of water at 20°C, while μ_1 is the viscosity of water at some other temperature. Different ways of defining the effective viscosity and their influence on the obtained results were considered in [5]. Using this method of defining the reduced Reynolds number, it is possible to plot a dependence of two fluids mixing efficiency on this number (see figure 3b). It is obvious that the transition occurs at reduced Reynolds number equal or close to 145 and the mixing efficiency is approximately the same for all studied cases.

Besides, the pressure drop between the micromixer inlet and mixing channel outlet was analyzed. In figure 4a shown the dependence of the pressure drop between the micromixer outlet and the inlet, where was supplied the water at a temperature of 20°C on Reynolds number. It is obvious that the differences between them are not very significant. When considering the pressure drop between the micromixer outlet and the inlet, where was supplied the water at the varied temperature depended on the Reynolds number, more substantial difference between results was obtained for different cases (figure 4b). It is apparent that these curves are arranged in order of increasing the initial inlet temperature of the fluid. This behaviour is easy to explain, since the pressure drop is proportional to the fluid viscosity and density, therefore the increase of the inlet fluid temperature leads to decrease in pressure drop.

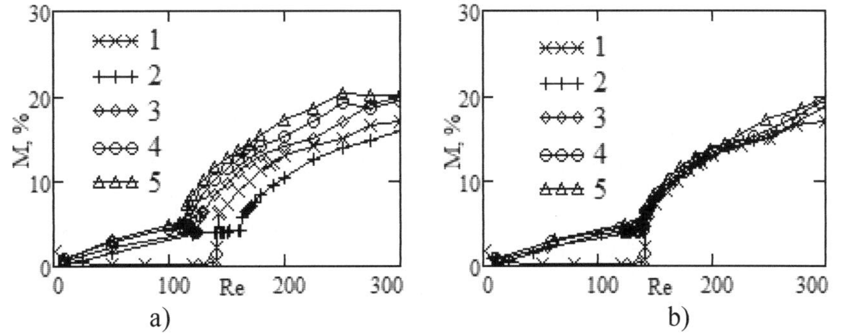

Figure 3. Mixing efficiency versus Reynolds number and reduced Reynolds number: Δt = 0°C (1); Δt = -10°C (2); Δt = 10°C (3); Δt = 20°C (4); Δt = 30°C (5)

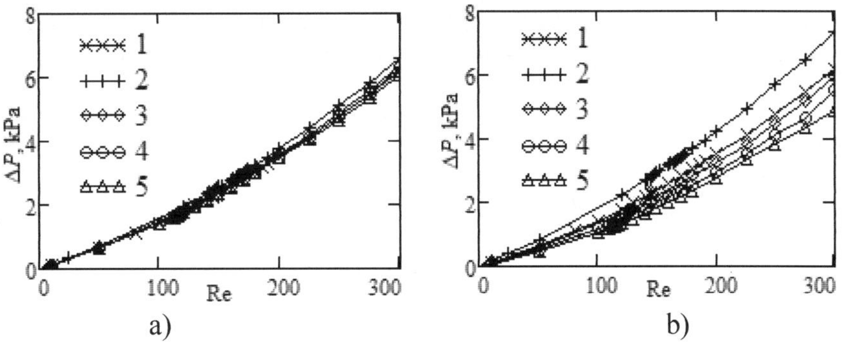

Figure 4. Pressure drop versus Reynolds number: Δt = 0°C (1); Δt = -10°C (2); Δt = 10°C (3); Δt = 20°C (4); Δt = 30°C (5)

4. Conclusions

The conducted numerical simulation allowed identifying following regimes for incompressible fluid flow in T-shaped micromixer: stationary vortex-free flow; stationary symmetric vortex flow; intermittent transition from symmetric to asymmetric flow (engulfment flow); and stationary asymmetric vortex flow. It is shown that the presence of a difference in initial temperatures of miscible fluids also leads to a shift of flow regimes. The simultaneous effect of differences in the viscosities and densities due to different initial temperatures of the fluids on the mixing efficiency was revealed. It was also shown that the flow and mixing of two fluids with different initial temperatures can be considered as self-similar pattern with regard to the reduced Reynolds number. Besides, the considerable effect of the difference in initial temperatures of miscible fluids on a pressure drop between the outlet and inlet of micromixer, through which water with variable temperature was supplied was revealed.

Acknowledgments

The work is performed at partial financial support of the projects funded by the Russian Foundation for Basic Research and Krasnoyarsk Regional Fund for Support of Scientific and Scientific-Technical Activities (Contract No. 16-48-243042) and by the Ministry of Education and Science of the Russian Federation Government contract with Siberian Federal University in 2017 (No. 16.8368.2017).

IOP Conf. Series: Journal of Physics: Conf. Series **899** (2017) 022010 doi:10.1088/1742-6596/899/2/022010

References

[1] Tabeling P 2005 *Introduction to Microfluidics* (Oxford: Oxford University Press)
[2] Karnidakis G, Beskok A and Aluru N 2005 *Microflows and Nanoflows. Interdisciplinary Appl. Math.* (Springer Science+Business Media Inc.), **29** p 817
[3] Karnik R 2008 *Microfluidic Mixing. Encyclopedia of Microfluidics and Nanofluidics* (Springer)
[4] Minakov A, Yagodnitsina A, Lobasov A, Rudyak V and Bilsky A 2013 Study of fluid flow in micro-mixer with symmetrical and asymmetrical inlet conditions *La Houille Blanche* **5** 12-21
[5] Lobasov A S, Minakov A V and Rudyak V Ya 2016 Viscosity effect on the flow modes in T-type micro-mixers *Fluid Dynamics* **51** 381-388
[6] Andreussi T, Galletti C, Mauri R, Camarri S and Salvetti M V 2015 Flow regimes in T-shaped micro-mixers *Comp. and Chem. Eng.* **76** 150-159
[7] Fani A, Camarri S and Salvetti M V 2013 Investigation of the steady engulfment regime in a three-dimensional T-mixer *Phys. Fluids* **25**
[8] Fani A, Camarri S and Salvetti M V 2014 Unsteady asymmetric engulfment regime in a T-mixer *Phys. Fluids* **26**
[9] Harr L, Gallagher J S and Kell G S 1984 *NBS/NRC Steam Tables* (Hemisphere Publ. Corp.)
[10] Marsh K N Ed. 1987 *Recommended Reference Materials for the Realization of Physicochemical Properties* (Oxford: Blackwell Scientific Publications)
[11] Sengers J V and Watson J T R 1986 Improved international formulations for the viscosity and thermal conductivity of water substance *J. Phys. Chem. Ref. Data* **15** 1291

IOP Conf. Series: Journal of Physics: Conf. Series **899** (2017) 022011 doi:10.1088/1742-6596/899/2/022011

Criterion of lifting force influence on the motion of buoyancy vortex ring

V V Nikulin[1]

[1]Lavrentyev Institute of Hydrodynamics SB RAS, Novosibirsk, Russia

Abstract. Criterion characterizing the lifting force influence on buoyancy vortex ring motion is worked out. If this criterion is small compared unity then lifting force influence on buoyancy vortex ring motion is weak, if criterion is large compared unity then such influence is strong. To check criterion experiments with buoyancy vortices moving in the direction of lifting force are carried out. Comparison of the experimental results with theoretical predictions shows their agreement.

1. Introduction

Several studies [1, 2] are devoted to investigations of the motion of homogeneous and inhomogeneous buoyant vortices in the gravity field. At the same time, the motion of buoyancy vortices having initial velocity has practically not been investigated. Such a situation is associated with the complexity of formation of the buoyancy vortex having initial velocity and containing a set amount of substance, which is more lightweight than the environmental one. In particular it is unknown the manner in which the lifting force impact on the vortex ring motion is varied with the initial velocity or corresponding Reynolds number. However, an understanding of it represents scientific and practical interest with respect to the problem of the dynamics of vorticity and the problems related to the processes of transfer and mixing. Experimental data are necessary for constructing theoretical models, which are adequate for observations.

One of the bright properties of vortex rings is the linearity of increasing their sizes in dependence on the distance traversed, at least at the initial stage of motion. This effect is characterized by the ring expansion coefficient α, which is equal to $\mathrm{d}R/\mathrm{d}z$, R is the ring radius, and z is the coordinate along which the motion proceeds. At the present time, it is established that the expansion coefficient of buoyant vortices in gravity field are an order of magnitude larger than that of vortices in a homogeneous liquid [2, 3]. So the buoyancy or lifting force has a great influence on the vortex ring dynamics. In this connection it is useful to work out criterion characterizing the lifting force influence on vortex ring motion.

2. Lifting force influence criterion

A vortex ring moving along z-axis parallel to the gravitational acceleration g in an unbounded viscous fluid of density ρ_0 at infinity is considered. From experiments it is known that a vortex ring during the motion retains its structure. Approximately it is composed of a toroidal core with a circular cross-section, containing a fluid of constant vorticity, and the atmosphere surrounding the core with zero vorticity [5]. For buoyancy vortex it is further assumed that all inhomogeneity is contained within the core. Then as the characteristic parameters of the flow are taken: radius of the vortex ring R (equal to the radius of the center line of the torus), radius of the cross section of the core a, velocity of the ring u, density ρ_0, kinematic viscosity ν and integral buoyancy force F. Other vortex parameters such as circulation Γ, momentum P can be obtained from characteristic one.

Criterion characterizing the buoyancy force influence on vortex ring motion must be constructed as a dimensionless complex containing characteristic parameters. So determination of the vortex characteristic parameters from its formation conditions leads to the criterion calculation and prediction of the buoyancy force influence on vortex ring motion.

To estimate buoyancy force influence we will compare expansion coefficients of homogeneous and buoyancy vortices. If these coefficients are distinguished significant that influence is significant.
In [4] it is presented analytical model describing the motion of homogeneous and buoyancy heterogeneous vortex rings. In particular in this work it is obtained formulae for calculation of initial expansion coefficient α_1, which may be written as:

$$\alpha_1 = \alpha_* + \frac{\beta F}{2\pi\rho_0 \nu^2 \mathrm{Re}_1^2} \tag{1}$$

$\beta = const \approx 0.3$, $\mathrm{Re}_1 = u_1 R_1/\nu$, R_1 and u_1 are initial radius and velocity of the vortex. Parameter α_* can be considered as an expansion coefficient of ring in homogeneous liquid because the case $F = 0$ is consistent with the motion in homogeneous liquid. If the motion occurs in the direction of lifting force then $F > 0$, if in opposite one then $F < 0$. From this equation it follows that lifting force increases the expansion coefficient if the motion occurs in its direction and decreases if the motion is the opposite. Let

$$k = \frac{\beta F}{2\pi\rho_0 \nu^2 \mathrm{Re}_1^2 \alpha_*} \tag{2}$$

Then equation (1) may be written as

$$\alpha_1 = \alpha_*(1+k) \tag{3}$$

From (3) it follows that distinction α_1 from α_* depends on the magnitude of k. So k may be adopted as criterion characterizing the buoyancy force influence on vortex ring motion. To final calculate k it must be given parameter α_*. Because α_* may be adopted as expansion coefficient of ring in homogeneous liquid then for it calculation we use empirical equation obtained in [3] for $\mathrm{Re}_1 > 3.5\cdot10^3$. This equation has the next form

$$\alpha_* = 3\cdot10^{-3} + \frac{50}{\mathrm{Re}_1 - 10^3} \tag{4}$$

Substituting (4) in (2) we obtain final expression for k.

$$k = \frac{\beta F}{2\pi\rho_0 \nu^2 \mathrm{Re}_1^2 (3\cdot10^{-3} + 50/(\mathrm{Re}_1 - 10^3))} \tag{5}$$

From (5) it follows that k is monotonically decreasing function of Re_1 if $F > 0$ and in contrast if $F < 0$. So the lifting forces influence on vortex ring motion decreases with increasing of Reynolds number and the peak distinguish between α_1 and α_* will be attained at minimum Reynolds number.

To establish the boundary value of k between strong and weak influence of lifting force we must prescribe corresponding value to k. Put that this boundary corresponds to $|k| = 0.5$. Note that from (3) it follows that for buoyancy vortex moving in the direction of lifting force ($F > 0$) this boundary corresponds to the case $\alpha_1 = 1.5\alpha_*$. After substitution of k = 0.5 in (5) we obtain equation for minimum lifting force at which their influence will be appreciable

$$F = \frac{\pi\rho_0 \nu^2 \mathrm{Re}_1^2}{\beta}\left(3\cdot 10^{-3} + \frac{50}{\mathrm{Re}_1 - 10^3}\right) \qquad (6)$$

3. Experimental results

Experiments with buoyancy vortex rings moving in water in the direction of lifting force have been carried out to check criterion characterizing the lifting force influence on vortex motion. In the experiments for creating the buoyancy, air bubbles are put in the vortex. In this case evidently $F = \rho_0 gV$, g is gravitational acceleration, V is volume of the air captured in vortex. The air bubbles are captured by a formed vortex, and, then, fall in its torus shaped core under the action of centrifugal force, where they are accumulated on the axial line of the torus displaced together with the vortex. The motion of the vortex ring is recorded by a high speed velocity video camera. The shadow transmission shooting is made perpendicularly to the direction of motion. The position of the vortex is visualized because of the presence air bubbles in it. We measure the external transverse size d of the region occupied with bubbles in dependence on the distance z travelled by the vortex.

In figure 1 the photograph of the air bubbles captured by the vortex is shown. In figure 2 the typical dependences $d(z)$ are shown by the points. The solid lines in figure 2 are the linear regressions drawn through the experimental points. From the half angle of the inclination of the linear regressions graphs, we find the expansion angle of the vortex ring (because the expansion angle is related to the radius instead of the diameter).

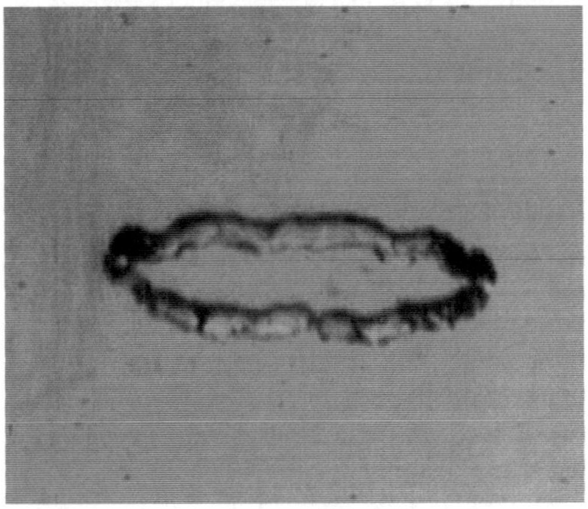

Figure 1. Air bubbles captured in the vortex_ring core.

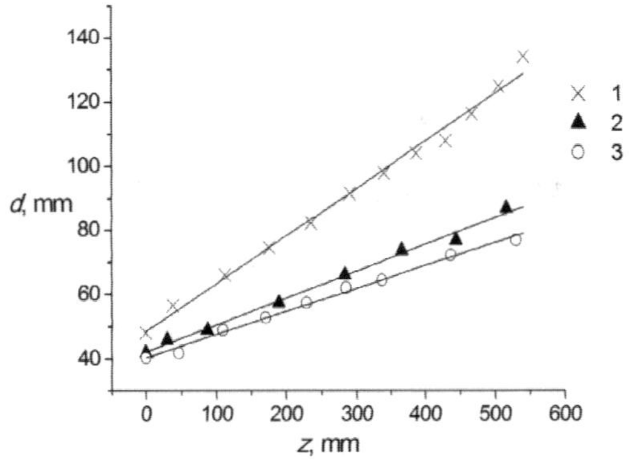

Figure 2. The external transverse size d of the region occupied with bubbles in dependence on the distance z travelled by the vortex. $V = 7.5$ cm^3, $\mathrm{Re}_1 = 5180$ (1), 6480 (2) and 7520 (3).

Using equation (6) we can estimate a minimum volume V_m of air bubbles captured by the vortex which is needed for influence of lifting force will be essential.

The right side of (6) is monotonically increasing function on Re_1. Then the minimum volume will be achieved at minimum Reynolds number. A minimum Reynolds number at which bubbles are held in the vortex core according to experiments is approximately equal to 3500. Substituting in (6) $F = \rho_0 gV$, $Re_1 = 3500$ we obtain $V_m \approx 0.3$cm^3. During the calculation we put $\nu = 10^{-2}$cm^2/s, $g = 980$cm/s^2, $\beta = 0.3$.

In the experiments both kinds of vortices are investigated, with small buoyancy, $V = 0.3\text{cm}^3$ and large buoyancy $- V = 7.5\text{cm}^3$; Reynolds numbers are varied from 4200 to 28500. For vortices with small buoyancy parameter k is varied from 0.4 to 0.04, with large buoyancy k is varied from 140 to 70. In figure 3 by the points it is shown expansion coefficients of vortices with small buoyancy, in figure 4 – expansion coefficients of vortices with small and large buoyancy. The solid line in the figures is empirical dependence (4) of expansion coefficient of vortices in homogeneous liquid, taken from [3].

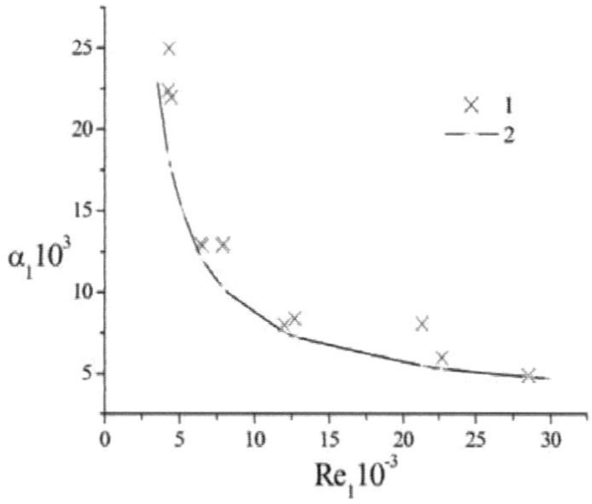

Figure 3. 1 – expansion coefficients of buoyancy vortex ring containing in the core air of volume $V = 0.3\text{cm}^3$ and 2 – empirical dependence (4).

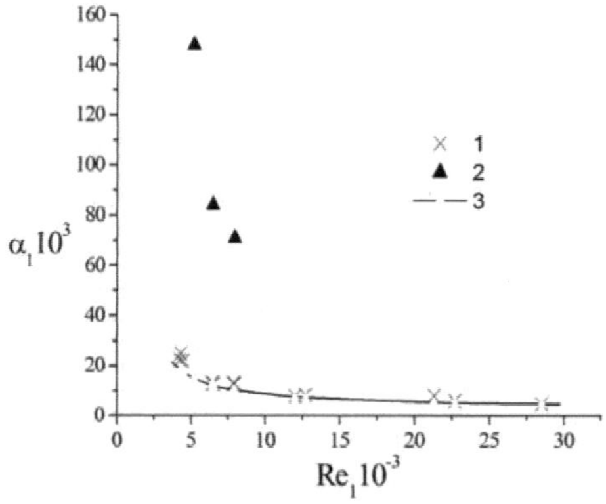

Figure 4. 1 – expansion coefficients of buoyancy vortex ring containing in the core air of volume $V = 0.3\text{cm}^3$, 2 – of volume $V = 7.5\text{cm}^3$ and 3 – empirical dependence (4).

4. Analysis of the results and conclusions.

From figures 3 and 4 it follows that for buoyancy vortex ring for which parameter $k >> 1$, also $\alpha_1 >> \alpha_*$. For buoyancy vortex ring for which parameter k is small compared 1 expansion coefficient is distinguished weakly from α_* – expansion coefficient of vortex in homogeneous liquid (figure 3). From this we conclude that parameter k can be considered as criterion characterizing lifting force influence on the motion of buoyancy vortex ring.

[1] Akhmetov D G 2009 *Vortex rings* (Berlin Heidelberg: Springer-Verlag)
[2] Zaslavskii B I and Sotnikov I M 1983 Experimental study of buoyant vortex rings *J. Appl. Mech.Tech. Phys*. **24** 16–21
[3] Tarasov V F 1973 *Dinamika Splosh. Sredy* **14** 120–7 (in Russian)
[4] Nikulin V V 2014 Analitical model of motion of turbulent vortex rings in an incompressible fluid *J. Appl. Mech. Tech. Phys*. **55** 558–64
[5] Lamb H. 1945 *Hydrodynamics* (New York: Dover)

Numerical and experimental study of the pressure pulsations at the free discharge of water through the turbine

D V Platonov[1,2]

[1]Siberian Federal University, 660074, Krasnoyarsk, Russia
[2]Institute of Thermophysics SB RAS, 630090, Novosibirsk, Russia

tov-andrey@yandex.ru

Abstract. The free discharge through the turbine is applied in the course of construction of hydro power plant or in case of excessive water inflow during floods or emergency situation. The experimental and numerical investigation of flow-induced pressure pulsation in hydraulic turbine draft tube at free discharge was performed.

1. Introduction

Low-frequency flow-induced pulsation occurs in Francis and Kaplan hydraulic turbines at off-design operation. Since early 20th century, oscillations of power of hydraulic turbines have been observed. In 1940, Rheingans [1] showed that the oscillations are associated with draft tube surges. In 1960, Deriaz [2] showed that the draft tube surges are caused by vortex rope precession. Generally, the pulsation is associated with the unsteady swirling flow in the draft tube of the hydraulic turbine.

One of the options for operating the hydraulic unit with the maximum pressure pulsation is a free discharge through the turbines. A free discharge through the hydraulic turbine unit is used in the construction of a power plant or in case of water discharge and is a complex technological process. In any cases of the blank pass, the flow around the main elements of the flow path structure will take place in an unclear mode and, consequently, will be accompanied by a significant increase in non-stationary processes and cavitation, which leads to additional wear and a shortened service life of the unit. There are not many systematic studies of free discharge for high head stations at the current time, so the study and optimization of the process is an urgent task.

2. Experimental technique and numerical algorithm

An experimental study was performed on an open aerodynamic stand. The configuration considered is a 61:1 scaled-down laboratory model of a Francis turbine (RO-230) of an LMZ (Leningradsky Metallichesky Zavod). Air supply in the draft tube model was carried out using a high performance air blower with the maximum flow rate Qmax =170 m3/h and the excess pressure of Δp=0.4 bar. An ultrasonic flow meter «IRVIS-Ultra» and a frequency converter of the blower with feedback controlled air flow rate were used. Figure 1 provides a photo of the experimental turbine model. The draft tube model, runner, spiral case and guide vanes geometry were provided by the company LMZ and were manufactured using the technology of 3D printing.

Servo drive SPSh10-3410 ensured precise rotation frequency of the runner in the range from 0 to 3000 rpm. A computer ensured the control of the experiment. Using the original software, it was possible to maintain the flow regime with an accuracy of 1.5% and 0.5% of the flow rate and the runner rotation frequency respectively.

The experiments included measurements of the pressure pulsations with special acoustic sensors. The acoustic sensors were based on high precision microphones attached with tiny tips for capturing pressure pulsations at the local points [3, 4]. Two of these sensors were used to conduct peak-to-peak measurements. The perturbation mode that corresponds to the precessing vortex core was identified by analysing the signal from the sensors located in diametrically opposite angular positions. Note that the possibility of using well-proven acoustic technology is another advantage of air model stand.

Pressure fluctuations were recorded using a microphone Bruel & Kjaer Type 2250. The signals from the microphones were digitized by an analog-to-digital converter L-card E-440 [5]. The experiments were conducted at the runner rotation frequency from f = 0 to 2000 rpm with a step 500 rpm and a flow rate, varing in the range from Q =100 to 150 m3/h with a step of 50 m3/h. At each point, the signal detected by the acoustic sensors was digitized for 5 seconds with a sampling rate of 2 kHz.

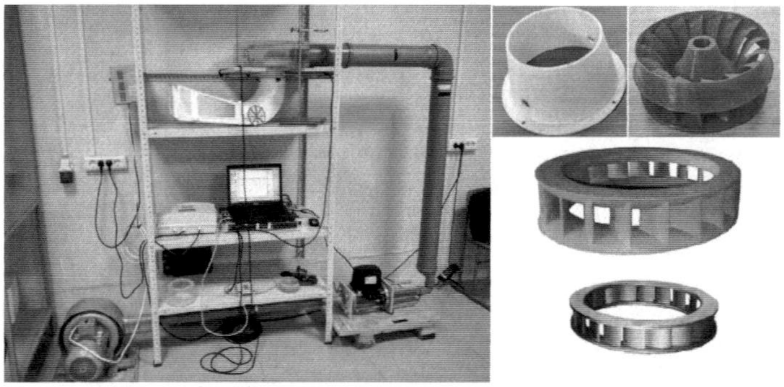

Figure 1. The photo of the experimental stand.

Describing the flows in water turbines numerically, one have to face several problems. The first challenge is related to modelling of turbulence in channels of complex geometric shapes and with strong swirl flow. In this paper, using the numerical technique for calculating of low-frequency pressure pulsations in a hydraulic turbine is based on the DES turbulence model (based on Menter k-ω SST model). Also in the simulation of hydraulic turbines, it is necessary to consider the rotation of the runner. There are many approaches for modelling flows with rotating bodies, they include: dynamic grid, sliding mesh, moving mesh and method based on a moving reference frame. In this work the modelling of the runner rotation was performed with rotated reference frame for runner zone. Numerous test calculations show the correctness of this approach, both in the description of integral characteristics of flow and pressure pulsation [6-8].

3. Results and discussion

In this paper, the results of a study of nonstationary processes at free discharge of water through a turbine are presented. The figures 2-4 show the results of experimental studies for three idle modes (without runner, "frozen" runner, free runner). During the experiment, pulsations of pressure in the diffuser of the draft tube were recorded and corresponding spectra of pulsations were constructed. It can be seen that the maximum pulsations are observed for the regime without runner.

IOP Conf. Series: Journal of Physics: Conf. Series **899** (2017) 022012 doi:10.1088/1742-6596/899/2/022012

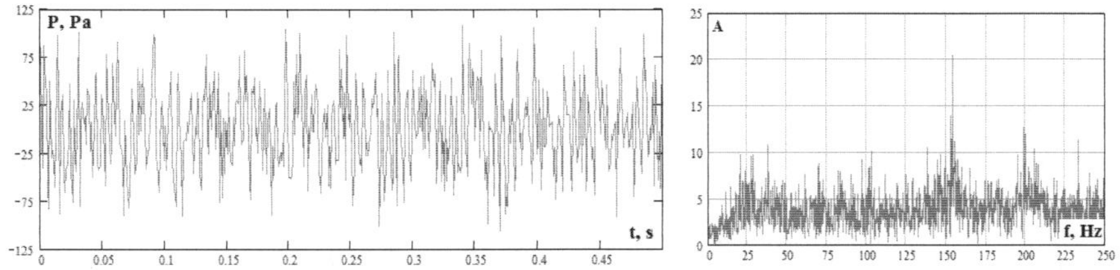

Figure 2. The pressure pulsations and their spectrum in the diffuser of the draft tube *(without runner)*.

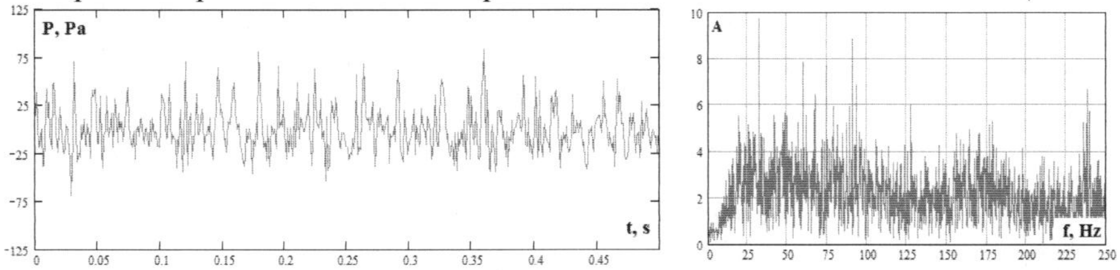

Figure 3. The pressure pulsations and their spectrum in the diffuser of the draft tube *("frozen" runner)*.

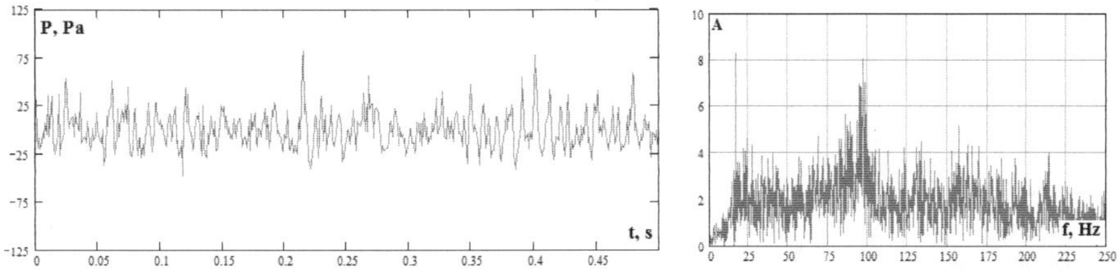

Figure 4. The pressure pulsations and their spectrum in the diffuser of the draft tube *(free runner)*.

Further, for the same regimes, a numerical study was carried out using the proposed numerical technique. Figure 5 shows the results of numerical simulation for the variant with the dismounted runner. The figures show that a structure is formed along the turbine cover, containing several precessing vortex rope. Analyzing the calculations, one can say that the vortex structure is very irregular, the vortices periodically break and form again.

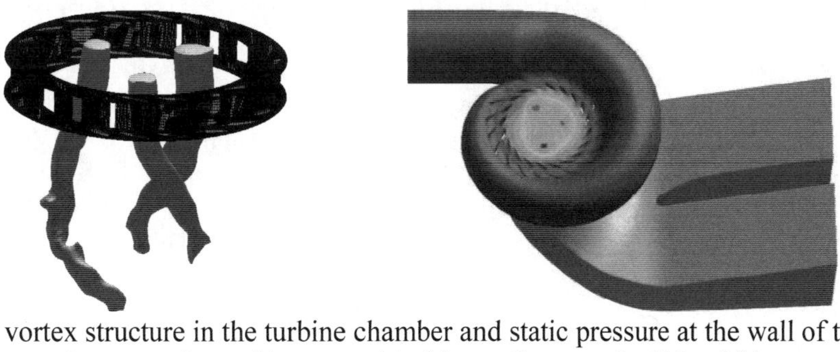

Figure 5. The vortex structure in the turbine chamber and static pressure at the wall of turbine (for regime: opening guide vanes A0=18mm, flowrate Q=130 m3/h).

Figure 6 presents a comparison of the numerical and experimental spectrum of pressure pulsations in the diffuser of the draft tube, for the mode without runner. The graph shows several expressed frequencies, which indicates a structure with several precessing vortex ropes. In general, the calculation and experiment are in good agreement with each other.

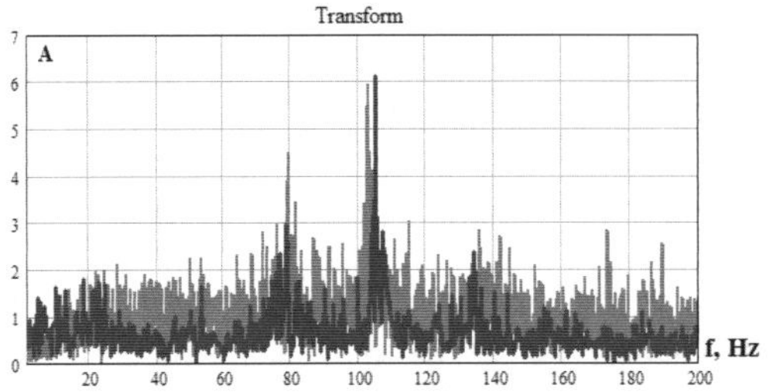

Figure 6. The spectrum pressure pulsations in the diffuser of the draft tube *(red color – experiment data, blue – calculation result).*

4. Conclusions

In the work, a numerical and experimental study of free discharge through a hydroelectric unit for high head hydro power stations was carried out. On the experimental stand, various options for free discharge were investigated. It is shown that the maximum pulsations are observed for the regime without runner. A numerical technique for simulating of free discharge was also proposed and a numerical study was performed. Comparison of calculated and experimental data is carried out, a fairly good agreement of the results is shown. These results may indicate the adequacy of the numerical technique and its applicability for further research.

Among the options considered for the free discharge as a whole, the lowest level of pressure pulsations was obtained in the cases of the pass with a free runner. The amplitude of pressure pulsations in the diffuser of the draft tube for one of the options is about 8% of the head, which is even slightly lower than the same value at the rated operating mode of the turbine.

Acknowledgments

The current work is performed at partial support of the projects funded by the Russian Foundation for Basic Research and Krasnoyarsk Regional Fund for Support of Scientific and Scientific-Technical Activities (16-48-240707).

References

[1] Rheingans W.J. 1940 Power swings in hydroelectric power plants. Transactions of the ASME. **62** pp 171-184

[2] Deriaz P. 1960 A contribution to the Understanding of Flow in Draft Tubes of Francis Turbines. France, Nice: *Hydraulic Machinery and Equipment Symposium*. p 13

[3] Minakov A.V., Platonov D.V., Litvinov I.V., Shtork S.I., Hanjalić K. 2017 Vortex ropes in draft tube of a laboratory Kaplan hydroturbine at low load: an experimental and LES scrutiny of RANS and DES computational models *Journal of Hydraulic Research*. In press, published online April 26, 2017

[4] Litvinov I.V., Mitryakov A.S., Shtork S.I. A 2016 Study of transient flow modes in a hydraulic turbine draft tube model. *Power Technology and Engineering*. **50** pp 1-5

[5] Litvinov I. V., Shtork S.I., Kuibin P.A., Alekseenko S. V., Hanjalic K., 2013, Experimental study and analytical reconstruction of precessing vortex in a tangential swirler *Int. J. Heat Fluid Flow*. **42** pp 251–264

[6] A.V. Minakov, A.V. Sentyabov, D.V. Platonov, A.A. Dekterev, A.V. Zakharov 2015 The numerical simulation of low frequency pressure pulsations in the high-head Francis turbine *Comp. fluids*. **111** pp 197-205

[7] A.V. Minakov, A.V. Sentyabov, D.V. Platonov, A.A. Dekterev, A.V. Zakharov 2015 The analysis of unsteady flow structure and low frequency pressure pulsations in the high-head Francis turbines International *J. Heat Fluid Flow*. **53** pp 183–194

[8] D. Platonov, A. Minakov, D. Dekterev, A. Sentyabov, A. Dekterev 2016 Numerical and experimental study of low-frequency pressure pulsations in hydraulic units with Francis turbine, *J. Physics: Conference Series* **754**

An experimental study of aperiodic pressure pulses in Francis turbine

D Platonov[1,2], D Dekterev[1,2], A Minakov[1,2] and A Maslennikova[1]

[1] Siberian Federal University, 79 Svobodny Ave., 660041 Krasnoyarsk, Russia
[2] Institute of Thermophysics, SB RAS, 1 Lavrentyev Ave., 630090, Novosibirsk, Russia
E-mail: platonov-08@yandex.ru

Abstract. The paper presents the results of experimental studies of the vortex rope reconnection in the flow of a Francis turbine. The studies were carried out on a mid-scale hydrodynamic set-up, with a model similar to the actual hydroelectric units. With the help of high-speed camera in a mode with a load of 82% a reconnection of the precessing vortex rope with detachment of the vortex rings was observed in the diffuser of the draft tube. Aperiodic pressure pulses associated with the vortex rope reconnection in diffuser in the draft tube.

1. Introduction

One of the reasons for the increased level of pressure and vibrations in water turbines of high-head hydro power plants is the phenomenon of a precessing vortex core. Precession of the vortex rope, which formed downstream the runner in the modes of partial or forced load of the hydraulic unit, is one of the main sources of the generation of low-frequency pulsations of the high-swirling flow [1-6].

Analysis of the experimental and numerical data shows [2-4] that the maximum pressure pulsations are observed in the mode of partial load on the hydraulic unit (40-50% of the nominal power of the turbine). However, in some works [1, 7, 8] the presence of aperiodic pressure pulses due to irregular and unstable forms of the vortex rope was observed in modes close to the optimum (70-80% of the nominal power). In [9], a phenomenon of reconnection of a precessing vortex rope with subsequent formation and detachment of the vortex ring was found on a set-up with a model simulating the conical diffuser of draft tube of water turbine. The process of reconnection is random, that gave grounds for formulating the scenario of the onset of aperiodic pulses caused by the passage of vortex rings near the wall [9].

Systematic experimental studies of the vortex reconnection phenomenon in real water turbines are currently unknown. This fact was the main motivation to perform current study on a hydrodynamic set-up with mid-scale model of a real Francis turbine.

2. Experimental methods

The set-up with a model water turbine is shown in Figure 1. The hydraulic unit consists of: 1 - a pressure tank performing the functions of the overflow, 2 - turbine flow path with a diameter of 400 mm with the electric valve 3. Water from the flow path 2 enters the spiral case 4, than goes through the guide vanes to the runner, draft tube 5, lower tank 6, drain tank 7, then through pipe 8 water goes to storage tank 9. From storage tank water is pumped back to the pressure tank by the pump 10. Thus, continuous circulation of water is ensured in the set-up. The maximum head of water at the set-up is

3.5 m, the diameter of the runner is D = 0.3 m. The diffuser of the draft tube is made of plexiglass for possibility of optical flow diagnostics and flow structure observing.

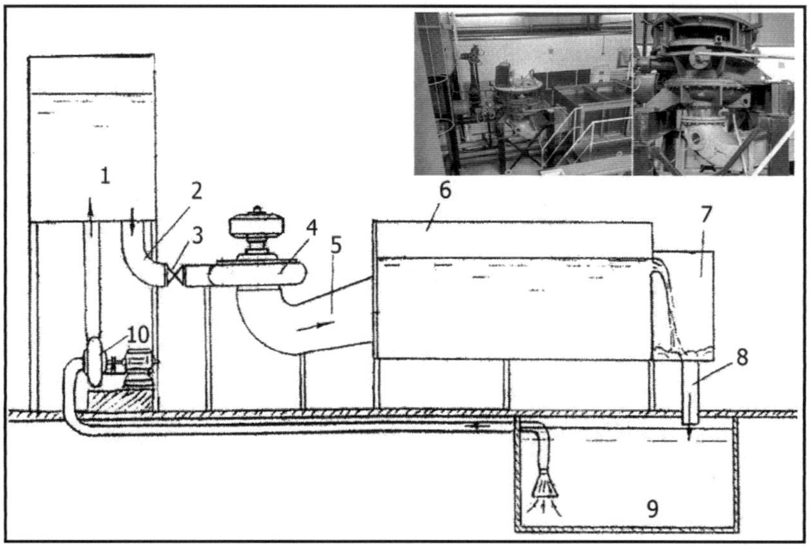

Figure 1. Scheme of the set-up

Recording of pressure pulsations was made by calibrated piezoelectric pressure sensors 014MT, which convert the rapid changes of pressure to electrical signal. Two sensors were mounted in the opposite walls of conical diffuser of draft tube. An external 4-channel charge amplifier LE-41 module was used to amplify the signal. The ADC E14-140 was used to record signals on the PC. In addition to the pressure pulsations, the water head, water flow, runner rotation frequency and the power produced by the turbine were measured.

Variation of operating modes is carried out by changing of the water head and the wideness of opening of the guide vanes. Pressure pulsations were measured and the flow structure was visualized in a wide range of operating modes of the hydraulic unit. The paper shows three successive operating modes of the hydraulic unit corresponding to 73%, 82% and 91% of the nominal power approximately.

3. Results

The 73% mode is characterized by the presence of a large vortex rope with occasional instability of the spiral structure. The vortex structure and the pressure sensor signal are shown in Figure 2. A uniform signal with constant amplitude can be observed.

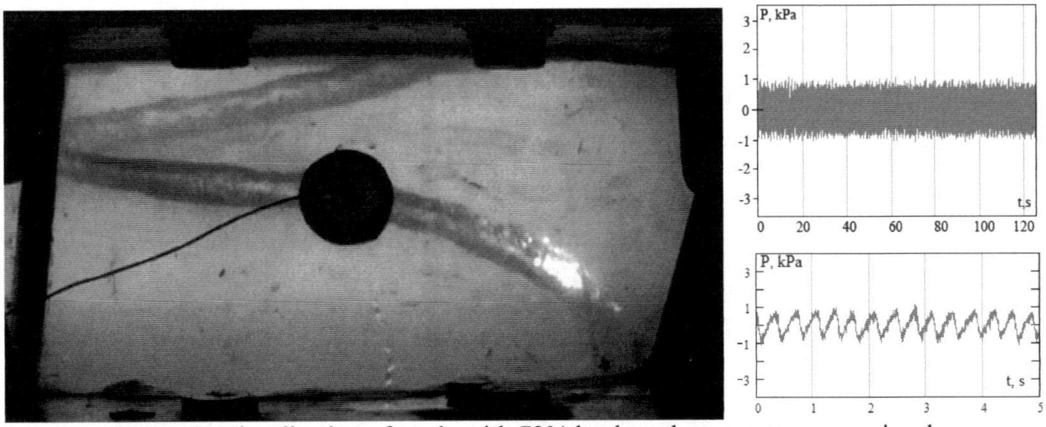

Figure 2. Visualization of mode with 73% loads and a pressure sensor signal

Due to the rather stable structure of the vortex the fast Fourier transform (FFT) of the signal makes it possible to accurately determine the precession frequency. In this case, the precession frequency is $f_{pvc} = 2.9$ Hz and the frequency of the runner rotation is $f_{rot} = 8$ Hz (Fig. 3)

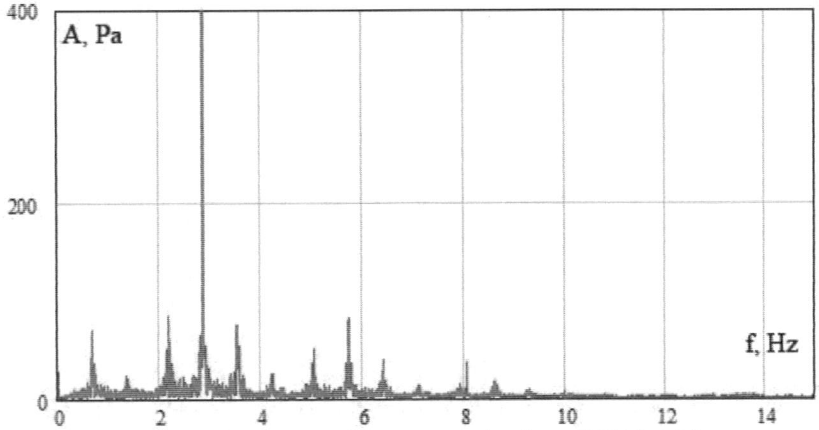

Figure 3. Spectrum of pressure pulsations (73% loads)

At 82% mode, a thin "wriggling" vortex with the reconnection effect and a subsequent formation and detachment of the vortex ring are observed. The vortex structure and synchronous signals from two pressure sensors (red line and blue line) are shown in Figure 4.

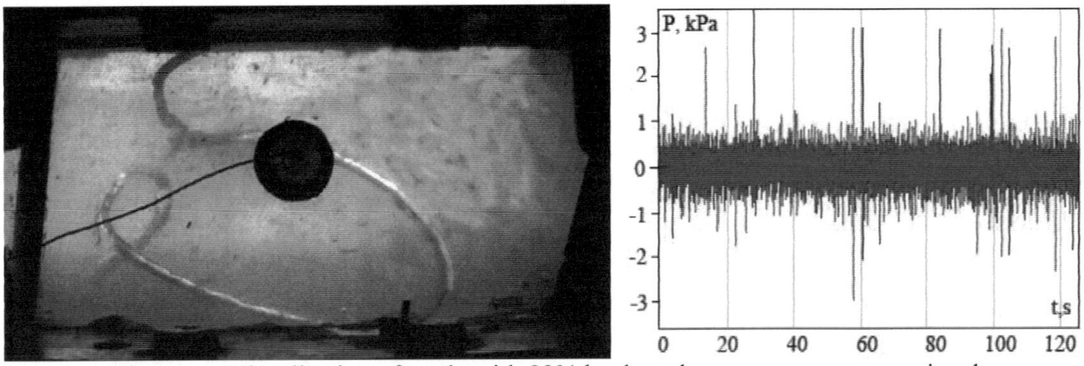

Figure 4. Visualization of mode with 82% loads and two pressure sensors signals

It can be seen that the overall level of pulsation decreases, but the signals have aperiodic pressure pulses with amplitude of 2-5 above the mean. It is clearly seen that the pressure pulses are random in time and different for both sensors.

Visual observations made it possible to determine that the formation of vortex rings occurs fairly regularly with frequency of about 0.45 Hz, it corresponds to 0.18 of the vortex precession frequency and 0.05 of the runner rotation frequency. However, from pressure sensor signals it can be seen that pressure pulses are irregular and occur much less frequently. The synchronization of high-speed video recording and the signal from the pressure sensor showed that pulses of pressure fluctuations are observed only when the vortex ring either hits directly to the sensor or passes very close to it.

The result of FFT is shown in Figure 5. In this case the frequency of precession is not so single-valued; nevertheless, a range of 2.4 Hz to 2.7 Hz can be distinguished. Also on the plot is clearly visible $f_{rot} = 8.3$Hz.

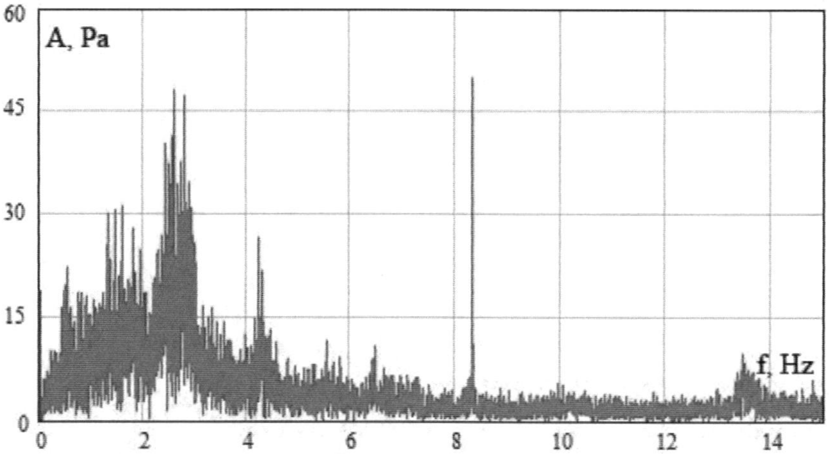

Figure 5. Spectrum of pressure pulsations (82% loads)

For a more detailed study of the operation mode with a reconnection of a vortex (83% load), signals with length of one hour for two pressure sensors were recorded and analyzed. The time between two neighboring pulses was determined to plot a histogram of the probability that a vortex ring hits a sensor or passes close to it. Probability of pressure pulses appearance in each time period T divided to the period of runner rotation T_{rot} shown in figure 6.

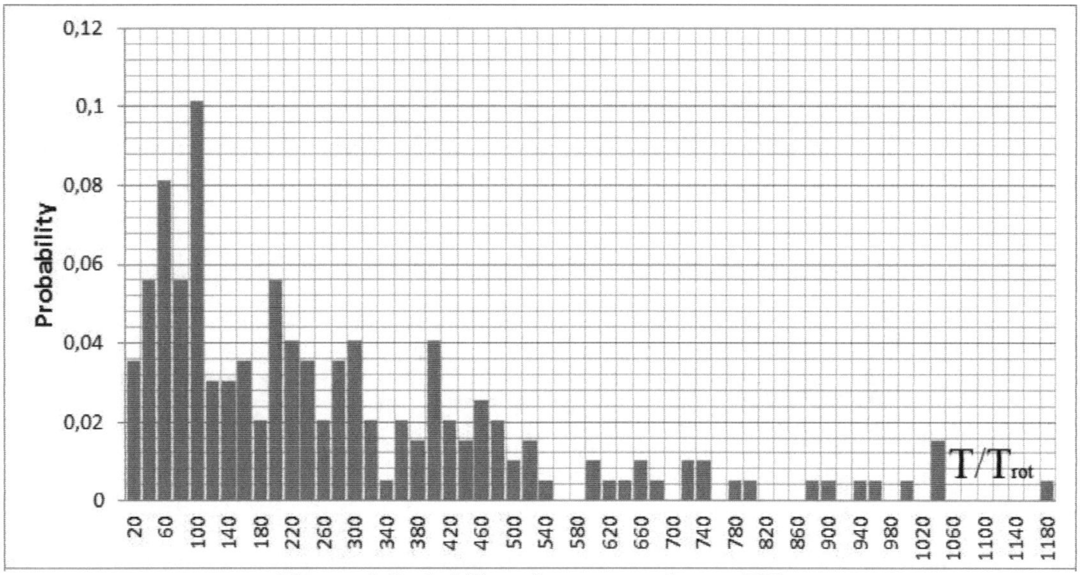

Figure 6. Probability of pressure pulses to **T/T$_{rot}$**

As can be seen from the distribution, the maximum probability of the pressure pulses occurs at 60-100 T_{rot}, although, the average period of pulses appearance is 303 T_{rot} (defined by the total time and number of pressure pulses). Table 1 summarizes the periods of main phenomena of flow with vortex reconnection with dependence to runner turn period.

Table 1.

Phenomena	Average period
Precession	**3.33 T_{rot}**
Ring separation	**18.5 T_{rot}**
Pressure pulse	**303 T_{rot}**

At 91% mode, vortex structures are not observed visually, the sensor signal also does not have sharp pressure pulses (Fig. 7).

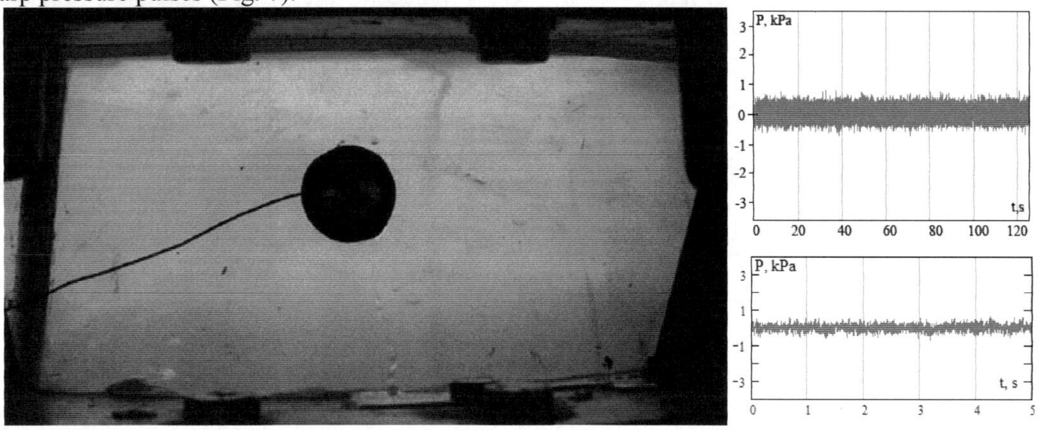

Figure 7. Visualization of mode 91% and a pressure sensor signal

The FFT also does not detect the presence of vortex structures with a constant frequency. Only the f_{rot} = 8.5 Hz is visible (Fig.8).

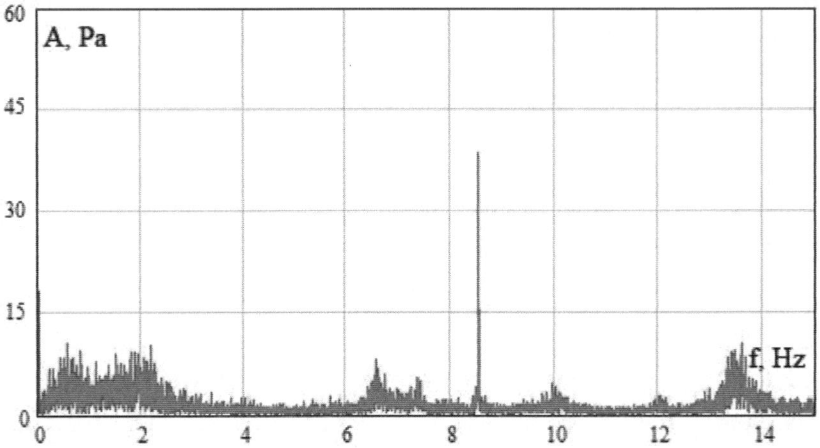

Figure 8. Spectrum of pressure pulsations (91% loads)

4. Conclusions

Thus, in this paper, on a hydraulic set-up with a Francis turbine constructed similar to real hydraulic units the operation mode with phenomenon of vortex rope reconnection was studied. Aperiodic pressure pulses connected with separation of vortex rings and their passage near the wall were observed experimentally. The following parameters of the flow with phenomenon of the vortex rope reconnection and separation of vortex rings were defined in first time and divided to period of runner rotation: Average periods of vortex rope precession, vortex rings separation, pressure pulses on the pressure sensor and most probable period of pressure pulses appearance.

Such phenomena can take place in the flow path of full-scale water turbines. Therefore, reconnection of the vortex and formation of vortex rings in water turbines is of great interest for fundamental and applied science and requires further study.

Acknowledgments

The work was performed at partial support of the projects funded by the Russian Foundation for Basic Research and Krasnoyarsk Regional Fund for Support of Scientific and Scientific-Technical Activities 16-41-243081.

References

[1] Dorfler P, Sick M, Coutu A 2013 Springer-Verlag London 242 p
[2] Minakov A, Sentyabov A, Platonov D, Dekterev A, Zakharov A 2015 Comp.and fluids p197-205
[3] Platonov D, Minakov A, Dekterev D, Sentyabov A, Dekterev A 2016 J.of Ph. Conf. Ser. V 754
[4] Minakov A, Platonov D, Dekterev A, Sentyabov A, Pylev I, Zakharov A. 2015 Power Techn. Eng. 49(**2**) 90-97.
[5] Vinokurov A, Shtork S, Alekseenko S. 2015 Letters JTF vol 41 (17) pp 61-67
[6] Skripkin S, Kuibin P, Shtork S. 2015 Letters JTF Vol 41 (13) pp 48-55
[7] Nishi M, Liu S. Int. J. Fluid Mach. Syst. 2013. Vol 6 No 1 P 33–48
[8] Nicolet C, Zobeiri A, Maruzewski P, Avellan F 2011 Int. J. Fluid Mach. Syst. Vol 4 No 1 P 179– 190
[9] Alekseenko S, Kuibin P, Shtork S, Skripkin S, Sonin V, Tsoy M, Ustimenko A 2016 IOP Conf. Series: Earth and Environmental Science Vol 49 Paper 082025, 8p

IOP Conf. Series: Journal of Physics: Conf. Series **899** (2017) 022014 doi:10.1088/1742-6596/899/2/022014

Large-eddy simulations of the near field of a turbulent annular jet

V. Ryzhenkov[1,2], **R. Mullyadzhanov**[1,2]

[1] Institute of Thermophysics SB RAS, Novosibirsk, Russia

[2] Novosibirsk State University, Novosibirsk, Russia

E-mail: v.o.ryzhenkov@gmail.com

Abstract. Using high-order spectral element method and large eddy simulations we study a turbulent annular jet at the Reynolds number of 8900, based on the bulk velocity in the annular channel and its half-width. The region of high turbulent kinetic energy is located inside the recirculation zone behind the bluff-body where the low-frequency motion is known to occur which is detected with energy spectra. Further analysis includes the azimuthal Fourier and Proper orthogonal decomposition (POD). It is shown that the main eigenmodes with low azimuthal wavenumbers m rotate around the symmetry axis in both directions providing the explanation for the precessing motion of the recirculating zone.

1. Introduction

Annular jets are produced by the flow issuing from the annular channel between two coaxial cylinders. Since the inner cylinder usually appears as a central bluff-body, the flow has a recirculation zone in the near field. This leads to inner and outer mixing layers where the first one is between the jet and recirculation zone and the second one is between the jet and the ambient fluid. A combination of features inherent to jets and separated flows makes an annular jet a complex object. The goal of the present study is to investigate the dynamics of the recirculation zone which is of utmost importance for a wide range of practical applications.

In first experiments Chigier and Beer [1] compared circular and annular jets studying the their similarities. Ko and Chan [2, 3, 4] considered different configuration focusing on the near field. Coaxial jets issuing from a double annular channel also attracted a lot of attention due the combustion applications [5]. While the dynamics of the outer shear layers is similar to that appearing in circular jets, the recirculation zone behind the bluff-body follow its rules. Berger et al. [6] studied the dynamics of the wake of a disk and a sphere at relatively high Reynolds numbers and found very low-frequency oscillations of the recirculating bubble. Similar observations were made for various bluff-body geometries [7, 8, 9, 10].

Recently Vanierschot and Van den Bulck [11] observed the precessing motion of the recirculating zone in annular jets. While Proper orthogonal decomposition (POD) is found useful to study the low-frequency dynamics [12], the analysis so far was two-dimensional due to experimental capabilities. In the present work we perform three-dimensional simulations and apply POD to volumetric data focusing on the vortical structures responsible for deformations of the recirculation zone.

IOP Conf. Series: Journal of Physics: Conf. Series **899** (2017) 022014 doi:10.1088/1742-6596/899/2/022014

2. Computational details and problem

To describe the motion of the incompressible fluid we consider the Navier–Stokes equations:

$$\frac{\partial \boldsymbol{u}}{\partial t} + (\boldsymbol{u}\nabla)\boldsymbol{u} = -\nabla p + \frac{1}{Re}\Delta \boldsymbol{u} - \nabla \cdot \tau, \qquad (1)$$

$$\nabla \cdot \boldsymbol{u} = 0, \qquad (2)$$

where $\boldsymbol{u}$, p are the dimensionless spatially filtered velocity field and pressure, respectively, τ represents the subgrid-scale stresses appearing in the large eddy simulations framework. The above described governing equations are implemented in the Nek5000 code [13] with a spatial discretisation based on the spectral-element method using Lagrange polynomials. The Navier–Stokes equations are cast in a weak form and discretised in space by means of Galerkin approximation using Nth-order Lagrange polynomial interpolants on the Gauss-Lobatto-Legendre points for the velocity field and pressure ($\mathbb{P}_N - \mathbb{P}_N$ formulation). In the present work the polynomial order N is set to 8. The semi-implicit time-stepping scheme is of the third-order accuracy. In the LES simulation the two highest modes are filtered with parabolic transfer function, and the filter amplitude was 5% in the last mode [14].

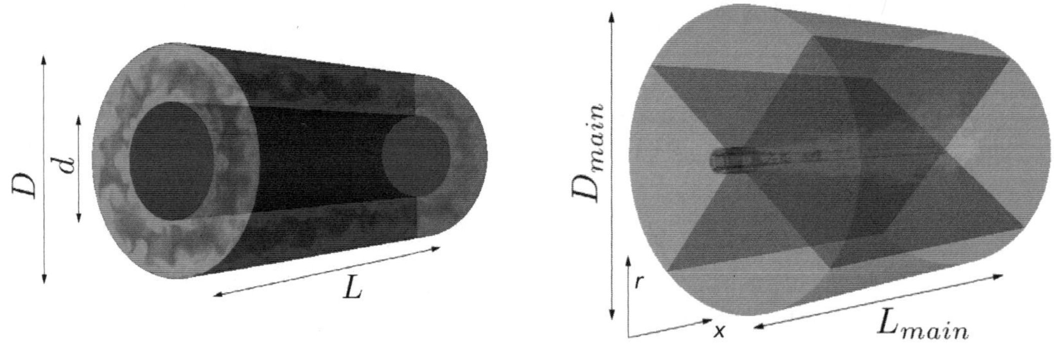

Figure 1. Left: annular channel with periodic boundary conditions to generate the inflow in the main simulation. Right: main computational domain of the jet flow and coordinate system.

We consider a turbulent annular jet of incompressible fluid with the Reynolds number $Re = 8900$, based on the bulk velocity U_b and outer diameter D, while $d = D/2$ is the inner diameter of the annular channel. Figure 1 shows the annular channel of length $L = 2.5D$ which is used to simulate with periodic boundary conditions a fully turbulent inflow velocity fields for the main computation of the annular jet. To validate this simulation performed with $8 \times 32 \times 20 = 5120$ spectral elements for r, θ and x directions, the results were compared with DNS by Chung et al. [15] with the same computational domain and flow parameters showing excellent agreement, see Fig. 3. The main region represents a cylinder with a diameter $D_{main} = 12D$ and length $L_{main} = 17D$, which includes a feeding annular channel with length equal $2D$. The thickness of the outer wall of the channel is $0.03D$. The velocity field in some $r - \theta$ plane from a periodic channel simulation was copied each time step to the annular channel of the main computation. The coflow in the main domain was set to $0.04U_b$. Zero derivative condition was set on the rest of the boundaries. The main area consisted of about 45 thousand spectral elements ($40 \times 32 \times 35$ elements in r, θ and x). Since we use polynomials of the eighth order, there are $9^3 = 729$ computational nodes inside each spectral element, so that the total number of computational nodes is about 32×10^6 for the main simulation and about 3.7×10^6 for the periodic annular channel. The distance between computational nodes at the outer wall inside the periodic channel is $\Delta r^+ = 0.6$, $(r\Delta\theta)^+ = 5$ and $\Delta x^+ = 50$ in radial, azimuthal and

IOP Conf. Series: Journal of Physics: Conf. Series **899** (2017) 022014 doi:10.1088/1742-6596/899/2/022014

longitudial direction, where the superscript '+' corresponds to the non-dimensionalization using the friction velocity and viscosity. The computational grid satisfies the near-wall well-resolved LES criteria [16]. Inside the recirculation zone we calculate a ratio between spatial grid step and the Kolmogorov scale, which does not exceed value of two Kolmogorov scales. This fact means that our calculation corresponds to the DNS accuracy.

3. Results

To get the impression on the main flow features we show the instantaneous and time-averaged axial velocity fields in $x - r$ plane, see Fig. 2. The outer and inner shear layers merge after around $3D$ leading to the formation of the jet profile. The recirculation bubble having the length of $0.52D$ has a complex dynamics which is addressed below. Figure 3 shows the radial profiles of the time-averaged axial velocity and its fluctuations at several axial stations along x direction. First profiles correspond to the periodic channel simulations where the results show excellent agreement with DNS calculation of Chung et al. [15]. This agreement confirms sufficient spatial

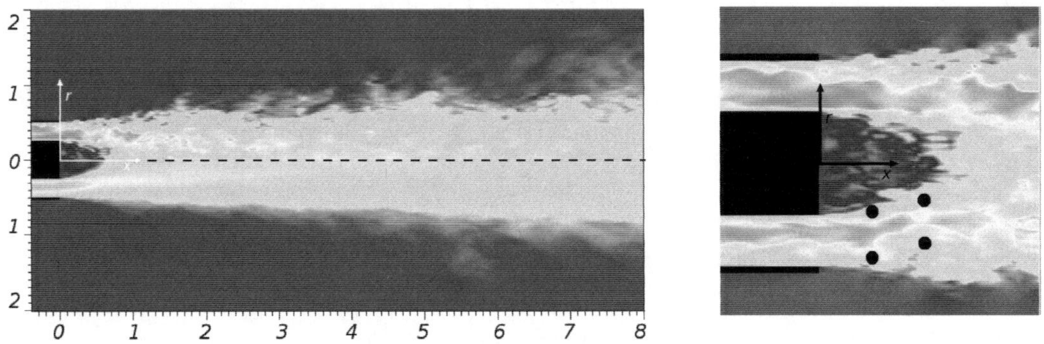

Figure 2. Left: instantaneous (top) and time-averaged (bottom) axial velocity fields and coordinate system. Right: The blow-up of the near-nozzle area and four points used to further analyze spectral characteristics.

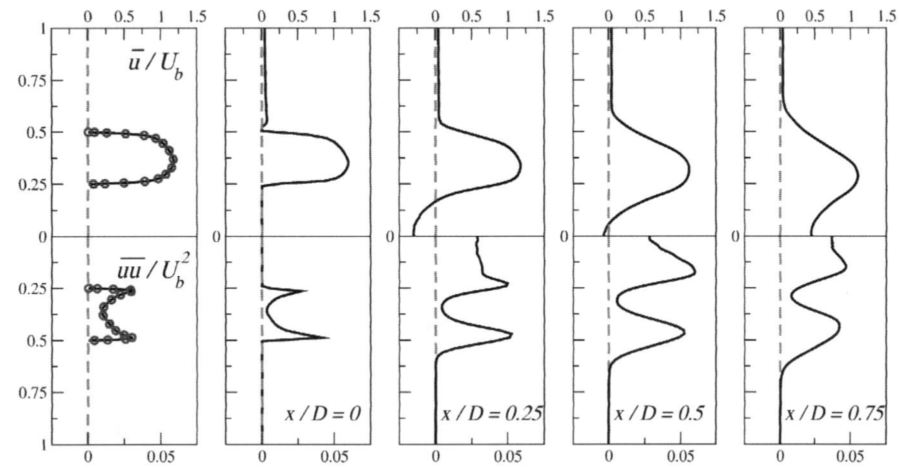

Figure 3. Upper rows: time-averaged radial profiles of axial velocity. Bottom rows: profiles of axial velocity fluctuations. In the first column the comparison of the periodic channel present simulation with the DNS [15] (symbols) is shown.

resolution of the computational mesh. Four more axial stations are shown, i.e. $x/D = 0$, 0.25, 0.5 and 0.75. The recirculation zone is characterized by the high level of fluctuations especially close to the stagnation point.

Further we investigate the time history and spectral characteristics at four points in the plane $x - r$ (see Fig. 2) where the velocity field was analyzed during $100tU_b/D$ time units. Two points were located in the inner mixing layer and two in the outer mixing layer at a distance of $0.25D$ and $0.5D$ from the nozzle edge. Figure 4 shows the energy espectra for individual velocity components. It is interesting to note that the Kelvin-Helmholtz instability is most clearly pronounced for the radial and azimuthal velocity components with a non-dimensional frequency $f_0 = fD/U_b \approx 1$. This peak is marked with the letter 'A' in the figure. A similar effect was observed in the mixing layer for a turbulent circular jet [17]. There is a small peak in the range of lower frequencies (marked as 'B') which will be investigated below with a more sophisticated method. Similar results were obtained for the Fourier transform from the autocorrelation functions (not shown here).

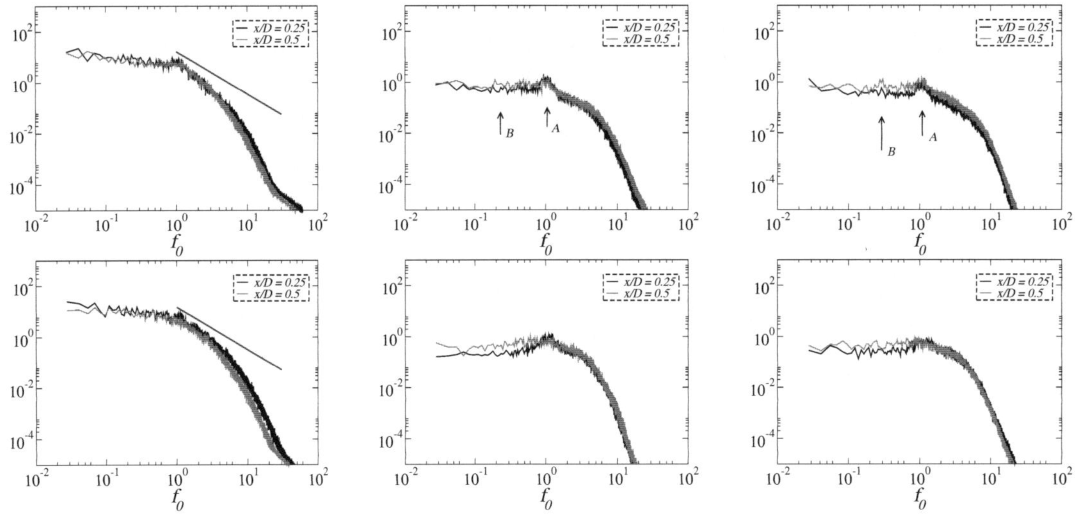

Figure 4. Energy spectra for separate velocity components (left: u, middle: v, right: w). Upper rows correspond to the inner mixing layer, bottom rows – to the outer mixing layer. Blue line illustrates the "$-5/3$" slope.

The round geometry suggests decomposition of the flow field into Fourier modes in the azimuthal direction ϕ. The instantaneous velocity field $\boldsymbol{u}$ is therefore decomposed into complex Fourier coefficients

$$\boldsymbol{u}(r,x,m,t) = \boldsymbol{u}^m = \frac{1}{2\pi}\int_0^{2\pi} \boldsymbol{u}(r,x,\phi,t)e^{im\phi}d\phi, \tag{3}$$

where m is the azimuthal wavenumber. We further apply snapshot Proper Orthogonal Decomposition [18] for each azimuthal wavenumber m and in a cylindrical subdomain located at a distance of $0.1D$ from a nozzle with a length of $1D$ and a radius of $1.5D$ to an ensemble of $N = 2660$ Fourier-decomposed fields $[\boldsymbol{u}_1^m, \boldsymbol{u}_2^m, ..., \boldsymbol{u}_N^m]$ corresponding to subsequent time instants $t = t_1, ..., t_N$ with $\Delta t = 10^{-2}D/U_b$:

$$\boldsymbol{u}_i^m(r,x) = \boldsymbol{u}(r,x,m,t_i) = \sum_{q=1}^{N} a_q^m(t_i)\lambda_q^m \boldsymbol{v}_q^m(r,x), \tag{4}$$

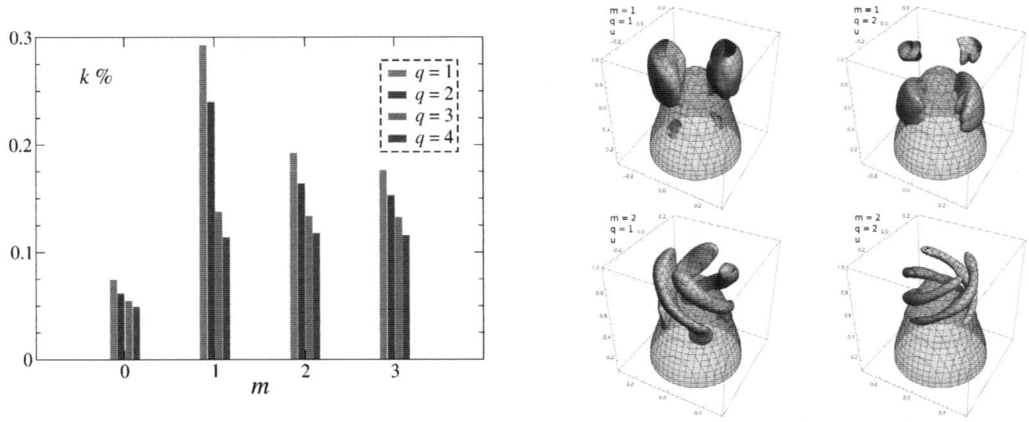

Figure 5. Left: energy distribution among most energetic POD modes for low wavenumbers $m = 0{-}3$. Right: 3D visualization of POD modes. A semitransparent axisymmetric figure is the contour of the time-averaged axial velocity corresponding to the value $\overline{u} = 0.5$, which represents closely the location of the recirculation zone. The blue and orange isosurfaces correspond to $u = \pm 1.5$, $v = \pm 0.5$ and $w = \pm 1.5$.

where $\boldsymbol{v}_q^m(r, x)$ and $a_q^m(t)$ are the non-dimensional complex-valued spatial eigenfunctions and temporal amplitudes satisfying orthonormal conditions

$$2\pi \int_V \boldsymbol{v}_i^m(r, x)\boldsymbol{v}_j^m(r, x) r dr dx = V\delta_{ij}, \tag{5}$$

$$\sum_{n=1}^{N} a_i^m(t_n) a_j^m(t_n) = N\delta_{ij}, \tag{6}$$

where λ_q^m are the real eigenvalues, δ_{ij} the Kronecker delta and V the volume of the considered domain (see [17] for details).

Figure 5 (left) shows the distribution of energy among different POD modes q for various m. Note that kinetic evergy of a particular mode integrated over a volume of interest is $(\lambda_q^m)^2/2$. The results suggest that the dynamics is governed by energy containing modes with low m. The most energetic mode has $m = 1, 2$ and $q = 1, 2$. Figure 5 (right) shows the three-dimensional visualization of these four modes. To investigate the dynamics of these structures the complex-valued temporal coefficient is decomposed as $a(t) = \beta(t)e^{2\pi i\gamma(t)}$. Figure 6 shows that behaviour of real-valued β and γ for $m = 1$ and $q = 1, 2$. While $\beta(t)$ oscillates, $\gamma(t)$ follows a linear function corresponding to the rotation of the coherent structure around the axis of symmetry with constant angular velocity. Note that the value of $d\gamma/dt$ is the frequency of rotation which is close to the one observed previously with energy spectra ($f_0 \approx 0.2$). The described framework of precessing motion will be further developed for this annluar jet flow.

4. Conclusion

Using large eddy simulations we studied the near field of the turbulent annular jet. The Reynolds number based on the bulk velocity in the feeding channel and the outer diameter was 8900. The assessment of statistical characteristics showed that the area around the stagnation point in the end of the recirculating region has a high level of turbulent kinetic energy. These fluctuations display a low-frequency coherent motion revealed by energy spectra. To study the spatial shape of these fluctuations we used Fourier and Proper orthogonal decomposition (POD). It is shown that the perturbations with Fourier harmonics $m \neq 0$ rotate around their own axis with low frequency, while the amplitude of the perturbation oscillates.

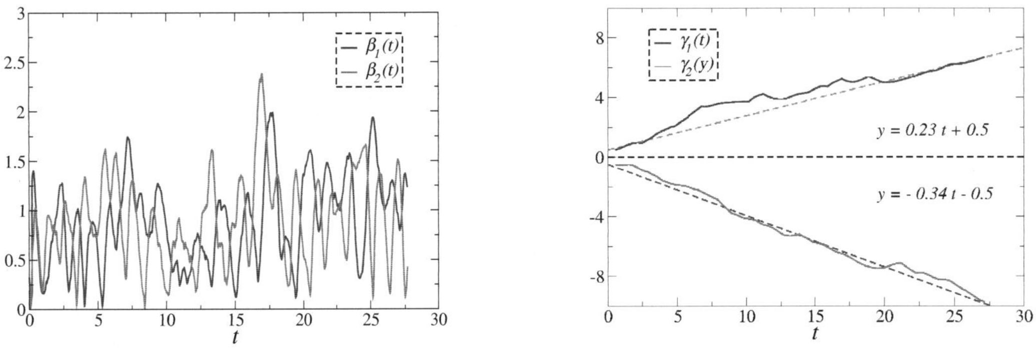

Figure 6. The time history of $\beta(t)$ and $\gamma(t)$ corresponding to $m = 1$ and $q = 1, 2$.

Acknowledgments

This work is funded by Russian Science Foundation grant No. 14-19-01685. The computational resources are provided by Siberian Supercomputer Center SB RAS (Novosibirsk) and Supercomputer Center of Novosibirsk State University.

References

[1] N. A. Chigier and J. M. Beer. The flow region near the nozzle in double concentric jet. J. Basic Engnrg. 86:797–804, 1964
[2] N. W. M. Ko and W. T. Chan. Similarity in the initial region of annular jets: three configurations. J. Fluid Mech. 84:641–656, 1978.
[3] N. W. M. Ko and W. T. Chan. Coherent structures in the outer mixing region of annular jets. J. Fluid Mech. 89:515–563, 1978.
[4] N. W. M. Ko and W. T. Chan. The inner region of annular jets. J. Fluid Mech. 93:549–584, 1979.
[5] F. H. Champagne and I. J. Wygnanski. An experimental investigation of coaxial turbulent jets. Int. J. Heat Mass Transf. 14:1445–1464, 1971.
[6] E. Berger, D. Scholz and M. Schumm. Coherent vortex structures in the wake of a sphere and a circular disk at rest and under forced vibrations. J. Fluid Struct. 4:231–257, 1990.
[7] F. M. Najjar and S. Balachandar. Low-frequency unsteadiness in the wake of a normal flat plate. J. Fluid Mech. 370:101–147, 1998.
[8] J. J. Miau, J. T. Wang, J. H. Chou and C. Y. Wei. Characteristics of the low-frequency variations embedded in vortex shedding process. J. Fluid Struct. 13:339–359, 1999.
[9] O. Lehmkuhl, I. Rodriguez, R. Borrell and A. Oliva. Low-frequency unsteadiness in the vortex formation region of a circular cylinder. Phys. Fluids 25:085109, 2013.
[10] E. Palkin, R. Mullyadzhanov, M. Hadziabdic and K. Hanjalic. Scrutinizing URANS in shedding flows: The case of cylinder in cross-flow in the subcritical regime. Flow Turbul. Combust. 97:1017–1046, 2016.
[11] M. Vanierschot and E. Van den Bulck. Experimental study of low precessing frequencies in the wake of a turbulent annular jets. Exp. Fluids, 50:189–200, 2011.
[12] B. Patte-Rouland, G. Lalizel, J. Moreau and E. Rouland. Flow analysis of an annular jet by particle image velocimetry and proper orthogonal decomposition. Meas. Sci. Tech. 12:1404–1412, 2001.
[13] S. G. Kerkemeier, P. F. Fischer, and J. W. Lottes. Nek5000: Open source spectral element cfd solver. available at http://nek5000.mcs.anl.gov, 2008.
[14] O. Marin and R. Vinuesa and A. V. Obabko and P. Schlatter. Characterization of the secondary flow in hexagonal ducts. Phys. Fluids, 28:1–26, 2016.
[15] S. Y. Chung, G. H. Rhee, and H. J. Sung. Direct numerical simulation of turbulent concentric annular pipe flow. Part 1: Flow field. Int. J. Heat Fliud Flow, 23:426–440, 2002.
[16] U. Piomelli and J. R. Chasnov. Turbulence and transition modeling. Large-Eddy Simulations: Theory and Applications. Springer-Science, 1995.
[17] R. Mullyadzhanov, S. Abdurakipov and K. Hanjalic. Helical structures in the near field of a turbulent pipe jet. Flow Turbul. Combust., 98:367–388, 2017.
[18] L. Sirovich. Turbulence and the dynamics of coherent structures. Part i: Coherent structures. Quart. Appl. Math. 45:561–571, 1987.

Intensity of vertical turbulent exchange in the Black Sea summer pycnocline around the Crimean peninsula

A S Samodurov[1] and A M Chukharev[1]

[1]Marine Hydrophysical Institute RAS, Sevastopol, Russia.

E-mail: anatol_samodurov@mail.ru

Abstract. When taking small-scale structure field measurements in the upper stratified layers the probe-turbulimeter with high sampling frequency is used. By the spectra of temperature fluctuations and velocity we are evaluated by virtue of the previously developed theory the characteristic scales of turbulent patches formed due to the internal waves breaking. Then the vertical turbulent diffusion coefficient dependence on buoyancy frequency for the specified layers is determined. It is shown that the spatial distribution of coefficient depends on the bottom topography; its maximum values are concentrated in the interface area of the shelf and continental slope, due to intensification of shear instability and internal waves breaking in this region.

1. Introduction

Turbulent exchange in the ocean plays an important role in the formation of heat, salt and other dissolved substances fluxes. Vertical turbulent exchange in stratified layers of the ocean is studied for decades. The fundamental works of Munk and Anderson [1], Thorpe [2, 3], Osborn [4], Gregg [5] and many others revealed main regularities of dynamic processes in these layers. Recent work among a large number of studies is experimental study of internal waves in the thermocline [6], where discovered the influence of Kelvin waves on the turbulence intensity. In [7] direct numerical simulation of the behavior of stably-stratified flow based on the dependence of liquid viscosity on temperature at different Reynolds numbers is performed. In the paper [8] based on observations in the Arctic Circumpolar Current the spatial distribution of the dissipation rate of turbulent energy and diapycnical mixing is studied, and it was shown that the dissipation rate is higher in areas with higher intensity of internal waves, and near the bottom irregularities.

In a stable stratified ocean away from all the sharp frontal zones, areas of evident appearance of double diffusion convection etc., the main source of turbulence is the shear instability mechanism in the quasi-inertial internal waves field [9]. Under the action of this mechanism, the so-called turbulent patches are formed, inside which the distributions of density and other physical characteristics are quasihomogeneous due to intensive turbulent mixing. Thus eventually the average vertical fluxes are generated. The evaluation of these fluxes is one of the most important tasks in modern oceanography. The goal of this work is specifying the dependence of the vertical exchange intensity on the local stratification on the basis of field measurements in the Black Sea upper summer stratified layer.

2. Field measurements

In-situ measurement of pulsating characteristics are conducted by using the measuring complex "Sigma-1" [10] designed to study the microstructure and turbulent exchange in the upper layer of the sea to a maximum depth of 300 m. The appearance of the complex is shown in figure 1. Probing was conducted in free fall at a speed of approximately 0.7 m/s from the surface to depth of 50 – 300 m depending on the location of station. In layers with a significant contribution of temperature to the

IOP Conf. Series: Journal of Physics: Conf. Series **899** (2017) 022015 doi:10.1088/1742-6596/899/2/022015

density gradient magnitude, the spectral analysis realized for a scalar value temperature fluctuations. In layers with small temperature gradients, on the contrary, the calculations were done mainly on the velocity fluctuations. The results of processing allow us to judge about the character of pulsation motion (wave or turbulent nature), about the supply level of small-scale turbulence, and finally, in conjunction with other parameters of the studied layer on the coefficients of vertical turbulent exchange. The typical profile of the buoyancy frequency with depth is shown in figure 2. The data collected were subjected to necessary pre-processing and validation, including cross-calibration with hydrological complex "Sea-Bird". To assess the extent of the turbulent patches in a stratified layer, we used the technique developed in [11]. On the density profiles the curve segment was selected with approximately constant gradient. On these segments vertical gradient spectra of temperature fluctuations versus wave number was built, figure 3. Effective vertical scale of the patches corresponds to the position of local minimum on the spectrum. Then one built the dependence of the average patches sizes versus buoyancy frequency $L \cong DN_c^{\alpha}$ at all profiles, where dimension of $[N_c]$ = cycle/hour, $[L]$ = m. By this dependence the values $\alpha \cong -1$ and $D \cong 1{,}4\,\mathrm{m}\cdot(\mathrm{cycle/hour})$ was determined; taking it into account the relationship for the vertical turbulent diffusion coefficient K was specified.

Figure 1. The probing measuring complex "Sigma-1"

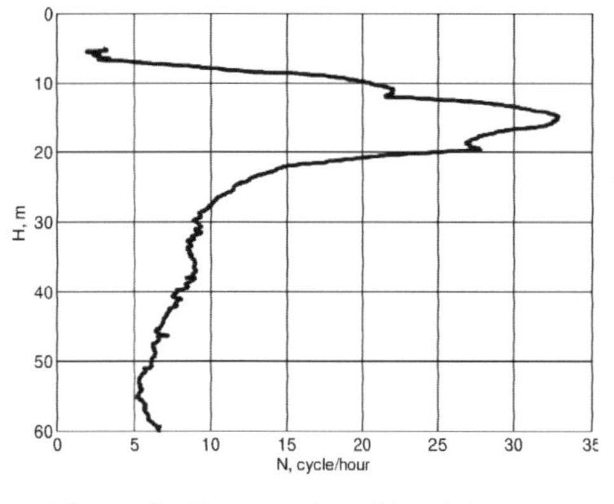

Figure 2. The typical profile of the buoyancy frequency N_c. R/V "Professor Vodyanitsky", 2016/07/03.

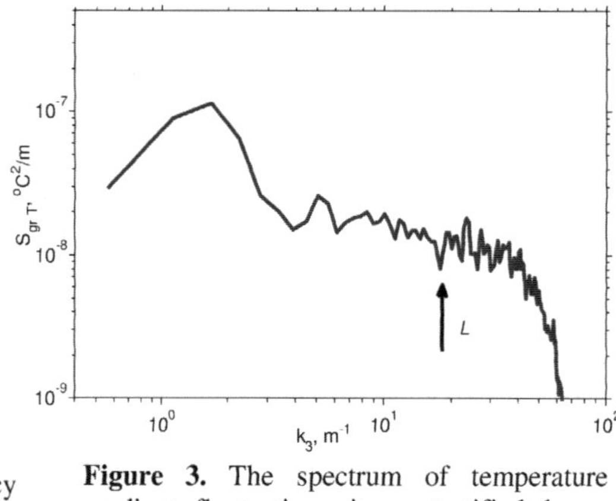

Figure 3. The spectrum of temperature gradient fluctuations in a stratified layer. 2016/07/01. R/V "Professor Vodyanitsky".

3. Analysis of measurement data

The main source of turbulence in a stratified layer in the study area is internal waves closed to the minimum of the inertial frequency. Due to shear instability the breaking of waves leads to the formation of turbulent patches. The occurrence of formation and size of these patches both determine the intensity of vertical turbulent diffusion and energy dissipation rate in stratified layers. In the investigated layer the contribution of temperature gradient to the vertical density distribution was crucial. That is why the calculation of the average patch thickness L was based on structure of the temperature first differences fluctuations spectra analysis. Previously it was shown that the effective vertical scale of natural turbulent patches corresponds to a vertical scale of stable minimum in the small-scale region of the spectrum [2], see figure 3. This approach allows finding the expression for the turbulent energy dissipation rate and turbulent diffusion coefficient in a wide interval of the N variability with involvement of semi-empirical model based on the energy balance in the system [11]. To the goal achievement during the 87-th cruise of R / V "Professor Vodyanitskii" (30.06.2016 – 18.07.2016), research are conducted to specify the dependence of the coefficient of vertical turbulent diffusion K from the frequency of buoyancy in the stratified layer from 20 to 80 m. The corresponding relationship for K on the top of main pycnocline (below the maximum of N) has the form [12]:

$$K(N_c) \cong 9{,}36 \cdot 10^{-5} (\tilde{N}_c)^{-1}\ \mathrm{m^2 s^{-1}}. \quad (1)$$

Here $\tilde{N}_c = N_c \cdot (\text{cycle/hour})^{-1}$ – dimensionless value, and N_c – as above cyclic buoyancy frequency. In figure 4 it is shown the spatial distribution of the coefficient K.

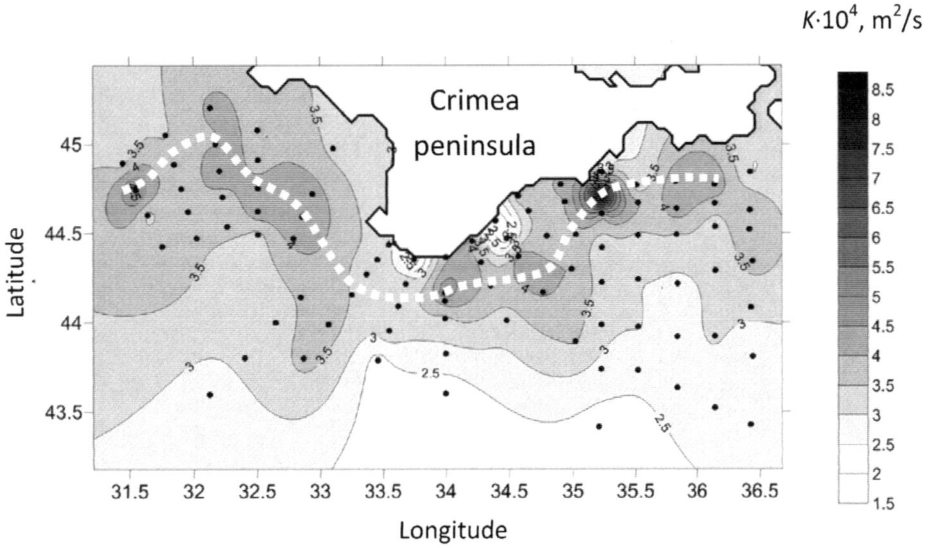

Figure 4. Horizontal distribution of vertical turbulent diffusion coefficient K in the layer of 20-80 m. The white dashed line is located close to the continental shelf and the continental slope interface area. Black dots mark the locations of stations.

Analysis of measurement data shows a non-trivial result: the maxima of the diffusion coefficient are located in the interface area of the shelf and the continental slope (dark background). In the shelf zone and in the deep region the values of K in the same layer are reduced. Formally, this means that the frequency of buoyancy in the interface area is on average smaller in comparison with the surrounding areas. To explain this fact we can use the results of [13]. There it is shown that the bottom topography in the interface area leads to significant increasing of breaking intensity for the reflected from the slope quasi-inertial waves in contrast to the surrounding areas. We assume that the energy inflow of the quasi-inertial waves is equal to the unit area for the entire study area. Here we consider quasi-inertial transverse internal waves ($\vec{u} \cdot \vec{k} = 0$, where $\vec{u}$ is the velocity vector of the flow in the wave and $\vec{k}$ is the wave vector), which are quasi-horizontal periodic flows. Then, when waves are reflected from the features of the bottom relief in the interface area of the continental shelf and the continental slope, the wave beam contracts, substantially reducing the vertical scale of the waves and, accordingly, increasing the amplitude. This transformation of the waves after reflection leads to a shear instability of the wave flows by the Richardson criterion and to wave breaking. It should be noted the fact that on an average quasi-inertial waves in the basin are in a mode closed to unsustainable. All of the above suggests that the turbulent regime in this area should be accompanied by increased dissipation of energy and the weakening of stratification. To take into consideration the results of the above analysis, it can be argued that ultimately this should lead to formation and maintenance of a balanced stripe with a weakened stratification and an intensification of the vertical turbulent exchange over the interface area. This exchange mode was discovered in the presented field experiment.

References

[1] Munk W H and Anderson E R 1948 J. Marine Research 7 3
[2] Gregg M C 1989 J. Geophys. Res. 94 C7 pp 9686 - 9698
[3] Thorpe S A 1973 J. Fluid Mech. 61 pp 731-751
[4] Thorpe S A 1987 J. Geophys. Res. 92 5231
[5] Osborn T R 1974 J. Phys. Oceanogr. 4 pp 109 -115
[6] Preusse M et al 2010 Limnol. Oceanogr. 55(6) pp 2353 - 2365
[7] Zonta F et al 2012 J. Fluid Mech. 697 pp 1 - 29
[8] Waterman S et al 2013 J. Phys. Oceanogr. 55 pp 259 - 282
[9] Wunsh C and Ferrari R 2004 Annu. Rev. Fluid Mech.. 36 pp 281 - 314
[10] Samodurov A S et al 2005 *Physical Oceanography* **15** 5 pp 311 - 322
[11] Samodurov A.S. et al 1995 *Physical Oceanography* **6** 3 pp 177-190.
[12] Samodurov A S 2016 *Physical Oceanography* **5** 6 pp 32-42.
http://physical-oceanography.ru/repository/2016/6/en_201606_04.pdf
[13] Samodurov A S et al 2013 *Fundament. and applied hydrophys.* **6** 2 pp 12-24 (in Russian)

Aperiodic pressure pulsation under non optimal hydraulic turbine regimes at low swirl number

Skripkin S.G.[1], Tsoy M.A.[1], Kuibin P.A.[1], Shtork S. I.[1]

[1]Kutateladze Institute of Thermophysics, SB RAS, Novosibirsk 630090, Russia

E-mail:skripkin.s.g@gmail.com

Abstract. Off-design operating conditions of hydraulic turbines is hindered by pressure fluctuations in the draft tube of the turbine. A precessing helical vortex rope develops, which imperils the mechanical structure and limits the operation flexibility of hydropower station. Understanding of the underlying instabilities of precessing vortex rope at low swirl number is incomplete. In this paper flow regimes with different residual swirl is analysed, particular attention is paid to the regime with a small swirl parameter. Study defines upper and low boundaries of regime where aperiodic pressure surge is observed. Flow field at the runner exit is investigated by Laser Doppler Velocimetry and high-speed visualizations, which are complemented draft tube wall pressure measurements.

1. Introduction

Nowadays, hydropower stations integrated in electrical grid have a very important task to provide flexible changes of electrical grid load. It is ensured by operation hydraulic turbine beyond designed conditions. Under part or over – load or starts-stops turbines the flow downstream the runner has a residual swirl which together with expanding part of draft tube lead to formation of recirculation zone and emergence of spiral vortex breakdown know as precessing vortex core (PVC). It is well recognized that such phenomenon has negative impact on hydraulic turbine operating stability and efficiency of hydraulic unit.

Detailed description and study of physical principle leading to PVC phenomenon are paramount to development flow control techniques, the main task of which to reduce PVC influence on hydraulic system while maintaining the efficiency of the turbine. This phenomenon in hydraulic turbine is widely described in literature; these are investigations of twin vortex rope [1,2], pulsations at upper part load [3], deep part load and low load investigations [4,5], turbine transient processes [6] etc. Due to abundance of various turbines designs and wide operating range in which turbines have to operate under grid regulating, the study of flow under turbine runner are still demanded.

Particular attention should be paid to regimes close to optimal at best efficiency point (BEP), commonly at such regimes whether part or over – load the swirl of flow is quite small and close to 0.2-0.5. Thus precessing vortex rope is formed weak and no intensive . Nevertheless, Dorfler [7] mentions regime near 0.85 Q_{BEP} when precessing vortex rope rotates irregularly, with a stable period no longer than 4 revolutions. At disruption of PVC pattern pressure spikes in the draft tube are observed that correspond to partial collapse of the vortex rope.

Recently it was found that the cause of pressure spikes in regime close to BEP is vortex rings formed due to reconnection of unstable spiral vortex [8,9]. The authors also note PVC instability each

3-4 of vortex rope revolutions in average. Due to induced motion vortex ring moves close to a draft tube wall and generates pressure spike or drop. At that draft tube wall pressure pulsations caused by PVC in observed regime are insignificant.

In that work explored regime is expanded compare to previous study [9] where regime with vortex rings formation was found at free rotating runner, rotational speed is proportional to flow rate. Here runner rotational speed is controlled and varied from 0 to 800 rpm. This allowed us to establish the boundaries of the unstable regime with aperiodic pulsations in terms of the swirl parameter. Experiments are conducted at fixed Reynolds number Re = 10^5 (constant flow rate) and inlet flow conditions are changed only by varying runner rotational speed.

2. Methodology

The study are conducted at the experimental set-up of the Kutateladze Institute of Thermophysics. The test rig is designed in a closed loop configuration driven centrifugal pump with a maximum flow rate of up to 150 m^3/h. As a hydro turbine prototype simplified model of hydraulic turbine with straight draft tube is used. The flow distribution close to the distribution downstream a Francis turbine runner operating at non-optimal conditions is provided by combination of stationary guide vanes and forcibly rotating runner. An electromotor regulates the rotating speed of the runner through magnetic couple.

Two piezo resistive pressure transmitters are mounted in a draft tube wall to detect pressure disturbance caused by precessing vortex rope. Flow rate is held constant to compare draft tube wall pressure pulsation at constant Reynolds number and different swirl. Various flow regimes are obtained by runner rotational speed regulation. Detailed description and scheme of experimental set-up can be found in works [2, 9].

Contactless technique laser Doppler anemometry (LDA) is used to obtain quantitative information about flow structure which is needed to understand the flow phenomena and verification of numerical simulation. The flow exits the runner into the Plexiglas draft tube cone with full optical access for LDA measurements and visualizations. Flat draft tube cone walls significantly decrease distortions related to cone curative, tangential velocity is specified according methodologic proposed by Z. Zhang [10]. Correction in 8% of measured tangential velocity is specified near central axis of draft tube cone.

The axial and tangential velocity components used for the calculation of the flow swirl in the inlet cross-sections of the draft tube cone are obtained by performing LDA measurements in a 2-D mode in regular steps between the cone wall and the centerline. The velocities are measured every 5 mm, completing each measurement with a last point located on the cone centerline. Based on mean axial and tangential velocity distribution integral swirl parameter characterized a flow regime is calculated as follows:

$$S = \frac{\int_0^R VUr^2 dr}{R\int_0^R U^2 r dr} \tag{1}$$

where V - mean tangential velocity, U – mean axial velocity, R is the radius of the cross-sectional area. High-speed visualization complements of velocity profiles that gives complete flow pattern.

3. Experimental results

Vortex rope patterns at different runner rotational speed are shown on figure 1. At N =800 rpm (swirl number S ~ 1.5) PVC represents a close interweaving of smaller vortices rotating as a unit. At N =480 rpm PVC has pronounced spiral pattern and shorter spatial wavelength. On figure 1 one can see separated vortex ring marked dashed line which moves toward draft tube wall.

IOP Conf. Series: Journal of Physics: Conf. Series **899** (2017) 022016 doi:10.1088/1742-6596/899/2/022016

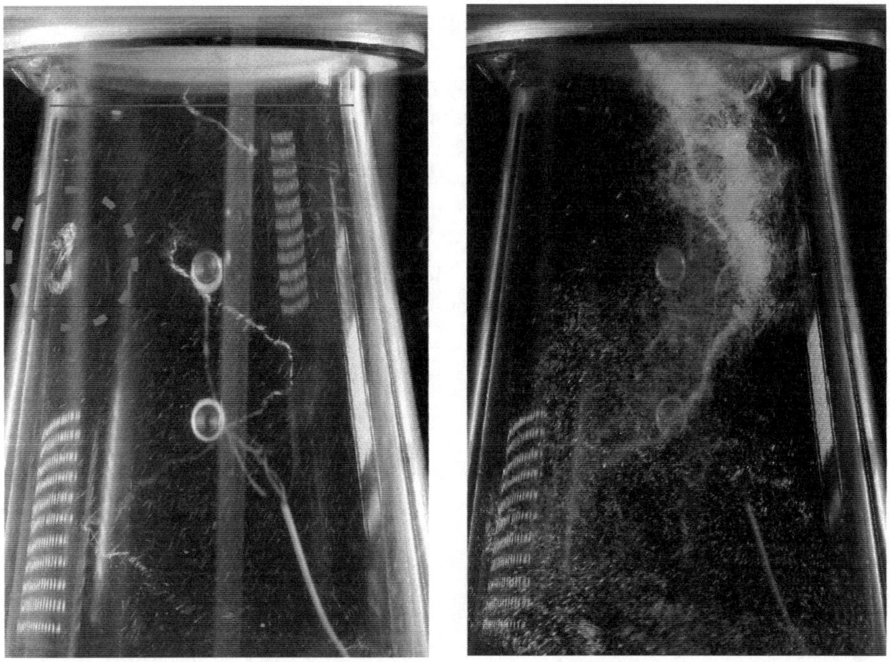

Figure 1. Visualization of vortex structure, N = 480 rpm (left), N = 800 rpm (right), dashed line marks separated vortex ring near wall, solid line indicates cross section for LDA measurements.

Pressure signal at N = 480 rpm and N = 800 rpm are presented on figure 2 (a, b). With runner rotational speed increasing i.e. at higher swirl number PVC radius is increased and peak-to-peak amplitude of wall pressure pulsation is increased too. Nevertheless, pressure spikes of the same order like at N = 800 are also observed in a regime near optimal when swirl of flow is low. It one can see at time 0.5 s on figure 2 (a). This pressure spike is caused by vortex ring and slightly differs from pressure drops related to vortex rope precession, the width in time of pressure drop detected by pressure transmitter in 3-5 times longer than pressure spikes.

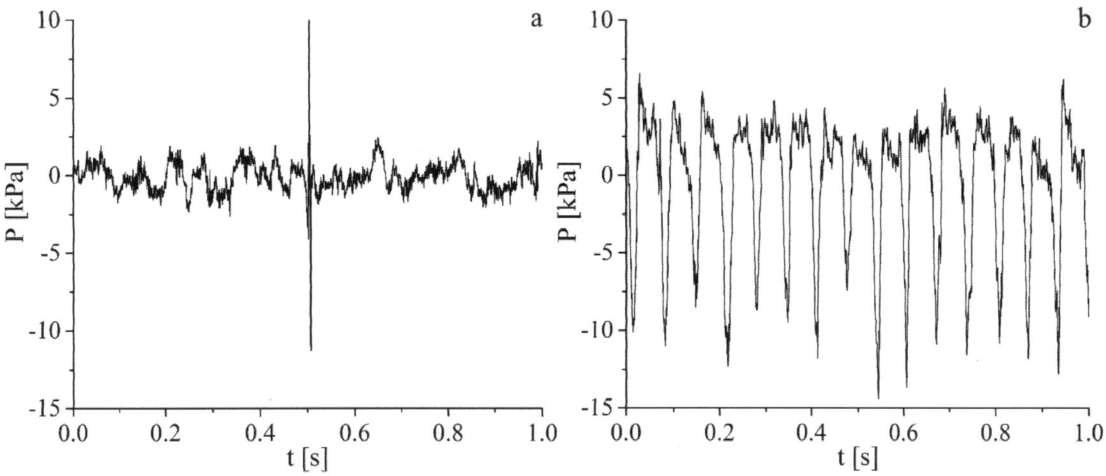

Figure 2. Dynamic pressure pulsations, zero on vertical axis correspond static pressure in measured cross section, N = 480 rpm (a), 800 rpm (b).

Upper and low boundaries in terms of runner rotational speed i.e. of swirl number are defined using high-speed visualization technique. As the speed of rotation of the runner decreases relative to the low boundary, the stagnation point is displaced downstream and the spiraling vortex structure ceases to exist, with an increase in the rotational speed above the upper boundary, the vortex spiral becomes stable and separation of the vortex rings is not observed.

Axial and tangential velocity distributions shown on figure 3 are measured in a throat part of draft tube conical diffuser marked solid line on figure 1 in boundaries regimes. One may see that the axial velocity distributions at upper and low boundary are the same, but recirculation zone is absent at a low boundary when column like vortex is formed. Dimensionless tangential distribution has a local minimum at r = 0.3R and in average in 1.8 times less than at upper boundary.

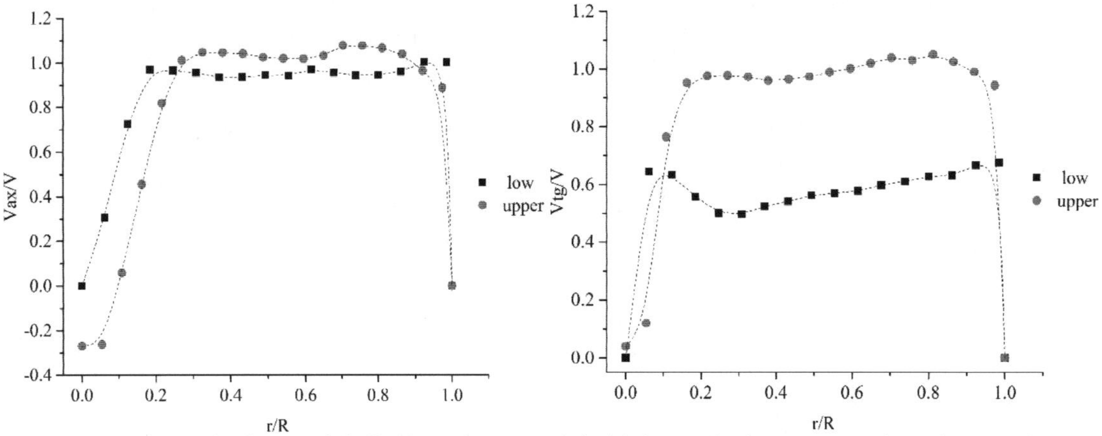

Figure 3. Dimensionless axial (left) and tangential (right) velocity measured at low and upper boundaries.

The dependence of integral swirl number on runner rotational speed is shown on the figure 4 at fixed flowrate Q = 70 m^3/h. As expected quicker runner rotation leads to increasing tangential velocity component and increasing swirl parameter. One can see that the dependence is well approximated by a linear function. Negative values of the swirl parameter correspond to the rotation of the flow counterclockwise twisted over the stationary guide vanes, in terms of hydraulic turbines this mode corresponds to overload. Dashed lines on figure 4 limits the unstable zone in which vortex ring formation is observed.

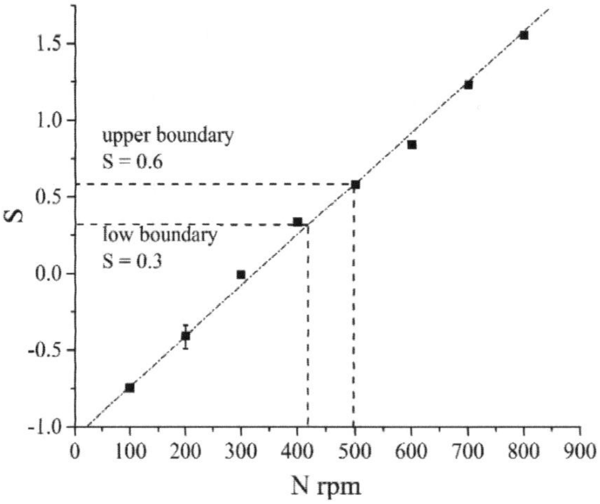

Figure 4. Dependence of swirl number on runner rotational speed.

Low threshold corresponds S = 0.3. Axisymmetric vortex column with breakdown point downstream is formed on the draft tube axis. With a slight increase in the runner rotational speed the breakdown point shifts upstream until it places on the tip of the cowl. Inside the unstable regime with increasing of swirl number the spatial period of the vortex rope and precession radius start to increase, which together leads to an increase in the local radius of curvature, and starting from S = 0.6 the vortex spiral becomes stable.

4. Conclusion

Study of the precessing vortex rope at low swirl parameter in the simplified draft tube model provides some new information regarding the vortex instabilities. Varying the rotation speed of runner at constant flow rate various flow patterns corresponding different swirl numbers are obtained. High-speed visualisation, LDA and pressure measurements provides insight into the physical understanding of phenomenon related to aperiodic pressure pulsations in hydraulic turbines. The boundaries of unstable regime accompanied by vortex ring formation are defined.

The most important is that near the optimal regime with a small swirl parameter S ~ 0.5, pressure shock caused by the vortex rings have the same amplitude with pulsations at the swirl parameter S ~ 1.5, which corresponds to a significant part load. The vortex ring formation in such complex system is very interesting from a practical and fundamental point of view, that requires further detailed investigations and explanations of principle leading to instability and vortex reconnection.

Acknowledgments

This work was supported by the Russian Science Foundation, grant No 16-19-00138.

References

[1] Wahl T L 1994 *Hydro Review* **13** 60–69.
[2] Skripkin S, Tsoy M, Shtork S, Hanjalic´ K *J. Hydraul. Res.* **54** 450-460.
[3] Nicolet C, Zobeiri A, Maruzewski P and Avellan F 2011 *International Journal of Fluid Machinery and Systems* **4** 179-190.
[4] Yamamoto K, Müller A, Favrel A, Landry C and Avellan F 2017 *Journal of Physics: Conference Series* **813** 012029.
[5] Minakov A, Platonov D, Litvinov I, Shtork S and Hanjalić K 2017 *J. Hydraul. Res.* 1-18.
[6] Trivedi C, Gandhi B, Michel C J 2013 J *Hydraul. Res.* **51** 121-132.
[7] Dörfler P, Sick M, Coutu A 2014 Flow-induced pulsation and vibration in hydroelectric machinery London:Springer.
[8] Alekseenko S V, Kuibin P A, Shtork S I, Skripkin S G and Tsoy M A 2016 *JETP letters* **103** 455-459.
[9] Skripkin S, Tsoy M, Kuibin P and Shtork S 2017 *J. Fluids Eng* **139** 081103
[10] Zhang Z 2004 *Exp. Fluids* **37** 29-39

IOP Conf. Series: Journal of Physics: Conf. Series **899** (2017) 022017 doi:10.1088/1742-6596/899/2/022017

Helicity sources in a rotating convection

A Teimurazov, A Sukhanovskii, A Evgrafova and R Stepanov

Institute of Continuous Media Mechanics, Academ. Korolyov, 1, Perm, 614013, Russia

E-mail: san@icmm.ru

Abstract. Convective flows from localized heater in a rotating layer were studied numerically in a three-dimensional non-stationary formulation. Distribution of helicity, its mean and fluctuating contributions were simulated for two specific regimes. In the first one the stable cyclonic vortex and intensive convective jet produce substantial amount of helicity near the axis of rotation. In the second one convective flow is more chaotic, cyclonic vortex appears at some distance from the center and as a result helicity is more dispersed in the lower layer of the fluid. Helicity in a described system is characterized by high level of pulsations. Spatial and temporal variations of helicity sources were analysed using equation for helicity balance. Time variations of viscous term and buoyancy term in helicity equation strongly exceed variations of other terms and the helicity time derivative. In a described system the buoyancy term is a source of helicity and viscous term is a sink. It was found that they are in antiphase and compensate each other.

1. Introduction

Nonlinear dynamics of fluids can lead to significant effects manifested in a particular flow structure. Explanation of these effects needs detailed consideration of physical mechanisms. Such analysis is getting especially complex when flow consists of structures of different scales. Their interactions can be local or nonlocal in the Fourier space and provide an effective transport of inviscid invariants up or down scale. Energy cascades (direct or inverse) in turbulent flows are well known examples. The theory of turbulence suggests a phenomenology which helps to consider a turbulent flow in the frame of simplified model, introducing eddy viscosity and simulate a mean flow. However a concept of homogeneous and isotropic turbulence is a very idealized view and should be used carefully because real turbulent flows are usually non-homogeneous and non-isotropic. Existence of boundaries leads to the different approaches for description of the bulk and boundary layer turbulence. In addition there are external volume forces like a buoyancy in the case of convection. This force may inject in flow not only kinetic energy but also helicity which is the second inviscid invariant. Helicity ($h = \mathbf{u} \cdot \boldsymbol{\omega}$, characterizes velocity-vorticity correlation) is observed in supercell of numerically simulated convective flow under rotation [1]. Also helicity can lead to a reduction of turbulent viscosity and energy accumulation in large scales [2, 3, 4] or play a significant role in the generation of large-scale helical vortex structures in the atmospheres of planets [5, 6, 7].

In the present paper we focused on nonlinear dynamics of the convective flow from localized heating source in the rotating layer. The general structure of the flow, its dependence on governing parameters were studied experimentally and numerically [8, 9, 10]. It contains global cyclone and anti-cyclone rotation and convective circulation. It was shown [11, 12] that described flow is characterized by strongly non-homogeneous distribution of helicity. Formation

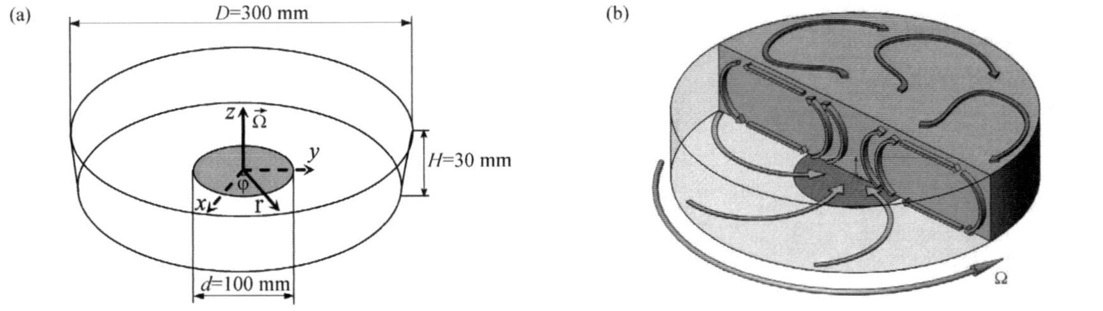

Figure 1. (a) – scheme of the computational domain, dimensions and location of the coordinate system; (b) – scheme of large-scale circulation.

of secondary flows in the shape of horizontal rolls and thermal plumes over localized heater [13] provides complex flow dynamics in the central part of the layer. All of these make this model flow a very interesting object to study a link between vortex formation and helicity. We aim to analyse helicity equation and define the dominant generation terms using numerical simulations.

2. Formulation of the problem and the mathematical model

The parameters of the numerical model were deliberately chosen similar to the experimental ones. The computational domain is a rotating cylinder with the the diameter $D = 300$ mm and height $H = 30$ mm (figure 1a). There is a heater with the diameter $d = 100$ mm in the central part of the bottom.

The problem is considered in a three-dimensional nonstationary formulation. The numerical code solves the Oberbeck-Boussinesq equations of thermal convection. All the simulations were run using the free and open source finite volume code OpenFOAM 4.1. The mesh has a regular block structure with a total of 3.5 million nodes with additional mesh refinement near the boundaries (mesh grid step is 0.25 mm near the boundaries). As a part of verification procedure numerical results were compared with experimental data [10] and numerical calculations [12]. Velocity profiles in figure 2 show good quantitative agreement between them. The PISO (Pressure Implicit with Splitting of Operators) algorithm based on the pressure correction procedure is used to solve the system of equations. The discretization of terms with time derivatives are done using an implicit Euler scheme. The convective terms are calculated by the "limitedlinear" TVD (Total Variation Diminishing) scheme [14]. The Courant number in our computations does not exceed 0.5. The equation for helicity evolution was implemented inside the solver in order to use the same numerical procedure for calculating its terms as for governing equations of thermal convection.

Silicone oil was used as a working fluid. Its physical properties at temperature 28°C, are: Prandtl number $Pr = 60$, kinematic viscosity $\nu = 5.35 \cdot 10^{-6}$ m^2/s, density $\rho = 904$ kg/m^3, and thermal expansion coefficient $\beta = 9 \cdot 10^{-4}$ K^{-1}. Rotating cylinder is considered thus the "rotating wall" boundary condition with angular velocity $\Omega = 0.0816$ 1/s and 0.170 1/s (which correspond to period of revolution 77 s and 37 s) is specified to bottom and sidewalls of the vessel. The free-slip velocity boundary condition is applied at the top surface. Fixed heat flux $Q = 17$ W is applied on the heater surface. Therefore, the heat flux density is $q_h = Q/S_h$. The equal amount of heat is removed from the top surface of the domain that is $q_t = -Q/S_t$. Here S_h and S_t are area of the heater and top surface of the model respectively. The sidewalls and the bottom except the area of the heater are heat insulated.

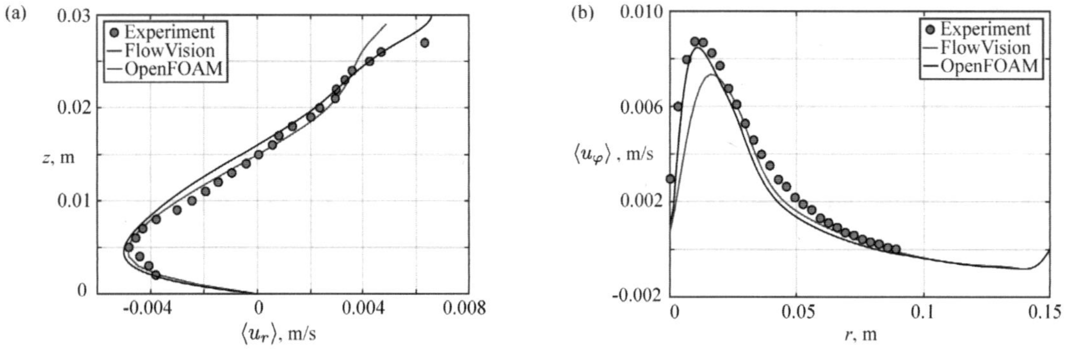

Figure 2. Time-averaged velocity profiles from experiment [10], numerical simulations using FlowVision [12] and OpenFOAM at $\Omega = 0.0816$ 1/s. (a) – radial velocity component $\langle u_r \rangle$ at the $r = 0.015$ m, (b) – azimuthal velocity component $\langle u_\varphi \rangle$ at $z = 0.015$ m (in a rotating frame).

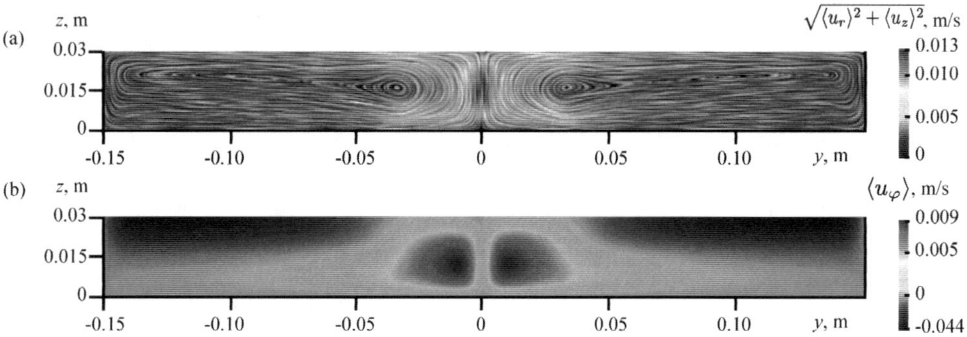

Figure 3. Time-averaged velocity field in the yOz plane, $\Omega = 0.0816$ 1/s: (a) – streamlines and magnitudes of meridional velocity (shown by color), (b) – azimuthal velocity component $\langle u_\varphi \rangle$ (in a rotating frame).

3. General structure of the flow

Detailed description of the flow formation can be found in [10] and here we present only brief description of the general structure of the large-scale flow. The heat flux in the central part of the bottom is a source of the intensive upward motion above the heater. Warm fluid cools at the free surface and moves toward the periphery where the cooled fluid moves downward along the side wall. Finally the large-scale advective flow occupies the whole vessel (figure 1b, figure 3a).

The cyclonic vortex formation in the laboratory system can be described by following scenario. Large-scale radial circulation leads to the angular momentum transport and the angular momentum exchange on the solid boundaries. Convergent flow in the lower layer brings the fluid parcels with large values of angular momentum from the periphery to the center and produces cyclonic motion (figure 1b, lower horizontal cross-section). In the upper layer situation is opposite – divergent flow takes the fluid with low values of angular momentum to the periphery resulting in anticyclonic motion (figure 1b, upper horizontal cross-section). Mean distribution of azimuthal velocity in a vertical cross-section is shown in figure 3b. The friction in the viscous boundary layers leads to the sink of angular momentum in the part of the bottom occupied by cyclonic flow and produces source of angular momentum on the sidewalls when anticyclonic flow comes to the periphery. Zero net angular momentum flux on the solid boundaries is the necessary condition for the steady-state regime [9].

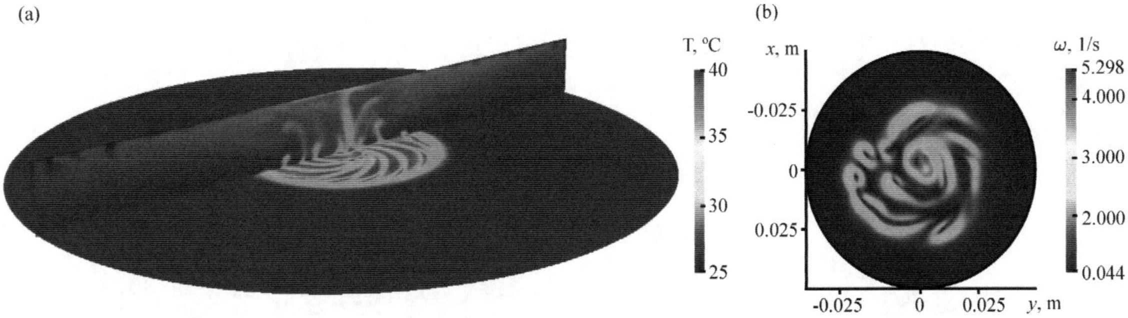

Figure 4. (a) – instant temperature field in the yOz plane at $x = 0$ and xOy plane at $z = 0.001$ m; (b) – magnitude of instant vorticity field in the xOy plane (only area above the heater is shown) at $z = 0.01$ m. $\Omega = 0.0816$ 1/s.

The large-scale advective flow in the lower part of the layer leads to the formation of boundary layer with unstable temperature stratification above the heater and generation of the horizontal rolls and thermal plumes (figure 4a). Horizontal rolls have spiral shape because of intensive cyclonic motion. One pair of rolls brings relatively cold fluid down and another pair takes warm fluid up. Periodically overheated parcels of fluid continue to float and produce thermal plumes moving to the center with the main flow.

Specific feature of the vortex formation in a rotating layer with localized heating is its strong dependence on the dimensionless rotation rate $Re = \Omega h^2/\nu$. Up to some values of Re (≈ 23) intensive cyclonic vortex is located in the center (figure 4b) and is quite robust (stable regime). At $Re > 23$ the vortex appears at larger radii and different azimuthal locations (unstable regime).

The main focus of the presented paper is a helicity generation in the described system. Instantaneous and mean helicity fields for the stable and unstable regimes are shown in figure 5. As we expected the change of the flow structure leads to the remarkable change of helicity distribution. Instantaneous helicity fields are strongly non-homogeneous as a result of small-scale structures dynamics (figure 5a,b). In the stable regime (I) mean helicity is concentrated in the central part where upgoing jet and intensive cyclonic vortex are located (figure 5c). In the unstable regime (II) mean helicity is spread in the lower part of the layer (figure 5d). The intensity of helicity pulsations can be characterised by the standard deviation (SD). Figure 5e,f shows the helicity SD fields. Helicity in a described system is characterized by high level of pulsations. For the regime I there are two distinct areas where peaks of pulsations are localized. Near the periphery of the heater where thermals are formed and near the center due to intensive upgoing jet fluctuations. The flow in regime II is less organized and more chaotic so pulsations are distributed more uniformly except periphery of the heater where secondary flows are absent. The level of pulsations is about half of the mean helicity values for the regime I and it is comparable with value of mean helicity in the regime II. It is useful to consider different contributions of helicity (parts of scalar product of velocity and vorticity vectors):

$$h = h_r + h_\varphi + h_z = u_r\omega_r + u_\varphi\omega_\varphi + u_z\omega_z. \quad (1)$$

Mean fields of helicity contributions are shown in figure 6. For the both regimes h_r is essentially weaker than h_z and h_φ.

In this section we focused on the qualitative description of helicity fields but understanding of helicity distribution and dynamics requires analysis of helicity sources which are considered in Section 4.

IOP Conf. Series: Journal of Physics: Conf. Series **899** (2017) 022017 doi:10.1088/1742-6596/899/2/022017

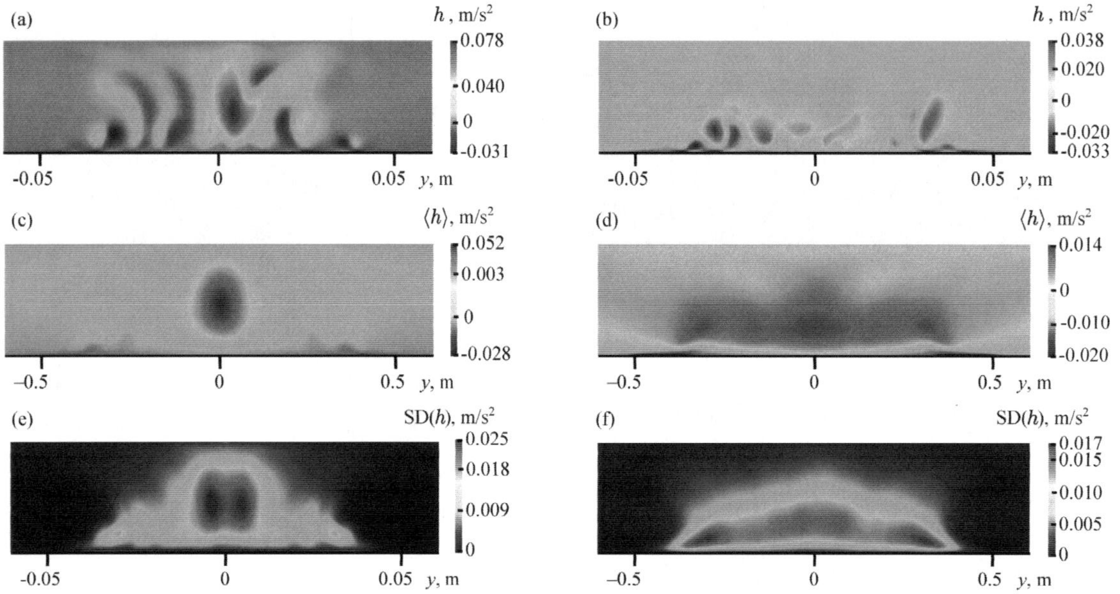

Figure 5. Distribution of helicity in the yOz plane (only central area is shown) at $x = 0$: (a), (b) – instant helicity field; (c), (d) – time-averaged helicity field; (e), (f) – standard deviation of the helicity field. Left column – $\Omega = 0.0816$ 1/s; right column – $\Omega = 0.170$ 1/s.

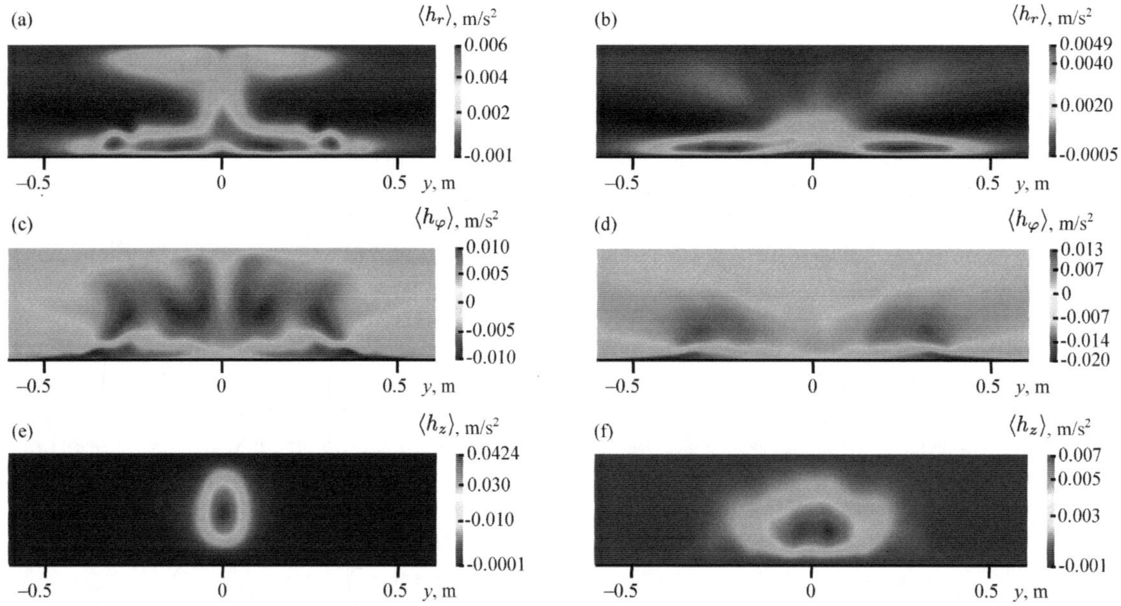

Figure 6. Distributions of different contributions of helicity in the yOz plane (only central area is shown) at $x = 0$: (a), (b) – time-averaged radial contribution of the helicity field $\langle h_r \rangle$; (c), (d) – time-averaged azimuthal contribution of the helicity field $\langle h_\varphi \rangle$; (e), (f) – time-averaged vertical contribution of the helicity field $\langle h_z \rangle$. Left column – $\Omega = 0.0816$ 1/s; right column – $\Omega = 0.170$ 1/s.

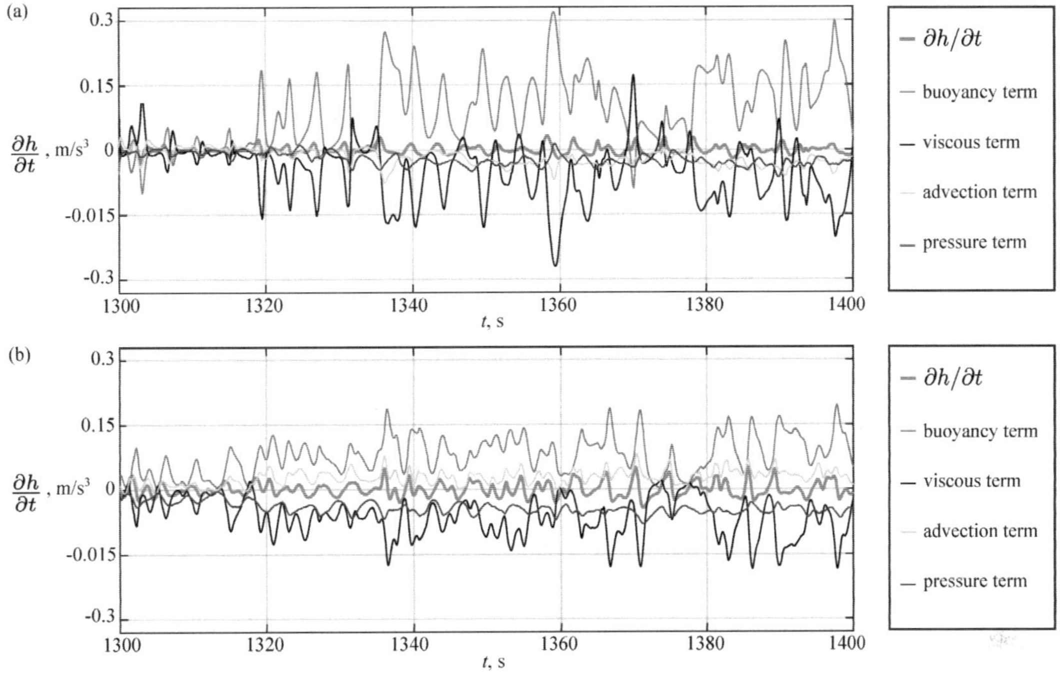

Figure 7. Time series for variations of the helicity time derivative and all the terms in the right side of the Eq. (2) at $\Omega = 0.0816$ 1/s for the points on the central axis of the cavity at $z = 0.003$ m (a) and $z = 0.015$ m (b).

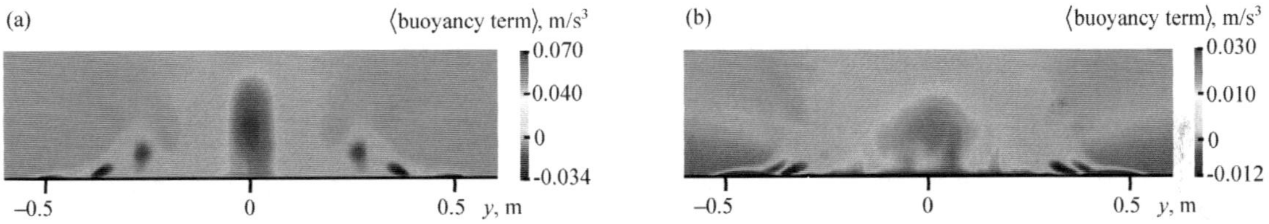

Figure 8. Time-averaged fields of the buoyancy term of Eq. (2) in the yOz plane, at $x = 0$, for two regimes: (a) – stable vortex ($\Omega = 0.0816$ 1/s), (b) – unstable vortex ($\Omega = 0.170$ 1/s).

4. Helicity balance

The governing equation for helicity evolution has the form [15]:

$$\frac{\partial h}{\partial t} = \mathbf{u}\cdot\nabla\times(\mathbf{u}\times\boldsymbol{\omega}) - \boldsymbol{\omega}\cdot(\mathbf{u}\cdot\nabla\mathbf{u}) - \boldsymbol{\omega}\cdot\nabla P/\rho + \nu(\boldsymbol{\omega}\cdot\nabla^2\mathbf{u} + \mathbf{u}\cdot\nabla^2\boldsymbol{\omega}) + \beta g(T\boldsymbol{\omega}\cdot\mathbf{e}_z + \mathbf{u}\cdot\nabla\times(T\mathbf{e}_z)), \quad (2)$$

where a gravity acceleration $\mathbf{g} = -g\mathbf{e}_z$. There are several terms which are responsible for helicity balance. First two terms at the right side of Eq. (2) correspond to the advection of helicity by the flow. The third term is related to the pressure gradient. The fourth term is responsible for viscous dissipation and the last one determines the buoyancy effects.

Due to complexity of helicity equation we decided to analyse helicity production at two specific points: at $z = 0.003$ m which is located in the boundary layer and in the middle of the layer at $z = 0.015$ m. Time series for variations of the helicity time derivative and all the terms in the right side of the helicity equation (buoyancy term, viscous term, advection term, pressure term) at $\Omega = 0.0816$ 1/s at different points on the central axis of the cavity are shown in figure 7. Time variations of viscous term and buoyancy term (two last terms in equation 2) strongly exceed variations of other terms and the helicity time derivative. The most interesting part is that they are in antiphase and compensate each other. In a described system the buoyancy term is a source of helicity and viscous term is a sink. The time evolution of helicity sources and sinks is very complex and its analyses requires long-time calculations. The spatial distributions of the buoyancy term for two regimes are shown in figure 8. In a stable regime the helicity generation by buoyancy term is concentrated in the center and in the unstable regime with more chaotic convective flows it is spread over larger area.

5. Conclusion

Convective flows from localized heater in a rotating layer were studied numerically using open source finite volume code OpenFOAM 4.1. Distribution of helicity, its contributions and pulsations were reconstructed for two specific regimes. In the first one the stable cyclonic vortex and intensive convective jet produce substantial amount of helicity near the axis of rotation. In the second one convective flow is more chaotic, cyclonic vortex appears at some distance from the center and as a result helicity is more dispersed in the lower layer of the fluid. Helicity in a described system is characterized by high level of pulsations. The level of pulsations is about half of the mean helicity values for the regime I and it is comparable with value of mean helicity in the regime II. Analysis of helicity sources using equation for helicity balance was carried out for two locations, in the boundary layer and in the middle of the layer. Time variations of viscous term and buoyancy term in helicity equation strongly exceed variations of other terms and the helicity time derivative. The amplitude of viscous and buoyancy terms are higher in the boundary layer but resulting helicity time derivative is larger in the middle of the layer. The buoyancy term produces helicity and viscous term efficiently devours most of it. This result is very interesting and deserves special attention. It is unclear how variation of dimensional parameters such as kinematic viscosity or coefficient of thermal expansion would change the ratio of buoyancy and viscous terms.

Acknowledgments

This work was supported by the Russian Science Foundation (grant RSF-16-41-02012). Computing resources of the supercomputer "Triton" were provided by Center of Shared Facilities (Institute of Continuous Media Mechanics UrB RAS).

References

[1] Marino R, Mininni P D, Rosenberg D and Pouquet A 2013 *Phys. Rev. E* **87** 033016 (*Preprint* 1211.3159)
[2] Yokoi N and Yoshizawa A 1993 *Phys. Fluids* **5** 464–477
[3] Stepanov R, Golbraikh E, Frick P and Shestakov A 2015 *Phys. Rev. Lett.* **115** 234501 (*Preprint* 1508.07236)
[4] Kessar M, Plunian F, Stepanov R and Balarac G 2015 *Phys. Rev. E* **92**(3) 031004
[5] Moiseev S S, Sagdeev R Z, Tur A V, Khomenko G A and Shukurov A M 1983 *Soviet Physics Doklady* **28** 926
[6] Ivanov M F, Gal'Burt V A and Fortov V E 1996 *JETP Lett.* **63** 813–817
[7] Lilly D K 1986 *J. Atmos. Sci.* **43** 126–140
[8] Bogatyrev G P 1990 *Soviet Journal of Experimental and Theoretical Physics Letters* **51** 630
[9] Batalov V, Sukhanovsky A and Frick P 2010 *Geophysical and Astrophysical Fluid Dynamics* **104** 349–368
[10] Sukhanovskii A, Evgrafova A and Popova E 2016 *Quarterly Journal of the Royal Meteorological Society* **142** 2214–2223
[11] Sukhanovskii A, Evgrafova A and Popova E 2016 *Journal of Physics: Conference Series* vol 754 (IOP Publishing) p 072005
[12] Sukhanovskii A, Evgrafova A and Popova E 2017 *arXiv:1705.00802 [physics.flu-dyn])*
[13] Sukhanovskii A, Evgrafova A and Popova E 2016 *Phys. D* **316** 23–33
[14] Versteeg H K and Malalasekera W 2007 *An Introduction to Computational Fluid Dynamics: The Finite Volume Method* (Pearson Education, Harlow)
[15] Kurgansky M V 2006 *Meteorologische Zeitschrift* **15** 409–416

IOP Conf. Series: Journal of Physics: Conf. Series **899** (2017) 022018 doi:10.1088/1742-6596/899/2/022018

Experimental study of shock-wave structure of supersonic underexpanded microjets

I Timofeev[1], V Aniskin[2] and A Maslov[2]

[1] Novosibirsk State University, 630090, Novosibirsk, Russian Federation
[2] Khristianovich institute of theoretical and applied mechanics, 630090, Novosibirsk, Russian Federation

e-mail: ivan.timofeev.koi@gmail.com

Abstract. The structure of supersonic underexpanded room temperature air microjets is experimentaly studied. Microjets escaping from the nozzles with characteristic heights of 22.3 and 83.3 μm and widths of 2410 and 3900 μm respactacly with the Reynolds numbers of 640 and 2390. At these Reynolds numbers hot-wire anemometer measurements of mass flow pulsations are performed. It is shown that low value of the mass flow pulsations are corresponding with the high supersonic core length.

1. Introduction

In view of the broad possibilities of application of microscopic gas and liquid devices (microfluidics), considerable research interest is presently devoted to studying fluid flows on the microscopic scale, including both the fluid flow in channels and jet emission from micron-sized orifices. Depending on the output pressure, a jet can be either subsonic or supersonic. Both subsonic and supersonic microjets used in macroscopic flow control, noise reduction, and jet cooling systems.

The number of reported experimental investigations devoted to supersonic microjets is not large [1–7]. Previously, we have studied in detail the structure of axisymmetric supersonic microjets [3, 4]. It was found that the relative shock cell size in micro and macrojets have no significant differences. However, at some values of the jet pressure ratio, the supersonic core length of microjets exhibits a significant increase as compared to that of macrojets. Here, the jet supersonic core length defined as the distance from the nozzle exit to a point on the jet axis where the flow velocity attains the local sound velocity value. A transition from the regime of a large supersonic core length to that characteristic of macrojets takes place at Reynolds numbers between 1100 and 2100 and related to the transition from laminar to turbulent flow.

There are numerous experimental and theoretical works devoted to studying the structure, noise, and interaction with obstacles for supersonic rectangular macrojets [8–10]. However, to the best of our knowledge, no published data are available on the supersonic core length of two-dimensional supersonic macrojets. Experimental investigations of the structure of supersonic microjets have not been reported except for our recent work [5], in which data were presented for the only one Reynolds number.

The goal of the present research was in getting deeper knowledge of the structure of the supersonic jet core. In this investigation hot-wire anemometer measurements were used to show the average signal and mass flow rate pulsations in the supersonic microjet.

Used nozzles were produced by the technology from [4]. Each nozzle had a cylindrical prechamber with a diameter of 4 mm and a wedge shaped narrowing taper part. In the present work, data are presented for jets escaping from nozzles with height of 83.3 μm and 22.3 μm.

2. Experiments

The nozzle dimensions and experimental parameters listed in the table 1. Here AR is the aspect ratio - ratio of width and height of the nozzle, JPR is the jet pressure ratio (defined as the ratio of static pressure at the nozzle exit to pressure in the surrounding space), Re is the Reynolds number calculated for the given nozzle height and gas parameters at the nozzle exit.

Table 1. Nozzle dimensions and flow parameters.

Height (μm)	Width, (μm)	AR	JPR	Re
83.3	3823	45.9	1.06	2390
22.3	2593	116	1.06	640

Previously supersonic flat underexpanded microjets studied using the method of flow visualization [7] and measuring the total pressure distribution along the jet axis [6]. Pitot tube measurements shows that for micronozzles with height below 50 μm realized a high supersonic core length effect.

Hot-wire anemometer experiment's scheme presented in figure 1. The working fluid was room temperature air, and the supersonic air microjet was escaping into atmosphere. The wire sensor, which had the wire diameter of 5 μm and the length of 200 μm, mounted in a holder that could be moved in three mutually perpendicular directions with the Narishige NT 88E micromanipulator and positioned in space to within an accuracy of ±1 μm. The sensor position monitored using the Nikon SMZ 1500 stereoscopic microscope.

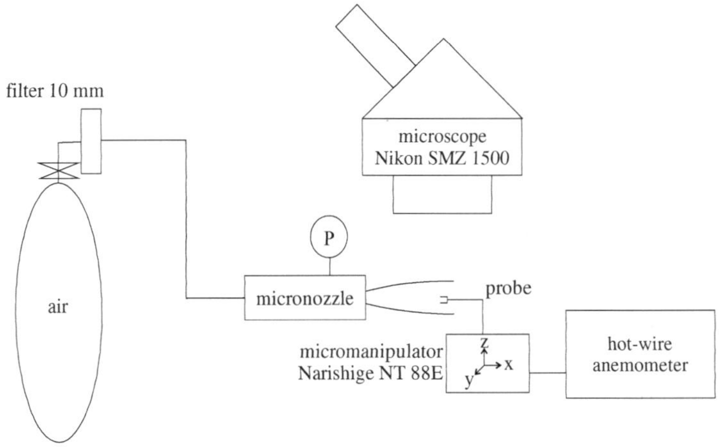

Figure 1. The hot wire measurements experiments scheme.

Figure 2(a) and figure 2(b) shows the typical microphotographs along the greater axis of the jets escaping from the nozzles with height of 83.3 μm and 22.3 μm consequently. Pressure in the prechamber equal to 2 atm.

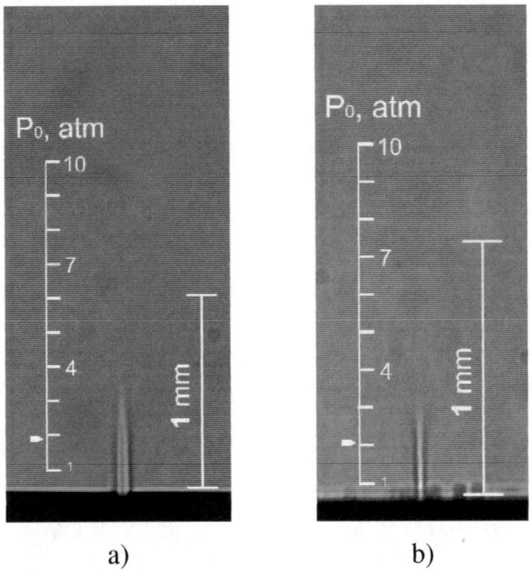

Figure 2. Photographs of air microjets escaping from the micronozzles at pressure in the prechamber equal 2 atm. Height of the micronozzle: (a) 83.3 μm, (b) 22.3 μm.

Figure 3 presents the longitudinal profile of the pressure measured by the Pitot tube in the jet escaping from the nozzle with height of 83.3 μm. The measurements performed along the line of intersection of the nozzle symmetry planes.

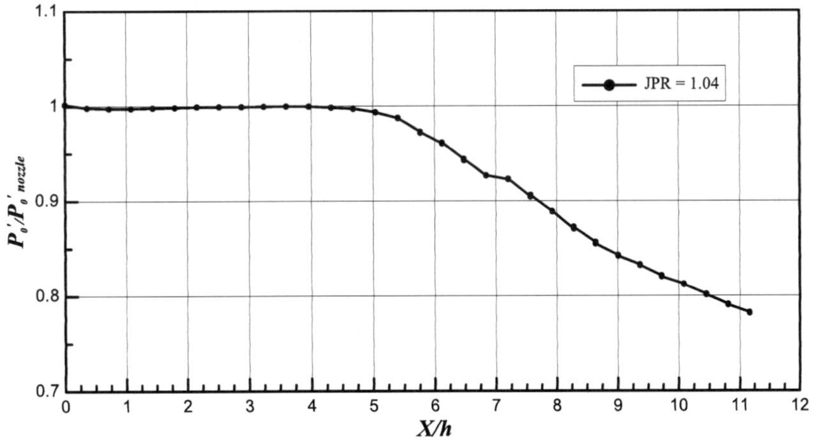

Figure 3. Longitudinal distribution of pressure along a jet escaping from the nozzle with h = 83.3 μm for various off-design parameters.

As can be seen, the P_0' value exhibits a smooth change of pressure along the microjet. The pressure value in the prechamber is not high enough that a shock-wave structure is realized in the jet.

The supersonic core length of the flat underexpanded microjets was determined from P_0' distributions along the line of intersection of the nozzle symmetry planes. The results presented in figure 4. However, the supersonic core length of the jet escaping from nozzle with h = 22.3 μm at JPR < 1.5 is greater than the values for other microjet, i.e. long supersonic core length regime. Analogous increase in the jet range observed for supersonic axisymmetric microjets [4]. Indeed, it was found in [4] that the supersonic core length of axisymmetric microjets escaping from nozzles with diameters above 60 μm corresponds to that of axisymmetric macrojets, while the microjets escaping from nozzles with diameters below 60 μm exhibit long supersonic core length.

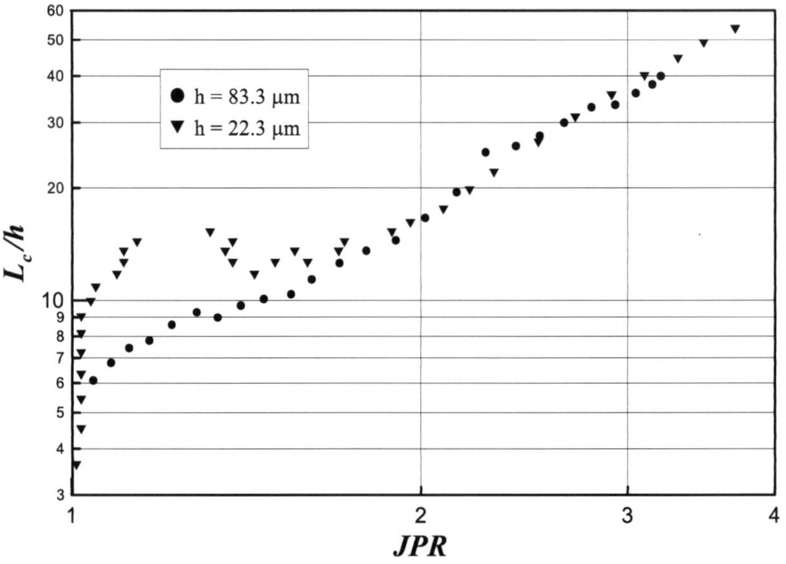

Figure 4. Supersonic core length vs JPR.

Figures 5 and figure 6 shows contours of the average signal (a) and the RMS mass-flow pulsations (b) for the supersonic jets escaping from micronozzles with the height of 83.3 μm and 22.3 μm reference. Pressure in the prechamber in both cases were 2 atm. As can be seen for the contours of microjet escaping from the nozzle with height of 22.3 μm (figure 6) at the distance of 10 nozzle sizes from nozzle plane, microjet starts to expand. It depends on the laminar-turbulent transition in the jet-mixing layer. Also we could notice that in the case of the jet escaping from the nozzle with height of 83.3 μm jet starts to expand closely to the nozzle exit.

Recently, we have demonstrated [11] that the long supersonic core length regime of supersonic axisymmetric microjets related to a laminar flow in the jet-mixing layer. It suggested that an increase in the supersonic core length of supersonic flat underexpanded microjets also related to a laminar flow in the mixing layer.

IOP Conf. Series: Journal of Physics: Conf. Series **899** (2017) 022018 doi:10.1088/1742-6596/899/2/022018

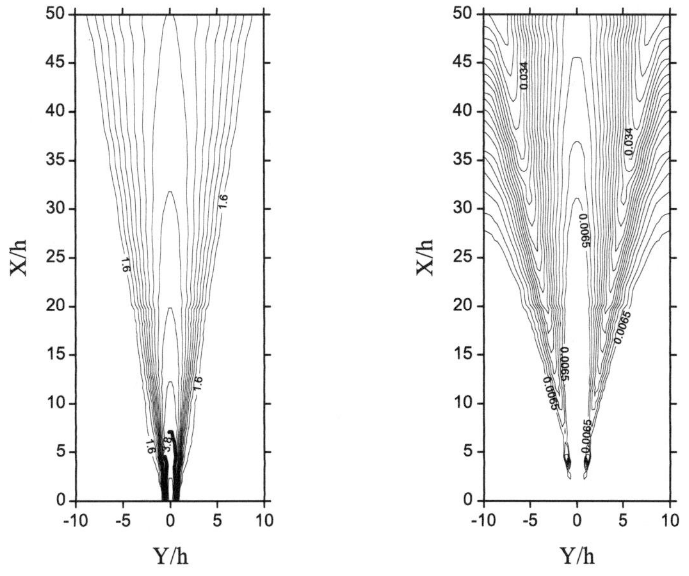

a) b)

Figure 5. The contours for microjet escaping from the nozzle with height of 83.3 μm: (a) average signal and (b) RMS signal.

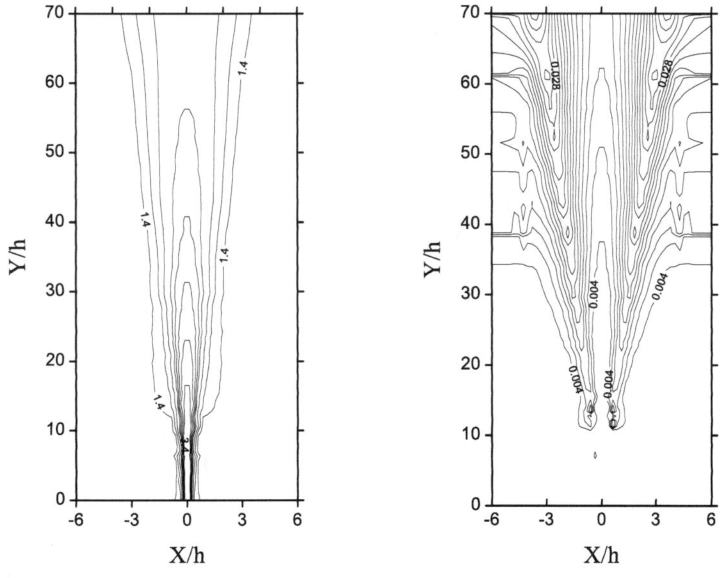

a) b)

Figure 6. The contours for microjet escaping from the nozzle with height of 22.3 μm: (a) average signal contour and (b) RMS signal.

As for the axisymmetric microjets, high supersonic core length effect is realized for the two-dimensional microjets. The growths of the supersonic core length are associated with the laminar flow in the jet-mixing layer. Supersonic core length of the jet escaping from the nozzle with height of 22.3 μm is equal to the $L_c/h = 10$.

In the jet's downstream the flow starts to expand. This effect could be seen in the plot of average and RMS signals. For the jet escaping from the nozzle with height of 83.3 μm, length of laminar mixing layer is commensurate with nozzle height. The turbulence of the jet-mixing layer occurs starts closer to the nozzle exit. Due to this effect jet starts to expand on the distance of X/h = 2.5 and supersonic core length is equal to $L_c/h = 5.5$.

3. Conclusion
Hot-wire anemometer measurements shows that flow regime of jet-mixing layer are affect to the supersonic core length. It is shown that laminar flow regime in the jet-mixing layer contribute to the length of the supersonic section increases.

Acknowledgments
This study was supported in part by the Russian Science Foundation 17-19-01157.

References
[1] Scroggs S D and Settles G S 1996 An experimental study of supersonic microjets *Experiments in Fluids* vol 21, pp 401-409
[2] Phalnicar K A, Kumar R and Alvi F S 2008 Experiments on free and impinging supersonic microjets *Experiments in Fluids* vol 44, pp 819-830
[3] Aniskin V M, Maslov A A and Mironov S G 2011 Effect of nozzle size on supersonic microjet length *Technical Physics Letters* vol 37(11), pp 1046-1048
[4] Aniskin V M, Mironov S G and Maslov A A 2013 Investigation of the structure of supersonic nitrogen microjets *Microfluid. Nanofluid* vol 14(3), pp 605-614
[5] Aniskin V M, Mironov S G and Maslov A A 2012 The structure of two-dimensional and axisymmetric microjets *International Journal of Microscale and Nanoscale Thermal and Fluid Transport Phenomena* vol 3(1), pp 49-59
[6] Aniskin V M, Maslov A A and Mironov S G 2013 Relaminarization in supersonic microjets at low Reynolds numbers *Technical Physics Letters* vol 39(8), pp 734-736
[7] Aniskin V M, Maslov A A, Tsirulnikov I S and Timofeev I V 2015 Visualization of supersonic axisymmetric and plane underexpanded microjets *Journal of Flow Visualisation and Image Processing* vol 22 (4), pp 213-227
[8] Sheeran W J and Dosanjh S 1968 Observations on jet flows from a two- dimensional, underexpanded, sonic nozzle *The American Institute of Aeronautics and Astronautics Journal* vol 6, pp 540-542
[9] Kashimura H, Masuda Y, Miyazato Y and Matsuo K 2011 Numerical analysis of turbulent sonic jets from two-dimensional convergent nozzles *Journal of Thermal Science* vol 20(2), pp 133-138
[10] Tam C K W 1988 The shock-cell structures and screech tone frequencies of rectangular and non-axisymmetric supersonic jets *Journal of Sound and Vibration* vol 121(1), pp 135-147
[11] Aniskin V M, Mironov S G, Maslov A A and Tsyryulnikov I S 2015 Supersonic axisymmetric microjets: structure and laminar–turbulent transition *Microfluidics and Nanofluidics* vol 19(3), pp 621–634

IOP Conf. Series: Journal of Physics: Conf. Series **899** (2017) 032001 doi:10.1088/1742-6596/899/3/032001

Nonlinear waves in a falling film with phase transition

S P Aktershev[1], S V Alekseenko[1,2]

[1] Kutateladze Institute of Thermophysics, Lavrentiev Ave. 1, Novosibirsk 630090, Russia
[2] Novosibirsk State University, Pirogova St. 2, Novosibirsk 630090, Russia
E-mail: sergey-aktershev@mail.ru

Abstract. Nonlinear wave formation and heat transfer in wavy film flowing over the isothermal wall in the present of phase transition are studied numerically. The integral-boundary-layer model, modified with account of the phase change at the interface has been used to describe the wave motion. For the first time the nonlinear evolution of forced two-dimensional waves was investigated, and wave effect on heat transfer was determined. It is shown that forced waves essentially intensify heat transfer within a certain range of frequencies as compared to the case of naturally occurring waves. Heat transfer enhancement by waves due to the predominant contribution of the thin residual layer between the peaks was demonstrated. It is shown that by applying the superimposed periodic oscillations, one can intensify heat transfer within a certain range of frequencies as compared to the case of naturally occurring waves.

1. Introduction

Intensive investigation of thin liquid films is caused by their wide application in engineering. Film condensation on the cooling wall and film evaporation on the heated wall are ones of the most important applications. The theoretical study of this problem began with the Nusselt's [1] pioneering work where the stationary laminar film condensation has been considered. In most practically important cases, the film flow is unstable and the liquid surface is covered by developing waves. It is known that the waves in liquid film intensify processes of heat and mass transfer even for laminar flow. Thus, in [2, 3] heat transfer coefficients at evaporation and condensation have been experimentally measured for a wave flow of laminar liquid film. The measurement results are 40-80% higher than the theoretical values for a smooth film. Wave generation and transport processes in falling liquid films without the phase transition were investigated in many studies; the results of these studies are summarized in [4]. The wave dynamics and heat transfer in non-isothermal films are much less studied. In [5] the linear stability of the condensate film is considered in the presence of vapor flow over the film surface. Dispersion dependences, curves of neutral stability, and characteristics of maximum growth waves were obtained.

The purpose of the presented research is to study the nonlinear evolution of waves in the film with a phase transition, and to determine the effect of waves on heat transfer. An integral-boundary-layer model [6] which takes into consideration phase transition is applied to describing nonlinear waves in a liquid film.

2. Theoretical model

Let us consider a vertical laminar film flow on a uniformly heated or cooled plate. Let us introduce Cartesian coordinate system *Oxy* with *Ox* axis directed downward, and *Oy* axis normal to the plate. Let us assume the following basic simplifications, acceptable for a wide range of practically important flow conditions. 1) Plate temperature $T_w = const$, the liquid surface is in contact with stagnant saturated vapor with temperature $T_s = const$. 2) The perturbation of the film surface is considered to be longwave (the typical length of the perturbation is much greater than the film thickness). 3) Density ρ, dynamic viscosity μ, liquid conductivity λ, surface tension σ, and latent heat of evaporation r are assumed constant. Let us introduce the distance scale l along the Ox axis and take film thickness h_m at $x = 0$ as the distance scale along the Oy axis. We introduce velocity scale $u_m = gh_m^2/3\nu$, time scale $t_m = l/u_m$, flow rate scale $q_m = u_m h_m$, temperature scale $\Delta T = |T_s - T_w|$, and move on to the dimensionless variables x/l, y/h_m, h/h_m, q/q_m, t/t_m, u/u_m, v/u_m keeping for them the same notations. In dimensionless variables the film flow with phase transition is described by the set of equations [7, 8] with respect to film thickness $h(x,t)$, flow rate $q(x,t)$ and temperature $\theta(x,\eta,t) = \pm(T - T_w)/\Delta T$:

$$\begin{aligned} &\frac{\partial q}{\partial t} + \frac{\partial}{\partial x}\left(\frac{6q^2}{5h}\right) = \frac{3}{\chi \operatorname{Re}_m}\left(h - \frac{q}{h^2}\right) + \chi^2 Weh\frac{\partial^3 h}{\partial x^3}, \\ &\frac{\partial h}{\partial t} + \frac{\partial q}{\partial x} = \pm\frac{A}{\chi \operatorname{Re}_m h}, \\ &\frac{\partial \theta}{\partial t} + u\frac{\partial \theta}{\partial x} + \frac{W}{h}\frac{\partial \theta}{\partial \eta} = \frac{1}{\chi \operatorname{Re}_m \Pr h^2}\frac{\partial^2 \theta}{\partial \eta^2}. \end{aligned} \tag{1}$$

Here and in other formulae sign "plus" corresponds to condensation and sign "minus" corresponds to evaporation, $A = \varepsilon \cdot \partial\theta/\partial\eta|_{\eta=1}$, $\varepsilon = c_p \Delta T/(r\,Pr)$ is a parameter characterizing intensity of phase transition, $Re_m = gh_m^3/3\nu^2$ is the Reynolds number at the inlet, $\chi = h_m/l$ is the ratio of scales, $We = (3Fi/Re_m^5)^{1/3}$, $Fi = \sigma^3/\rho^3 g\nu^4$ is the Kapitsa number, Pr is the Prandtl number,

$$W = \left(\eta - \frac{3\eta^2}{2} + \frac{\eta^3}{2}\right)\frac{\partial q}{\partial x} - \frac{3(\eta^2 - \eta^3)}{2}\frac{\partial}{\partial x}\left(\frac{qA}{4 + A}\right) - \frac{A\eta}{\chi\,Re_m\,h}.$$

For steady flow the effect of surface tension can be neglected. Under a condition $\varepsilon << 1$ the temperature profile is closed to linear $\theta = \eta$, and the solution to the first two equations (1) will be as follows [7, 8]:

$$h_{steady} = \left(1 \pm \frac{4\varepsilon \cdot x}{3\chi\,Re_m}\right)^{1/4}, \quad q_{steady} = h^3. \tag{2}$$

Wave flow regimes are obtained by numerical solution to equations (1) using the finite-difference method described in [7]. The waves were generated by a small artificial perturbation of the flow rate at the inlet $q(0,t) = q_0(1 + Q_a \sin 2\pi f t)$. Here q_0 is the undisturbed flow rate, Q_a is small amplitude, and f is flow rate oscillation frequency. The constant film thickness $h(0,t) = 1$ and the linear temperature profile $\theta(0,\eta,t) = \eta$ were set at the inlet. Linear temperature profile was set at time $t = 0$ for the all region of calculation; distributions of $h(x,0)$ and $q(x,0)$ were set by formulas (2). Boundary conditions for the energy equation are $\theta|_{\eta=1} = 1$, $\theta|_{\eta=0} = 0$. Region of computation was large enough to be able to trace the development of the waves downstream. Duration of calculations was also chosen sufficiently large to assure that the wave regime is developed over the all computation region.

For the computation of heat transfer from the liquid to the isothermal plate, time averaged local Nusselt number was calculated as $\langle Nu_x \rangle = \frac{1}{t}\int_0^t Nu(x,t)dt$. Time averaged local Reynolds number was calculated at the same time: $\langle Re_x \rangle = \frac{1}{t}\int_0^t Re(x,t)dt$. Here $Nu(x,t) = \frac{1}{(3Re_m)^{1/3} h(x,t)} \frac{\partial \theta}{\partial \eta}\Big|_{\eta=0}$ and $Re(x,t) = Re_m\, q(x,t)$ are instantaneous values of the local Nusselt and Reynolds numbers.

3. Simulation results

All calculations were carried out for water at the temperature of 373 K ($Pr = 1.75$, $Fi^{1/3} = 14700$) for $Q_a = 0.01$, $\varepsilon = 0.005$. Reynolds number at the inlet $Re_m = 1$ in a case of condensation and $Re_m = 40$ in a case of evaporation. The natural waves caused by instability of film flow appear in the absence of artificial perturbations at the inlet. In a case of condensation the perturbations of the film surface become obvious in the narrow zone near $x = 0.4$ m; above this zone the film remains smooth. At distance 0.4 –0.45 m from the inlet perturbations develop fast into the nonlinear waves of high amplitude. Similar natural waves were observed in [8] at numerical simulation of the R11 condensate film flow.

3.1. Forced waves

The forced waves imply the waves, which develop in the liquid film due to periodic external action. In this case the wave pattern over a certain frequency range takes the regular 2D character. This allows the detailed investigation of wave dynamics and identification of the mechanism of wave effect on transfer processes that is the main goal of the current research. Usually, in experiments with forced waves the pulsations of liquid flow rate are used. As applied to the isothermal films, the forced waves are described in detail in [4]. The wave surface of condensate film at forcing frequency of 18 Hz, which slightly differs from frequency of developed natural waves, is shown in Fig 1. The forced waves have the regular character and develop at a distance of 0.25–0.3 m from the inlet, i.e. earlier than in case of natural waves. Upstream the wave amplitude is so small that perturbations are not visible in Fig 1. Time period $1/f$ corresponds to the distance between the peaks indicated in Fig. 1. We should note that the term “wavelength” here is conventional. Since the thickness of undisturbed film and the average film flow velocity increase downstream, there is no strict spatial periodicity of the waves. We note also that despite the amplitude of developed wave increases downstream, the residual layer thickness between the peaks does not depend on coordinate. The film surface at lower frequency of 6 Hz is shown in Fig. 2. Here, in contrast to the case with $f = 18$ Hz, we can observe spatial irregularity, i.e. existence of lower peaks between the high peaks. Appearance of additional peaks caused by the growth of higher harmonics in the region of nonlinear wave development occurs at frequencies significantly lesser than 18 Hz.

On Fig. 3 the wave surface of an evaporating film for frequency of 18 Hz is shown. Like the condensation case, the forced waves are regular and develop much earlier, than natural ones (on

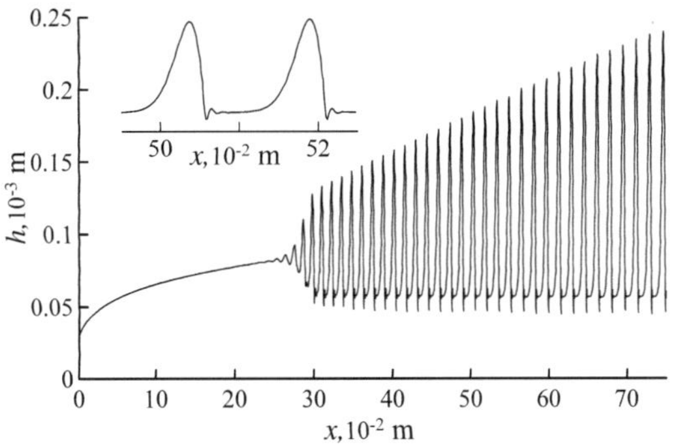

Figure 1. Forced waves in a condensate film; $f = 18$ Hz; Detail structure of the wave is shown in the upper insertion

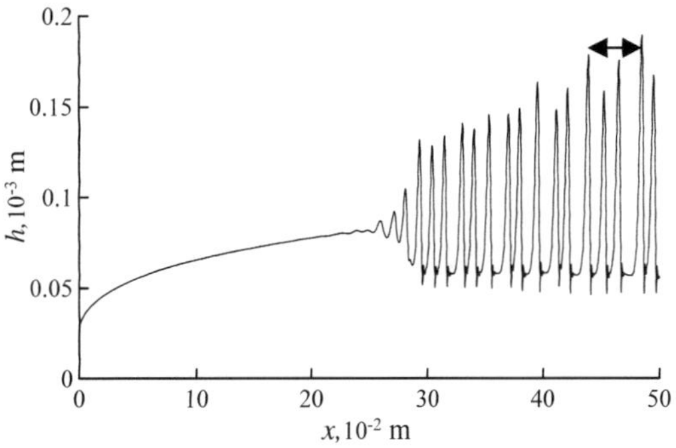

Figure 2. Forced waves in a condensate film; $f = 6$ Hz. The length with the arrows shows the "wave length" corresponding to 6 Hz.

IOP Conf. Series: Journal of Physics: Conf. Series **899** (2017) 032001 doi:10.1088/1742-6596/899/3/032001

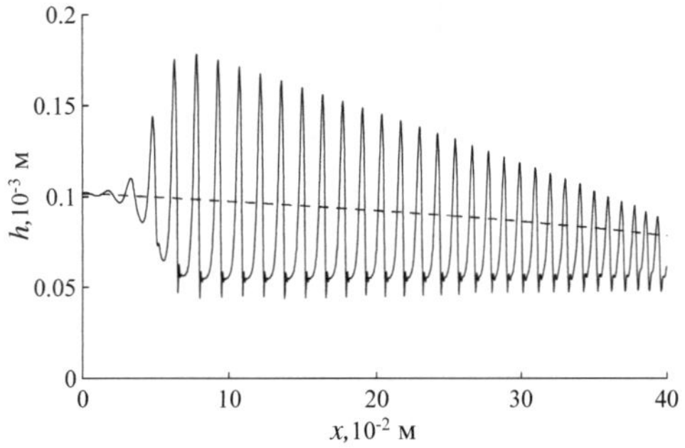

Figure 3. Forced waves, f = 18 Hz (evaporating film).

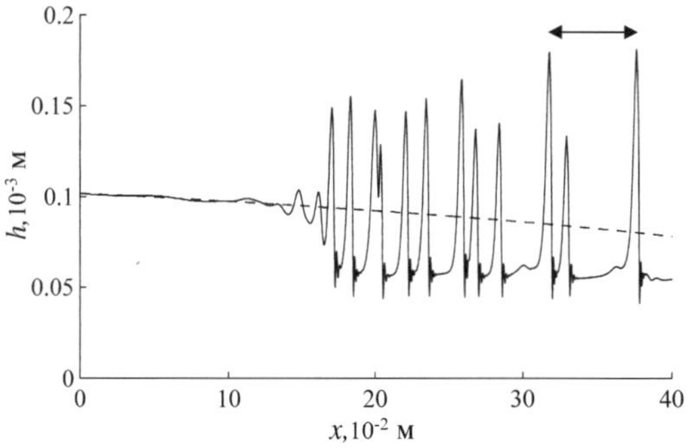

Figure 4. Forced waves, f =5 Hz (evaporating film). The length with the arrows shows the "wave length".

distance of 4–5 wavelengths from an inlet). Apparently from a figure, the amplitude of the developed waves in evaporating film smoothly decreases with distance, unlike a condensate film for which the amplitude of the forced waves increases downwards on a stream. As well as in case of condensation, the time period $1/f$ corresponds to distance between adjacent peaks. As an undisturbed film thickness decreases with distance, here there is no strict space periodicity of a wave. So for frequency of 18 Hz the space interval between peaks smoothly decreases from 1,5 cm on an input to 0,9 cm on distance of 0,4 m owing to decreasing of movement velocity of peaks. The wavy film surface for frequency of 5 Hz is shown on Fig. 4. Here intermediate peaks like on Fig. 2, appear already at a linear stage simultaneously with the main peaks. On an interval in one "wavelength" are formed two additional peaks. As a result of merging of intermediate peaks the amplitude of the main peaks considerably grows. As a whole, for an evaporating film as well as for a condensate film the same features of evolution of the forced waves, are observed. At frequency comparable to ones of natural waves the steady wave mode takes place with monotonously varying amplitude downwards on a stream. At frequency essentially smaller, than ones of natural waves, intermediate peaks are formed already at a linear stage. Unlike a condensation case, the wave amplitude decreases with distance, but thickness of a residual layer increases downstream a little.

3.2. The effect of waves on heat transfer

The temperature profiles in a condensate film are shown in Fig. 5 for forcing frequency of 9 Hz in different film cross-sections. According to the figure, in the region of residual layer (cross-section 1) the temperature profile does not differ from the linear one, and in the peak area (cross-sections 2, 3, 4) it differs significantly from linear trend. This proves a considerable contribution of convection into heat transfer in the peak area. This fact is also proved by Fig. 6, where the streamlines are shown in a

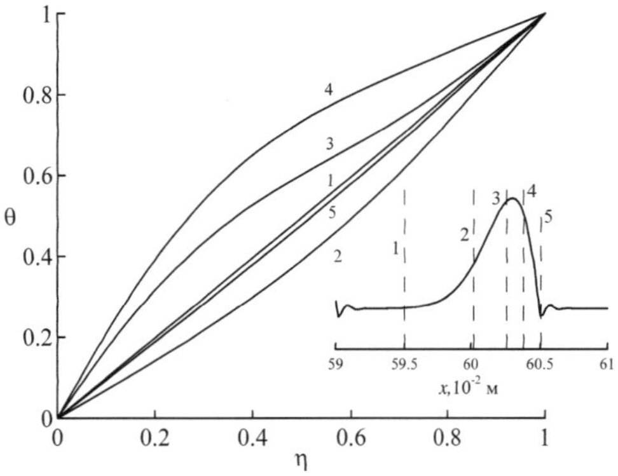

Figure 5. Temperature profiles in the cross-sections 1-5; condensate film; forced waves.

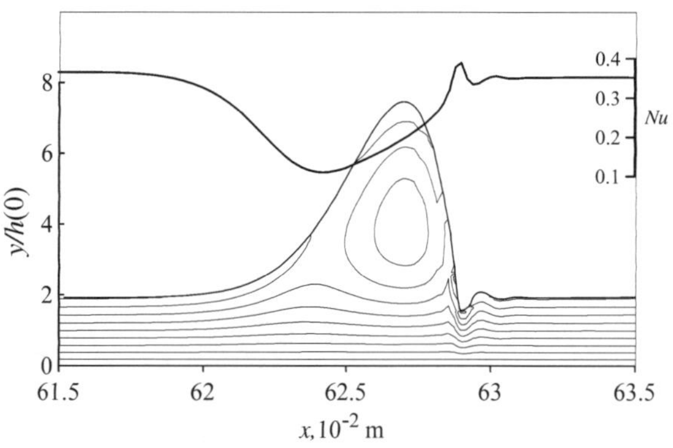

Figure 6. The streamlines in a reference frame related to a wave (bottom) and local Nusselt number (top).

reference frame moving with the speed of the wave. As it follows from the figure, there is a zone of circulation near the maximum, where convection should enhance heat transfer in transverse direction. The zone of circulation was observed repeatedly at numerical simulation of large waves.

Dependences of time-averaged local Nusselt number $\langle Nu_x \rangle$ on coordinate are shown in Figures 7 and 8 for different values of forcing frequency. In the region of smooth film calculation coincides with theoretical dependence, but in the region of nonlinear wave development the value of $\langle Nu_x \rangle$ increases rapidly. The coordinate, at which the nonlinear wave starts to develop, depends on frequency, but in all cases the forced waves develop earlier than the natural ones. Calculations show that heat transfer enhancement by the waves occurs mainly because of a decrease in the film thickness between the peaks. The local Nusselt number is determined by the length of the region between the peaks and depends on the coordinate and the forcing frequency. The distance between the peaks increases because of wave merging, and value of $\langle Nu_x \rangle$ increases significantly owing to absorbing of lower peaks by the higher ones.

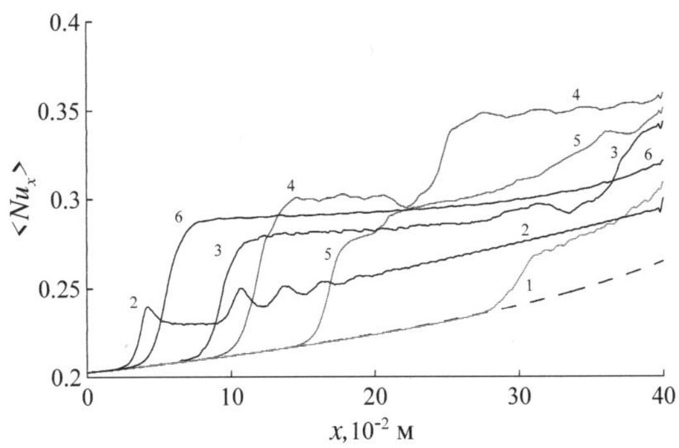

Figure 7. The Nusselt number versus coordinate for evaporating film; 1 – natural waves; forced waves: 2 – 30Hz, 3 – 10Hz, 4 – 8Hz, 5 – 5Hz, 6 – 18Hz; dashed line: the theoretical value for smooth film.

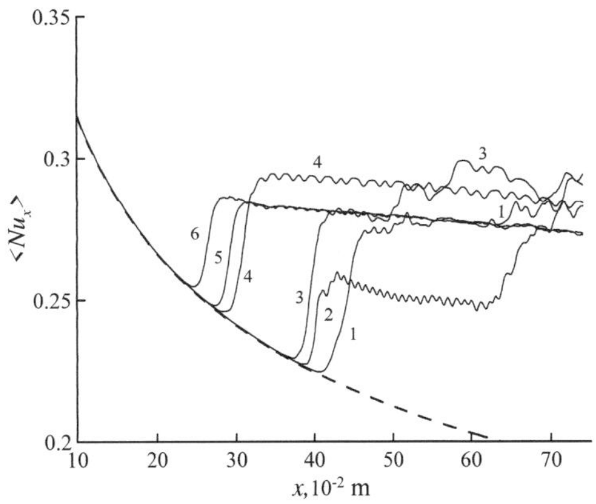

Figure 8. Nusselt number versus coordinate for condensate film; 1 – natural waves; forced waves: 2 – 25Hz, 3 – 3Hz, 4 – 5Hz, 5 – 18Hz, 6 – 9Hz; dashed line: the theoretical value for smooth film.

4. Conclusions

In the presented paper, a numerical method is used to study nonlinear wave formation and heat transfer in the film of water flowing down a vertical wall in the presence of a phase transition. For description of the wave motion of the liquid film the integral-boundary-layer model modified with considering the phase transition is applied. The equations were solved using a finite difference method for the spatial statement of the problem. The evolution of both natural waves arising due to the instability of the flow, and forced waves generated by small periodic perturbation of flow rate at the inlet, was investigated. Natural waves appear at some distance from the inlet and very quickly transform into large-amplitude nonlinear waves with a thin residual layer between them. Because of stochastic behavior of the natural waves, their parameters at the nonlinear development stage change randomly, that leads, in particular, to the merging of the individual waves.

The application of small periodic forced perturbation within a certain frequency range results in appearance of regular waves with stable characteristics. For the first time the evolution of forced waves in a film with phase transition is studied in a wide range of frequencies. The coordinate, at which the nonlinear wave starts to develop, depends on frequency, but in all cases the forced waves develop earlier than the natural ones. At sufficiently low forcing frequency, the intermediate peaks (generation of higher harmonics) are observed. This results in the wave interaction with the subsequent merge of the individual peaks. Sufficiently high forcing frequency results in appearance of subharmonic instability, which also leads to the interaction and merging of the waves.

Heat transfer calculations show that at the appearance of waves (both natural and forced) heat transfer coefficient abruptly increases up to a certain level, depending on the frequency of the waves. The analysis of the calculation results shows that the heat transfer coefficient in the wavy film is described rather accurately in the approximation of a linear temperature profile, i.e. the local Nusselt number is proportional to the value of $1/h$, where h is local film thickness. Thin residual layer between peaks gives main contribution to the heat transfer intensification. The vortex zone observed in nonlinear waves also leads to the heat transfer intensification, though this enhancement is insignificant as compared to the contribution of the residual layer. Wave interaction plays an important role in heat transfer; it manifests itself at low and high forcing frequencies, as well as in case of natural waves. This leads to merging of the waves and increasing the zone of residual layer, i.e. results in heat transfer enhancement. From a practical point of view, the results presented can be used to select the operation modes of industrial condensers and evaporators.

Acknowledgments
The work was supported by the Russian Science Foundation (grant No. 14-29-00093).

References

[1] Nusselt W 1916 *ZVDI.* **60**, 541-546
[2] Hirshburg R I and Florschuetz L W 1982 *Trans. ASME J. Heat Transfer.* **104**, 459-464
[3] Gross U, Storch Th, Philipp Ch and Doeg A 2009 *Int. J. Multiphase Flow.* **35**, 398-409
[4] Alekseenko S V, Nakoryakov V E and Pokusaev B G 1994 *Wave Flow of Liquid Films* (New York: Begell House 313 p)
[5] Aktershev S P and Alekseenko S V 2005 *Int. J. Heat Mass Transfer.* **48**, 1039-1052
[6] Shkadov V Ya 1968 *Fluid Dynamics.* **3**, 12-15
[7] Aktershev S P and Alekseenko S V 2013 *Physics of Fluids.* **25**, 083602
[8] Aktershev S P and Alekseenko S V 2014 *Bulletin of Tomsk Polytech. University.* **324**, 6-14
[9] Miyara A 2001 *Trans. ASME, J. Heat Transfer.* **123,** 492-500

IOP Conf. Series: Journal of Physics: Conf. Series **899** (2017) 032002 doi:10.1088/1742-6596/899/3/032002

Applying the relaxation model of interfacial heat transfer to calculate the liquid outflow with supercritical initial parameters

M V Alekseev[1], I S Vozhakov[1,2], S I Lezhnin[1,2] and N A Pribaturin[1]

[1]Kutateladze Institute of Termophysics SB RAS, Lavrentev Ave. 1, 630090, Novosibirsk, Russia

[2]Novosibirsk State University, Pirogov Str. 2, 630090, Novosibirsk, Russia

E-mail: alekseev@itp.nsc.ru

Abstract. A comparative numerical simulation of the supercritical fluid outflow on the thermodynamic equilibrium and non-equilibrium relaxation models of phase transition for different times of relaxation has been performed. The model for the fixed relaxation time based on the experimentally determined radius of liquid droplets was compared with the model of dynamically changing relaxation time, calculated by the formula (7) and depending on local parameters. It is shown that the relaxation time varies significantly depending on the thermodynamic conditions of the two-phase medium in the course of outflowing. The application of the proposed model with dynamic relaxation time leads to qualitatively correct results. The model can be used for both vaporization and condensation processes. It is shown that the model can be improved on the basis of processing experimental data on the distribution of the droplet sizes formed during the breaking up of the liquid jet.

1. Introduction

De-pressurization of vessels and pipelines in power plants, filled with water under high pressure, leads to a release of superheated water with a drastic pressure decrease in the area of rupture, formation of compression waves in the surrounding medium and further formation of the jet of the boiling-up coolant [1]. If thermodynamic parameters of the coolant (pressure, temperature) are in the supercritical region, there is a case when parameters of the fluid at outflowing pass not only through the curve of saturation and «boiling», but through the curve of «condensation» as well [2]. The process of adiabatic expansion, as a rule, is non-equilibrium and its description frequently requires a thermodynamic non-equilibrium two-temperature model of steam-water mixture motion [3]. In the simulation of vaporization in real conditions it is necessary either to set the parameters of the heterogeneous mechanism of bubbles formation (e.g. [4]), or to use a different (phenomenological) approach (e.g., relaxation [5]). In particular, in [4] it is proposed to set the number of vaporization centres added with asymptotic models of bubbles growth as an empirical parameter. Due to the use of a bubble structure to describe the two-phase medium, this approach loses its adequacy at large ($\alpha > 0.3 - 0.5$) volumetric steam qualities.

Earlier, authors of [6] based on the performed analysis proposed a simple relaxation model for the description of thermal and dynamic processes, involving water boiling at de-pressurization of high pressure vessel. The model assumes droplet structure of the liquid-vapor flow, which is typical at high differential pressure «vessel – atmosphere». This paper is an attempt to generalize this model to the case of non-equilibrium condensation.

There are many experimental studies of the features of the stationary jet outflow of superheated steam and boiling coolant on the obstacle under high pressure. For example, authors of [7] measured the impact of the steam jet on the obstacle and the pressure distribution over the obstacle radius. They showed similarity of the stationary outflow of the superheated steam and that of the gas jet. Experimental study of the stationary outflow of boiling coolant from the pipe end on the barrier in an industrial scale was carried out in [8]. An adequate calculation of the characteristics of the wave action on the barrier at small times is of great interest in modelling the flow of the coolant both with subcritical and supercritical parameters.

2. Governing equations

The present study considers an axisymmetric problem of water coolant outflow from high pressure vessel with the aim to obtain the characteristics of wave impact on the outer barrier and other parameters of the outflow. Numerical simulation is carried out in the absence of the phase slip, and given that the vapor and liquid phases can be in a metastable state. The system of model equations includes the Euler equations in the axisymmetric approximation:

$$\frac{\partial}{\partial t}(\rho r) + \frac{\partial}{\partial z}(\rho r u) + \frac{\partial}{\partial r}(\rho r v) = 0 \tag{1}$$

$$\frac{\partial}{\partial t}(\rho r u) + \frac{\partial}{\partial z}\left(\rho r u^2\right) + \frac{\partial}{\partial r}(\rho r u v) = -r\frac{\partial p}{\partial z} \tag{2}$$

$$\frac{\partial}{\partial t}(\rho r v) + \frac{\partial}{\partial z}(\rho r u v) + \frac{\partial}{\partial r}\left(\rho r v^2\right) = -r\frac{\partial p}{\partial r} \tag{3}$$

$$\frac{\partial}{\partial t}(Er) + \frac{\partial}{\partial z}(ur(E+p)) + \frac{\partial}{\partial r}(vr(E+p)) = 0 \tag{4}$$

added with the equation for interphase mass transfer:

$$\frac{\partial(\rho r X)}{\partial t} + \frac{\partial(\rho r u X)}{\partial z} + \frac{\partial(\rho r v X)}{\partial r} = \Gamma \tag{5}$$

$$\Gamma = \frac{\rho(X_S - X)}{\theta} \tag{6}$$

Here, z, r are the axial and radial coordinates, u, v are the axial and radial components of velocity, ρ is the density, p is the pressure, E is the specific internal energy, X is the steam quality and X_s is the equilibrium steam quality (at given p and E), Γ is the rate of steam generation, and θ is the relaxation time. The system of equations (1)-(6) is closed by the equation of state (relationship between p, E, ρ, X).

Calculation of properties of steam and water in sub- and supercritical condition was performed using the software package TTSE [9]. The program code was developed to solve a wide class of problems on de-pressurization of the vessels or pipelines with superheated water coolant with the use of computer complex LCPFCT [10].

Due to low density of steam the droplet structure of the two-phase medium is typical even at small steam qualities (but moderate and large volumetric ones) [11], and can be used in the modelling of intensive evaporation of liquid. To determine the relaxation time in (6), an analytic relaxation model [6] is used (as mentioned above), constructed on the basis of the asymptotic solution of the thermal problem in superheated water droplets of the same size:

$$\theta = \frac{R^2}{\pi^2 a} \tag{7}$$

In a real situation, the droplet size distribution takes place [12]. Therefore, according to the existing spectrum, it is necessary to calculate an effective droplet size. The droplet size is easy determined from the simplified analysis of instability of «water-steam» interface: the droplet size corresponds to the length of waves, the fastest destroying the interfacial surface, and is determined by critical Weber number We_{cr} approximately equals to 2π (water, liquid metals) [11, 13]. According to the model, at the micro-level of drops the heat required for vaporization is supplied from the superheated liquid. The temperature of the droplet surface is equal to the steam temperature [14]. To calculate the relaxation time we proposed the formula [6].

$$\theta = \begin{cases} \frac{\sigma^2 \rho_L^2}{4a_L \rho_V^2 (P-P_S)^2}, \ P - P_S > \frac{\rho_L L^2}{2\left(\frac{\rho_L}{\rho_V}-1\right)^2 C_{P,L} T} \\ \frac{\sigma^2 \left(\frac{\rho_L}{\rho_V}-1\right)^4 C_{P,L}^2 T^2}{a_L \rho_V^2 L^4}, \ P - P_S \leq \frac{\rho_L L^2}{2\left(\frac{\rho_L}{\rho_V}-1\right)^2 C_{P,L} T} \end{cases} \quad (8)$$

Here, σ is the surface tension coefficient, a is the thermal diffusivity, $C_{P,L}$ is specific heat of liquid and L is the enthalpy of phase transition. To use the relaxation model not only for evaporation, but condensation as well, we substantially modified the equation of state of two-phase medium, which closes the system of equations (1)-(6) to the case of existence of not only saturated, but supercooled steam. Although the size of drops formed in the condensation may change from the case of jet disintegration [11, 13], the formula (7) was used for test calculations. Simulation was carried out in two versions: a) with fixation of the relaxation time; b) taking into account its dependence on locally varying parameters (dynamic relaxation model, as well as in the paper [5]).

The relaxation time θ (7) is proportional to the square of the drop size, that is very sensitive to the value of this parameter. In addition, the characteristic size of the drop, according to the model, is determined by the magnitude of the critical Weber number Wecr and formally must reach this value asymptotically for long times. Therefore, the correct choice of the characteristic (effective) size remains an important problem. The most reasonable approach is to analyze the real distribution of droplets in sizes obtained from experiments on fragmentation of boiling-up jet upon its outflow from a pressure vessel (see, for example, the review [15]). Figure 1 presents a concrete example of such a droplet size distribution [12]. The right maximum of the distribution curve is associated with the classical jet fragmentation, which is determined by the We_{cr} and exists both for the hot and cold (no boiling up) liquid outflow. The left maximum is explained by the fragmentation of large droplets, caused by their explosive effervescence [12, 16] and for a cold liquid is not observed.

Applying the procedure for deriving the relaxation equation for a monodisperse medium [6] to a polydisperse discrete distribution, we obtain the following model evolution equations for the mass content of the liquid drops $X_{drop} = \sum_i X_i$:

$$\delta X_{drop} \approx \frac{C_{P,L}}{L} \sum_i X_i \delta T_i \quad (9)$$

$$\frac{dX_{drop}}{dt} = -\frac{C_{P,L}}{L} \sum_i X_i \frac{\delta T_i}{\theta_i} \quad (10)$$

where $\theta = R_i^2/\pi^2 a, \delta T_i = (T_0 - T_s(P))exp(-t/\theta_i)$

The procedure for choosing the effective (characteristic) droplet size is proposed as follows: the solution of equation (9) is compared with the exponent $exp(-t/\tau_{eff})$, where the parameter τ_{eff} is chosen so that this exponent, simultaneously with the exact solution (9), reaches a dimensionless sensation, for example, 0.01. This is equivalent to the generally accepted definition of boundary layer thickness in fluids.

The analysis shows that in a wide range of design parameters for water the equilibrium model is adequate with sufficient accuracy at θ less than 10^{-6} – 10^{-5} s. Therefore, it was logical to "cut off" droplets of small size, for which, for example, θ less than 10^{-6} s (for the experiment (Figure 1), this corresponds to drops with R less than 1.3 μm). Then we can assume that the liquid in small droplets instantly cools, forming a thermodynamically equilibrium vapor-liquid medium together with the vapor. In this case, relaxation takes place only on larger drops.

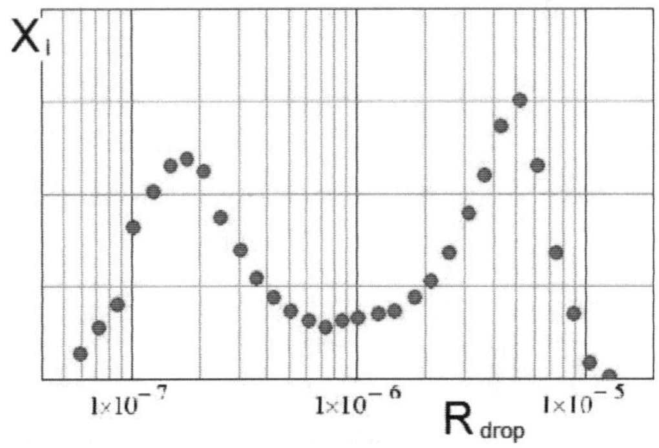

Figure 1. Dependence of the volume content of the liquid on the droplet radius at a distance of 60 mm from the cut of the nozzle at a water temperature of 473K and a pressure of 8MPa [12]

The characteristic radius of the droplets for the distribution in Fig. 1 calculated according to the proposed procedure with «cut-off» of small drops, is equal to 9.8 μm, which corresponds to a relaxation time value of $5.8 \cdot 10^{-5}$ s. The use of an exact solution with a full spectrum of drops substantially reduces the characteristic relaxation time, and it is not practical to use it.

For this reason, it is clearly inappropriate to use the so-called sauter mean size as a characteristic size, which is acceptable when the character of the heat exchange on droplets of different sizes is the same. Unfortunately, in many experiments to investigate the fragmentation of drops, namely this parameter is given (for example [16]).

In our opinion, for the logical completion of the relaxation model construction, it is necessary to carefully analyse all available experimental data on the distribution of droplets in the jet of boiling liquid, depending on the initial pressure and temperature. Then, based on the algorithm proposed above, we get a reasonable empirical formula for the effective droplet radius and relaxation time. Further comparison with a verified more complete program or an appropriate precision experiments will make the final model choice.

3. The calculation results

Earlier in [2] it was shown that a small change of initial parameters leads to a significant change in the nature of a supercritical fluid outflow, in particular, to a change in direction of phase transition. The present work provides calculations of the outflow of water in the supercritical state in a wide range of parameters at different initial parameters and different geometry of the computational domain.

Both equilibrium and non-equilibrium relaxation models of the phase transition were used. The thermodynamic equilibrium model of two-phase medium was, in fact, obtained from the general system (1)-(4) with its closing by the equation of state of the equilibrium two-phase mixture. A fixed relaxation time $\theta = 10^{-4}$ was chosen in accordance with the relaxation model (7) for the experimentally determined droplet sizes of 10^{-5} m [12].

In the numerical calculation, a dynamic relaxation model was also used. The relaxation time was determined from expression (8), which includes the local parameters of the mixture. It turned out that both at evaporation and condensation, the relaxation time varies in the range from 10^{-6}s to 10^{-3} s, and the main change occurs in the range from 10^{-5}s to 10^{-4}s. These values are in good agreement with the experimental data on relaxation time during steam condensation in Laval nozzles. In particular, in [16] it was shown that the characteristic relaxation time is of the order of 10^{-5}s.

Figure 2 shows the calculation results of spatial distribution of normalized pressure gradient and steam quality for the case of the boiling up water jet outflow and its impact on the barrier. Profiles are calculated at time of 0.3 s after de-pressurization for the dynamic relaxation model. Two pressure wave fronts are observed, as in the case of the outflow of a coolant located at subcritical parameters [17]. The first pressure front corresponds to the compression wave, the second front to the rarefaction wave. It can be noted that the difference in the fields of the pressure gradient for boiling and condensation processes is inessential . The amplitudes of the pressure waves for these cases are also close. A significant difference can be observed for fields of mass vapor content (Fig. 2, bottom). Inside the channel,both the boiling front (Fig. 2a) and the condensation one (Fig. 2b) propagate. The velocity of the condensation front is higher than the velocity of the boiling one. In both cases, the regions of sharp changes in the mass vapor content correspond to the regions of the fronts of the sudden change in pressure.

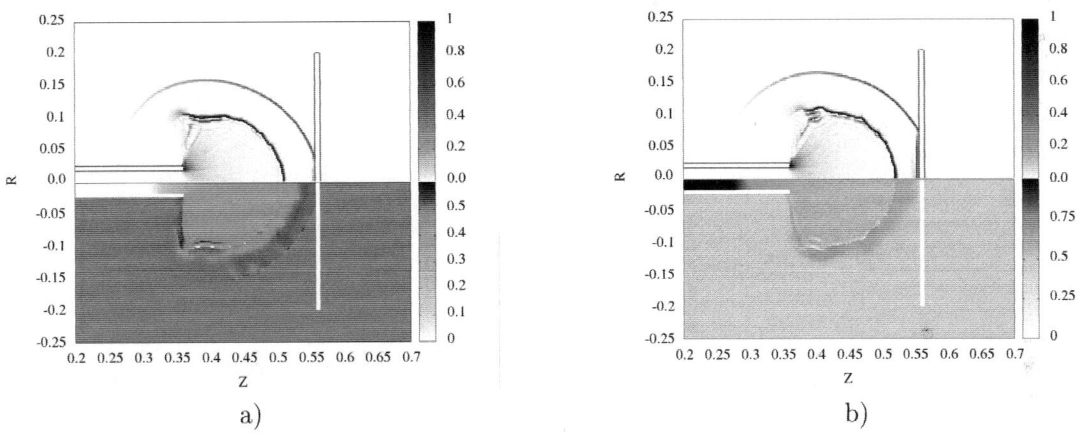

Figure 2. Spatial distribution of normalized pressure gradient (top) and steam quality (bottom) at time t =300 μs for initial conditions $P_0 = 23$ MPa, $T_0 = 370^0$C (a) and $T_0 = 380^0$C (b)

Fig. 3, 4 shows axial distributions of steam quality calculated by three models at different points in time. It is seen that inside the channel the calculation results obtained by the equilibrium model and for the dynamic relaxation time coincide with graphical accuracy. The small value of the relaxation time (8) is most likely due to the close to zero value of the surface tension σ in the near-critical region. At a greater distance from the nozzle the relaxation time increases to values of $\theta \approx 10^{-4}$s. This leads to an increasing deviation of the results of the calculation on the dynamic model from the equilibrium model and the approximation of the results with a fixed relaxation time.

4. Conclusion

We have performed a comparative numerical simulation of the supercritical fluid outflow on the thermodynamic equilibrium and non-equilibrium relaxation models of phase transition for different times of relaxation: fixed, based on the experimentally determined radii of the drops, and dynamically changing, depending on local parameters. The application of the proposed model with dynamic relaxation time leads to qualitatively correct results. The model can be

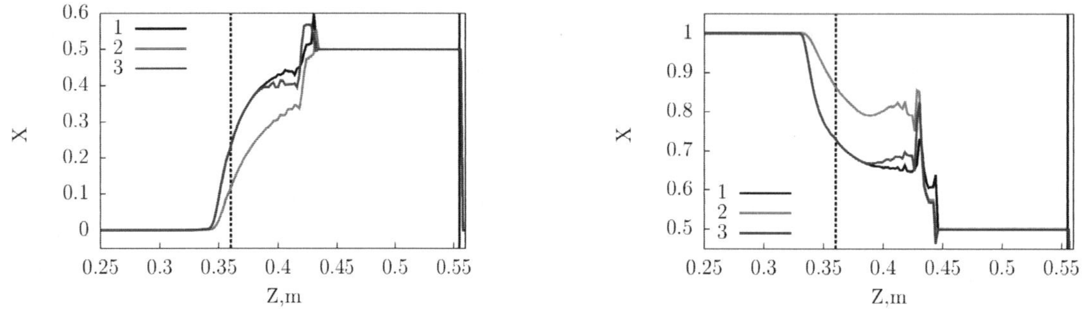

a) b)

Figure 3. Axial distribution of steam quality for the initial conditions $P_0 = 23$ MPa, $T_0 = 365^0$C (a) and $T_0 = 385^0$C (b), $t = 90\mu$s (1 – equilibrium model; 2 – fixed relaxation time $\theta = 10^{-4}$s; 3 – dynamic relaxation model).

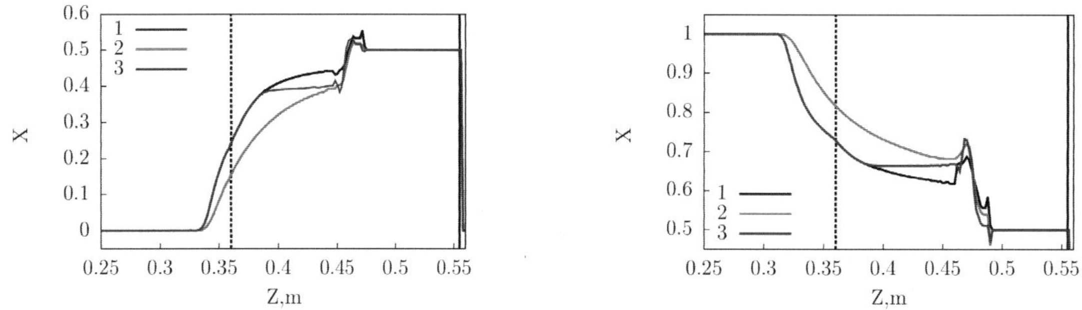

a) b)

Figure 4. Axial distribution of steam quality for the initial conditions $P_0 = 23$ MPa, $T_0 = 365^0$C (a) and $T_0 = 385^0$C (b), $t = 150\mu$s (1 – equilibrium model; 2 – fixed relaxation time $\theta = 10^{-4}$s; 3 – dynamic relaxation model).

used for both vaporization and condensation processes. It is shown that the model can be improved on the basis of processing experimental data on the distribution of the droplet sizes formed during the breaking up of the liquid jet.

Acknowledgments

The work was financially supported by the Russian Science Foundation (grant No. 14-29-00093).

References

[1] Alekseev M V, Vozhakov I S, Lezhnin S I and Pribaturin N A 2016 *Thermophysics and Aeromechanics* **23**(6) 905 – 914

[2] Vozhakov I S, Alekseev M V, Lezhnin S I and Pribaturin N A 2016 *Journal of Physics: Conference Series* **754** 032021

[3] Artemov V, Minko K and Yan'kov G G 2016 *Thermal Engineering* **62**(12) 897 – 905

[4] Bolotnova R and Buzina V 2014 *Computational Continuum Mechanics* **7** 343–352

[5] Zapolski P D, Bilicky Z, Bolle L and Franco J 1996 *Int. J. Multiphase Flow* **22**(3) 473 – 483

[6] Lezhnin S I, Vozhakov I S, Alekseev M V and Bogomolov A 2016 *Vestnik of KusSTU* **116**(5) 93 – 101 (in Russian)

[7] Masuda F and et al 1982 *Nuclear Engineering and Design* **67**(2) 273 – 286

[8] Isozaki T and Miyazono S 1986 *Nuclear Engineering and Design* **96** 1–9
[9] 2003 *Guideline on the Tabular Taylor Series Expansion (TTSE) Method for Calculation of Thermodynamic Properties of Water and Steam Applied to IAPWS-95 as an Example. The International Association for the Properties of Water and Steam* Vejle, Denmark
[10] Boris J P, Landsberg A M, Oran E S and Garder J H LCPFCT - Flux-Corrected Transport Algorithm for Solving Generalized Continuity Equations. NRL/MR/6410-93-7192
[11] Nigmatulin B I and Soplenkov K I 1980 *High Temperature* **18**(1) 118 – 131 (in Russian)
[12] Batenin V M, Zalkind V I, Zeigarnik Y A and Nizovskiĭ V L 2010 *Doklady Physics* **55** 115–118
[13] Levich V G 1959 *Physico-chemical hydrodynamics* (Moscow: State publishing house of physico-mathematical literature) (in Russian)
[14] Landau L D and Lifshitz E M 1986 *Hydrodynamics (Theoretical Physics, Vol. 6)* (Moscow: Nauka) (in Russian)
[15] Polanco G, Holdo A E and Munday G 2010 *Journal of hazardous materials* **173** 2 – 18
[16] Saltanov G A 1972 *Supersonic two phase flows* (Minsk: High school) (in Russian)
[17] Alekseev M V, Lezhnin S I, Pribaturin N A and Sorokin A L 2014 *Thermophysics and Aeromechanics* **21**(6) 763–766

IOP Conf. Series: Journal of Physics: Conf. Series **899** (2017) 032003 doi:10.1088/1742-6596/899/3/032003

Surfactat's Impact on the Evaporation Intensity and a Vapor Embryos Generation Kinetics within the Water Droplets

M P Anisimov[1, 3]**, V I Terekhov**[2] **and N E Shishkin**[1, 2]

[1]Aerosol Technology Research Laboratory, Novosibirsk State Technical University, 630073, 20, K. Marks Street, Novosibirsk, Russia
[2]Institute of Thermophysics, Siberian Branch of the Russian Academy of Sciences, 1 Ac. Lavrentyev Prospect, 630090, Novosibirsk, Russia
[3]Nanoaerosol Research Laboratory, FSBSI Technological Design Institute of Scientific Instrument Engineering SB RAS, 630058, 41 Russkaya Street, Novosibirsk, Russia

Abstract. The research of water-surfactant droplets evaporation was provided. The sodium dodecylsulfate was taken as a surfactant (SAS). The initial SAS mass contain was ~ 0,15%. Water and water-surfactant solution droplets were evaporated under the dry air flow blowing at temperatures $T_0 \sim (20 \div 90)^0$C and air linear velocity $V_O \sim 2 \div 5$ m/s. The droplet initial diameter (d_0) was equal to ~ 2 mm, Reynolds number was within the interval of Re = 500 ÷ 2000.
The droplet evaporation rate and their surface temperature were measured by an Infrared Microscope. The measurements display that the temperature heterogeneity on the droplet surface is initiated by the vapor fluctuating bursts at the local centers. A surfactant admixture into the base liquid (water) has a considerable impact on the vapor centre generation kinetics as well as on the time behavior of the droplets evaporating surfaces. The environmental conditions have the weak impact on the vapor formation, but the inner conditions, i.e. droplet's temperature and pressure in it and physical properties of the base liquid have the considerable impact on the vapor generation kinetics. A surfactant adding makes surface tension higher. That depresses a heat-mass transfer and rises an average temperature of droplets.

1. Introduction

The attention to evaporation of the droplets and the liquid films is initiated by the attractive ability to manage by surfactant the surface tension of the interphase boundary and therefore to change the boundary heat mass transfer. Authors [1] think that SAS additives have impacts on the heat emission during the liquid films sliding along the vertical walls is initiated by froth formation instead of the surface tension changing. Situation is complicate in case of the droplets in comparison with a film flow. It seems that a droplet size gets smaller during the evaporation. A circulation movement and composition as well as a droplet surface tension are resulted by changing. At some time a monolayer of the unsolvable substances can be formed from SAS or can be sorbet from the ambient air. The experimental and theoretical results [2-7] clearly show the considerable dropping of the evaporation rate in comparison with the pure liquid surface. The droplet surface temperature is changing with time qualitatively in the result of SAS additive [6, 7]. The heat-mass transfer and droplet evaporation is considerably changed due the different compositions SAS additive as well as at different ambient gas temperatures. That is shown in article [8] where PVP surfactant is raising the intensity of water droplets evaporation. That effect is interpreted as the surface tension changing in the result of the surfactant additive [9]. The evaporation rate is strongly dependent on the SAS concentration, which is shown clearly under SAS concentration variation [6]. It was shown by the multiple systematic authors' measurements that the droplet surface temperatures are irregular for the all tested liquids. The non-uniform temperatures are produced by the inner processes in droplets. It was independent on the blown air flow velocity. The present research authors did not found in the scientific literature the detail explanation of the non-uniform droplet surface temperatures. The thermal measurements display that the temperature heterogeneity is initiated by the vapor fluctuating bursts at the local centers on the droplet surfaces. The falls of local temperatures on a droplet surface are produced by the heat consumption by boiling areas on the droplet surfaces. The idea of the nucleation rate measurements for a vapor production on the surface of evaporated suspended droplet is developing in the present research. A number of the boiling centers and dynamics of their appearance along the time was calculated by the special computer code on the base of the infrared scanner pictures.

2. The measuring system

The basic elements of the measuring system are shown schematically in Figure 1. Air of low humidity (less that 2%) was blowing the water droplets with the linear velocity in the interval 1 ÷ 5 m/s, a turbulence level was around ~ 5,5 %, the air flow temperature were varied from the experimental set to set in the interval ~ 20 ÷ 100 ^{0}C. The air flow was directed up (against the droplet weight force) to avoid the droplet upset from the supporting glass fiber. The initial droplet diameter is equal to ~ 2 mm. The droplet surface temperatures and each droplet size were measured for pure water and SAS-water droplets by the infrared imager of Thermo Tracer TH7102MV. The infrared images were analyzed by computer using the home written computer code. We have got the vapor embryos nucleation rates as the result of our measurements.

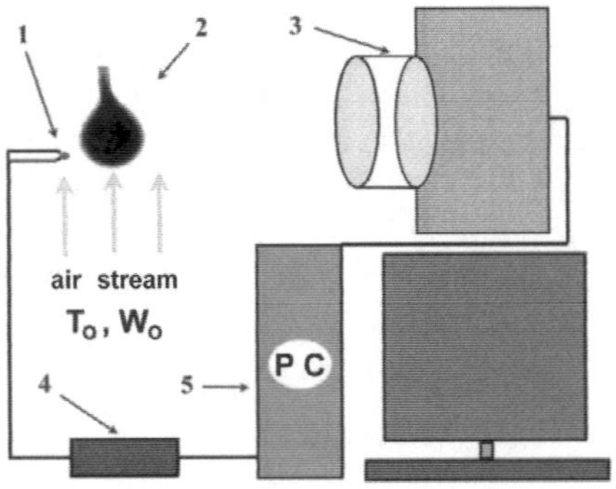

Figure 1. The schematically view of measuring system.
1 – Thermocouple, 2 - A droplet under investigation, 3 – An infrared imaging set up,
4 – Digitizer and thermocouple intermediate data collector, 5 – Computer.

Syringe was used to set a droplet sample for each measuring cycle. The glass fiber diameter was not bigger of 250 micrometer, which was enough restriction for heat flow along the fiber. The sodium dodecylsulfate was taken as a surfactant (SAS). It is treated as anionic SAS [10]. Its chemical formula can be presented such as C_{12} H_{25} O S O_3^- Na^+. The SAS is not toxic and is used, for example, in toothpaste, as well as in the food industry, washing solutions *etc.* It is an excellent penetrating agent and frothier. SAS of C_{12} H_{25} O S O_3 Na was used as solution in the distillated water with a SAS contains [5 M / l] $\cdot 10^{-3}$, which is presenting the mass concentration 0.15 weight %. More careful description of the experiment procedure and the experiment accuracy are described in articles [11, 12].

3. Research results

The temperature measurements reveal a considerable impact of SAS on the droplet surface temperature distribution as the result of the local liquid evaporations. The average temperature for the pure water droplets as well as its minimal temperature are stabilizing and keeping constant with time. The sodium dodecylsulfate additive initiates the temperature rise in 1.5 – 2 times (on 4-5 centigrade) as it is shown in Figure 2. One can assume that rise of a SAS contains on the droplet surface is decreasing the liquid evaporation and initiates the temperature dropping which is suppressed as consequence.

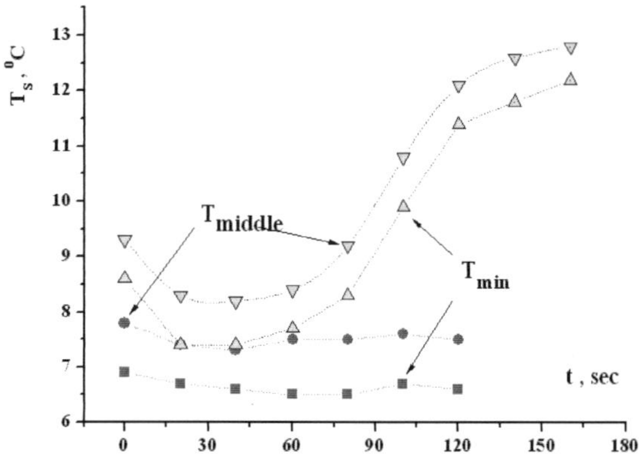

Figure 2. Surface temperature of the drops: green points – pure water, blue points – water at 0.15 % of surfactant; T_c = 19.8 ^{0}C, V_c = 3.2 m/s.

The time dependent droplet surface temperature behaviors of the distilled water and SAS-water solutions are shown in Figure 2. Results on the water droplet diameter measurements are presented in Figure 3. The relative decrease of a droplet diameter (d/d_o) is shown versus time of droplet evaporation, where d_o is the initial and d is a current diameter of droplet. T is temperature and V is a linear velocity of the air flow. It is easy to see that within the interval of parameters variation of an initial droplet's diameter and the velocities of the air flow both have the negligible impact on the droplet diameter in comparison with the blown air temperature.

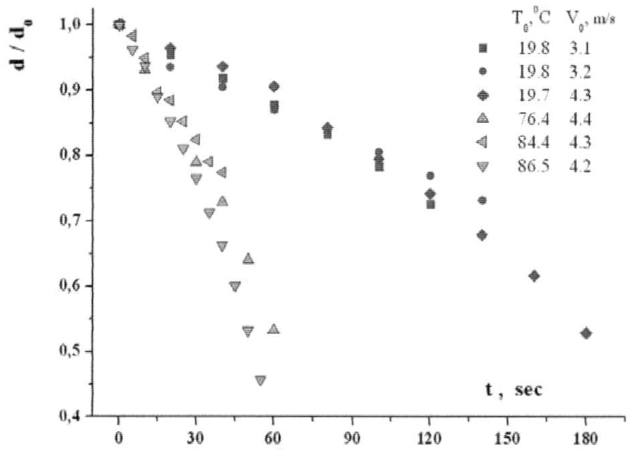

Figure 3. Results on the water droplet diameter, d/d_o, measurements are presented ($d_0 \sim 2.0 \div 2.3$ mm).

The same results are collected for SAS+water solutions (Figure 4). The air flow temperature has the strongest impact on the droplet rate of evaporation. The direct comparison of the data for the pure water and water and surfactant solutions is shown in Figure 5, where the experimental data are compared for the same or close experimental conditions. The slowed SAS+water droplet evaporation gives rise of the droplet surface temperature (Figure 2) and recession of the droplet evaporation rate. The diameter changing due the evaporation was close to the linear one in the all cases.

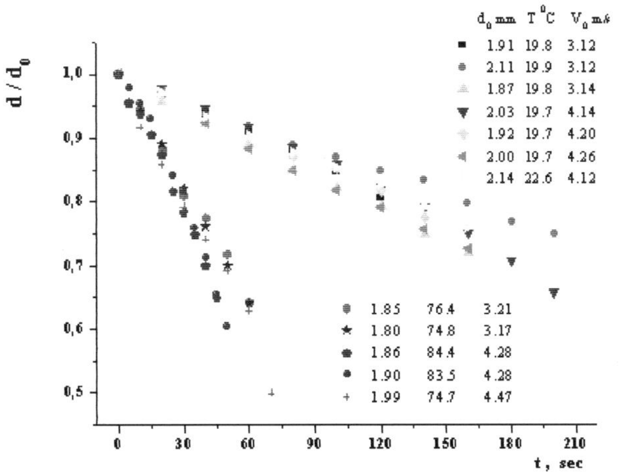

Figure 4. The initial parameters impact on the decrease of SAS droplets diameter, d/d_o.

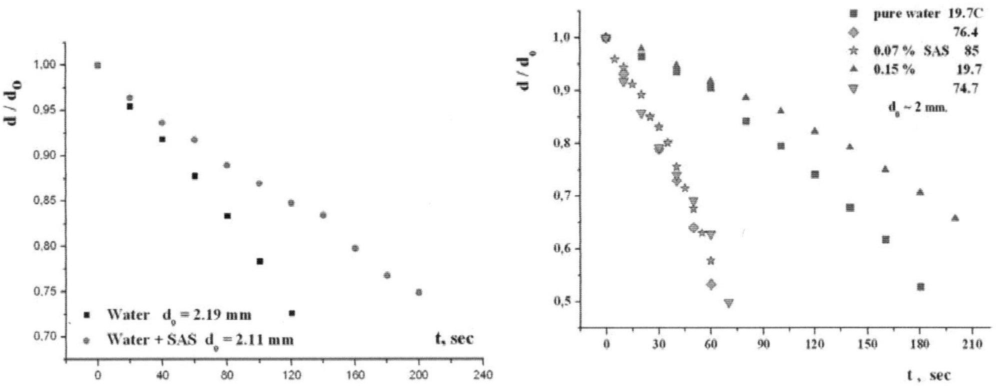

Figure 5. Influence of substance and temperature on the droplet diameter decrease:
a) $V_0 \sim 3.1$ m/s, $T_0 = 19.8\ ^0C$; b) $V_0 \sim 4.3$ m/s.

The empirical results on the evaporation rate of droplets are described by relation (1) which is published in articles [5,7]:

$$(d / d_0)^{3/2} = 1 - A \cdot \frac{Re_0^{0.5} \cdot Fo}{Ku}, \qquad (1)$$

where $Fo = t \cdot a_o / d_o^2$ is Fourier number, $Ku = r / [c_{po} \cdot (T_o - T_s)]$ is criteria of phase transformation by Kutateladze, r and c_{po} are latent heat of evaporation and heat capacity of gas, T_s и T_0 are vapor-gas mixture temperature on the droplet surface (saturation condition line) and the ambient air temperature accordingly. The experimental data can be described by equation (1) as two separate linear functions (Figure 6a for pure water and 6b for the water-SAS solutions). The infrared imager of Thermo Tracer TH7102MV has no ability to register each boiling event; therefore the number of the boiling centers was estimated by counting the quantity of the low temperature regions on the infrared image of each droplet. It is supposed that that the biggest temperature gradients (and quantities of the boiling centers) are appeared within the upper layer of droplets. The dipper volume of the droplets is keeping the near constant temperature of droplet smoothing the temperature fluctuations within the upper layer. The special estimations of the temperature gradients were not performed. It was assumed that the boiling centers distribution was uniform to calculate the nucleation rates on the droplet surface. The upper layer of droplet was taken as a volume where nucleation takes place. That layer is easy detectable by infrared imager.

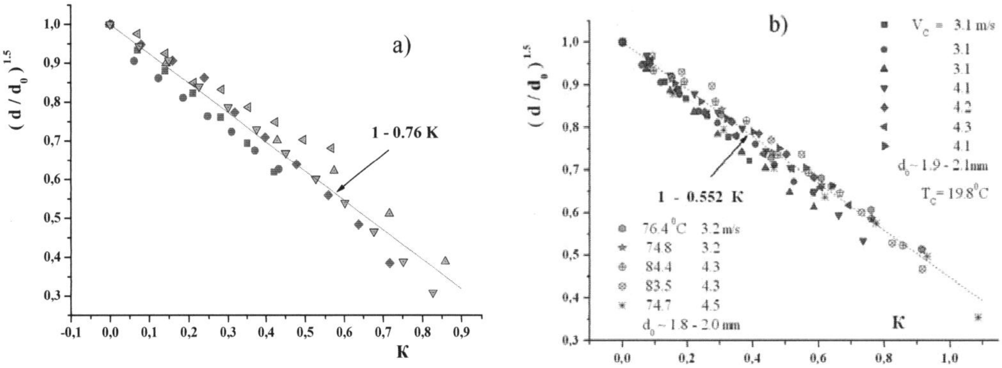

Figure 6. Generalization of experimental data on evaporation of a) water; b) surfactant droplets.

An equation $\omega = N{\cdot}k/(\tau{\cdot}V)$ was used to calculate the nucleation rate, ω (number of boiling centers within the unit volume of 1 sm^3 per unit time of 1 second). The number of boiling events (N) was calculated on the selected peace of the infrared image which was exposed in the time interval τ. In our case τ = 0,25 seconds and k (form factor) is used to estimate the total number of boiling centers on the droplet surface in the sphere approximation. The selected peace on the droplet is less that a total droplet surface then a form factor is bigger of unity. V is the volume of the upper layer of droplet, i.e. a boiling layer on the droplet. It is equal to the product of droplet surface and sickness of the boiling layer (sickness of the boiling layer is taken as to 2 pixels on the droplet's image in the present case). Results of the nucleation rate measurements on the droplet surface are presented in Figure 7 (for data of Figure 6).

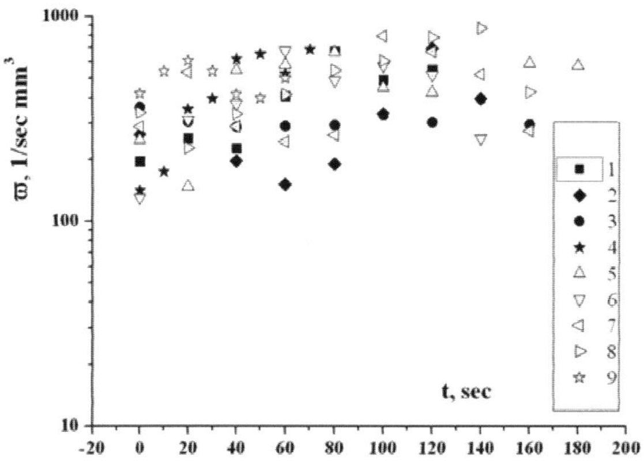

Figure 7. The vapor nucleation rates on the droplet surface: pure water is presented by dark points (1-4); SAS-water droplets are shown by light points (5 – 9). The air flow temperature was 19,8 °C, except for points (4, 9) where temperature was T ~ 80 °C

One can see the deviations which are looking like considerable. We have understanding the origin of some problems such as the insufficient statistics of the random nucleation events and nonobjective fixation of the number of the lowered temperature events. We hope to make better algorithm for data analysis of the nucleation events in the nearest future. The ambient temperature can have the influence on the droplet evaporation rate. That effect was studied at the special measurements where the air flow temperature was varied. The number of boiling centers within the droplet surface layer is shown in Figure 7 (the flow velocity is within interval of 3 ÷ 4 m/s). One can find the weak influence on the vapor nucleation rates in relation of the outside temperature T_0 and the surfactant additives.

The inner droplet parameters such as temperature, pressure, physical properties of the base liquid have the main influence on the vapor nucleation rates in the upper layer of droplets. SAS additives are raising the surface tension of the blown droplet. It makes the mass exchange less intensive and the average droplet temperature is rising in that case. The vapor embryos generation is under governing of the interfacial conditions.

4. Conclusions

Results of the empirical research of the ambient temperature and the evaporation rate of the liquid droplets at the variable contain of SAS (sodium dodecylsulfate) are presented and discussed. The initial concentration of SAS was constant at the single experiment, beginning from 0.15 wt %. Experiments were provided at the broad temperature conditions of the dry air ($T_0 \sim 20 \div 100\ ^0C$) and air velocities. The experimental research of the evaporation rates and temperature of the droplet surface were provided with help of infrared microscope. It was found that adding of SAS into the base liquid (water) has a considerable impact on the heat- and mass transfer. The temperature measurements of droplet surface shows that the vapor embryos generation is defined by the internal conditions mostly, and SAS additive has impact on the mass surfactant carry-over intensity only. The presented original and new results show the complexity and multitask of the research problem. The results of the present research can be used as basis for the new detailed and comprehensive researches which are under variation of the base liquids, SAS, their concentrations as well as model of the droplet heat-mass transfer with phase transitions verifications.

Acknowledgments

This work was supported by a grant from the Russian Ministry of Science and Education under Contract № 14.Z50.31.0041 which was issued at 13th February 2017 (MA&PH). TV & ShN acknowledge the Russian Academy of Sciences (RAS) for financial support through the DEMMPC Program (grant № 26).

References

[1] Basit H. Shan and Ron Darby. The effect of evaporative heat transfer in vertical film flow // Int. J. Heat and Mass Transfer. – 1973. – V. 16. – P. 1889 – 1903.
[2] Bradley R.S. The rate of evaporation of micro-drops in the presence of insoluble monolayers // J. Colloid. Sci. – 1955. – V. 10. – P. 571 – 575.
[3] Deryagin B.V., Bakanov S. P., Kurgin Yu.S. On influence of monolayers on the dropletsevaporation // DAN USSR. – 1960. – V. 135, № 6. – P. 1417–1420.
[4] Borzilov V.A., Klepikova N.V., Merkulovich V.M. The Surfactant influence on the droplet growth and evaporation // Meteorology and Hydrology. – 1979. – № 5. – P. 41 – 48.
[5] Deryagin B.V., Kurgin Yu. S. Nonstationary evaporation of droplets and adsorption layer coatings // DAN USSR. – 1964. – T. 155, № 3. – P. 644 – 646.
[6] Sadd P.A., Lamb J.A., Clift R. The effect of surfactants on heat and mass transfer to water drops in air // Chemical Engineering Science. – 1992. – V. 47. No 17/18. – P. 4415 – 4424.
[7] Mohsen A. Hashem. Effect of surfactants on mass transfer coefficients of single liquid drops // Alexandria Engineering Journal. – 2005. – V. 44. – P. 477 – 486.
[8] Chen R.-H., Phuoc T. X., Martello D. Effect of nanoparticles on nanofluid droplet evaporation // Int. J. Heat and Mass Transfer. – 2010. – V.53. – P. 3677 – 3682.
[9] Chen R.-H., Phuoc T. X., Martello D. Surface tension of evaporating nanofluid droplets // Int. J. Heat and Mass Transfer. – 2011. – V.54. – P. 2459 – 2466.
[10] Surfactants and detergents. // Handbook under editing by Abramzon A.A., Moscow. 1993. 270 p.
[11] Terekhov V.I., Terekhov V.V., Shishkin N.E., Q.C.Bi .Experimental and numerical research of non-stationary evaporation of drops of a liquid. Journal of Engineering Physics and Trermophysics, 2010. – Vol. 83. Is 5. P. 883 – 890.
[12] Terekhov V.I., Shishkin N.E. Evaporation of Water Droplets Containing Carbon Nanotubes. Technical Physics Letters. – 2012. – Vol 38. No.1. – P. 25 – 28.

New trends in the nucleation research

M. P. Anisimov[1,3] and P. K. Hopke[1,2]

[1] Aerosol Technology Research Laboratory, Novosibirsk State Technical University, 20 K. Marks str., Novosibirsk, 630073, Russia
[2] Clarkson University, Box 5708, Potsdam, NY 13699-5708
[3] Nanoaerosol Research Laboratory, FSBSI Technological Design Institute of Scientific Instrument Engineering SB RAS, 41 Russkaya str., Novosibirsk, 630058, Russia

E-mail: anisimovmp@mail.ru

Abstract. During the last half of century the most of efforts have been directed towards small molecule system modeling using intermolecular potentials. Summarizing the nucleation theory, it can be concluded that the nowadays theory is far from complete. The vapor-gas nucleation theory can produce values that deviate from the experimental results by several orders of magnitude currently. Experiments on the vapor-gas nucleation rate measurements using different devices show significant inconsistencies in the measured rates as well. Theoretical results generally are quite reasonable for sufficiently low vapor nucleation rates where the capillary approximation is applicable. In the present research the advantages and current problems of the vapor-gas nucleation experiments are discussed briefly and a view of the future studies is presented. Using the brake points of the first derivative for the nucleation rate surface as markers of the critical embryos phase change is fresh idea to show the gas-pressure effect for the nucleating vapor-gas systems. To test the accuracy of experimental techniques, it is important to have a standard system that can be measured over a range of nucleation conditions. Several results illustrate that high-pressure techniques are needed to study multi-channel nucleation. In practical applications, parametric theories can be used for the systems of interest. However, experimental measurements are still the best source of information on nucleation rates. Experiments are labor intensive and costly, and thus, it is useful to extend the value of limited experimental measurements to a broader range of nucleation conditions. Only limited experimental data one needs for use in normalizing the slopes of the linearized nucleation rate surfaces. The nucleation rate surface is described in terms of steady-state nucleation rates. It is supposed that several new measuring systems, such as High Pressure Flow Diffusion Chamber for pressure limit up to 150 bar will be created soon in the frame of the Russian Ministry of Science & Education project under Contract № 14.Z50.31.0041 issued by February 13th of 2017. The Project will provide the nucleation studies for basic problems of theoretical and practical applications. Published under licence in Journal Title by IOP Publishing Ltd.

1. Introduction

The nucleation rate measurements data have the key impotence for the nucleation theory development, experimental data on nucleation interpretation, a material production, etc. The first nucleation experiment can be considered to be associated with measurements of liquid and crystals supercooling done by Fahrenheit [1]. The nucleation of bubbles in gas saturated solutions was observed and the concept of critical embryos of a new phase was introduced in nucleation science during the second half of 19th century [2]. The quality of vapor/liquid nucleation rate results has improved substantially beginning in 1980s because of the development of new measurement systems. For example, the first prototype of the Flow Diffusion Chamber for vapor nucleation rate measurements was developed by Anisimov et al. [3, 4]. As a rule, theoretical and experimental data on nucleation rate are not in a good agreement over a range of temperatures and/or pressures. It appears that there may be problems in both the experiments and theory and deficiencies can be identified in all versions of nucleation theories and practically all of the reported experimental results. Current theories correspond mostly to various modifications of Classical Nucleation Theory [5]. The theoretical results look quite reasonable for sufficiently low vapor nucleation rates where the droplet approximation is applicable [6]. At the present time, vapor-gas nucleation theory can produce values that deviate from the experimental results by up to several orders of magnitude [9, 10]. However, nucleation experiments using different devises also show significant inconsistencies in the measured rates, for example [10]. Both problems produce difficulties in establishing standard vapor/liquid nucleation systems that could be used to test accuracy of vapor-gas nucleation rate measurements. The advantages and current problems of the vapor-gas nucleation experiments are discussed below and a view of the future studies is presented based on the assessment of vapor-gas/liquid nucleation experimental results.

One can find in scientific literature a pretty fresh idea of the nucleation rate surfaces design over the diagrams of phase equilibria [6]. That idea looks like profitable for the nucleation theory development and for various practical applications where predictions of theory have no high enough accuracy for today. The usual way of any theory is to reduce the many-body problem to one body in some field. The features of one body and the field are adjusted usually to get the reasonable compliance to the empirical results. Relation of the theoretical and empirical results is not good enough in case of the Nucleation Theory.

2. The empirical methods

Initial measurements of vapor nucleation were made by Aitken [12] and Coulier [13, 14] using the rapid adiabatic expansion of air saturated with water vapor to provide the necessary supersaturation. Allen and Kassner [15] modified the experimental procedure using an expansion/compression cycle in a Wilson-type chamber where compression is introduced after the adiabatic expansion of a vapor-gas mixture. The resulting vapor supersaturation permits the growth of the nucleated clusters to optically detectable sizes. The pressure change, intensity of transmitted/scattered laser light, and other parameters are recorded. Temperatures are measured with uncertainties of the order of several tenths of a degree Kelvin. The shock tube is a version of the expansion technique for the nucleation research (for example, Peters & Paikert [18]). It has been thought that wave tube measurements are less accurate "because pressure pulses cannot always be accurately reproduced" in comparison with two-piston chamber (Holten et al. [19]). Supersonic jets have been used to produce the same adiabatic expansion scheme where cluster generation and growth are decoupled (Kim et al. [20]). The expansion techniques cover a total nucleation rate interval from 103 up to 1017 cm-3s-1. The available nucleation data involve the nucleation pressures from 100 kPa to 7 MPa.

3. Diffusion chambers

Langsdorf [21] created the first static diffusion chamber (SDC). The SDC is described in detail by Katz [22], Kacker and Heist [23], etc. The static diffusion chamber consists of two wet plates maintained at different temperatures. Vapor diffusion and temperature gradient produce vapor supersaturation and nucleation in the space between the plates such that clusters of the new phase are formed. Clusters then grow. Particles can move toward the cold plate direction by temperature and vapor concentration gradients. A particle drops when its mass gets sufficiently large. A special SDC design was used for vapor-gas nucleation measurements at elevated pressures up to 4.0 MPa (Heist et al.[24]). Idea of SDC is used for the Flow Diffusion Chamber (FDC) design. Initial versions of the FDC were described by Anisimov et al. [3, 4]. The FDC scheme uses a hot laminar vapor-gas flow within cold boundary conditions. Hot vapor-gas flow enters the chilled tube (condenser). The vapor-gas velocity distribution at the condenser beginning is parabolic one. The distributions of temperature, T, vapor supersaturation ratio, S, and nucleation rate, J, along the tube axis can be revealed as the result of the Navier-Stokes equations solution. Embryos form in the nucleation volume and subsequently grow in the supersaturated vapor. A flow laminator is used to produce fully developed initial laminar flow. Parameters of that flow are used as boundary conditions for the stationary heat-mass transfer problem.

An algorithm for estimation of the average nucleation rate over FDC nucleation volume was suggested by Anisimov et al. [25]. The maximum experimental value of the nucleation rate, Jmax, in diffusion chambers can be measured using an obvious relation Jtheor/Ntheor=Jmax/Nexp, where Jtheor is maximum theoretical nucleation rate; Ntheor and Nexp are the theoretical and experimental particle concentrations respectively (Hyvarinen et al. [26]; Wagner and Anisimov [27]). That relationship can be used also for Jmax measurements using a supersonic nozzle (Streletzky et al. [28]). The current FDC scheme has been used for vapor nucleation rate measurements at total pressures from 0.03 to 0.50 MPa. The FDC data span over six orders of magnitude in nucleation rate that can be measured in a single experimental system. Nucleation temperatures from 230 to 400 K can be obtained in these systems now. Versions of FDC have built in the USA (Warren et al. [31]; Brock et al. [32]; Brock et al. [33]) as well as in Finland (Anisimov et al. [29]; Lihavainen & Viisanen [30]), Czech Republic (Brus et al. [10]), Nguyen et al. [37] have studied homogeneous and heterogeneous nucleation of a single vapor, etc. Several groups have used rapid turbulent mixing to measure nucleation rates of the single and binary vapors (Kogan & Burnasheva [34]; Sutugin & Fuchs [35]; Okuyama et al. [36]; etc). Most studies have only measured the critical supersaturation for two-component vapor systems. The problem of a turbulent mixing scheme is associated with wide spectrum of nucleation conditions. Fluctuations of vapor supersaturation and nucleation temperature can be large enough in the pre-nucleation zone to generate some pre-particles before system achieves homogeneity. These pre-particles initiate a heterogeneous nucleation. Thus, turbulent flow systems are rarely used for homogeneous nucleation measurements. However, they have proved useful for heterogeneous (Lee et al. [43]; Mavliev et al. [44]) and ion-induced (Seto et al. [45]; Gamero-Castano, et al. [46]; etc) nucleation measurements.

During the past five decades, several research groups have examined the effects of pressure and carrier gas composition on homogeneous nucleation. Classical nucleation theory assumes that the only role of the carrier gas is to maintain the temperature equilibrium of the clusters with the ambient media. Frank and Hertz [47] made the first observations of a gas-pressure effect. The result was reproduced in several other measurements (Katz et al. [48]; Brus and Zdimal [49], etc). Other experimental results have supported a dependence of the nucleation rate on the nature and total pressure of the carrier gas (Luijten and van Dongen [53]; Luijten et al. [54]; Anisimov et al. [42]; Lihavainen and Viisanen [55]. Gas-pressure effects were detected by van Remoortere et al. [56] although most of the measurements from this research group (i.e., Viisanen et al. [57]; Viisanen and Strey [58]) did not observe carrier gas influences on nucleation rate. Fladerer and Strey [9] did attempt to measure supersaturated argon nucleation using a cryogenic nucleation pulse chamber. Experimental results of different experimental schemes used for nucleation rate measurements at cryogenic temperatures are still inconsistent (Fladerer and Strey [9]) because of low accuracy of the experimental data. Several research groups have made comparative measurements of nucleation rates (Anisimov et al., [29, 59]; Wilck et al. [60]; Brus et al. [10], etc.) The nucleation rates of n-butanol in helium using both an FDC (Hyvarinen et al. [61]) and an expansion chamber (Strey et al. [16]) illustrate the data sets deviation on 4-5 orders of magnitude. The origin of these data discrepancies can be better understood through the consideration of nucleation rate surfaces (Anisimov et al. [62, 63]). In the most cases, the gas is treated as an inert media to absorb the heat released from the phase transitions. Inconsistencies among the experimentally measured values from the different experimental schemes are a major problem for current vapor-gas nucleation experiments. Consideration of vapor-gas nucleation as a binary system is a reasonable way to resolve the data inconsistencies. It is plausible to think that different experimental systems have the inconsistent trajectories along the nucleation rate surface when the vapor-gas system is treated as binary system. Several results, for example, Anisimov et al. [41, 63] illustrate that high pressure measuring techniques need to be designed to study multi-channel nucleation.

4. Reference system for nucleation rate measurements

A In order to test the accuracy of an experimental system, it is important to have a standard system that can be measured over a range of nucleation conditions. The n-pentanol–helium system was proposed in Prague, 1995 for such measurements. The available results from several research groups were collected and compared (Anisimov et al. [11]). They proposed a reference equation for nucleation rates of n-pentanol–helium as a practical test of any experimental measurement system for total pressures from 0.10 to 0.30 MPa. The problem of a nucleation standard can only be solved when consistent results have been obtained by independent groups that use the different experimental schemes.

5. Nucleation rate surfaces design

Nucleation experiment can be provided in very local nucleation conditions even the nucleation takes place from the critical conditions down to the absolute zero temperature limit and from zero nucleation rates at phase equilibria up to the spinodal conditions. Theory predictions have low reliability as a rule. It is well known that any phase diagram has several lines of phase equilibria. It is easy to show that each line of phase equilibria generates the nucleation rate surface in space of nucleation process parameters. It means that one has multi sheet nucleation rate surfaces in the common case. Each nucleation rate surface is related to single phase state generation, or it is related to a one channel of nucleation. Semiempirical design of the nucleation rate surfaces over diagrams of phase equilibria have a potential ability to provide a reasonable quality information on nucleation rate for each channel of nucleation. One reason for nucleation theory problem is the application of inconsistent assumptions that are used to interpret the experimental results. The most experimental data on vapor nucleation are interpreted as a one-surface (one nucleation channel) approximation of a nucleation theory. However, two or more nucleation channels are really present in most systems as it was reported by Anisimov et al [6]. To empirically detect a single nucleation channel one needs to have the appropriate measuring system, but these kinds of systems are not currently under active development. The semiempirical approach permits the prediction of a realistic topology for the nucleation rate surface. That topology can be designed over the phase equilibria diagrams [62, 66, etc]. One needs at least several experimental points on the nucleation rate surface and the phase diagram to effectively develop these surfaces [66].

A short history of semiempirical design of the nucleation rate surfaces over phase equilibrium diagrams can be discussed briefly. This approach involves a simple idea developed by Anisimov et al. [62, 65, etc], that nucleation rate surfaces arise up from the phase equilibria lines. It means that each line of the real and metastable phase equilibria produces two nucleation rate surfaces both surfaces reflect the nucleation kinetics for each of two metastable phases that can be joined by an equilibrium line. Experimental detection of nucleation rate for each of two individual vapor nucleation rate channels, which are generating two single nucleation rate surfaces for vicinity of the triple point, one can see in article [67]. That result is unique to the present time.

Theory involves a number of assumptions to describe small clusters [68]. Further, when the size dependencies of the surface tension and density of nuclei were taken into account [6] and the inherent degrees of freedom were used to calculate the statistical sum for a nascent cluster [69], agreement between theoretical and experimental results get worse. Semiempirical design over diagram of phase equilibria is applied for metastable volume construction [65-68]. That way has a considerable potential in the nucleation rate surface description. A major problem in detection of the nucleation rate surface singularities is the random error inherent in any experimental data. This problem exists even one has experimental results of relatively high accuracy as it is shown by Anisimov [6]. A conceptual problem

in vapor-gas to liquid (or to solid phase) nucleation is its treatment of the nucleation as a single component problem instead as a two component system.
Results of the present research illustrate clearly that gases can generate a phase transitions in a condensed state of any vapor-gas system. A direct mass spectrometric measurements show the presence of the carrier gas in the critical cluster under some nucleation conditions, for example, that result was shown in case the monosilane-argon system [71].

6. Conclusions

In the present review, advantages and problems related to aerosol generation experiments are discussed. It can be concluded that the development of accurate experimental techniques for vapor-gas nucleation research are still in progress. The current recommendations for vapor-gas nucleation rate measurements can be summarized such as: vapor nucleation and cluster growth events should be decoupled to allow the embryos to grow before light scattering detection. Theory independent algorithms are needed to estimate experimental nucleation rates based on FDC and others flow systems experimental data. Homogeneous nucleation rates should be measured at the sufficiently high vapor supersaturation [4] to reach conditions of the vapor homogeneous nucleation. Impurities in the vapor substance must not exceed 0.2 % for the present time. This level of impurities shifts the nucleation rate values within one order of magnitude as shown by Anisimov et al. [39], Strey et al. [17], etc.

Vapor-gas systems are strongly recommended to be considered as binary nucleation systems. The problem of experimental data inconsistencies should be resolved to create a nucleation rate standard. Aerosol size distribution measurements are strongly recommended to identify the possibility of two and more channels for homogeneous and heterogeneous nucleation. It can be hoped that the uncontrolled parameter(s) will be identified in the near future and permit consistent nucleation rate data to be derived from different research methods. Introduction of the nucleation standard(s) is a major current problem. Success in the nucleation standard development and its introduction in nucleation research practice is a key issue for current nucleation experiments.

Empirical results illustrate clearly that gases can generate a phase transitions in a condensed state of matter in the critical line vicinity. A direct mass spectrometric measurements [71] show the presence of the carrier gas in the critical cluster under some nucleation conditions. Several contemporary achievements are collected in the present review. It is clear that idea of a semiempirical design of the nucleation rate surfaces is promising tool for the nucleation theory development. The data base file and the digital information on nucleation rates, equations of states, phase equilibria diagrams, binodal and spinodal conditions, etc will be intensively collected without doubt. Time when soft to the nucleation rate surface design will be available for each scientist and engineer is approaching and the knowledge on nucleation will get position of powerful tool which is effective in scientific research and industrial applications.

7. Acknowledgement

Research is supported by grant of the Russian Ministry of Science and Education under Contract № 14.Z50.31.0041 at 13 February of 2017. Academician Howard Reiss is gratefully acknowledged for the idea valuable discussion of nucleation rate surfaces design and their relations to the nucleation rate measurements. Academician Howard Reiss was lost tragically due the car accident at December of 2015, unfortunately. The present research is dedicated to academician Howard Reiss.

References

[1] Ostwald, W. 1896-1903, Lehrbuch der Allgemeinen Chemie. W. Engelmann: Leipzig.
[2] Volmer, M. 1939, Kinetik der Phasenbildung. T. Steinkopf: Dresden-Leipzig.
[3] Anisimov, M.P., Costrovskii, V.G., & Shtein, M.S. 1978, Colloidniy Journal (Russian), 40 (1), 116-121.
[4] Anisimov, M.P. and Cherevko, A.G. 1985, J.Aerosol Sci., 16(2), 97-107.

[5] Frenkel, Ja.I. 1975, Kinetic theory of liquids, Nauka: Leningrad.
[6] Anisimov, M.P. 2003, Russian Chemical Reviews, 72, 591 – 628.
[7] Baydakov, V.G. 1995, Superheating of cryogenics liquids. Ural Branch RAS: Ekaterinburg.
[8] Protsenko, S.P., Baydakov, V.G., *et al.* 2006, Ural Branch RAS, Ekaterinburg, 8, 171 – 202.
[9] Fladerer, A. and Strey, R. 2006, Journal of Chemical Physics, 124, 164710 – 164715.
[10] Brus, D., Hyvarinen, A. Zdimal, V., & Lihavainen, H. 2005, JCP, 122, 214506 – 21514.
[11] Anisimov, M.P., Hopke, P.K., & Shvets, I. 2000, Journal of Chemical Physics, 113, 1971 – 1975.
[12] Aitken, J. 1888, Transactions of the Royal Society of Edinburgh, 35, 1-9.
[13] Coulier, P. J. 1875, J. de Pharmacie et de Chimie, Paris, Ser. 4, 22:165-173.
[14] Coulier, P. J. 1875, J. de Pharmacie et de Chimie, Paris, Ser. 4, 22:254-255.
[15] Allen, L.B. & Kassner, J.L., Jr. 1969, Journal of Colloid and Interface Science, 30, 81 – 93.
[16] Strey, R., Wagner, P.E., & Schmeling, T. 1986, JCP, 84(4), 2325 – 2335.
[17] Strey, R., Wagner, P. & Viisanen, Y. 1994, Journal of Physical Chemistry, 98, 7748 – 7758.
[18] Peters, F. & Paikert, B. 1989, Journal of Chemical Physics, 91, 5672 – 5678.
[19] Holten, V., Labetski, D., & van Dongen, M.E. 2005, JCP, 123, 104505 – 104509.
[20] Kim, Y.J., Wislouzil, B.E., Wilemski G, Wölk J, Strey, R. 2004, JCP, 108(A), 4365 – 4377.
[21] Langsdorf, A. 1939, Review of Scientific Instruments, 10, 91 – 103.
[22] Katz, J.L. 1970, Journal of Chemical Physics, 52, 4733 – 4748.
[23] Kacker, A. & Heist, R.H. 1985, Journal of Chemical Physics, 82, 2734 – 2744.
[24] Heist, R.H., Bertelsmann, A., Martinez, D., & Chan, Y.F. 2003, Atmospheric Research, 65, 189 – 209.
[25] Anisimov, M.P., Costrovskii, V.G., Mikheev, V.B. 1980, Colloid J. (Russian), 42, 941 – 945.
[26] Hyvarinen, A.P., Brus, D., Zdimal, V., Smolík, J., Kulmala, M., 2006, JCP, 124, 224304 - 224311.
[27] Wagner, P.E. & Anisimov, M.P. 1993, Journal of Aerosol Science, 24, 1, 103 – 104.
[28] Streletzky, K.A., Wyslouzil, & Strey, R. 2002, Journal of Chemical Physics, 116, 4058 – 4070.
[29] Anisimov, M. P., Hameri, K., Kulmala, M. 1993, Report Series in Aerosol Scince, 23, 19 – 24.
[30] Lihavainen, H., & Viisanen, Y. 1998, Report series in aerosol science No. 41, Eds by J.Hienola and K.Hameri (pp. 381 – 382). Helsinki.
[31] Warren, D.R., Okuyama, K. et al. 1987, J. Colloid &Interface Sci., 116, 563-581.
[32] Brock, J.R., Kuhn, P.J. & Zenavi, D. 1986, Journal of Aerosol Science, 17, 11 – 22.
[33] Brock, J.R., Kuhn, P.J. & Zenavi, D. 1988, Journal of Aerosol Science, 19, 413 – 424.
[34] Kogan, Ja. I. & Burnasheva, Z.A. 1960, Journal of Physical Chemistry (Russian), 34, 2630 – 2639.
[35] Sutugin, A.G. and Fuchs, N.A. 1970, Colloidniy Journal (Russian), 32, 255 – 260.
[36] Okuyama, K., Kousaka, Y., Flagan, R.C., & Seinfeld, J.H. 1988, JCP, 89, 6442 – 6453.
[37] Nguyen, H. V., Okuyama, K., Mimura, T., Kousaka, Y., Flagan, R.C., & Seinfeld, J. H. 1987, Journal of Colloid and Interface Science, 119, 491 – 504.
[38] Kodenev, G.G., Baldin, M.N. & Vaganov, V.S. 1987, Cluster Physics, Eds A. A. Vostrikiv and A. K. Rebrov, (pp.110 – 115). ITP Novosibirsk.
[39] Anisimov, M.P., Vershinin S.N., Aksenov A.A., Sgonnov A.M., Semin G.L. 1987, Colloidniy Journal (Russian), 49, 842 – 846.
[40] Kashchiev, D. 1982, Journal of Chemical Physics,76, 5098 – 5102.
[41] Anisimov, M.P., Koropchak, J.A., Nasibulin, A.G., & Timoshina, L.V. 1998, JCP, 109, 10004 – 10010.
[42] Anisimov, M.P., Koropchak, J.A., & Timoshina, L.V. 2000, JCP,112, 9917 – 9928.
[43] Lee, D.-W., Hopke, P.K., Rasmussen, D.H., Mavliev, R. 2003, J. Phys. Chem. B 107, 13813-13822.
[44] Mavliev, R., Hopke, P.K., Wang, H.-C., Lee, D.-W. 2004, J. Phys. Chem. B 108, 4558-4564.
[45] Seto, T., Okuyama, K., de Juan, L., Fernández de la Mora, J. 1997, JCP 107, 1576-1585.

[46] Gamero-Castano, M., Fernandez de la Mora, J. 2002, Journal of Chemical Physics 117, 3345-3353.
[47] Frank, J. & Hertz, H. 1956, Zeitschrift fur physik, 143, 559 – 590.
[48] Katz, J. L., Hung, C. H. & Krasnopoler, M. 1988, Lecture Notes in Physics V.309, 356 – 359.
[49] Brus, D., Zdimal, V., & Stratmann, F. 2006, Journal of Chemical Physics, 124, 164306 – 1634310.
[50] Anisimov, M.P. & Vershinin, S.N. 1990, Journal of Aerosol Science, 21 (1), 11 – 14.
[51] Heist, R.H., Ahmed, J. & Janujua, M. 1994, Journal of Physical Chemistry, 98, 4443 – 4453.
[52] Heist, R.H., Ahmed, J. & Janujua, M. 1995, Journal of Physical Chemistry, 99, 375 – 383.
[53] Luijten, C.C.M, & van-Dongen, M.E.H. 1999, Journal of Physical Chemistry, 111, 8524 – 8534.
[54] Luijten, C.C.M, Peeters, P. & van-Dongen, M.E.H. 1999, Journal of Chemical Physics, 111, 8535 – 8544.
[55] Lihavainen, H. & Viisanen, Y. 2001, Journal of Physical Chemistry, 105, 11619 – 11629.
[56] Van Remoortere, P., Heath, C., Wagner, P., & Strey, R. 1996, Proceedings of the 14-th ICNAA, Helsinki, Eds by M. Kulmala and P. Wagner, Pergamon: Oxford, 256 – 259.
[57] Viisanen, Y., Strey, R., & Reiss, H. 1993, Journal of Chemical Physics, 99, 4680 – 4692.
[58] Viisanen, Y. & Strey, R. 1994, Journal of Chemical Physics, 101, 7835 – 7843.
[59] Anisimov, M.P., Hameri, K., & Kulmala, M. 1994, Journal of Aerosol Science, 25 (1), 23 – 32.
[60] Wilck, M., Stratmann, F., & Kulmala, M. 1998, Journal of Aerosol Science, 29, 899 – 911.
[61] Hyvarinen, A.-P., Lihavainen, H., & Kulmala, M. 2004, Journal of Chemical Physics, 120, 11621 – 11633.
[62] Anisimov, M.P., Hopke, P.K; Rasmussen, D.H., & Pinaev, V.A. 1998, JCP, 109, 1435 – 1444.
[63] Anisimov, M.P. Nasibulin, A.G., & Shandakov, S.D. 2000, Journal of Chemical Physics, 112, 2348 – 2354.
[64] Vortisch H., Kramer, B., Weidinger, I., Wöste, L., Leisner, T., Schwell, M., Baumgärtelb, H. and Rühlc, E. 2000, Phys. Chem. Chem. Phys., 2, 1407-1413.
[65] Anisimov, M. P., J. Aerosol Sci. 1990, 21, 1, 23-25.
[66] Anisimova, L., Anisimov, M, Semin, G, Hopke, P. K. 2005, J. Colloid Interface Sci., 290, 107-116.
[67] Anisimova, L., Hopke, P. K., and Terry, J. 2001, J. Chem. Phys., 114(22), 9852-9855.
[68] Anisimov, M. P., Fominykha, E. G., Akimova, S.V., Hopke, P.K. J. Aerosol Sci., 2009, 40, 733-746.
[69] Lothe, J. and Pound, G. 1966, J. Chem. Phys., 45, 630-634.
[70] Reid, R. C., Prausnitz, J. M., and Sherwood, T. K., 1977, The Properties of Gases and Liquids, McGraw-Hill, New York.
[71] Sharafutdinov, R. G., Zarvin, A. E., *et al.*, 1999, Technical Physics Letters, 25(21), 47.

IOP Conf. Series: Journal of Physics: Conf. Series **899** (2017) 032005 doi:10.1088/1742-6596/899/3/032005

Simulating interaction of nonlinear spatial waves on a free surface of the shallow viscous liquid layer

D G Arkhipov[1,2], **G A Khabakhpashev**[1,2], **N S Safarova**[2]

[1] Physics Dept, Novosibirsk State University, 2 Pirogov Street, Novosibirsk, 630090, Russia
[2] Dept of Physical Hydrodynamics, Institute of Thermophysics, 1 Acad. Lavrentyev Avenue, Novosibirsk, 630090, Russia

E-mail: theory@itp.nsc.ru

Abstract. This paper deals with the combined approach to describing the evolution of weakly nonlinear three-dimensional moderately long perturbations of free surface of viscous liquid. The initial system of hydrodynamic equations is reduced to the novel model system of equations. The first of them is integro-differential equation for nonlinear perturbation of the free surface, taking into account non-stationary shear stress on a weakly sloping bottom. Another equation is an auxiliary linear equation for determining the liquid horizontal velocity vector, averaged over the layer depth. This vector is present in the main equation only in the term of the second order of smallness. The proposed model is suitable for finite-amplitude waves, traveling in different directions in the horizontal plane. Some problems of interactions and collisions of such perturbations over the horizontal and weakly sloping bottom are solved numerically.

1. Introduction

Gravity waves in fluids are among the classical objects of the nonlinear phenomena mechanics. In particularly, Korteweg and de Vries derived their famous equation precisely for perturbations of the free surface of a shallow water layer (see, for example, [1]). Kadomtsev – Petviashvili equation [2] was surely the next major step in the development of evolutionary equations for perturbations of small but finite amplitude. The model assumes that the horizontal scale of the wave processes under consideration is appreciably smaller in the longitudinal rather than in the transverse coordinate.

In recent decades, much attention of researchers has been attracted to essentially three-dimensional finite-amplitude perturbations of the water surface. However, almost all simplified models (e. g., articles [3–5]) are applicable only to nonlinear waves, propagating mainly in one direction (therein lies the fundamental restriction of the above-mentioned equations resulting from the fact that the equations of motion involve convective terms). Only in these cases, the fluid velocity can be eliminated from these terms, and the problem may be reduced to one equation for the perturbation of the free surface. For this reason, finite-amplitude waves traveling simultaneously in different directions (at arbitrary angles to each other) can be described only by the systems of equations incorporating both the perturbation of the free boundary and the liquid velocity. In the systems proposed before (for example, [6–8]), even the linear terms of all equations involve terms depending on the fluid velocity.

The equation of the second order in time, assuming head-on collision of plane nonlinear waves, was suggested in the article [9]. Its author managed to avoid the liquid velocity in the first order

IOP Conf. Series: Journal of Physics: Conf. Series **899** (2017) 032005 doi:10.1088/1742-6596/899/3/032005

of smallness and as a result to use the linearized equation of continuity for its determination. This idea served as a basis for the present work.

The purpose of this research is to propose a system of equations, which would be convenient for analysis of the spatial nonlinear waves interactions. The first version of such system was suggested in the report [10], where the results of only two simple calculations for an invisid liquid with the horizontal bottom were given.

2. Problem statement and model equations

Since the propagation rate of gravity perturbations is much lower than the speed of sound in liquids, we may use the incompressible liquid approximation. Moreover, we assume that
(i) the stationary components of the liquid flow equal zero;
(ii) the typical horizontal scale of the free surface perturbations l_w is noticeably larger than the typical equilibrium depth of the liquid layer h_0 ($h_0/l_w \sim \sqrt{\varepsilon}$, where ε is a small parameter);
(iii) the perturbation amplitude η_a is considerably small ($\eta_a/h_0 \sim \varepsilon$);
(iv) the bottom of the pool is gently sloping ($\boldsymbol{\nabla} h \sim \eta_a/l_w$, where $\boldsymbol{\nabla} = (\partial/\partial x,\, \partial/\partial y)$, x and y are the horizontal coordinates, $h(x,y)$ is the undisturbed depth of the liquid layer);
(v) the capillary effects are moderate (Bond number $\mathrm{Bo} = \rho g h_0^2/\sigma \sim 1$, where ρ is the liquid density, g is the acceleration of free fall, σ is the surface tension);
(vi) the emerging boundary layers remain thin, that is, the time of expansion of boundary layers over the entire thickness of liquid is much greater than the typical time of perturbation propagation over any considered point of the pool t_w (the hydrodynamical homochronicity number $\mathrm{Ho}_\nu = \nu t_w/h_0^2 \sim \varepsilon^2$, where ν is the kinematic viscosity). We note that $t_w = l_w/c$, where c is the propagation rate of very long linear gravitational waves. Hence, the emerging flow should be potential everywhere except for the thin boundary layer.

By analogy with the approach used in [10] for an inviscid liquid, we reduced the initial system of the continuity equation and the Stokes equations as well as the conventional kinematic and dynamic boundary conditions on the free surface and the no-slip boundary condition on the fixed bottom to the basic evolution equation of the wave type in the following form:

$$\frac{\partial^2\eta}{\partial t^2} - gh\nabla^2\eta - \frac{g}{2}\nabla^2(\eta^2) - h\nabla^2\left(\mathbf{u}^2\right) - \frac{h^2}{3}\nabla^2\frac{\partial^2\eta}{\partial t^2} + \frac{\sigma h}{\rho}\nabla^4\eta -$$
$$- g\boldsymbol{\nabla} h\cdot\boldsymbol{\nabla}\eta + g\sqrt{\frac{\nu}{\pi}}\int_0^t \frac{\nabla^2\eta\; dt'}{\sqrt{t-t'}} = \frac{1}{h}\sqrt{\frac{\nu}{\pi t}}\;\boldsymbol{\nabla}\cdot\mathbf{u}_0\,, \qquad (1)$$

where $\mathbf{u}$ is the vector of the horizontal component of liquid velocity, averaged over the layer depth, the subscript 0 marks the initial value. The equation (1) looks unusual (it contains initial data). It is explained by the fact that in the initial moment the profile of liquid velocity is not yet related to previous dynamics and has to be set explicitly. It is seen that in the course of time, the right hand side of the equation asymptotically tends to null, and its contribution in the perturbation evolution is negligible [11, 12]. The specific feature of Eq. (1) is that the average values of horizontal components of the liquid velocity are included only in the second-order term. Thereby we can use a simple approximate linear equation to determine this velocity $\mathbf{u} = \boldsymbol{\nabla}\varphi$:

$$\boldsymbol{\nabla}^2\varphi = -\frac{1}{h}\frac{\partial\eta}{\partial t}\,, \qquad (2)$$

where φ is the potential of the liquid horizontal velocity in the irrotational case. Equation (2) is actually the law of conservation of mass in the first approximation.

We emphasize that the derived model (Eqs. (1) and (2)) is applicable to waves traveling at arbitrary angles to each other. It is essentially easier for analysis than the known systems of differential equations (see, e.g., [12]) in which all equations involve liquid velocity in both linear and nonlinear terms.

3. Numerical modelling

The work considered solutions to the system of Eqs. (1) and (2) that are periodical on spatial coordinates. The pool depth $h(x, y)$ was supposed to have respective symmetry. Two functions $\eta_0(x, y)$ and $\partial\eta/\partial t|_0(x, y)$ should be set as initial condition, since according to Eq. (2) the velocity vector $\mathbf{u}_0$ is determined by the time derivative of free surface perturbation. Numerical solution of the obtained evolution problem was found with the use of Fourier method. The unknown functions η and φ, as well as the pool depth h were presented in series:

$$[\eta,\, \varphi,\, \mathbf{u},\, h]\,(x, y, t) = \sum_{n=-\infty}^{\infty}\sum_{m=-\infty}^{\infty}\left[\hat{\eta}_{n,m}(t),\, \hat{\varphi}_{n,m}(t),\, \hat{\mathbf{u}}_{n,m}(t),\, \hat{h}_{n,m}\right] e^{i\alpha n x + i\beta m y} \tag{3}$$

where $\alpha = 2\pi/L_x$, $\beta = 2\pi/L_y$, L_x, L_y are the dimensions of the computing area in certain directions. Complex amplitudes of spatial harmonics are restricted by the condition of functions to be real: $\hat{\eta}_{-n,-m} = \hat{\eta}^*_{n,m}$, $\hat{\varphi}_{-n,-m} = \hat{\varphi}^*_{n,m}$, $\hat{\mathbf{u}}_{-n,-m} = \hat{\mathbf{u}}^*_{n,m}$, $\hat{h}_{-n,-m} = \hat{h}^*_{n,m}$, here the superscript asterisk indicates operation of complex conjugation.

Substituting expressions (3) into equation (1) and equating coefficients at each exponent lead to an infinite system of ordinary differential equations for the evolution of amplitudes $\hat{\eta}_{n,m}(t)$. Confining ourselves to some rather large number of the first spatial harmonics, we obtain its finite-dimension analogue. Equation (2) is an algebraic expression for the amplitudes of potential $\hat{\varphi}_{n,m}(t)$ at a fixed time moment. Fourier coefficients of liquid velocity vector components are determined from the condition: $\hat{\mathbf{u}}_{n,m}(t) = (i\alpha k\, \hat{\varphi}_{n,m}(t),\, i\beta l\, \hat{\varphi}_{n,m}(t))$. An integral in equation (1) corresponding to viscous effects was calculated by trapezoidal method. The resulting system of ODEs was solved by the Runge–Kutta method of the fourth order of accuracy.

We will first consider the case of evolution of initial peakless periodic cross-type perturbations above the horizontal bottom. Fig. 1 shows one fourth of the spatial period for transparency: the

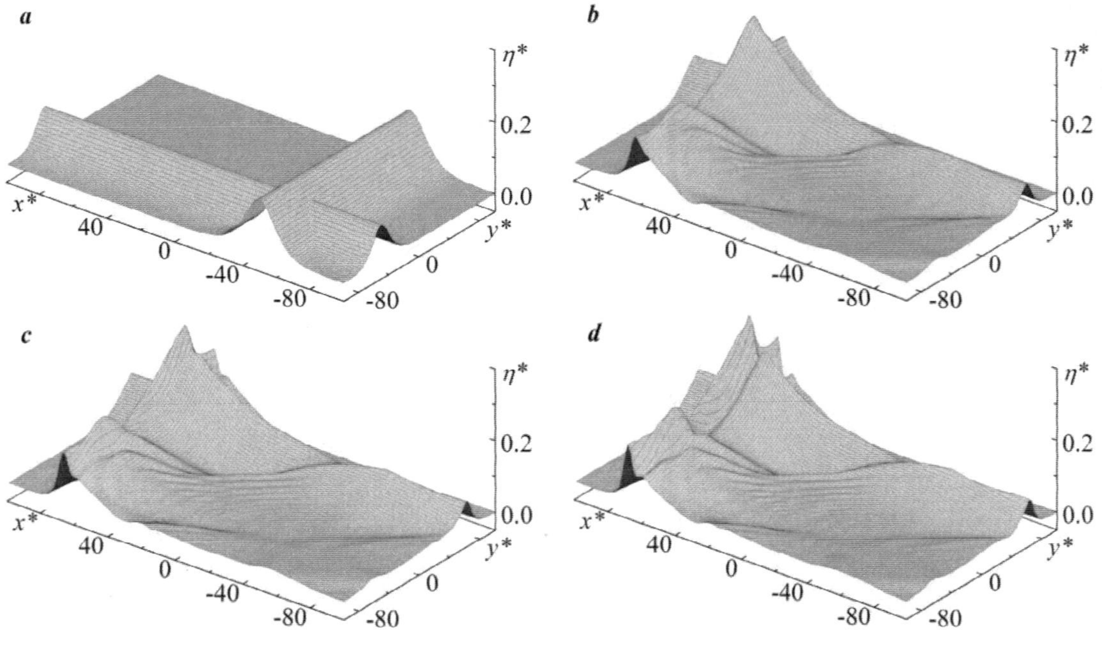

Figure 1. Shapes of free surface ($\eta^* = \eta/h$) at $t = 0$ (a) and $t^* = t\,(g/h)^{1/2} = 100$ ($b-d$) when two initial plane waves run along the x- and y-axis ($x^* = x/h$, $y^* = y/h$) above the horizontal bottom: calculations are performed with KP equation (b), 2DB equation (c) and our model (d).

full calculated area is obtained by mirror reflection, first relative to the plane $x^* = 100$, and then relative to $y^* = 100$. This perturbation is the superposition of four plane waves, propagating in line with growing x-axis or y-axis. Namely, at initial moment of time there is $\eta = \eta_x$ if $\eta_x > \eta_y$ and $\eta = \eta_y$ if $\eta_y > \eta_x$, where $\eta_j = \eta_{ja}/\cosh^2[(j - U_j t)/4l_j]$, $U_j = \sqrt{gh(1 + \eta_{ja}/3h)/(1 - 2\eta_{ja}/3h)}$, $l_j = 2h\sqrt{1/3 + h/\eta_{ja}}$, $j = x, y$ (U_j and l_j are the velocity and the length of plane soliton solutions). In this case, with time we observe a peak that appears at the cross of two wave fronts. In particular, at $t = 100\sqrt{h/g}$ we have $\eta_{\max} = 0.312\,h$ if $\eta_{xa} = 0.15\,h$ and $\eta_{ya} = 0.1\,h$, that is $\eta_{\max} = 1.25\,(\eta_{xa} + \eta_{ya})$ ($\eta_{\max} = 0.6\,(\eta_{xa} + \eta_{ya})$ for $t = 0$). Calculating this problem using

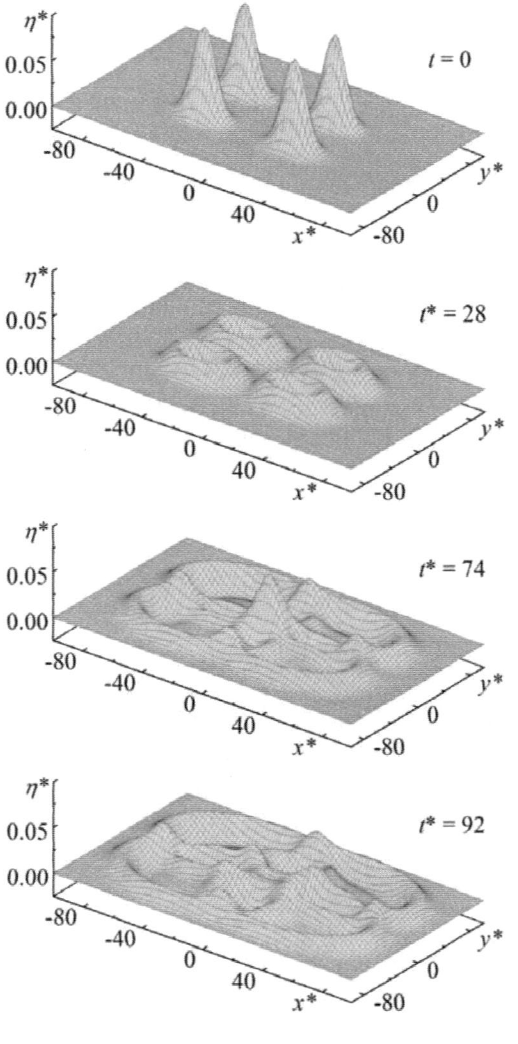

Figure 2. Shapes of free surface ($\eta^* = \eta/h_\infty$) at 5 points in time ($t^* = t\,(g/h_\infty)^{1/2}$) when 4 initial three-dimensional solitary bell-type perturbations are at rest above the horizontal bottom ($x^* = x/h_\infty$, $y^* = y/h_\infty$).

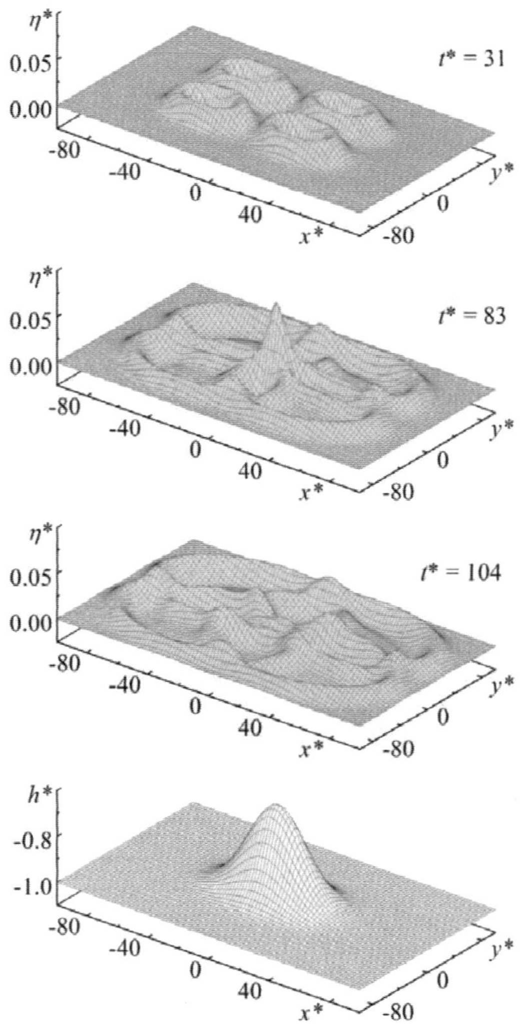

Figure 3. Shapes of free surface at 4 points in time when 4 initial three-dimensional solitary bell-type perturbations are at rest; and shape of the bottom ($h^* = -h/h_\infty$).

IOP Conf. Series: Journal of Physics: Conf. Series **899** (2017) 032005 doi:10.1088/1742-6596/899/3/032005

the two-dimensional Boussinesq (2DB) equation, we obtained that $\eta_{\max} = 1.15\,(\eta_{xa} + \eta_{ya})$. And the use of Kadomtsev–Petviashvili (KP) equation (the cross-type perturbation running approximately along the bisector of the angle $x0y$) has resulted in $\eta_{\max} = 1.07\,(\eta_{xa} + \eta_{ya})$. From the mathematical standpoint, these equations allow studying interactions of perturbations, propagating at any angle. However, 2DB and KP equations were derived with the assumption that nonlinear perturbations travel mainly in one direction. Besides, KP model supposes that in such wave processes a characteristic horizontal scale along the longitudinal coordinate is less than along the transversal one. Therefore for this problem, the KP model correctly describes

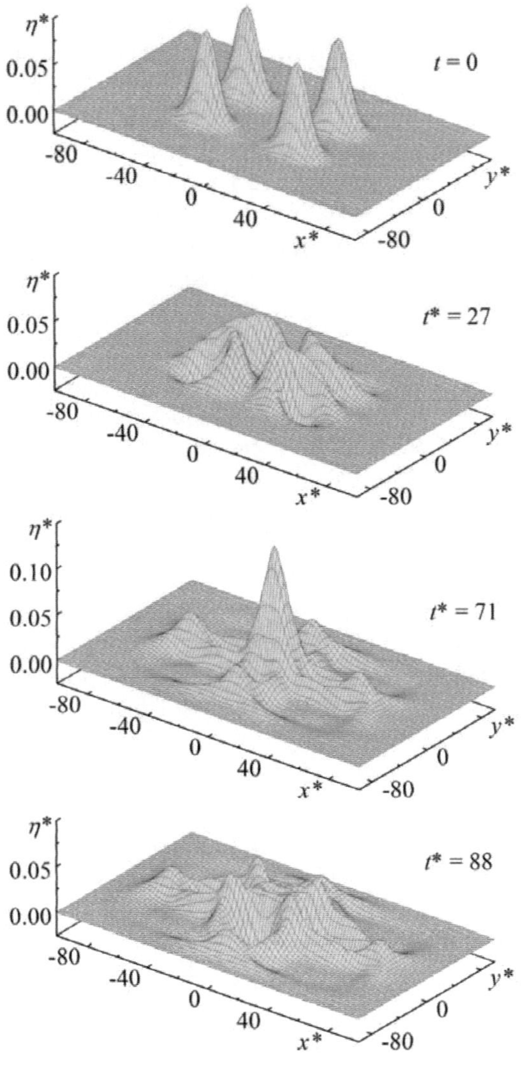

Figure 4. Shapes of free surface ($\eta^* = \eta/h_\infty$) at 5 points in time ($t^* = t\,(g/h_\infty)^{1/2}$) when 4 initial three-dimensional solitary bell-type perturbations are characterized by centripetal motion above the horizontal bottom ($x^* = x/h_\infty$, $y^* = y/h_\infty$).

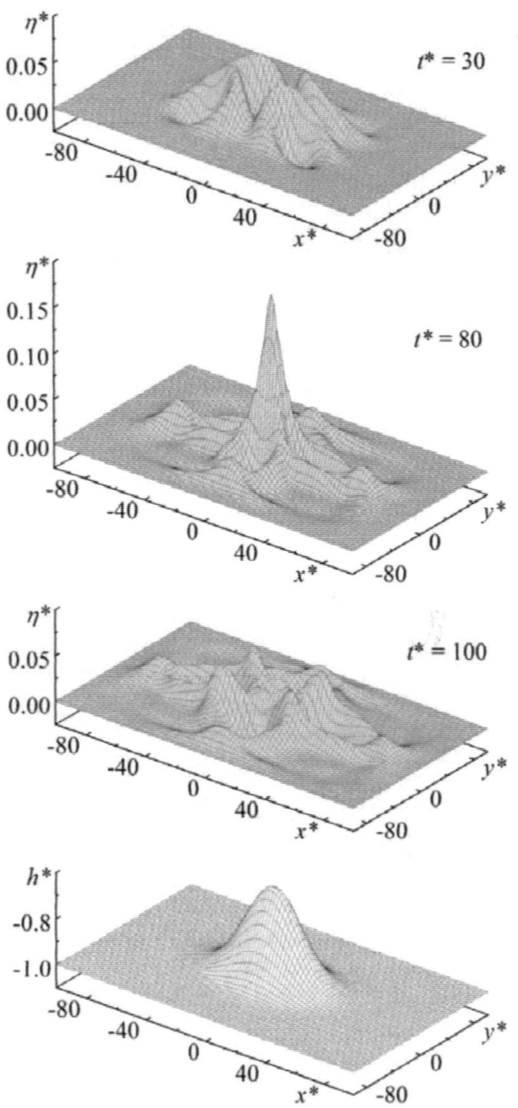

Figure 5. Shapes of free surface at 4 points in time when 4 initial three-dimensional solitary bell-type perturbations are characterized by centripetal motion; and shape of the bottom ($h^* = -h/h_\infty$).

neither nonlinearity nor dispersion.

Below are the results of calculations of the evolution of four bell-type perturbations $\eta_k(x,y) = 0.1\,h_\infty \exp\left(-[(x-x_k)^2+(y-y_k)^2]/(20\,h_\infty)^2\right)$ that were initially at rest. Here and below h_∞ is the liquid depth at $x^* = 100$ and $y^* = 100$. Figure 2 demonstrates the interaction of perturbations above the horizontal bottom, and Fig. 3 shows such interaction above the underwater mountain $h(x,y) = h_\infty\left[1 - 0.4\exp\left(-(x^2+y^2)/(50\,h_\infty)^2\right)\right]$. Maximum dimensionless deviations of the free surface in the center of the pool are equal to 0.063 in the first case, and 0.082 in the second case. They take place for $t^* = t\sqrt{g/h_\infty} = 74$ in the first case, and for $t^* = 83$ in the second case. This, of course, is due to the fact that the waves run slower, but their nonlinearity is higher above a submarine mountain than in the pool with constant depth h_∞.

One can see that if these perturbations initially move to the center of the pool their interaction is more intensive, and the diverging waves are noticeably weaker (see Figs. 4 and 5) than in previous cases. At time moments of maximal interaction of such perturbations ($t^* = 71$ for Fig. 4 and $t^* = 80$ for Fig. 5) the errors of calculations, based on 2DB equation, equal approximately 6 % for Fig. 4 and 12 % for Fig. 5. Emphasize that in the case of uneven bottom, a nonlinearity of perturbations is determined by the local ratio $\eta\,(x,y)/\,h\,(x,y)$. In particular, the local nonlinearity of the perturbation equals approximately 0.3 in the center of the pool at $t^* = 80$ for the case in Fig. 5.

4. Conclusion

The new combined approach to the description of three-dimensional perturbation evolution on a free surface of a viscous liquid is proposed. The model consists of one basic nonlinear equation and one linear auxiliary equation for determining the horizontal liquid velocity vector. This vector is present in the main equation only in the term of the second order of smallness. The method is suitable for moderately long waves of small but finite amplitude, simultaneously running at any angle. This approach is, in essence, easier than the known systems of equations in which all equations contain both linear and nonlinear terms.

Validity of this approach for a number of planar problems of nonlinear disturbance transformation is shown by means of numerical experiments. It is important to note, that at uneven bottom (Figs. 3 and 5) the topography stronger affects the change of coefficients of the main linear and dispersive terms, compared to the term, containing the gradient of liquid depth, which is inessential.

5. Acknowledgments

The study was financed by the Russian Science Foundation (grant No.14-22-00174).

6. References

[1] Miles J W 1981 *J. Fluid Mech.* **106** 131–147
[2] Kadomtsev B B and Petviashvili V I 1970 *Sov. Phys. Dokl.* **15** 539–542
[3] Kim K Y, Reid R O and Whitaker R E 1988 *J. Comput. Phys.* **76** 327–348
[4] Pelinovsky D E and Stepanyants Yu A 1994 *JETP* **79** 105–112
[5] Khabakhpashev G A 1997 *Comput. Technol.* (in Russian) **2** 94–101
[6] Peregrine D H 1967 *J. Fluid Mech.* **27** 815–827
[7] Karpman V I 1975 *Nonlinear Waves in Dispersive Media* (New York: Pergamon Press)
[8] Green A E and Naghdi P M 1986 *J. Fluid Mech.* **78** 237–246
[9] Johnson R S 1996 *J. Fluid Mech.* **323** 65–78
[10] Arkhipov D G and Khabakhpashev G A 2006 *Doklady Physics* **51** 418–422
[11] Khabakhpashev G A 1987 *Fluid Dyn.* **22** 430–437
[12] Liu P and Orfila A 2004 *J. Fluid Mech.* **520** 83–92

Study of the convective fluid flows with evaporation on the basis of the exact solution in a three-dimensional infinite channel

V B Bekezhanova[1] **and O N Goncharova**[2,3]

[1] Institute of Computational Modeling SB RAS, Academgorodok 50/44, Krasnoyarsk, Russia
[2] Institute of Thermophysics SB RAS, Ac. Lavrentieva ave 1, Novosibirsk, Russia
[3] Altai State University, pr Lenina 61, Barnaul, Russia

E-mail: vbek@icm.krasn.ru, gon@math.asu.ru

Abstract. The solution of special type of the Boussinesq approximation of the Navier – Stokes equations is used to simulate the two-layer evaporative fluid flows. This solution is the 3D generalization of the Ostroumov – Birikh solution of the equations of free convection. Modeling of the 3D fluid flows is performed in an infinite channel of the rectangular cross section without assumption of the axis-symmetrical character of the flows. Influence of gravity and evaporation on the dynamic and thermal phenomena in the system is studied. The fluid flow patterns are determined by various thermal, mechanical and structural effects. Numerical investigations are performed for the liquid – gas system like ethanol – nitrogen and HFE-7100 – nitrogen under conditions of normal and low gravity. The solution allows one to describe a formation of the thermocapillary rolls and multi-vortex structures in the system. Alteration of topology and character of the flows takes place with change of the intensity of the applied thermal load, thermophysical properties of working media and gravity action. Flows with translational, translational-rotational or partially reverse motion can be formed in the system.

1. Introduction

The fluid flows with an interface being under action of the co-current gas fluxes and accompanied by evaporation or condensation have been the subject of extensive theoretical and experimental investigations in the last few decades (for a review, see [1]). The flows are applied in systems of the fluidic cooling or thermal controlling of highly efficient semiconductor equipment, in membrane evaporators, distillers, thermal coating applications technologies, etc. Precisely forecasting the fluid dynamics in the processes requires comprehensive analysis based on modeling the two-layer flows with evaporation. Results of the theoretical study can help one to clarify physical aspects of evaporative convection and phenomena in the liquid media caused by the gas flow. One way of modeling real fluid flows is obtaining the exact solutions. It allows one effectively and rapidly to get some evaluation characteristics or to forecast outcome of experiments on preliminary stages of working-out [2]. Finally, in multiparameter problems the exact solution certainly gives information on degree and character of influence of various factors and provides possibility to make more precise a mathematical model.

In present work the character and structure of the joint flow of evaporating liquid and co-current laminar gas flux are investigated on the basis of new exact solution of the Boussinesq approximation of the Navier – Stokes equations. Obtained solution is a generalization of the well-known Ostroumov – Birikh solution [3, 4] for three dimensional case. The group nature of the Birikh solutions [5] allowed one to generalize the solutions both for 3D convection problem in the non-axis-symmetrical case [6] and for problems of evaporative convection in the domains with internal interfaces admitting phase transition [7]. The invariant and partially invariant solutions are used to study the fundamental and secondary features of physical processes described by the convection equations. They imply the natural properties of space-time symmetry and symmetry of spatial fluid motion. It should be noted that the asymptotic character of the Birikh type solution has been justified in [8], where the thermocapillary gravitational convection in a long horizontal cavity has been studied experimentally and numerically on the basis of a 2D problem. Complete investigation of properties of the constructed solutions expects the study of the stability characteristics and their perturbation spectrum. In [9, 10] the stability of the two-dimensional convective flows with evaporation was investigated. Influence of intensity of the thermal loads on the external channel walls, gravity and linear thicknesses of the working media layers on the stability characteristics have been studied in the framework of the problem statement without taking into account the Soret effect. The aim of the paper is to carry out mathematical modeling of the three-dimensional convective fluid flows with evaporation as well as to analyze the influence of the gravity and external thermal load on the flow patterns. The study is performed in the framework Oberbeck – Boussinesq model taking into account the Soret and Dufour effects in vapor – gas phase.

2. Generalization of the Ostroumov – Birikh solution for 3D flows with evaporation at interface

Let the Cartesian coordinate system be chosen so that the gravity acceleration vector $\mathbf{g}$ is directed opposite to the Ox axis ($\mathbf{g} = -g\mathbf{i}$, $\mathbf{i}$ is the unit vector of Ox). Two fluid layers are separated by the thermocapillary interface Γ given here by the equation $x = 0$ (see fig. 1). Let the linear size of the flow domain in the y-direction h be the characteristic length. The characteristic values for the coupled problem of the liquid-gas flows are introduced on the basis of the characteristics of the liquid so that u_*, T_* and $p_* = \rho_1 u_*^2$ are the characteristic velocity, temperature drop and pressure, respectively. Viscous incompressible fluids (liquid and gas – vapor mixture) fulfill the infinite horizontal layers Ω_1 and Ω_2 (fig. 1). The boundaries of the domains Ω_1, Ω_2 are the fixed impermeable walls. The stationary three-dimensional convective flows of j-th medium (here and subsequently $j = 1, 2$ relate to the liquid and gas – vapor mixture, respectively) are described by the Oberbeck – Boussinesq approximation of the Navier – Stokes equations [7, 11].

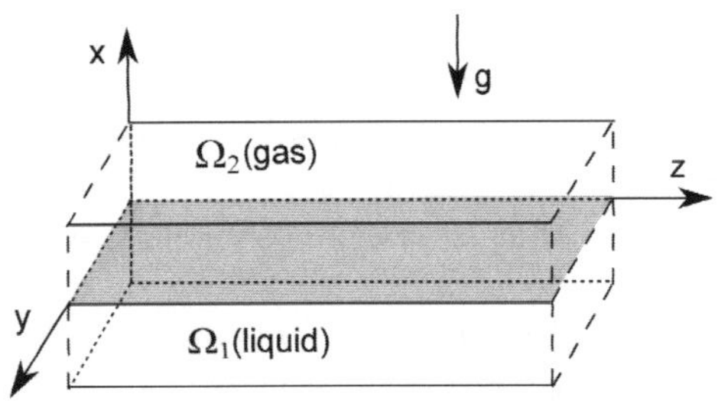

Figure 1. Flow domain

The Dufour and Soret effects (or the effects of diffusive thermal conductivity and thermodiffusion [12, 13, 14]) are taken into account in the gas phase. We suppose that the vapor is a passive admixture, the vapor diffusion in the gas phase is described by the diffusion equation. We construct the exact solution of the Oberbeck – Boussinesq, which is characterized by dependence of the components of the liquid $\mathbf{v}_1 = (u_1, v_1, w_1)$ and gas velocity $\mathbf{v}_2 = (u_2, v_2, w_2)$ vectors on the transverse coordinates (x, y) (see [7]). The temperature functions T_j, pressure p_j and vapor concentration C have the terms Θ_j, q_j, Φ similarly depending on the transverse coordinates (x, y):

$$u_j = u_j(x,y), v_j = v_j(x,y), w_j = w_j(x,y),\ T_j = -Az + \Theta_j(x,y),\ C = Bz + \Phi(x,y),$$

$$p_1 = -A\frac{Gr}{Re^2}xz + q_1(x,y),\ p_2 = -A\bar{\rho}\bar{\beta}\frac{Gr}{Re^2}xz + B\bar{\rho}\gamma\frac{Ga}{Re^2}xz + q_2(x,y).$$

Here $Re = u_*h/\nu_1$ is the Reynolds number, $Gr = \beta_1 T_* gh^3/\nu_1^2$ is the Grashof number, $Ga = gh^3/\nu_1^2$ is the Galilei number, $\bar{\rho} = \rho_2/\rho_1$, $\bar{\beta} = \beta_2/\beta_1$ are the ratios of the densities ρ_j and coefficients of thermal expansion β_j of the gas and liquid, respectively; ν_j is the coefficient of kinematic viscosity, γ is the concentration coefficient of the gas density. The coefficients A and B determine the constant longitudinal temperature and concentration gradients along the interface. If A_* and B_* are the dimensional longitudinal gradients of temperature and concentration functions then $A = A_*h/T_*$, $B = B_*h$. The following relation between A and B will take place $B = -C_*\bar{\varepsilon}A$ because of the interface condition for saturated vapor concentration. (Here C_* is the saturated vapor concentration at $T_2 = T_0$; $\bar{\varepsilon} = \varepsilon T_*$, $\varepsilon = \lambda\mu/(R^*T_0^2)$, λ is the latent heat of evaporation, μ is the molar mass of the evaporating liquid, R^* is the universal gas constant; for details and choice of T_0 see [2, 9].)

The interface boundary conditions are formulated on the basis of the conservation laws and some additional assumptions [6, 7, 11]. The kinematic and dynamic conditions (projection on the tangential and normal vectors to the interface) should be fulfilled on the interface $x = 0$ [6, 7]. Conditions of continuity of tangential velocities and temperature are assumed to be fulfilled on the thermocapillary interface $x = 0$. At $x = 0$ the heat transfer condition with respect to the diffusive mass flux due to evaporation and the vapor mass balance equation are formulated. These relations take into account the Dufour and Soret effects. The linearized form of an equation for saturated vapor concentration at interface is used as a condition for the vapor concentration function at interface [9]. It is a consequence of the Clapeyron – Clausius equation and the Mendeleev – Clapeyron equation for an ideal gas. On the fixed impermeable walls of the channel the no-slip conditions for velocity fields and the conditions of thermal insulating of the lateral walls are imposed. The case of absence of vapor flux on the upper and lateral rigid boundaries is studied in the present statement. Both conditions provide a fulfillment of the conditions for full heat flux and for full mass flux with respect to the Dufour and Soret effects, respectively, on the fixed walls [11, 15].

3. Numerical investigations

The analytical calculations for construction of the exact solutions of the three-dimensional convection problem with evaporation at the interface is complemented by the numerical investigations. The numerical algorithm of solving of each 2D problems is based on the longitudinal transverse finite difference scheme known as the method of alternating directions [6, 11, 16].

Numerical investigations are performed for the liquid – gas system like ethanol – nitrogen and HFE-7100 – nitrogen with the physicochemical properties described in [17] (see also [2, 7, 9]). The values of other parameters used by simulations including the coefficients characterized the Soret and Dufour effects in the gas – vapor layer are chosen similarly to that described in [2, 7, 9]. We investigate the flow topology computed with following values of the Grashof number $Gr = \{47000, 470\}$ for the ethanol – nitrogen flows and with the values $Gr = \{1220000, 12000\}$ for the flows of the HFE-7100 – nitrogen system. These values of the Grashof number correspond to the conditions of normal ($g = g_0 = 9.81$ m/s^2) and low gravity ($10^{-2}g_0$ m/s^2), respectively, when the characteristic temperature drop and characteristic length are equal to $T_* = 10$ K and $h = 10^{-2}$ m. The characteristic velocity is chosen equal to the velocity of viscous stresses relaxation ν/h so that $u_* = 0.15\cdot 10^{-5}$ (m/s) for ethanol and $u_* = 0.38\cdot 10^{-4}$ (m/s) for HFE-7100.

We interpret constructed solution as a solution describing the flow on the working area $[0, z_0]$ in a sufficiently long cavity. The examples of flow structure (temperature distribution and projections of stream lines in $z = 0.5z_0$ cross-section) are presented in figs. 2, 3 for ethanol – nitrogen and figs. 4, 5 for HFE-7100 – nitrogen systems. In the numerical experiments the non-dimensional longitudinal temperature gradient A has the values $A = 0.1$ and $A = 0.3$, which correspond to the dimensional values equal to $A_* = 100$ K/m and $A_* = 300$ K/m, respectively. The flows are characterized as the translational motion and progressively rotational flow and can be realized in various forms.

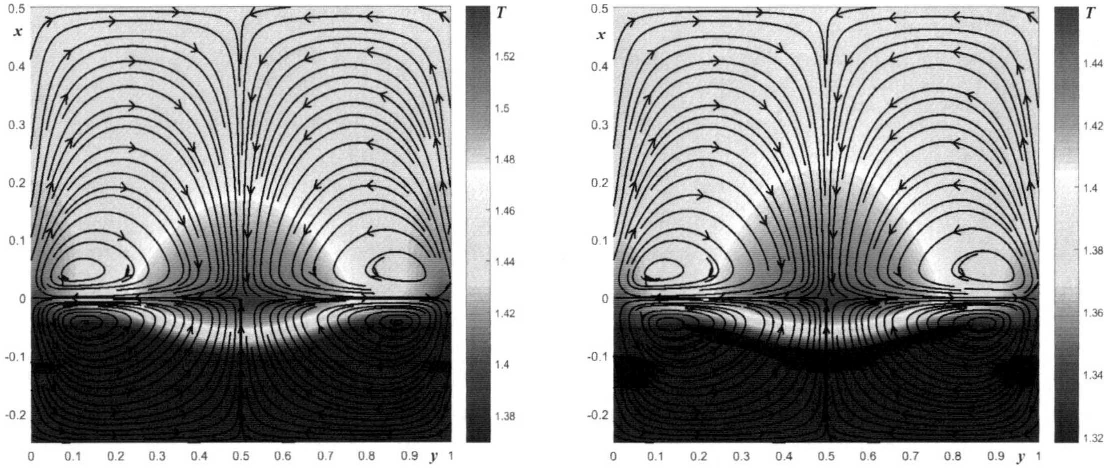

Figure 2. Flow pattern. Ethanol – nitrogen system; $A = 0.1$: $Gr = 470$ (left); $Gr = 47000$ (right).

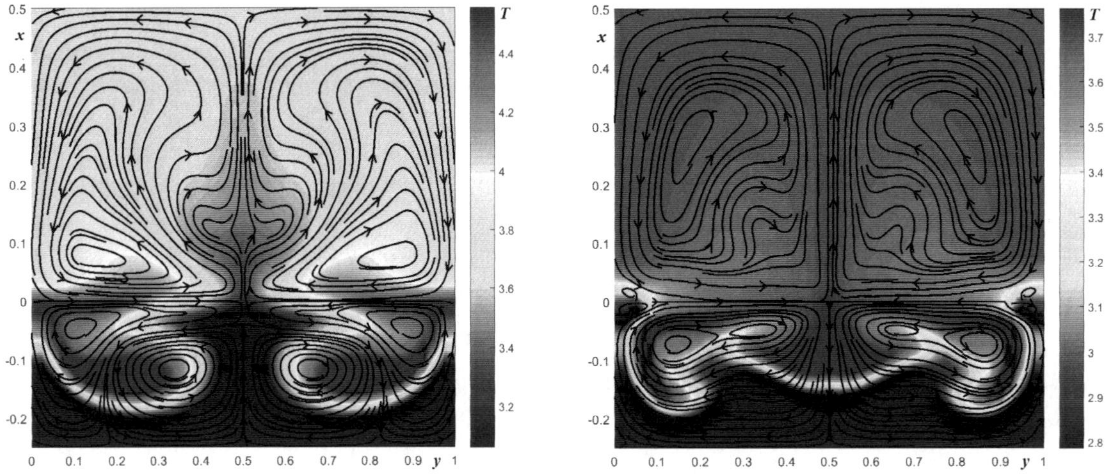

Figure 3. Flow pattern. Ethanol – nitrogen system; $A = 0.3$: $Gr = 470$ (left); $Gr = 47000$ (right).

Flows in both phases in fig. 2 are characterized by clearly marked symmetric two-vortex structures by $A = 0.1$. Vortexes in the liquid have the convective nature and vortex centers are located close to the interface in the "corners" (fig. 2). At the same time the bright qualitative differences are found by simulation of the HFE-7100 – nitrogen flows (fig. 4). By $A = 0.1$ we observe the two-vortex structure in the gas layer and "quadruple" flow in the liquid layer if $Gr = 12200$. Two additional small corner vortices appear in the liquid layer in the case of normal gravity $Gr = 1220000$. Formation of the extra vortexes is explained by weaker viscous effects in HFE-7100 in comparison with ethanol.

More complicated vortex patterns in fluids are observed in the case with more intensive longitudinal temperature regimes (at $A = 0.3$, see figs. 3, 5). Splitting two big vortex into four

IOP Conf. Series: Journal of Physics: Conf. Series **899** (2017) 032006 doi:10.1088/1742-6596/899/3/032006

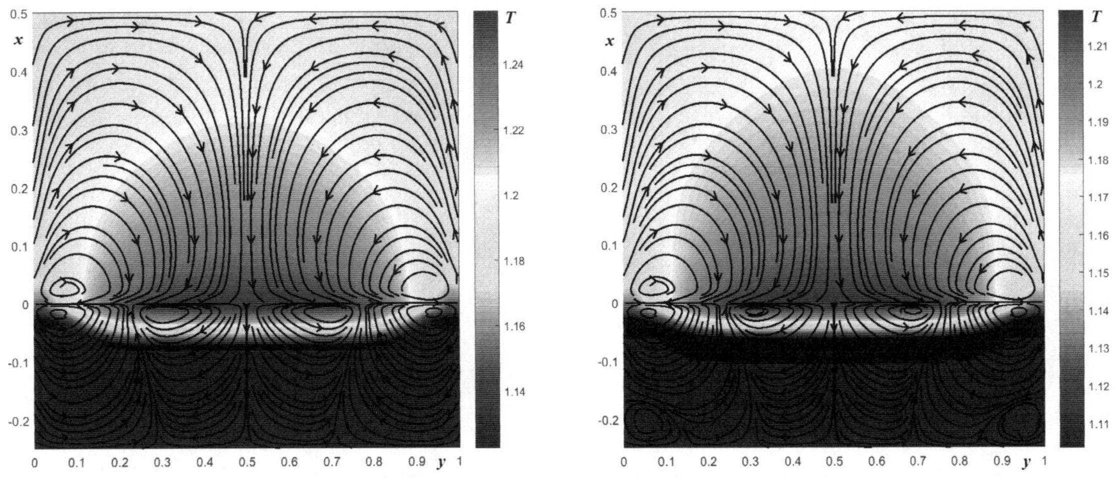

Figure 4. Flow pattern. HFE-7100 – nitrogen system; $A = 0.1$: $Gr = 12200$ (left); $Gr = 1220000$ (right).

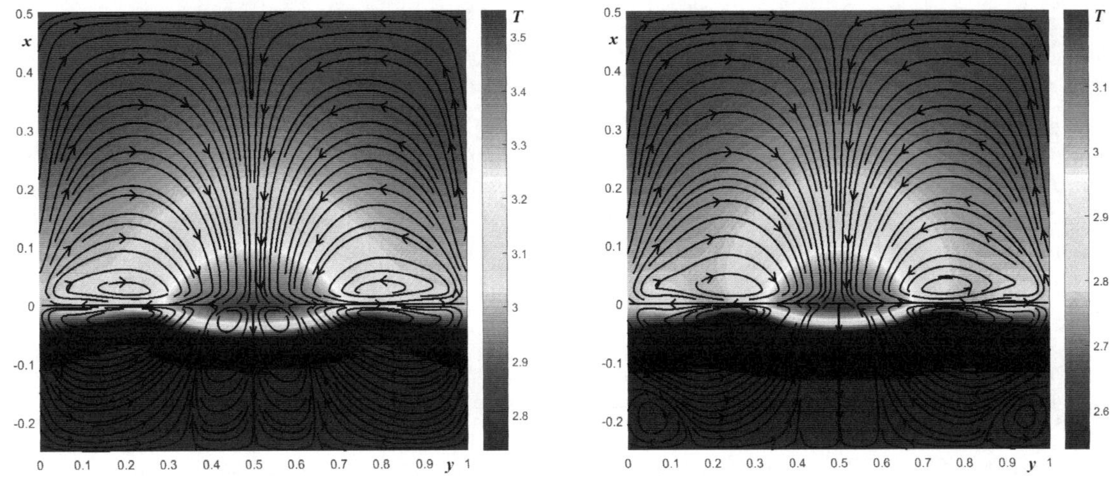

Figure 5. Flow pattern. HFE-7100 – nitrogen system; $A = 0.3$: $Gr = 12200$ (left); $Gr = 1220000$ (right).

asymmetrical swirls in the liquid layer occurs in the ethanol – nitrogen system in microgravity conditions. Two of the vortices are deformed into "angled" vortexes, but their cores stay close to the interface in "corners" near the lateral walls. Liquid particles from the hot spot move along the interface due to the Marangoni effect and go down near the lateral walls because of unstable thermal stratification. Thus, the upper part of the "angled" vortex is generated by thermocapillary effect, but the side part appears due to convective mechanism. In HFE-7100 – nitrogen system in the case with more intensive longitudinal temperature regimes (see fig. 5) the liquid flows are characterized by four separate vortices under microgravity and six swirls in the terrestrial conditions. For the system the flow regimes differ only in the liquid layers (compare fig. 4 and 5), whereas the fluid flow structures are rebuilt in both upper and lower layers in ethanol – nitrogen system (fig. 2 and 3). In all the cases flow patterns are symmetric with respect to the plane $y = 0.5$ in both phases. Thus, we have different flow topology for the various gas – liquid systems. More intensive rotational flow has been observed by normal gravity (here $Gr = 47000$ for ethanol – nitrogen, $Gr = 1220000$ for HFE-7100 – nitrogen systems). Some minor quantitative differences are observed by comparison of the flows under conditions of normal gravity and microgravity for both liquids. However, it should be pointed out that there are more greater quantitative differences between liquid and gas flows in the flows of the ethanol – nitrogen system, than in the HFE-7100 – nitrogen system.

4. Conclusions

The solution of special type of the 3D stationary coupled problem of the gravitational and thermocapillary convection with respect to evaporation is used to describe the convective flows in an infinite channel of a rectilinear cross section without assumption of axial symmetry of the flow domains. This solution has the group nature and is the analogue of the Ostroumov – Birikh solution of the convection equations which additionally include the Dufour and Soret effects in the gas phase. Considering the infinite channel under action of the constant longitudinal temperature gradient we study a model problem of evaporative convection and interpret the constructed solution as an one that describes the flow on the working area being a sufficiently long cavity.

The flows of both fluids (of a liquid and gas – vapor mixture) modeled with the help of the exact solution can be characterized as a translational motion and progressively rotational flows and realized in the various forms. The qualitative and quantitative differences are confirmed for the flows of various working fluids (ethanol – nitrogen and HFE-7100 – nitrogen systems). The numerical investigations allow one to analyze the possible flow structure with respect to the intensity of the gravitation field and longitudinal temperature gradients created on the interface. The intensity of the liquid flows depends on intensity of the gravitation field and on interface temperature regime. Topologically various structures of flows are formed due to combined influence of the thermocapillary and convective mechanisms and evaporation/condensation process, affecting the thermal pattern of the flows.

Acknowledgments

This work is supported by the Russian Science Foundation (project RSF 15-19-20049).

References

[1] Bekezhanova V B and Goncharova O N 2017 Problems of the evaporative convection (Review) *Journal of Applied Mathematics and Mechanics* (accepted for publication)
[2] Goncharova O N, Rezanova E V, Lyulin Yu V and Kabov O A 2015 Modeling of two-layer liquid-gas flow with account for evaporation *Thermophysics and Aeromechanics* **22(5)** 655–661
[3] Ostroumov G A 1952 *Free convection under the conditions of an internal problem* (Moscow-Leningrad: Gostekhizdat Press)
[4] Birikh R V 1969 Thermocapillary convection in a horizontal layer of liquid *Journal of Applied Mechanics and Technical Physics* **3** 43–45
[5] Pukhnachov V V 2000 Group-theoretical nature of the Birikh's solution and its generalizations (Krasnoyarsk: Book of Proc. Symmetry and differential equations) 180–183
[6] Goncharova O N, Kabov O A and Pukhnachov V V 2012 Solutions of special type describing the three dimensional thermocapillary flows with an interface *Int. J. Heat Mass Transfer* **55** 715–725
[7] Goncharova O N and Kabov O A 2016 Investigation of the two-layer fluid flows with evaporation at interface on the basis of the exact solutions of the 3D problems of convection *Journal of Physics: Conference Series* **754 (032008)** 1–6
[8] Kirdyashkin A G, Polezhaev V I, Fedyushkin A I 1983 Thermal convection in a horizontal layer with lateral heating *Journal of Applied Mechanics and Technical Physics* **24(6)** 876–882
[9] Bekezhanova V B and Goncharova O N 2016 Stability of the exact solutions describing the two-layer flows with evaporation at interface *Fluid Dynamics Research Journal* **48(6)** 061408
[10] Bekezhanova V B, Goncharova O N, Rezanova E V and Shefer I A 2017 Stability of two-layer fluid flows with evaporation at the interface *Fluid Dynamics* **52 (2)** 189–200

[11] Andreev V K, Gaponenko Yu A, Goncharova O N, Pukhnachov V V 2012 *Mathematical models of convection (de Gruyter Studies in Mathematical Physics)* (Berlin/Boston: De Gruyter)
[12] De Groot S R and Mazur P 1984 *Thermodynamics of irreversible processes* (London: Dover 1984)
[13] Prigogine I and Dufay R 1954 *Chemical thermodynamics* (London, New York: Longmans, Green)
[14] Gebhart B, Jaluria Y, Mahajan R L and Sammakia B 1988 *Bouyancy-induced flows and transport* (Berlin, Heidelberg, New York, London, Paris, Tokyo: Springer Verlag)
[15] Landau L D and Lifshitz E M 1987 *Fluid Mechanics Vol. 6 (2nd ed.)* (Butterworth – Heinemann)
[16] Roache P J 1976 *Computational Fluid Dynamics* (Albuquerque: Hermosa Publishers)
[17] Weast C R C 1979 *Handbook of Chemistry and Physics* (Boca Raton, Florida: CRC Press Inc)

Formation of a cavitation cluster in the vicinity of a quasi-empty rupture

E S Bol'shakova[1] and V K Kedrinskiy[1]

[1]Lavrentyev Institute of Hydrodynamics, Siberian Branch of Russian Academy of Sciences, Lavrentyevprosp.15, Novosibirsk, 630090, Russia

E-mail: kedr@hydro.nsc.ru

Abstract. The presentation deals with one of the experimental and numerical models of a quasi-empty rupture in the magma melt. This rupture is formed in the liquid layer of a distilled cavitating fluid under shock loading within the framework of the problem formulation with a small electromagnetic hydrodynamic shock tube. It is demonstrated that the rupture is shaped as a spherical segment, which retains its topology during the entire process of its evolution and collapsing. The dynamic behavior of the quasi-empty rupture is analyzed, and the growth of cavitating nuclei in the form of the boundary layer near the entire rupture interface is found. It is shown that rupture implosion is accompanied by the transformation of the bubble boundary layer to a cavitating cluster, which takes the form of a ring-shaped vortex floating upward to the free surface of the liquid layer. A *p-κ* mathematical model is formulated, and calculations are performed to investigate the implosion of a quasi-empty spherical cavity in the cavitating liquid, generation of a shock wave by this cavity, and dynamics of the bubble density growth in the cavitating cluster by five orders of magnitude.

1.Introduction.

Explosive volcanic processes have been intensely investigated in both experiments and simulations [1, 2]. Geophysical data show that the magma melt is saturated by a high-pressure gas. The decrease in pressure induced by conduit opening violates the thermodynamic equilibrium state. Rarefaction waves are generated, diffusion processes behind the fronts of these waves arise, and intense development of bubble cavitation is observed, which can lead to magma melt rupture. The formation of such rupture can be considered as a hydrodynamic mechanism of cyclic eruptions. An experimental model of rupture formation in a liquid layer was proposed in [3]. Those experiments were performed in electromagnetic hydrodynamic shock tubes (EM HSTs) with the diapason of storage energies up to 0.8 – 1.2 kJ [3]. It was shown that the shock wave generated in the liquid layer by a membrane under the action of a pulsed magnetic field initiates the development of intense cavitation processes in the liquid layer after SW reflection from the free surface. In the presented problem (for small EM HST with storage energies up to 100 J) the pulsed motion of the membrane leads to the formation of a quasi-empty rupture already (in 10 μs) on the motionless membrane. The dynamics and structure of this rupture can be considered as a qualitative model of the above-mentioned process. Distilled water is a multiphase medium with very small gas bubbles, solid microparticles, and their combinations with sizes of about 1.5 μm and densities of the order of 10^6 cm^{-3} [4]. New experimental data on the dynamics of the state of the bubble boundary layer on the rupture interface and on its transformation to a cavitating cluster at the rupture implosion "point" are reported below. A two-phase mathematical model of the implosion of a quasi-empty spherical bubble under conditions of an intensely developing cavitating cluster is considered.

2. Statement of experiments

The test section of the small experimental setup (EM HST) includes a transparent cuvette 4 cm in diameter, which is filled with distilled water [4]. The bottom of the cuvette is a conducting Duralumin membrane, which is placed directly onto a flat helical coil connected via an electronic switch to a

IOP Conf. Series: Journal of Physics: Conf. Series **899** (2017) 032007 doi:10.1088/1742-6596/899/3/032007

high-voltage capacitor bank. When the switch is closed, a pulsed magnetic field is generated on the coil, which pushes the membrane due to the skin effect, thus, generating a powerful SW in the liquid (the SW amplitude is of the order of 10 MPa). The process is recorded by a high-speed video camera with a frequency of 10^5 frames/**s.**

3.Experimental results: dynamics of the state of the quasi-empty rupture and bubble cluster

When the electronic switch is closed, an SW is generated in the liquid layer. This SW is reflected from the free surface of the layer in the form of a rarefaction wave and leads to the formation of an intense cavitation region (**Fig. 1, a**). In 0.5 ms, a rupture is formed on the membrane (**Fig. 1, b**). A bubble boundary layer on the surface, which covers the entire rupture interface, is clearly visible. The next two frames (**Figs. 1, c and 1, d**) demonstrate the dynamics of the collapsing rupture and SW radiation.

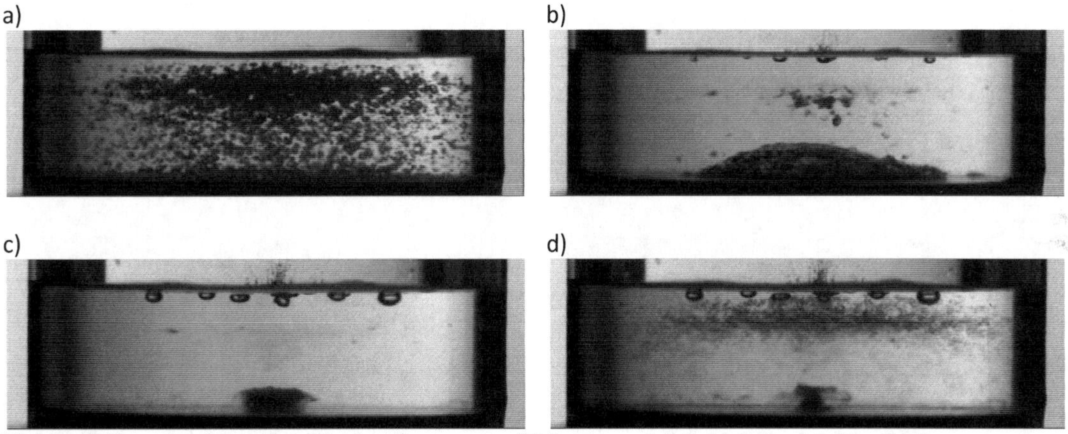

Figure 1. Rupture dynamics. The loading energy 32 J, at t = a) 0.1; b) 0.5; c) 1.04; d)1.1ms.

The analysis of the rupture state dynamics shows that it acquires a shape close to a spherical segment. It is known that an implosion of an empty cavity in a homogeneous liquid near a solid wall is accompanied by the formation of a cumulative jet. In the present case, however, such a situation is not observed: the rupture implodes symmetrically, without changing its topology. It is of interest that the liquid flow on the membrane in the course of the rupture implosion forms a ring-shaped flow of a homogeneous liquid, which is closed under the rupture with a high velocity (**Fig. 1, c**). At the end of the rupture implosion, a dense core is formed, and the bubble boundary layer transforms to a bubble cluster surrounding the rupture The state of the rupture surrounded by a dense bubble cluster at the instant of the implosion can be defined as a state with a high intenal energy. This is confirmed by generation of a sufficiently powerful secondary SW after the rupture implosion (**Fig. 1. d**). Propagation of this SW and its interaction with the free surface leads to the formation of the secondary cavitation region (**Fig. 1d**). It was demonstrated earlier that the rupture retains its topology, i.e., a shape similar to a spherical segment. Based on this fact, we use the dimensional method, assuming that the basic parameters of the liquid state are the hydrostatic pressure p_0 and density ρ of the liquid, the storage energy E, and the coefficient α characterizing the ratio of the maximum potential energy of the rupture to the loading energy Q/E. Let us find a combination of parameters that yields the time characteristics of the examined process:

$$T = \alpha^{1/3} E^{1/3} \rho^{1/2} . p_0{}^{-5/6} \quad (1)$$

The formulas of the rupture dynamics have the form:

$$r \approx 13.4 + 6.4 \cdot t - 19 \cdot t^2., \qquad h \approx 0.3 + 14 \cdot t - 12 \cdot t^2.$$

IOP Conf. Series: Journal of Physics: Conf. Series **899** (2017) 032007 doi:10.1088/1742-6596/899/3/032007

The value of α predicted by Eq. (1) is equal to the value for the maximum potential energy: $\alpha = 0.0032$ or $Q \approx 0.1$ J. It is of interest that the experimental results obtained for identical storage energy, but for different heights of the water layer in the cavity are close to each other. Thus, it is confirmed that the dynamics of cluster formation depends to a greater extent on the loading energy and is almost independent of the water layer height. **Fig.2** illustrates the cluster dynamics after the rupture implosion and generation of the secondary SW. It is seen that the cluster emerges to the water layer surface and possesses a sufficient amount of the initially stored energy to move to the liquid layer surface almost uniformly (1. 5 cm in 2 ms). The character of the bubble cluster motion and the fact that a ring-shaped flow along the membrane is formed due to the rupture implosion allow us to assume that the bubble boundary layer transforms after the closure of the quasi-empty rupture to a ring-shaped cluster (vortex).

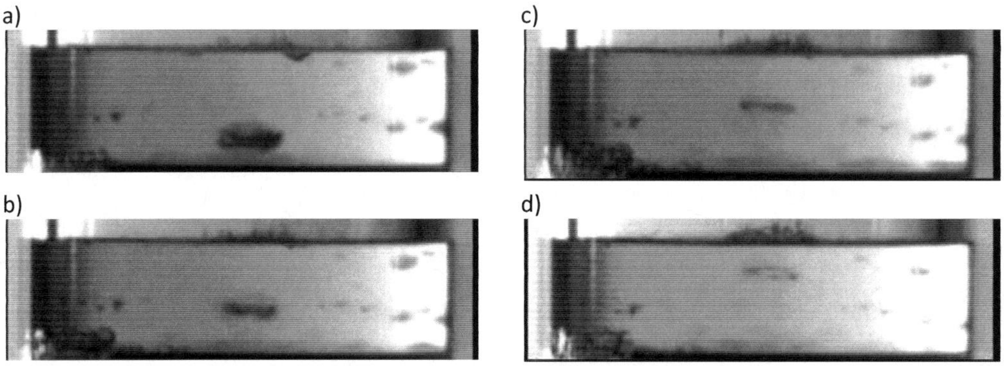

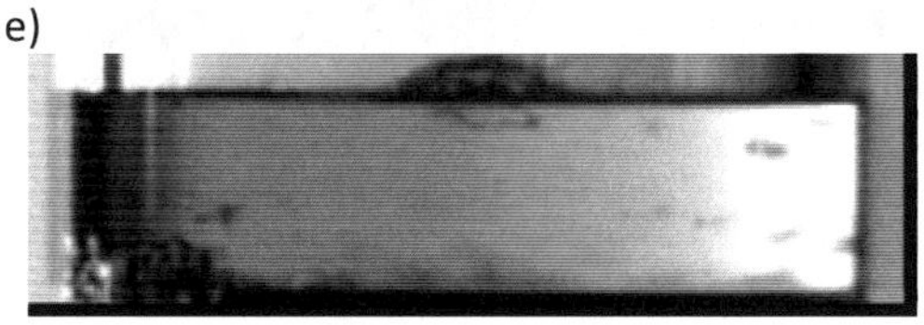

Figure 2. Dynamics of the bubble cluster. The loading energy is 32 J; the time instants are 1.3 (a), 1.7 (b), 2.1 (c), 2.7 (d), and 3.1 ms (e).

The results of the experimental analysis for the cavity width of 4 cm, water layer height of 1.5 cm, and loading energy of 18, 24.5, and 32 J are shown in **Fig. 3**. It is seen that the plot of motion is an almost straight line for the greatest energy examined in this study. As the energy decreases, the plot becomes curved, and the emergence velocity becomes smaller. The bubble cluster velocity is approximately equal to 7.5 m/s for the storage energy of 32 J.

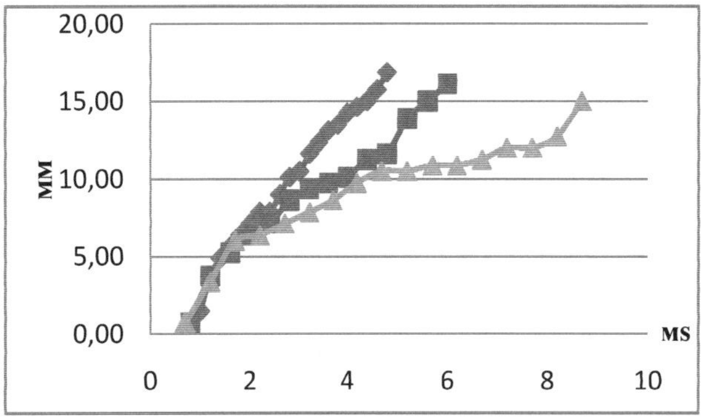

Figure3. Dynamics of bubble cluster emergence for different loading energies. The water layer height is 1.5 cm. The loading energy is 32 (blue), 24.5 (red), and 18 J (green).

4 Mathematical model of the imploding rupture in the intense cavitation region.

Formulation of the problem. A spherical cavity with a radius of 1 cm is placed into a space willed with a liquid, which is distilled water with known parameters of its heterogeneous state (radius of free gas microbubbles 1.5 μm, p_0= 1 atm, and number density of these nuclei 10^6 cm^{-3} for the volume concentration of the gas phase $k_0 = 10^{-6}$). The pressure in the cavity (R radius) and microbubbles is equal to the hydrostatic pressure p_0= 1 atm. The system is in the equilibrium state. At the time $t = 0$, the pressure inside the cavity instantaneously decreases to an extremely low value: $p_0=10^{-8}$ atm. A cavitation process starts to develop near the cavity boundary under the action of the initial pressure in the bubbles.

Dynamics of the state of the cavitating liquid. By introducing a new function $\zeta = p - p\_ k^{-\gamma}$ and a variable $\eta = r\,\alpha\, k^{1/6}$ as a new independent variable, the system of the *pk-model* can be converted to the following form within the framework of some approximations underthe condition of an incompressible liquid component

$$\frac{d^2\zeta}{d\eta^2} + \frac{\nu}{\eta}\left(\frac{d\zeta}{d\eta}\right) - \zeta = 0, \qquad \frac{\partial^2 k}{\partial t^2} = -\frac{3}{\rho_0 r_0{}^2} k^{\frac{1}{3}}\,\zeta + \frac{1}{6k}\left(\frac{\partial k}{\partial t}\right)^2,$$

where $\nu = 0, 1, 2$ defines the type of the flow symmetry. In the spherical formulation, the first equation has the solution $\zeta = \eta^{-\frac{1}{2}}\left[AI_{\frac{1}{2}}(\eta) + BK_{\frac{1}{2}}(\eta)\right]$, where A=0 because the solution is limited at infinity, and, $K_{\frac{1}{2}}(\eta) = (\pi/2\eta)^{-1/2}e^{-\eta}$ is a modified Bessel function.

The boundary conditions for $r = R$:

$$\zeta|_{r=R} = p_0\left(\frac{R}{R_0}\right)^{-3\gamma} - p_\infty k^{-\gamma} = \zeta_R, \quad \eta|_{r=R} = R\alpha k^{1/6}.$$

$$\zeta|_{r=R} = \frac{B}{\eta_R}\sqrt{\frac{\pi}{2}}e^{-\eta_R}, \qquad B = \zeta_R \eta_R \sqrt{\frac{2}{\pi}}e^{\eta_R}.$$

Thus, we obtain $\zeta = \eta^{-\frac{1}{2}}\zeta_R\eta_R\sqrt{\frac{2}{\pi}}e^{\eta_R}\left(\frac{\pi}{2\eta}\right)^{\frac{1}{2}}e^{-\eta} = \frac{\zeta_R\eta_R}{\eta}e^{-(\eta-\eta_R)}.$

Let us now return to the variables *p-k*:

$$p(t) - p_\infty k^{-\gamma} = \left(p_0\left(\frac{R}{R_0}\right)^{-3\gamma} - p_\infty k^{-\gamma}\right)\frac{R}{r}e^{-\alpha k^{1/6}(r-R)}$$

Substituting the resultant expression into the equation for k, we obtain

$$\ddot{k} = \frac{3k^{\frac{1}{3}}}{\rho_0 r_0{}^2}\frac{R}{r}\left(p_\infty k^{-\gamma} - p_0\left(\frac{R}{R_0}\right)^{-3\gamma}\right)e^{-\alpha k^{1/6}(r-R)} + \frac{1}{6k}\left(\frac{\partial k}{\partial t}\right)^2.$$

Dynamics of the cavity in the cavitating liquid [5]. The dynamics of the spherical cavity can be described by using an analog of the Herring equation in a bubbly liquid

$$R\ddot{R} + \frac{3}{2}\dot{R}^2 = H + \frac{\dot{R}}{c_b}\frac{dH}{dt}.$$

In this equation, H is the enthalpy and c_b is the velocity of sound in the bubbly liquid, which is a function of the gas phase concentration in the cavitating liquid: $c_b{}^2 = \gamma p_\infty[\rho_0 k(1-k)]^{-1}$ The equation of state of the bubbly liquid with an incompressible component is $pk/(1-k) = a$.

The enthalpy $H = \int\frac{dp}{\rho_b} = \frac{P_g - p_\infty}{\rho_0} + \frac{a}{\rho_0}\ln\frac{P_g}{p_\infty}$, $\frac{dH}{dt} = -\frac{3\gamma}{\rho_0}p_0\left(\frac{R}{R_0}\right)^{-3\gamma}\frac{\dot{R}}{R}\left(1 + \frac{a}{p_0}\left(\frac{R}{R_0}\right)^{3\gamma}\right)$ – derivative determines the acoustic loss. Then the equation for a spherical cavity of radius R takes the form

IOP Conf. Series: Journal of Physics: Conf. Series **899** (2017) 032007 doi:10.1088/1742-6596/899/3/032007

$$R\ddot{R}+\frac{3}{2}\dot{R}^2=\frac{P_g-p_\infty}{\rho_0}-\frac{3\gamma}{\rho_0}p_0\left(\frac{R}{R_0}\right)^{-3\gamma}\frac{\dot{R}}{c_b}\left(1+\frac{a}{p_0}\left(\frac{R}{R_0}\right)^{3\gamma}\right)$$

Thus, the dynamics of the implosion of a quasi-empty cavity in a cavitating medium with a dynamically changing concentration of the bubbly liquid is described by the system of equations

$$\begin{cases} \ddot{k}=\dfrac{3k^{\frac{1}{3}}}{\rho_0 r_0{}^2}\dfrac{R}{r}\left(p_\infty k^{-\gamma}-p_0\left(\dfrac{R}{R_0}\right)^{-3\gamma}\right)e^{-\alpha k^{1/6}(r-R)}+\dfrac{1}{6k}\left(\dfrac{\partial k}{\partial t}\right)^2. \\ R\ddot{R}+\dfrac{3}{2}\dot{R}^2=\dfrac{p_0\left(\frac{R}{R_0}\right)^{-3\gamma}-p_\infty}{\rho_0}-\dfrac{3\gamma}{\rho_0}p_0\left(\dfrac{R}{R_0}\right)^{-3\gamma}\dot{R}\left(\dfrac{\rho_0 k(1-k)}{\gamma p_\infty}\right)^{1/2}\left(1+\dfrac{a}{p_0}\left(\dfrac{R}{R_0}\right)^{3\gamma}\right) \end{cases}$$

The numerical solution of this system under the initial conditions $k_0=10^{-6}$ and $p_0=10^{-8}$ yields the following result (**Figs. 4 and 5**). Let us confine our consideration to the data on k at the cavity boundary r = R.

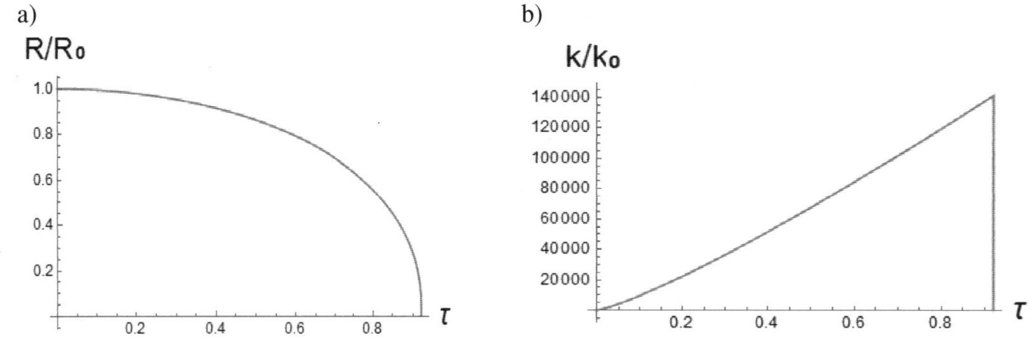

Figure 4. a) Dynamics of the spherical cavity; b) dynamics of the gas phase concentration.

It is seen that the concentration of the gas phase surrounding the imploding cavity increases under the action of the initial internal pressure in microbubbles. However, at the cavity implosion instant, the pressure in the cavity increases and the concentration rapidly decreases. **Fig.5,a** illustrates the dynamics of the acoustic losses in the cavity. It is seen that the cavity implosion is accompanied by pulsed release of the SW energy in a narrow region near the cavity implosion point. The integral acoustic loss of energy (**Fig.5,b**) is described by the formula

$$I=\int_{R_0}^{R}\frac{3\gamma}{\rho_0}p_0\left(\frac{R}{R_0}\right)^{-3\gamma}\dot{R}\left(\frac{\rho_0 k(1-k)}{\gamma p_\infty}\right)^{1/2}\left(1+\frac{a}{p_0}\left(\frac{R}{R_0}\right)^{3\gamma}\right)2R^2dR$$

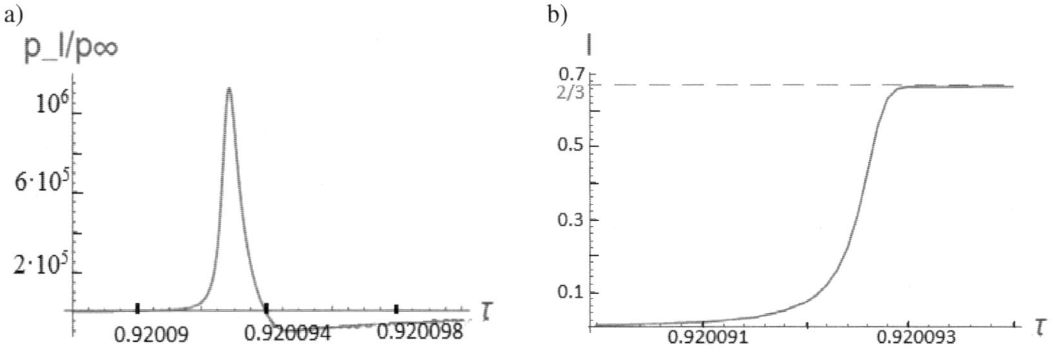

Figure 5. a) Dynamics of acoustic losses; the signal width is $\Delta t=1.2*10^{-8}$ s; b) integral acoustic loss of energy.

In dimensionless variables, the initial potential energy of the quasi-empty cavity is 2/3. If the initial internal pressure in the cavity is extremely low, the entire energy transforms to pulsed radiation (**Fig. 5, b**).

5.Conclusions

The analysis of the experimental data shows that shock wave loading of the liquid layer leads to the rupture development on the bottom of the cuvette. The rupture shape is close to a spherical segment, and there is a bubble boundary layer on the rupture interface. After the implosion, this boundary layer transforms to a bubble cluster. The rupture implosion generates the secondary shock wave, which is confirmed by the formation of the secondary cavitation region and which forms the cluster structure. After shock wave generation, the bubble cluster emerges to the liquid layer surface. The character of the cluster dynamics indicates the formation of a ring-shaped vortex moving upward to the liquid surface. It is shown that the cluster motion is almost independent of the liquid layer height and is mainly determined by the loading energy. A *p-κ* mathematical mode is formulated, and calculations are performed to investigate the implosion of a full quasi-empty spherical cavity in the cavitating liquid, generation of a shock wave by this cavity, and dynamics of the bubble density growth in the cavitating bubbly cluster by five orders of magnitude. It is of interest that the integral acoustic loss of energy is equal to the initial maximal potential energy of cavity.

Acknowledgement

This work was supported by the Russian Foundation for Basic Research (Grant No. 15-05-03336a) and partly by Siberian Branch of the Russian Academy of Sciences (Project No. III.22.3.1).

References

[1] Woods W.: The dynamics of explosive volcanic eruptions// Reviews of geophysics, V. 33, No. 4, p. 495–530, (1995)
[2] Gonnermann, H.M., Manga M. The fluid mechanics inside a volcano Annu. Rev. Fluid Mech.2007 ,39 P. 321–56
[3] Bolshakova E. S., Kedrinskiy V. K., Jornal of Physics: Conference Series – 2016.– V. 754. № 4, P. 042003
[4] BesovA. S., KedrinskiyV. K., PalchikovE. I. Letters in JTPh. –1984. –V. 10, № 4. P. 67–70
[5] Kedrinskiy V. K, Bolshakova E. S. The Journal of the Acoustical Society of America . – 2015. – V. 138. №3, Pt. 2, P. 1829.

IOP Conf. Series: Journal of Physics: Conf. Series **899** (2017) 032008 doi:10.1088/1742-6596/899/3/032008

Study of development of disturbance waves in annular gas-liquid flow.

Andrey V Cherdantsev[1,2], Mikhail V Cherdantsev[1], Sergey V Isaenkov[1] and Dmitriy M Markovich[1,2]

[1]Kutateladze Institute of Thermophysics, 1 Lavrentiev ave, 630090 Novosibirsk, Russia

[2]Novosibirsk State University, 2 Pirogov str, 630090 Novosibirsk, Russia

cherdantsev@itp.nsc.ru

Abstract. Downstream development of disturbance waves properties in annular regime of gas-liquid flow was conducted in adiabatic air-water downwards flow in a vertical pipe with inner diameter of 11.7 mm. The measurements were conducted using brightness-based laser-induced fluorescence technique. Instantaneous distributions of local thickness of liquid film along one longitudinal section of the duct over the first 45 cm from the inlet were obtained with sampling frequency of 10 kHz. Based on these spatiotemporal plots, dependence of local average velocity of disturbance waves on downstream distance was obtained for a wide range of gas and liquid flow rates. Three main stages of flow development were identified: a stage prior to formation of disturbance waves, a stage of constant acceleration of disturbance waves and a stage of deceleration nearly compensating the initial acceleration. Transitions to both second and third stages occur closer to the inlet at higher gas velocities and lower liquid flow rates. The initial acceleration is defined by the effect of the gas shear; it grows in parabolic manner with superficial gas velocity and shows weak dependence on liquid flow rate. The deceleration is supposed to occur due to entrainment of liquid from disturbance waves.

1. Introduction

In annular regime of gas-liquid flow high-velocity gas stream flows through the central part of the duct; the walls of the duct are covered by liquid film. The liquid can be entrained from film surface into the gas core in form of droplets. The liquid film is covered by large-scale disturbance waves separated by thin residual layer and small-scale ripple waves covering both the residual layer and the disturbance waves. The disturbance waves play important role in annular flow: they carry the major fraction of liquid, affect shear stress and heat and mass transfer in the flow and generate the ripples on their rear slopes (Alekseenko et al., 2008). The ripples, in turn, contribute to the shear stress by increasing the interface roughness (Schubring & Shedd, 2011) and, being torn into droplets by the gas stream at the tops of the disturbance waves, contribute to liquid entrainment. In order to create physically based models predicting pressure drop and liquid entrainment in annular flow it is necessary to accurately predict the properties of the disturbance waves. Such modelling requires basic assumptions on waves properties and behaviour which may be obtained via comprehensive experiments.

Large amount of experimental data on average velocity, height and frequency of disturbance waves at different flow rates and physical properties of the phases was collected in previous years. The majority of these experiments were conducted at large distances from the inlet, where the flow is believed to be stabilised. In particular, it was shown (Hall Taylor et al., 1963) that far from the inlet the disturbance waves travel with constant speed over long distances. But the disturbance waves are formed close to the inlet (Zhao et al., 2013) and their properties change significantly with downstream distance, x, at the first meters of the duct length. In particular, their frequency decreases with x due to numerous acts of irreversible coalescence (Hall Taylor & Nedderman, 1968) and their velocity undergoes complex evolution with strong initial acceleration and weaker subsequent deceleration (Wolf et al., 2001). The number of experimental works devoted to studying disturbance waves evolution is small and, as a rule, the measurements were conducted by pointwise methods in a small number of downstream positions of the probes. The disturbance waves should be modelled starting from their appearance and subsequent development, since these processes define their properties downstream.

In the recent work by Alekseenko et al. (2015) the process of formation of disturbance waves was studied using field measurements of film thickness at the first 100 mm of the pipe length. It was found that the disturbance waves are formed due to coalescence of high-frequency initial waves. The disturbance waves were found to accelerate after formation, but studying their further development was restricted by the short length of the region of interrogation. In the present work, the investigated longitudinal section of the pipe was expanded to 450 mm from the inlet to study the further stages of waves development.

2. Experimental setup and measurement technique

The experiments were conducted in adiabatic air-water flow at near-atmospheric conditions. Downward annular flow was organised in a vertical acrylic resin pipe with inner diameter of 11.7 mm and length of 650 mm. The scheme of the experimental setup is shown in Fig. 1(a).

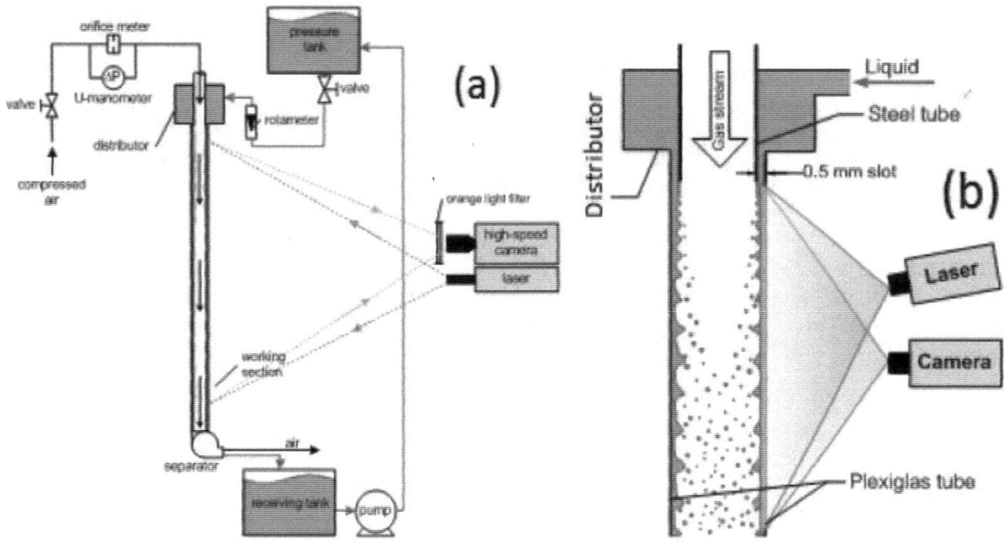

Figure 1. (a) Scheme of the experimental setup. (b) Scheme of the inlet.

Liquid circulated in a closed hydrodynamic loop. Prior to the experiments it was pumped from the receiving tank into the pressure tank located 3 metres above the pipe entrance. During the experiments the pump was switched off and liquid went into the distributor under hydrostatic pressure. The distributor ensures uniform wetting of the pipe walls by the working liquid, which is introduced as a film through a ring-shaped 0.5 mm thickness slot between the inner surface of the pipe and the outer surface of gas-feeding tube (Fig. 1b). Liquid flow rate was controlled with a float rotameter. The gas was introduced from the centralised line of compressed air through the thin-walled steel tube mounted coaxially with the main pipe. Gas flow rate was controlled by an orifice meter connected to U-manometer. The gas-liquid mixture leaving the pipe entered swirl separator. After separation the liquid phase entered the receiving tank and the air went to the atmosphere.

Liquid Reynolds number, Re_L, was defined as volumetric liquid flow rate divided by wetted perimeter of the pipe and by kinematic viscosity of the liquid. Re_L values of 140, 220, 300 and 400 were used. Superficial gas velocity, V_g, defined as volumetric gas flow rate per cross-section of the pipe varied in the range from 22 m/s to a maximum of 100 m/s at Re_L=140. Due to restrictions of the rig pressure, maximum V_g was 90 m/s for Re_L=220, 80 m/s for Re_L=300 and 57 m/s for Re_L=400. According to Webb & Hewitt (1975) and Alekseenko et al. (2015) in this range of Re_L the selected range of Vg ensures presence of disturbance waves.

Film thickness was measured using brightness-based laser-induced fluorescence technique. Fluorescent dye Rhodamine 6G was dissolved in working liquid in small (15 mg/l) concentration. For excitation of fluorescence a continuous 2W laser with wavelength 532 nm was used. The laser beam is converted into a narrow vertical light sheet illuminating a longitudinal section of the pipe with length of 450 mm starting from the inlet. Rhodamine 6G absorbs the laser radiation and re-emits it in red spectral domain. The brightness of the re-emitted light depends on the thickness of liquid layer. This brightness is measured by high-speed camera equipped with orange optical filter to avoid registration of reflected and scattered laser light. The camera sampling frequency was either 10 or 20 kHz depending on gas velocity. Exposure time was 40-60 μs. A line of 1000 camera pixels was "seeing" the region of interrogation, which yielded spatial resolution of 0.45 mm per pixel. For each combination of gas and liquid flow rates a record with duration of 2 seconds was obtained. Brightness of re-emitted light, J, is related to film thickness, h, as:

$$J(x) = C(x)(1 - e^{-\alpha h(x)})(1 + Ke^{-\alpha h(x)})$$

Here $C(x)$ is a compensation matrix, α is the absorption coefficient, K is coefficient of reflection from water-air interface, which is equal 0.02. α is measured separately by comparing brightness of fluorescent light emitted by liquid in two slots of known thicknesses. $C(x)$ can be constructed based on in situ measurements of time-averaged brightness of liquid layer with known average thickness. A simple example of such reference signal is thin (Re_L=40) film falling under action of gravity; thickness of such film is described well by the Nusselt solution.

3. Results and discussion

An example of spatiotemporal film thickness matrix, $h(t,x)$, is shown in Fig. 2 in graphical representation. Local brightness of each pixel of this image is linearly proportional to local film thickness measured at given $\{x,t\}$-point. White corresponds to film thickness equal to or more than 1 mm. Each string of such a matrix represents an instantaneous distribution of film thickness along the downstream coordinate, x. Each column represents a temporal record of film thickness obtained at fixed downstream coordinate. Waves are visible in this image as bright inclined lines, with local slope of a line to the t-axis corresponding to the instantaneous velocity of the wave.

At the very inlet (the leftmost part of the image) regular small-amplitude waves of very high frequency can be observed. Due to coalescence of these initial waves the disturbance waves are formed (Alekseenko et al., 2015). The disturbance waves are characterised by large longitudinal size, amplitude, speed and lifetime; they are separated by thin base film layer; their crests are covered with ripples moving faster than the waves themselves. Spatiotemporal trajectories of the disturbance waves are curved upwards in the beginning, which means that the speed of the waves grows downstream. In the rightmost part of the image the trajectories are almost straight, which indicates nearly constant speed. The number of disturbance waves gradually decreases with x due to waves coalescence. Acts of coalescence can also be observed in Fig. 2 as intersections of two waves' trajectories after which only one wave travels further.

Using spatiotemporally resolved data, average velocity of waves can be easily measured at every x value by calculating the cross-correlation function (CCF) for neighbouring temporal records of film thickness, $h(t,x\pm\delta x)$, where δx is the length "seen" by one camera pixel; this length is equal to 0.45 mm. The velocity is then estimated as $2\delta x/\delta t$, where δt is the temporal delay corresponding to maximum of CCF. To increase accuracy of the measurements, the vicinity of the CCF maximum was interpolated by a parabola, and location of its maximum was used as a more precise estimation of δt. The velocity measured this way corresponds to the velocity of initial waves at the very beginning. Afterwards it corresponds to the average velocity of the disturbance waves: though small-scale ripples coexist with disturbance waves on film surface, they are of much smaller amplitude and their contribution into CCF is negligible.

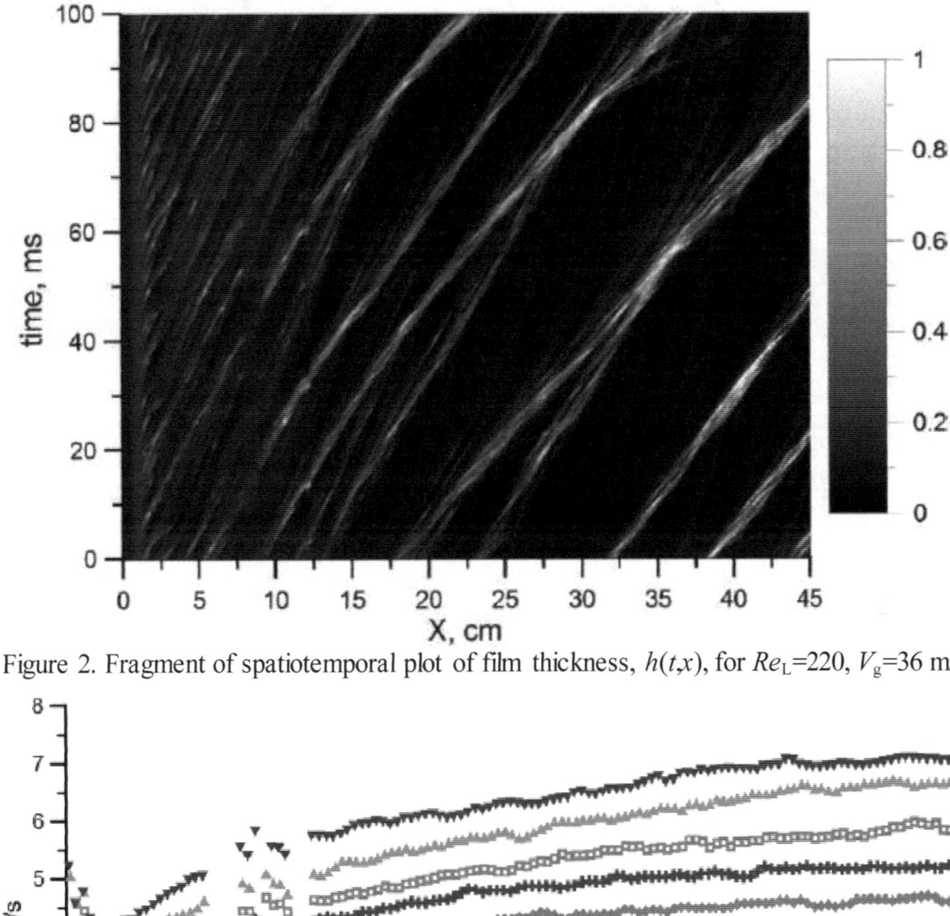

Figure 2. Fragment of spatiotemporal plot of film thickness, $h(t,x)$, for Re_L=220, V_g=36 m/s.

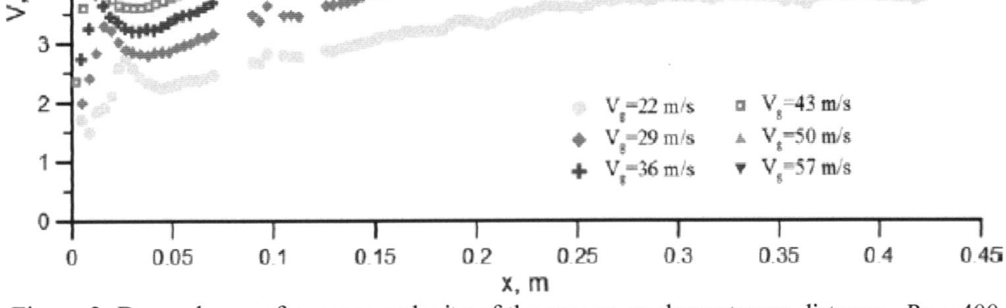

Figure 3. Dependence of average velocity of the waves on downstream distance. Re_L=400.

The obtained dependencies $V(x)$ for one Re_L are shown in Fig. 3. At the initial stage a short high-amplitude peak in the velocity plots can be observed. This peak describes the velocity of the initial waves prior to formation of disturbance waves. According to Alekseenko et al. (2016), the initial waves are two-dimensional at first, but after some distance they undergo fragmentation into localised 3D waves. The latter coalesce to form quasi two-dimensional disturbance waves. Thus, at the 2D stage the initial waves undergo acceleration (the growing part of the peak); the fragmentation into 3D waves is accompanied by decrease in the speed of the waves.

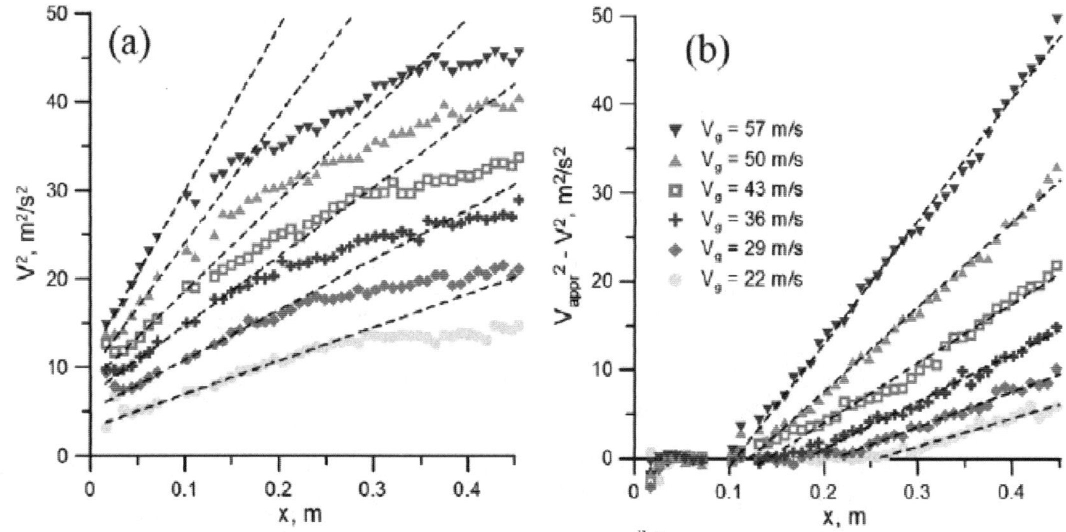

Figure 4. (a) Search for linear stage in the dependence $V^2(x)$. (b) Difference between the linear extrapolation shown by the dashed lines in Fig. (a) and the measured squared velocity. Re_L=400.

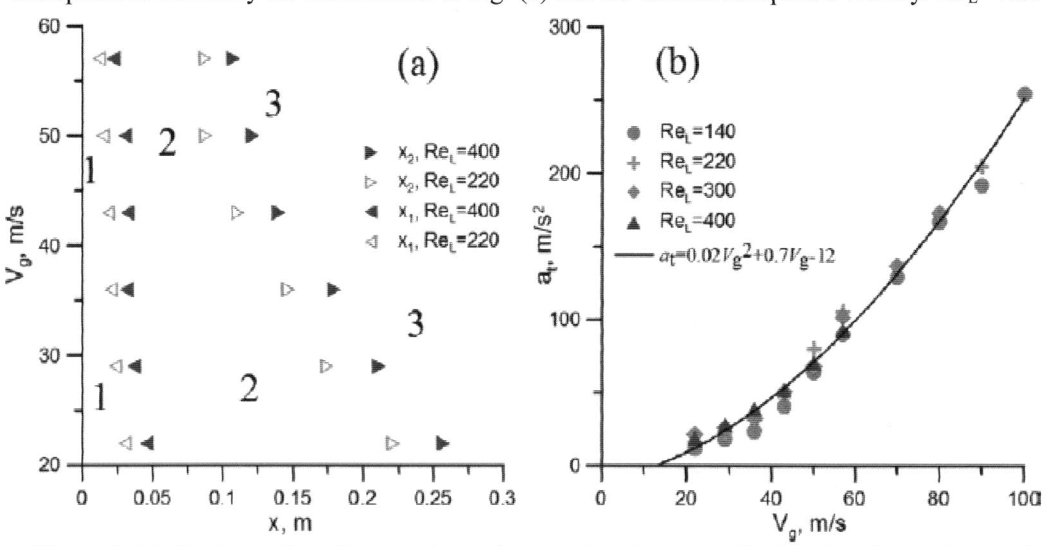

Figure 5. (a) Borders of the three regions of waves development. (b) Acceleration in time at the stage of constant initial acceleration.

After the stage of constant acceleration, waves' speed continues to grow, but at significantly lower rate. Figure 4(b) shows the difference between the constant-acceleration prediction and the measured values of $V^2(x)$. This difference also linearly grows with x; at higher gas velocities the growth starts closer to the inlet and the rate of growth is larger. Wolf et al. (2001) who observed deceleration of the disturbance waves relatively far from the inlet, related it to liquid entrainment, which leads to thinning of liquid film. This idea seems reasonable since the dependencies shown in Fig. 4(b) are very similar to the dependencies of entrained fraction on x (see, e.g., equation 23 in Kataoka et al., 2000). Nonetheless, we suppose that the way entrainment affects the waves speed is more complex than just integral thinning of liquid film. Since the entrained droplets are "extracted" directly from the disturbance waves, the entrainment may lead to decrease in the area of interaction between the gas stream and a disturbance wave, and thus reducing the effective drag force.

To summarise, three stages of waves development can be identified based on present experimental data: (1) the region of initial waves prior to formation of the disturbance waves; (2) the region of constant acceleration of disturbance waves; (3) the region of reduced acceleration due to liquid entrainment. The borders of the three regions are shown in Fig. 5(a). Both regions (1) and (2) shorten at higher gas velocities and lower liquid flow rates. The measured values of constant acceleration at in the region (2) for the whole investigated range of gas and liquid flow rates are shown in Fig. 5(b). at grows with V_g in a parabolic manner, and it does not show any strong dependence on Re_L.

4. Conclusions

Development of waves in downward annular gas-liquid flow was studied using brightness-based laser-induced fluorescence technique. Spatiotemporal evolution of film thickness and local velocity of waves was studied over first 45 cm of a vertical pipe with inner diameter 11.7 mm. Three stages of waves evolution were identified. At the first stage, initial high-frequency waves appear on film surface. These waves undergo strong acceleration followed by deceleration. Based on previous observations in three-dimensional space (Alekseenko et al., 2016) we suppose that the velocity increases while the initial waves conserve two-dimensionality and decreases after the 2D-waves are fragmented into small 3D-waves.At the second stage the large-scale disturbance waves appear and move with constant acceleration, which grows in parabolic manner with gas velocity without strong dependency on liquid flow rate. At the third stage the acceleration suddenly drops to very small values. The reason of such a sharp decrease in acceleration is supposedly related to start of liquid entrainment from the tops of disturbance waves leading to decrease in waves amplitudes and hence the effective drag force exerted on waves by the gas stream. The length of each stage decreases with gas velocity and increases with liquid flow rate. It is very likely that downstream the region of interrogation used in the present work one more stage of waves development can be observed, related to start of deposition of liquid droplets from the gas core onto film surface and consequent stabilisation of waves velocity. This assumption requires experimental validation through spatiotemporal measurements of film thickness conducted farther from the inlet.

References

[1] Alekseenko S V, Antipin V A, Cherdantsev AV, Kharlamov S M and Markovich D M 2008 *Microgravity Sci. Technol.* **20** 271
[2] Alekseenko S V, Cherdantsev A V, Cherdantsev M V, Isaenkov S V and Markovich D M 2015 *Int. J. Mult. Flow* **77** 65
[3] Alekseenko S, Cherdantsev A, Cherdantsev M, Isaenkov S and Markovich D 2016 *MATEC Web Conf.* **84** 00001
[4] Kataoka I, Ishii M and Nakayama A 2000 *Int. J. Heat Mass Transfer* **43** 1573
[5] Hall Taylor N S, Hewitt G F and Lacey P M C 1963 *Chem. Eng. Sci.* **18** 537
[6] Hall Taylor N S and Nedderman R M 1968 *Chem. Eng. Sci.* **23** 551
[7] Schubring D and Shedd T A 2011 *Int. J. Heat Fluid Flow* **32** 730
[8] Webb D R and Hewitt G F 1975 *Int. J. Mult. Flow* **2(1)** 35
[9] Wolf A, Jayanti S and Hewitt G F 2001 *Chem. Eng. Sci.* **56** 3221
[10] Zhao Y, Markides C N, Matar O K and Hewitt G F 2013 *Int. J. Mult. Flow* **55** 111

Acknowledgments

The work was supported by Russian Science Foundation (project RSF 16-19-10449).

Investigation of the evolution of the bubble size distribution in the ascending and descending flows

A S Chernyshev and A A Schmidt

Computational Physics Laboratory, Ioffe Institute, Politekhnicheskaya 26, Saint Petersburg, 194021, Russia

E-mail: alexander.tchernyshev@mail.ioffe.ru

Abstract. Numerical simulations of the liquid-bubble ascending (air-lift) and descending flows were carried out with account of polydispersity. Simulations were conducted in the framework of the previously developed mathematical model which is based on the Euler-Euler approach and includes multiphase momentum transfer and k-ω-SST turbulence model with bubble induced turbulence generation and bubble path dispersion. Polydispersity is taken into account by the multi-size group model. Bubbly flows were studied in vertical pipes and the evolution of the bubble size distribution were investigated. The effect of the polydispersity was shown and results were compared with the cases with monodisperse bubbles, also the relative concentration of bubbles of different classes were inspected at the locations close to the inlet and outlet boundaries.

1. Introduction

Polydisperse phenomenon is widely encountered in the multiphase and, particularly, in bubbly flows, and frequently occurs in nature and industrial applications (like coolant systems, pipelines, chemical reactors and so on). Bubble size distribution is one of the main characteristics of polydisperse flows and that distribution can be altered due to the relative motion of the bubbles and interphase interactions between bubbles and carrier liquid. Such an alteration of the bubbly flow pattern can modify the entire flow and, thus, can play significant role in the prediction of the flow structure inside the flow domain. Because of that, investigations of the polydisperse bubbly flow are of high importance.

Both experimental and numerical techniques are used recently to study such flows (see, for example, [1, 2]) while numerical simulations became a robust tool for flow pattern predictions in the cases when exact flow structure can not be precisely measured in experiments.

Numerical studies of polydisperse flows are widely presented in the literature [3-5], different approaches are used to describe multiphase medium and polydispersity. Frequently used the Euler-Lagrange description of multiphase flows [5] can be easily applied to the simulation of polydisperse particles, but at high concentrations it becomes resource demanding. The Euler-Euler approach [3, 6] is suitable for almost any concentrations of disperse particles with the same continuity equations as for single phase flows for all phases, but special technique required to capture multiple size species. A multi-size-group (MUSIG) [6] model will be used in the simulations as one of the common polydisperse models utilized in the investigations for a long time and proved their reliability, and because that model can be simply implemented in the monodisperse Euler-Euler model. Moreover, that model provides straightforward description of the non-homogeneous flow regime characterized by significant difference of the velocity of different groups of bubbles which is crucial in the present work.

Investigations by the means of numerical simulations were performed to analyze the impact of polydispersity on the flow structure inside the vertical pipes and to study the evolution of the bubble distribution by their sizes due to relative bubble motion. Simulations were carried out in the framework of the proposed model which was previously tested [7] and good correlation with experimental data was obtained. Under the current setup only the influence of the interphase forces were studied and bubble coalescence and break-up were omitted.

2. Mathematical model and governing equations

Previously developed and verified mathematical model for the polydisperse bubbly flows is used which is capable of description of gravity-driven flows with account of turbulence and interphase momentum transfer [7]. The model is based upon the Euler-Euler approach of the description of the multiphase flows, under current approach it is assumed that each phase is continuous one and fills up entire domain with specified volume fraction α in each point. The number of bubble classes M in the framework of the MUSIG model are introduced, thus volume fraction of the liquid can be expressed as:

$$\alpha_l = 1 - \sum_{i=1}^{M} \alpha_{ib}, \tag{1}$$

here subscript i refers to the i-th class of the bubbles, i=1,...,M. The Navier-Stokes equations for all phases under the current assumptions will have the following form:

$$\begin{gathered}
\frac{\partial \alpha_{ib} \rho_{ib}^{0}}{\partial t} + \operatorname{div}\left(\alpha_{ib} \rho_{ib}^{0} \vec{V}_{ib}\right) = 0, \\
\frac{\partial \alpha_{ib} \rho_{ib}^{0} \vec{V}_{ib}}{\partial t} + \operatorname{div}\left(\alpha_{ib} \rho_{ib}^{0} \vec{V}_{ib} \vec{V}_{ib} + p \bar{E}\right) = S_{ib}, \\
\frac{\partial \alpha_{l} \rho_{l}^{0}}{\partial t} + \operatorname{div}\left(\alpha_{l} \rho_{l}^{0} \vec{V}_{l}\right) = 0, \\
\frac{\partial \alpha_{l} \rho_{l}^{0} \vec{V}_{l}}{\partial t} + \operatorname{div}\left(\alpha_{l} \rho_{l}^{0} \vec{V}_{l} \vec{V}_{l} + p \bar{E} - \bar{\tau}_{l}\right) = S_{l}.
\end{gathered} \tag{2}$$

Additionally, the density of the bubbles can vary due to pressure changes inside the liquid. Such variations can be described by equation of state derived from Rayleigh-Plesset equation using bubble pressure:

$$\rho_{ib}^{0} = \frac{p + 2\Sigma / R_{ib}}{\Re_b T_0}, \tag{3}$$

here Σ is the surface tension for the liquid and unless otherwise specified is equal to the 0.072 N/m which is close to the value for water at 297 K. Thus, the temperature working conditions T_0 is set to the 297 K also. $\Re_b$ is the gas constant for the specified gas in the bubbles.

This system should be accomplished by conservation equations of number densities per each class since the bubble radius is not constant and can change with the variation of the bubble gas density:

$$\frac{\partial N_{ib}}{\partial t} + \operatorname{div}\left(N_{ib} \vec{V}_{ib}\right) = 0. \tag{4}$$

The initial distribution of bubbles by their sizes in the bulk liquid, or distribution of bubbles at inlet boundaries will be taken either uniform or described by log-normal distribution:

$$\frac{1}{R\sigma\sqrt{2\pi}}\cdot\exp\left(-\left(\ln(R)-\mu\right)^2\cdot 0.5\sigma^{-2}\right). \quad (5)$$

Mathematical expectations in that distributions are taken equal to most probable bubble size, which can be estimated from the problem set-up parameters or taken from the experimental data. The number of classes M is usually in the range from 5 to 10, and selected on the criteria of converged solution and invariance of the solution from M.

The k-ω-SST model of Menter [8] is used to describe turbulence. The model is expanded to take into account turbulence generation near the bubble surfaces due to the non-zero slip velocity and the work of the drag force. Such modifications are described in the previous paper [9] and based on the work of Troshko & Hassan [10] for the k-ε models and adapted to the k-ω model. Source terms for k and ω equations with modifications according to the bubble turbulence generation and dissipation terms:

$$S_{ibk}=\frac{3}{8}\frac{C_{iD}}{R_{ib}}\alpha_{ib}\rho_l\left|\vec{V}_{irel}\right|^3,\quad S_{ib\omega}=0.8\frac{k}{\mu_{turb}}S_{ibk}. \quad (6)$$

Source terms in the right-hand side of the momentum conservation equations (2) comprise the Buoyancy force and interphase forces:

$$S_{ib}=\alpha_{ib}\rho_b^0\vec{g}+\vec{F}_{iD}+\vec{F}_{iL}+\vec{F}_{iWL}+\vec{F}_{iTD},\quad S_l=\alpha_l\rho_l^0\vec{g}-\sum_{i=1}^{M}\left(\vec{F}_{iD}+\vec{F}_{iL}+\vec{F}_{iWL}+\vec{F}_{iTD}\right). \quad (7)$$

Previous model was extended by inclusion of the wall lubrication force F_{iWL} , which repels bubbles from walls. Expressions for the Stokesian drag force F_{iD}, lift force F_{iL}, wall lubrication force F_{iWL} and turbulent dispersion force F_{iTD} are presented below:

$$\vec{F}_{iD}=\frac{3\rho_l}{8R_{ib}}C_{iD}\vec{V}_{irel}\left|\vec{V}_{irel}\right|,\vec{V}_{irel}=\vec{V}_l-\vec{V}_{ib}, \quad (8)$$

$$\vec{F}_{iL}=C_{iL}\alpha_{ib}\rho_l\vec{V}_{irel}\times\operatorname{rot}\vec{V}_l, \quad (9)$$

$$\vec{F}_{iWL}=-C_{iWL}\alpha_{ib}\rho_l\vec{V}_{irel}^2\vec{n}_W, \quad (10)$$

$$\vec{F}_{iTD}=-1.2C_{iD}\frac{3}{8}\frac{\left|\vec{V}_{irel}\right|}{R_{ib}}\mu^{eff}\nabla\alpha_{ib}, \quad (11)$$

expressions for C_{iD}, C_{iL}, C_{iWL} were taken from [11], [12] and [13], respectively.

3. Numerical method

The numerical method is based on the unstructured grids, finite-volume approach and pseudo-transient scheme for iterative process. Pressure and velocity fields are found in cell centers by separate fashion in the framework of the PC-SIMPLE algorithm which is based on the original SIMPLE proposed by Patankar [14]. The Rhie-Chow correction applied to prevent checker-boarding which arises from using collocated storage of variables. TVD schemes with low dissipative limiters are used to provide second order accuracy over spatial coordinates.

4. Results

Two different cases in the current work were considered. First one is the air-lift flow in a cylindrical bubble column. Gravity-driven flow in an axisymmetric column with the sparger installed in its bottom was considered and setup was taken from the paper [11]. The cylindrical column has radius R=0.07 m and height H=0.65 m. Gas is supplied to the column by the circular sparger with the radius of 0.05 m and volumetric flow rate of 15 l/h, which is equivalent to the superficial velocity of the gas V_G=0.53 mm/s. At the top of the column the outlet is located which is represented as free surface with free outlet for the gas and slip conditions for the liquid phase. The most probable bubble radius was taken equal to the 0.245 mm, the bubbles were injected across the inlet area with the same distribution at each location at the inlet. There were taken 7 bubble classes, which have corresponding bubble radii (from smaller to bigger): 0.126 mm, 0.168 mm, 0.215 mm, 0.263 mm, 0.313 mm, 0.362 mm, 0.415 mm. Experiments were carried out at normal conditions with temperature T=297 K and pressure P=101350 Pa.

Second case represents descending bubbly flow inside a pipe, the setup of the current problem was taken close to that described in the paper [15]. The straight vertical pipe had the diameter D=25.4 mm and length L=20D. Water with immersed air bubbles is supplied to the top end. At the inlet, the superficial liquid velocity V_L=1.1 m/s, superficial gas velocity V_G=6.85 mm/s and total gas void fraction for all classes α=0.00685. Bubble size distributions and operating conditions are the same as in the setup for the upward flow in the bubble column.

Simulations were carried out for monodisperse and polydisperse flows, in the case of monodisperse flow the most probable bubbles were taken with radius R=0.245 mm, for polydisperse cases the log-normal distribution was applied. Liquid velocity along the radial coordinate was compared for both cases and results presented in figures 1 and 2. The cross-section for the first case located at 0.325 m above the bottom wall, for the second case at the outlet, at the distance of 20D from the inlet location. It can be seen that polydispersity alters the flow structure and velocity profile, especially at the axis of the column and the pipe due to the fact that bubbles were dragged towards the center of the domain by wall force and drag force due to liquid convection (in the first case) and both wall and lift forces (in the second case).

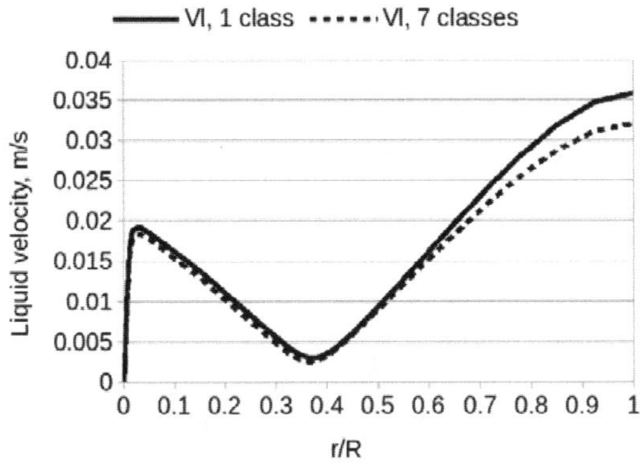

Figure 1. Comparison of the radial distribution of the liquid velocity in the monodisperse (1 class) and polydisperse cases for the air-lift flow problem.

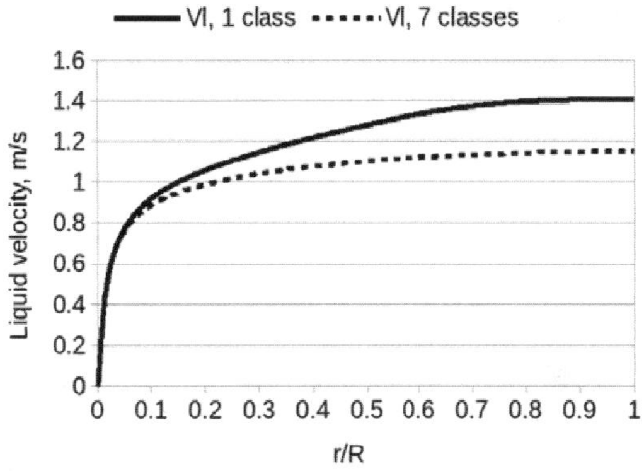

Figure 2. Comparison of the radial distribution of the liquid velocity in the monodisperse (1 class) and polydisperse cases for the descending flow problem.

In the case of polydisperse flows the evolution of the bubble distributions was investigated for 2 types of initial bubble distributions: constant and log-normal. The later one was selected as a distribution that is in the correlation with the real distributions of the particles or bubbles observed in the experiment (see, for example, [11]). The data obtained in the simulations are represented in terms of relative volume fraction per each bubble class, which is the ratio of volume fraction of i-th class to the total sum of volume fractions of all classes. The simulated data are sampled in 3 points: Point A and Point B located at the axis of the symmetry, Point A at 10 mm above the sparger for the first case, or 10 mm below the inlet boundary for the second, Point B at 10 mm from the outlet boundary. Point C located at 10 mm from the outlet boundary and at half of the pipe or the column radius from the axis of symmetry.

The data for the air-lift flow case are presented in figures 3 and 4. The toroidal vortex liquid structure is produced inside the column due to the central sparger mounting, such vortex effectively drags light bubbles towards the center of the column which can be seen from the comparison of the initial distribution and distribution in Point A. The bubbles redistributed across the column by turbulent dispersion as they moved towards the outlet and the resulted distribution at points B and C became closer to the initial one, with increased concentration of the smaller bubbles towards the column center. These effects are more pronounced in the case of constant distribution since in that situation there are more small bubbles.

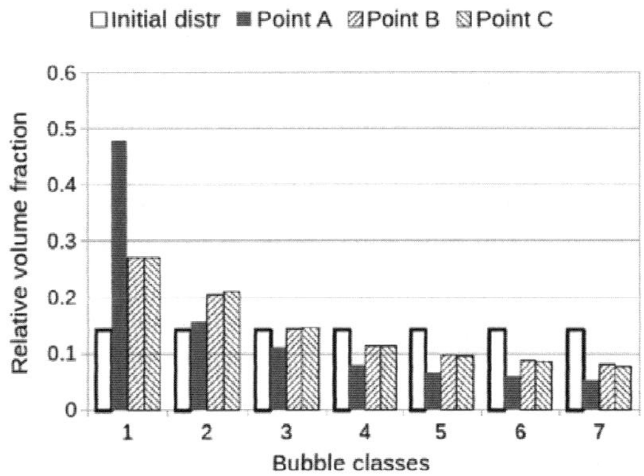

Figure 3. Distribution of the bubbles by their sizes at 3 points for air-lift flow and constant distribution at the inlet.

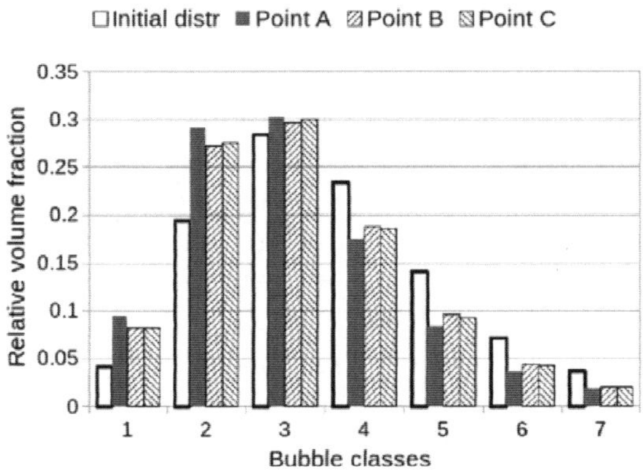

Figure 4. Distribution of the bubbles by their sizes at 3 points for air-lift flow and log-normal distribution at the inlet.

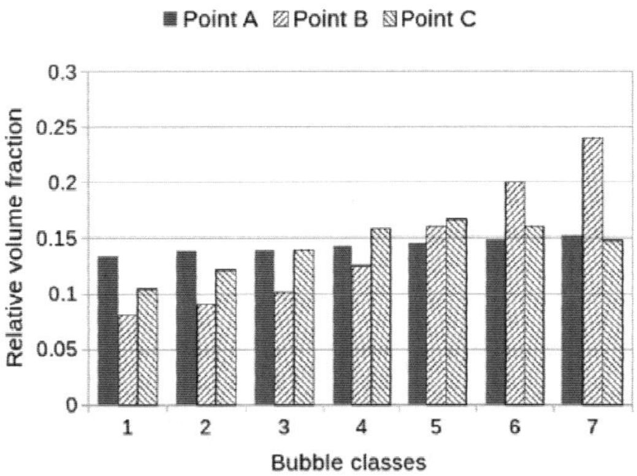

Figure 5. Distribution of the bubbles by their sizes at 3 points for descending flow and constant distribution at the inlet.

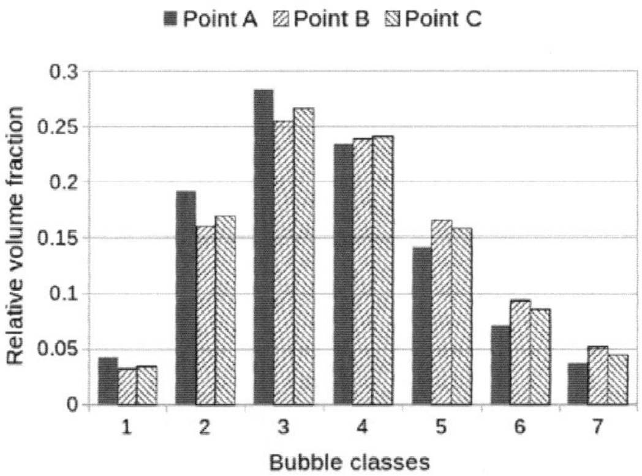

Figure 6. Distribution of the bubbles by their sizes at 3 points for descending flow and log-normal distribution at the inlet.

As for the case of descending flow (figures 5 and 6) the initial distribution is almost unaffected at the initial stages of the flow (close to the inlet) and indistinguishable with that measured at the Point A, but as bubbles move along the pipe they became more affected by transverse forces. That resulted in the significant shift of the distribution at the axis of the pipe towards the bigger bubbles. Smaller bubbles can be more uniformly distributed across the pipe due to turbulent dispersion force. The effect of re-distribution is less pronounced for the middle-sized bubbles, thus in the case of the log-normal distribution with prevailing bubbles of the size close to the 0.25 mm the distribution is less affected.

IOP Conf. Series: Journal of Physics: Conf. Series **899** (2017) 032009 doi:10.1088/1742-6596/899/3/032009

In the both cases the log-normal distribution seems more stable and preserves its form during the flow. Deviations from the initial distribution at the inlet can be addressed to the absence of the coalescence and break-up models, thus future investigations will be addressed to the inclusion that effects into simulations.

5. Conclusion

Simulations of the polydisperse bubbly flows inside the cylindrical bubble column and vertical pipe were carried out and the evolution of the bubble size distributions were investigated. The bubbles were supplied to the domain of interest with 2 distributions: same volume fractions for all classes (constant distribution) and log-normal distribution as one which describes the distribution of the particles . The breakup/coalescence models were omitted in the current formulations thus only effect of lateral bubble motion was investigated.

The impact of the polydispersity on the flow structure is noticeable and mainly observable at the axis of the column and the pipe due to the fact that bubbles are dragged from the walls by interphase forces and that impact differs for bubbles of various sizes.

It was shown that interphase forces can significantly alter the initial distribution of the bubbles by their sizes. This effect can be addressed to the influence of the specific flow structure, like in the case of the bubble column, or to the cumulative and prolonged action of the transverse forces, like in the case of the descending flow in the pipe. The log-normal distribution became less affected by the interphase interactions, but for detailed investigation of polydisperse flows the inclusion of coalescence and break-up models required.

References

[1] Kashinsky O N, Lobanov P, Pakhomov M A, Randin V V and Terekhov V I 2006 *Int. J. of Heat and Mass Transfer* **49**(19) 3717-27
[2] Chernyshev A and Schmidt A 2013 *AIP Conf. Proc.* **1558** 1071–4
[3] Bannari R, Kerdouss F, Selma B, Bannari A and Proulx P 2008 *Comp. and Chem. Eng.* **32** 3224–37
[4] Morel C, Ruyer P, Seiler N and Laviéville J M 2010 *Int. J. of Mult. Flow* **36** 25–39
[5] Lain S, Broder D and Sommerfeld M 1999 *Chem. Eng. Sci.* **54** 4913-20
[6] Krepper E, Lucas D, Frank T, Prasser H-M and Zwart P J 2008 *Nucl. Eng. and Design* **238** 1690–702
[7] Chernyshev A S and Schmidt A A 2016 *J. of Phys.: Conf. Ser.* **754**(3) 032005
[8] Menter F R, Kuntz M and Langtry R 2003 *Turb. Heat and Mass Tran.* **4** 625-32
[9] Chernyshev A S and Schmidt A A 2013 *Tech. Phys. Lett.* **39**(6) 548–51
[10] Troshko A A and Hassan Y A 2001 *Int. J. Multiphase Flow* **27** 1965-2000
[11] Lain S, Sommerfeld M, Broder D and Goz M 2002 *Int. J. Multiph. Flow* **28** 1381–407
[12] Tomiyama A, Tamai H, Zun I and Hosokawa S 2002 *Chem. Eng. Sci.* **57** 1849–58
[13] Frank Th, Zwart P J, Krepper E, Prasser H-M and Lucas D 2008 *Nucl. Eng. and Design* **238** 647-59
[14] Patankar S V 1980 *Numerical Heat Transfer and Fluid Flow* (New York: McGraw-Hill)
[15] Ishii M, Paranjape S, Kim S and Sun X 2004 *Int. J. Multiph. Flow* **30** 779–801

IOP Conf. Series: Journal of Physics: Conf. Series **899** (2017) 032010 doi:10.1088/1742-6596/899/3/032010

Experimental research of 134a hydrate formation in cyclic boiling-condensation process with variation of working volume

Elistratov D.S., Meleshkin A.V., Chernov A.A., Pilnik A.A.

Kutateladze Institute of Thermophysics SB RAS, Novosibirsk, Russia

Novosibirsk State University, Novosibirsk 630090, Russia

E-mail: mav@itp.nsc.ru

Abstract. On the basis of researches on examination of gas hydrates previously made in Kutateladze Institute of Thermophysics SB RAS the new method was developed. That method is based on cyclic boiling-condensation process of gas-hydrating agent in enclosed volume. This method compares favorably with previously developed ones by its simplicity. There is no need in constant gas supply and mixing device.

1. Introduction

All the researches in the field of gas-hydrates can be roughly divided into fundamental and applicative. They examine general conditions as well as the most important lines. Those lines include the research of the structure of gas-hydrates, their physical-chemical, thermophysical, mechanical and other properties as well as general conditions and patterns of their formation, schemes and types of growth of the hydrate mass etc.[1–3]. A big attention is paid to methods of research of artificially made as well as natural gas-hydrates [4–7]. A big part of works is dedicated to experimental and mathematical simulation of processes of formation and dissolution of gas-hydrates [8–12].

Before the experimental researches of making of gas-hydrates of different gases by different ways were made. All those researches were mainly based on shock action on gas-liquid mediums. Their outstanding feature was a high speed of hydrate formation that can be explained by the essential intensification of heat and mass exchanging processes.

The logical extension of those researches was the creation of the boil-off method: the freon gas under the pressure is fed into the hermetic vessel halfway filled by the water cooled to 2-4°C. Freon creates the pressure, its temperature goes down to the environment temperature and it becomes liquideous. Since the liquefied freon is significantly heavier than water, it goes down and creates a layer on the bottom of the vessel. When freon is on the bottom of the vessel depressurizing begins by opening of the valve. As the result the liquefied freon on the bottom of the vessel explosively boils and mixes with the water. The boiling process is accompanied by creation of shock waves that crushes gas bubbles and creates a developed interfacial area. Bubbles move with respect to water but the water around them is significantly cooler. This combination of factors leads to a big formation of freon hydrate in the whole volume of water. It should be noted that the speed of formation of hydrate depends on the heat transmission and is not limited by diffusion of the gas into liquid. Along with the crushing of the gas phase a significant importance has a heat transfer from interfacial area on which the gas hydrate forms. According to researches carried out by Japanese scientists the enthalpy of hydrate formation of freon R-134A is 124 kJ/mole. Therethrough when 1 kg of freon gas converts into gas hydrate creates 296 kcal of heat energy.

This piece of work describes the method of creation of gas-hydrate freon based on boiling-condensation of gas-hydrating agent in enclosed volume. This article shows the influence of height of water column on the process of formation of hydrates. This method is not so fast as the one described above but has the advantage of its simplicity and low energy consumption.

2. Experimental setup

The working area constituting a hermetic vessel of 740 mm height and with 150X150 mm section (fig. 1). Two vessel walls are made of optical transparent material that allows to observe processes taking place inside the working area. Maintenance of required temperature inside the unit is provided by the built in cooling system (water jacket). The bottom of the unit has no heat insulation and gets heat from external environment. Measurement of temperature was taken by the sensor ДТС204-PT100, measurement of pressure was taken by the sensor ПД-100. The unit allows to make the research in the wide range of temperatures (from -10 to 60°C) and pressure (from 0.1 to 10MPa).

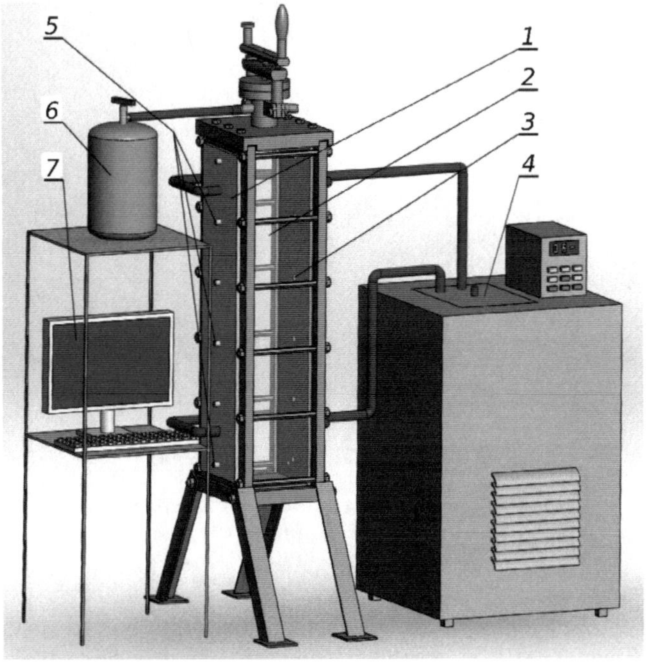

Figure 1. The schematics of the unit: 1 — Side walls; 2 — Working area; 3 — temperature sensor; 4 — cryostat; 5 — pressure sensors; 6 — gas tank; 7 — PC.

Figure 2. Gas-hydrate gained in the experiment

3. Method and results

The experiment: the reaction vessel is filled with water from 2 to 10 liters. The water is cooled down to 7 °C. Then 0,3 kg of R-134A freon is fed into the chamber. Since the walls of the vessel have subzero temperature freon liquefies and goes down to the bottom. After the delivery of the gas into the working area is cut off the pressure alines and boiling on the bottom of the vessel begins. Due to intensive boiling a big number of freon bubbles appear. When the gas goes up it cools down and it

IOP Conf. Series: Journal of Physics: Conf. Series **899** (2017) 032010 doi:10.1088/1742-6596/899/3/032010

creates conditions for formation of freon hydrates. The hydrate film becomes to appear on walls of bubbles. When bubbles reach the free liquid surface they destroy and leave hydrate flakes. Freon gas which wasn't converted into hydrate is condensed on walls and then goes down to the bottom of the vessel and takes part in boiling process again. As you can see from the description the process is cyclic and goes until all gas is converted into hydrate (fig. 2). Intensity of the process depends on heat exchange on side walls and heating of the bottom of the vessel as well as on the distance that bubbles go during the boiling on every cycle. The fig. 3 shows dependence of growth of gas-hydrate "hat" from the distance that bubbles go from the bottom of the unit to the water surface. As you can see from the chart the more distance bubbles go the slower gas-hydrates forms. It can be explained in the following way: the longer bubbles go to the surface the more gas-hydrate film forms and the harder gas-hydrates contacts with water. Also the boiling-condensation process takes more time.

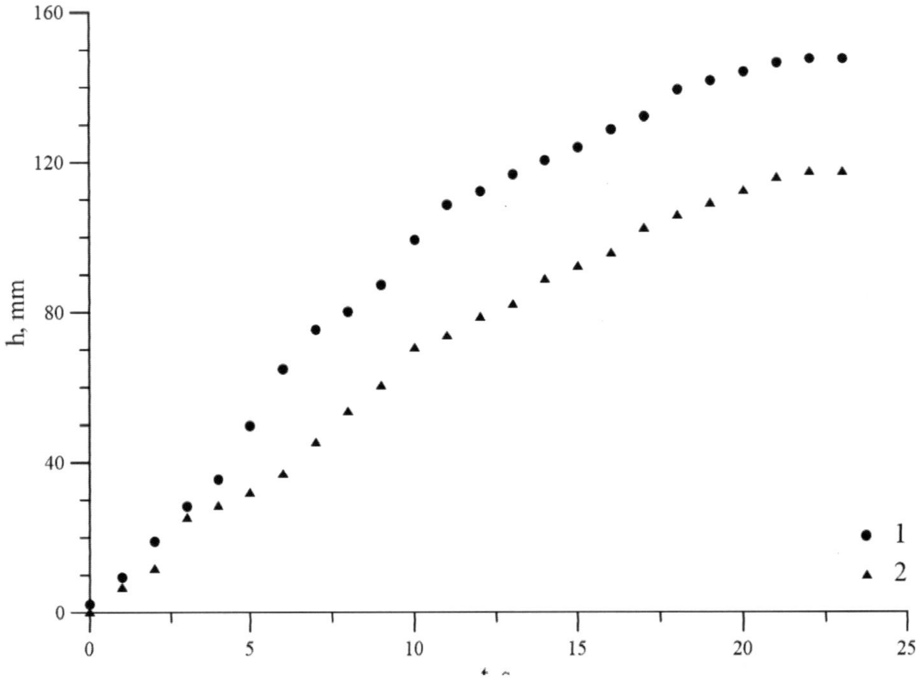

Figure 3. Chart of growth of hydrate mass R-134A in the vessel depending on the level of water: 1 – Height of water column is 80 mm, 2 - Height of water column is 420 mm.

4. Conclusion

A method for obtaining a gas hydrate of Freon 134a was described, which is based on the cyclic boiling-condensation process of the gas-hydrate-forming agent in a closed water volume. The influence of the height of the water column on the hydrate formation process is shown. This method of hydrate formation is characterized by its simplicity and low energy consumption.

Acknowledgments
This work was supported by the Russian Science Foundation, project №15-19-10025. We express our gratitude to the Novosibirsk State University for 145 giving us an access to the library.

References
[1] Makagon Y F 1997 *Hydrates of Hydrocarbons* (Tulsa, Oklahoma. Pennwell publishing company)

[2] Yakushev V S, Istomin V A 1993 Gas-hydrates self-preservation effect *Physics and Chemistry of Ice* pp 136-139

[3] Sloan E D, Koh C A 2008 Clathrate hydrates of natural gases 3rd ed *Chemical industries series* p 721

[4] Nakoryakov, V.E., Mezentsev, I.V., Meleshkin, A.V., Elistratov, D.S., Manakov, A.Y. 2015 Experimental investigation of gas-hydrate formation by underwater boiling of a condensed gas layer *Journal of Engineering Thermophysics* Vol. 24 № 4 pp. 335-337

[5] Nakoryakov, V.E., Mezentsev, I.V., Meleshkin, A.V., Elistratov, D.S. 2015 Visualization of physical processes occurring on liquid nitrogen injection into water *Journal of Engineering Thermophysics* Vol. 24 № 4, pp. 322-329

[6] Chernov A.A., Pil'nik A.A., Elistratov D.S., Mezentsev I. V., Meleshkin A.V., Bartashevich M.V., Vlasenko M.G. 2017 New hydrate formation methods in a liquid-gas medium *Scientific Reports* Vol. 7, 40809

[7] Chernov A.A., Elistratov D.S., Mezentsev I.V., Meleshkin A.V., Pil'nik A.A. 2017 Hydrate formation in the cyclic process of refrigerant boiling-condensation in a water volume *International Journal of Heat and Mass Transfer* Vol. 108 Part B P. 1320–1323

[8] Nakoryakov V E, Dontsov V E, Chernov A A 2006 Gas hydrate formation in a gas-liquid mixture behind a shock wave *Doklady Physic* Vol. 51 № 11 P 621-624

[9] Dontsov V E,. Chernov A A 2009 Dilution and hydrate forming process in shock waves *Int. J. Heat Mass Transfer* Vol. 52 № 21-22 P 4919-4928

[10] Chernov A A, Dontsov V E 2011 The processes of dissolution and hydrate forming behind the shock wave in the gas–liquid medium with gas mixture bubbles *Int. J. Heat Mass Transfer* Vol 54 № 19-20 P 4307-4316

[11] Misura S Y 2013 Effect of Heat Transfer on the Kinetics of Methan hydrate dissociation *Chem. Phys. Lett* Vol. 583 pp 34-37

[12] Nakoryakov V.E., Misyura S.Y., Elistratov S.L. 2013 Methane combustion in hydrate systems: water-hydrate and water-hydrate-isopropanol *Journal of Engineering Thermophysics* Vol. 22 №3 P. 169-173.

[13] Misyura SY. 2016 The influence of porosity and structural parameters on different kinds of gas hydrate dissociation *Scientific Reports* Vol. 6, 30324

[14] Misyura S.Y., Donskoy I.G. 2016 Dissociation of natural and artificial gas hydrate *Chemical Engineering Science* Vol. 148 P. 65-77

[15] Shagapov V.Sh., Musakaev N.G., Khasanov M.K. Formation of gas hydrates in a porous medium during an injection of cold gas // Int. J. of Heat and Mass Transfer. 2015. V. 84. P.1030-1039.

[16] Shagapov V.Sh., Khasanov M.K., Musakaev N.G., Ngoc Hai Duong Theoretical research of the gas hydrate deposits development using the injection of carbon dioxide // Int. J. of Heat and Mass Transfer. 2017. V.107. P. 347-357.

[17] V. Sh. Shagapov, G. R. Rafikova, M. K. Khasanov On the theory of formation of gas hydrate in partially water-saturated porous medium when injecting methane // High Temperature. 2016. V.54, №6. P. 858–866.

[18] M. K. Khasanov, I. K. Gimaltdinov, M. V. Stolpovsky Specific features of the formation of gas hydrates during the injection of a cold gas into a porous medium saturated with a gas and water // Theoretical Foundations of Chemical Engineering. 2010. V.44, №4. P. 424–431.

Waves in porous media saturated with bubbly liquid

A A Gubaidullin, O Yu Boldyreva and D N Dudko

Tyumen Branch of Khristianovich Institute of Theoretical and Applied Mechanics SB RAS, Tyumen State University, Tyumen, Russia

E-mail: gubai@tmn.ru

Abstract. Wave processes in a porous medium saturated with bubbly liquid, taking into account the nonlinearity of the oscillations of the bubbles, are numerically investigated. The nonlinearity is accounted in the equation of gas state and the equation of Rayleigh - Lamb. Numerical study in the framework of nonlinear model is performed for the first time. The dispersion relations for linear waves in a porous medium containing bubbly liquid, as well as for the main acoustic mode in a cylindrical cavity in such a medium, are obtained. Frequency dependences of velocity and attenuation of harmonic waves in a cylindrical waveguide are calculated. The influence properties of the bubbly liquid saturated porous medium on the wave velocity and attenuation and the penetration depth of perturbation into the surrounding porous medium are analyzed. The passage of the step wave from fluid into a porous medium saturated with bubbly liquid is investigated. The influence of porous medium parameters and intensity of the incident wave on the wave evolution in bubbly liquid saturated porous medium is studied. The calculation results show good qualitative agreement with the experimental data of other authors.

1. Introduction

The investigation of waves in porous medium saturated with bubbly liquid is of particular interest, because often the oil extraction process is accompanied by formation of gas bubbles in the reservoir. For example, when the pressure decreases in the reservoir, formation of bubbles from dissolved gas, or gas separation from gas condensate, or the decomposition of gas hydrates may occur. Gas is also formed during the acidizing of low-permeability zones.

The study of the properties of acoustic waves in a cylindrical waveguide is important in connection with numerous applications, for example, determination of the elastic properties of the surrounding space [1]. In particular, in geophysical research the acoustic waves in a cylindrical cavity are used to determine the characteristics of water-bearing and hydrocarbon reservoirs. The basic acoustic mode of such waves is usually called tube waves.

In traditional acoustic well studies, the rock porosity is determined by the first entry of the acoustic pulse. One of the next arrivals, the Stonely wave (also called the tube wave), is used to determine permeability, and the mechanical properties are determined by the whole wave train. For correct interpretation of acoustic logs, a deep understanding of the influence of rock properties (porosity, permeability, lithology, hydrocarbon saturation) on wave propagation in a cylindrical waveguide (borehole), is required.

In the present time in the scientific literature there is a number of experimental studies [2-5] for the wave propagation in porous media saturated with liquid with gas bubbles. The theoretical studies mainly focus on the dispersion relations for waves in such media and the propagation of perturbations in the framework of linear theory [2, 5, 6]. Insufficiently investigated there remains the problem of evolution of pressure perturbations in a porous medium saturated with bubbly liquid. In this work, we numerically investigated wave processes in a porous medium saturated with bubbly liquid, in the framework of nonlinear model. The nonlinearity is taken into account in the equation of gas state and the equation of Rayleigh - Lamb.

2. Basic Equations

To study wave propagation in a porous medium containing a bubbly liquid, two velocity, two stress tensors model of saturated porous medium is used [7, 8]. The pore space is filled two-phase mixture of liquid with gas bubbles, bubble size is calculated using the equation of Rayleigh-Lamb for a gas bubble in a porous medium.

The equations of mass and momentum balance are in the form:

$$\frac{\partial \rho_{l+g}}{\partial t}+\nabla^i\left(\rho_{l+g} v_l^i\right)=0, \quad \frac{\partial \rho_s}{\partial t}+\nabla^i\left(\rho_s v_s^i\right)=0, \quad \frac{\partial n_b}{\partial t}+\nabla^i\left(n_b v_l^i\right)=0 \tag{1}$$

$$\rho_{l+g} \frac{d_l v_l^i}{dt}=-\alpha_{l+g} \nabla^i p_l-F^i, \quad \rho_s \frac{d_s v_s^i}{dt}=-\alpha_s \nabla^i p_l+\nabla^j \sigma_{s*}^{ji}+F^i, \tag{2}$$

where ρ_j, v_j, α_j are the density, velocity, volumetric content of the j - th phase, the subscripts $j = s, l, l{+}g$ relate to the skeleton of the porous medium, liquid or mixture of liquid and bubbles; σ_{s*}, p_l are effective stress in the skeleton and the pressure in the liquid, respectively, n_b is the number of bubbles per unit volume.

The skeleton of the porous medium is assumed to be elastic with effective moduli of elasticity λ_{s*}, μ_{s*}:

$$\sigma_{s*}^{ij}=\alpha_s\left(\lambda_{s*}\delta^{ij}\varepsilon_s^{mm}+2\mu_{s*}\varepsilon_s^{ij}+\gamma_{s*}\delta^{ij} p_l\right), \quad \gamma_{s*}=\beta_s\left(\lambda_{s*}+2/3\,\mu_{s*}\right), \tag{3}$$

$$\frac{d_s \varepsilon_s^{ij}}{dt}=\frac{1}{2}\left(\nabla^i v_s^j+\nabla^j v_s^i\right),$$

where ε_s is strain of the solid phase, γ_{s*} is the ratio of bulk elastic moduli of the skeleton and the solid.

The state equations for solid and liquid phases are given in the acoustic approximation:

$$p_s-p_{s0}=C_s^2\left(\rho_s^\circ-\rho_{s0}^\circ\right), \quad p_l-p_{l0}=C_l^2\left(\rho_l^\circ-\rho_{l0}^\circ\right),$$

here subscript 0 indicates the unperturbed value, ρ°_j is the true density, C_j is the sound velocity in the material of the j - th phase, p_s is true pressure of the solid phase.

Interphase interaction is a sum of viscous friction force F_μ and added masses force F_m:

$$F=F_m+F_\mu, \quad F_m=\frac{1}{2}\eta_m\alpha_s\rho_{l+g}\left(\frac{d_l v_l}{dt}-\frac{d_s v_s}{dt}\right), \quad F_\mu=\eta_\mu\alpha_s\alpha_{l+g}\mu_l a_{s*}^{-2}\left(v_l-v_s\right), \tag{4}$$

where a_{s*} is the characteristic size of the grains of the skeleton, μ_l is viscosity of the fluid, η_m, η_μ are the coefficients of interaction between phases dependent on the structure of the pore space.

The gas in the bubbles is adiabatically compressed (γ is the ratio of specific heats):

$$\frac{p_g}{p_{g0}} = \left(\frac{\overset{\circ}{\rho}_g}{\overset{\circ}{\rho}_{g0}}\right)^{\gamma} = \left(\frac{a_{b0}}{a_b}\right)^{3\gamma}, \tag{5}$$

bubble radius a_b changes according to the equation of Rayleigh-Lamb for the bubble in a porous medium [2]:

$$\frac{da_b}{dt} = w_R + w_A, \qquad w_A = \frac{p_g - p_l - \dfrac{2\Sigma}{a_b}}{\overset{\circ}{\rho}_{l0} C_l x_g^{1/3}}, \tag{6}$$

$$\overset{\circ}{\rho}_{l0}\left(\dot{w}_R a_b + \frac{3}{2} w_R^2\right) = p_g - p_l - \frac{2\Sigma}{a_b} - 4\mu_l \frac{w_R}{a_b}\left(1 + \frac{1}{4}\alpha_s \eta_\mu \left(\frac{a_b}{a_{s*}}\right)^2\right)$$

here x_g is the gas volume fraction in the bubbly liquid, Σ is the coefficient of surface tension at the liquid and gas boundary, w_A takes into account the compressibility of the liquid [9].

To close the system of equations the relations between the true pressures p_l, p_s in the phases and the effective pressure p_{s*} in the skeleton are used

$$\rho_s = \alpha_s \overset{\circ}{\rho}_s, \quad \rho_{l+g} = \alpha_l \overset{\circ}{\rho}_l + \alpha_g \overset{\circ}{\rho}_g, \quad \alpha_s + \alpha_l + \alpha_g = 1, \quad \alpha_g = \frac{4}{3}\pi a_b^3 n_b, \tag{7}$$

$$p_{s*} = \alpha_s (p_s - p_l), \qquad p_{s*} = -\frac{1}{3}\sigma_{s*}^{mm}.$$

3. Dispersion Relations

After linearization the above system of equations the dispersion relation was obtained, and the velocity and the damping of linear waves in porous media saturated with bubbly liquid were calculated.

When considering the wave propagation along a cylindrical cavity in a porous medium saturated with bubbly liquid it was assumed that the cavity is also filled with bubbly liquid. The boundary conditions at the interface between of the cavity and a porous medium are the continuity of the liquid flow, pressure in a liquid and total stress. The cases of open pores (when fluid flow between the cavity and the porous medium is free) or closed pores (there is no flow at the interface) at the boundary are considered. The equations of motion are written in cylindrical polar coordinates, the system of equations is linearized, and we seek a solution in the form of a harmonic wave corresponding to the propagation of perturbations along the cylindrical cavity. In this case the perturbation in the porous medium is a combination of heterogeneous bulk longitudinal and transverse waves. The dispersion relation for tube waves is derived from the conditions at the boundary of the cavity and the porous medium.

In figure 1 the phase velocity and the linear damping decrement of the deformational (1) and filtrational (2) waves in bubbly liquid saturated porous medium (black lines) are presented. For comparison, the velocity and attenuation of waves in bubbly liquid (red lines), and liquid saturated porous medium (blue lines) are also shown.

The equilibrium parameters of the medium used in calculations and presented in figure 1, are the following: bubbly liquid is water + methane, the equilibrium pressure is 50 bar (ρ°_g = 34.36 kg/m3), the volume fraction of gas in the bubbly liquid is 0.01, the radius of the bubbles is 1 mm, the skeleton of the porous medium is quartz, with porosity of 0.4 and a typical pore radius of 0.1 mm; effective elastic moduli of skeleton are $\lambda_{s*} = \mu_{s*} = 2\cdot 10^9$ Pa.

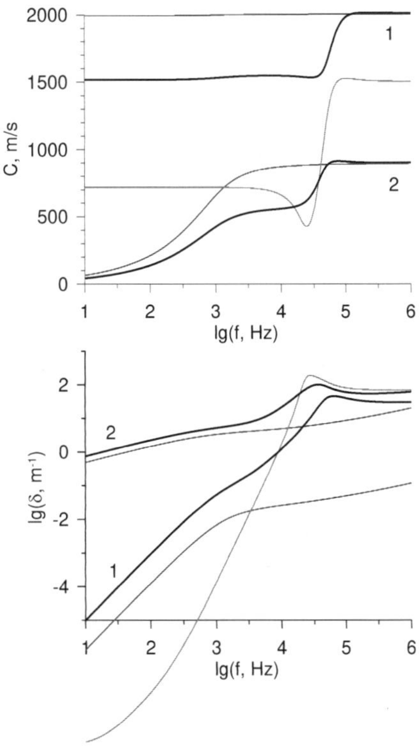

Figure 1. The phase velocity and the linear damping decrement of the deformational (1) and filtrational (2) waves in bubbly liquid saturated porous medium (blue lines), in a bubbly liquid (red lines), and liquid saturated porous medium (black lines).

Minimum velocity of wave propagation in bubbly fluid (red line) corresponds to the natural frequency of the bubbles which is equal to 22.4 kHz. At these and lower frequencies the velocity of the fast and slow waves in porous medium saturated with bubbly liquid is lower and the attenuation is higher than that in a porous medium saturated with liquid without gas bubbles. With increasing frequency the velocities reach the same values as in a porous medium with a liquid without gas bubbles.

Analysis of the obtained numerical results showed that the increase in the equilibrium pressure leads to increase of the velocity of deformational and filtrational waves at frequencies below the natural frequency of the bubbles. Damping of waves in this frequency range is somewhat reduced.

4. Propagation of stepwise wave

Methodology for calculating the motion of a porous medium containing liquid with gas bubbles based on the method of Lax – Wendroff was developed, and the numerical study of pressure wave propagation in such medium was made. Perturbation in the porous medium ($x>0$) is generated by a wave incident from the pure liquid ($x<0$).

We have investigated the influence of medium parameters and of the initial pulse on the wave evolution in a porous medium containing the bubbly liquid.

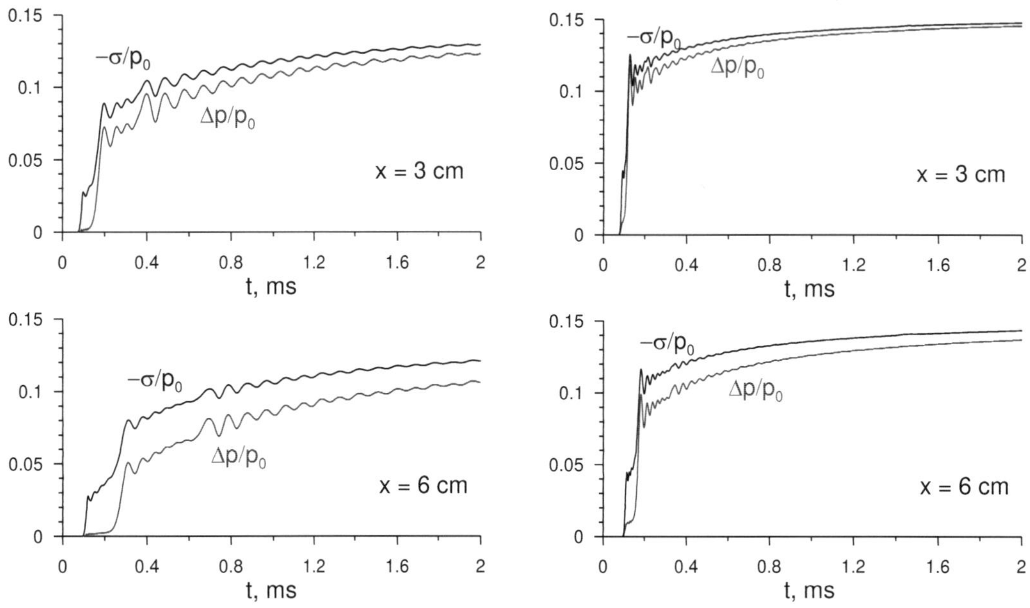

Figure 2. The pressure in the liquid and total stress in the propagation of the step wave of the amplitude equal to $0.1\cdot p_0$ from a liquid into a porous medium at a distance of 3 cm and 6 cm from the boundary. Equilibrium pressure is 10 bar (left) and 50 bar (right)

In figure 2 shows the variation of the pressure in the liquid p and the total stress $\sigma = \sigma_{s*} - p$ in the step wave of the amplitude equal to $0.1\cdot p_0$ passing from the liquid to the bubbly liquid saturated porous medium. Equilibrium pressure is p_0 =10 bar (left) and p_0 =50 bar (right). The other parameters of the porous medium and babbly liquid are the same as in figure 1.

When entering a porous medium initial pulse splits into fast (deformational) and slow (filtrational) waves. At the fast wave passage the abrupt increase in full stress σ is observed (black line), and the arrival of the slow wave is characterized by a gradual increase in pore pressure and subsequent damped oscillations associated with pulsations of the bubbles. The observed period of oscillations coincides with the period of natural oscillations of the bubbles (in this case about 0.10 ms (left) and 0.045 ms (right)). If you increase the amplitude of the original incident wave, pressure variations become less regular. In a medium with a higher equilibrium pressure, there is a high propagation velocity of both fast (deformational) and slow (filtrational) waves.

With decreasing bubble size the amplitude and period of oscillations decrease. With increasing gas content in the bubbly liquid decreases the velocity of waves in a porous medium containing bubbly liquid, and its acoustic stiffness. As a result, with increase in gas content the pressure increasing in the wave passed into the porous medium is slower.

The calculation results show good qualitative agreement with the experimental data on the passage of step waves from water to porous medium saturated with water with air bubbles [2, 3]. A quantitative comparison is not possible, as in these works are not all the medium parameters are specified.

This work was supported by RFBR (Project No. 17-51-540001) and by the President of Russian Federation in Support of the Leading Scientific Schools (Project No. NSH-6987.2016.1).

References

[1] White J E 1983 *Underground sound. Application of seismic waves* (Elsevier: Amsterdam –

Oxford – New York)
[2] Dontsov V E, Kuznetsov V V, Nakoryakov V E 1987 *Izv. USSR Academy of Sciences, MZhG* **4** 85
[3] Grinten J G M van der 1987 *An experimental study of shock-induced wave propagation in dry, water-saturated, and partially saturated porous media (Proefschrift Eindhoven)*
[4] Smeulders D M J 1992 *On wave propagation in saturated and partially saturated porous media (Proefschrift Eindhoven)*
[5] Smeulders D M J and van Dongen M E H 1997 *J. Fluid Mech.* **343** 351
[6] Dunin S Z, Mikhailov D N and Nikolaevskii V N 2006 *J. Appl. Math. Mech.* **70(2)** 251
[7] Nigmatulin R I 1978 *Fundamentals of Mechanics of Heterogeneous Media* (Moscow: Nauka)
[8] Nigmatulin R I 1990 *Dynamics of Multiphase Media. Part 1* (New York: Hemisphere Publ)
[9] Nigmatulin R I, Shagapov V S and Vakhitova N K 1989 *USSR Academy of Sciences Doklady* **304** 1077

IOP Conf. Series: Journal of Physics: Conf. Series **899** (2017) 032012 doi:10.1088/1742-6596/899/3/032012

Capillary hydrodynamics and transport processes during phase change in microscale systems

V V Kuznetsov[1,2]
[1] Institute of Thermophysics Siberian Branch of Russian Academy of Sciences
1, Academician Lavrentyev Av., Novosibirsk, 630090, Russian Federation
[2] Novosibirsk State University, 2, Pirogova Str., Novosibirsk, 630090, Russian Federation

E-mail: vladkuz@itp.nsc.ru

Abstract. The characteristics of two-phase gas-liquid flow and heat transfer during flow boiling and condensing in micro-scale heat exchangers are discussed in this paper. The results of numerical simulation of the evaporating liquid film flowing downward in rectangular minichannel of the two-phase compact heat exchanger are presented and the peculiarities of microscale heat transport in annular flow with phase changes are discussed. Presented model accounts the capillarity induced transverse flow of liquid and predicts the microscale heat transport processes when the nucleate boiling becomes suppressed. The simultaneous influence of the forced convection, nucleate boiling and liquid film evaporation during flow boiling in plate-fin heat exchangers is considered. The equation for prediction of the flow boiling heat transfer at low flux conditions is presented and verified using experimental data.

1. Introduction

Due to a rapid growth of the applications that require transport of the large amount of heat in the limited space, more and more attention has recently been paid to the development of micro-scale heat exchangers. These heat exchangers are characterized by extremely large ratio of surface area to volume and traditionally classified as compact heat exchangers. Two-phase compact heat exchangers offer ample opportunities for engineering applications such as high compact generators/condensers of power and cryogenic units and natural gas liquefaction units. The mechanisms of microscale heat transfer during flow boiling and condensation in mini/micro-channels were investigated, for example, in [1, 2] and are fairly inconsistent especially for flow boiling. It has been found in a number of studies that heat transfer during flow boiling in mini- and microchannels is determined by the heat flux and depends only weakly on the mass flux [3]. Experimental study of heat transfer during boiling and condensing of cryogenic fluids, hydrocarbons and refrigerants in the plate-fin heat exchangers with minichannels showed weak dependence of heat transfer coefficient on the heat flux as it shown in [4]. Liquid film flow, which provides large heat flux at small temperature difference, deserves special attention in the analysis of heat transport in microscale systems [5]. Published methods for predicting the heat transfer with flow boiling in micro-scale heat exchangers are based either on certain modifications of models that characterize boiling in conventional tubes [1] or on specially developed correlations [3], but their application in various heat-exchange regimes was hardly a success. In particular, this applies to cryogenic compact heat exchangers operated at low mass and heat fluxes.

The aim of the present study is revealing the mechanisms of fluid flow and microscale heat transfer for boiling in downward flow in the plate-fin heat exchanger with minichannels when the value of heat flux is not sufficient to support active nucleate boiling. The characteristics of gas-liquid flow and heat transport during flow boiling of refrigerant R-21, resulting from the numerical studies, are discussed. This refrigerant has similar thermophysical properties as the cryogenic liquids at low temperatures. The results of numerical modelling of the liquid evaporation in rectangular minichannel are presented for annular flow. The equation for prediction of flow boiling heat transfer in plate-fin heat exchanger at low flux conditions that accounts simultaneously the influence of the forced convection, nucleate boiling and liquid evaporation is presented. The results for the calculation of heat transfer during flow boiling are compared with the experimental data.

2. Capillary hydrodynamics and heat transfer during evaporation in the annular flow

Identification of the characteristics of gas-liquid (vapor-liquid) flow and heat transport during flow boiling and condensing plays a key role in justification of the micro-scale or microstructured heat exchangers. In these heat exchangers the fluids flow in the channels with typical lateral dimensions below the capillary constant. The gas-liquid flow in these channels is characterized by considerable influence of the capillary forces on flow patterns and the gas bubbles are elongated due to the restriction of channel wall. In plate-fin heat exchangers, the form of the mini- and microchannels usually is closed to a rectangle. Three basic flow patterns such as transition flow with elongated bubbles, annular flow with disturbance waves and annular flow with liquid meniscus near the channel corners were observed in [6] during adiabatic experiments in plate-fin heat exchanger with downward flow of refrigerant R-21. For relatively small heat fluxes, liquid evaporation is one of the most promising heat transfer mechanisms. For this case, the heat transport from the wall in annular flow is determined by liquid film thickness distribution along the channel perimeter. The gas and liquid superficial velocities when the transition to annular flow occurs can be determined from the model of equivalent film thickness [7]. In this model, it is assumed that the transition to the annular flow occurs if the film thickness δ_{eb} in elongated bubble δ_{eb} or transition flow is equaled to the film thickness in the annular flow δ_{film}

$$\delta_{eb} = \delta_{film} \quad (1)$$

This means that the gas velocity in a channel becomes sufficiently high to move all liquid in the film under the action of shear stress and the formation of liquid plugs is not occurs. For elongated bubble flow, the residual film thickness δ_{eb} can be determined using the “Taylor” law from [8] as follows:

$$\delta_{eb}/D = 0.67Ca^{2/3}/\left(1+3.35Ca^{2/3}\right) \quad (2)$$

where D is the tube diameter, $Ca=U_b\mu_{liq}/\sigma$ is the capillary number defined by bubble velocity U_b, liquid viscosity μ_{liq} and surface tension σ. For rectangular channel, the hydraulic channel diameter D_h can be used in the equation (2) instead of D as it was shown in [9]. Corresponding film thickness in the annular flow δ_{film} can be calculated using the annular flow model of Asali et al. [10] without liquid entrainment.

Let's consider the gas-liquid flow and heat transfer during liquid evaporation in the rectangular minichannel of the plate-fin heat exchanger. In annular flow, the capillary forces are crucial for liquid flow distribution and move liquid into the corners of the channel forming a thin film on the walls. The model of the creeping flow for annular flow in rectangular channel [11] is based on the allocation of liquid flow in two zones of the flow: flow in a corner of the channel bounded by meniscus and thin film flow on the channel wall; and matching solutions for these zones with the conditions of their conjugation. In approximation of a thin layer there is a small parameter $\varepsilon = \delta_0/a << 1$, where δ_0 is the initial film thickness and a is the half width of the long side of the channel. For this approximation, the Navier-Stokes equation for the downward flow can be reduced to the dimensionless equation for the local thickness of curved film with a last term which takes into account the evaporation of liquid

$$(\gamma \mathrm{m}^3 + 1.5\frac{\kappa}{\varepsilon}\mathrm{m}^2)_x + (\mathrm{m}^3\mathrm{m}_{yyy})_y = \frac{3}{\varepsilon^4}(\mathrm{Ga}(\frac{\mathrm{m}_y}{\mathrm{m}})_y - \frac{\mathrm{G}_0\Theta_{w,i}}{\mathrm{m}\varepsilon}) \quad (3)$$

Here $\gamma = 1-\left(dp_{gas}/dx\right)/\rho_{liq}g$, $\kappa = \tau/\rho_{liq}ga$, $\mathrm{Bo} = \rho_{liq}\mathrm{g}a^2/\sigma$, $\mathrm{Ga} = \mathrm{A}_0/(6\pi a^2\sigma)$, the dimensionless film thickness is scaled via initial film thickness $m=\delta/\delta_0$, longitudinal x and transverse y coordinates are scaled via aBo/ε and a respectively, τ is shear stress at the interface, A_0 is the Gamaker's constant, $G_0 = \lambda_{liq} T_* \nu_f / (h_{fg} \sigma a)$ and $\Theta_{w,i}=(T_{w,i}-T_{sat})/T_*$ are determined by temperature of the inner wall $T_{w,i}$, saturation temperature T_{sat} and characteristic temperature $T_* = H \cdot a^2 / \lambda_{liq}$ for the volume heat density H in the wall area, h_{fg} is the latent heat of vaporization.

Equation (3) is solved together with the equations for liquid flow in the meniscus and total liquid flow rate conservation. The boundary conditions are the symmetry conditions at the center of the channel and conditions $m_y=0$ и $m_{yy}=1/(r\varepsilon)$ at the point of conjunction of interface solutions for the film and the meniscus, where $r=R/a$. The singularity due to the fact that thickness of the film streams toward zero in the vicinity of the contact line is eliminated accounting the disjoining pressure. The inspection of the rivulet half-width for the current liquid flow is required to calculate the shape of the interface. When the rivulet half-width becomes less than the distance from the center of the channel to the meniscus, the film rupture is supposed and new configuration of the flow consisting of rivulet, menisci and dry spots between them is established. When the interface shape is obtained, the Fourier equations for liquid and wall areas are solved together to determine the local heat transfer coefficient.

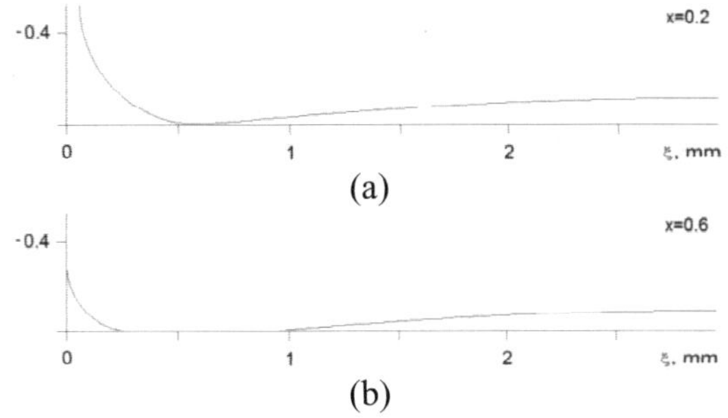

Figure 1. Liquid interface for x=0.2 (a) and x=0.6 (b) for R-21 evaporation in the 1.05x6 mm channel with q_w=1.7 kW/m^2 and G=50 kg/m^{2}s.

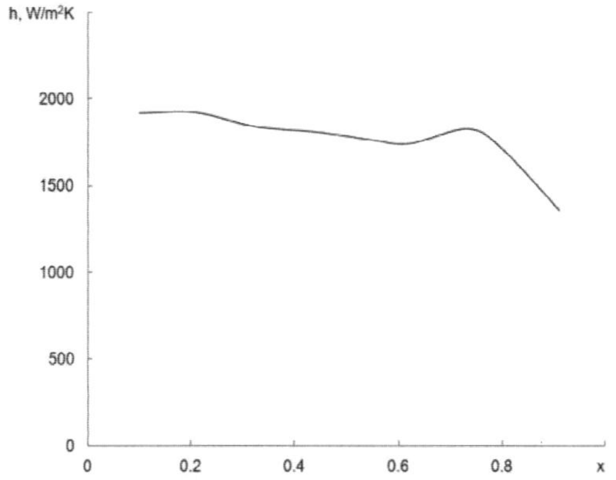

Figure 2. Dependence of average heat transfer coefficient on vapor quality for R-21 evaporation with q_w=1.7 kW/m^2, G=50 kg/m^{2}s.

Figure 1 shows the results of calculation of the liquid interface during R-21 evaporation in the 1.05x6 mm channel with supplied heat flux q_w=1.7 kW/m^2 and mass flux G=50 kg/m^{2}s. The liquid interface is presented for two distances from channel inlet that correspond to vapor quality of x=0.2 and x=0.6. Formation of dry spots and increasing local heat flux in the vicinity of contact line are typical for evaporation in minichannels. Heat flux in this area is limited by the roughness of the wall equaled to 7.5 μm. The condensation provokes alignment of the film thickness near the meniscus, and lowers the coefficient of heat transfer in comparison with the ruptured film which is typical for the evaporation. Figure 2 presents the dependence of average heat transfer coefficient on vapor quality for R-21 liquid film evaporation in the 1.05x6 mm rectangular channel with q_w=1.7 kW/m^2 and G=50 kg/m^{2}s. The calculation results show that the mechanism of heat transport during evaporation in micro-scale heat exchangers is conductive heat transfer through curved liquid film and in the vicinity of liquid-vapor-wall contact line. The heat transfer deterioration occurs for vapor quality higher then 0.7, when liquid in meniscus will be evaporated.

3. Flow boiling heat transfer in micro-scale heat exchangers at low flux conditions

The mechanism of flow boiling in plate-fin heat exchangers at low heat and mass fluxes is more complex then only evaporation. The reliable model for prediction of heat transfer should account for the determining mechanisms of heat transfer selected in [12] as follows: the nucleate boiling, nucleate boiling suppression, forced convection, evaporative heat transfer in the annular and elongated bubble flows. The physical basis of the approach proposed in this paper consists in integration of all these mechanisms in one general equation. To account the contribution of nucleate boiling, forced convection and liquid film evaporative heat transfer on the two-phase heat transfer coefficient h_{tp} it is suggested to use the method proposed in [12] as follows:

$$h_{tp}^2 = \left(h_{con}F\right)^2 + \left(h_{boil}\Psi_{sup}S\right)^2 + \left(E\varphi h_{ev}\right)^2 \quad (4)$$

Here h_{con}, h_{boil} and h_{ev} are the forced convection, nucleate boiling and liquid film evaporative heat transfer coefficients, accordingly; F and S are the factors of forced convection enhancement and nucleate boiling suppression, proposed by Liu & Winterton [13]

$$F = \left(1 + x\,\mathrm{Pr}_{liq}\left(\rho_{liq}/\rho_{gas} - 1\right)\right)^{0.35}$$
$$S = \left(1 + 0.055F^{0.1}\,\mathrm{Re}_{liq0}^{0.16}\right)^{-0.1} \quad (5)$$

and Ψ_{sup} is the boiling suppression factor. Evaporation factor E corresponds to the relative cross-sectional areas along the flow occupied by the continuous gas core. For the elongated bubble flow and transition flow, evaporation factor can be estimated as vapor volume fraction for homogeneous flow model $E = \left(1 + \rho_{gas}(1-x)/\rho_{liq}x\right)^{-1}$, and for annular flow, the gas core is continuous and $E = 1$. Parameter φ equals the part of the channel perimeter, where the liquid film exists.

To account the effect of thermo-physical properties of the working fluid and channel dimensions on forced convection heat transfer, the suitable correlation for heat transfer coefficient is used depending on all-liquid flow Reynolds number. For the laminar all-liquid flow, corresponding convective heat transfer coefficient h_{con} should be calculated accounting the channel geometry and developing flow conditions. For the rectangular channel, the correlating equations of Shah & London [14] for thermally developing laminar flow is used as follows:

$$\mathrm{h}_{con} = \frac{\lambda_{liq}\mathrm{Nu}_z}{\mathrm{D}_h}, \quad \mathrm{Nu}_z = \mathrm{Nu}_{dev} + 8.68\cdot\left(10^3 z^*\right)^{-0.506} e^{-41 z^*} \quad (6)$$

Here $z^* = (z/\mathrm{D}_h)/\mathrm{Re}_{liq0}\mathrm{Pr}_{liq}$, z is the distance from the channel inlet, Nu_{dev} is the thermally developed heat transfer coefficient. For the rectangular channel, Nu_{dev} is determined as follows $Nu_{dev} = 8.235\cdot\left(1 - 2.042\gamma + 3.085\gamma^2 - 2.4765\gamma^3 + 1.0578\gamma^4 - 0.1861\gamma^5\right)$, where γ is the channel aspect ratio [14].

The correct selection of nucleate boiling model h_{boil} in the equation (4) is very important for reliable heat transfer prediction, especially for the case of dominant nucleate boiling. For nucleate boiling in plate-fin heat exchanger, the Cooper equation [15] is used

$$h_{boil} = 55M^{-0.5}p_r^{0.12}\left[-\log(p_r)\right]^{-0.55} R_p^{-0.2\log(p_r)} q_w^{0.666} \quad (7)$$

Here R_p is the surface average roughness and $p_r = p/p_{cr}$ is the reduced pressure. Nucleate boiling suppression for a thin liquid film is considered in [12] using boiling suppression factor Ψ_{sup} as the multiplier for nucleate boiling suppression term S in the equation (4). For rectangular channels, it is necessary to take into account the absence of nucleate boiling suppression in the menisci area as follows:

$$\Psi_{sup} = \left(\alpha_m + (1-\alpha_m)\Psi_{sup0}\right) \quad (8)$$

where $\alpha_m = 6R_m/(2a+b)$, R_m is the meniscus radius for the channel corner, b and a are the length of short and long sides of the channel, accordingly,

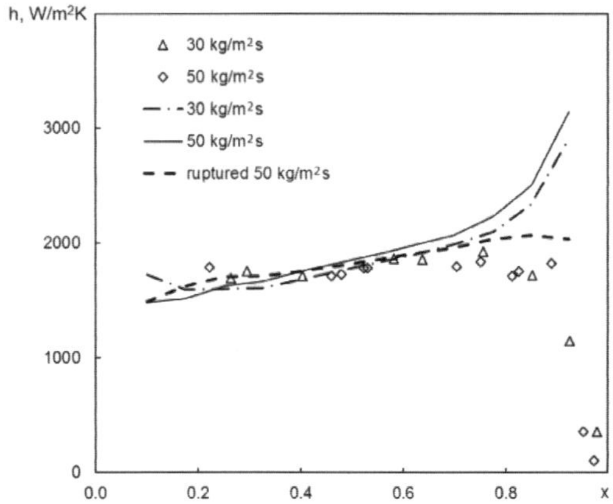

Figure 3. The dependence of heat transfer coefficient on vapor quality for G=30 kg/m^{2}s and G=50 kg/m^{2}s. Points are the data [6] and lines are calculation results for q_w =1.7 kW/m^2.

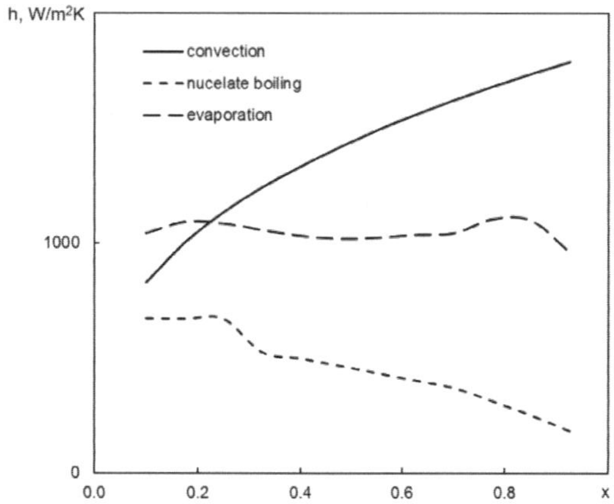

Figure 4. The values of forced convection, nucleate boiling and evaporation terms during R-21 flow boiling at G=50 kg/m^{2}s and q_w=1.7 kW/m^2.

$\Psi_{\sup 0} = \tanh^2\left(2.5\cdot 10^{-3}\theta_{\sup}^2\right)$, $\theta_{\sup} = \left(y_{vis}/d_{\tan}\right)^{0.6} / \left(Bo_x^{0.4} We_{liq0}^{-0.08} \Pr_{liq}^{1/3}\right)$ is the boiling suppression parameter introduced in [12], $Bo_x = q_w / \left(G\cdot x\cdot h_{fg}\right)$ is the Boiling number based on the local liquid flow rate.

In a mini- and microchannels, the liquid film becomes very thin due to the high surface area, and the contribution of evaporation can exceed the contribution of nucleate boiling. For the elongated bubble flow, liquid surface is controlled by capillary forces providing a thin liquid film inside a bubble. In this particular case, the heat transfer coefficient for evaporating liquid film can be calculated as follows:

$$h_{ev} = \left(\lambda_{liq}/\delta_{eb}\right),\ \ \delta_{eb}/D_h = 0.67Ca^{2/3}/\left(1+3.35Ca^{2/3}\right) \qquad (9)$$

Here, liquid film thickness δ_{eb} is determined in [8]. To determine the heat transfer coefficient for the evaporating wavy film in annular flow, the model of [16] is used with calculation of the shear stress according to model [10] without entrainment.

The solid line in figure 3 shows the results of calculation of heat transfer coefficient according to the equations (4)–(9) for mass fluxes of G=30 kg/m^{2}s and G=50 kg/m^{2}s, and heat flux of 1.7 kW/m^2 in comparison with experimental data [6] for boiling of R-21 in downward flow in plate-fin heat exchanger. The channels in this heat exchanger have not strictly rectangular cross-section, therefore, the arithmetic mean of the Nusselt numbers for rectangular $Nu_{dev/rec}$ and sine $Nu_{dev/sin}$ ducts from [14] $\bar{N}u_{dev} = \left(Nu_{dev/rec} + Nu_{dev/\sin}\right)/2$ is used to perform calculations. The calculation results are in an agreement with experimental data up to vapor quality of 0.6 where heat transfer deterioration occurs. The better agreement with the data is observed when heat transfer model is completed by the results of numerical simulation of evaporative heat transfer with liquid film rupture presented in figure 2 instead of uniform film thickness model. Figure 4 presents the values of forced convection, nucleate boiling and evaporation terms in equation (4) during R-21 flow boiling with G=50 kg/m^{2}s and q_w=1.7 kW/m^2. As can be seen in figure 4, the nucleate boiling is not determining mechanism of heat transfer in plate-fin heat exchanger at low flux conditions.

4. Conclusions

Identification of the characteristics of gas-liquid flows and heat transport during flow boiling and condensing plays a key role in justification of the micro-scale heat exchangers. The gas-liquid (vapor-liquid) flow in these exchangers is characterized by considerable influence of the capillary forces on flow patterns. In rectangular channels, the gas bubbles are elongated due to the restriction of channel wall and a liquid is accumulated near the channel corners both for elongated bubble, transition and annular flows forming thin curved liquid film on the walls. For relatively small heat and mass fluxes, the evaporation of liquid is one of the most promising heat transfer mechanisms for micro-scale heat exchangers and the heat transport from the wall is determined by liquid film thickness distribution along the channel perimeter. The calculations according to model of liquid film evaporation in annular flow show that increasing of local heat flux in the vicinity of contact line for the case of dry spot formation is typical for evaporation in minichannels. In contrarily, the condensation provokes alignment of the film thickness near the meniscus, and lowers the coefficient of heat transfer. Nevertheless, the mechanism of flow boiling in plate-fin heat exchangers at low fluxes is more complex then only evaporation. The reliable model for prediction of heat transfer should account for the determining mechanisms of heat transfer including nucleate boiling heat transfer, nucleate boiling suppression, forced convection heat transfer and evaporative heat transfer for elongated bubble, transition and annular flows. It is confirmed by comparison of the results of calculation according to the equation that units of all these mechanisms in one general equation with experimental data. The better agreement with experimental data is observed when heat transfer model is completed by the results of numerical simulation of the evaporative heat transfer with liquid film rupture instead of uniform liquid film thickness model.

Acknowledgments
This work performed in the Kutateladze Institute of Thermophysics SB RAS by a grant from the Russian Science Foundation (project No 14-49-00010).

References

[1] Kandlikar S G 2010 *Heat Transfer Eng.* **31** 159
[2] Kim S M and Mudawar I 2012 *Int. J. Heat Mass Transfer* **55** 984
[3] Bertsch S S, Groll E A and Garimella S V 2009 *Int. J. Multiphase Flow* **35** 142
[4] Kuznetsov V V, Safonov S A and Shamirzaev A S 2015 *Tech. Phys. Lett.* **41** 1124
[5] Pavlenko A N, Li X, Li H, Gao X, Volodin O A, Surtaev A S and Serdyukov V S 2015 *Tech. Phys. Lett.* **41** 774
[6] Kuznetsov V V, Dimov S V and Shamirzaev A S 2017 *J. Eng. Thermophys.* **26** 353
[7] Kuznetsov V V, Kozulin I A and Shamirzaev A S 2012 *J. Phys. Conf. Ser.* **395** 012093
[8] Aussillous P and Quere D 2000 *Phys. Fluids* **12** 2367
[9] Bartkus G V and Kuznetsov V V 2016 *MATEC Web Conf.* **84** 00005.
[10] Asali J C, Hanratty T J and Andreussi P 1985 *AlChE J.* **31** 886
[11] Kuznetsov V V and Safonov S A 2013 Fluid flow and heat transfer with phase change in minichannels and microchannels *Heat Pipes and Solid Sorption Transformations: Fundamentals and Practical Applications*, eds. L L Vasiliev and S Kakac (Boca Raton: CRC Press) pp 465–496
[12] Kuznetsov V V and Shamirzaev A S 2016 *Heat Transf. Eng.* **37** 1105
[13] Liu Z and Winterton R H S 1991 *Int. J. Heat Mass Transfer* **34** 2759
[14] Shah R K and London A L 1978 Laminar flow forced convection in ducts *Advances in Heat Transfer* (New York: Academic Press) Suppl. 1
[15] Cooper M G 1984 *Advances in heat transfer* **16** 157
[16] Butterworth D 1974 *Int. J. Multiphas. Flow* **1** 671

IOP Conf. Series: Journal of Physics: Conf. Series **899** (2017) 032013 doi:10.1088/1742-6596/899/3/032013

The influence of various factors on the droplet desorption

Misyura S.Y., Morozov V.S.

Kutateladze Institute of Thermophysics SB RAS, Novosibirsk, Russia

E-mail: misura@itp.nsc.ru

Abstract. Experimental data on sessile droplet desorption of aqueous salt solution of LiBr on a heated wall were implemented. High-temperature desorption of water-salt solutions in air atmosphere leads to significant difficulties at modeling heat and mass transfer. In this case, the evaporation rate multiply decreases with time and the diffusion coefficient, the desorption heat and the salt concentration change significantly. With the growth of salt concentration in solution from 10 % to 65 %, the steam partial pressure at the interface falls by dozens of times. In this study, we performed experiments in a wide range of salt concentrations and proposed a simple estimated method for calculating the mass flow. The resulting technique can predict the droplet solution behavior with a significant change in the partial vapor pressure on the droplet interphase with time.

1. Introduction

The Spray cooling are used in heat exchangers and small devices to improve the efficiency of heat transfer. The problems of combustion and dissociation are also closely related to gas-droplet flows. For example, burning of fuel droplets in a moving flow [1] and the formation of water droplets, vapor and gas methane during the combustion of methane hydrate [2, 3]. The size of constructions and energy costs depend on cooling efficiency. Droplet evaporation of single-component liquids on the hydrophilic surfaces was investigated in Refs. [4-8]. The droplet behavior on hydrophobic surfaces is studied in Refs. [9, 10]. Boiling of the gas-drop flow is considered in Refs. [11]. At brine evaporation, salt concentration in liquid increases. Desorption processes are accompanied by additional dissolution heat because concentrations of components in solution change. Desorption of aqueous salt solutions is used in desorbers of thermotransformers. One of the way to increase the heat pump efficiency is increasing wall superheating, which allows a reduction in desorption time and an increase in heat transfer coefficient. The diffusion equation is important for salt solutions modeling. If we neglect diffusion, this can lead to overstating the absorption rate more than twice [12]. Thermodynamic and thermal-physical properties of salt solutions are considered in Ref. [13, 14]. An increase in salt concentration leads to an increase in desorption heat and repeated decrease in the desorption rate. There is a strong non-linearity of the density of steam mass flux vs. time [15]. An increase in the heat flux in a salt solution droplet leads to a transitional regime of boiling crisis [16]. Formation of crystalline hydrates in technical devices can lead to crystal plugs and cessation of the mixture flow in the channel. The high-temperature flow of salt solution in minichannels results in formation of emulsion multiphase flow. The crystalline film is formed on the wall. This flow is unstable and the flow is oscillatory in nature [17, 18]. In contrast to the single-component liquids, the Marangoni flow in the salt solution droplets is formed due to the concentration gradient, which affects the bubble microlayer dynamics [19]. The combined effect of boiling and free convection is considered in Ref. [20]. In the last decades, the attention to the construction of the models, taking into account the combined effect of the natural convection in liquid and gas, has increased. An increase in the droplet diameter and Rayleigh number leads to a multiple increase in the evaporation rate for liquids [21, 22].

Thus, droplets desorption of aqueous salt solutions is characterized by continuous decrease in the desorption rate and the partial vapor pressure. The purpose of this study is to carry out the experimental research for the desorption of aqueous salt solutions of LiBr in a wide range of a change in salt concentration with time from 10 to 60 %.

2. Experimental method and analysis

The droplet of aqueous salt solutions of LiBr was located on the metallic horizontal heated cylinder. The cylinder was made from titanium (Fig. 1). All experiments were carried out at the air temperature of 21 °C and air pressure of 1 atm. The initial droplet temperature was equal to the ambient air temperature (21 °C). Relative air humidity was 30-40 %. The cylinder diameter was 80 mm and its height was 65 mm. The wall temperature under the droplet was measured by a thermocouple and this temperature was kept constant automatically. The liquid surface temperature was determined by the thermal imager. The values of initial mass salt concentrations of C_{01} were determined by the densimeters. Current mass salt concentration C_{1i} was determined using the weight methods (Fig. 1). The water mass concentration is C_2=1-C_1.

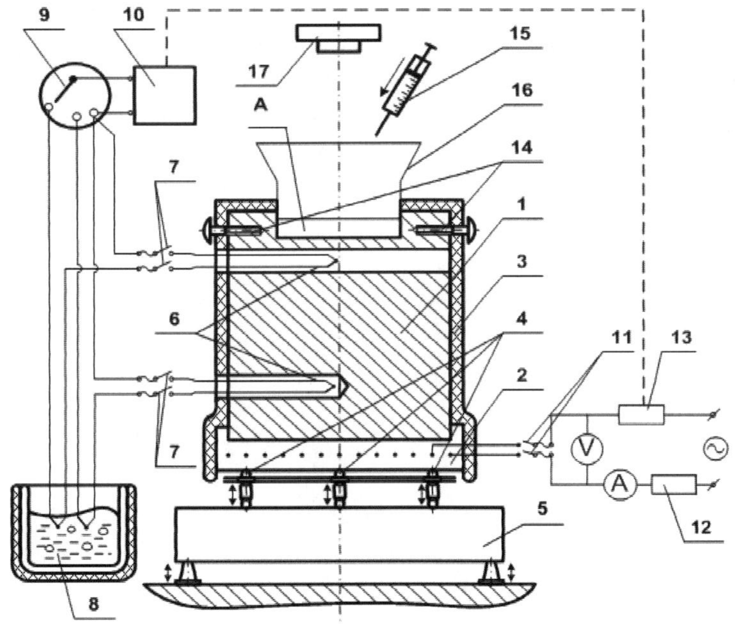

Figure 1. Principle scheme of setup:
A – working heating surface; 1 – titanium cylinder; 2 – electric heater; 3 – multilayer screen thermal insulation; 4 – adjustable supports; 5 – analytical balance; 6 – thermocouples; 7 – flexible detachable element of thermocouple; 8 – Dewar vessel with ice; 9 – thermocouple switch; 10 – millivoltmeter; 11- flexible detachable element of electric wiring; 12 – laboratory thermotrnasformer; 13 – temperature regulator; V and A – voltmeter and ammeter; 14 – puller ear; 15 – metering device; 16 – protecting screen; 17 – video and shooting cameras, thermal imager

The maximum relative error of the mass flow measurements was in the range of 8-10%. With time a change in the interfacial temperature will lead to a shift of the point on the equilibrium curve from (C_0, T_{s0}) to point (C_{si}, T_{si}), at that, the driving force of absorption (desorption) will decrease $\Delta C_{nisos}=C_{si}-C_{s0}$, and ΔC_{nisos} for the non-isothermal case will be significantly less than ΔC_{isos} for the isothermal problem. For the case of isothermal desorption, when the Stefan flow is taken into account (interface moves with velocity V), the expression for the water mass flow density j_2 will take form (1) [12].

$$j_2 = -\frac{\rho D}{1-C_{02}}\left(\frac{\partial C}{\partial y}\right)_{y=0} = \frac{\rho\sqrt{D}(C_{02}-C_{so2})}{\sqrt{\pi t}}\frac{1}{(1-C_{02})} \quad (1)$$

For the infinitely large layer of solution at y → ∞, $T = T_0 = T_w$, $C = C_{02}$. It is also necessary to write down the interface conditions. Desorption heat is spent for liquid heating $r_d j_v = -\lambda\frac{\partial T}{\partial y}$. At that, free convection in gas and liquid is neglected. In a small alteration range of equilibrium water concentration C_{2i} for the interface, temperature dependence of concentration can be presented in the linear form $C_{2si} = k_1\text{-}k_2T_{si}$, where coefficients k are functions of pressure (p_s). Expression (2) for consideration of for non-isothermal nature is proposed in Ref. [12]

$$j_2 = -\frac{\rho D}{1-C_{02}}\left(\frac{\partial C}{\partial y}\right)_{y=0} = \frac{\rho\sqrt{D}(C_{02}-C_{so2})}{\sqrt{\pi t}}f_1, \quad (2)$$

where f_1 is function taking into account non-isothermal nature (desorption heat r), $f_1 = \frac{1}{1+(C_{02}-C_{s02})K\sqrt{Le}}$, $K = \frac{r}{C_p(T_w-T_{s0})}$, $Le = \frac{D}{a}$, where D, a are the coefficients of diffusion and thermal diffusivity in solution, C_p is heat capacity of liquid. The average temperature vs. time for the droplet surface is approximately $\overline{T_s} = \frac{T_{s1}+T_{scr}}{2} = \frac{(T_w-15)+(T_w-7)}{2} = T_w - 11$. The average partial pressure is determined as $\overline{p_{si}} = \frac{p_{s0}+p_{scr}}{2}$, where p_{s0} is initial equilibrium vapor pressure near the liquid surface, corresponding to initial concentration, p_{scr} is vapor pressure before crystallization beginning. Then, C_{2si} corresponds to point $(\overline{T_{si}}, \overline{p_{si}})$. Thus, the nonstationary problem of the equilibrium pressure is reduced to the problem with stationary boundary conditions (p_{si}=const, T_{si}=const). Moreover, this problem can be considered as the quasi-isothermal because T_s increases with time slightly for most of the desorption time.

The mass change is determined by time integration of the flow density $G = F\int J_{i2}\,dt = 2J_2Ft$, F is average area of the droplet free surface, $F=(F_{in}+F_{fin})/2$, F_{in} is initial area of the liquid free surface after droplet placement on the wall, F_{fin} is final area before crystallization. Data for the water mass change for $(\overline{T_{si}}, \overline{p_{si}})$ are shown in Fig. 2.

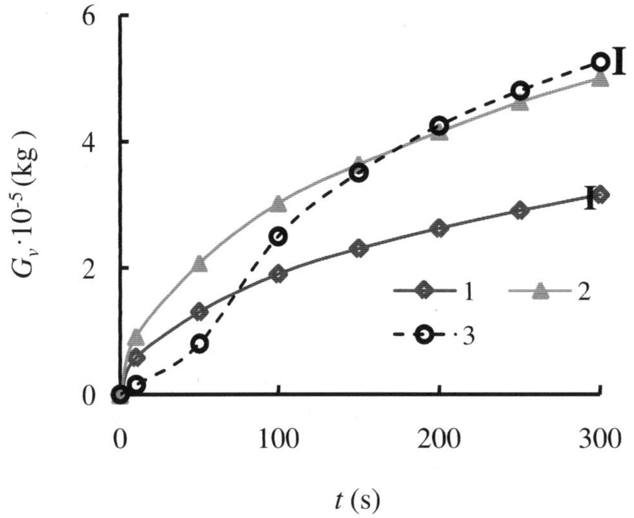

Figure 2. The water mass change (the initial mass salt concentration of LiBr is C_{01} = 55 %, T_w = 85 °C, V_0 = 250 μl): 1 – simulation by (1) without interface transportation (V = 0); 2 – simulation with consideration of V by (1); 3 – experimental data

As it can be seen from Fig. 2, the largest deviation of simulation from the experiment corresponds to the period of droplet heating and with constant concentration $C_{i1} = C_{01} = \text{const}$. This model is not designed to calculate the initial warm-up period. For the next quasi-isothermal interval, the calculation curve (2) is in satisfactory agreement with the experimental data (3).

Conclusion

Thus, the mass flow can be estimated by the simplified procedure, when an unknown point on the equilibrium curve is defined as the average value of the equilibrium vapor pressure. The mass flow, taking into account the velocity of free surface transportation, shows a satisfactory agreement with the experimental data, except the initial period of heating.

Acknowledgments

This work was carried out at the Kutateladze Institute of Thermophysics SB RAS and financially supported by the Russian Science Foundation (project number 15-19-10025).

References

[1] Volkov RS, Kuznetsov GV, Strizhak PA 2016 *Int. J. Heat Mass Transfer* **96** 20-28.
[2] Misyura SY, Nakoryakov VE 2013 *Energy and Fuels* **27** 7089.
[3] Nakoryakov VE *et al* 2013 *J. Eng. Thermophys.* **22** (3) 169.
[4] Brutin D *et al* 2011 *Exp. Therm. Fluid Sci.* **35** 521.
[5] Misyura SY 2016 *Exp. Therm. Fluid Sci.* **70** 389.
[6] Log T 2016 *Appl. Therm. Eng.* **93** 214.
[7] Misyura SY 2017 *Appl. Surf. Sci.* **414** 188.
[8] Misyura SY 2017 *Appl. Therm. Eng.* **113** 472.
[9] Misyura SY 2016 *Exp. Therm. Fluid Sci.* **75** 43.
[10] Feoktistov DV *et al* 2015 *MATEC Web of Conferenses* **23** 01054.
[11] Chernov AA *et al* 2017 *Scientific Reports* **7** 40809.
[12] Nakoryakov VE, Grigoryeva NI Nonisothermal absorption in thermotransformers: Novosibirsk, Nauka, 2010.
[13] Lower H Thermodynamishe und physikalische eigenschaften der wassrigen lifhum-bromid losung: Dissertation, Karlsruhe, 1960.
[14] Conde MR 2004 *Int. J. Therm. Sci.* **43** 367.
[15] Nakoryakov VE *et al* 2012 *Therm. Sci.* **16** 997.
[16] Nakoryakov VE *et al* 2013 *J. Eng. Thermophys.* **22** 1.
[17] Nakoryakov VE *et al* 2011 *J. Eng. Thermophys.* **20** 338.
[18] Nakoryakov VE *et al* 2014 *J. Eng. Thermophys.* **23** 1.
[19] Misyura SYa 2015 *Int. J. Therm. Sci.* **92** 34.
[20] Pavlenko AN, Tairov EA, Zhukov VE, Levin AA, Moiseev MI 2014 *J. Eng. Thermophys.* **23** 173.
[21] Carle F *et al* 2016 *Int. J. Therm. Sci.* **101** 35.
[22] Carrier O *et al* 2016 *J. Fluid Mech.* **798** 774.

IOP Conf. Series: Journal of Physics: Conf. Series **899** (2017) 032014 doi:10.1088/1742-6596/899/3/032014

Dissociation of methane hydrate granules

Misyura S.Y., Donskoy I.G., Morozov V.S.

1. Kutateladze Institute of Thermophysics of the Siberian Branch of the Russian Academy of Sciences, 1. Lavrentiev Ave., Novosibirsk, 630090, Russia,
2. National Research Tomsk Polytechnic University, pr. Lenina 30, Tomsk 634050, Russia

E-mail: misura@itp.nsc.ru

Abstract. The methane hydrate dissociation at negative temperatures and under external pressure of 1 bar is studied experimentally. It is shown that the dissociation rate of the gas hydrate depends on the granule diameter and heat transfer. The dissociation curve has an extremum. The dissociation rate initially increases due to the temperature increase and reaches the maximum value and then sharply falls due to the curvature of the granules. When describing dissociation kinetics of the spherical granules, it is important to take into account the granule size and their composition.

1. Introduction

Currently, increasingly greater attention worldwide is given to investigates on gas hydrates formation and dissociation. The growth and dissociation of gas hydrates [1, 2] are determined by many key factors: the external pressure and temperature, subcooling degree of pellets, structural characteristics of gas hydrate surface and sample defects. There are several main structures of gas hydrates which are most often found in studies: cubic structure (sI); cubic structure (sII); hexagonal structure (sH) [1]. Methane gas hydrates have structure (sI). Gas hydrate deposits contain huge reserves of natural gas and will be promising alternative sources in the coming decades [2]. Theoretical and practical studies on generation and dissociation of artificial methane hydrates are being carried out intensively. Much attention is paid to the problems of transportation and storage [3] of methane granule hydrates in large containers [2, 4]. The tankers carry liquefied natural gas at the temperatures below 111 K. Pelleted gas hydrates can be stored for a long period due to the self-preservation phenomenon at the temperatures of about 253 K. Besides, the decay rates depend on the grain size and the thickness of powder layer. The dissociation at different heating rates were studied in Ref. [5]. It is shown that a slow heating can significantly affect the dissociation intensity. When changing rate of heating from 0.2 to 1 K/h, the dissociation rate may differ by tens of percent. In the literature, there is practically no data on the methane hydrate dissociation at heat flux variations by tens times. High heat fluxes and high dissociation rates are realized by burning methane hydrate [6, 7]. Experimental investigation of flame spreading over pure methane hydrate in a laminar boundary layer were studied in Refs. [8, 9]. Of course, for a turbulent incoming stream, it is necessary to evaluate the effect of external turbulence and the mechanisms of turbulent transport, which are considered in [10,11]. The efficiency of burning the methane hydrate depends on the organization scheme of the oxidant flow. The best method corresponds to the combination of the longitudinal flow and the transverse one through the granule layer [12]. High rates of the decomposition and the gas hydrates formation are realized by the explosive boiling of liquid gases in water at atmospheric pressure. High rates of evaporation and hydration growth are caused by the large area of small gas bubbles formed during boiling [13, 14]. The decomposition kinetics of the methane hydrate granules taking into account the influence of porosity is considered in Refs. [15, 16]. The porosity can be formed at both destruction and formation of gas hydrates [17]. Gas hydrates dissociation of the porous granule layers are considered in Ref. [18]. The technique for obtaining thermal imaging measurements with multiple magnification, used in this paper, was considered in Ref. [19, 20].

Lately, the increased attention is paid to the phenomenon of "self-preservation" [21-24] when abnormally low rates of dissociation is achieved. The phenomenon is important for storage and transportation of methane contained in the hydrates. To date, the self-preservation mechanism and its descriptions are still far from fundamental understanding. Most of the works associate the abnormally stable properties of gas hydrates with formation of a solid ice crust and the appearance of characteristic structures during annealing [25]. It is shown, based on Microfocus X-ray Computed Tomography, that the ice crust increases in the direction from the outer shell to the granule center.

Thus, the study of the influence of porosity and granule size is important both for the development of storage and transportation technologies of methane hydrate and for combustion technologies of environmentally friendly raw materials.

2. Experimental method and analysis

The experimental setup and is presented in [6, 7]. The external pressure was 1 atm. The setup was located in a closed shell, and the outer temperature was kept constant. In the experiment, initial methane mass concentration in the methane hydrate synthesized under the laboratory conditions corresponded to 12.1-12.4 wt.%. Methane hydrate was consistent with sI structure. Formula of the elementary cell is 2D·6T·46H2O, or taking into account the number of edges and faces 2(512)+6(51262). The cell consists of 46 water molecules, two small and six large areas. Characteristics of methane hydrate powder were determined by gravimetric and volumetric methods. The methane concentration values closely match with the weight measurements by gas hydrate dissociation.

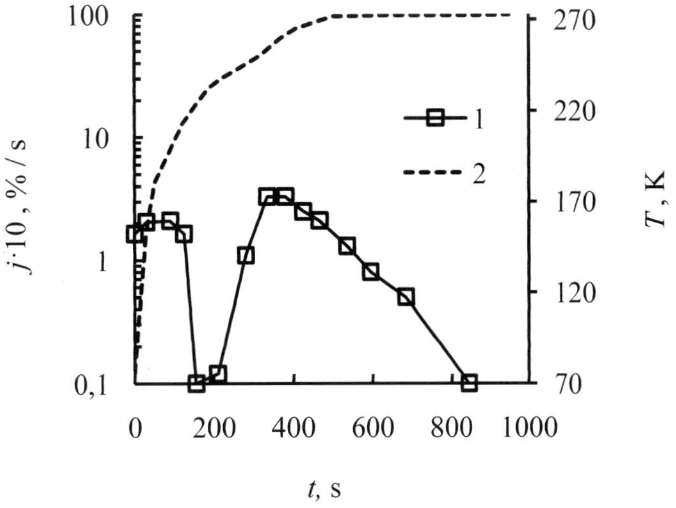

Figure 1. Dissociation rate of methane hydrate. Curve 1 – Experimental data for dissociation rate; Curve 2 – temperature variations of powder surface

Figure 1 shows the methane gas hydrate dissociation rate ($j = \Delta V_i/V_0 \cdot 100\%$) depending on time. Despite the fact that the temperature of the granules continuously increases, the dissociation curve has an extremum. The dissociation rate initially increases due to the temperature increase and reaches an extremum. With increasing temperature, the deviation of the thermodynamic system from an equilibrium increases. Reduction in dissociation rate in the range where $t > 400$ s is associated with the curvature of the granules. As the granule decays, and when the granules (sphere) approach the center, the surface area of the reaction (decay) decreases sharply. Therefore, the amount of methane released from the powder is sharply reduced. Since the decay rate of the methane powder depends on the curvature, i.e. it depends from the average diameter of the granules, it is necessary to take into account the granular composition in the modeling of dissociation.

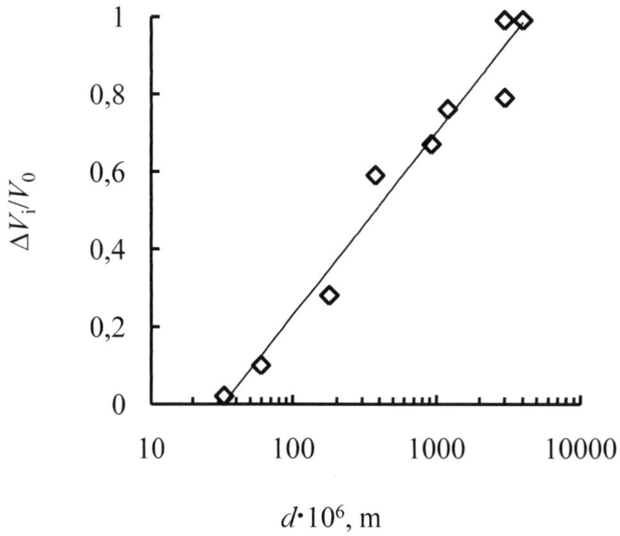

Figure 2. The change in the dissociation rate as a function of the diameter of methane hydrate granules

Fig. 2 represents experimental data of the dissociation rate of methane hydrate versus the granules diameter and for powder temperature $T = 225$K. The decomposition of the gas hydrate begins at a temperature of 193 K. The temperature of the powder continuously increases and when $T = 225$K the total amount of methane removed from the granules is measured by the weight method. The average conditional dissociation rate is characterized by the ratio $\Delta V_i/V_0$, where ΔV_i is the amount of gas released from the powder at T=225K, V_0 is the total initial amount of methane in the powder before the dissociation beginning. As can be seen from the figure, the rate of distraction decreases substantially with the growth of granules diameter. The strong influence of the diameter is probably determined by two main reasons: 1) the effect of heat transfer (as the diameter of the granule grows, the mass grows as R^3, and the heat trsnsfe area as R^2). 2) With increasing diameter, the effect of self-preservation is more pronounced. The phenomenon of "self-preservation" is observed at annealing temperatures ($T > 235$K) when observed abnormally low rates of dissociation.

Conclusion

Thus, the experimental studies of nonisothermal dissociation of methane hydrate were implemented. The significant influence of the granules diameter on the decay rate during dissociation was established. Despite the continuous increase in the temperature of the powder and the increase in the deviation of the methane hydrate from the equilibrium state, the decay curve is essentially nonlinear. The dissociation curve has an extremum. The dissociation rate initially increases due to the temperature increase and reaches the maximum value. Reduction in dissociation rate in the range where $t > 400$ s is associated with the curvature of the granules. As the granule decays, and when the granules approach the center, the decay surface area decreases sharply. Therefore, the amount of methane released from the powder is sharply reduced. The distraction rate of methane hydrate substantially decreases with the growth of granules diameter. Hereby, it is necessary to take into account the granular composition in the modeling of dissociation.

Acknowledgments
Experiments were carried out at the Kutateladze Institute of Thermophysics SB RAS and financially supported by the Russian Science Foundation (project number 15-19-10025). The methodology of calculation and using the software and library programs with respect to researches of the processes of heat exchange intensification at the interphase under the conditions of phase transformations and chemical reactions was carried out within the framework of the development program of the National Research Tomsk Polytechnic University in the project of the leading universities of the world 5-100.

References
[1] Istomin VA, Yakushev VS Gas hydrates in nature - Moscow: Nedra, 1992.
[2] Sum AK *et al* 2009 *Ind. Eng. Chem. Res.* **48** 7457.
[3] Xie Y *et al* 2010 *Appl Energy* **87** 3340.
[4] Javanmardi J *et al* 2005 *Appl. Therm. Eng.* **25** 1708.
[5] Sato H. *et al* 2013 *Chemical Engineering Science* **91** 86.
[6] Nakoryakov VE *et al* 2013 *J. Eng. Thermophys.* **22** (3) 169.
[7] NakoryakoVE *et al* 2013 *J. Eng. Thermophys.* **22** (2) 87.
[8] Maruyama Y *et al* 2013 *Proceedings of the Combustion Institute* **34** 2131.
[9] Nakamura Y *et al* 2009 *Energy & Fuels* **23** 1445.
[10] Lebedev VP *et al* 1993 *Fluid dynamics* **28** 624.
[11] Misyura SY, Nakoryakov VE 2013 *Energy and Fuels* **27** 7089.
[12] Misyura SY 2016 *Energy* **103** 430.
[13] Chernov AA *et al* 2017 *Int. J. Heat Mass Transfer*, **108** 1320.
[14] Chernov A.A *et al* 2017 *Scientific Reports* **7** 40809.
[15] Misyura SY 2016 *Scientific Reports* **6** 30324.
[16] Misyura SY, Donskoy IG 2016 *Chem. Eng. Sci.* **148** 65.
[17] Kuhs WF *et al* 2000 *Geophys. Res. Lett.* **27** 2929.
[18] Misyura SY, Donskoy IG 2017 *Fuel Processing Technol.* **158** 154.
[19] Misyura SY, 2016 *J. of Heat Transfer ASME.* **138**(11) 111501.
[20] Nakoryakov VE *et al* 2012 *Thermal Science* 16(4) 997.
[21] Zhang G, Rogers RE 2008 *Chem. Eng. Sci.* **63** 2066.
[22] Kuhs WF *et al* 2004 *Phys.Chem.Chem.Phys.* **6** 4917.
[23] Takeya S *et al* 2002 *J. of Crystal Growth* **237-239** 379.
[24] Stern LA *et al* 2001 *J. Phys. Chem. B* **105** 1756.
[25] Falenty A, Kuhs WF 2009 *J. Phys. Chem. B* **113** 15975.

IOP Conf. Series: Journal of Physics: Conf. Series **899** (2017) 032015 doi:10.1088/1742-6596/899/3/032015

Hysteresis in a swirling two-fluid flow

I.V. Naumov[1], B.R. Sharifullin[1] and V.N. Shtern[2]

[1]Kutateladze Institute of Thermophysics, SB RAS, Novosibirsk 630090, Russia
[2]Shtern Research and Consulting, Houston, Texas 77096, USA

E-mail: naumov@itp.nsc.ru

Abstract. This experimental study describes a hysteresis—a vivid manifestation of strongly nonlinear flow physics. A sealed vertical cylindrical container of radius 45 mm and height 90 mm is filled with water and sunflower oil. The rotating lid drives swirl and the meridional circulation of both fluids. As the rotation strength Re increases, the oil-water interface rises near the axis, touches the lid at $Re = Re_1$, and moves toward the container sidewall. Then as Re decreases, the interface returns to the axis and separates from the lid at $Re = Re_2 < Re_1$. At each Re from the range, $Re_2 < Re < Re_1$, two different stable steady flow states are observed that is typical of hysteresis. The hysteresis only occurs if a volume fraction of oil is small. The hysteresis disappears as the oil fraction exceeds a threshold, which is around 0.4.

1. Introduction

Swirling, two-fluid flows recently attracted the attention of researchers due to applications in aerial vortex bioreactors, a rapidly developing technology [1-3]. These flows have a number of intriguing features of fundamental interest: numerous topological metamorphoses [4], multiple eddies [5], and thin circulation layers [6-8]. Fujimoto and Takeda [9] visualized a flow of silicon oil and water driven by the rotating lid in a sealed vertical cylindrical container. As the rotation strength Re increases, the interface takes shapes named by the authors as "hump", "cusp", "Mt. Fuji" and "bell". Analogous results were obtained for a flow of soybean oil and water by Tsai et al. [10], who also detected vortex breakdown in the oil part of flow.

Our study explores one more strongly nonlinear phenomenon—hysteresis—in a confined two-fluid flow. A hysteresis can be related to an inverse bifurcation, e.g., in a plane channel flow [11]. A hysteresis, composed by fold bifurcations, is described by the analytical solutions [12-14] of the Navier-Stokes equations. These solutions model jump transitions observed in tornadoes [15]. Hysteretic transitions are typical of strongly swirling flows. The experiment [16] revealed a hysteresis in the appearance and disappearance of time oscillations in a single-fluid flow driven by a rotating disk in a sealed container. The experimental [17] and numerical [18] studies described hysteresis in annular swirling jets.

A hysteresis also can be a capillary phenomenon [19]. However, we have found no study of hysteresis in a two-fluid swirling flow in the literature. This work partially fills this gap and describes a hysteresis related to the capillary interaction of the sunflower- oil-water interface with the rotating lid in a vertical sealed cylindrical container.

2. Experimental set up

Figure 1(a) is a schematic of experimental set up showing the cylindrical container (1), CCD camera (2), stepping motor (3), LED 250W lamp (4) and reflecting white screen (5). The apparatus was housed in a darkened room, with lighting provided by a LED 250W lamp reflected off a white sheet of paper. The photos of interface were shot by CCD 6-megapixel camera CASIO EX-F1. The cylindrical container is made of transparent optical glass with the inner radius 45.75 mm and height 126 mm. The lid of radius R = 45 mm can be positioned at a varying height h and is set here at h = 2R = 90 mm. The container is filled with tap water ($0 < z < h_w$, dark in figure 1) and sunflower oil ($h_w < z < h$, light in figure 1); z is the distance from the container bottom and $h=h_w+h_o$.

At the room temperature 22.6°C, the fluid densities and kinematic viscosities are $\rho_w = 1000$ kg/m^3, $\rho_o = 920$ kg/m^3, $\nu_w = 1$ cSt, and $\nu_o = 55$ cSt. Hereafter subscripts "w" and "o" denote "water" and "oil". The oil-water interface tension is $\sigma_{ow} = 26$ mN/m. The container is still except the lid which can rotate with angular velocity ω driven by the stepping motor. The rotation strength is characterized by the Reynolds number $Re = \omega R^2/\nu_o$; Re = 100 corresponds to about 0.5 RPS. For the Re range here considered, the variation in ω during a revolution is less than 0.3%.

The interface height at the axis was measured by a laser beam and a precise scale as figure 1(b) illustrates. A semiconductor laser (with power 250 mW and wavelength of 684 nm) was fixed on the coordinate device providing movement along z-axis with a step of 0.1 mm. The collimated laser beam was established horizontally. The beam diameter is 0.1 mm at the container axis providing the measuring accuracy about 0.1 mm. As the beam meets the interface it reflects. The laser beam was shifted up starting from the bottom along z-axis with a step of 0.1 mm until the reflection of laser beam from the interface terminates and the beam becomes tangent to the interface, touching the interface top.

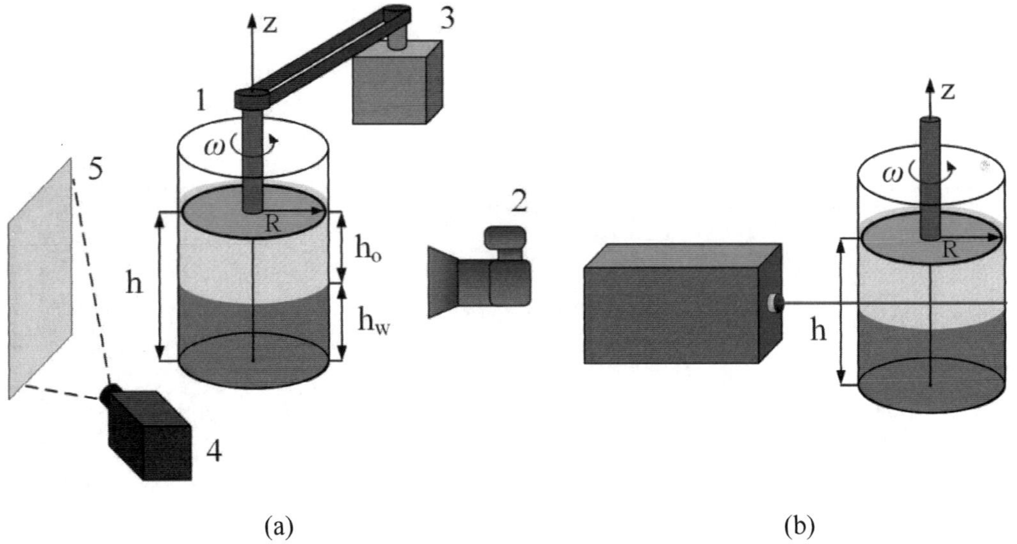

Figure 1. Schematic of experimental set up: (a) visualization, (b) laser measuring of the interface height at the axis.

3. Visual hysteresis observation

The experimental results are presented here at $h_o/R = 0.1, 0.25, 0.5$ and 0.75. For larger h_o, the interface does not reach the rotating lid because the flow becomes time-oscillating, the interface breaks, and water mixes with oil.

At rest, the interface is flat and horizontal. As angular velocity ω increases, the oil-water interface deforms up near the axis and down near the sidewall as typical of water spouts. Then, as ω further increases the interface touches the lid and water spreads over the near-axis part of lid. As ω further increases, the interface breaks and water mixes with oil. In order to detect hysteresis phenomena, the measurements were repeated by decreasing Re. As we started to decrease ω before the interface breaks, the water separates from the lid at ω value being less than that at which water touches the lid as Re increases.

The experimental results for the dependence of characteristic Re values on volume fraction of the sunflower oil h_o/R are presented in table 1. The Reynolds number, at which the interface touches the rotating lid at the axis, is denoted as Re_1. The Reynolds number, at which the flow becomes unsteady, is denoted as Re_u. If Reynolds number decreases before Re_u, the Reynolds number, at which the interface detaches from the lid, is denoted as Re_2. The hysteresis range is $\Delta Re = Re_1 - Re_2$. In the case, $h_o = 0.75R$, the hysteresis range is small and uncertainty in the difference ΔRe is around the step in Re. The steady state loses its stability for sufficiently high Re except for $h_o \geq 0.75R$ where the flow becomes time-oscillating since Re_u is smaller than Re_1 and Re_2. For this reasons this experimental investigation focuses on the volume fraction of the sunflower $h_o/R \leq 0.75$.

Table 1. Dependence of Re values, characterizing hysteresis, on sunflower oil volume fraction h_o/R.

h_o/R	Re_1	Re_2	ΔRe	Re_u
0.1	170	115	55	440
0.25	266	226	40	456
0.5	465	445	20	570
0.75	675	670	5	660

Figure 2 shows photos illustrating the interface shapes at $h_o = 0.75$. As Re increases, first the hump shape develops as figure. 2(a) illustrates at Re = 400. The bell shape develops as figure. 2(b) illustrates at Re = 575. This development is similar to that described in the literature [9, 10]. The difference starts at Re = 675 as the interface touches the rotating lid at the axis as figure 2(c) illustrates. Then the water spreads over the lid and the interface breaks.

(a) (b) (c)

Figure 2. Interface shapes at $h_o = 0.75R$ and Re = 400 (a), Re = 575 (b), Re = 675 (c).

Figure 3 shows photos illustrating the interface shapes at Re = 300. Figures 3(a), 3(b) illustrate the hump shape at $h_o = 0.75$ and 0.5. This development is similar to that described in the literature [9, 10]. The difference starts at $h_o = 0.25$ when the interface touches the rotating lid at the axis and the water spreads over the lid as figure 3(c) illustrates. This interface is close to the "Mt. Fuji" shape [9], but here its flattop part is attached to the lid. At $h_o = 0.1R$, the interface shape is close to that at $h_o = 0.5R$.

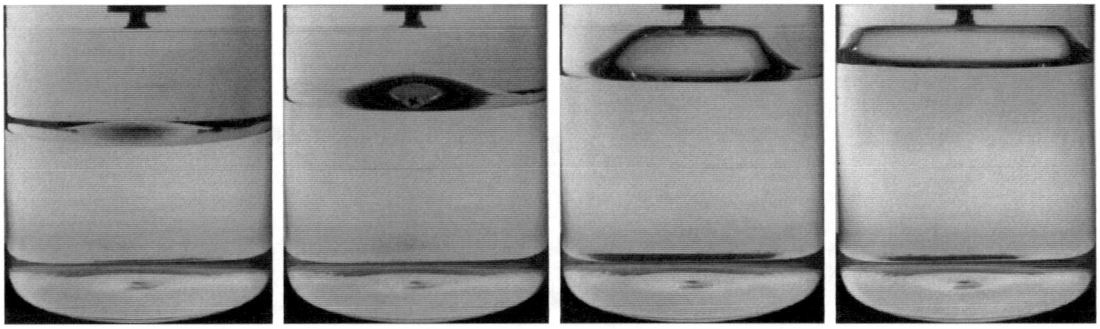

(a) (b) (c) (d)

Figure 3. Interface shapes at Re = 300 and h_o = 0.75 (a), 0.5 (b), 0.25 (c), 0.1 (d).

Figure 4 presents photos illustrating the moment when the interface touches the rotating lid and the established steady state just after the detachment. Figure 4(a), 4(b) shows the interface at h_o = 0.5R and Re = 465 as Re increases, Re = 445 as Re decreases. Figure 4(c), 4(d) shows the interface at h_o = 0.1R and Re = 170 as Re increases, Re = 115 as Re decreases.

(a) (b) (c) (d)

Figure 4. Interface shapes at h_o = 0.5R and Re_1 = 465 (a), Re_2 = 445 (b) and at h_o = 0.1R and Re_1 = 170 (c), Re_2 = 115 (d).

4. Summarized hysteresis results

Figure 5 characterizes the hysteresis by presenting the dependence of the interface height at the axis, h_i, on the rotation speed, Re. First, Re was increased from 0 (the rest state) with an increment ≤ 25. The results are presented by filled symbols in figure 5. Next, Re was decreases with the step down being $\leq$ 25. The results are presented by empty symbols in figure 5.

Figure 6 depicts the dependence of hysteresis characteristics on the oil volume fraction, h_o/h, where $h_o = h - h_w$ is the depth of oil layer at rest; see figure 1. The symbols depict the measured values and the curves are fitting approximations. In figure 6(a), the solid curves show the upper (Re_1) and lower (Re_2) hysteresis values of the Reynolds number while the dashed curve corresponds to the stability boundary: the flow becomes unsteady above this curve. It is important that the stability and hysteresis curves intersects near h_o/h = 0.375 (h_o = 0.75R) in figure 7(a). For h_o/h > 0.375, Re_u is smaller than Re_1 and Re_2.

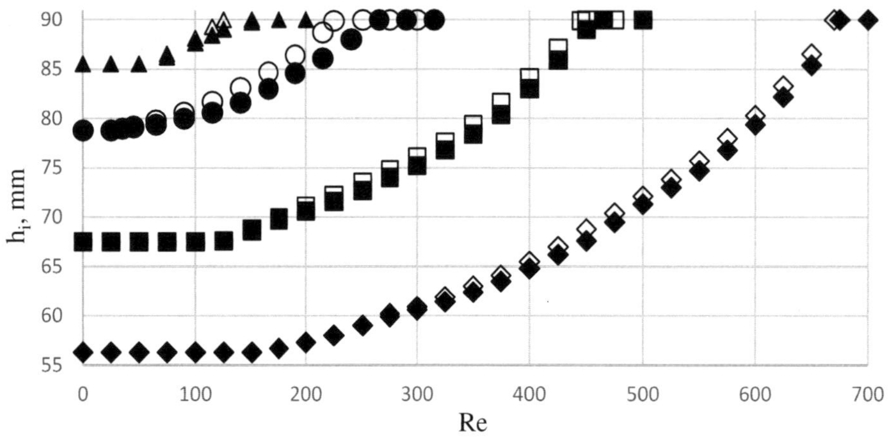

Figure 5. Dependence of interface height h_i on the rotation speed Re as Re increases (filled symbols) and decreases (empty symbols) at $h_o = 0.75R$ (rhombuses), 0.5R (squares), 0.25R (circles) and 0.1R (triangles).

Figure 6(b) depicts the width of hysteresis range ΔRe, where the square symbols present the measurement results and the curve depicts the fitting parabolic relation, $\Delta Re = 132(h_o/h)^2\text{-}215(h_o/h)+65$. This relation helps predict important hysteresis characteristics which are not easy to measure: the maximal hysteresis range, $\Delta Re_{max} = 65$, as h_o/h tends to zero, and the oil fraction, $h_o/h = 0.402$ (or h_o= 0.804R), at which the hysteresis disappears (ΔRe diminishes down to zero).

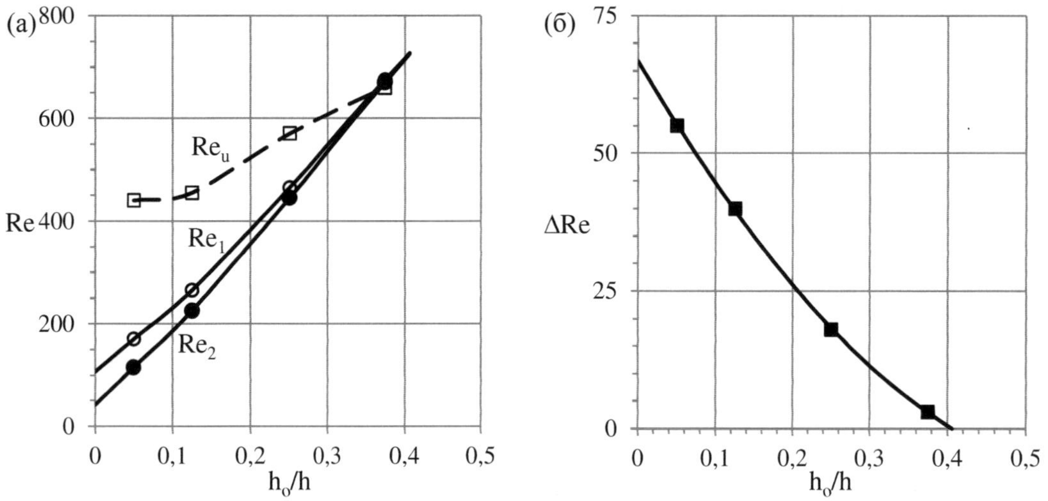

Figure 6. Dependence on the oil volume fraction (h_o/h) of (a) the critical (Re_u) and the hysteresis upper (Re_1) and lower (Re_2) values of the Reynolds number and (b) the hysteresis range ΔRe.

5. Conclusion - Physical reason of hysteresis

This experimental study of a hysteresis was performed in a sealed vertical cylindrical container filled with water and sunflower oil. The rotating lid drives swirl and the meridional circulation of both fluids. As the rotation strength Re increases, the oil-water interface rises near the axis, touches the lid and moves toward the container sidewall. Then as Re decreases, the interface returns to the axis and separates from the lid at a smaller Re value. At each Re two different stable steady flow states are observed that is typical of hysteresis. The hysteresis only occurs if a volume fraction of oil is small.

The physical reason of the hysteresis is the capillary effect. To detach the water from the lid requires a remarkable effort similar to the sufficient weight required for detaching water droplets from a wet ceiling. Here the rotating lid generates the centrifugal force, which pushes the oil toward the sidewall near the lid thus developing the oil meridional circulation in addition to the swirl. The oil converges toward the axis near the interface and goes upward near the axis. This oil motion entrains the adjacent water, acts against gravity, and raises the interface near the axis — the development typical of water spouts [6, 20]. As the rotation speeds up, this effect becomes sufficiently strong for the interface to reach the lid at the axis. Then the centrifugal force helps water to spread over the lid.

To detach the interface from the lid in the reversed process, the water wetting requires a weaker rotation and therefore stronger gravity contribution than those in the direct process. This results in the hysteresis. The hysteretic range of rotation strength decreases as the oil volume fraction increases because the oil motion decays as the distance from the lid increases. Therefore, the entrainment effect becomes weaker while the gravity action becomes stronger at the same rotation speed. This explains why the hysteresis disappears as the oil volume fraction exceeds a threshold which is 0.402 here.

6. Acknowledgments

Research was supported by the Russian Science Foundation (Project № 14-29-00093).

References

[1] Liow K Y S, Thouas G, Tan B T, Thompson M C and Hourigan K 2008 *IFMBE Proceedings* **23** pp.1672-75
[2] Liow K Y S, Tan B T, Thouas G and Thompson M C 2009 *Modern Physics Letters* B **23** pp 121-7
[3] Lo Jacono D, Nazarinia M and Brøns M 2009 *Phys. Fluids* **21** p 111704
[4] Balci A, Brøns M, Herrada M A and Shtern V N 2015 *Fluid Dyn. Res.* **47** p 065503
[5] Herrada M A and Shtern V N 2014 *J. Fluid Mech.* **744** pp 65-88
[6] Herrada M A, Shtern V N and Lopez-Herrera J 2013 *Phys. Fluids* **25** p 093604
[7] Brady P T, Herrman M and Lopez J M 2012 *Phys. Rev.* E **85** p 016308
[8] Brady P T, Herrman M and Lopez J M 2012 *Phys. Rev.* E **85** p 067301
[9] Fujimoto S and Takeda Ya 2009 *Phys. Rev.* E **80** p 015304(R)
[10] Tsai J C, Tao C Y, Sun Y C, Lai C Y, Huang K H, Juan W T and Huang J R 2015 *Phys. Rev.* E **92** p 031002(R)
[11] Chen T C, Joseph D D 1973 *J. Fluid Mech.* **58** pp 337-51
[12] Shtern V and Hussain F 1993 *Phys. Fluids* A **5** pp 2183-95
[13] Shtern V and Hussain F 1996 *J. Fluid Mech.* **309** pp 1-44
[14] Shtern V 2012 *Counterflows: Paradoxical Fluid Mechanics Phenomena* (New York: Cambridge University Press)
[15] Burggraf O R and Foster M R 1977 *J. Fluid Mech.* **80** pp 685-704
[16] Sorensen J N, Naumov I V and Mikkelsen R 2006 *Exp. Fluids* **41** pp 425-40
[17] Vanierschot M and Van den Bulck E 2007 *Exp. Thermal Fluid Sci.* **31** pp 513-24
[18] Vanierschot M and Van den Bulck E 2007 *Int. J. Numer. Meth. Fluids* **54** pp 313-24
[19] Eral H B,'t Mannetje D J C M and Oh J M 2013 *Colloid & Polymer Sci.* **291** pp 247-60
[20] Carrión L, Herrada M A and Shtern V N 2017 *Phys. Rev. Fluids* **2** p 024702

Bubbly free and impinging jets under forced flow conditions: experimental study by means of PIV/PFBI

Mikhail Yu Nichik[1], **Konstantin S Pervunin**[1,2] **and Dmitriy M. Markovich**[1,2,3]

[1] Department of Physics, Novosibirsk State University, 2, Pirogov Str., Novosibirsk, 630090, Russia
[2] Kutateladze Institute of Thermophysics, Siberian Branch of the Russian Academy of Sciences, 1, Lavrentyev Ave., Novosibirsk, 630090, Russia
[3] Institute of Power Engineering, Tomsk Polytechnic University, 30, Lenin Ave., Tomsk, 634050, Russia

pervunin@itp.nsc.ru

Abstract. Jet flows are extensively used in various practical applications. Presently, the development of technical equipment where jets are employed is connected with the improvement and optimization of different methods of flow control. In the paper, an experimental investigation of the turbulent structure of forced bubbly free and impinging jets was carried out by means of PIV and PFBI techniques. PIV was applied to measure velocity distributions and turbulent characteristics in the continuous phase, while PFBI approach was applied to visualize bubbles in the flow and evaluate their sizes. The flow was studied at the Reynolds number of 12,500 and three void fractions $\beta = 0$, 1 and 2% for forced conditions St = 0.5. The mean air bubble diameter was estimated to be roughly 0.8 mm for all β. It was revealed that in the free jet the air bubbles and flow pulsations reduces substantially the longitudinal dimension of the jet core. In two-phase flow with forcing distribution of turbulence kinetic energy was similar to one-phase case but maximum value was two and a half times higher then for one-phase unforced jet. In the impinging jet flow, the bubbles produced a maximum of the turbulence kinetic energy near the wall, which increased two and a half times in forced conditions.

1. Introduction

Turbulent shear flows including jets are an inherent attribute of the modern world. They are met in everyday life and are used in many technological processes due to their relative simplicity of implementation and high efficiency. The range of application of jet flows is very wide. Free jets are found in air- and spacecraft engines, water propulsors, air-atomizing nozzles, evaporators and other industrial equipment and have been already studied in detail. Impact jets are primary utilized in heat exchangers and are less studied today. At present, the most promising approach to increase the performance of various devices where the jets are employed is application of different methods of flow control. This, in turn, requires the elaboration of these methods and determination of optimal parameters of manipulation, which is a quite challenging issue.

All of the methods are conventionally divided into two main classes: passive and active. Passive methods are commonly based on changing the flow geometry and do not require energy supply from the outside to the system. Distinctive features of the passive methods are the absence of forced phase transformations in the considered flow region, the conservation of mass of the operating liquid at the inlet and outlet of the test section and the equilibrium distribution of the dispersed phase in the supply duct when the operating liquid is two-phase. The main idea of all active methods is to manage the flow by violating one of the above mentioned conditions, for which an external energy is necessary to be added into the system.

One of the most well-known methods of the active jet control is imposition of periodic flow pulsations. If this is the case, when the jet escapes into the free space, a mixing layer characterized by high transverse velocity gradients in which large-scale vortices, so-called coherent structures, develop and distributions of the hydrodynamic flow characteristics change substantially [1]. One of the passive methods that can be used to manipulate jet flows is an injection of a dispersed gas phase. Variation of the dispersed phase parameters (the volume gas fraction and its size distribution) can significantly modify the hydrodynamics of the whole flow. Interaction of the continuous and dispersed phases is, as a rule, of a complex nonlinear character because of their mutual influence: the local nonequilibrium turbulence in liquid affects the bubbles trajectories; on the other hand, the bubbles dynamics significantly impacts the turbulent flow structure.

In the paper, bubbly free and impinging round jets are studied experimentally at the Reynolds number Re = 12,500 under forced conditions with the Strouhal number St = 0.5 for several volume gas fractions β = 0, 1, 2% and the same mean bubble size of approximately D_B = 0.8 mm by means of a combination of the time-resolved particle image velocimetry (PIV) and planar fluorescence for bubbles imaging (PFBI) approaches. We report here on PIV measurements of instantaneous velocity fields, with a special emphasis put on the effect of the dispersed phase and produced flow-rate pulsations on distributions of turbulence kinetic energy.

2. Experimental conditions and measurement techniques

The bubbly free and impinging round jet flows were reproduced in the hydrodynamic rig in Kutateladze Institute of Thermophysics SB RAS. The experimental setup is of closed type and continuous operation. Its test section is made of organic glass with the following dimensions (HxWxL): 425x300x300 mm (see details in [2]). The water temperature was maintained constant at 30±0.1 °C in the test section by means of a thermostatic regulator. The jets were formed by a round contraction nozzle with the inner diameter of D_N = 15 mm at its outlet. In case of the impinging jet, the distance between the nozzle edge and the impingement surface was $H = 3D_N$. The Reynolds number based on D_N and the superficial mixture velocity $V_0 = 4\cdot(Q_W+Q_A)/(\pi\cdot D_N^2) = 0.663$ m/s, where Q_W and Q_A are the volume flow rates of the operating liquid (distilled water) and air respectively, equaled to Re = 12,500 in case of the one-phase flow. The water flow rate was measured by an ultrasonic flowmeter Kamstrup MULTICAL 402 with the measurement error of 2%. The air temperature, pressure and flow rate were measured by a TSI thermal mass flowmeter (model 4140) with the measurement uncertainties of 0.1 °C, 0.1 kPa and 2%, respectively. In order to introduce bubbles of a close-to-monodisperse size distribution into the flow, a bubble generator of a special construction [3] was used. Different gas-water flow-rate ratios $\beta = Q_A/(Q_W+Q_A)$ were achieved by changing Q_A using a needle valve. In experiments, β possessed three values: 0, 1 and 2%. The mean diameter of the air bubbles D_B was approximately 0.8 mm for all β. In order to impose external pulsations on the water flow rate, an electromagnetic vibrator was utilized. The vibrator was connected to the water supplying duct by a sylphon bellows upstream of the bubble generator. The pulsation frequency was set by a Berkeley Nucleonics Corporation pulse/delay generator (model 575) and equaled to f = 22 Hz (Strouhal number St = fD_N/V_0 = 0.5). As found in previous tests, this frequency is of the best jet receptivity. The pulsation amplitude was changed using an amplifier and equaled to $0.05Q_W$ as it must be small relative to the water flow rate.

In order to implement a planar fluorescence for bubbles imaging (PFBI) approach [4, 5], the operating liquid was mixed with Rhodamine 6G as a fluorescent dye, the concentration of which was quite low (about 20 μg/l). In order to visualize the bubbly flow and measure the flow velocity, a PIV-system with a high temporal resolution consisting of a pulsed Nd:YAG Photonics Industries DM-532-50 laser (wavelength 532 nm, repetition rate 15 kHz, pulse duration 190 ns, pulse energy 7 mJ), Photron FASTCAM SA5 CMOS-camera (digit capacity 12 bits, resolution 1,024x1,024 pixels, acquisition rate 7 kHz) equipped with a Nikon AF Micro-Nikkor 60 mm f/2.8D lens and a low-pass optical filter (transmission edge at 570 nm) and Berkeley Nucleonics Corporation pulse/delay generator (model 575) for external synchronization was employed. Fluorescent seeding particles filled with Rhodamine B (fraction 1-20 μm) produced by Microparticles GmbH of polymethyl methacrylate were added into the operating liquid for PIV measurements. Raw data (a series of 10,000 snapshots for each regime) gathered continuously during 5 seconds were processed using a PC with "ActualFlow" software. Velocity fields were calculated using an iterative cross-correlation algorithm with a continuous window shift, deformation and 75% overlap of interrogation windows. Details of the data processing procedures used can be found in [2].

3. Results and discussion

In this section, some selected results of the experimental study are presented for the regimes listed in Table 1. Below, we demonstrate an original instantaneous image from the high-speed PFBI visualization followed by a discussion of distributions of the mean streamwise velocity and turbulent kinetic energy (TKE) for the continuous phase extracted from the PIV measurements. As the jet flows are axisymmetric, we use the cylindrical coordinate system for the data presentation. The full TKE of the bubbly axisymmetric jets was estimated on the assumption that $\langle \upsilon_r^2 \rangle \approx \langle \upsilon_\varphi^2 \rangle$:

$$K = (\rho/2)\left(\langle \upsilon_z^2 \rangle + \langle \upsilon_r^2 \rangle + \langle \upsilon_\varphi^2 \rangle\right) = \rho\left(\langle \upsilon_z^2 \rangle/2 + \langle \upsilon_r^2 \rangle\right),$$

where υ_z, υ_r and υ_φ are the axial, radial and azimuthal components of the fluctuating velocity, respectively. This assumption is valid as the jet flows under consideration have a preferential direction along their axes.

Table 1. Reference designation of the flow regimes considered in the research

β	free jet		impinging jet	
	unforced	forced	unforced	forced
0%	F0U	F0F	I0U	I0F
1%	F1U	F1F	I1U	I1F
2%	F2U	F2F	I2U	I2F

3.1. *Bubbly free jet*

Figure 1 shows an example of the PFBI visualization of the bubbly free jet under the forced conditions. As seen, most of the bubbles are collected in the cores of toroid-shaped coherent structures (CS) – large-scale vortices – that originate regularly due to the periodical flow excitation. Patterns of these bubbles are highly overlapped even if the volume gas concentration is relatively low and it is extremely difficult to distinguish individual bubbles (Figure 1). That is why we do not go beyond an analysis of the hydrodynamic structure of the jet flows without taking into consideration the dynamics of the dispersed phase.

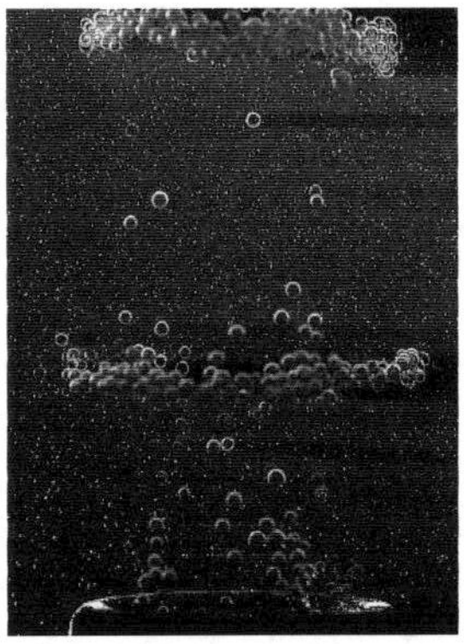

Figure 1. An instantaneous photograph of the initial region of the flow captured by the PFBI technique for Case F1F. The nozzle outlet is visible at the bottom. The flow is directed upward. Small bright spots in the image are the fluorescent tracers employed for the PIV approach.

Downstream evolution of the mean velocity of the carrier phase in the forced jet flow is presented in Figure 2 in comparison with the one for the unforced flow conditions. In the single-phase flow (Cases F0U and F0F, Figure 2-a), the mean velocity is distributed uniformly when the jet escapes the nozzle ($z/D_N = 0$). Downstream, the profiles slope, the flow velocity drops and the jet expands radially. This process is found to occur faster under the forced conditions. When the dispersed phase is injected, the mean velocity at the nozzle edge increases by 4% and 6% for F1U and F2U, respectively (see Figure 2-b and 2-c). In the two-phase flow, the trend of the downstream jet widening and velocity redistribution is more pronounced. Unlike the unforced jet that broadens out gradually downstream with a certain rate almost independent of β so that the velocity profile bend in the mixing layer eventually disappears, the flow excitation leads to that the jet grows quickly in the radial direction right after leaving the nozzle up to the distance of $1D_N$ from its edge (Case F0F) and $0.5D_N$ when the flow is two-phase (Cases F1F and F2F). Downstream, once this threshold cross-section is overcome, the forced jet starts to entrain the ambient (initially stagnant) liquid up to the upper border of the measurement region, with the velocity profile bend retained (Figure 2-b and 2-c).

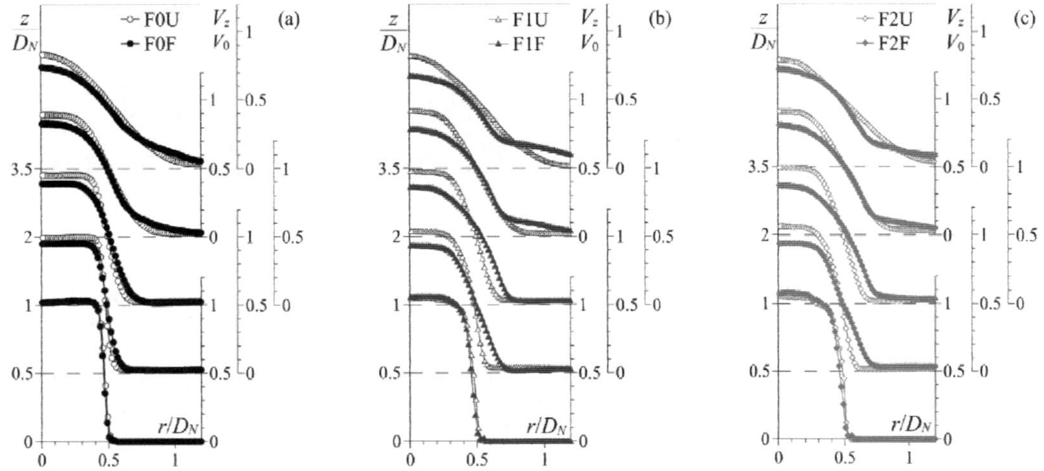

Figure 2. Profiles of the streamwise component of the mean velocity of the continuous phase in five cross-sections for the forced and unforced conditions and different gas-water flow-rate ratios β: (a) 0%, (b) 1% and (c) 2%. The reference point ($z/D_N = 0$, $r/D_N = 0$) is at the centre of the nozzle outlet.

In Cases F1U and F2U, the jet core appears to be substantially reduced as compared to Case F0U (Figure 2). For instance, at $z/D_N = 0.5$ its width does not exceed $0.6D_N$ for the both two-phase flows whereas, in the single-phase jet, it equals to $0.9D_N$. This occurs because the bubbles induce the flow turbulization and cause its slowdown over the jet perimeter where their concentration is relatively high. When the forcing is applied (Case F0F, Figure 2-a), the jet core breaks down beyond the measurement region at $z/D_N \approx 5.5$ but decays faster than in Case F0U ($z/D_N \approx 6$). This is likely linked with that the CSs which become more intensive when the jet is excited increase the mixing rate, bringing larger amounts of the slowed ambient liquid into the flow and throwing away the accelerated liquid from the jet core. In Cases F1F and F2F, the jet core vanishes completely at $z/D_N = 0.7$ as a result of CS extension due to the bubbles.

As seen in Figure 3, the TKE fields depend considerably on both governing parameters: the dispersed phase concentration and external excitation. These parameters have an opposite influence on the TKE distributions. Forcing leads to an intensification of the velocity fluctuations in the mixing layer, with the TKE maximum shifted to the cross-section $z/D_N = 1$ and increased to $K = 0.09V_0^2$ (cf. Cases F0U and F0F, Figure 3-a, 3-b and 3-e). Moreover, in the excited jet (Case F0F) TKE starts to grow from the very nozzle edge while it preserves its minimal values ($K \approx 0.01V_0^2$) until $z/D_N \approx 0.3$ when no forcing is imposed (Case F0U), afterwards it increases until $z/D_N \approx 1$ ($K = 0.045V_0^2$) and then slowly drops downstream (Figure 3-e). Injection of the dispersed phase results in a suppression of the turbulent fluctuations (cf. Cases F0U and F2U, Figures 3-a and 3-c). In the two-phase jet (Case F2U), the maximum of the TKE distribution moves towards the jet axis to the point ($r/D_N = 0.4$, $z/D_N = 0.35$) where $K = 0.04V_0^2$ (Figure 3-c) that is half the one for Case F0U (Figure 3-e). Downstream ($z/D_N > 1$), it remains almost constant (Figure 3-e). In Case F2F, the TKE maximum turns out to be in the outer region of the mixing layer in the point ($r/D_N = 0.57$, $z/D_N = 0.86$) and becomes by 30% higher ($K = 0.12V_0^2$) compared to Case F0F and 3 times higher than in Case F2U. The rise of the turbulent fluctuations in the latter case can be probably connected with that when the flow rate pulsations are applied the bubbles size can also oscillate due to alternations of the local pressure field, which, in turn, can act as an additional source of the TKE growth.

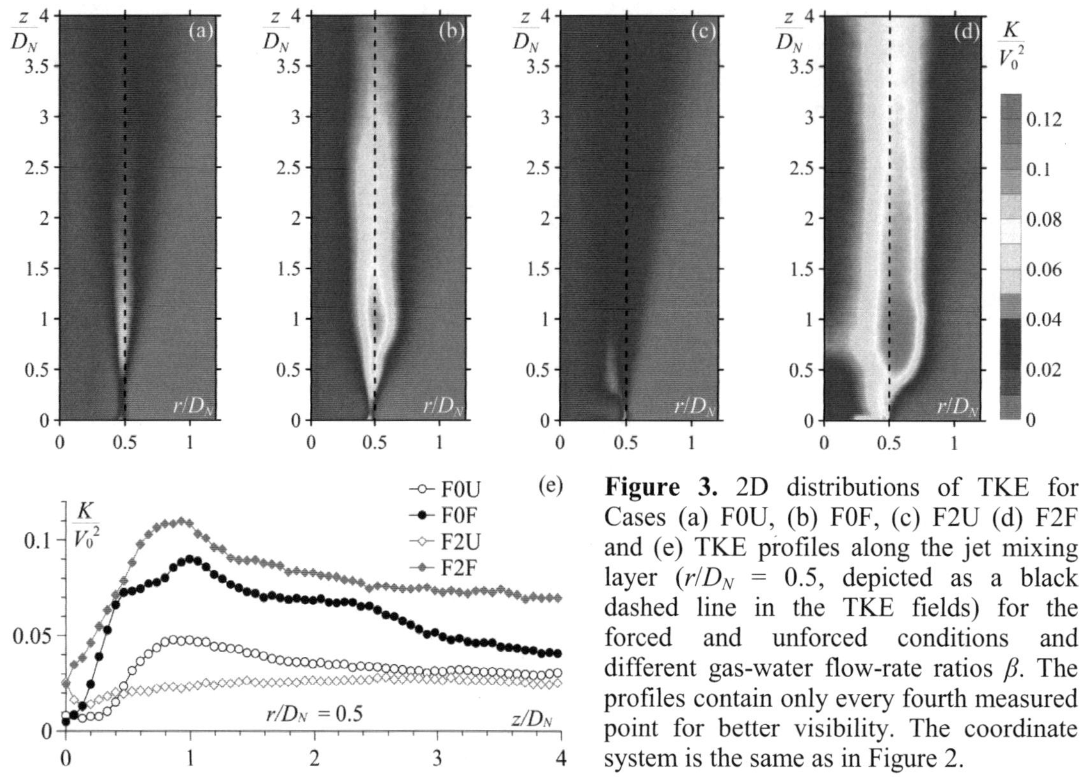

Figure 3. 2D distributions of TKE for Cases (a) F0U, (b) F0F, (c) F2U (d) F2F and (e) TKE profiles along the jet mixing layer (r/D_N = 0.5, depicted as a black dashed line in the TKE fields) for the forced and unforced conditions and different gas-water flow-rate ratios β. The profiles contain only every fourth measured point for better visibility. The coordinate system is the same as in Figure 2.

3.2. Bubbly impinging jet

In the initial region of the impinging jet, the TKE fields qualitatively look similar to those for the free jet (cf. Figures 3 and 4 for the same flow conditions when $z/D_N < 2$). When approaching the wall ($2 < z/D_N < 3$), the TKE distributions noticeably differ for the two test arrangements – free and impinging jets. Along the impingement surface, the mean velocity in Case I0U initially increases progressively with r/D_N (not shown) because of a higher pressure near the stagnation point but then reduces gradually downstream (for $r/D_N > 0.8$) by the law of $1/r$ due to the flow expansion. The turbulent fluctuations start to grow in the boundary layer at the wall and in the mixing layer below the surface (Figure 4-a). When the single-phase jet is excited (Case I0F), TKE rises up to $K = 0.055V_0^2$ at r/D_N = 1.35 and decreases downstream, reaching the value of $K = 0.037V_0^2$ at $r/D_N = 2$, which is close to the one in Case I0U (Figure 4-b and 4-e). The main difference between the single- and two-phase impinging jets (Cases I0U and I2F) is that the bubbles cause an appearance of the maximum in the TKE fields at $r/D_N = 0.7$ ($K = 0.057V_0^2$) (Figure 4-c and 4-e). Downstream, the turbulent fluctuations decay so that, for $r/D_N > 1.2$, no considerable discrepancy between the one- and two-phase jets is observed. In Case I2F, the TKE amplitude significantly grows (up to $0.13V_0^2$ at $r/D_N = 0.85$) due to the same reason as in the initial region of the free jet (Case F2F). Downstream, TKE diminishes to $K = 0.05V_0^2$ at $r/D_N = 2$ (Figure 4-e), which is by 40% higher than in Case I0U.

4. Summary

Periodically excited (St = 0.5) bubbly free and impinging jets with volume gas fractions up to $\beta = 2\%$ and the mean bubble diameter of 0.8 mm were studied by means of the PIV and PFBI methods. In case of the two-phase free jet, the application of the flow rate pulsations was revealed to reduce substantially the streamwise dimension of the jet core and cause an intensification of the turbulent fluctuations in the mixing layer up to about three times as compared to the single-phase unforced flow. In the two-phase impinging jet, the bubbles produce a maximum in the distributions of turbulent kinetic energy close to the wall at the distance of 70% of the nozzle diameter from the stagnation point. When the pulsations are imposed, this maximum rises about two times and somewhat shifts downstream from the stagnation point.

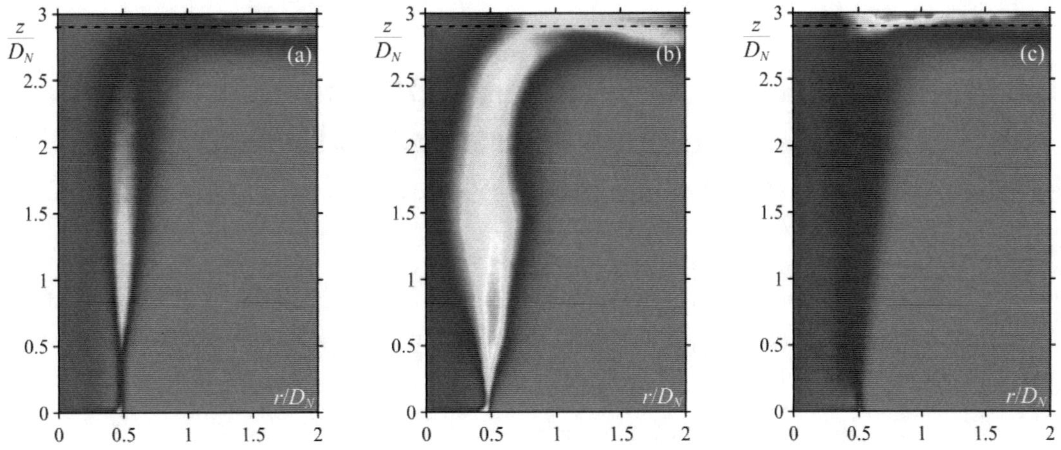

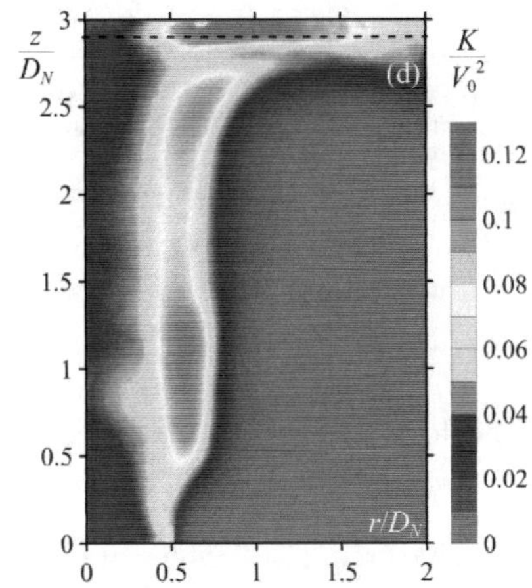

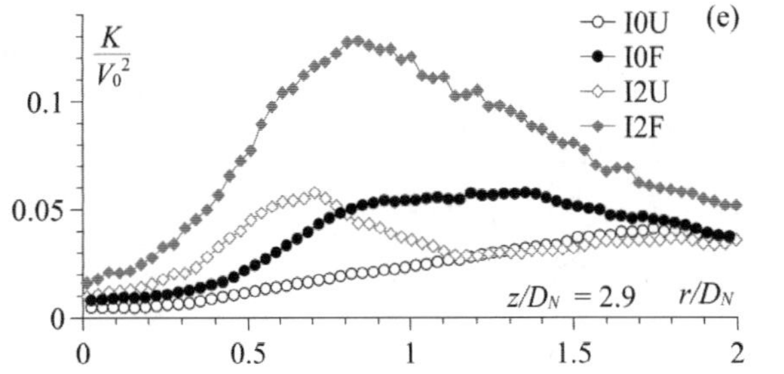

Figure 4. 2D distributions of TKE for Cases (a) I0U, (b) I0F, (c) I2U (d) I2F and (e) TKE profiles along the impingement surface (z/D_N = 2.9, depicted as a black dashed line in the TKE fields) for the forced and unforced conditions and different gas-water flow-rate ratios β. The profiles contain only every second measured point for better visibility. The coordinate system is the same as in Figure 2.

Acknowledgments
The research was funded by a grant from the Russian Foundation for Basic Research (Project No. 15-08-03909-A supervised by PhD K. Sharov) and a grant of the President of the Russian Federation for the state support of leading scientific schools (Project No. NSh-10179.2016.8 supervised by Prof. D. Markovich).

References
[1] Alekseenko S V, Markovich D M and Semenov V I 1999 Suppression of large-scale structures in a gas-saturated impact jet *Tech. Phys. Lett.* **25** 374–376
[2] Pervunin K S and Timoshevskiy M V 2016 Bubbly free and impinging jets: experimental study by means of PIV and PFBI *J. Phys. Conf. Ser.* **754** (032016)–6
[3] Timkin L S and Gorelik R S 2009 The problems of the upward monodispersed microbubble flow in a vertical tube *J. Eng. Thermophys.* **18(1)** 69–71
[4] Akhmetbekov Ye K, Alekseenko S V, Dulin V M, Markovich D M and Pervunin K S 2010 Planar fluorescence for round bubble imaging and its application for the study of an axisymmetric two-phase jet *Exp. Fluids* **48** 615–629
[5] Dulin V M, Markovich D M and Pervunin K S 2012 The optical principles of PFBI approach *AIP Conf. Proc.* **1428** 217–224

IOP Conf. Series: Journal of Physics: Conf. Series **899** (2017) 032017 doi:10.1088/1742-6596/899/3/032017

Distribution of local flow parameters at separation of liquid mixture under distillation on the structured packing at a periodic change in the drip point pattern

A N Pavlenko, V E Zhukov, N I Pecherkin, A D Nazarov, X Li, H Sui, and H Li

Institute of Thermophysics SB RAS, 630090, Lavrentyev av.1, Novosibirsk, Russia

pavl@itp.nsc.ru; nazarov@itp.nsc.ru

Abstract. The effect of non-stationary irrigation of a structured packing on distribution of flow parameters over the cross-section and height of a distillation column was studied using a controlled liquid distributor. We have determined the characteristic time of switching the valves of the liquid distributor at periodic sequential irrigation of two zones of the column cross-section, corresponding to the most effective suppression of stable large-scale maldistribution of the flow parameters in the packing layers, typical of stationary uniform irrigation. It is shown that application of dynamically controlled liquid distributor in the studied regimes of structured packing irrigation allows an improvement of the efficiency of mixture separation by 20%.

1. Introduction

Today, rectification is used in various areas of chemical technology throughout the world, for example, in the industries of organic synthesis, production of isotopes, polymers, products of cryogenic air separation, and other high purity substances.

Stationary irrigation of the structured packing in large-scale industrial distillation columns is usually performed by the liquid distributors with a fixed drip point density over the column cross-section. For stationary setting of irrigation points under the conditions of the liquid film flow over the packing and the counterflow vapor flow, large-scale maldistributions of the flow density develop over the packing cross-section and height [1]. Maldistribution significantly reduces the efficiency of mixture separation in distillation columns of the large diameter.

Deterioration of separation efficiency of mixtures in the large-scale mass exchange apparatuses of chemical technology, as a rule, has a hydrodynamic nature [2], i.e. it is a consequence of development of liquid and vapor maldistribution over the cross-section and height of the structured packing. Flow maldistribution along the column height can also occur during mixture separation, when the vapor density in the upper part of the column becomes significantly higher than in the lower part [1]. The reverse flows of vapor along with longitudinal mixing [3] also cause deterioration in the efficiency of mixture separation.

One of the methods for suppressing the transverse maldistribution of flow parameters is the creation of an organized and controlled flow structure. The use of the modern structured packing allows much more even distribution of the liquid film over the packing surface [4]. The pulsating supply of the liquid flow [5] makes it possible to create phase distribution in the packed column, when

*Corresponding author: nazarov@itp.nsc.ru

the height of transfer unit becomes significantly less than that under the stationary conditions of irrigation. In the experiments presented below, the idea of non-stationary influence on the large-scale stationary structures in the column cross-section with the purpose of their destruction was implemented using a dynamically controlled liquid distributor.

2. Results

The experiments were carried out on the large-scale research setup, whose description is given in [1]. A structured Mellapak 350Y packing was mounted in the distillation column with the diameter of 0.9 m. The experiments were carried out with the packing height of 2.1 and 4 m, respectively. The mixture of R-114 and R-21 Freons was used as a working mixture. All experiments were carried out at pressure P = 3 bar at complete reflux. To irrigate the packing, a specially designed controlled liquid distributor was used [6]. In the liquid distributor, there were 128 solenoid valves, overlapping the holes with the diameter of 5 mm. Each valve was controlled by a PC according to any predetermined algorithm. The column response to the action of the liquid distributor was observed in real time by the readings of three groups of thermometers [6]. The thermometers were installed in the lower (two layers from the bottom of the packing), middle and upper (two layers from the top of the packing) cross-sections. In a binary mixture there is a definite relationship between concentration and temperature at a constant pressure. The displayed information made it possible to estimate both the structure of the large-scale maldistribution of temperature and mixture composition, and the magnitude of this maldistribution. An estimate of the degree of temperature maldistribution, which indirectly characterizes the efficiency of mixture separation, is the value of standard deviation of temperature over the packing cross-section.

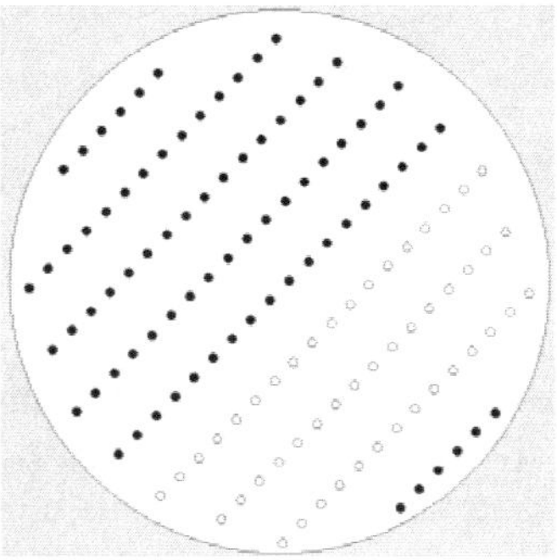

a.

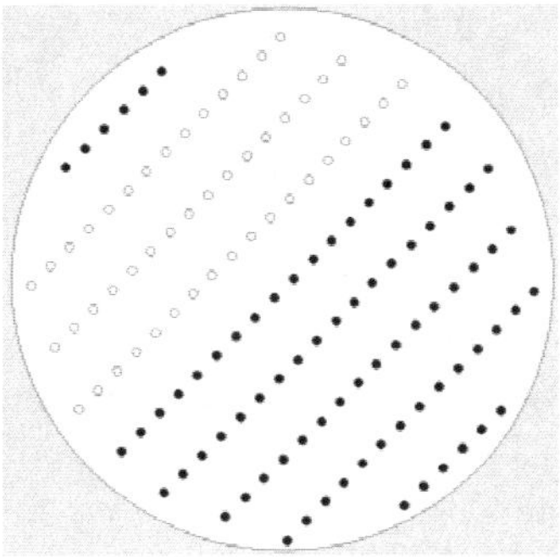

b.

Figure 1. Periodical irrigation (Half Column). Packing height is 4 m. a – state of distributor t_1; b – state t_2.

One of the alternatives of the dynamic effect on the stationary structures is periodic alternating irrigation of two halves of the packing cross-section (*HC* pattern (*Half_Column*)). During time t_1, the holes in one half of the distributor are opened, the holes in the second half of the distributor remain closed (figure 1, a). Then, the valves are switched and the irrigated and non-irrigated zones are changed for time t_2 (figure 1, b). After this, the cycle is repeated with period $T = t_1 + t_2$. The main data array for the *HC* pattern was obtained for equal times $t_1 = t_2$ in the range from 2.5 to 120 s.

It is obvious that for large values of t_1 and t_2, irrigation of a half the column cross-section is substantially non-uniform, herewith in the case of stationary irrigation t_1, $t_2 \rightarrow \infty$, the efficiency of mixture separation will be much lower than at uniform packing irrigation. In the case of periodic irrigation of diametrically located regions of the packing cross-section with a period, commensurable with the time of liquid stay in the packing, a tendency to formation of a new large-scale temperature distribution structure is observed in different cross-sections of the column.

Since each switching causes formation of large-scale maldistribution with diametrically opposite location, then for the switching periods, commensurable with times of formation of large-scale maldistribution, a pulsating change in the degree of maldistribution is observed. More efficient mixture separation corresponds to this regime of dynamic packing irrigation. Figure 2 shows the comparison of a time change in the standard deviation of temperature in the lower packing layer for the stationary irrigation and HC pattern at $t_1 = t_2 = 20$ s, 90 s, and 120 s, respectively.

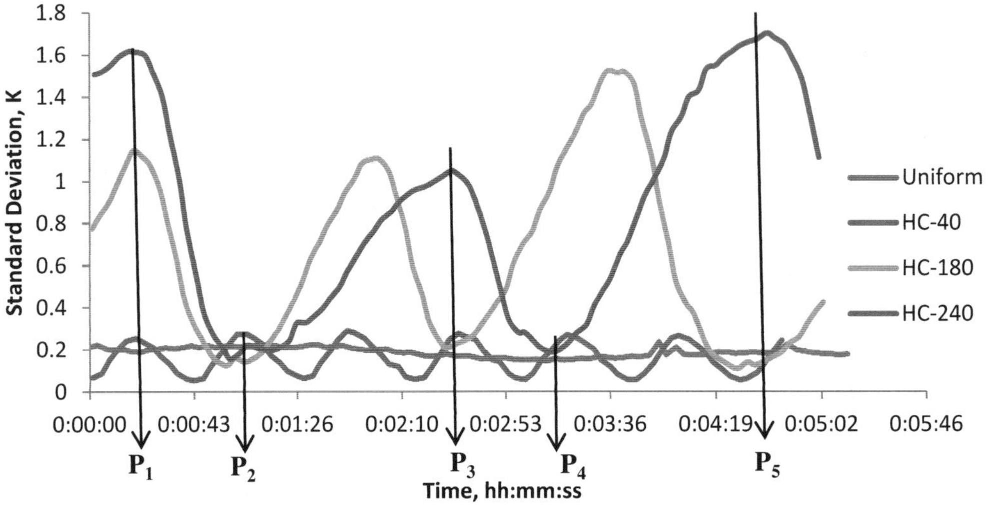

Figure 2. A change in the value of standard deviation of temperature in the lower packing layer with time. Initial moment – moment of distributor valve switching. Drip point pattern: HC-40 – $t_1 = t_2 = 20$ s, HC-180 – $t_1 = t_2 = 90$ s, HC-240 – $t_1 = t_2 = 120$ s. $P_1 - P_5$ – extreme points of the HC-240 curve. Reduced vapor velocity $K_v = 0.028$ m/s. Packing height – 4 m.

The periodic character of curves indicates the irrigation sequence of two halves of the packing. According to data presented, for $t_1 = t_2$ equal to 90 s and 120 s, a stable large maldistribution (the maximal values of temperature deviation) can form along the entire height of the packing. More uniform distribution of the flows, corresponding to the minimal standard deviations of temperature, is associated with reformation of the flows along the cross-section and height of the structured packing during periodic irrigation of two semidiameters of the upper cross-section of the mass exchange surface. As it can be seen from the data presented, under the regime with time of valve switching $t_1 = t_2 = 20$ s, the suppression of stable stationary large-scale maldistributions in the packing layers is most effective as compared to other *HC* regimes. This regime is more effective than steady-state uniform irrigation for the given packing height and specified liquid and vapor flow rates.

Diagrams of local temperature distribution over the packing cross-section at the bottom of the column are presented in figure 3 for period of switching two halves of the distributor at the points corresponding to the extremes of the standard deviation curve (see figure 2) and they are compared with temperature distribution at *Uniform* pattern in the same packing cross-section. Time moments corresponding to points P_1, P_3 and P_5 (figure 3, a, c, e, respectively) are the moments of completion of irrigation of one and beginning of irrigation of another half of the packing. Time interval from P_1 to P_3 is $t_1 = 120$ s. Accordingly, time interval $P_3 - P_5$ is $t_2 = 120$ s. The isotherms in the diagrams are drawn in 0.2 K. The range of absolute temperature changes in 5 ÷ 6°C (figure 3, a, c, e, respectively) indicates significant maldistribution of flows along the height and cross-section of the packing. The diagrams of local temperature distribution of time moments P_2 and P_4 (figure 3, b and d, respectively) reflect the process of flow reformation along the height and cross-section of the structured packing from one steady distribution, corresponding to irrigation of one packing half, to the state of the flows, corresponding to irrigation of another half. The range of temperature variation for time moments P_2 and P_4 indicates more even distribution of flows over the height and cross-section of the packing, even in comparison with the *Uniform* drip point pattern (figure 3, f).

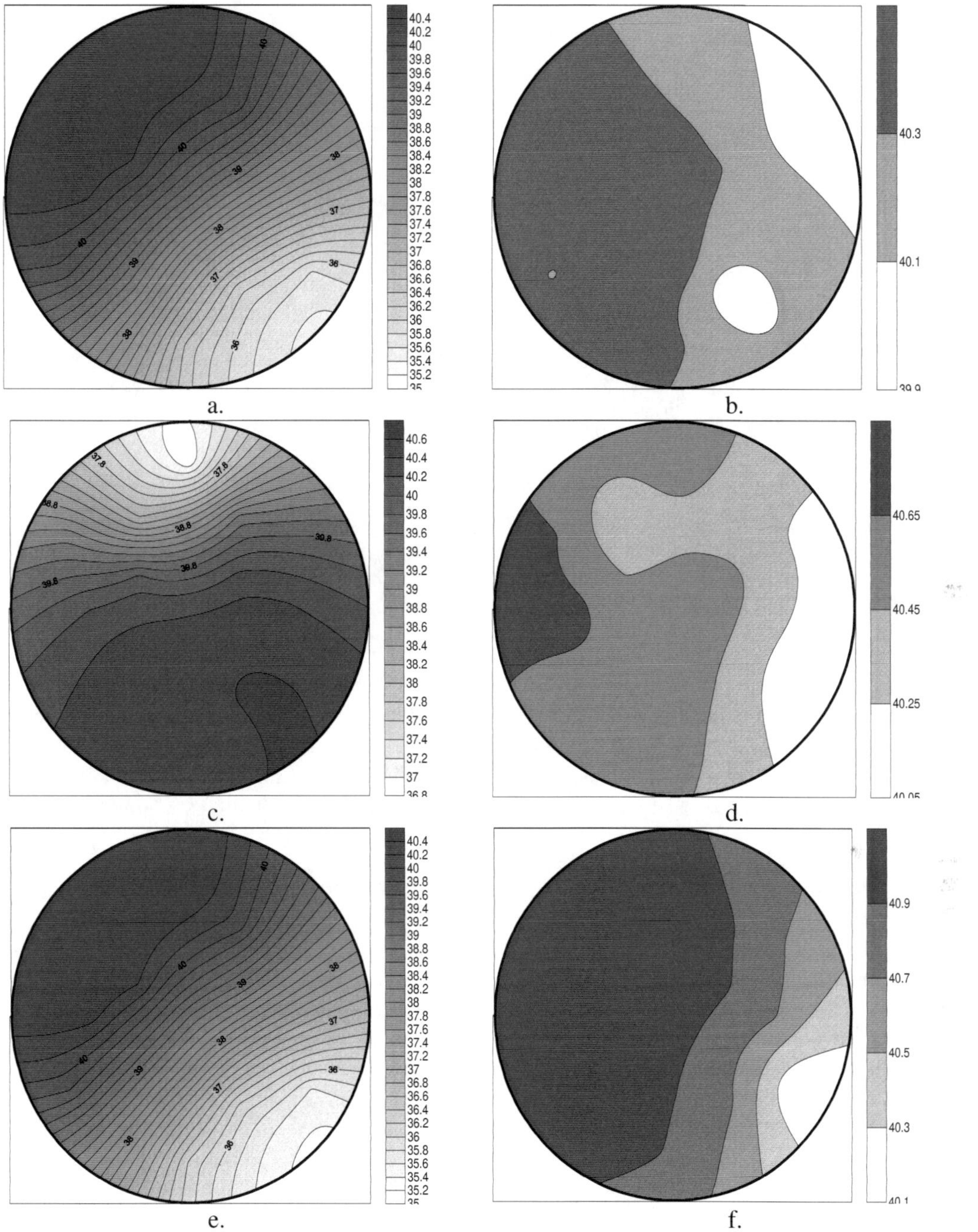

Figure 3. Diagrams of local temperature distribution in the lower packing layer. *a, b, c, d, e* – time moments P_1, P_2, P_3, P_4, P_5 (Fig. 2), respectively; *f* – pattern *Uniform*. Duration of half packing irrigation: $t_1 = t_2 = 120$ s. $K_v = 0.028$ m/s. Packing height – 4 m.

The results of experiment on investigation of separation efficiency of mixtures for different periods of non-stationary periodic irrigation of the packing half-diameters are shown in figure 4. Based on the analysis of experimental data, we have determined the flow characteristics of distillation column operation, when periodic alternating irrigation of two halves of the packing cross-section provides an increase in separation efficiency as compared to the traditional uniform drip point pattern.

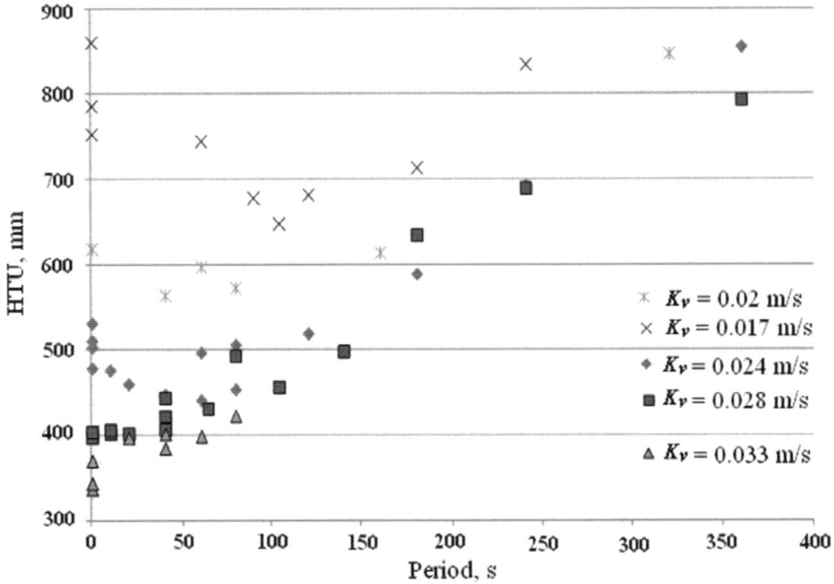

Figure 4. Efficiency of mixture separation (*HTU*) depending on switching period for *HC* pattern. Packing height – 4 m.

3. Conclusion

Periodic irrigation of the specified zones of the structured packing allows increasing the efficiency of column separation within 20 % for the switching periods, commensurable with intrinsic characteristic times of formation of large-scale maldistribution of the flow parameters over the cross-section and height of the packing. Thus, when operating the large-scale industrial distillation columns, there is a real opportunity to use new liquid distributors with a controlled drip point pattern to improve the efficiency of mixture separation in such mass exchanging apparatuses.

Acknowledgments

The study was performed at IT SB RAS thanks to the grant of the Russian Science Foundation (project No. 14-49-00010)

References

[1] Pavlenko A N, Zhukov V E, Pecherkin N I, Volodin O A, Shilkin A and Grossmann C 2014 *AIChE J.* **60** 690–705
[2] Rozen A M, Kostanyan A E 2002 *Theoretical Foundations of Chemical Engineering* **36** 339–46
[3] Lapteva E A, Laptev A G 2015 (Russia, Kazan, Pechat-Servis XXI vek) 236
[4] Pavlenko A N, Zeng J, Pecherkin N I, Zhukov V E, Volodin O A 2016 Journal of Engineering Thermophysics **25** 1–14
[5] Memedlyaev Z N, Kulov N N, Moskalik V M 1994 *Theoretical Foundations of Chemical Engineering* **28** 483
[6] Pavlenko A N, Zhukov V E, Pecherkin N I, Nazarov A D, Serov A F, Mis'kiv N B, Li X, Jiang B, Sui H, Li H, Gao X 2017 *Optoelectronics, Instrumentation and Data Processing* **53** 15–20

IOP Conf. Series: Journal of Physics: Conf. Series **899** (2017) 032018 doi:10.1088/1742-6596/899/3/032018

Controlled irrigation of a structured packing as a method for increasing the efficiency of liquid mixture separation in the distillation column

A N Pavlenko, V E Zhukov, N I Pecherkin, A D Nazarov, X Li, H Li, X Gao, and H Sui

Kutateladze Institute of Thermophysics SB RAS, Novosibirsk, Russia

E-mail: pavl@itp.nsc.ru, zhukov@itp.nsc.ru

Abstract. The use of modern structured packing in the distillation columns allows much more even distribution of the liquid film over the packing surface, but it does not completely solve the problem of uniform distribution of flow parameters over the entire height of the packing. Negative stratification of vapor along the packing height caused by different densities of vapor mixture components and higher temperature in the lower part of the column leads to formation of large-scale maldistributions of temperature and mixture composition over the column cross-section even under the conditions of uniform irrigation of packing with liquid. In these experiments, the idea of compensatory action of liquid distributor on the large-scale maldistribution of mixture composition over the column cross-section was implemented. The experiments were carried out in the distillation column with the diameter of 0.9 m on 10 layers of the Mellapak 350Y packing with the total height of 2.1 m. The mixture of R-21 and R-114 was used as the working mixture. To irrigate the packing, the liquid distributorr with 126 independently controlled solenoid valves overlapping the holes with the diameter of 5 mm, specially designed by the authors, was used. Response of the column to the action of liquid distributor was observed in real time according to the indications of 3 groups of thermometers mounted in 3 different cross-sections of the column. The experiments showed that the minimal correction of the drip point pattern in the controlled liquid distributor can significantly affect the pattern of flow parameter distribution over the cross-section and height of the mass transfer surface and increase separation efficiency of the column within 20%.

1. Introduction

Distillation columns are widely used in the oil refining, chemical, and food industries to separate the mixtures of various liquids. The main requirement for efficient operation of the distillation column is maximal uniformity of the distributions of the liquid film on the packing surface and counter-current flow of vapor over its cross-section. To achieve this goal, the packing consisting of metal sheets with applied structures of different scale, as well as highly permeable porous or mesh packing [1-10] have been developed. The use of modern structured packing allows much more even distribution of the liquid film over the packing surface, but does not completely solve the problem of uniform distribution of flow parameters along the entire height of the packing [1, 3, 6]. Negative stratification of vapor along the packing height caused by different densities of vapor mixture components and higher temperature in the lower part of the column leads to formation of large-scale maldistributions of temperature and mixture composition over the column cross-section even under the conditions of uniform irrigation of packing with liquid. In these experiments, the idea of compensatory action of liquid distributor on the large-scale maldistribution of mixture composition over the column cross-section was implemented.

2. Set-up and method description

The experiments were carried out on a large-scale research installation "Large Freon Column", whose detailed description is given in [3]. The structured Mellapak 350Y packing was mounted in the distillation column with the diameter of 0.9 m. The experiments were carried out on 10 packing layers with the total height of 2.1 m. The mixture of freons R-21 and R-114 was used as the working mixture, which was selected due to its characteristics to simulate separation of cryogenic mixtures. To irrigate the packing, the liquid distributor specially designed by the authors was used; its description is given in [11]. The distributor had 126 solenoid valves, overlapping the holes with the 5-mm diameter. Each valve was controlled independently by a PC through specially developed software allowing implementation of both the manual control of a single valve or an arbitrary group of valves, and programmable switching of valves according to any predetermined algorithm with time sampling of 1 s. The column response to the action of liquid distributor was observed in real time according to indications of 3 groups of thermometers installed in 3 different cross-sections of the column. Each group consisted of 16 thermometers, evenly distributed over the packing cross-section. Their readings were displayed on the computer monitor in the form of topographs. The groups of thermometers were installed in the lower (2 layers from the packing bottom), middle and upper (2 layers from the packing top) cross-sections of the packing. Information displayed on the monitor made it possible to evaluate both the structure of the large-scale temperature maldistribution and magnitude of this maldistribution.

The experiments were carried out with the fixed number of drip points in the liquid distributor. The initial drip point pattern corresponded to the uniform packing irrigation. When the column achieved the quasi-stationary operation and the structure and degree of large-scale temperature maldistribution over the column cross-section were registered, a minimal correction of the drip point pattern in the liquid distribution was made. This correction was made by opening one hole in the liquid distributor in the region of projection of the large-scale zone of maximal temperatures with simultaneous closing one hole in the region of projection of the minimal temperature zone. As a result, the structure and degree of large-scale temperature maldistribution across the column cross-section and, consequently, the efficiency of mixture separation were changed. By means of two or three above iterations, a minimal degree of temperature maldistribution in the column cross-sections was achieved and, as a consequence, the best efficiency of mixture separation. An example of described correction of drip point irrigation is shown in figure 1.

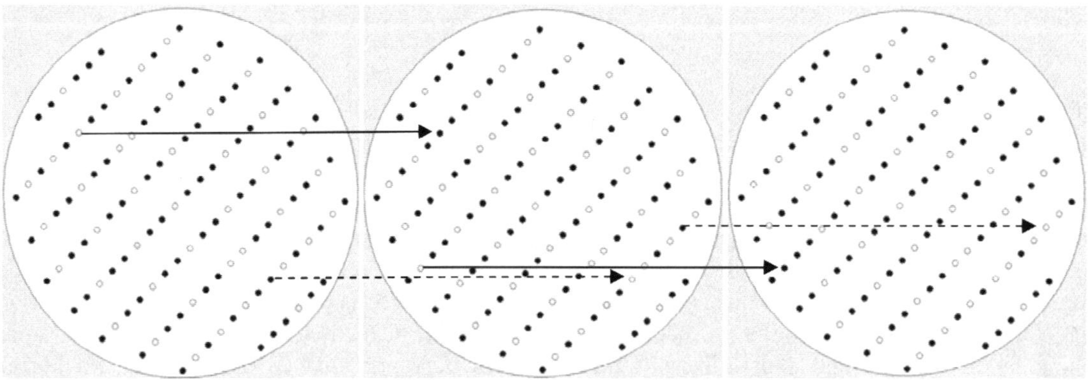

a) b) c)

Figure 1. Correction of drip point pattern. Solid arrow – valve opening, dashed arrow – valve closing. Pattern indications: *a – Uniform*; *b – Step 1*; *c – Step 2*.

IOP Conf. Series: Journal of Physics: Conf. Series **899** (2017) 032018 doi:10.1088/1742-6596/899/3/032018

3. Results and discussions

The experiments were performed under pressure P = 3 bar at complete reflux in the range of the reduced vapor velocity 0.017 $<K_v<$0.045 m/s. Under the condition of uniform packing irrigation (*Uniform* pattern, figure 1a), large-scale maldistribution of mixture concentration and, consequently, temperature was formed throughout the packing cross-section along the entire height of the column. As a rule, the presence of diametrically located poles of high and low temperature is characteristic of the structural non-uniformity revealed in the experiments. The topograms of temperature distribution in three different packing cross-sections along the column height are shown in figure 2 for *Uniform* drip point pattern at reduced vapor velocity K_v = 0.03 m/s. For each topogram, the value of standard deviation *StDev* is shown by indications of all thermometers mounted in the given packing cross-section.

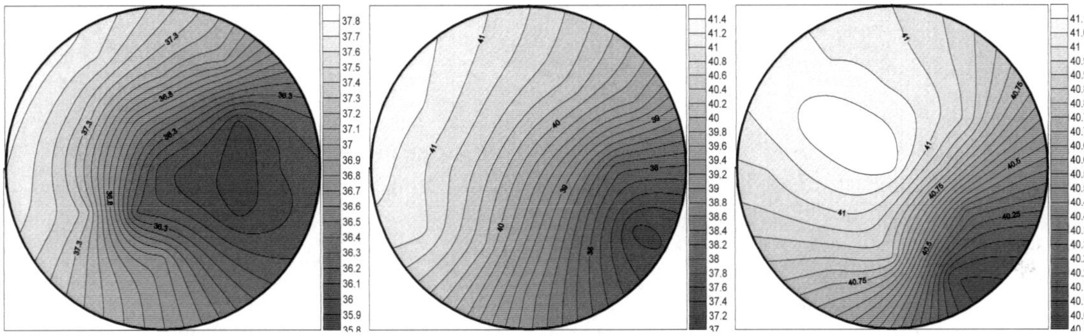

a) b) c)

Figure 2. Topograms of temperature distribution (°C) in different packing cross-sections h along the height (from the bottom edge). Drip point pattern *Uniform*; K_v = 0.03 m/s; *HTU* = 299 mm: *a* – layer 8, h = 1.68 m, *StDev* = 0.64 K; *b* – layer 5, h = 1.05, *StDev* = 1.20; *c* – layer 2, h = 0.42, *StDev* = 0.32.

As it follows from the analysis of presented topograms, large-scale maldistribution spreads over the entire height of the packing and has the areas of increased and decreased temperature in approximately the same parts of the column cross-section. The region of decreased temperature (dark area in topograms) corresponds to the region with higher concentration of volatile component and, as a rule, higher density of liquid flow rate in the given cross-sectional projection. The effect of packing irrigation on this projection of cross-section (pattern *Step 1*, figure1b) led to a change in the character and degree of temperature maldistribution in the packing cross-section (figure 3).

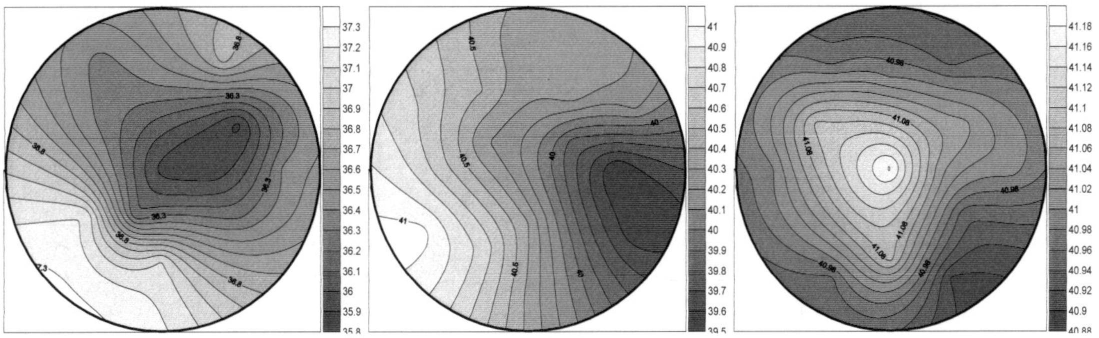

a) b) c)

Figure 3. Topograms of temperature distribution (°C) in different packing cross-section h along the height (from the bottom edge). Drip point pattern *Step 1*; K_v = 0.03 m/s; *HTU* = 270 mm: *a* – layer 8, h = 1.68 m, *StDev* = 0.41 K; *b* – layer 5, h = 1.05, *StDev* = 0.44; *c* – layer 2, h = 0.42, *StDev* = 0.08.

The following effect of packing irrigation (drip point pattern *Step 2,* figure 1c) led to a following change in the character and degree of temperature maldistribution in the packing cross-section (figure 4).

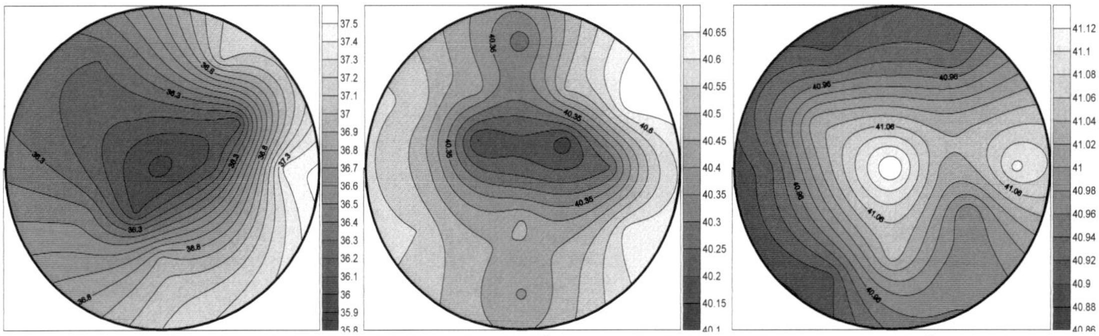

a) b) c)

Figure 4. Topograms of temperature distribution (°C) in different packing cross-section *h* along the height (from the bottom edge). Drip point pattern *Step 2;* K_v = 0.03 m/s; *HTU* = 262 mm: *a* – layer 8, *h* = 1.68 m, *StDev* = 0.49 K; *b* – layer 5, *h* = 1.05, *StDev* = 0.17; *c* – layer 2, *h* = 0.42, *StDev* = 0.08.

As it follows from figures 2-4, the step-by-step change in the drip point pattern led to a change in the structure of large-scale temperature maldistribution in the column cross-section, as well as to a decrease in the standard deviation of temperature distribution in the packing cross-sections. Distribution topograms of the liquid flow rate density in the column cross-section directly below the bottom edge of the packing are shown in figure 5. According to the figure, for the *Uniform* drip point pattern, maldistribution of liquid flow rate density over the column cross-section is higher than for *Step 2* drip point pattern. The achieved reduction in maldistribution of the flow parameter after the correction procedure led to an increase in the efficiency of mixture separation: the *HTU* value decreased 299 mm to 262 mm, respectively.

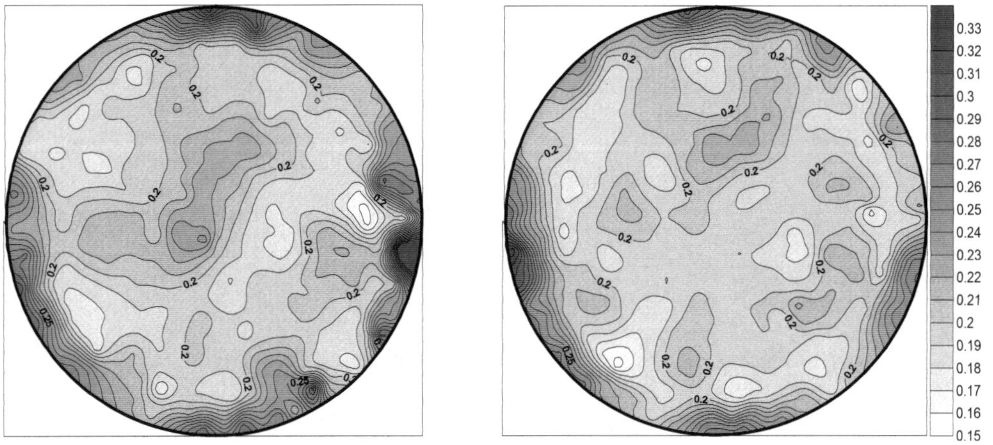

a) b)

Figure 5. Distribution topograms of liquid flow rate density (ml/cm^{2}c) in the column cross-section at the packing outlet. K_v = 0.03 m/s: *a* – *Uniform*, *StDev* = 14.5 %, *HTU* = 299 mm; *b* – *Step 2*, *StDev* = 11.7 %, *HTU* = 262 mm.

Dependences of the height of transfer unit on the reduced vapor velocity for *Uniform* drip point pattern and for the pattern obtained as a result of the correcting the drip point patterns are shown in figure 6.

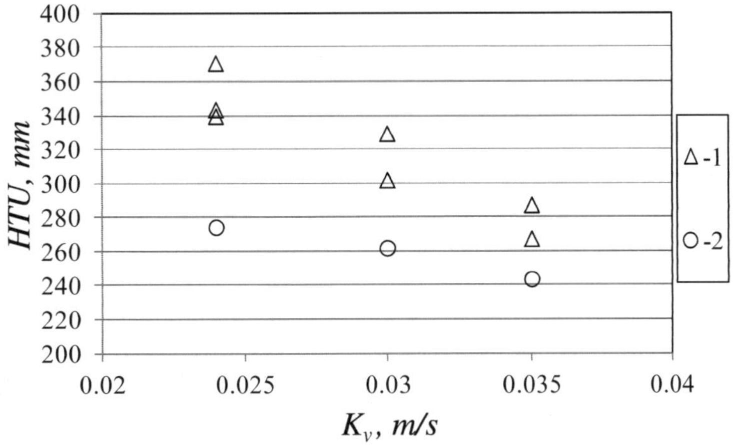

Figure 6. Separation efficiency of the packing for *Uniform* drip point pattern (*1*) and after final correction (*2*).

As it can be seen from the diagram, application of the above procedure for correcting the drip point pattern in the entire range of vapor velocities resulted in a significant reduction in the height of transfer unit (10-20%), and this corresponds to more efficient separation of mixture.

4. Conclusions

The experiments showed that minimal correction of the drip point pattern of the liquid distribution can significantly affect distribution of the flow parameters over the cross-section and height of the mass-transfer surface and increase the separation efficiency of the column within 20%.

Acknowledgment

This research has been performed at IT SB RAS with the support of the Russian Science Foundation (project No. 14-49-00010).

References

[1] Pavlenko A N, Pecherkin N I, Chekhovich V Yu, Zhukov V E, Sunder S and Houghton P 2009 *Theoretical Foundations of Chemical Engineering* **43** 1–11
[2] Pavlenko A N, Zhukov V E, Pecherkin N I, Chekhovich V Yu, Sunder S and Houghton P 2010 *Theoretical Foundations of Chemical Engineering* **44** 869–76
[3] Pavlenko A N, Zhukov V E, Pecherkin N I, Volodin O A, Shilkin A and Grossmann C 2014 *AIChE J.* **60** 690–705
[4] Soulaine C, Horgue P, Franc J and Quintard M 2014 *AIChE J.* **60** 3665–74
[5] Ehler C and Fieg G 2014 *AIChE J.* **60** 3833–47
[6] Pavlenko A N, Li X, Zhukov V E, Pecherkin N I, Volodin O A, Surtaev A S, Gao X, Zhang L, Sui H and Li H 2015 *Journal of Engineering Thermophysics* **24** 210–21
[7] Pavlenko A N, Zeng J, Pecherkin N I, Zhukov V E and Volodin O A 2016 *Journal of Engineering Thermophysics* **25** 1–14
[8] Oluji´c Z, Rietfort T, Jansen H and Zich E 2015 *Chemical Engineering Research and Design*

99 14–19
[9] Yao Y, Pavlenko A N and Volodin O A 2015 *Journal of Engineering Thermophysics* **24** 222–36
[10] Stepykin A, Sidyagin A and Ulyanov V 2015 *Transactions TSTU* **21** 445–52
[11] Pavlenko A N, Zhukov V E, Pecherkin N I, Nazarov A D, Serov A F, Miskiv N B, Li X, Jiang B, Sui H, Li H and Gao X 2017 *Optoelectronics, Instrumentation and Data Processing* **53** 19–25

IOP Conf. Series: Journal of Physics: Conf. Series **899** (2017) 032019 doi:10.1088/1742-6596/899/3/032019

Exact solution of the problem of dissolved gas segregation by the plane crystallization front

Andrey Pil'nik, Andrey Chernov

Kutateladze Institute of Thermophysics, SB RAS, Novosibirsk, Russia
Novosibirsk State University, Novosibirsk, Russia

E-mail: `pilnik@itp.nsc.ru`

Abstract. Analytical solution of the segregation problem was found for the arbitrary crystal growth law using the quasi-steady-state approximation. The segregation is caused by the displacement of dissolved gas by moving plane crystallization front. The effect of solidification shrinkage on the crystallization process was taken into account. It is shown that in the case of "equilibrium crystallization" (when the growth rate is in inverse ratio to time) the solution of the problem becomes self-similar. In this case gas concentration at the crystallization front stays the same during the whole process while the diffusion layer thickness increases with time.

1. Introduction

The major part of modern materials are being created by the way of solidification from melts. Depending on the conditions one can create various forms of solid matter: crystalline, amorphous, clathrate compound and others [1–6]. All processing technologies of producing solid materials are in one form or another associated with minimizing or eliminating two factors: shrinkage and gas porosity. Thus, creating fully consistent theory allowing to account for these seems to be the relevant task.

Ultra-rapid quenching from melt causes extreme cooling rates allowing to transfer the melt into deeply metastable state. Under such conditions the melt solidification is a result of fluctuation nucleation and the growth of new phase nuclei [1, 7]. It is noted in a number of articles on the experimental research of such a processes that the micropores and cracks in solidified material can be caused by the shrinkage of substance in solidification process [8–12]. In [13] it is shown that the formation of even a small fraction of crystal phase leads to significant tension stress in uncrystallized melt which leads to intensive cavitation. The results show that the size of crystal grains and cavities in solidified material depend very heavily on the rate of melt cooling.

Another mechanism of pore formation during crystallization of gas-saturated melts is the gas segregation in the process of crystallization [14]. The solubility of gas in the solid is substantially less than solubility of gas in the liquid, so the gas dissolved in the melt will be forced out by the moving crystallization fronts. As a result the gas will "accumulate" near the phase transition front. At the moment when gas concentration exceeds the critical one (corresponding to gas solubility in the melt under given conditions) nucleation and growth of gas bubbles occur. Further on crystallization front captures bubbles and the solidified material has a porous structure as a result. Number of articles are devoted to this problem [15–21]. Alas, the majority of them are dealing with the case of low crystal growth rate and take no account for the shrinkage of the substance.

2. Statement of the problem

Let us consider the problem of segregation of gas dissolved in the melt caused by moving crystallization front in the case of plane crystallization.

The dependence of dissolved gas concentration field on time during crystallization is described by the diffusion equation:

$$\frac{\partial C}{\partial t} + v_l \frac{\partial C}{\partial x} = \frac{\partial}{\partial x}\left(D\frac{\partial C}{\partial x}\right), \tag{1}$$

where t — time; x — the coordinate with the origin at the crystallization center; C — mass fraction of the gas dissolved in the melt; D — diffusion coefficient for the gas in the melt.

Let us note that eq. (1) has a convective term caused by the shrinkage of the melt during solidification. For incompressible liquid: $v_l = -k\dot{X}$. Where X — coordinate of crystallization front (initially $X = 0$); $k = \rho_s/\rho_l - 1$ — shrinkage coefficient; ρ_s and ρ_l — densities of solid and liquid phases respectively.

The dependence of the coordinate of crystallization front X is obtained by solving the problem of crystal growth in supercooled melt. Without getting into specifics let us note that the crystallization processes can be divided into two categories [2, 22, 23]: non-equilibrium and equilibrium ones.

We will assume that initially the dissolved gas concentration in the melt is homogeneous and equal to C_0:

$$(C)_{t=0} = C_0, \tag{2}$$

moreover $C_s < C_0 < C_l$, where C_s and C_l — solubility of gas in solid phase and solubility in liquid phase respectively.

Boundary conditions consist of following. At large distance from the crystal concentration of dissolved gas is equal to initial one:

$$(C)_{x\to\infty} = C_0; \tag{3}$$

Material balance equation hold true at the interphase:

$$\rho_l (C)_{x=X}\,(dX + kdX) = \rho_s C_s dX - \rho_l D\left(\frac{\partial C}{\partial x}\right)_{x=X} dt,$$

or

$$-D\left(\frac{\partial C}{\partial x}\right)_{x=X} = (1+k)\dot{X}\left\{(C)_{x=X} - C_s\right\}. \tag{4}$$

Boundary problem (1)–(4) fully describes the segregation process during crystallization. Introducing following non-dimensional variables: $\bar{C} = \dfrac{C - C_0}{C_0 - C_s}$; $\tau = t/t_0$; $\bar{x} = x/x_0$; $\bar{X} = X/x_0$; $\acute{\bar{X}} = d\bar{X}/d\tau = \dot{X}/v_0$; $\chi = \bar{x}/\bar{X}$, where $t_0 = D/v_0^2$ and $x_0 = D/v_0$ — characteristic time of the process and characteristic size respectively. As a characteristic velocity of the process v_0 we will use the characteristic growth rate of the crystal.

By using the variables τ and χ we transform the boundary value problem (1)–(4) into the problem with fixed boundaries. Similar approach was also used in [23, 24]. Using these new variables we can transform the problem to the following form:

— diffusion equation:

$$\frac{\partial \bar{C}}{\partial \tau} = \frac{\acute{\bar{X}}}{\bar{X}}(k+\chi)\frac{\partial \bar{C}}{\partial \chi} + \frac{1}{\bar{X}^2}\frac{\partial^2 \bar{C}}{\partial \chi^2}; \tag{5}$$

— initial condition and boundary conditions:

$$(C)_{\chi\to\infty} = 0; \quad -\left(\frac{\partial \bar{C}}{\partial \chi}\right)_{\chi=1} = (1+k)\bar{X}\acute{X}\left\{(\bar{C})_{\chi=1} + 1\right\}. \tag{6}$$

This boundary problem in the general form can only be solved numerically. But, considering that the concentration of dissolved gas in the process of crystallization is a slowly varying function of time we can assume $\partial \bar{C}/\partial \tau \approx 0$. It will be shown below that this assumption allows us to obtain the quasi-stationary solution of the problem. It is in good agreement with the numerical solution and in the case of equilibrium crystallization represent the exact solution.

3. Analytical solution of the problem

Integrating the diffusion equation (5) and using boundary conditions the quasi-stationary solution of the problem for plane crystallization front can be found:

$$\bar{C}(\tau,\chi) = \bar{C}(\tau,1)\,\frac{\operatorname{erfc}\left\{\sqrt{\bar{X}\acute{X}/2}\;(\chi+k)\right\}}{\operatorname{erfc}\left\{\sqrt{\bar{X}\acute{X}/2}\;(1+k)\right\}}, \tag{7}$$

where $\bar{C}(\tau,1)$ — gas concentration at the crystallization front. It is a sought-for function and can be found from the material balance equation (6):

$$\bar{C}(\tau,1) = \frac{\sqrt{\pi}(1+k)\sqrt{\bar{X}\acute{X}/2}\,\operatorname{erfc}\left\{(1+k)\sqrt{\bar{X}\acute{X}/2}\right\}}{\exp\left\{-(1+k)^2\,\bar{X}\acute{X}/2\right\} - \sqrt{\pi}(1+k)\sqrt{\bar{X}\acute{X}/2}\,\operatorname{erfc}\left\{(1+k)\sqrt{\bar{X}\acute{X}/2}\right\}}. \tag{8}$$

In the case of equilibrium crystallization of the melt the function $\bar{X}\acute{X} = 2\beta^2 = \text{const}$. It means that the concentration field in this case is stationary. Thus, the solution of the boundary problem becomes a self-similar one (with self-similar variable χ) and the exact one at that because no approximations were used to obtain it:

$$\bar{C}(\chi) = \frac{\sqrt{\pi}\,\beta\,(1+k)\operatorname{erfc}\{\beta\,(\chi+k)\}}{\exp\left\{-\beta^2\,(1+k)^2\right\} - \sqrt{\pi}\,\beta\,(1+k)\operatorname{erfc}\{\beta\,(1+k)\}}. \tag{9}$$

In the case of no shrinkage ($k = 0$) similar solution was obtained in [14] using different technique.

For the cases of small and big values of the parameter β following asymptotic approximations can be found.

For $\beta \ll 1$ (close to the crystallization front):

$$\bar{C}(\chi) \approx \bar{C}(1)\left\{1 - (2\beta/\sqrt{\pi})(\chi-1)\right\}, \quad \bar{C}(1) \approx \sqrt{\pi}\beta(1+k); \tag{10}$$

For $\beta \gg 1$:

$$\bar{C}(\chi) \approx \bar{C}(1)\,\frac{1+k}{\chi+k}\,e^{-\beta^2\left\{(\chi+k)^2-(1+k)^2\right\}}, \quad \bar{C}(1) \approx 2\beta^2(1+k)^2 + 2; \tag{11}$$

4. Conclusions

The problem of dissolved gas segregation in melt by the plane crystallization front is formulated taking account of the solidification shrinkage. Analytical solution of the problem was found using the quasi-steady-state approximation. The solution is applicable for the arbitrary crystal growth law (the obtained approximate solution shows good fit with the exact solutions for extreme cases). It is found that in the case of "equilibrium crystallization" (when the growth rate is in inverse ratio to the square root of time) the solution of the problem becomes self-similar and exact. In this case gas concentration at the crystallization front instantly increases and stays at the value that depends on the crystal growth rate. At the same time the diffusion layer thickness increases proportionally to the square root of time. The effect of solidification shrinkage on the process at hand is shown.

Acknowledgements

This work was supported by the Russian Science Foundation, project # 15-19-10025. We express our gratitude to the Novosibirsk State University for giving us an access to the library.

References

[1] Fedorchenko A I and Chernov A A 2003 *International Journal of Heat and Mass Transfer* **46** 921 – 929 ISSN 0017-9310
[2] Chernov A A, Pil'nik A A and Islamov D R 2016 *Journal of Crystal Growth* **450** 45 – 49 ISSN 0022-0248
[3] Dontsov V E and Chernov A A 2009 *International Journal of Heat and Mass Transfer* **52** 4919 – 4928 ISSN 0017-9310
[4] Chernov A A and Dontsov V E 2011 *International Journal of Heat and Mass Transfer* **54** 4307 – 4316 ISSN 0017-9310
[5] Chernov A A, Pil'nik A A, Elistratov D S, Mezentsev I V, Meleshkin A V, Bartashevich M V and Vlasenko M G 2017 *Scientific Reports* **7** 40809
[6] Chernov A, Elistratov D, Mezentsev I, Meleshkin A and Pil'nik A 2017 *International Journal of Heat and Mass Transfer* **108, Part B** 1320 – 1323 ISSN 0017-9310
[7] Chernov A A 2009 *International Journal of Heat and Mass Transfer* **52** 3290 – 3296 ISSN 0017-9310
[8] Zhang X, Xu B, Xuan F, Tu S, Wang H and Wu Y 2009 *Applied Surface Science* **255** 4362 – 4371 ISSN 0169-4332
[9] Zhang X, Xu B, Xuan F, Wang H and Wu Y 2009 *Journal of Alloys and Compounds* **473** 145 – 151 ISSN 0925-8388
[10] Salimijazi H, Raessi M, Mostaghimi J and Coyle T 2007 *Surface and Coatings Technology* **201** 7924 – 7931 ISSN 0257-8972
[11] Bengtsson P and Johannesson T 1995 *Journal of Thermal Spray Technology* **4** 245–251 ISSN 1544-1016
[12] Kuroda S and Clyne T 1991 *Thin Solid Films* **200** 49 – 66 ISSN 0040-6090
[13] Chernov A A and Pil'nik A A 2012 *International Journal of Heat and Mass Transfer* **55** 294 – 301 ISSN 0017-9310
[14] Fedorchenko A I and Chernov A A 2003 *International Journal of Heat and Mass Transfer* **46** 915 – 919 ISSN 0017-9310
[15] Geguzin Y and Dzuba A 1981 *Journal of Crystal Growth* **52** 337 – 344 ISSN 0022-0248
[16] Bianchi M V and Viskanta R 1997 *International Journal of Heat and Mass Transfer* **40** 2035 – 2043 ISSN 0017-9310
[17] Wei P and Hsiao S 2012 *International Journal of Heat and Mass Transfer* **55** 8129 – 8138 ISSN 0017-9310
[18] Poon G G and Peters B 2013 *Crystal Growth & Design* **13** 4642–4647
[19] Wei P and Hsiao S 2016 *International Journal of Heat and Mass Transfer* **103** 920 – 930 ISSN 0017-9310
[20] Wei P and Hsiao S 2016 *International Journal of Heat and Mass Transfer* **103** 931 – 939 ISSN 0017-9310
[21] Wei P and Hsiao S 2017 *Journal of Crystal Growth* **460** 126 – 133 ISSN 0022-0248
[22] Chernov A A 2009 *Journal of Engineering Thermophysics* **18** 238–240 ISSN 1990-5432
[23] Chernov A A and Pil'nik A A 2015 *JETP Letters* **102** 526–529 ISSN 1090-6487
[24] Chernov A A, Kedrinsky V K and Pil'nik A A 2014 *Physics of Fluids* **26** 116602

Study of transport coefficients of nanodiamond nanofluids

M I Pryazhnikov[1,2], A V Minakov[1,2], D V Guzei[1,2]

[1] Siberian Federal University, 79 Svobodny pr., Krasnoyarsk, 660041, Russia
[2] Institute of Thermophysics SB RAS, 1 Acad. Lavrentiev pr., Novosibirsk, 630090, Russia

Abstract. Experimental data on the thermal conductivity coefficient and viscosity coefficient of nanodiamond nanofluids are presented. Distilled water and ethylene glycol were used as the base fluid. Dependences of transport coefficients on concentration are obtained. It was shown that the thermal conductivity coefficient increases with increasing nanodiamonds concentration. It was shown that base fluids properties and nanodiamonds concentration affect on the rheology of nanofluids.

1. Introduction

Nanoparticle colloidal suspensions, referred to as 'nanofluids' [1], have potential application in many heat transfer areas because of their intriguing properties such as a considerable increase in thermal conductivity, long-term stability and prevention of clogging in micro-channels [2–3]. Recently extensive efforts have been made to prepare thermal performance enhanced nanofluids [4–5]. Diamond is one of the materials with very high thermal conductivity, and diamond particles are often used as filler in mixtures for upgrading the performance of a matrix [6].In some studies nanodiamonds are used as particles that improve the thermal conductivity of heat-transfer liquids, such as water, oil and ethylene glycol just as particles of metals (gold, copper), metal oxides (aluminum and copper oxides) and silicon carbide [7-11]. It is reasonable to expect that the addition of diamond nanoparticles would lead to thermal performance enhancement in a base fluid. However, available information about them in the open literature is very scant.

Torii et al. [12] conducted experiments in nanodiamond-water nanofluids. The authors of work [12] investigated the coefficients of viscosity and thermal conductivity of nanodiamond-water nanofluids. Particle size ranged from 2 to 10 nm were used. Viscosity coefficient of nanofluid with a particle volume concentration of 2% is 55% higher the viscosity coefficient of water. Thermal conductivity coefficient of nanofluid with a particle volume concentration of 5% is 15% higher the thermal conductivity of water.

In work [13], nanodiamonds were chosen to form nanofluids with mixture base fluid composed of distilled water with a volume fraction of 0.55 and ethylene glycol with a volume fraction of 0.45. The thermal transport properties, including the thermal conductivity, the viscosity and the convective heat transfer coefficient, were investigated. The thermal conductivity enhancement of nanodiamond nanofluids increases with nanodiamonds loading and the thermal conductivity enhancement is more than 18.0% for a nanofluid at a nanodiamonds volume fraction of 0.02. Viscosity measurements show that the nanodiamond nanofluids demonstrate Newtonian behaviour, and the viscosity significantly decreases with temperature.

Yeganeh et al. [14] measured thermal conductivity enhancements of nanodiamond particles suspended in pure deionized with different volume fractions in the range from 0.8% to 3%. The highest observed enhancement in the thermal conductivity is 7.2% for a volume fraction of 3% at a temperature of 30°C. The thermal conductivity increases by about 9.8% as the temperature rises to 50°C.

This work presents an experimental study of thermal conductivity coefficient and viscosity coefficient of nanodiamond nanofluids. Distilled water and ethylene glycol were used as the base fluid.

2. Preparation of nanofluids

The powder of nanodiamonds was produced by Joint Stock Company Federal Research & Production Center ALTAI.

Properties of nanodiamonds. The size of primary particles is 4-6 nm, it was defined by X-ray phase analysis. The aggregate size is 20-2500 nm and it is obtained by using a scanning electron microscope (see figure 1). Density 3.0±0.1 g/cm^3 was measured by the pycnometric method. The specific surface is 280±60 m^2/g and is measured by the BET method. The chemical analysis showed a diamond phase content not less than 91.0%. The mass fraction of non-combustible impurities in the solid phase which was defined by the burning method is not more than 5.0%. Chemical impurities are O, N, H (elemental analysis).

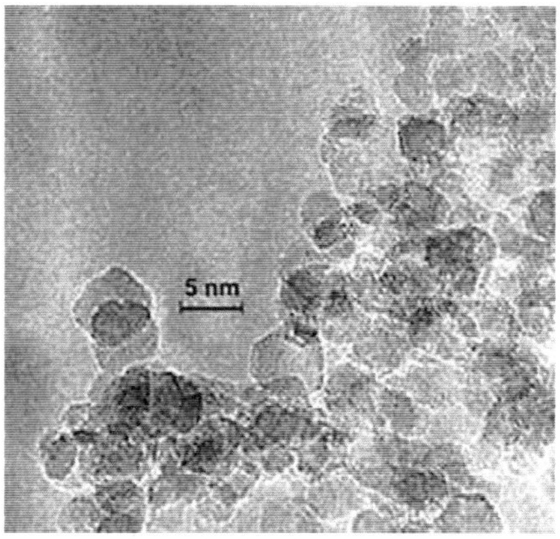

Figure 1. SEM image of nanodiamonds.

The volume concentration of nanoparticles varied from 0.25 to 2%. Preparation of nanofluids was based on standard twostep process. At first we have added the required amount of nanopowder to the base fluid, then the nanofluid was careful mixed mechanically, then it was placed into an ultrasonic disperser 'Sapphire TC-10338' for a half-hour to destruct conglomerates of particles.

3. Dependence of thermal conductivity on particle concentration

Thermal conductivity measurements were performed by nonstationary hot-wire method. Detailed description of the test bench and testing technique is given in [15]. The resultant relative measurement error of fluid thermal conductivity coefficient does not exceed 3%.

Two series of measurements of thermal conductivity were carried out. The first series was carried out for a nanofluid based on distilled water. In the second series, ethylene glycol was used as the base fluid. Dependences of relative coefficient of thermal conductivity of nanofluids (the ratio of thermal conductivity of nanofluid to coefficient of thermal conductivity of base fluid) on nanodiamonds concentration were obtained (see figure 2-3).

Figures 2-3 show that the coefficient of thermal conductivity of nanofluids increases with increasing nanodiamond volumetric concentration φ.

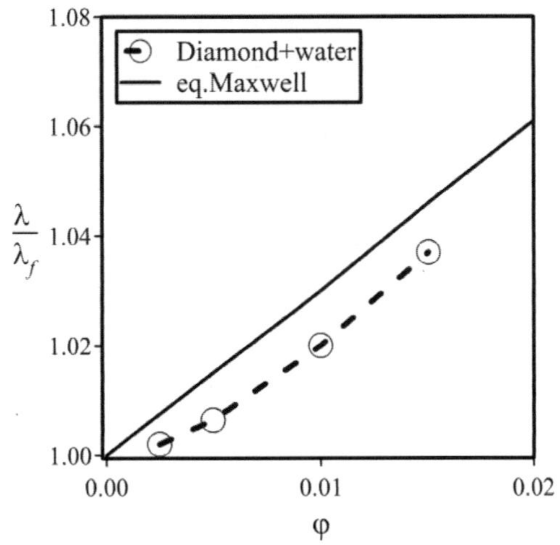

Figure 2. Relative thermal conductivity coefficient of water based nanofluids depending on diamond particle concentration.

Furthermore, figures 2-3 show the dependences of the thermal conductivity of nanofluids predicted by the Maxwell theory (solid line):

$$\lambda = \lambda_f \frac{\lambda_p + 2\lambda_f + 2\varphi(\lambda_p - \lambda_f)}{\lambda_p + 2\lambda_f - \varphi(\lambda_p - \lambda_f)} \tag{1}$$

where λ_p, λ_f are the thermal conductivity coefficient of particle material and base fluid.

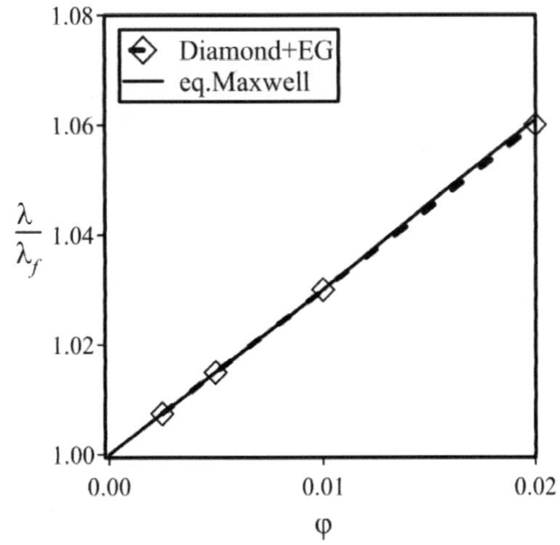

Figure 3. Relative thermal conductivity coefficient of ethylene glycol based nanofluids depending on diamond particle concentration.

It is seen that the experimental coefficient of thermal conductivity of ethylene glycol-based nanofluid is well described by Maxwell's theory. The experimental thermal conductivity coefficient of water-based nanofluid is slightly lower than the coefficient predicted by Maxwell's theory.

IOP Conf. Series: Journal of Physics: Conf. Series **899** (2017) 032020 doi:10.1088/1742-6596/899/3/032020

4. Dependence of viscosity on particle concentration

The viscosity of nanofluids was measured by a rotational viscosimeter Brookfield DV2T with the adapter ULA(0) for a low viscosity. The installations description and its approbation for the viscosity measurement of nanosuspensions are given in [16]. The accuracy of viscosity's measurement is not lower than 2 %. All the data below are obtained at the temperature of 25°C.

Two series of measurements were also carried out, as is the case with the study of the thermal conductivity of nanodiamond nanofluids. The first series of measurements was carried out for water-based nanofluids. Figure 4 shows the dependences of the relative viscosity coefficient on the spindle speed by different nanodiamond concentrations. It is seen that viscosity coefficient increases with increasing nanodiamond concentration. Furthermore, the dependence of viscosity on the spindle speed is well seen. This demonstrates the non-Newtonian properties of nanofluids. The obtained data are well described by a power law:

$$\frac{\mu}{\mu_f} = K \cdot \dot{\gamma}^{\,n-1} \tag{2}$$

where K is flow consistency index (Pa·s^{-n}), $\dot{\gamma}$ is the shear rate (s^{-1}); n is the flow behavior index of nanofluids.

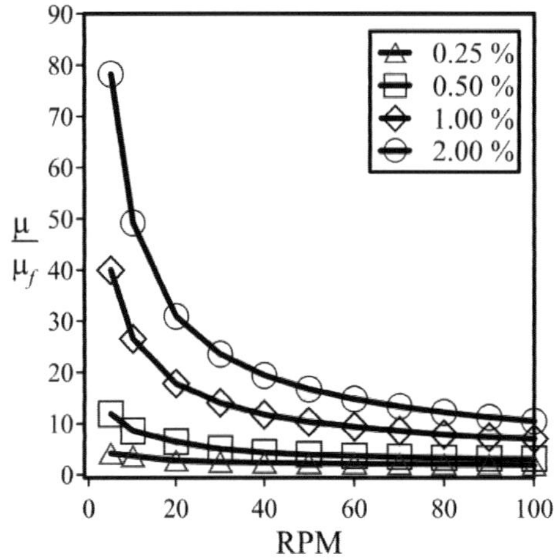

Figure 4. Relative viscosity coefficient of water based nanofluids depending on RPM.

The flow behavior index and the flow consistency index of investigated nanofluids are represented in table 1. It demonstrates that the flow behavior index decreased and the flow consistency index increased with increasing particle concentration. Thus, it was shown that non-Newtonian properties increase with increasing particle concentration.

Table 1. Flow behavior index and flow consistency index depending on particle concentration

φ	n	K (mPa·s^{-n})
0.0025	0.84	6.37
0.0050	0.55	24.0
0.0100	0.42	101
0.0200	0.33	230

Similar studies of ethylene glycol-based nanofluids did not reveal non-Newtonian properties. The nanofluid remained Newtonian at all considered concentrations. The dependence of relative viscosity coefficient of ethylene glycol-based nanofluid on nanodiamond concentration is shown in figure 5. Figure 5 shows the dependence of the coefficient of viscosity on nanodiamond concentration predicted by Einstein's theory. Also figure 5 represent that the relative viscosity coefficient of nanofluid is much higher than the coefficient predicted by Einstein's formula:

$$\frac{\mu}{\mu_f} = 1 + \frac{5}{2}\varphi \quad (3)$$

where μ_f is the viscosity coefficient of base fluids.

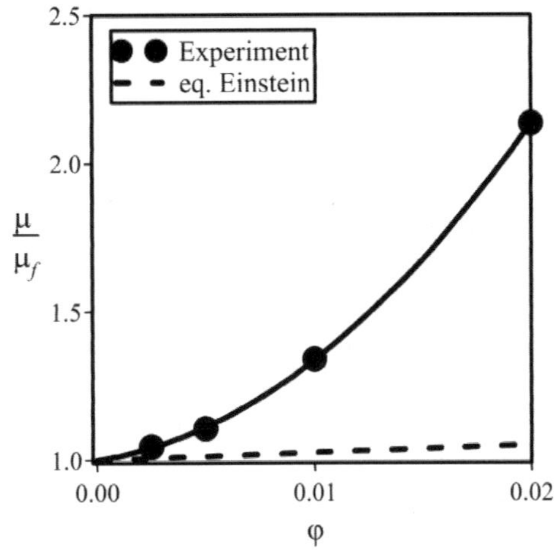

Figure 5. Relative viscosity coefficient of ethylene glycol based nanofluids depending on particle concentration.

5. Conclusion

Complex experimental study of the coefficient of thermal conductivity and the viscosity coefficient of nanodiamond nanofluids was carried out.

It was obtained that the coefficient of thermal conductivity of nanofluids increases with increasing nanodiamond concentration in all considered cases. It is shown that the increase in the thermal conductivity coefficient is greater for fluids with lower thermal conductivity. The experimental value of thermal conductivity of nanodiamond nanofluid is well described by the Maxwell formula (1).

It is shown that viscosity coefficient increases with increasing nanodiamond concentration. The experimental value is much higher than the value predicted by the Einstein formula. The strong influence of base fluid properties on the rheology of nanofluids is noteworthy. It is established that water-based nanofluids become non-Newtonian with increasing nanodiamonds concentration. The obtained data are well described by a power law (see formula (2)). The flow behavior index and the flow consistency index of the investigated nanofluids were determined (presented in table 1). However, the ethylene glycol-based nanofluids remain Newtonian.

Acknowledgments

The work is supported by the Ministry of Education and Science of the Russian Federation Government contract with Siberian Federal University in 2017 (No. 16.8368.2017).

References

[1] Choi U S 1995 *Development and Applications of Non-Newtonian Flows* ed Siginer D A and

IOP Conf. Series: Journal of Physics: Conf. Series **899** (2017) 032020 doi:10.1088/1742-6596/899/3/032020

Wang H P FED-vol **231**/MD-vol **66** (New York: ASME) p 99.

[2] Yu W, France DM and Choi S U S and Routbort J L 2007 Review and assessment of nanofluid technology for transportation and other applications ANL/ESD/07-9.

[3] Wang X Q and Mujumdar A S 2007 *Int. J. Therm. Sci.* **46** 1-19.

[4] Xie H, Wang J, Xi T and Ai F 2002 *J. Appl. Phys.* **91** 45684572.

[5] Eastman J A, Choi S U S, Li S, Yu W and Thompson L 2001 *J. Appl. Phys. Lett.* **78** 718-720.

[6] Tyler T O, Shenderova G, Cunningham J, Walsh J and Drobnik G 2006 *Diam. Relat. Mater.* **15** 2078-2081.

[7] Chopkar M, Das P K and Manna I 2006 *Scripta Mater.* **55**(6) 549-552.

[8] Chen G, Yu W and Singh D 2008 *J. Nanopart. Res.* **10** 1109-1114.

[9] Mintsa H A, Roy G, Nguyen C T and Doucet D 2009 *J. Therm. Sci.* **48** 363.

[10] Eastman J A, Choi S U S, Li S, Yu W and Thompson L J 2001 *J. Appl. Phys. Lett.* **78** 718.

[11] Pryazhnikov M I, Minakov A V, Rudyak V Ya and Guzei D V 2017 *Int. J. Heat Mass Transfer.* **104** 1275–1282.

[12] Torii S and Yang W J 2009 *J. Heat Transfer.* **131** (043203) 1-5.

[13] Xie H, Yu W and Li Y 2009 *J. Phys. D: Appl. Phys.* **42** (095413) 1-5.

[14] Yeganeh M, Shahtahmasebi N, Kompany A, Goharshadi A K, Youssefi A and Šiller L 2010 *Int. J. Heat Mass Transfer.* **53** 3186-3192.

[15] Minakov A V, Guzei D V, Pryazhnikov M I, Lobasov A S and Rudyak V Ya 2015 *J. eng. phys. thermophys.* **88** 149-162.

[16] Rudyak V Ya, Minakov A V, Smetanina M S and Pryazhnikov M I 2016 *Dokl. Phys.* **61**(3) 152-154.

IOP Conf. Series: Journal of Physics: Conf. Series **899** (2017) 032021 doi:10.1088/1742-6596/899/3/032021

Generation of a wall jet to control unsteady cavitation over a 2D hydrofoil: visualization and hydroacoustic signal analysis

Mikhail V Timoshevskiy[1] and Ivan I Zapryagaev[1]

[1] Kutateladze Institute of Thermophysics of SB RAS, 1, Lavrentyev Ave., Novosibirsk, 630090, Russia

timoshevskiy.mv@gmail.com

Abstract. In the paper, the possibility of active manipulation of unsteady cavitating flow over a 2D hydrofoil, a scaled-down model of Francis turbine guide vane (GV), was tested. The flow manipulation was implemented by generating of a wall jet at various flow rates through a spanwise slot nozzle in the foil surface. The experiments were carried out at two attack angles of 3° and 9°. Different cavitation conditions were reached by varying the cavitation number and injection velocity. Temporal and spatial cavity characteristics were studied by means of a high speed visualization. Hydroacoustic measurements were performed in order to investigate pressure pulsations spectral characteristics. It was found that the wall jet generation technique appears to be effective in suppressing cavity unsteady behavior or at least to reduce the corresponding pressure pulsations at low inclination angle and less efficient at high angles of attack, allowing only minor decrease of pressure pulsations.

1. Introduction

Cavitation occurs widely in hydraulic machinery and cause flow disruptions such as a pressure and flow pulsations, erosion and efficiency decrease. The suppression or control on development of cavitation process is one of the approaches in order to improve hydraulic machinery efficiency and safety. One of the promising techniques of cavitating flow control is a generation of a wall jet along hydrofoil surface in order to input additional momentum in to boundary layer and to overcome adverse pressure gradient [1]. However, the literature lacks on the experimental studies of cavitation manipulation by generation of wall jet, which are obviously necessary to improve the method and facilitate its implementation within the real scale hydraulic equipment. The paper aims at experimental verification of possibility to control the intensity and dynamics of partial cavities at the unsteady regimes, by generating a wall jet along the suction side of a scaled-down model of guide vanes of a Francis turbine through a spanwise slot channel in the surface and assessment of the effectiveness of the considered approach of flow manipulation. In order to investigate the wall jet generation method efficiency several experimental technique were applied, namely, high speed visualisation to analyse spatial and temporal evolution of the attached cavities and hydroacoustic measurements to evaluate pressure pulsation intensity. The experiments were carried out at attack angles of 3° and 9° and various values of the cavitation number.

2. Experimental setup and measurement techniques

The experiments were carried out in the cavitation tunnel in Kutateladze Institute of Thermophysics SB RAS. Description of the test rig, experimental conditions and measurement techniques used can be found in [2]. The test object was a scaled-down model of guide vanes (GV) of a Francis turbine which geometry was modified in such a way to create a spanwise slot channel in its upper surface to generate a wall jet so that the hydrofoil remained two-dimensional but became asymmetric. The detailed information about the test object is given in [3] and [4]. The investigation was performed for the attack angle α of 3° and 9° at the unsteady cavitating regimes with the cavitation numbers $\sigma = (p_{in}-p_V)/(\rho U_0^2/2)$ of 0.87 and 0.82 for α = 3° and σ = 1.56 for α = 9°, where p_{in} is the static pressure at the inlet of the test channel, p_V is the saturation vapor pressure of the operating liquid (distilled water), ρ is the liquid density, by changing the static pressure and/or the mean flow velocity U_0 measured by PIV in the central longitudinal section close to the test channel inlet.

3. High-speed visualization and hydroacoustic measurements

3.1. Low attack angle

At the attack angle of 3° cavitation is initiated right behind the leading edge at the distance of 0.11*C* on the suction side of the GV profile in form of travelling bubbles merging into a single attached cavity (not presented, the detailed information about the inception of cavitation can be found in previous paper [4]). At σ = 0.87, the attached cavity becomes unstable and starts to oscillate at the reduced frequency $\mathrm{St} = fL_c^{max}/U_0 = 0.08$ with reaching its maximal length L_c^{max} of 0.51*C* (Fig. 1-a.1).

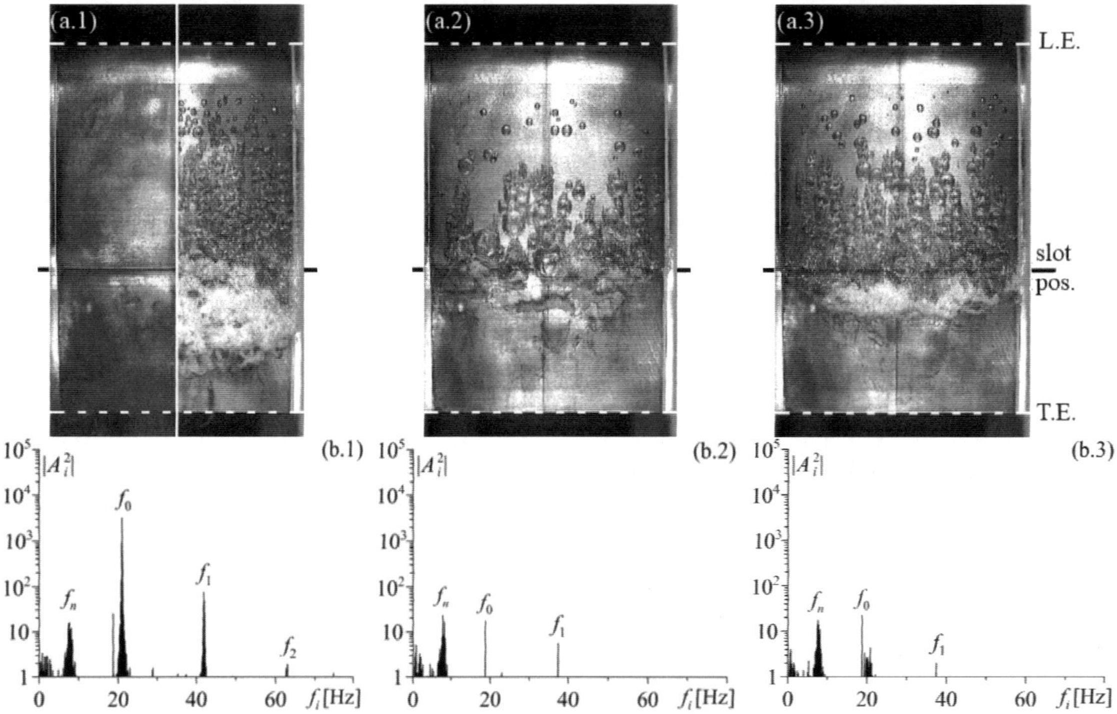

Figure 1. (a) Photographs of the partial cavities (top view) occurring on the suction side of the GV at α = 3° and (b) pressure pulsations power spectra for the following flow conditions: σ = 0.87 when (1) $U_{inj}/U_0 = 0$, (2) $U_{inj}/U_0 = 0.73$ and (3) $U_{inj}/U_0 = 1.28$. Half images indicating two phases of the attached cavity evolution when it has (left) the shortest (at the moment whereupon the cavity starts to grow) and (right) longest length (just before the cavity interface breakup) are shown together in the same picture. The flow direction is from the top. At the pressure pulsations power spectra the peaks frequencies are marked as f_n – the natural frequency of the rig, f_0 – fundamental signal harmonic and f_1 and f_2 – are multiple harmonics.

Once the injection is started (U_{inj}/U_0 = 0.24), the cavity becomes stable, changing the cavitation regime to the transitional one and decreasing the cavitation area to 0.46*C* (not shown). The increase of the injection rate to U_{inj}/U_0 = 0.49 causes the cavitation area to grow up to 0.51*C*, however the cavity dynamic remains stable. Figure 1-a.2 presents a typical cavitation patter for so called transitional cavity at U_{inj}/U_0 = 0.49. The following increase of the wall jet velocity $U_{inj} > 0.49U_0$ results in a gradual cavitation area growth, reaching the its maxima of 0.59*C* at U_{inj}/U_0 = 1.28 without any changes in cavitation pattern despite the increased cavitation area streamwise dimension (Fig. 1-a.3).

This wall jet impact on cavity dynamic is clearly observable on the wake pressure pulsations power spectra where the peak height of pressure pulsations corresponding to the fundamental harmonic f_0 at the St = 0.08 (f_0 = 21Hz) on Figure 1-b.1 (completely corresponding to the cavity oscillation frequency extracted from the analysis of visual data) decreases as the injection rate rises. At the wall jet velocity of $U_{inj} = 0.24U_0$ the dimensionless amplitude of the peak decreases by 70% (comparing to the unforced case) and at the attached cavity becomes stable. At $U_{inj} > 0.24U_0$ the peak amplitude diminishes and becomes comparable to noise level (see Figure 1-b.2, b.3 and Figure 3).

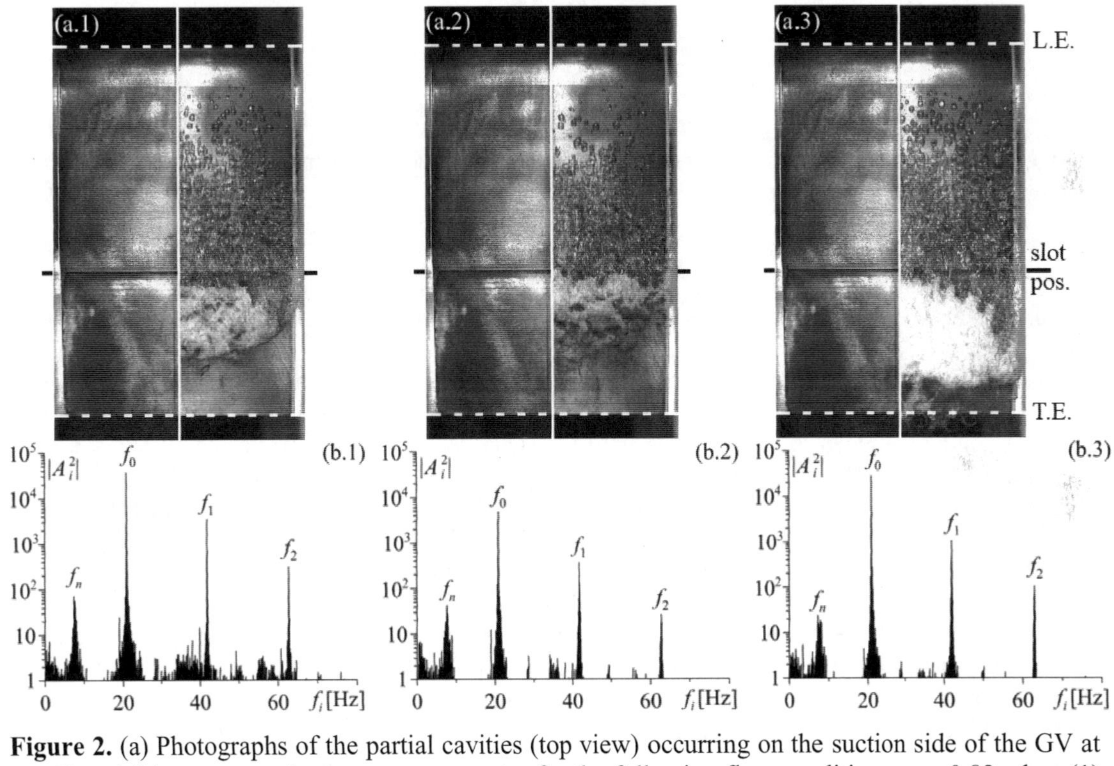

Figure 2. (a) Photographs of the partial cavities (top view) occurring on the suction side of the GV at α = 3° and (b) pressure pulsations power spectra for the following flow conditions: σ = 0.82 when (1) $U_{inj}/U_0 = 0$, (2) $U_{inj}/U_0 = 0.75$ and (3) $U_{inj}/U_0 = 1.25$. Half images indicating two phases of the attached cavity evolution when it has (left) the shortest (at the moment whereupon the cavity starts to grow) and (right) longest length (just before the cavity interface breakup) are shown together in the same picture. The flow direction is from the top. At the pressure pulsations power spectra the peaks frequencies are marked as f_n – the natural frequency of the rig, f_0 – fundamental signal harmonic and f_1 and f_2 – are multiple harmonics.

A minor decrease in cavitation number down to 0.82 leads to that the wall jet cannot suppress the cavity unsteadiness but it is still capable to affect the cavity length. Starting with cavity length of 0.58*C* (Fig. 2-a.1) it reduces it to 0.52*C* at $U_{inj} = 0.25U_0$ and consequently increases up to 0.7*C* at U_{inj} = $1.25U_0$ (Fig. 2-a.3). Likewise the liquid injection impacts on the pressure pulsations decreasing the amplitude by 61% for the injection rate of U_{inj}/U_0 = 0.25 and reaching minimal value of the amplitude reduction of 87% at $U_{inj}/U_0 = 0.75$ (Fig. 2-b.2). The following wall jet velocity increase causes the pressure pulsations amplitude to grow gradually up to the amplitude reduction of 25% at $U_{inj}/U_0 = 1.25$ (see Figure 3). Thus, despite the fact, that the cavity dynamic remains unsteady the wall jet generation appears to be effective at suppressing the amplitude of pressure pulsations at the unsteady regime.

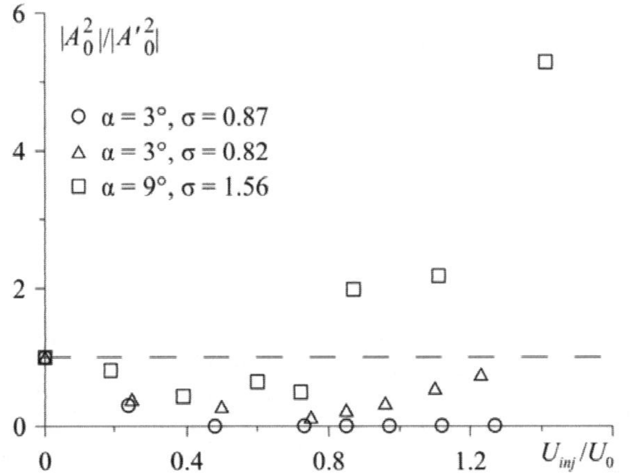

Figure 3. The dependence of the pressure pulsations amplitude corresponding to the fundamental harmonic f_0 normalized by the amplitude for zero injection rate vs the wall jet velocity for attack angles α of 3° when cavitation numbers 0.87, 0.82 and α = 9° when σ = 1.56.

3.2. High attack angle

At higher inclination at α = 9° the attached cavity is formed behind the leading edge at the distance of $0.01C$ in form of a sheet cavity (see details in previous paper [4]). At σ = 1.56 the attached cavity becomes unstable and starts to oscillate (Fig. 4-a.1). However, the periodicity of cavity detachment appears to be disturbed as there is no narrow band of detachment frequencies. The cavity length and the corresponding reduced frequency (measured for each cavity detachment) vary within $0.3 < L_c^{max} < 0.67$ and $0.14 < \mathrm{St} < 0.5$ respectively. Moreover, the pressure pulsations power spectra shows that there is a narrow peak at St = 0.07 although the cavity detachment frequencies are widely dispersed (Fig. 4-b.1) (cavity detachment also causes a pressure pulsations, however their amplitudes appears to be significantly lower comparing to the amplitudes of pressure waves caused by cavitation surge and could be identified on the power spectra as a higher noise level and wider peaks). The injection appears to be less efficient on this angle of attack slightly reducing the cavity length variation to $0.4 < L_c^{max} < 0.65$ and narrowing the detachment frequencies band to $0.15 < \mathrm{St} < 0.45$ for $U_{inj}/U_0 = 0.6$. At higher injection ratios the cavity length variation increases to $0.4 < L_c^{max} < 0.67$ for $U_{inj}/U_0 = 0.73$ (Fig. 4-a,b.2) and to $0.35 < L_c^{max} < 0.75$ for $U_{inj}/U_0 = 1.42$ (Fig. 4-a,b.3), increasing the detachment frequencies band to $0.12 < \mathrm{St} < 0.48$ for $U_{inj}/U_0 = 0.73$ and narrowing it to $0.11 < \mathrm{St} < 0.3$ for $U_{inj}/U_0 = 1.42$. The peak on the pressure pulsations power spectra has a similar to the 3 degree case dependency reducing the amplitude by 57% at wall jet velocity of $U_{inj} = 0.39U_0$, and significantly increasing the amplitude by more than 500% at $U_{inj} = 1.42U_0$ (see Figure 3), although the amplitude values remain lesser comparing to the 3 degree case.

Such cavity behavior at α = 9° is conditioned by the following. As was mentioned previously there are two main mechanisms of cavity unsteadiness – cavitation surge and re-entrant jet instability. Cavitation surge occurs for relatively long and thin cavities when they become susceptible to pressure fluctuations with usual dimensionless frequencies St ≈ 0.1 and producing pressure waves propagation within the duct. Cloud cavitation appears for a relatively short and thick cavities when the adverse pressure gradient is sufficient enough to form the re-entrant jet separating the cavity from the foil with the dimensionless frequencies St = 0.4 - 0.5. As the GV maximal thickness is locates at $x/C = 0.44C$ so the flow on the GV suction side remains contracted in wide area which in turn hinder the re-entrant jet formation. Also slot channel forms a backwards facing step which act as an obstacle for re-entrant jet. As a consequence, the cavity appears to be long enough to become susceptible for pressure fluctuations and cause cavitation surge unsteadiness. Thus, the cavity becomes exposed to both instability mechanisms simultaneously resulting in such intermittent unsteady cavitation. In this case, the injection causes not only the pressure rise, slightly decreasing the evaporation rate, on the GV suction side as in 3 degree case but also it hampers the re-entrant jet formation and shifting the unsteady intermittent cavitation regime to cavitation surge.

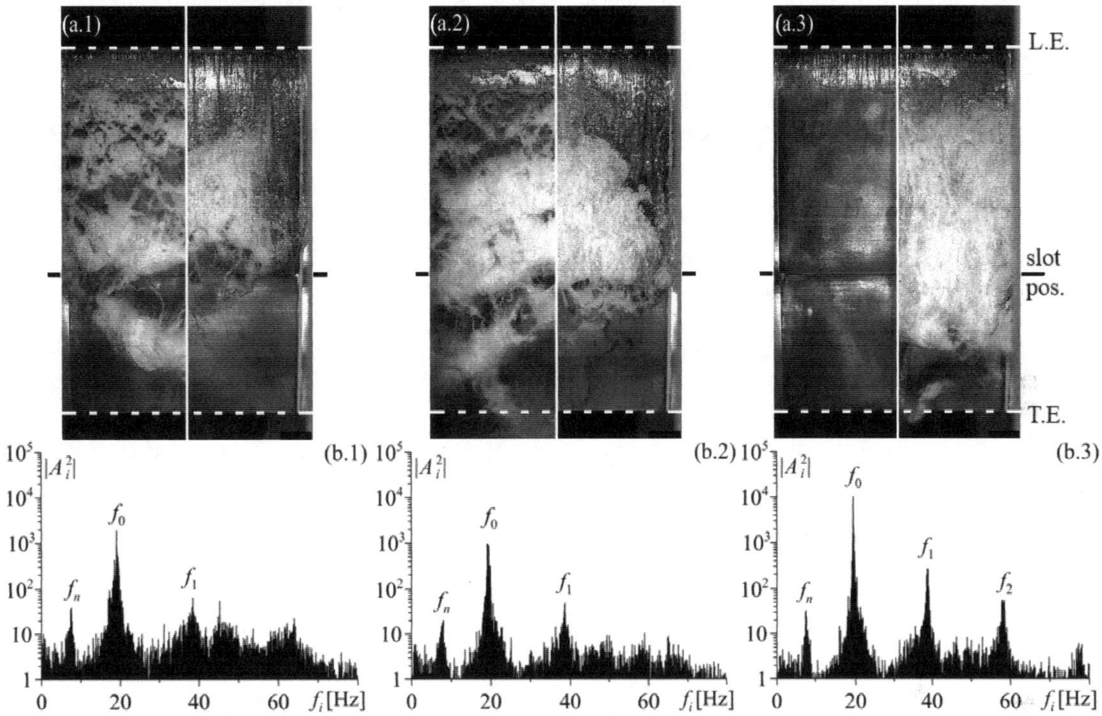

Figure 4. (a) Photographs of the partial cavities (top view) occurring on the suction side of the GV at $\alpha = 9°$ and (b) pressure pulsations power spectra for the following flow conditions: $\sigma = 1.56$ when (1) $U_{inj}/U_0 = 0$, (2) $U_{inj}/U_0 = 0.73$ and (3) $U_{inj}/U_0 = 1.42$. Half images indicating two phases of the attached cavity evolution when it has (left) the shortest (at the moment whereupon the cavity starts to grow) and (right) longest length (just before the cavity interface breakup) are shown together in the same picture. The flow direction is from the top. At the pressure pulsations power spectra the peaks frequencies are marked as f_n – the natural frequency of the rig, f_0 – fundamental signal harmonic and f_1 and f_2 – are multiple harmonics.

4. Conclusions

The wall jet generation on the suction side of the Francis turbine GV has proved to be an effective tool for cavitating flow manipulation at the injection rate $U_{inj}/U_0 < 1$, especially on low attack angles. On the contrary, the application of this method with wall jet velocity could be undesirable from the viewpoint of cavitation manipulation. The observed intermittent unsteady cavitation regime on the tested hydrofoil appears to be less susceptible to control by means of liquid injection, however, due to the widespread frequencies of cavity oscillations the possibility of resonant phenomena is significantly lower comparing to the case of typical unsteady cavitation. Combining this phenomena with the effect of pressure pulsations reducing and the simplicity of wall jet implementation on real scale hydro power plants may give a good economical advantage.

Acknowledgments

The research was funded by a grant from the Russian Scientific Fund (Project No. 16-19-00138 supervised by Dr. P.A. Kuibin in IT SB RAS).

References

[1] Kozhukharov P G, Hadjimikhalev V H, Mikuta V I and Maltzev L I 1985 *ASME Fluids Eng. Div.* **31** 67–74
[2] Kravtsova A Yu, Markovich D M, Pervunin K S., Timoshevskii M V and Hanjalić K 2014 *Thermal Engineering* **61** No. 14 1003-1010
[3] Timoshevskiy M V, Churkin S A, Kravtsova A Yu, Pervunin K S, Markovich D M and Hanjalić, K 2016 *International Journal of Multiphase Flow* **78** 75-87
[4] Timoshevskiy M V, Zapryagaev I I, Pervunin K S and Markovich D M 2016 Proc. Int Symp. And School of Young Scientist «Interfacial Phenomena and Heat transfer » vol 84 (Les Ulis Cedex: EDP Sciences - Web of Conferences) pp (00039)-4
[5] Timoshevskiy M V, Zapryagaev I I, Pervunin K S and Markovich D M 2016 Proc. Int. Conf. on Methods of Aerophysical Research (Perm) vol 1770 (New York: American Institute of Physics) pp (030026)-8

Controlling the coarse droplets streams by blade profile shape modification in the last stages of steam turbines

V Tishchenko[1], R Alekseev[1], I Gavrilov[1], V Gribin[1], A Tishchenko[1]

[1]National Research University "Moscow Power Engineering Institute", Krasnokazarmennaya 14, Moscow, 111250 Russia

AlexeevRA@mpei.ru

Abstract. This work is devoted to investigation of wet steam flow in steam turbines blade passages. The methodology of flow laser diagnostic system application in order to study the trajectories and parameters distributions of coarse droplets in wet steam flow is presented. Obtained data was used to fit mathematical model of coarse droplets generation (as a result of droplet interaction with the blade surfaces) to experimental results. The calculation performed with this model allowed to optimize the shape of stator blade profile in order to control the streams of coarse droplets and minimize their diameters downstream the cascade.

1. Introduction

Presence of a liquid phase in the steam flow is the feature of the work of steam turbine last stages. It imposes strict requirements on the reliability and efficiency of the last stages operation conditions. The generation of discrete liquid phase intensifies a number of negative processes that affect the parameters of the steam flow. Detailed experimental and numeric studies have shown that the efficiency of the last stages is determined by the thermodynamic processes of condensation on the surfaces of fine droplets (the diameter of which are less than 1 μm) [1]. These phenomena are well studied [2, 3] and may be modeled using modern CFD codes. At the same time, the reliability of the rotor blades is affected by erosion damage caused by coarse droplets (with diameter more than 10 μm) which are formed upstream in stator blades passages. Currently, the extraction of liquid films from blade surfaces and erosion-resistant materials are used to minimize the negative effects connected with this process. However, the experience of steam turbines operation shows, despite the effectiveness of these methods, erosion damage remains one of the most important problems that must be solved in order to design efficient steam turbine flow path [4]. The profiling of stator blades shape with the aim of controlling the coarse droplets streams in blade passages seems to be a very effective method in the combination with the traditional ways of erosion damage reduction. But in order to implement this technique the understanding of coarse droplets development processes in blade cascades is needed.

The generation of erosion-hazardous drops is a complex process, which is realized due to several mechanisms. At the moment, there is only indirect knowledge about the nature of the formation and motion of coarse drops in the blade passages of steam turbines. It is connected with the difficulties of experimental investigations. It imposes a number of limitations on the possibility of minimizing erosion damage and improving the reliability of the blade. The generalized behavior of coarse droplets streams in a blade passage is shown in figure 1 (according to [5]).

It should be noted that the formation and movement of the liquid phase particles (except the primary droplets) is virtually impossible to model, because a lot of factors must be taken into consideration for its mathematical description [6]. But from another side the droplets interaction with dry and wetted solid surfaces is studied well in idealized laboratory conditions. And there are a lot of statistical models that describe the processes of droplet impact on surfaces [7] with the further effects: rebounding, splashing, depositing. The experience of these models applications has shown that their free parameters can be fitted to real engineering systems [8]. The application of modern experimental approaches, based on flow laser diagnostics systems, allows studying the features of motion of erosion-hazardous droplets in more detail [9]. On the basis of obtained experimental data, parameters of statistical models can be determined for the conditions of wet steam flow in steam turbines. Formulated by this way mathematical model of motion and generation of coarse droplets, can be used to optimize the shape of stator blade profile in order to control the streams of coarse droplets and minimize the erosion damage of rotor blades.

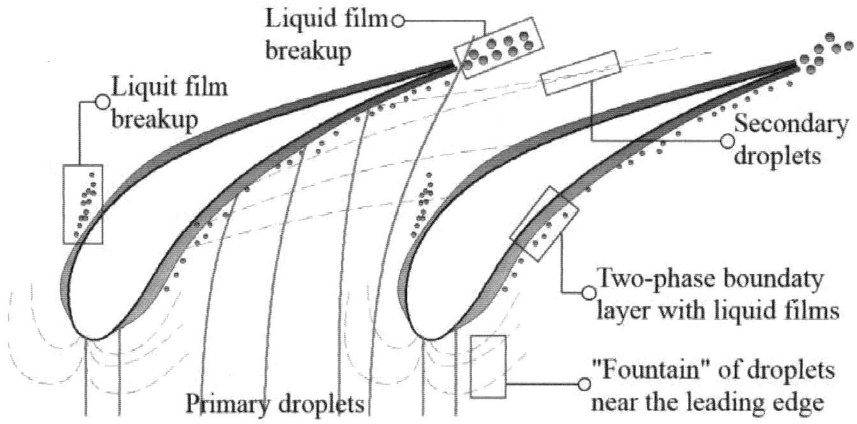

Figure 1. The scheme of movement of liquid phase particles in the nozzle blade cascade.

2. The experimental facility and object of study

In order to obtain the behavior of droplets motion in the nozzle blade cascade, the experimental studies were performed. The investigations were carried out in the experimental facility Wet Steam Circuit (WSC) in the turbine laboratory of the Moscow Power Engineering Institute (MPEI). This experimental plant is used to study the flow of superheated, saturated and wet steam in stationary channels. The main feature of this rig – steam for experimental investigations is extracted from the operating steam turbine. It was designed in order to study the polydispersed coarse droplets movement in wet steam flow.

The object of study for current investigation was the flat nozzle blade cascade. The chord of the blades was determined on the basis of Reynolds number, and the number of blades is limited by the experimental facility dimensions. The geometry of the flat cascade, which consists of 5 blades, is shown in figure 2a.

In order to obtain the characteristics of liquid phase in the blade passage, laser diagnostic system “POLIS” was used. It implements the PTV (particle tracking velocimetry) method that allows to obtain instantaneous velocity vectors for each droplet detected by the method in studied flow domain. The optical scheme of laser diagnostic system is shown in figure 2b. The wet steam flow in the blade passage is illuminated by a plane laser knife formed by a dual impulse laser. It is directed through the endoscope into working part and illuminates droplets moving in the inter-blade channel. The high-speed PIV camera takes photos of them. The obtained droplet flow photos are used as initial data for the PTV method. This technique obtains irregular vector field for each pair of photos. In order to increase statistical significance of the results, 1000 photos were made for each studied conditions. This allows detecting total amount of droplets approximately equal 20e6.

Operation conditions were controlled by measurement of total pressure (p_0), total temperature (T_0) and initial steam wetness (y_0) upstream the experimental cascade and average static pressure (p_1) downstream the blade cascade. Different operating conditions have been considered. In this article, we show results only for one of them: total pressure $p_0 = 60000$ Pa; theoretical Mach number downstream the nozzle blade cascade $M_{1t} = 0.8$; initial steam wetness $y_0 = 3\%$; the liquid to vapor phase density ratio upstream the blade cascade $\bar{\rho}= 2684.7$. The $\bar{\rho}$ parameter is determined as follows:

$$\bar{\rho} = \frac{\rho_l}{\rho_g}, \tag{1}$$

where ρ_l-initial liquid density, ρ_g- steam density.

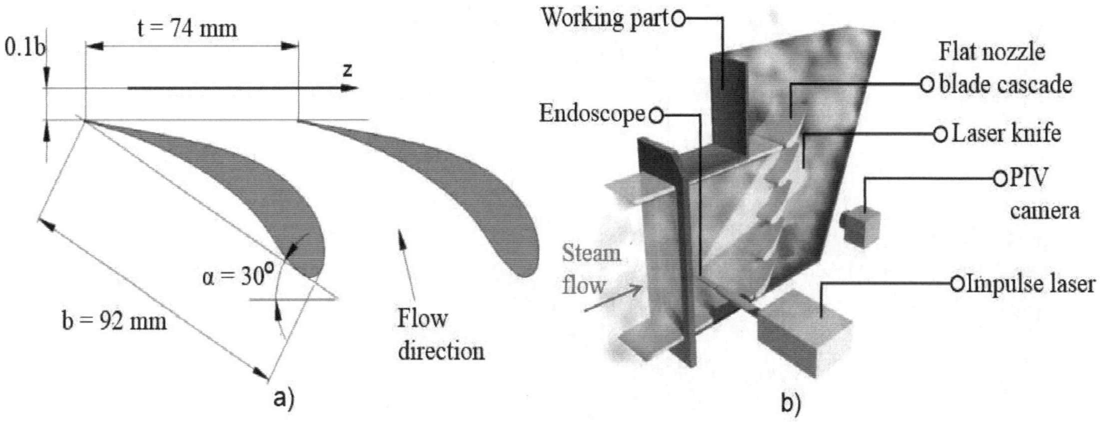

Figure 2. Geometry of the nozzle blade cascade (a); optical scheme of the laser diagnostic system (b).

3. Droplets motion structure in the studied blade passage

Here we present a brief analysis of experimental data. A more detailed analysis is presented in [10]. Experimental data obtained by PTV method was used to investigate the behavior of droplets streams in the studied blade passage. The laser diagnostic system provides the visualization of the discreate-phase flow structure in the blade passage. The photo of illuminated coarse droplets is presented in figure 3a. These data confirm the theoretical conception about the processes that take place in the studied flow domain (see figure 1). There is a "fountain" of secondary particles of liquid phase (areas 1 and 2) at the inlet of the nozzle blade cascade.The maximum concentration of liquid particles is observed near the blade pressure side (zone 3). They form the two-phase boundary layer. In addition, the area near the blade suction side should be noted (zone 4). It is situated in location, which represents the "aerodynamic shadow" for most of primary and secondary coarse droplets – the concentration of liquid particles is minimal for this zone.

The instantaneous velocity field of droplets obtained by PTV method in the studied blade passage is shown in figure 3b. Processing a large number of these fields allows to analyze parameters of droplets in each point of the blade passage. The obtained experimental data shows that the secondary droplets originated from the blade pressure side move along the surface and form a "two-phase boundary layer" – the ensemble of steam boundary layer, liquid film and secondary droplets. The distribution of liquid particles parameters in this area differs considerably from the zone which is situated in flow core. It can be clearly seen from the comparison of the droplets parameters distributions shown in figure 3c. In this figure velocities and angles of each droplet detected by PTV method in the point 1 (flow core) and 2 ("two-phase boundary layer") are presented. As one can see in the "two-phase boundary layer" (point 2) liquid particles move in a broad range of angles (-30°...120°) and basically have low velocity in comparison with the flow core (point 1).

IOP Conf. Series: Journal of Physics: Conf. Series **899** (2017) 032022 doi:10.1088/1742-6596/899/3/032022

4. Mathematical model validation

In the current study we attempted to simulate the formation processes of secondary coarse droplets which are originated from the blade surfaces. The trajectories of coarse droplets were computed under the assumption that only aerodynamic force from the steam flow acts on the liquid particle. So the equation of motion is following:

$$\frac{1}{2}A_d C_x \rho_g \left|\vec{c_g} - \vec{c_d}\right| \left(\vec{c_g} - \vec{c_d}\right) = \vec{a_d}, \quad (2)$$

where A_d– cross-sectional area of the droplet; C_x - aerodynamic drag coefficient of spherical particles; c_g- velocity of the steam flow; c_d- droplet velocity; a_d- droplet acceleration; m_d- droplet mass.

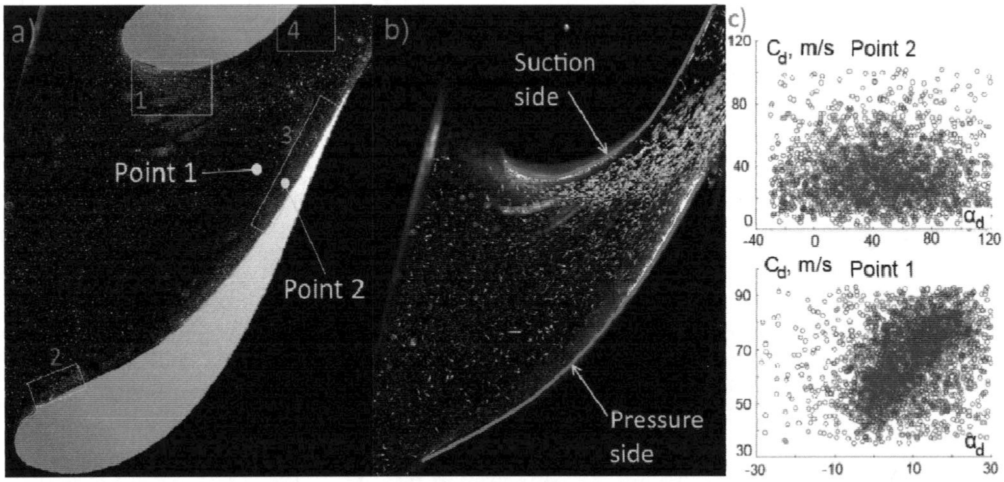

Figure 3. Photo of illuminated droplets(a); instantaneous vector droplets velocity field (b); distribution of droplets velocities and angles in flow domain area (c).

Steam flow parameters were obtained by CFD code. In current study the reverse impact of liquid phase on main flow was not considered. According to flow operation conditions one can mark out 2 scenarios of droplets interaction with the blade surfaces [7]. They are presented in figure 4. The first one is a full splashing process: after the impact on the surface the primary liquid particle disintegrates and polydisperse stream of secondary droplets leaves the wall. At the second scenario a fraction of the droplet mass deposits on the wall and forms the liquid film.

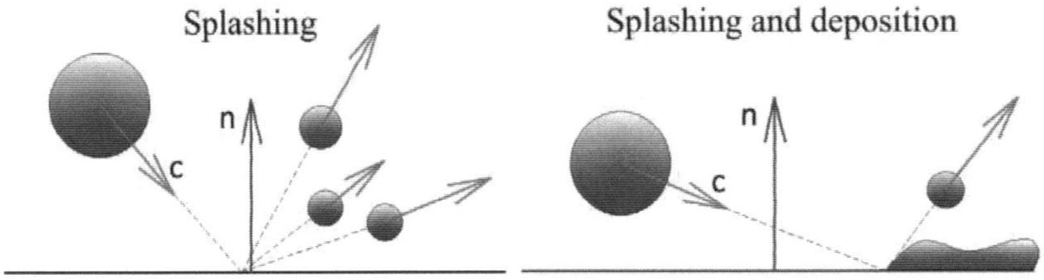

Figure 4. Scenarios of droplet interaction with the surface.

In order to take into account these processes and calculate the interaction of primary droplets with the blade surfacesthe Mundo model [11] was used. As input parameters for this model the initial velocity of the primary droplets (obtained by PTV method) and initial distribution of the primary droplets diameters (obtained by fingerprint probe) were used. The parameters of this model were fitted to the results obtained by PTV method. They were determined by comparing the parameters distributions of splashed droplets obtained by the experiment and the mathematical simulation. It is important to note that using such splashing model in wet steam flow conditions is a rough approximation. But this can help to estimate the behavior of the coarse droplets streams in turbines blade passages.

The results of droplets motion simulation for studied condition are presented in figure 5a. The primary droplets correspond to the red trajectories and the secondary droplets correspond to green trajectories. The structure of the liquid phase motion in the blade passage obtained by numerical investigation is in a good agreement with the experimental data (see figure 3a). As an example, the comparisons of experimental (obtained by PTV method) and numerical liquid particles parameters (velocity and angle) distributions in the area, marked as blue square in figure 5a, are shown in figures 5b and 5c. As one can see they don`t match but have the same behavior.

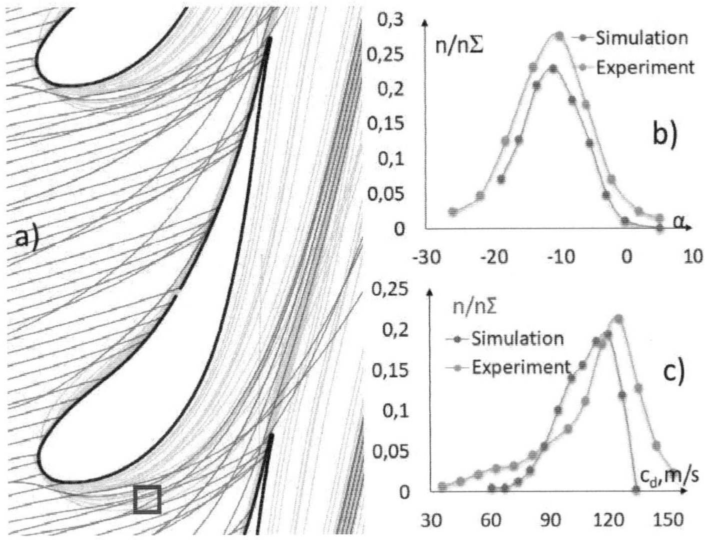

Figure 5. Results of simulation of droplets in the studied blade passage (a); comparison of droplets angles distribution (b); comparison of droplets velocities distribution (c).

5. Strategy of the blade modification

Based on the experimental and numerical data described in sections 3 and 4, one can pick out the following features of coarse secondary droplets movement in the blade passage.

Droplets in "fountain" which originate from the leading edge near the suction side leave the blade passage without interaction with the surfaces. So, they cannot be evacuated from turbine flow path by means of separation system.

Secondary droplets formed due to the splashing process on the blade pressure side, move inside the "two-phase boundary layer" and doesn`t leave it. The experiment investigation carried out in [12] has shown that separation slots on the blade surface effectively extract the liquid film but they are not able to evacuate these coarse droplets moving above the liquid film surface.

So, in the nozzle blade passage the coarse erosion-hazard droplets are generated and they cannot be influenced by the steam turbine liquid phase separation system. In order to minimize the erosion damage of rotor blades caused by these droplets we have to decrease the size of secondary droplets and increase the fraction of the primary droplet mass which deposits on the wall and forms the liquid film. It can be achieved by changing the shape of the blade profile. As was derived in [13], the part of primary droplet which remains on the wall after splashing process can be termed as:

$$\frac{M_{spl}}{M_0} = min\left[2.9 \cdot 10^{-4}\sqrt{Re_n}(We - We_c), 0.75\right], \quad (3)$$

where M_0 - primary droplet mass; M_{spl} – mass of secondary droplets generated by splashing process; Re_n – droplet normal Reynolds number; We - Weber number of droplet; We_c - the critical Weber number, defines the minimum droplet impact energy which is needed for intensification of splashing process. As one can see from the equation 3, in order to minimize the total mass of secondary droplets, the normal component of the primary liquid particle velocity vector should be decreased. This can be achieved by decreasing the angle between the surface normal and droplet velocity vector (see figure 4). This statement was used to optimize the geometry of nozzle blade profile. Modification of the baseline blade profile (see figure 2a) has been carried out by the blade parameterization method, described in [14]. Using the obtained in this study model of coarse droplets movement it is necessary to achieve the reduction of droplets diameters downstream the blade cascade. In figure 6 the modified profiles which meet this requirement are shown. These two modifications have a pronounced inlet area where the curvature of suction and pressure sides varies insignificantly (compared with the baseline geometry). The angle of the tangent to the surface in this region is close to average inlet angle of the primary droplets. The shape of the pressure sides for the modified profiles has a more uniform distribution of curvature in comparison with the baseline geometry. And the change in the tangent angle is closest to the change in the angle of the droplet trajectories in the channel. As one can see from figure 6 the suction side shape of the modified profiles differs considerably. It was done in order to experimentally study the impact of the main flow pressure distribution on the processes of liquid film formation. It is important to note that in order to confirm the efficiency of the chosen approach the experimental investigations for these new profiles geometries are needed. It will be done in the future.

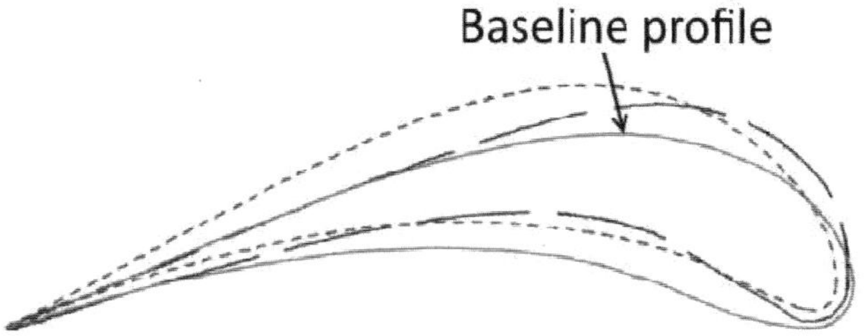

Figure 6. Geometry of the modified profiles.

6. Conclusions

The following conclusions can be made:

1. The new method of experimental investigation of coarse droplets movement in steam turbines blade passages has been presented in this study.
2. Model which describes the interaction between droplets and solid surfaces has been fitted to experimental results with a good agreement.
3. The possibility of controlling streams of coarse droplets by optimizing the profile geometry is shown.

Acknowledgements

This study was supported by Russian Science Foundation (project N 16-19-10484).

References

[1] Petr V, Kolovratnik M 2014 *Proc. IMechE: J.Power and Energy* A **228** 206-15
[2] Starzmann J, Casey MM, Mayer JF, Sieverding F 2014 *Proc. IMechE: J.Power and Energy* A **228** 216-31
[3] Starzmann J et al 2016 Results of the international wet steam modeling project *Proc. Wet Steam Conference (Prague)* (Prague: Cesce vysoke uceni technicke v Praze) p 146
[4] Cai X, Ma L, Tian C, Li J, Ning D. and Xu W 2012 *Proc. Baumann Centenary Conf.* (Cambridge)
[5] Deitch ME 1996 Gasdynamics of turbine cascades (Moscow: Energoatomizdat)
[6] Filippov G, Gribin V, Tischenko A, Gavrilov I, Tischenko V 2014 *Proc. IMechE: J.Power and*

Energy A **228** 168-77
[7] Rosa N G, Villedieu P, Dewitte J, Lavergne G 2006 *Proc. 10th Int. Conf. liquid atomization and spray systems* (Tokyo) p 29
[8] Jaegle F, Senoner J.-M, Garica M, Bismes F, Lecourt R, Cuenot B, Poinsot T 2011 *Proc. of the Combustion Institute* **33** 2099-107
[9] Gavrilov I, PopovV, Sorokin I, Tishchenko V, Khomyakov S 2014 *Thermal Engineering* **61** 577-84
[10] Sorokin I, Gribin V, Gavrilov I, Tishchenko A, Tishchenko V, Popov V, Khomyakov S, Alexeev R 2016 *Proc. Wet Steam Conference(Prague)* (Prague: Cesce vysoke uceni technicke v Praze) p 201
[11] Mundo C, Sommerfeld M, Tropea C 1995 *Int J. Multiphase Flow* **21(2)** 151–73
[12] Gribin V, Tishchenko A, Gavrilov I, Popov V, Sorokin I, Tishchenko V, Khomyakov S 2016 *Power Technology and Engineering* **50** 180-187
[13] Yrin A, Weiss D 1995 *J. Fluid Mechanics* **283** 141-73
[14] Gribin V, Tishchenko A, Gavrilov I, Tishchenko V, Sorokin I, Alexeev R 2017 *Proc. 12th Conf. Turbomachinery Fluid dynamics and Thermodynamics* (Stokholm) paper ID: ETC2017-195

Simulating nonlinear steady-state traveling waves on the falling liquid film entrained by a gas flow

O Yu Tsvelodub[1, 2] A A Bocharov[3]

[1] Kutateladze Institute of Thermophysics SB RAS, Lavrentyev prosp. 1, Novosibirsk, 630090, Russia
[2] Novosibirsk State University, Pirogova Str. 2, Novosibirsk, 630090, Russia
[3] Rzhanov Institute of Semiconductor Physics SB RAS, pr. Lavrentieva 13, Novosibirsk, Russia, 630090
E-mail: tsvel@itp.nsc.ru

Abstract. The article is devoted to the simulation of nonlinear waves on a liquid film flowing under gravity in the known stress field at the interface. The paper studies nonlinear waves on a liquid film, flowing under the action of gravity in a known stress field at the interface. In the case of small Reynolds numbers the problem is reduced to the consideration of solutions of the nonlinear integral-differential equation for film thickness deviation from the undisturbed level. The periodic and soliton steady-state traveling solutions of this equation have been numerically found. The analysis of branching of new families of steady-state traveling solutions has been performed. In particular, it is shown that this model equation has solutions in the form of solitons-humps.

1. Introduction and problem statement

The joint flow of liquid and gas is a classical problem of hydrodynamics. Flows of thin liquid films in the presence of counter- or co-current gas flows often occur in various technological applications. Two stages of simulation are often distinguished: determining gas stresses on the film surface and further calculating the evolution of waves in the liquid. The possibility of the problem staging is justified in particular in [1, 2].

In this paper, we consider the dynamics of nonlinear waves on a fluid film entrained by the cocurrent gas flow and falling under the action of gravity along the vertical plane in a known stress field at the interface. In [2] for the system of hydrodynamic equations written in tensor form, invariant to coordinate systems, for the considered flow in the case of small flow rates ($\mathrm{Re} \sim 1$) an evolution equation for the film thickness h was obtained:

$$\frac{\partial h}{\partial t}+\frac{\mathrm{Re}}{\mathrm{Fr}}h^2\frac{\partial h}{\partial x}+\mathrm{Re}\,\tau_0 h\frac{\partial h}{\partial x}$$

$$+\varepsilon\frac{\partial}{\partial x}\left(\frac{1}{3}\varepsilon \mathrm{ReW}h^3\frac{\partial^3 h}{\partial x^3}+\frac{2}{15}\frac{\mathrm{Re}^3}{\mathrm{Fr}^2}h^5\frac{\partial h}{\partial x}(h+\tau_0\mathrm{Fr})+\frac{1}{2}\mathrm{Re}h^2\tau_0\int \hat{h}_k k\tau(k)e^{ikx}\mathrm{d}k\right)=0. \quad (1.1)$$

Here, $\mathrm{Re}=\rho h_0 u_0/\mu$ is Reynolds number, $\mathrm{Fr}=u_0^2/gh_0$ is Froude number, $\mathrm{W}=\sigma/\rho l_0 u_0^2$ is Weber number and $\varepsilon=h_0/l_0$ is relation of specific film thickness h_0 to specific wave length l_0. Direction of coordinate x coincides with the direction of gravity vector. In addition, in the equation (1.1) and in the dimensionless complexes we use the characteristic scales of velocity u_0 and time l_0/u_0. Here, σ is the surface tension, ρ is the density, $\mu\mu$ is the dynamic viscosity of liquid, g is the acceleration of gravity, τ_0 is undisturbed component of gas shear stress on the film surface, $\tau(k)=\tau_r(k)+i\tau_{im}(k)$

are the Fourier components of tangential stresses of gas due to the boundary curvilinearity, and $\hat{h}(k,t)$ are the Fourier components of the surface form expansion:

$$\hat{h}(k,t) = \frac{1}{2\pi}\int_{-\infty}^{+\infty} h(x,t)e^{-ikx}\mathrm{d}x.$$

Here it was assumed that $W\varepsilon \sim 1$. Equation (1.1) accurate to notations coincides with the equation obtained in [1]. Some differences in the coefficients are due to the fact that it is the countercurrent flow of gas that was considered in [1].

Restricting to perturbations of small but finite amplitude, using the transformation: $h = 1 + \varepsilon h_1$, and introducing slow ($t_0 = t$) and fast ($t_1 = \varepsilon t$) times into consideration in [2] it was shown that in the first approximation (fast times) perturbations of small but finite amplitude propagate with a characteristic constant velocity $c_0 = \mathrm{Re}/\mathrm{Fr}(1+\mathrm{Fr}\tau_0)$. The equation describing the nonlinear evolution of the perturbations at large (slow) times ($t_1 = \varepsilon t$) has the form [2]:

$$\frac{\partial h_1}{\partial t_1} + \frac{\mathrm{Re}}{\mathrm{Fr}}(2+\mathrm{Fr}\tau_0)h_1\frac{\partial h_1}{\partial x} + \frac{\mathrm{W\,Re}\,\varepsilon}{3}\frac{\partial^4 h_1}{\partial x^4}$$
$$+\frac{2}{15}\frac{\mathrm{Re}^3}{\mathrm{Fr}^2}\frac{\partial^2 h_1}{\partial x^2}(1+\tau_0\mathrm{Fr}) + \frac{1}{2}\mathrm{Re}\,\tau_0\int i\hat{h}_{1k}k^2\tau(k)e^{ikx}\mathrm{d}k = 0. \quad (1.2)$$

After replacement $t = bt_1$, $h_1 = AH$, $b = \mathrm{WRe}\varepsilon/3$, $\varepsilon = 0{,}4\mathrm{Re}^2(1+Fr\tau_0)/(W\,Fr^2)$, $A = 2\mathrm{Fr}b/(\mathrm{Re}(2-\mathrm{Fr}\tau_0))$ the equation (1.2) takes the form [2]:

$$\frac{\partial H}{\partial t} + 2H\frac{\partial H}{\partial x} + \frac{\partial^2 H}{\partial x^2} + \frac{\partial^4 H}{\partial x^4} + B\int_{-\infty}^{\infty} ik^2\,\tau(k)\,\hat{H}(k,t)e^{ikx}\mathrm{d}k = 0. \quad (1.3)$$

Here $B = \dfrac{\mathrm{Re}\,\tau_0}{2b} \equiv \dfrac{3\tau_0}{2\mathrm{W}\,\varepsilon}$.

The aim of this work is to find steady-state traveling periodic and soliton solutions of equation (1.3).

It is easy to study the linear stability of the unperturbed solution $H = 0$. Indeed, neglecting the nonlinear term in (1.3) for linear solutions $H \sim \exp(ik(x-ct))$, we obtain the dispersion relation:

$$c \equiv c_r + ic_i = Bk\tau(k) + i(k - k^3).$$

For unstable perturbations, the imaginary part of the phase velocity c_i will be greater than zero. They are long-wave disturbances. Their wave numbers are smaller than the neutral wave number k_n satisfying the equation:

$$1 - k_n^2 + B\,\tau_{im}(k_n) = 0. \quad (1.4)$$

As it can be seen from (1.4), for the freely falling film ($B = 0$) the neutral wave number $k_n = 1$. We will choose the parameters of the undisturbed flow, so that the neutral wave number k_n different from the unit. At the same time, it is required for this value of k_n to correspond to a definite value of τ_{im}. Then for the value of the parameter B from (1.4) we have:

$$B = \frac{k_n^2 - 1}{\tau_{im}(k_n)}.$$

Obtaining results presented below, we used data on friction pulsations obtained by the model of boundary conditions transfer to the unperturbed level in the paper [2].

In the paper [3] it was shown that in the vicinity of the neutral wave number k_n there is the soft type of branching: the wave numbers of the periodic steady-state traveling regimes of small but finite amplitude lie in the area of linear instability. For steady-state traveling solutions

$$H(x,t)=H(\xi),\ \xi=x-ct$$

and equation (1.3) is written in the form

$$-c\frac{dH}{d\xi}+2H\frac{dH}{d\xi}+\frac{d^2H}{d\xi^2}+\frac{d^4H}{d\xi^4}+B\int_{-\infty}^{\infty} ik^2\tau(k)\hat{H}(k)e^{ik\xi}dk=0. \tag{1.5}$$

This equation was solved numerically.

Equation (1.5) is invariant relative to the transform

$$H(\xi)\to H(\xi)+const, \qquad c=c+2const.$$

Taking this into account, we will further restrict to consider the periodic solutions, where

$$\langle H(\xi)\rangle \equiv \int_0^{2\pi/k} H(\xi)d\xi=0. \tag{1.6}$$

The use of solutions with normalization (1.6) means that the characteristic thickness scale is the average layer thickness in the wave flow regime.

2. Results of nonlinear wave calculation on the model equation

For finding solutions of equation (1.3) periodic on ξ the function H is represented as a spatial Fourier series:

$$H(\xi)=\sum_n H_n \exp(ikn\xi). \tag{2.1}$$

Since H is real function, then $\bar{H}_{-n}=H_n$. The overbar means the operation of complex conjugation. After substituting (2.1) into equation (1.5), given that in this case the integral term in (1.3) is transformed to the corresponding series, obtain an infinite system of algebraic equations for Fourier harmonics H_n. Supposing that all $H_n(t)$ with indices $|n|\geq N$ are equal to zero, obtain its finite-dimensional analogue.

For finding new families, we use the method of stability theory. The study of stability of the steady-state periodic traveling solutions $H_0(\xi)$ with respect to linear perturbations allows finding branching points for new solutions. The method of stability investigation is similar to those described in [4]. Below is its brief description.

Let $H_0(\xi)$ is the periodic solution of equation (1.5) with wave number *k*. To study its stability, we substitute $H(\xi)=H_0(\xi)+h(\xi,t)$ in (1.3) and linearize it on the perturbation $h(\xi,t)$. As a result, we obtain a linear equation with periodic coefficients. Taking into account that the variable t does not enter the equation explicitly, its solution has the form:

$$h(\xi,t)=\exp(-\gamma t)h_1(\xi)+C.C.$$

Here $C.C.$ is complex conjugated expression. From the Floquet theorem it follows that the physically reasonable solutions $h_1(\xi)$ bounded at infinities have the following form:

$$h_1(\xi)=\exp(iQk\xi)\varphi(\xi),$$

where $\varphi(\xi)$ is the periodic function of the same period as $H_0(\xi)$, and Q is the real parameter. Therefore, the study of the stability of steady-state traveling solutions of equation (1.3) is reduced to investigation of the spectrum of eigenvalues γ at various values Q. For these eigenvalues the resulting equation has periodic solutions $\varphi(\xi)$ of the same period as $H_0(\xi)$. It is sufficient to consider only the changes of Q in a unit interval, for example $0 \le Q \le 1$. The wave is stable if for any values of Q, in all γ the real part $\gamma_r > 0$. New steady-state traveling modes bifurcate from the solution $H_0(\xi)$, if at some point (k,Q) one of the eigenvalues γ is equal to zero:

$$\gamma(k,Q) = 0.$$

If $Q = p/r$ is the rational number, then there is a mode periodic on ξ with a new wave number $k_{new} = k/r$, and if $Q = 0$, then the new regime with the same period as $H_0(\xi)$ can branch off.

In the case of $B = 0$, the equation (1.3) is transformed into equation which is now called "Kuramoto-Sivashinski equation (K-S)". For the falling film it was derived in [5]. It is well known that the family (family I) of wavy steady-state traveling solutions branches from the trivial one in the point $k_n = 1$ (in our normalization). It can be continued into the region of instability to the value $k_* = 0.4979$. At this point every odd-numbered harmonics is equal to zero, and the solution with the wave number $k = 2k_* = 0.9958$ is obtained. Thus, this family is closed on itself and cannot be continued to the region of smaller wave numbers. For all values of wave numbers the phase velocity of the waves of the family I $c = 0$. At the point $k_{**} = 0.554$ there is a bifurcation with $Q = 0$, and from the first family two new families with the same wavelength and $c \ne 0$ branch [4]. These families in the limit $k \to 0$ transform into the localized solitary waves (Depending on what is larger in the value, the value of the maximum or minimum, these solitons are often called positive or negative, respectively). For the first time these solitons were found in [6]. The solution with $c > 0$ transforms to a known positive soliton (soliton-hump) with the characteristic oscillations in the forefront.

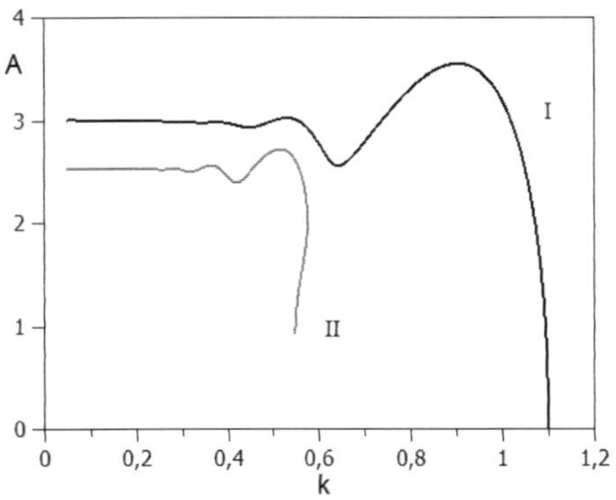

Figure 1. Dependences of the amplitudes of steady-state traveling solutions of the families on the wave number k: curve I - family I and curve II - family II. Neutral wave number $k_n = 1.1$, $B = 0.0037$.

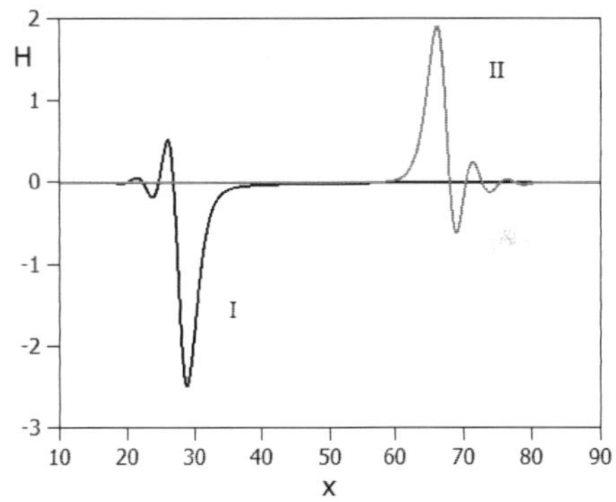

Figure 2. Surface profiles for the solitons: curve I - family I and curve II - family II. Neutral wave number $k_n = 1.1$, $B = 0.0037$.

The emergence of a new term with $B \ne 0$ leads to a change in the bifurcation picture. Thus, for small values of the parameter B the first family extends to the region of small wave numbers and in the limit becomes a negative soliton. Family II branches from the first family with period doubling ($Q = 0.5$). The limit of this family at $k \to 0$ is a positive soliton (soliton-elevation). The example of this case is presented in Figure 1, which shows the behavior of amplitudes ($A = H_{max} - H_{min}$) for the first two families of steady-state traveling solutions depending on the wave number. Here, the value of the parameter $B = 0.0037$, i.e. in this case the influence of the gas flow is weak. The first family branches off from the trivial solution $H_0 = 0$ in the neutral point $k_n = 1.1$. The new family II branches off from the first family at the point $k_1 = 0.5472$. The limits of these families at $k \to 0$ are a negative soliton with velocity $c=-1.53$ and a positive one with velocity $c=1.87$, respectively. The profiles of these solitons are shown in Figure 2. This result is fairly predictable. However, with further increase of the parameter B the pattern of branches is changing more radically.

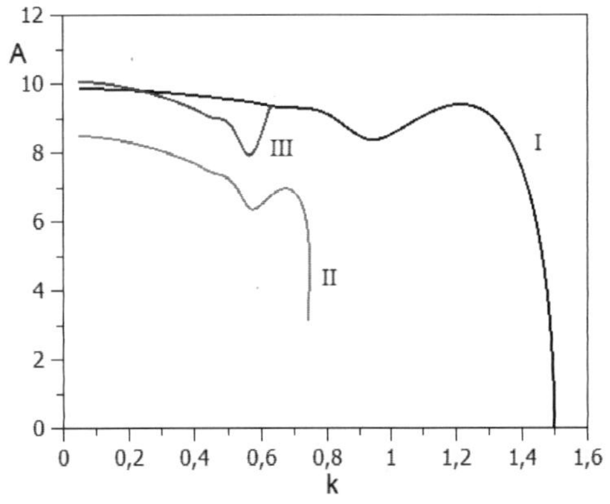

Figure 3. Dependences of the amplitudes of steady-state traveling solutions of the families on the wave number k: curve I - family I, curve II - family II and curve III - family III. Neutral wave number $k_n=1.5$, $B=0.222$.

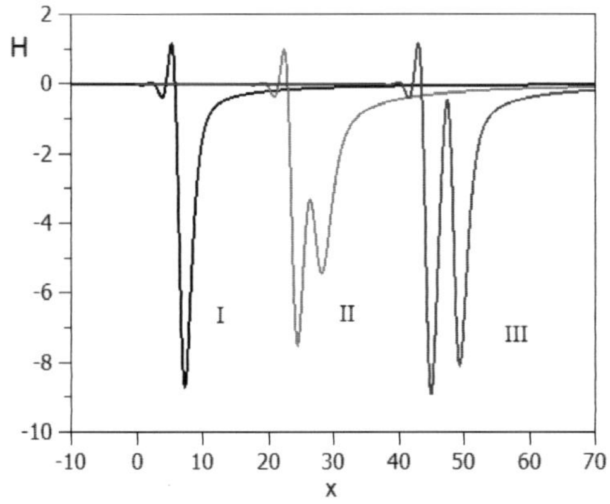

Figure 4. Surface profiles for the solitons: curve I - family I, curve II - family II and curve III - family III. Neutral wave number $k_n=1.5$, $B=0.222$.

As an example, this paper presents the results, obtained at a sufficiently large value of the parameter $B=0.222$. In this case, the impact of the gas flow blowing the film is significant.

Figure 3 shows the behavior of amplitudes for the first three families of steady-state traveling solutions depending on the wave number. The first family branches off from the trivial solution in the neutral point $k_n=1.5$. Families II and III arise from the first family at the points $k_1=0.7433$ and $k_2=0.6304$, respectively. They bifurcate as a result of the fact that at points $k_{1bif}=1.4866$ and $k_{2bif}=1.2608$ in the solutions of the first family at the parameter value $Q=0.5$ one of the eigenvalues

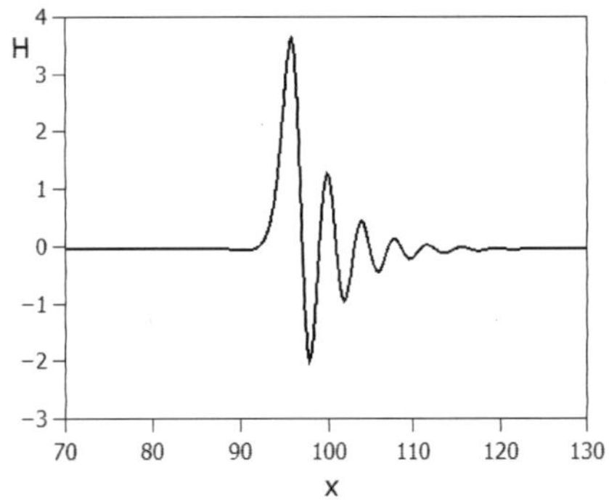

Figure 5. Surface profile for the one-hump positive soliton at $B=0.222$.

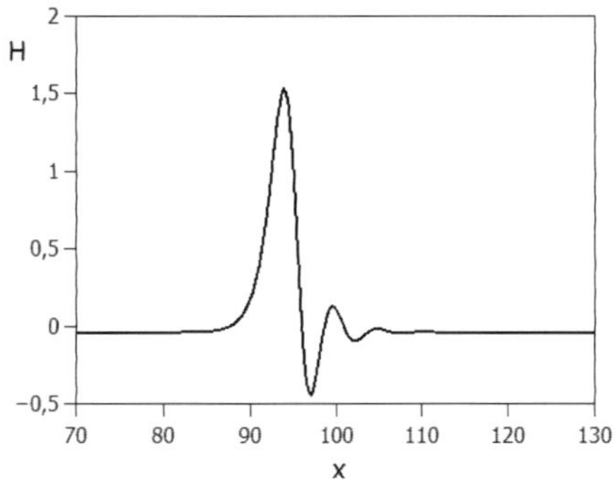

Figure 6. Surface profile for the one-hump positive soliton at $B=0$ (solution to the K-S equation).

$\gamma(k_{i\,bif}, Q) = 0$. All three families transform to soliton solutions, but all these solitons are negative. Family I transforms into the negative one-hump soliton, and families II and III transform into different two-hump solitons, respectively. Their profiles are shown in Figure 4. In contrast to the case of small B, for a given fixed value of the parameter $B = 0.222$ a family of steady-state traveling solutions, which limit is a positive one-humped soliton, can not be found by successive bifurcations. The corresponding families of solutions are found, moving along the continuity from small to large values of the parameter *B*. This soliton is shown in figure 5. For comparison, figure 6 shows the one-humped soliton being a solution to the K-S equation.

3. Conclusion

Nonlinear waves on the liquid film flowing under the action of gravity in the known stress field at the interface have been considered. In the case of small Reynolds numbers, the problem is reduced to considering solutions of a nonlinear integro-differential equation for the film thickness deviation from the undisturbed level. Steady-state traveling periodic and soliton solutions to this equation have been considered. Using the results of the stability analysis of the steady-state periodic traveling solutions we have found new families of waves, including those that transform into positive solitons in the limit of small wave numbers.

Acknowledgements
The work was performed with financial support of the Russian Science Foundation (grant No 16-19-10449).

References
[1] Tseluiko D and Kalliadasis S 2011 J. *Fluid Mech.* **673** 19–59
[2] Vozhakov I S, Arkhipov D G and Tsvelodub O Yu 2015 *Thermophysics and Aeromechanics* **22** 191-202
[3] Tsvelodub O Yu 2016 *J. Phys.: Conf. Series* **754** 032020
[4] Tsvelodub O Yu and Trifonov Yu Ya 1989 *Physica D* **39** 336–351.
[5] Nepomnyashchy A A 1974 *Fluid Dynamics* No. **3** 354–359
[6] Tsvelodub O Yu 1980 *Fluid Dynamics* No. **4** 591–594

IOP Conf. Series: Journal of Physics: Conf. Series **899** (2017) 032024 doi:10.1088/1742-6596/899/3/032024

Numerical simulation of gas volume motion during the gas injection into liquid metal coolant

E V Usov[1], P D Lobanov[2], N A Pribaturin[1,2], V I Chuhno[1], A E Kutlimetov[1], A I Svetonosov[2]

[1]Nuclear Safety Institute of the Russian Academy of Sciences, Novosibirsk Branch, Novosibirsk, 630090, Lavrentieva Street, 1, Russia
[2]Institute of Thermophysics, Siberian Branch of Russian Academy of Sciences, Novosibirsk, 630090, Lavrentieva Street, 1, Russia

Abstract. Detailed description of relations and numerical approaches to simulate transport of gas phase in a vertical liquid column is presented in a current paper. These approaches are important to calculate phenomena that take place during steam generator tube rapture in fast reactors with liquid metal coolant. Presented relations determine interphase friction between gas and fluid in different flow regimes of two-phase flow. It is shown that correct definition of interphase friction coefficients determines the correct value of bubble velocity that is very important to simulate two-phase flow in steam generator and reactor core. The paper also contains numerical algorithm to calculate motion of gas volume in fluid flow. Especial attention is paid to describe the algorithms for simulating two-phase flow with sharp edges between phases that are character for slug flow regime. Also some experimental results are presented in the paper. Comparison between experimental data and calculation results has been provided.

1. Introduction

The active studies of processes in two-phase flow of liquid metals been providing in the middle of the 20th century due to the development of project of fast reactors. Unfortunately, due to the inherent technological problems in operation with liquid metals, many countries refused to further develop this technology, with the exception of Russia, which is still developing projects for liquid metal-cooled reactors. There are projects of lead, lead-bismuth and sodium cooled reactors. Accidents caused by the steam generator tubes rupture and the coolant boiling due to insufficient of heat removal from the core is determined by the behaviour of the gaseous phase in liquid. The motion of bubbles in the core, pulsations of theirs volume strongly influence to the behaviour of the reactor due to the presence of a positive feedback between the core neutron power and the gas volume. Knowledge about the specific of gas bubble motions into liquid metal coolant is also important for metallurgy to calculate of boiling or blown steel-melting baths.

There is a lack of information about the processes in two-phase liquid metal flow. So it is important to provide numerical and experimental investigations of the motion of bubbles in liquid metals. During the sodium boiling in core or during the leakage in steam generator slug regime is the main two-phase [1, 2]. The feature of slug flow regime is the existence of sharp boundary between phases that make some difficulties to numerical simulating especially in one dimensional approximation.

2. Description of the models to calculate processes in two-phase flow

Common ways to simulate two-phase flow are based on two-liquid model [3]. To describe two-phase flow in frame of two-liquid model, system of mass, momentum and energy equations must be solved. The system of mass, momentum and energy equations is presented below:

$$\frac{\partial}{\partial t}\left(\varphi_j\rho_j\right)+\frac{\partial}{\partial z}\left(\varphi_j\rho_j v_j\right)=0, \tag{1}$$

$$\frac{\partial(\varphi_j\rho_j v_j)}{\partial t}+\frac{\partial(\varphi_j\rho_j v_j v_j)}{\partial z}+\varphi_j\frac{\partial}{\partial z}P=\tau_{\mathrm{ij}}+\tau_{wj}-\varphi_j\rho_j g\cdot\sin\vartheta\,. \tag{2}$$

$$\frac{\partial(\varphi_j\rho_j h_j)}{\partial t}+\frac{\partial}{\partial z}(\varphi_j\rho_j h_j v_j)-\varphi_j\frac{dP}{dt}=Q_{\mathrm{ij}}+Q_{wj}\,. \tag{3}$$

The first (1) equation is the mass conservation law; the second (2) is the momentum conservation law; the third (3) is the energy conservation equaton.
The index "j" corresponds to the number of phase: j=g for gas, j=f for fluid. In the equations φ is j-phase volume fraction, ρ is a density of the phase (kg·m^{-3}), v is a velocity (m·s^{-1}) of the phase, h is an enthalpy of the phase. The values on the right hand of momentum equation are the interphase friction (Pa·m^{-1}) per unit length, wall friction (Pa·m^{-1}) per unit length and the gravitational forces. It should be noted that the sum of the volume fractions must be equal to unity. The sum of the interphase friction and interphase heat transfer must be equal to zero:

$$\varphi_g+\varphi_f=1,\ \tau_{\mathrm{ig}}+\tau_{\mathrm{if}}=0,\ Q_{\mathrm{ig}}+Q_{\mathrm{if}}=0\,. \tag{4}$$

The relation to calculate interphase friction depends on flow regime. The flow regime map is shown in table 1.

Table 1. Flow regime map for vertical channel

№	Upper boundary of gas volume fraction	Regime
1	0.3	Bubble
2	0.99	Slug
3	1	Annular-mist

The upper boundary of gas volume fraction has been chosen according to work [1, 2]. To calculate the interphase friction coefficients, the following relation are used:

$$\tau_{ig}=\chi\left|v_g C_1-v_f C_0\right|\left(v_g C_1-v_f C_0\right). \tag{5}$$

C_0 is a distribution parameter.

$$\chi=0.75\cdot C_b\varphi_g\rho_f/D_b\,,C_1=\left(1-\varphi_g C_0\right)/\left(1-\varphi_g\right). \tag{6}$$

For bubble regime in vertical channel:

$$C_0=1.2-0.2\sqrt{\rho_g/\rho_f}\,. \tag{7}$$

For others flow regime C_0 and C_1 are equal to unity. Friction coefficient C_b for bubble in liquid can be calculated for wide range of parameter with help of [5]:

$$C_b=C_{\mathrm{b},We\to 0}+\tanh(0.021\cdot We^{1.6})(C_{\mathrm{b},We\to\infty}-C_{\mathrm{b},We\to 0})\,, \tag{8}$$

$$C_{\mathrm{b},We\to 0}=\frac{16}{\mathrm{Re}_b}(1+[\frac{8}{\mathrm{Re}_b}+\frac{1}{2}(1+\frac{3.315}{\sqrt{\mathrm{Re}_b}})]^{-1}),\ C_{\mathrm{b},We\to\infty}=\frac{8}{3}+\frac{24}{\mathrm{Re}_b}(\frac{2+3\mu^*}{3+3\mu^*})\,. \tag{9}$$

In the relations (8) – (9) Re_b is a Reynolds number for bubble, We is a Weber number, $\mu^* = \mu_f/\mu_g$ is a relative dynamic viscosity.

A slug flow regime is one of the main flow regimes that takes place during sodium boiling or steam generator tube rapture in the fast reactors. Different relations have been analysed for calculating friction coefficient for slugs. The comparison of different relations is shown in the figure 1 [6].

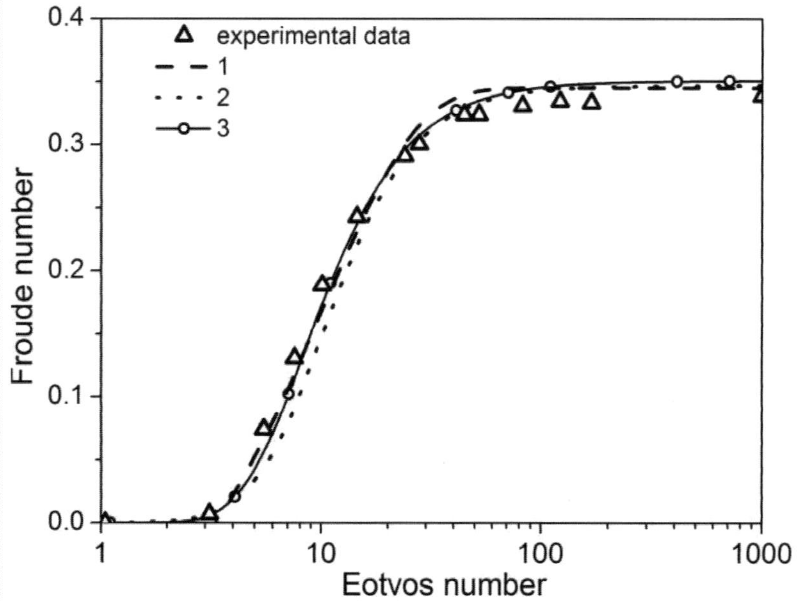

Figure 1. The comparison of different relations for slug friction coefficient. 1 – [7], 2 – [8], 3 – [9], experimental data – [10]

There is a good agreement between all of the relations and experimental data. To simplify the calculation procedure correlation from [7] has been chosen:

$$C_D = \frac{4}{3}\frac{d_S}{Dk^2\left(N_f, Eo\right)},\ k = 0.345\left(1 - e^{\frac{-0.01N_f}{0.345}}\right)\left(1 - e^{\frac{3.37-Eo}{m}}\right), \tag{10}$$

$$m = \begin{cases} 10; & N_f > 250 \\ 69N_f^{-0.35}; & 18 < N_f < 250 \\ 25; & N_f < 18 \end{cases}. \tag{11}$$

3. Numerical scheme to simulate two-phase flow

To numerical simulate, the system of equations (1) – (2) should be represented in finite-difference form. Finite difference form of mass and momentum conservation equation is shown below:

$$\frac{\varphi_k^{n+1} - \varphi_k^n}{\tau}\rho_k + \varphi_k^n \frac{\partial\rho}{\partial P}\frac{P_k^{n+1} - P_k^n}{\tau} + \frac{\left(\varphi_{k+1/2}^n \rho_{k+1/2}^n\right)v_{k+1/2}^{n+1} - \left(\varphi_{k-1/2}^n \rho_{k-1/2}^n\right)v_{k-1/2}^{n+1}}{\Delta z} = 0, \tag{12}$$

$$\begin{aligned} &\varphi_{k+1/2}^n \rho_{k+1/2}\frac{v_{k+1/2}^{n+1} - v_{k+1/2}^n}{\tau} + \varphi_{k+1/2}^n \rho_{k+1/2} v_{k+1/2}^n \frac{v_{k+3/2}^{n+1} - v_{k+1/2}^n}{\Delta z} + \varphi_{k+1/2}^n \frac{P_{k+3/2}^{n+1} - P_{k+1/2}^{n+1}}{\Delta z} = \\ &= \left(\tau_{ij}\right)_{k+1/2} + \left(\tau_{wj}\right)_{k+1/2} - \varphi_{k+1/2}^n \rho_{k+1/2} g \cdot \sin\vartheta \end{aligned}, \tag{13}$$

IOP Conf. Series: Journal of Physics: Conf. Series **899** (2017) 032024 doi:10.1088/1742-6596/899/3/032024

"k" is a number of calculation point, "n" is a number of time level. An "upwind" scheme is used to approximate convective terms. The shortcoming of the "upwind" scheme is a large numerical diffusion. A "Kabare" scheme [11] has been realized to overcome the shortcoming of the "upwind" scheme. A comparison between the "Kabare" scheme, "upwind" scheme and analytical solution for gas transfer in channel are presented in figure 2 and figure 3. As shown in figures, simulations with help of the "Kabare" scheme more close to analytical solution than calculation with "upwind" scheme. The diffusion of the "upwind" scheme also provides a spreading of sharp boundary, especially on large times.

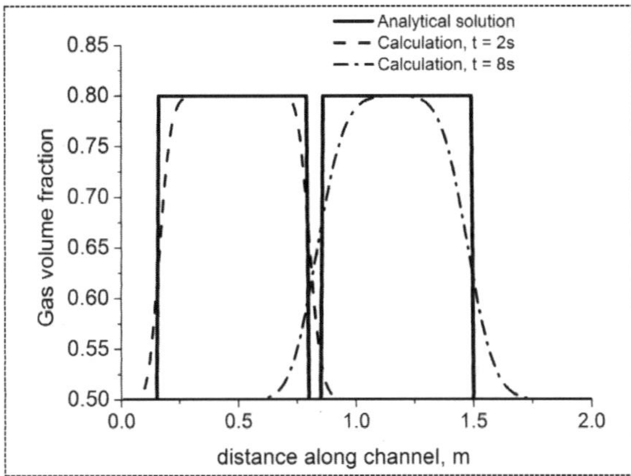

Figure 2. Simulation of gas volume transfer with help of "upwind" scheme.

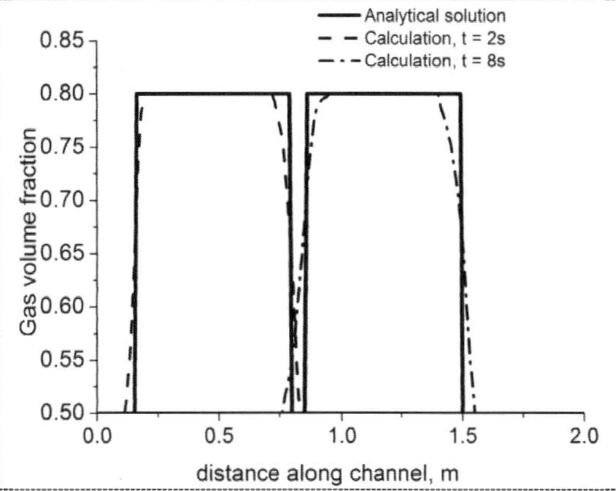

Figure 3. Simulation of gas volume transfer with help of "Kabare" scheme.

4. Validation of numerical approaches

To validate developed numerical model, experiments with bubble and slug motion in stagnant liquid metal have been provided. Experimental setup are shown in figure 4. Vertical tube with Rose's alloy has been used in the experiments. Tube diameter was equal 48 mm. Initial level of the alloy was equal to 40 sm. Its temperature was 423 K. Argon has being injected into the Rose's alloy. Injection has been made with help of round tube. Its diameter was equal to 4 mm. Volume flow rate was equal to 84, 120, 146, 180, 188, 206, 222, 240 ml/s. A conductivity sensor was mounted at height 33sm. It measured the value of gas volume fraction. The measuring error of gas volume fraction was equal to 15%.

The results of the experiments are shown in figure 5. To simulate the experiment, the calculation scheme has been made. There were two vertical tubes. The first tube was filled with Rose's alloy. The gas was injected through the second tube. The end of the second tube was been connected to the beginning of the first tube.

There is a good agreement between the calculation results and the experimental data with considering of experimental error. The presented experiment allows validating the models of motion of gas bubble in tubes. As shown in [4], gas volume fraction depends on the volume flow rate and the velocity of gas bubbles. The value of the velocity depends on the interphase friction. So the good agreement between the experimental data and the calculation results verifies chosen models to simulate motion of gas bubbles and slugs.

5. Conclusion

The approaches of simulation gas bubbles and slug motions in liquid metal are presented in the current paper. The approaches are based on two-liquid model with closing relations that determine the value of the interphase friction.

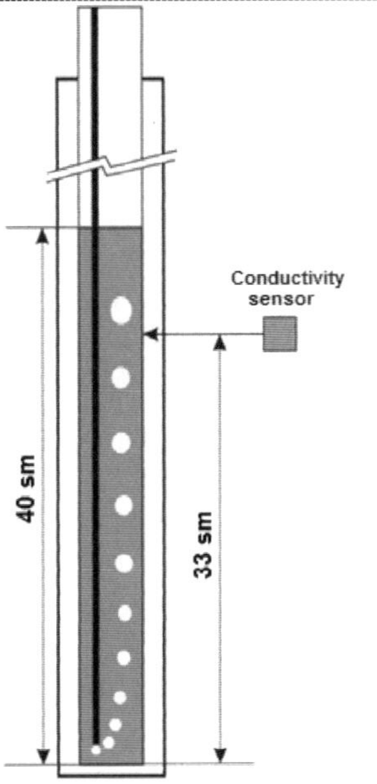

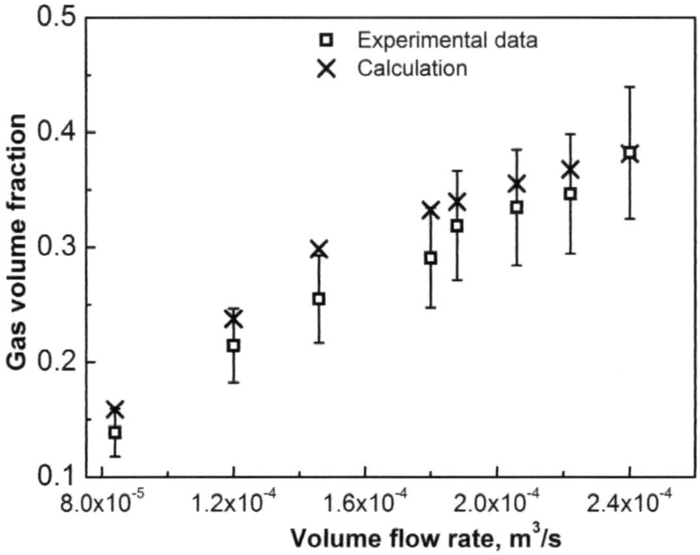

Figure 4. The scheme of the experiments.

Figure 5. Comparison the results of calculations and the experimental data.

The experiments with gas injection into the Rose's alloy have been provided. The experiments allowed validating numerical approaches that have been chosen. The difference between calculation results and experimental data is equal to 12% for gas volume fraction.

References

[1] Kikuchi Y, Haga K, Takahashi 1975 *Journal of Nuclear Science and Technology* **12** (2) 83
[2] Marten K, Baumgartner E, Khan M, Hoffman H 1982 Impact of Gas in Sodium Flow on the Temperature Variation of an LMFBR Rod Bundle *KFK* (3253) 47
[3] Nigmatulin R I 1978 Fundamentals of Mechanics of Heterogeneous Media (in Russian) (Moscow: Nauka)
[4] Ishii M One-dimensional drift-flux model and constitutive equations for relative motion between phases in various two-phase flow regimes, *ANL* (77-47)
[5] Loth E 2008 *Int. J. of Multiphase Flow* **34** 523
[6] Morgado A O, Miranda J M, Araújo J D 2016 *International Journal of Multiphase Flow* **85** 348.
[7] Wallis G B 1969 One-dimensional two-phase flow (New York: McGraw Hill)
[8] Hayashi K, Kurimoto R, Tomiyama A 2010 *Multiphase Science and Technol*ogy **22** (3) 197
[9] Kurimoto R, Hayashi K, Tomiyama A 2013 *Int. J. of Multiphase Flow* **49** 8
[10] White E T, Beardmore R H 1962 *Chem. Eng. Sci.* **17** 351
[11] Goloviznin V M 2006 *Mathematical modeling* **18** (11) 14 (in Russian)

IOP Conf. Series: Journal of Physics: Conf. Series **899** (2017) 032025 doi:10.1088/1742-6596/899/3/032025

Studying evolution of periodic wave regimes on a falling down liquid film

I S Vozhakov, D G Arkhipov and O Yu Tsvelodub

Kutateladze Institute of Thermophysics SB RAS, Lavrentev Ave. 1, 630090, Novosibirsk, Russia

Novosibirsk State University, Pirogov Str. 2, 630090, Novosibirsk, Russia

E-mail: vozhakov@gmail.com

Abstract. The numerical simulation of wavy regimes of the freely-falling liquid film was realized using the previously discovered symmetry of model equations. The calculation results are presented for the dynamics of amplitudes of significant harmonics and for the wave surface evolution.The paper has resulted in finding a specific regime, characterized by an alternate predominance of the first and second harmonics. At certain values of parameters this limiting cycle may have a very large period, during which the solution is attracted to the quasi-steady-state traveling regime, then suddenly transforms, and again tends to the quasi-steady-state with a phase shift of a half wave length.

1. Introduction

The flow of a thin layer of viscous liquid (film) over an inclined solid plane has been studied for over a hundred years [1]. Until the middle of the last century, the interest in the falling films was almost exclusively utilitarian. In 1941, Friedman and Miller [2] noted: "In the general field of diffusional processes, including absorption, extraction, heat transfer, humidification, and distillation, the flow of thin liquid films is often encountered". In accordance with practical problems, reserches were mainly experimental in nature and were focused on a turbulent flow regime. Precision measurements of the fluid velocity on the surface of very thin laminar liquid layers freely flowing over a vertical tube [2] have revealed its significant exceedence over the velocity calculated with the trivial Nusselt solution for a smooth film. A qualitative explanation of this consisted in the necessity to take into account wave formation on the liquid surface.

The impetus to a theoretical study of wave flows of liquid films was given in a series of brilliant study of Kapitza [3]. Notable photos of the film surface showed the existence of many regular wave flow regimes that were much more stable than the plane-parallel Nusselt flow. The sustained efforts of subsequent researchers to describe such regimes have been rewarded by the profound results that reveal the general nature of the hydrodynamic instability and have direct application in various fields of physics of continuous media. It is suffice to mention the widely known equation of Kuramoto-Sivashinskii, derived for the first time to simulate the very wave evolution on the liquid film surface [4]. The interaction of effects of long-wave pumping, short-wave dissipation and perturbation nonlinearity (each of them is expressed by only one term of the K-S equation) leads to an unusually complex pattern of steady-state traveling wave modes, formed due to the instability of the trivial solution, as well as to an extremely complex dynamics of perturbations, demonstrating main features of the spatial-temporal deterministic chaos.

The full formulation of the problem of the wavy film flow includes the Navier – Stokes and the continuity equations with appropriate kinematic and dynamic boundary conditions. The major problem is the uncertainty of the moving boundary position, which is determined in the course of solving. If the effects of droplet entrainment and solid surface drying are excluded from consideration, the fluid flow area turns out to be simply connected. The presence of surface tension ensures the absence of sharp edges on the film surface. Under these conditions, the function determining the surface position is often unambiguous. Then, there exists a continuously differentiable coordinate transformation, mapping the area of the fluid flow on a band of constant thickness [5]:

$$x = x, \quad \eta = y/h(x,t) - 1, \quad t = t. \tag{1}$$

Governing equations of fluid dynamics were written in new coordinates in the following dimensionless form [6]:

$$\begin{gathered} \frac{\partial Q}{\partial t} + \frac{\partial}{\partial x}\left(\frac{Q^2}{h}\right) + \frac{\partial}{\partial \eta}\left(\frac{QV}{h}\right) = \frac{1}{\varepsilon \mathrm{Re} h^2}\frac{\partial^2 Q}{\partial \eta^2} + \frac{3h}{\varepsilon \mathrm{Re}} + \frac{18}{5}h\frac{\partial^3 h}{\partial x^3}, \\ \frac{\partial h}{\partial t} + \frac{\partial Q}{\partial x} + \frac{\partial V}{\partial \eta} = 0, \\ Q = V = 0, \quad \text{at} \quad \eta = -1, \\ \frac{\partial Q}{\partial \eta} = 0, \quad V = 0, \quad \text{at} \quad \eta = 0. \end{gathered} \tag{2}$$

Here, h is a dimensionless film thickness, $Q = hu$, $V = hv$, where u and v are dimensionless longitudinal and transversal contravariant components of velocity vector, $\mathrm{Re} = gh_0^3/3\nu^2$ is the Reynolds number, and $\varepsilon = h_0/l_0$ is the wavelength parameter. In derived equations, perturbations are assumed to have long wavelengths $\varepsilon \ll 1$, and liquid flow rates are supposed to be moderate $\varepsilon\mathrm{Re} \sim 1$. The selected characteristic scales are: the average thickness h_0 and the superficial velocity of a waveless Nusselt flow $u_0 = gh_0^2/3\nu$. The scales of the length l_0 and time l_0/u_0 are determined from the relationship:

$$\frac{h_0}{l_0} = \varepsilon = \sqrt{\frac{18}{5}\frac{\mathrm{Re}^{5/3}}{3^{1/3}\mathrm{Fi}^{1/3}}},$$

where $\mathrm{Fi} = \sigma^3/\rho^3 g\nu^4$ is the film number, characterizing liquid properties. As it is shown in [7], this choice provides the value of neutral wave number $\alpha_n \approx 1$.

2. Calculation method

The problem (2) is solved by the pseudospectral (collocation) method. Functions depending on transverse coordinate η are expanded in a series on Chebyshev polynomials T_i. The paper [7] showed that at moderate values of Reynolds numbers the steady-state traveling solutions of the problem (2) in the area extended over the transverse coordinate $\eta \in [-1, 1]$ are characterized by the following symmetry:

$$u(x,\eta,t) = u(x,-\eta,t), \; v(x,\eta,t) = -v(x,-\eta,t).$$

Therefore, in this range one may choose only even Chebyshev polynomials for the function

IOP Conf. Series: Journal of Physics: Conf. Series **899** (2017) 032025 doi:10.1088/1742-6596/899/3/032025

Q, and only odd ones for the function V:

$$Q(x,\eta,t) = \sum_{j=1}^{M} Q_{2j}(x,t)\,(T_{2j}(\eta) - 1) \equiv \sum_{j=1}^{M} Q_{2j}(x,t)\,\tilde{T}_{2j}(\eta),$$
$$V(x,\eta,t) = \sum_{j=1}^{M} V_{2j+1}(x,t)\,(T_{2j+1}(\eta) - \eta) \equiv \sum_{j=1}^{M} V_{2j+1}(x,t)\,\tilde{T}_{2j+1}(\eta). \quad (3)$$

Here, M is a number of Chebyshev nodes.

The choice (3) provides an automatic satisfaction of boundary conditions on the free surface and on the solid wall.

So, for a set of Chebyshev nodes:

$$\eta_l = -\cos\left(\frac{l\pi}{2(M+1)}\right),\ l = 1, ..., M,$$

we obtain the set of equations:

$$\sum_{j=1}^{M} \frac{\partial Q_{2j}}{\partial t}\tilde{T}_{2j}(\eta_l) = \frac{1}{\varepsilon \mathrm{Re} h^2}\sum_{j=1}^{M} Q_{2j}\tilde{T}''_{2j}(\eta_l) + \frac{18}{5}h\frac{\partial^3 h}{\partial x^3} + \frac{3}{\varepsilon \mathrm{Re}}h -$$
$$-\sum_{j=1}^{M}\sum_{k=1}^{M}\frac{\partial}{\partial x}\left(\frac{Q_{2j}Q_{2k}}{h}\right)\tilde{T}_{2j}(\eta_l)\tilde{T}_{2k}(\eta_l) - \frac{1}{h}\sum_{j=1}^{M}\sum_{k=1}^{M} Q_{2j}V_{2k+1}\left(\tilde{T}_{2j}\tilde{T}_{2k+1}\right)'(\eta_l), \quad (4)$$

$$\frac{\partial h}{\partial t} + \sum_{j=1}^{M}\frac{\partial Q_{2j}}{\partial x}\tilde{T}_{2j}(\eta_l) + \sum_{j=1}^{M} V_{2j+1}\tilde{T}'_{2j+1}(\eta_l) = 0$$

Here apostrophe denotes a derivative with respect to η. Functions that depend on the longitudinal coordinate were expanded into spatial Fourier series:

$$[Q_{2j}, V_{2j+1}, h](x,t) = \sum_{k=-\infty}^{\infty} [Q_{2j,k}, V_{2j+1,k}, h_k](t)\, e^{i\alpha k x}. \quad (5)$$

Here α is the wave number of the periodic solution, i is the imaginary unit. The functions Q, V and H are real-valued, so $Q_{2j,-k} = \overline{Q}_{2j,k}$, $V_{2j+1,-k} = \overline{V}_{2j+1,k}$, $H_{-k} = \overline{H}_k$. The overline denotes the operation of complex conjugation.

Substituting representations (5) in (4) and limiting to finite numbers of harmonics we obtain the system of ordinary differential equations which was solved by the Runge-Kutta method of the 4-th order.

3. The calculation results

As it is known, the plane-parallel Nusselt flow is unstable relative to linear periodic perturbations with wave number $\alpha < \alpha_n \approx 1$. The dynamics of small perturbations with $\alpha > \alpha_n$ is trivial: they exponentially attenuate with time. Amplitudes of unstable perturbations with wave numbers close to a neutral wave number first exponentially grow. The first harmonic of such perturbations obtains energy from the main flow, and due to nonlinear effects passes it to the second and higher harmonics, which wave number is located in the stable area and in such a way providing energy dissipation. As a result, the steady-state traveling mode is usually formed quickly. According

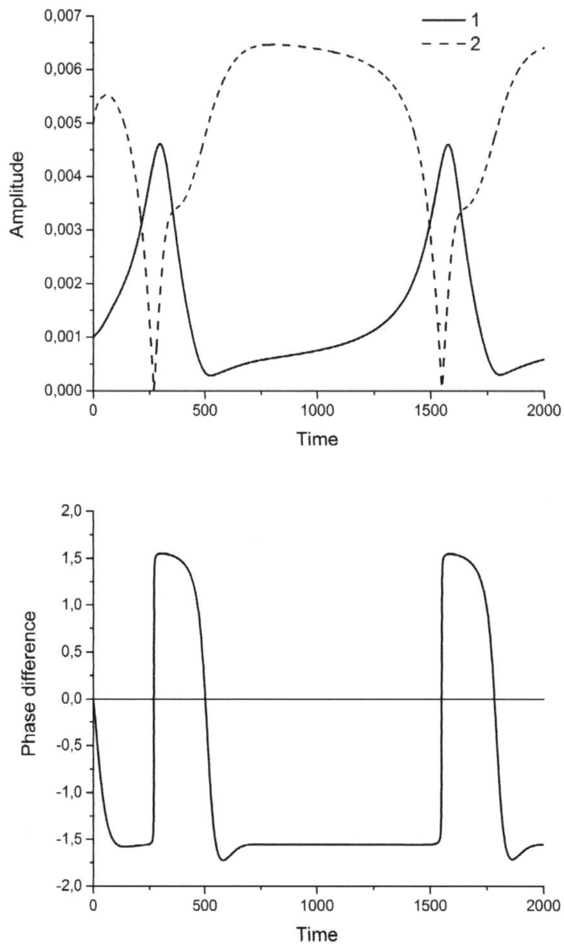

Figure 1. Evolution of the modulus of harmonics' amplitudes (upper) and phase difference (lower) at $\alpha = 0.42$.

to numerical experiments, such a behavior is specific for the majority of perturbations from the range $0.5 < \alpha < 1$. Farther to the area of instability, the second harmonic (its wave number is 2α) also turns out to be unstable. This means that at certain conditions during evolution there is a possibility of appearance of quasi-steady-state states with predominance of only even harmonics. Such a state may be interpreted as a steady-state-traveling mode with a double wave number 2α, with imposed perturbations with nonzero odd harmonics (such perturbations have a double spatial period and a wave number α).The research shows that at $0.41 < \alpha < 0.5$ this mode (the wave number 2α) is linearly unstable relative to such perturbations, so the evolution turns out to be more complicated.

The research numerically considers the dynamcis of periodic perturbations of free surface ($|H_1| \neq 0$ and $|H_2| \neq 0$) in the range of wave numbers $0.41 < \alpha < 0.5$ and Reynolds numbers $1 < \mathrm{Re} < 30$. Typical scenarios of the evolution of perturbations with small initial amplitude are presented in figure 1. Figure 1 presents the evolution of amplitudes of the first two harmonics ($H_1 = |H_1|e^{i\varphi}, H_2 = |H_2|e^{i\psi}$) at the value $\alpha = 0.42$. Here, there are dependences of moduli of harmonics' amplitudes (fig. 1a) and the "phase difference" $\phi = 2\varphi - \psi$ (fig. 1b). It is seen that

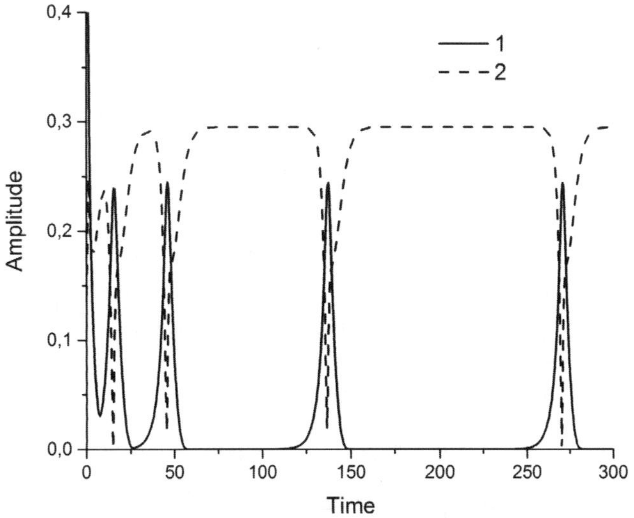

Figure 2. Evolution of the amplitude of the first (line 1) and second (line 2) harmonics

these functions change periodically. During a substantial part of the period, the amplitude of the second (and as the calculation has shown, the first-born even harmonics) are close to the values that correspond to the steady-state traveling solution with a double wave number. At that, the amplitude of the first and other odd harmonics exponentially grows. At the same time,there is an increase in nonlinear effects, the foremost of which turns out to be the energy transfer from the first to the second harmonic. As a result, the amplitude of the second harmonic passes through null (and its phase immediately changes to π), soon afterthat the growth of the first harmonic stops, and now the second harmonic grows for the account of the first harmonic.

As it is seen from fig. 1b, during the growth of the first harmonic the "phase difference" is negative and with good precision is equal to $-\pi/2$. Until $\phi > 0$ the first harmonic rapidly decreases, and the second one grows.

To better understand such pulsation regimes we will carry out their analysis on the basis of the simplified model of a film flow at small Reynolds numbers (equation of Nepomnyashchy):

$$H_t + 4HH_x + H_{xx} + H_{xxxx} = 0 \tag{6}$$

For these equations an interesting class of solutions was found in [8]. In them, such a "pulsating" character is more pronounced than in solutions of the equations (2). In case of initial perturbations evolution with $H_1 \neq 0$, $H_2 \neq 0$ and dimensionless wave numbers $0.41 < \alpha < 0.5$, an analogous quasiperiodical regime is set. An example of the evolution of moduli of the first two harmonics' amplitudes is given in Fig. 2($\alpha = 0.45$). It may be seen that after the onset the amplitude of the first harmonic tends to null, whereas the amplitude of the second harmonic reaches some constant value. The system remains in such a state for quite a long time (time range 50 - 120). Then, for a short period of time, the flow parameters drastically change (time range 120 - 150), and again a quasi-stationary regime becomes established.

Presenting the solution to equation (6) as Fourrier series, obtain an infinite system of ordinary differential equations for Fourrier harmonics.

$$\dot{H}_k - (\alpha^2 - \alpha^4)H_k + 2i\alpha k \sum H_{k'}H_{k-k'} = 0 \tag{7}$$

Supposing that all harmoinics except for H_1, H_{-1}, H_2, H_{-2} equal null, obtain a low-dimensional approximation of the equations (7):

$$\begin{aligned} \dot{H}_1 - (\alpha^2 - \alpha^4)H_1 + 2i\alpha H_2 H_{-1} = 0 \\ \dot{H}_{-1} - (\alpha^2 - \alpha^4)H_{-1} - 2i\alpha H_1 H_{-2} = 0 \end{aligned} \tag{8}$$

Seeking for the solution to the equations (8) in the form:

$$H_1 = |H_1(t)|e^{i\varphi(t)}, \quad H_2 = H_{-2} = \text{const} > 0 \tag{9}$$

In this case, the "phase difference" is $\phi = 2\varphi$, i.e. coincides with the double phase of the first harmonic. The equations (8) taking into account (9) is presented in the form:

$$|\dot{H}_1| - (\alpha^2 - \alpha^4)|H_1| + 2\alpha H_2 |H_1| sin(\phi) = 0 \tag{10}$$

$$\dot{\phi} + 4\alpha H_2 \cos(\phi) = 0 \tag{11}$$

From the equation (10) it is seen that at small values of H_2 the first harmonic H_1 will always be growing. At rather large value of H_2, the function $H_1(t)$ may become decreasing. From equation (11) it may be obvious that at large times the "phase difference" $\phi \to -\pi/2$, and such a regime will be unstable that is seen from equation (10). The change in sign H_2 will lead to $\phi \to \pi/2$. The formed quasi-steady regime differs in phase from the previous one by π. This means that wave maxima have changed to minima and vice versa.

4. Conclusion

The numerical simulation of wavy regimes of the freely-falling liquid film was realized using the previously discovered symmetry of model equations. The calculation results are presented for the dynamics of amplitudes of significant harmonics and for the wave surface evolution.The study has resulted in finding a specific regime, characterized by an alternate predominance of the first and second harmonics.At certain values of parameters this limiting cycle may have a very large period, during which the solution tends to the quasi-steady-state traveling regime, then suddenly transforms, and again is attracted to the quasi-steady-state with a phase shift of a half wave length.

Acknowledgments

The research was financially supported by the Russian Science Foundation (grant No. 14-22-00174).

References

[1] Nusselt W 1916 *Zetrschr. Ver. Deutch. Ing.* **60** 541-6
[2] Friedman S J and Miller C O 1941 *Industrial Engineering Chemistry* **33** 885-91
[3] Kapitza P L 1948 *Zhurnal Eksperimentalnoi i Teoreticheskoi Fiziki* **18** 3-18
[4] Nepomnyashchii A A 1974 *Fluid Dynamics* **9** 354-9
[5] Alekseenko S V, Arkhipov D G and Tsvelodub O Yu 2011 *Doklady physics* **56** 22-5
[6] Vozhakov I, Arkhipov D and Tsvelodub O 2016 *Journal of Physics: CS* **754** 032022
[7] Arkhipov D, Vozhakov I, Markovich D and Tsvelodub O 2016 *European Journal of Mechanics*-B/Fluids **59** 52-6
[8] Tsvelodub O Yu 2012 *Thermophysics and Aeromechanics* **19(1)** 91-9

IOP Conf. Series: Journal of Physics: Conf. Series **899** (2017) 032026 doi:10.1088/1742-6596/899/3/032026

Experimental study of ionic liquid-water flow in T-shaped microchannels with different aspect ratios

A A Yagodnitsyna[1,2], A V Kovalev[1,2] and A V Bilsky[1,2]

[1]Kutateladze Institute of Thermophysics, 1, Lavrentev Ave., Novosibirsk, 630090, Russia
[2]Novosibirsk State University, Novosibirsk, 2, Pirogova str., Novosibirsk, 630090, Russia

E-mail: yagodnitsinaaa@gmail.com

Abstract. Flow regimes of immiscible ionic liquid - water flow in T-shaped microchannels with 160 um hydraulic diameter and 1:2 and 1:4 aspect ratios are experimentally studied in the present work. Plug length and velocity were measured using high-speed visualization of the flow. Flow pattern maps were drawn for two channels. Parallel flow was shown to prevail for 1:4 aspect ratio channel in comparison to 1:2.

1. Introduction

In the recent years, one of the most promising immiscible liquid-liquid systems for industry applications is ionic liquid – water. Ionic liquids are salts with relatively low melting point (usually < 100°C), and many of them are liquids at room temperature. Unique properties of ionic liquids like resistance to high temperatures and radiation, low evaporation rate and vapor pressure make them very attractive for using in power plants and nuclear waste recycling [1-3]. The main limitation for ionic liquids is a high cost, thus development of efficient low-loss systems for handling liquid flow is essential. For this purpose promising microfluidic technology can be adopted. Microfluidic systems can strongly enhance heat and mass transfer because of high surface to volume ratio. In addition, using of microchannels instead of conventional devices reduces consumption of expensive chemicals or fluids [4,5].

However, there is a lack of fundamental knowledge about liquid – liquid flows on microscale and particularly about ionic liquid – water flows. To properly design and optimize microfluidic devices more data on hydrodynamics and fluid flows in microchannels is needed. Nowadays only a few research articles on immiscible liquid flows with ionic liquid as one of the phases are available in the literature.

Novak et al. performed a study of extraction reaction for ionic liquid based two-phase system in the case of parallel flow pattern when liquids occupy equal volumes [6]. A model for mass transfer in this particular case was proposed and compared with experimental results. Experimental research of flow structure inside water plugs during ionic liquid – water flow was done by Dore et al. [7]. Presence of two recirculation areas in plugs was shown and circulation time was calculated from velocity fields and plotted as function of plug velocity and length. Flow pattern maps for microchannels made of three different materials were obtained in [8]. Flow pattern and pressure drop were strongly depend on the liquid which first entered a channel. Influence of channel diameter and flow velocity on extraction process in plug flow regime was reported by Li et al. [9]. Extraction efficiency tends to decrease with increasing of flow velocity or channel diameter. As one can see from existing studies, hydrodynamic properties play important role in chemical reaction efficiency. But there is still not enough data about influence of channel surface wettability, fluids viscosities, channel geometry etc. on flow pattern map and properties of particular flow patterns such as plug flow.

In this work ionic liquid – water flow in T-shaped rectangular microchannels with different aspect ratios and equal hydraulic diameters was studied by means of high speed visualization. In order to determine influence of aspect ratio on liquid – liquid flow properties flow pattern maps for different channels were plotted and plug length and velocities were measured.

2. Experimental setup

In the present work 1-Butyl-3-methylimidazolium bis(trifluoromethylsulfonyl)imide ionic liquid and distilled water were used for flow regimes visualization in T-shaped microchannels. Liquid properties are noted in Table 1. Ionic liquid was held in contact with water long enough to rich a saturation limit before using. Corresponding properties in Table 1 are taken from work [8] for water saturated ionic liquid. The T-shaped microchannels used are made of SU-8. Two channels with different aspect ratios were used in the experiments. A, b and c dimensions are shown in figure 1 (right) were equal to 120 um, 240 um and 120 um for the first channel and 100 um, 400 um and 200 um for the second one, which corresponds to 1:2 and 1:4 aspect ratios, respectively. The length of inlet and outlet channels is 11.5 and 22.5 mm, respectively which is sufficient to get steady flow. The photo of experimental setup for flow visualization is presented in figure 1. For flow visualization pco.1200 hs high speed camera and inverted Zeiss Axio Observer.Z1 microscope with 5x magnification lens were used. Frame rate varied from 1 to 1000 Hz. Halogen lamp was used for flow illumination. Liquids were injected in the channel by KDS Gemini 88 double syringe pump. Flow visualization was done in two different channel areas: T-zone and at the end of the channel, which is 56 channel hydraulic diameters far from T-zone.

Table 1. Physical properties of liquids

Physical properties	Ionic liquid (water saturated)	Water
Density, kg/m^3	1420	997
Viscosity, mPa•s	41	0.894
Interfacial tension, mN/m	12.3	

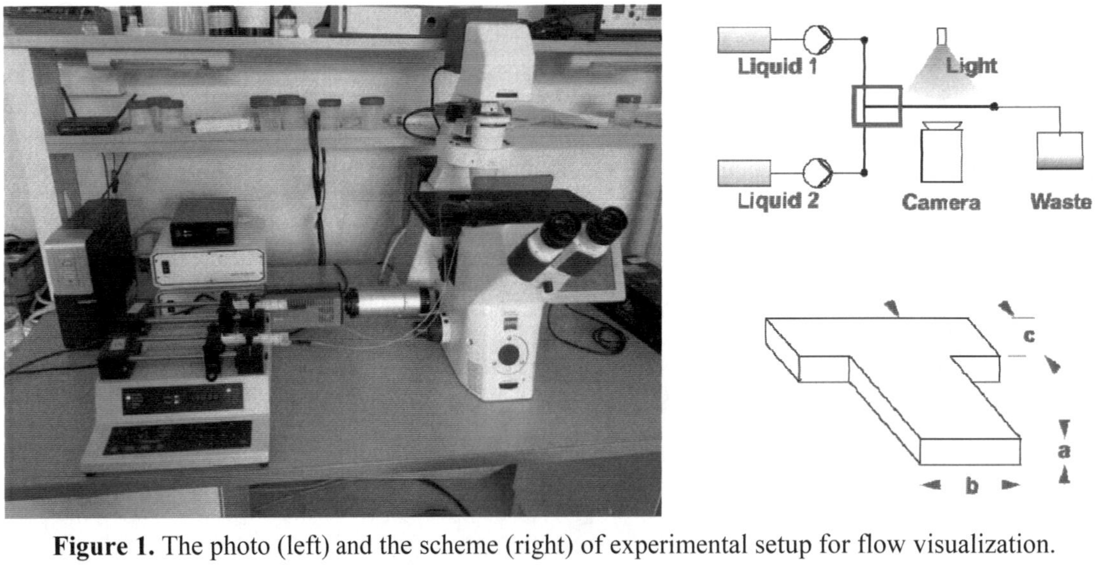

Figure 1. The photo (left) and the scheme (right) of experimental setup for flow visualization.

IOP Conf. Series: Journal of Physics: Conf. Series **899** (2017) 032026 doi:10.1088/1742-6596/899/3/032026

3. Results and discussion

3.1. Flow patterns

Typical flow patterns for two types of microchannels with 120x240 um and 100x400 um outlets are presented in figure 2. The water was a dispersed phase in this ionic liquid -water system inside SU8 channel because of wettability properties. It was found that flow patterns are independent from the phase which penetrates the channel first. Four basic flow patterns viz. plug, droplet, slug and parallel were observed for both microchannel sizes. In addition to those flow regimes there were throat-annular and rivulet flows for 120x240 um channel.

A plug flow corresponds to low superficial velocities, U_{il} = 0.3 mm/s and U_w = 0.15 mm/s for the case depicted in figure 2 a. Interfacial tension and adhesion forces dominate in this type of flow and tend water to form plugs with contact line on the channel wall. With increase of superficial velocities contact line is breaking and plugs are flowing surrounded by continuous phase film as presented in figure 2 b for U_{il} = 5.8 mm/s, U_w = 5.8 mm/s. The transition from plugs with contact line to non-contact line plugs is defined by dynamic contact angle dependence on contact line velocity. Usually only one type of plugs is observed in experiments [10], thus two different types for one set of liquids in this work indicate strong contact angle dependence from contact line velocity. Further increasing in superficial velocity of ionic liquid leads to droplet flow shown in figure 2 c for U_{il} = 14.5 mm/s, U_w = 2.9 mm/s. In this type of flow the water forms almost spherical drops that smaller than channel diameter. The mechanism of droplet formation is known in literature as the "dripping" mode in contrast to the "squeezing" mode inherent for the plug flow [11]. The slug flow was observed for high water to ionic liquid superficial velocities ratio, for instance U_{il} = 0.06 mm/s and U_w = 115.7 mm/s in figure 2 d. The parallel flow occurred when inertia force was strong enough to prevent plug formation. Typical pictures of this flow are represented in figure 2 e for U_{il} = 14.5 mm/s and U_w = 290 mm/s.

Throat-annular and rivulet flows were found only for 120x240 um channel. Throat-annular pattern also known as liquid ring was mentioned before in gas - liquid [12] and ionic liquid – water [8] flows and looks like core-annular flow with waves propagating over the interface. This flow pattern was realized in very narrow area of phase superficial velocities values, in figure 2 f U_{il} = 14.5 mm/s and U_w = 28.9 mm/s. In the rivulet flow a thin thread of ionic liquid flows upon one of channel walls. Rivulet flow is shown in figure 2 g for U_{il} = 1.16 mm/s and U_w = 580 mm/s.

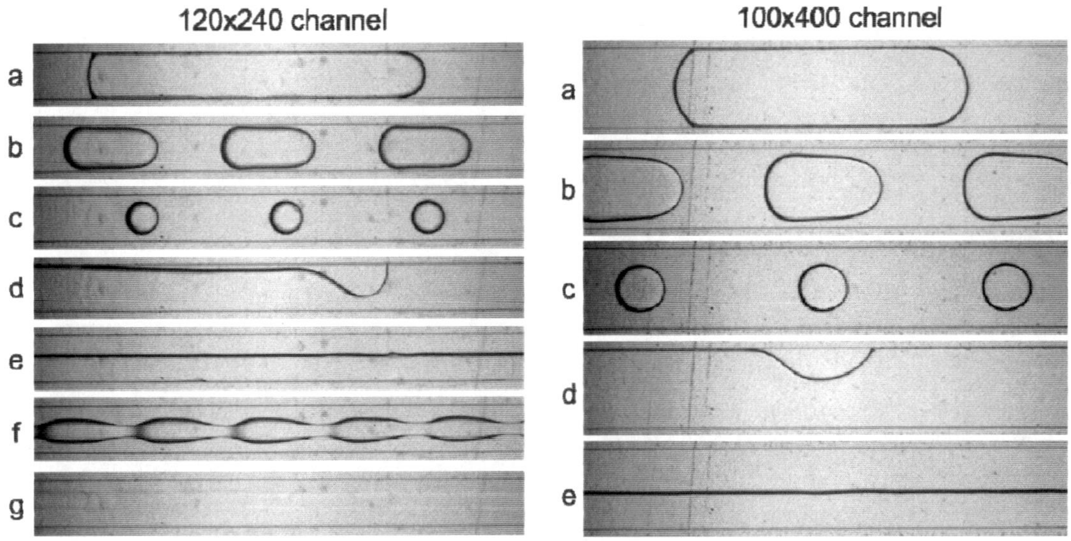

Figure 2. Typical flow patterns in 120x240 um (left column) and 100x400 um (right column) microchannels: a) plug flow, b) plug flow (no contact line), c) droplet, d) slug, e) parallel, f) throat-annular, g) rivulet.

IOP Conf. Series: Journal of Physics: Conf. Series **899** (2017) 032026 doi:10.1088/1742-6596/899/3/032026

3.2. Flow pattern maps

Based on visualization results flow pattern maps for both microchannel sizes were constructed. The dimensionless parameter proposed in [13] was used to plot flow pattern maps as diagrams presented in figure 3. It should be noted that almost every experimental point in the map for 120x240 um michochannel has corresponding point obtained for same superficial velocity in the map for 100x400 um microchannel. Consequently, dimensionless parameters such as Capillary number, Reynolds number and Weber number (We) written in terms of superficial velocity and hydraulic diameter have the same values in these couples of points. This fact allows us to specify aspect ratio influence on flow pattern maps keeping other parameters constant.

Physical significance of the using parameter Weber number multiplied by Ohnesorge number (Oh) is a ratio of forces that contribute to interface formation – capillary forces – to inertia forces and viscous dissipation acting in opposite direction. According to this definition and experimental results from [13] flow pattern maps are divided into several regions where different forces are prevail. There is a good match between experimental results and proposed regions borders. The plug flow points correspond to low WeOh values for both phases. Plug flow without contact line and droplet flow points hit the region with low WeOh of water and high WeOh of ionic liquid. There is parallel flow in the central region and the region of high values. Slug and rivulet points are fall in regions with high WeOh of water and low or medium WeOh values of ionic liquid respectively.

As one can note the main difference between flow pattern maps is that parallel flow in 100x400 um channel (figure 3 b) takes more area than in 120x240 um channel (figure 3 a). It spreads into plug and droplet regions in comparison to figure 3 a, and completely replaces rivulet flow. Probably, decrease in aspect ratio from 1:2 to 1:4 can enhance the role of viscous dissipation and make parallel flow more preferential. Therefore, we can conclude that channel with smaller aspect ratio is more appropriate for parallel flow applications such as various continuous chemical reactions.

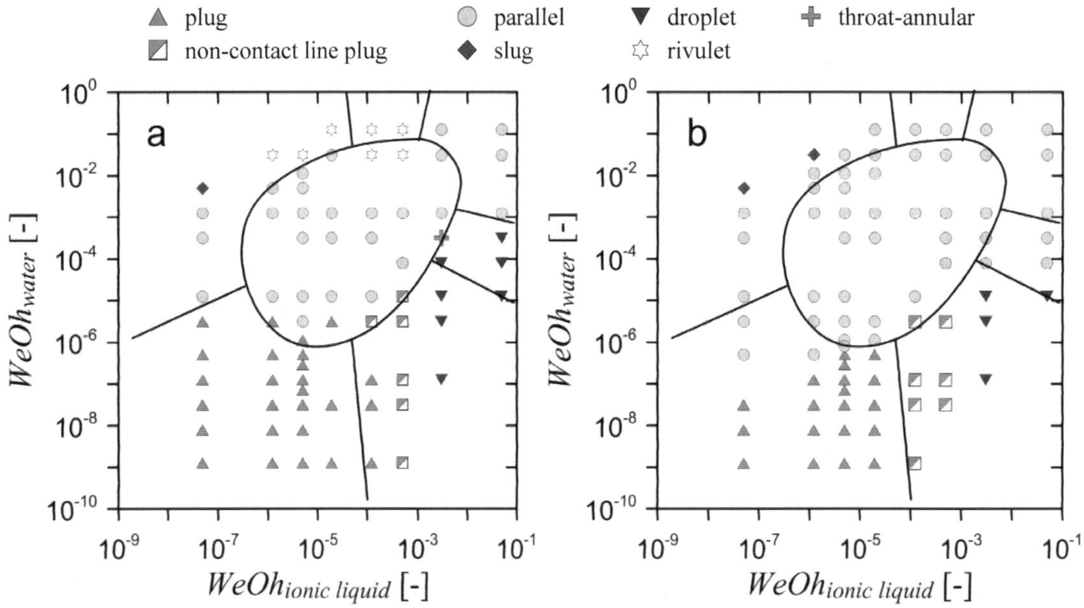

Figure 3. Flow pattern maps in We·Oh coordinates: a) 120x240 um channel outlet, b) 100x400 um channel outlet. Solid lines show transition between flow regimes obtained for several liquid sets are taken form [13].

3.3. Plug flow properties

For plug length and velocity measurements special software developed in the Kutateladze Institute of Thermophysics was used. Uncertainty of plug length measurements was about 2 pixels, which corresponds to 4 um. Plug velocity was calculated using plug leading and trailing edge positions for different time steps. The averaging over a number of plugs was done for every value of superficial velocity. In figure 4 (left) a plot of plug velocity versus bulk velocity is shown for both microchannels. Standard deviation of plug velocity is shown by error bars with 3 times higher magnitude in order to bring more clarity. It was found that for both channels equation (1) fits the experimental plug velocity data well with coefficient of determination R-squared = 0.992.

$$\boldsymbol{u_{\mathrm{plug}} = 2.34\, U_{\mathrm{bulk}}^{1.087}} \quad (1)$$

The power fitting function corresponds to the dependence of dynamic contact angle on contact line velocity for ionic liquid-water system. Analysis of plug length for different flow rates showed that plug length increases with the flow rate ratio increase and decreases with bulk flow increase. Therefore in figure 4 (right) non-dimensional plug length divided by channel hydraulic diameter is presented as a function of the ratio of bulk velocity to flow rate ratio for both channels. Similar trend can be easily seen for both channels with different coefficients in fitting functions.

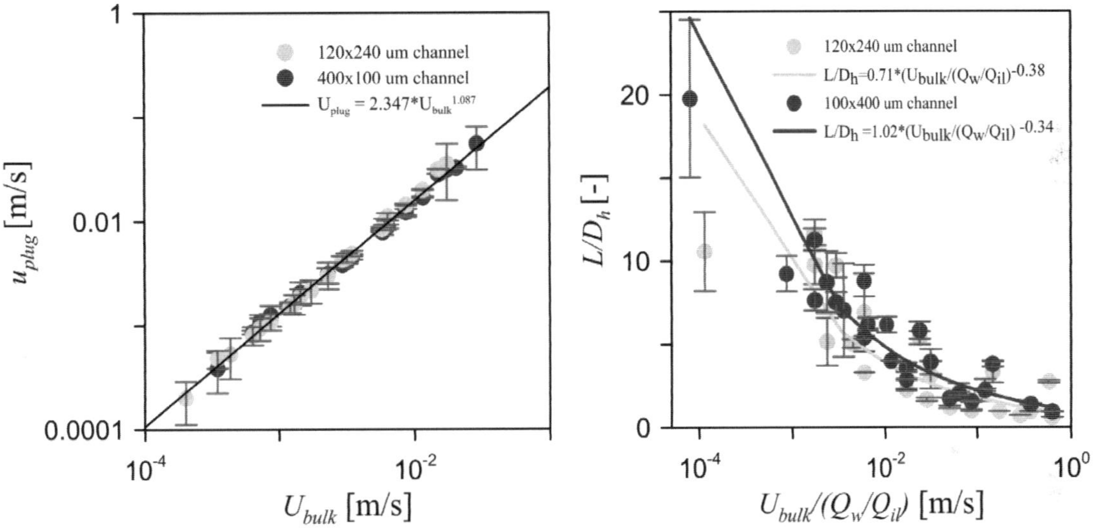

Figure 4. Left: plug velocity versus bulk velocity for 120x240 um and 100x400 um microchannels. Right: plug length as a function of the ratio of bulk velocity to flow rate ratio.

4. Conclusion

Experimental study of ionic liquid–water flow in T-shaped microchannels with different aspect ratios was done. The following flow regimes were obtained during flow visualization: plug, slug, parallel, throat-annular, rivulet and droplet flow. Two types of plugs, i.e. plugs with contact line and without one were observed which point out to the strong dependence of dynamic contact angle on contact line velocity. Flow maps were plotted for both microchannels in We*Oh numbers. It was shown that parallel flow in 100x400 um replaces rivulet and part of plug and droplet flows in contrast to 120x240 um channel. This fact is probably explained by aspect ratio difference and should be studied in detail more precisely. Plug length and velocity measurements were performed for a wide range of superficial velocities. For both microchannels plug length can be described as a function of bulk velocity very well and does not depend on the aspect ratio of the channels. Plug length can be fitted by power functions of bulk velocity and flow rate ratio with different coefficients for both channels.

Acknowledgements

This work performed in the Kutateladze Institute of Thermophysics SB RAS is supported by a grant from the Russian Science Foundation (project No 16-19-10519).

IOP Conf. Series: Journal of Physics: Conf. Series **899** (2017) 032026 doi:10.1088/1742-6596/899/3/032026

References

[1] Liu S, Fukuyama T, Sato M and Ryu I 2004 *Organic Process Research and Developmen* **8** 477–81
[2] Wishart J F and Shkrob I A 2009 *Ionic Liquids: From Knowledge to Application ACS Symposium Series* **1030** 119-34
[3] Wishart J F 2009 *Energy and Environmental Science* **2** 956–961
[4] Dummann G, Quittmann U, Gröschel L, Agar D W, Wörz O and Morgenschweis K 2003 *Catal. Today* **79–80** 433–9
[5] Burns J R and Ramshaw C 2001 *Lab Chip* **1** 10–5
[6] Novak U, Pohar A, Plazl I and Žnidaršič-Plazl P 2012 *Sep. Purif. Technol.* **97** 172–8
[7] Dore V, Tsaoulidis D and Angeli P 2012 *Chem. Eng. Sci.* **80** 334–41
[8] Tsaoulidis D, Dore V, Angeli P, Plechkova N V and Seddon K R 2013 *Int. J. Multiph. Flow* **54** 1–10
[9] Li Q and Angeli P 2016 *Chem. Eng. Sci.* **143** 276–86
[10] Kovalev A V, Yagodnitsyna A A and Bilsky A V 2016 *J. Phys. Conf. Ser.* **754**
[11] Fu T, Ma Y, Funfschilling D, Zhu C and Li H Z 2010 *Chem. Eng. Sci.* **65** 3739–48
[12] Serizawa A, Feng Z and Kawara Z 2002 *Exp. Therm. Fluid Sci.* **26** 703–14
[13] Yagodnitsyna A A, Kovalev A V and Bilsky A V 2016 *Chem. Eng. J.* **303** 547–54

Tip-leakage cavitation in the clearance of a 2D hydrofoil with fillets: high-speed visualization and PIV/PTV measurements

Ivan I Zapryagaev[1], Mikhail V Timoshevskiy[1] and Konstantin S Pervunin[1,2]

[1] Kutateladze Institute of Thermophysics, Siberian Branch of the Russian Academy of Sciences, 1, Lavrentyev Ave., Novosibirsk, 630090, Russia
[2] Department of Physics, Novosibirsk State University, 2, Pirogov Str., Novosibirsk, 630090, Russia

ivanzapryagaev@gmail.com

Abstract. Tip-clearance cavitation is one of the most aggressive forms of cavitation as it can cause surface erosion of hydraulic machinery elements and, as a result, their fatigue damage and disturb designed operating conditions. At present, the literature lacks for detailed experimental data on the inception and development of this type of cavitation at various flow conditions. In the paper, a tip-leakage cavitation occurring in the clearance between an end face of a 2D hydrofoil (a scaled-down model of guide vanes (GV) of a Francis turbine) and a transparent wall of the test section was studied. The experiments were carried out for different cavitating regimes on the cavitation number and two attack angles of 3° and 9°, with the gap size (tip clearance width) varied in the range from 0.4 to 0.8 mm. In order to determine the cavitation inception conditions and investigate the dynamics of the tip-leakage cavitation, a high-speed visualization was applied. A modified PIV/PTV technique with a diverging laser beam instead of a laser light sheet was used to measure the mean velocity distributions within the gap. It was shown that the cavitation pattern on the suction side of the GV model impacts the dynamics of the leakage flow in the gap but does not affect the sheet cavity formed close to the foil leading edge in the clearance as well as its size and dynamics. When the gap size is increased, the tip-leakage cavitation initiates at higher cavitation numbers or, in other words, conditions for the cavitation occurrence become more favorable.

1. Introduction

Tip-clearance cavitation occurs when a liquid leaks due to a high pressure gradient through narrow gaps between tip edges of rotor blades or end faces of guide vanes and stator housing when the clearance flow separates from the lateral surface of a hydrofoil. It is one of the most widespread forms of cavitation as it initiates at higher cavitation numbers, before any other type of cavitation appears [1]. General parameters of the tip-leakage flow that accompanies the tip-clearance cavitation are as follows: small gap thickness compared to its length and height, high pressure gradients and strong shear stresses. Thus, this type of cavitation has several recognizable features that make it difficult to study such flow both experimentally and numerically. According to the literature, the tip-clearance cavitation is commonly divided into three parts: gap cavitation itself occurring in the clearance, blade-end cavitation taking place at the leading edge due to stretching of the boundary layer vorticity when the gap size is small and the leakage vortex cavitation caused by interaction of the leakage flow with the primary flow and usually occurring when the gap size is relatively large [2]. In hydraulic

IOP Conf. Series: Journal of Physics: Conf. Series **899** (2017) 032027 doi:10.1088/1742-6596/899/3/032027

machinery, the presence of all these cavitation processes can cause erosion of surface material, decrease of equipment efficiency, increased noise and vibration level and pressure pulsations, which can eventually lead to the system failure.

In most research papers that can be found in the literature, the phenomena of the tip-leakage [3-5] and tip-vortex cavitation [6-9] are investigated numerically. The number of experimental studies is substantially lower [1, 10], which is linked with the technical difficulty of implementation of the specified flow geometry and application of measurement techniques under such conditions. Moreover, measurement results of velocity distributions in the clearance that are necessary to quantitatively compare numerical and experimental data and validate existing mathematical models are unavailable in literature nowadays. The research aims at experimental investigation of the tip-leakage flow, cavitation inception and development, assessment of the effect of the gap cavitation on velocity distributions in the clearance and evaluation of the influence of the gap size on all these processes. The study was carried out for the attack angle of 3° and 9° and various cavitation regimes on the cavitation number by means of a high-speed imaging and PIV/PTV technique adapted for velocity measurements in the clearance.

2. Experimental setup and measurement techniques

The experiments were performed in the cavitation tunnel in Kutateladze Institute of Thermophysics SB RAS. Description of the test rig can be found in [11]. The test object was a scaled-down model of guide vanes (GV) of a Francis turbine, which is a symmetrical hydrofoil equipped with fillets to reproduce exactly the geometry of original guide vanes (Figure 1). The foil chord length C equals to 100 mm, its maximum thickness of $h_{max}/C = 0.223$ is at $x/C = 0.44$. The GV leading edge is shaped with a rounding radius of $0.02C$ and its trailing edge was blunt with the height of $0.013C$. The detailed information about the foil profile is given in Table 1. The foil rotation axis with the diameter of 25 mm in the clearance h_1 between the GV end face and transparent wall of the test channel where the observations and measurements were conducted (see Figure 1-b) is located at the distance of $0.54C$ from its leading edge. The fillets nearby the gap h_1 are rounded along the foil chord with the radii of $0.176C$ and $0.143C$ from the suction and pressure sides, respectively, and curved in the transversal direction with the radius of $0.113C$ from both sides (Figure 1). In the experiment, h_1 was changed in the range between 0.4 mm and 0.8 mm. The attack angle α was set to be 3° and 9° and the cavitation number $\sigma = (p_{in}-p_V)/(\rho U_0^2/2)$, where p_{in} is the static pressure at the inlet of the test channel, p_V is the saturation vapor pressure of the distilled water (operating liquid), ρ is the water density, U_0 is the mean flow velocity measured by PIV in the central longitudinal section close to the test channel inlet, was varied in the range from 1.66 to 0.74 for $\alpha = 3°$ and from 3.58 to 1.66 for $\alpha = 9°$ by changing the static pressure and/or U_0. The liquid temperature was stabilized at 30±0.1 °C.

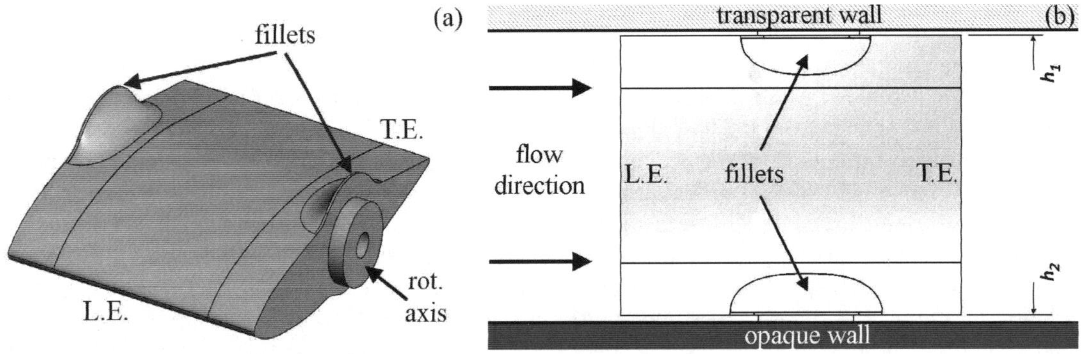

Figure 1. (a) 3D model of the guide vane equipped with fillets and an axis of rotation at both sides and (b) schematic of its disposition in the test channel (top view).

In order to visualize the cavitation patterns in the clearance, a high-speed visualization was applied. For this, a high-speed Photron FASTCAM SA5 CMOS-camera (digit capacity 12 bits, resolution 1024x1024 pix., acquisition rate 7 kHz) with Nikon AF Nikkor 50 mm f/1.4D lens was used. To illuminate the region of cavitation occurrence uniformly, three halogen lamps with the total power of 2.5 kW were utilized as a continuous incoherent light source. In the experiment, the camera was inclined at the angle of 25° to horizon so it was able to register both cavitation structures on the GV suction side and gas-vapor cavities in the clearance. A PIV-system consisting of a CCD-camera ImperX GEV B2020 (pixel depth 14 bit, matrix resolution 2048x2048, acquisition rate 6 Hz) with a Nikon micro-Nikkor 60 mm f/2.8D lens and a low-pass optical filter (transmission edge at 570 nm), a Nd:YAG laser Quantel EVG00200 (emission wavelength 532 nm, pulse repetition rate 15 Hz, pulse duration 10 ns, pulse energy 200 mJ) and a synchronizing processor POLIS (8 TTL channels, repetition rate of pulse pairs 10 Hz, minimum pulse delay 100 ns) and operated through a PC using "ActualFlow" software was used to measure velocity distributions in the gap. The camera was installed perpendicular to the transparent test channel wall. The gap was illuminated by a laser beam expanded using an Edmund Optics 2-8X 532 nm beam expander to cover the whole GV end face. The laser beam was directed at the angle of nearly 25° to the camera optical axis in the horizontal plane. PMMA microparticles of 20-50 μm size dyed with Rhodamine B were applied as tracers to avoid overexposure and glares in images. The description of the image processing algorithms is given in [11].

Table 1. Points of the generatrix of the GV model at different cross-sections x/C. $y^{\pm}$ are the transversal coordinates of the points on the upper and lower foil sides with respect to its chord (zero line). The GV shape is a result of a cubic spline approximation of these values.

x/C	$y^{\pm}/C$
0	0
0.1	±0.0658
0.2	±0.0898
0.3	±0.1045
0.4	±0.1098
0.5	±0.1087
0.6	±0.0981
0.7	±0.0792
0.8	±0.0544
0.9	±0.0312
1	±0.0065

3. High-speed visualization

3.1. Small gap size (0.4 mm)

At the lower attack angle of 3°, the cavitation first initiates over the GV fillets on the suction side in form of transient bubbles at $\sigma = 0.97$ (not shown). When the cavitation number is decreased to 0.92, the transient bubbles also appear in the central part of the hydrofoil and travel roughly from the GV leading edge to its midsection, where GV attains its maximum thickness ($x/C \approx 0.44$). Downstream of the midsection, the bubbles collapse due to a local growth of the static pressure. At these regimes, no tip-leakage cavitation was registered. A further decrease of σ to 0.84 makes the cavity to grow and leads to its unsteadiness (not shown). The primary cavity in the central part of the GV model merges the lateral cavities on both GV fillets. In the gap, cavitation appears only in the aft part of the GV model right behind the rotation axis as a bubbly flow similar to that shown in Figure 2-a ($\sigma = 0.74$).

IOP Conf. Series: Journal of Physics: Conf. Series **899** (2017) 032027 doi:10.1088/1742-6596/899/3/032027

The regimes at $\sigma = 0.84$ and 0.74 do not differ essentially but cavitation in the second case is more intensive so below we consider only the latter regime. The size of cavitation area in the clearance is governed by the dynamics of the primary cavity and alternates in phase with its length pulsations as a result of variation of the local pressure field due to the unsteady flow behavior. In addition, the traveling bubbles are periodically entrained into the gap from the suction or pressure sides of the hydrofoil and then collapse or escape to the same or opposite side, depending on the phase of the primary cavity evolution (Figure 2-a).

At the higher incidence of 9°, cavitation occurs both on the GV suction side and in the clearance right behind the leading edge in form of a sheet cavity at $\sigma = 2.34$ (not shown). The sheet cavity in the gap has a streaky structure similar to the one in Figure 2-b. This is likely connected with a high level of turbulence in this flow region, which can be a reason for the bypass transition. At $\sigma = 2.15$, dimensions of both cavities increase but the cavitation patterns remain the same (not presented). When the cavitation number is decreased down to $\sigma = 1.65$, both cavities expand downstream and the flow regime modifies to an unsteady one – so-called cloud cavitation is observed over the foil suction side (Figure 2-b). Despite the primary cavity is unstable, the sheet cavity in the clearance appears to be stable, with the cavitation pattern unchanged (Figure 2-b). As for the inclination of 3°, the transient bubbles are also periodically entrained into the gap right behind the GV axis from its both sides. However, the number of the transient bubbles occurs to be noticeably lower as compared to the case of $\alpha = 3°$.

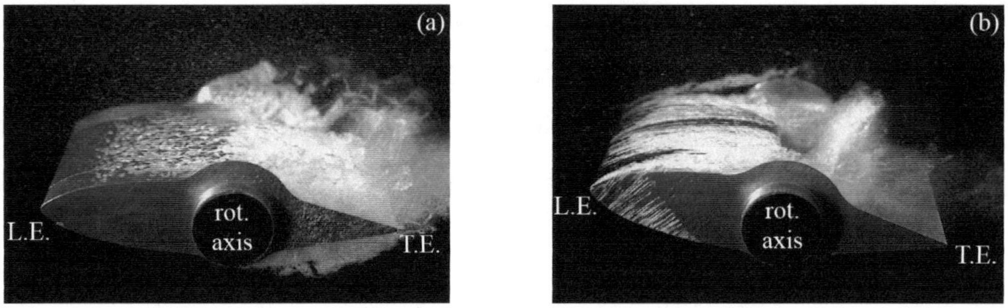

Figure 2. Instantaneous images of tip-clearance cavitation along with partial cavities on the suction side of the GV section for $h_l = 0.4$ mm when (a) $\alpha = 3°$, $\sigma = 0.74$ and (b) $\alpha = 9°$, $\sigma = 1.65$.

3.2. Large gap size (0.8 mm)

An increase of the gap width leads to noticeable changes in behavior of the tip-leakage flow. However, the primary cavity pattern and its dynamics change with a decrease of the cavitation number in the same way as it was for $h_l = 0.4$ mm. At $\alpha = 3°$, cavitation in the clearance initiates at $\sigma = 1.65$ right behind the foil leading edge as a sheet cavity similarly to the case presented in Figure 2-b but, for the smaller gap size, it is absent (see Figure 2-a). The sheet cavity in the gap has also a streaky structure and its dynamics remains stable even if the primary flow is unsteady (e.g. see the case of $\sigma = 0.74$ in Figure 3-a) as it was in the case shown in Figure 2-b. At the attack angle of $\alpha = 9°$, the sheet cavity initiates in the gap at higher cavitation numbers (about $\sigma = 3.58$) in comparison with the case of h_l = 0.4 mm for which the incipience cavitation number is $\sigma = 2.34$. The transition to cloud cavitation at σ = 1.66 does not modifies the tip-leakage cavity behavior like in the case of $h_l = 0.4$ mm. In general, the larger the gap size is, the larger the cavitation area in the clearance occurs to be. For instance, it is twice larger for $h_l = 0.8$ mm as compared to $h_l = 0.4$ mm.

4. PIV/PTV measurements

PIV/PTV measurements were performed only for $\alpha = 9°$ and $h_l = 0.8$ mm at the cavitation incipience ($\sigma = 2.37$) and cloud cavitation ($\sigma = 1.66$) regimes. The mean velocity distributions in the clearance are presented in Figure 4. The velocity distributions for both cases appear to be nonuniform and are influenced by the cavity. For example, the flow around the GV leading edge reaches the velocities up to $1.5U_0$ (see Area I in Figure 4). As seen (cf. Figures 3-b and 4-b), the cavity size almost completely corresponds to Area I. The sheet cavity located in this region contracts the flow cross-section, which leads to its acceleration. In Area II (Figure 4), the flow moves upward and eventually escapes the gap to the GV suction side. This occurs because of the existing pressure gradient between the pressure and suction sides. Above the GV rotation axis, there exists a recirculation zone marked as Area III in Figure 4, where the flow rises from the fillet base to its top due to an additional contraction of the primary flow by each fillet. Behind the GV rotation axis ($x/C > 0.65$), the flow is directed downward from the suction side to the pressure one (see Area IV in Figure 4), which is the opposite case to the one in Area II as the pressure and suction sides switch their places downstream of the axis when the attack angle is nonzero. However, the flow velocity in this region is significantly lower than the bulk velocity and is only about $0.2U_0$ since the pressure gradient behind the axis is lower in comparison with the one in front of it. This result is in good agreement with the acquired visual data for the unsteady regime (see Figure 3), where the direction of the traveling bubble motion periodically changes, so the mean flow velocity is in general very low.

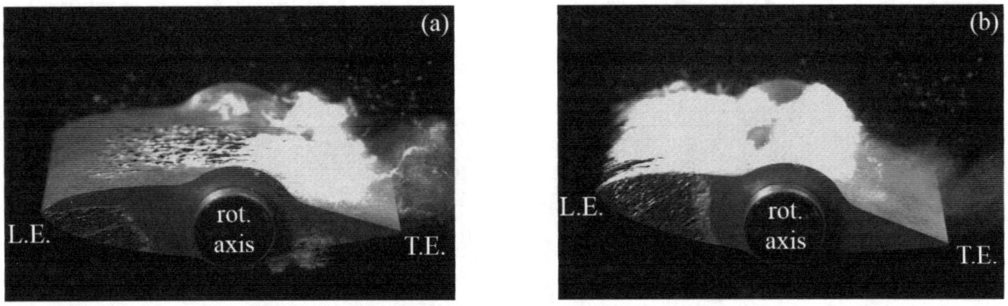

Figure 3. Instantaneous images of tip-clearance cavitation along with partial cavities on the suction side of the GV section for $h_1 = 0.8$ mm when (a) $\alpha = 3°$, $\sigma = 0.74$ and (b) $\alpha = 9°$, $\sigma = 1.66$.

5. Conclusions

The tip-leakage cavitation was successfully studied by means of a high-speed visualization and PIV/PTV approach. It was demonstrated that the cavity formation and evolution in the clearance is strongly dependent on the gap size for both attack angles. The larger the gap is, the more developed the cavity appears to be. The sheet cavity, in turn, influences the velocity distributions within the gap. At unsteady regimes, oscillations of the primary cavity over the foil suction side make the flow direction in the clearance to alternate periodically depending on the phase of the main cavity evolution, especially at small incidences. The GV fillets were shown to produce recirculation zone above the rotation axis where the flow rises from the fillet base to its top.

Acknowledgments

The research was funded by a grant from the Russian Foundation for Basic Research (Project No. 17-08-01199-A supervised by K. Pervunin) and a grant of the President of the Russian Federation for the state support of leading scientific schools (Project No. NSh-10179.2016.8 supervised by Prof. D. Markovich).

References

[1] Gearhart W S 1966 Tip clearance cavitation in shrouded underwater propulsors *J. Aircraft* **3(2)** 185–192
[2] Farrell K J and Billet M L 1994 A correlation of leakage cavitation in axial-flow pumps *J. Fluids Eng.* **116** 551–557
[3] You D, Wang M, Moin P and Mittal R 2006 Effects of tip-gap size on the tip-leakage flow in a turbomachinery cascade *Phys. Fluids* **18** 105102(14)

[4] Zhao Y, Wang G, Jiang Y and Huang B 2016 Numerical analysis of developed tip leakage cavitating flows using a new transport-based model *Int. Comm. Heat Mass Transf.* **78** 39–47
[5] Harwood C M and Young Y L 2014 A physics-based gap-flow model for potential flow solvers *Ocean Eng.* **88** 578–587
[6] Murayama M, Yoshida Y and Tsujimoto Y 2006 Unsteady tip leakage vortex cavitation originating from the tip clearance of an oscillating hydrofoil *J. Fluids Eng.* **128** 421–429
[7] You D, Wang M, Moin P and Mittal R 2007 Vortex dynamics and low-pressure fluctuations in the tip-clearance flow *J. Fluids Eng.* **129** 1002–1014
[8] Watanabe S, Seki H, Higashi S, Yokota K and Tsujimoto Y 2001 Modeling of 2-D leakage jet cavitation as a basic study of tip leakage vortex cavitation *J. Fluids Eng.* **123** 50–56
[9] Hsiao C T and Pauley L L 1998 Numerical study of the steady state tip vortex flow over a finite-span hydrofoil *J. Fluids Eng.* **120** 345–353
[10] Laborde R, Chantrel P and Mory M 1997 Tip clearance and tip vortex cavitation in an axial flow pump *J. Fluids Eng.* **119** 680–685
[11] Kravtsova A Yu, Markovich D M, Pervunin K S, Timoshevskii M V and Hanjalić K 2014 Cavitation on a semicircular leading-edge plate and NACA0015 hydrofoil: visualization and velocity measurement *Therm. Engng.* **61(14)** 1003–1010

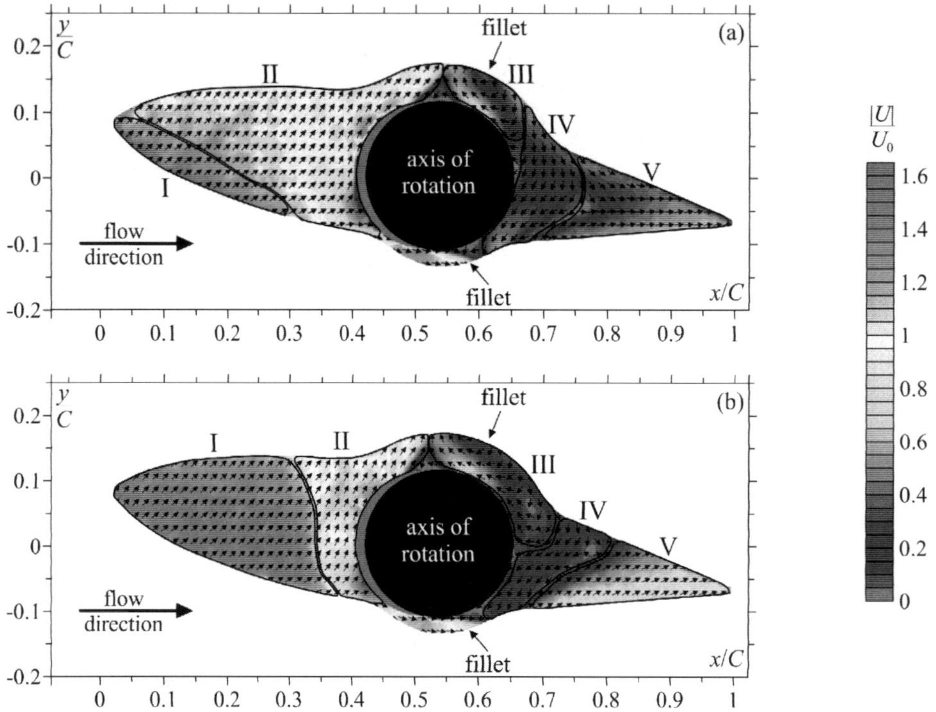

Figure 4. Plane distributions of the mean velocity (absolute values) in the gap between the GV end face and test channel wall in case of h_l = 0.8 mm at α = 9° for (a) σ = 2.37 (tip-leakage cavitation inception) and (b) σ = 1.66 (cloud cavitation). Black arrows show the flow direction. The black circle depicts the GV rotation axis and the grey area indicates its shadow. The velocity fields are separated into several distinct zones corresponding to: I – tip-leakage cavitation, II – preaxial upward flow, III – recirculating flow above the rotation axis, IV – downward flow behind the rotation axis and V – flow in the foil aft part.

Continuous spin detonation of a syngas-air mixture in a plane-radial vortex combustor

F A Bykovskii[1], S A Zhdan[1], E F Vedernikov[1]

[1] Lavrentyev Institute of Hydrodynamics SB RAS, Lavrentyev Prosp. 15, 630090, Novosibirsk, Russia

E-mail: bykovskii@hydro.nsc.ru

Abstract. Continuous spin detonation in syngas-air mixtures with three different syngas compositions [CO]/[H_2] = 1/3, 1/2, and 1/1 is experimentally studied in a flow-type radial vortex combustor 500 mm in diameter. It is found that all these mixtures with three syngas compositions can be effectively burned in air in the detonation regime. Transverse detonation waves of identical structure are detected. The limits of existence of continuous detonation in terms of the specific flow rates of the mixtures (minimum values) are determined.

1. Introduction

Intense investigations of burning various fuels in a continuously running transverse detonation wave (TDW) (according to Voitsekhovskii's concept [1]) are currently performed. Of interest for practice is continuous spin detonation (CSD) of fuel-air mixtures (FAMs) in flow-type combustors, which are used in air-breathing engines and power engineering facilities. Anthracite and lignite particles were burned in a detonation manner in a flow-type plane-radial vortex combustor 500 mm in diameter (PDK-500) located at the Lavrentyev Institute of Hydrodynamics of the Siberian Branch of the Russian Academy of Sciences [2]. The coal particles were transported to the combustor by a syngas flow with syngas compositions [CO]/[H_2] = 1/3, 1/2, and 1/1, which simultaneously served as a promoter of the chemical reaction in the detonation wave. Of independent scientific interest is burning a mixture of air with syngas containing no coal particles in the PDK-500, which is particularly important because a number of advanced technologies of syngas production have been developed. Depending on the method of syngas production, the ratio [CO]/[H_2] may have different values. Thus, gasification of coal with water vapor ($C + H_2O = CO + H_2$) allows syngas production with [CO]/[H_2] = 1/1; partial oxidation of saturated hydrocarbons (C_nH_{2n+2} + $1/2nO_2$ = nCO + (n+1)H_2) ensures syngas production with [CO]/[H_2] = 1/1 ÷1/2; finally conversion of saturated hydrocarbons with water vapor (C_nH_{2n+2} + nH_2O = nCO + (2n+1)H_2) allows syngas production with [CO]/[H_2] = 1/2÷ 1/3 [3]. As the number of carbon atoms in a hydrocarbon molecule decreases, the amount of released hydrogen increases (the greatest amount of hydrogen is released in the case of methane conversion: 1/3). The goal of the present paper is to obtain and study continuous detonation burning of FAMs based on syngas with the molar fractions [CO]/[H_2] = 1/1, 1/2, and 1/3 in the PDK-500 flow-type combustor.

2. Experimental

The experimental setup and the radial vortex combustor with a diameter d_{c1} = 500 mm and exhaustion toward the center are schematically shown in Fig. 1.

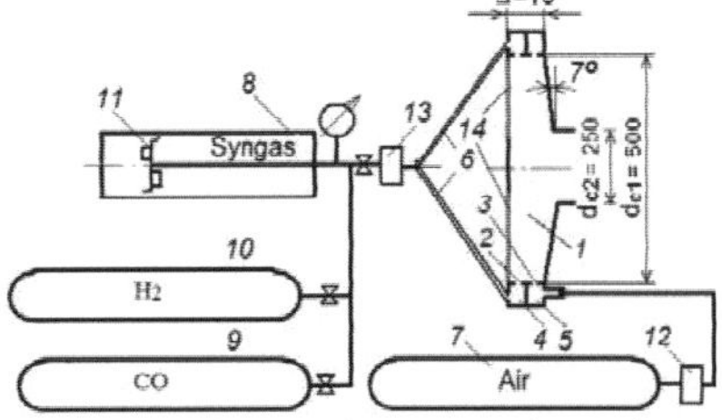

Fig. 1. Experimental setup and combustor.

The experimental combustor *1* is a semi-closed volume bounded by a cylindrical wall and two side walls (one flat wall and one conical wall with an angle $\alpha \approx 7^o$ from the base plane). The chosen geometry of the combustor *1* ensured an approximately constant cross-sectional area along the radius $S_c = \pi \cdot d_{c1} \cdot \Delta = 235.6$ cm^2, where $\Delta = 15$ mm is the distance between the walls near the cylindrical surface of the combustor. The reaction products escaped from the combustor through an orifice with a diameter $d_{c2} = 250$ mm, which was located in the conical wall.

The syngas was fed to the combustor *1* through an injector *2*, which had 150 orifices with a cross-sectional area of 0.5x1 mm^2 uniformly distributed over the cylindrical wall and directed along the radius. Air was supplied to the combustor *1* through 50 channels *3* formed in an annular slot of width $\delta = 5$ mm by ribs of thickness $\delta_r = 1$ mm inclined to the radius at an angle $\beta = 75^o$. The air flow was swirled in the combustor by guiding ribs with the minimum distance between them being 10 mm. The free cross-sectional area of the air slot at the combustor entrance was $S_\delta = (\pi \cdot d_{c1} - 50\ \delta_r) \cdot \delta = 76.04$ cm^2, and the combustor expansion ratio was $K_s = S_c/S_\delta = 3.1$.

Air and syngas were distributed over the orifices by annular manifolds *4* and *5,* respectively. In addition, a more uniform distribution of syngas over the manifold *5* was provided by using 24 tubes *6*. Air was supplied to the manifold *5* from a receiver *7* with a volume $V_{r,a} = 43$ liters, and syngas was supplied to the manifold *5* from a receiver *8* with a volume $V_{r,f} = 13.8$ liters, where it was prepared. Carbon monoxide from a gas holder *9* and hydrogen from a gas holder *10* were fed to the receiver *7,* where they were mixed. First the receiver was filled with carbon monoxide and then hydrogen was added. The gases in the receiver were mixed with the use of the Segner wheel *11* with blades on both arms. Hydrogen formed jets and rotated the wheel, while the blades provided the motion of the mixture in the axial direction. Thus, intense mixing of the gases was ensured. The time of the mixture formation (from the beginning of its preparation to its injection to the combustor) was 5 minute or more; therefore, the combustible components were well mixed over the entire receiver volume. An independent criterion of good mixing was the flow rate coefficient *c* [4], which had the same value for a particular mixture in different experiments. Air and syngas were injected into the combustor after opening of valves *12* and *13,* whose actuation time was defined by a specially programmed device on the control panel.

As the pressures in the receivers decreased, the flow rates of air and syngas also decreased: $G_a = 7.8 \rightarrow 0.76$ kg/s, $G_f = 1.2 \rightarrow 0.8$, $0.67 \rightarrow 0.076$, and $0.45 \rightarrow 0.05$ kg/s for the syngas compositions $[CO]/[H_2] = 1/1$, 1/2, and 1/3, respectively. The specific flow rates of air and fuel through the combustor cross section stayed within the interval $g_\Sigma = (G_a + G_f)/S_c = 340 \div 40$ kg/(s·m^2). The fuel-to-air equivalence ratio was almost constant: $\phi = 0.62$, 0.57, and 0.47 ± 0.02 for the syngas compositions $[CO]/[H_2] = 1/1$, 1/2, and 1/3, respectively. CSD was initiated in the combustor by means of burning an aluminum foil strip by electric current with an energy of about 5 J. The combustion products escaped into the atmosphere with an ambient pressure $p_a = 10^5$ Pa.

The process was photographed through two windows *14* located diametrically opposite along the combustor radius by a Photron FASTCAM SA5 type 775K-M2 high-speed video camera operating at 420000 frames per second. From each frame consisting of 16 pixels over the width of the window, only one pixel was chosen with the use of a special code written in the C^{++} programming language, and this pixel was then combined with single pixels chosen from other frames. As a result, the flow in the wave-fitted coordinate system was constructed. However, to obtain a real image of the TDW structure, the linear pattern should be rolled into a ring with retaining the scale of the examined object.

The pressures in the air and fuel receivers and manifolds ($p_{r,a}$ and $p_{r,f}$, $p_{m,a}$ and $p_{m,f}$, respectively), and also the pressures in the combustor (p_{ci}, where *i* is the number of the sensor located at a distance of 5, 47, 87, and 126 m from the cylindrical surface) were recorded by pressure sensors produced by the Wika company (Germany). The signals from the pressure sensors were recorded and processed on a computer.

3. Results of experiments

Stable CSD was observed in all experiments with burning FAMs consisting of air and syngas with three different compositions. A fragment of the photographic records of CSD for the CO + H_2 – air mixture is shown in Fig. 2. The linear sizes of the waves in the tangential direction near the cylindrical surface are corrected to the combustor diameter.

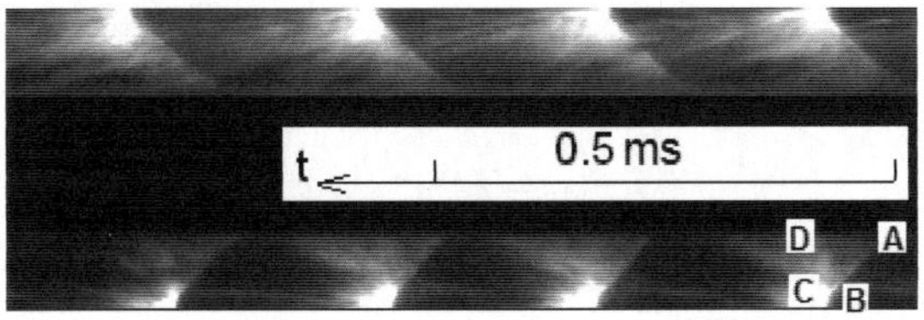

Fig. 2. CSD in the (CO+H_2) – air mixture (fragment of the photographic records), g_Σ = 294 kg/(s·m^2), ϕ = 0.62, $p_{m,a}$ =13.1·10^5 Pa, p_{c1} = 3.4·10^5 Pa, D = 1.67 km/s, and n = 4.

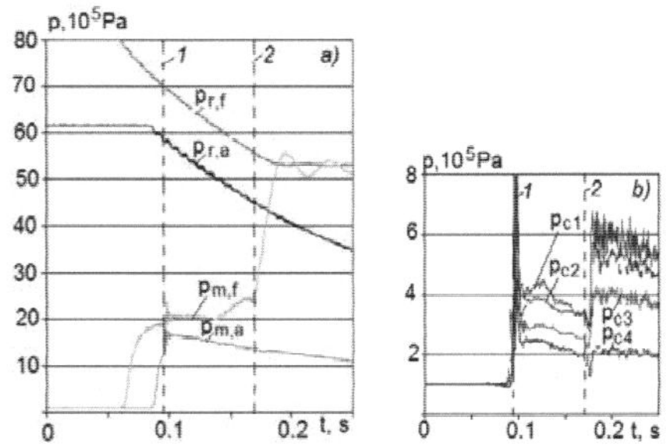

Fig. 3. Pressure oscillograms in the system of injection of the fuel components (a) and in the combustor (b).

The mixture is burned in four (n = 4) TDWs moving with a velocity D = 1.67 km/s. These waves move from left to right in the lower window and in the opposite direction in the upper window. The detonation front BC is the result of irregular reflection of the precursor AB from the cylindrical wall (Mach stem). A tail CD (shock wave in the products) emanates from the triple point C. A similar structure was observed in the case of CSD in homogeneous and two-phase mixtures in combustors of similar geometry [4]. A similar wave structure was also detected in FAMs containing syngas with the compositions CO + 2H_2 and CO + 3H_2. The ratio of the height of the front BC (h) to the distance between the waves (l) was h/l ≈ 1/15.

The oscillograms of the pressures in the system of injection of the fuel components and in the combustor for the experiment illustrated in Fig. 2 are shown in Fig. 3. In this experiment, the orifices of the fuel injectors were choked. As a result, the pressure sensors measured the pressures before the instant of initiation (*1*), in the CSD regime (time interval *1-2*), and during exhaustion of cold air (after the time instant marked by *2*). The injection system displayed only a minor response to CSD initiation, and the pressure in the combustor first increased due to exhaustion of the cold mixture and then decreased. Another increase in pressure was observed after injector choking. The pressures decrease toward the combustor center. The processes in the injection system and in the combustor at the instant of CSD initiation and existence for other syngas compositions were similar to the situation described above. The main parameters of the CSD regimes are listed in Table 1.

Table 1. Basic parameters of the CSD regimes for three syngas compositions.

syngas	G_A, kg/s	G_f, kg/s	g_Σ, kg/(s·m^2)	ϕ	$p_{m,a}$ ·10^5 Pa	p_{c1}	D, km/s	n
[CO]/[H_2]	7.8-5.98	1.0-0.63	364-294	0.61-0.63	17-13.0-	4.2-3.4	1.86-1.67	3-4
					13.1	6.4	-	-
[CO]/[2H_2]	7.8-0.87	0.67-0.08	354-40.3	0.57	16.7-2.5	4.6-1.6	1.87-1.97	3-1
[CO]/[3H_2]	7.6-0.9	0.45-0.05	341-41	0.48-0.46	16.3-2.2	4.4-1.07	1.88-1.55	3-2

For the [CO]/[H_2] composition, the flow parameters after CSD failure are also given.

4.Analysis of results

For similar specific flow rates of the combustible mixture, the CSD regime was found to be weakly dependent on the syngas composition. For close values of the equivalence ratio, an identical number of waves (n =3 or n = 2) and similar detonation velocities were obtained in both more and less chemically active syngas compositions. Apparently, the key role is played by acoustic phenomena in the combustor because the flow in the radial direction is subsonic. Figure 4 shows the detonation velocity (D) and the number of TDWs ПДВ (n) as functions of the specific flow rate of the fuel (g_Σ).

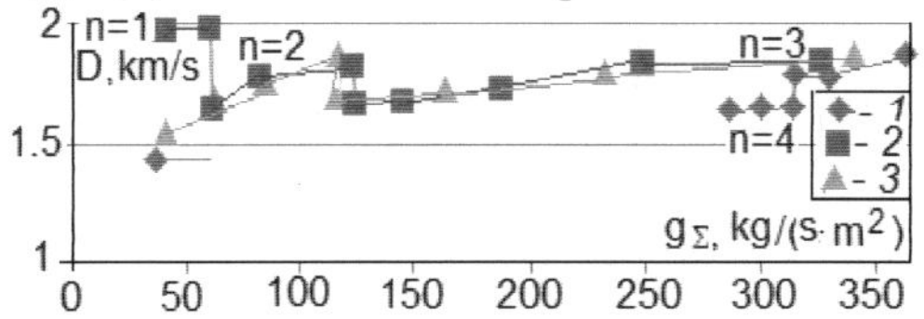

Fig. 4. Detonation velocity and number of TDWs versus the specific flow rate of the fuel for three syngas compositions: *1* -[CO]/[H_2], *2* - [CO]/[2H_2], *3* - [CO]/[3H_2].

During the experiments, the value of g_Σ decreased with decreasing flow rates of the FAM components. At high values of g_Σ, elevated detonation velocities and the greatest number of TDWs were observed. As previously [4], the detonation velocity decreased if the number of TDWs was fixed and the specific flow rate decreased. Moreover, when a certain value of g_Σ was reached, the number of TDWs also decreased (which was accompanied by a drastic increase in velocity). This tendency was clearly observed for the syngas compositions *2* and *3,* whereas the process for the syngas composition *1* displayed an anomalous behavior in the region $g_\Sigma \approx 300$ kg/(s·m^2): the number of waves increased from three to four. Further investigations are needed to explain this phenomenon. Nevertheless, the number of TDWs regularly decreased by the end of the experiment (when the lower limit in terms of g_Σ was approached); the extreme left point for the syngas composition *1* is taken from a different experiment. The minimum values of $g_{\Sigma min}$ were found below which the CSD process ceased (see also Table 1).

As previously [4], the pressures near the cylindrical surface of the radial combustor in experiments without CSD for all examined FAMs were higher than the pressures in the CSD regime with the same specific flow rates g_Σ (approximately by a factor of 2). The reason is the greater value of the centrifugal force in the case of cold gas exhaustion as compared to exhaustion of less heavy hot CSD products. Based on the pressure difference in the air channels ($p_{m,a}/p_{c1}$), it is seen (see Fig. 2 and Table 1) that critical exhaustion of air in the combustor occurs. Further investigations are required to reduce the pressure difference with retaining stable CSD.

Thus, multiwave CSD regimes in mixtures of air with the syngas compositions [CO]/[H_2] = 1/3, 1/2, and 1/1 were obtained for the first time in a plane-radial vortex combustion 500 mm in diameter. The structure of the detonation waves and the heights of the detonation fronts in the examined FAMs are similar to those observed previously for gaseous (methane-air) and heterogeneous kerosene-air and coal-air) mixtures [4].

Acknowledgment

This work was supported by the Russian Foundation for Basic Research (Grant No. 14-08-00080a).

5. References

[1] Voitsekhovskii B V 1959 Steady detonation *Dokl. Akad. Nauk SSSR* **129** (6) pp 1254-1256 [in Russian]

[2] Bykovskii F A Zhdan S A Vedernikov E F and Zholobov Yu A 2016 Detonation Burning of Anthracite and Lignite Particles in a Flow-Type Radial Combustor *Combustion, Explosion, and Shock Waves* **52** (6) pp 703-712

[3] Kozyukov E A Krylova A Yu and Krylov M V 2006 *Chemical Processing of Natural Gas* (Moscow: Izd. MAI) p 63 [in Russian]

[4] Bykovskii F A and Zhdan S A 2013 *Continuous spin detonation* (Novosibirsk: Izd. Sib. Otd. Ross. Akad. Nauk) p 423

IOP Conf. Series: Journal of Physics: Conf. Series **899** (2017) 042002 doi:10.1088/1742-6596/899/4/042002

Continuous spin detonation of poorly detonable fuel-air mixtures in annular combustors

F A Bykovskii[1], S A Zhdan[1]

[1] Lavrentyev Institute of Hydrodynamics SB RAS, Lavrentyev Prosp. 15, 630090, Novosibirsk, Russia

E-mail: zhdan@hydro.nsc.ru

Abstract. This paper reports on the results of experimental investigations of continuous spin detonation of three fuel-air mixtures (syngas-air, CH_4/H_2-air, and kerosene/H_2–air in a flow-type annular cylindrical combustor 503 mm in diameter. The limits of existence of continuous detonation in terms of the specific flow rates of the mixtures (minimum values) are determined. It is found that all gas mixtures, including the least detonable methane-air mixture, with addition of hydrogen can be burned in the continuous spin detonation regime.

1. Introduction

Results of experimental and theoretical investigations of continuous spin detonation (CSD) based on Voitsekhovskii's scheme [1], which are of interest for researchers all over the world, are summarized in the book [2]. The current status of experimental investigations of CSD in flow-type annular combustors (as a variant of an air-breathing engine) for some fuel-air mixtures (FAMs) is analyzed in [3]. It is known that the annular detonation chamber diameter d_c should be greater than a certain minimum value ($d_c > d_{c,min}$) for CSD to occur. Obviously, the minimum critical diameter $d_{c,min}$ for the least detonable methane-air mixture is very large. The critical diameter for CSD can be reduced by means of special fuel preparation. A well-known method [4] is incomplete thermal oxidation of methane ($CH_4 + 1/2O_2 = CO + 2H_2$) to a synthesis gas (syngas), which is a mixture of gaseous carbon monoxide (CO) and hydrogen (H_2) with the ratio of their molar fractions 1:2. The first results on CSD in a syngas-air mixture performed in the DK-300 combustor were reported in [5]. Of particular scientific interest is obtaining CSD in a gaseous methane-air mixture and heterogeneous kerosene-air mixture. Detonation burning of these mixtures in plane-radial rocket-type combustor with a diameter d_{c1} = 204 mm was studied in [6].

The goal of the present work is to obtain, investigate, and analyze the CSD process in a flow-type annular cylindrical combustor 503 mm in diameter (DK-500) for three FAMs: syngas-air, CH_4/H_2-air, and kerosene/H_2 – air.

2. Experimental

The DK-500 detonation chamber is a coaxial channel with an outer diameter d_c = 503 mm, length L_c = 490 or 540 mm, and gap Δ = 18 mm (Fig. 1). The free cross-sectional area of the combustor channel was $S_\Delta = \pi(d_c - \ )\cdot \ = 274.1$ cm^2. Air was fed to the combustor from two receivers *4* over an annular manifold *3* through an annular slot *2* of width δ = 3.5 mm (the free cross-sectional area was $S_\delta = \pi(d_c - \delta)\cdot\delta = 54.9$ cm^2). The DK-500 expansion ratio was $K_S = S_\Delta/S_\delta = 5.0$.

In experiments with gaseous FAMs, the two-component fuel ($CO + 2H_2$, $CH_4 + 8H_2$, or $CH_4 + 4H_2$) was fed from a receiver *7* over an annular manifold *6* through an injector *5*. The injector had 600 orifices with a cross-sectional area 0.35x1.45 mm^2, which were pairwise aligned at an angle of 90° to the combustor axis. The fuel components were mixed directly in the receiver *7*.

In experiments with heterogeneous FAMs, the liquid fuel (aviation kerosene TS-1) was displaced by hydrogen from a flowmeter *13* through a diaphragm *14* into the manifold *6* and then through the injectors *5* into the combustor (right part of Fig. 1). The pressure on the piston was provided by gaseous hydrogen incoming from a receiver *9*. Some part of hydrogen through a pipeline and diaphragm *15* was fed to a mixer *16*, where bubbling of liquid kerosene was ensured before its injection into the manifold *6*. The piston path L_p in the flowmeter was measured by a rheostat probe *18*.

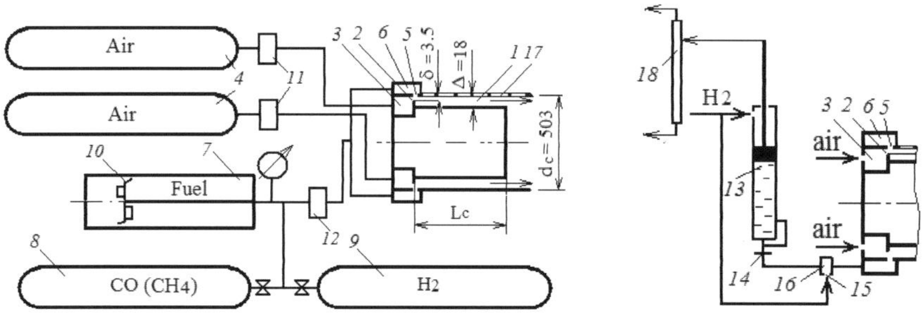

Figure 1. Schematic of the experimental setup.

The initial pressures of air in two air receivers were $p_{r,a0} = 30 \div 60\ 10^5$ Pa. The initial flow rates of air were varied in the interval $G_{a0} = 6.5 \div 34.5$ kg/s, and their initial flow rates G_a decreased approximately in proportion to the pressures in the receivers. The initial flow rat6es of the fuel were varied in the following ranges: a) for $CO + 2H_2$, $G_{f0} \approx 1.2$ kg/s; b) for CH_4/H_2, $G_{f0} = 0.75 \div 1.4$ kg/s; c) for kerosene, $G_f = (0.3 \div 1)$ kg/s and for H_2, $G_H = (0.06 \div 0.35)$ kg/s. The current specific flow rates of the FAMs $g_\Sigma = (G_a + G_f)/S_\Delta$ through the cross section of the DK-500 combustor and the fuel-to-air equivalence ratios ϕ varied in the following intervals: a) $g_\Sigma = 34 \div 234$ kg/(s m^2) and $\phi = 0.81 \div 1.29$; b) $g_\Sigma = 64 \div 850$ kg/(s m^2) and $\phi = 0.78 \div 1.02$; c) $g_\Sigma = 230 \div 565$ kg/(s m^2), $\phi_f = 0.17 \div 1.3$, $\phi_H = 0.23 \div 0.51$, and $m_H = 8.4 \div 42$ %. Here ϕ_f and ϕ_H are the equivalence ratios for kerosene and hydrogen, respectively, and $m_H = G_H/(G_f + G_H)$ is the mass fraction of H_2 in the two-phase fuel. The detonation products exhausted into the ambient atmosphere with an ambient pressure $p_a = 10^5$ Pa.

The process was photographed by a Photron Fastcam SA1.1 675K-M3 high-speed camera in the regime with 400000 frames/s through longitudinal Plexiglas windows *17* aligned one after the other along the combustor wall (see Fig. 1). The width and length of each window were 20 and 93 mm, respectively, and the distance between the windows was 24 mm. The method of determining the frequency *f*, the number of transverse detonation waves (TDWs) *n*, and the CSD velocity *D* was similar to that used in [2]. The pressures in the air receivers (p_{ra1} and p_{ra2}), receivers of the two-component fuel (p_{rf}) and hydrogen (p_{rH}), corresponding manifolds (p_{mf} and p_{ma}), and also the static (p_{c1}, p_{c2}, and p_{c3}) and total (p_{c30}) pressures were registered at distances of 15 and 100 mm from the combustor entrance (p_{c1} and p_{c2}) and near the combustor exit (p_{c3} и p_{c30}) by S-10 pressure sensors produced by the VIKA company.

3. Results of experiments

3.1. CSD in syngas-air mixtures

The DK-500 combustor with the length $L_c = 540$ mm allowed us to obtain multiwave ($n = 2 \div 3$) CSD regimes with the TDW rotation frequency $f = 1.51 \div 2.73$ kHz and detonation velocity $D = 1.13 \div 1.42$ km/s for the $CO + 2H_2$ + air mixture. Figure 2 shows a fragment of the photographic records of the CSD propagation process at the instant of the transition $n = 3 \rightarrow 2$.

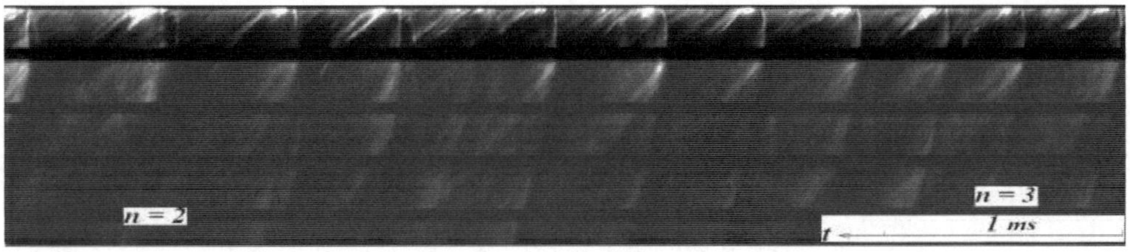

Figure 2. Photographic records (fragment) of CSD in the CO + $2H_2$ + air mixture: g_Σ = 180 kg/(s·m^2), ϕ = 1.25, n = 3: f = 2.58 kHz, D = 1.31 km/s; n = 2: f = 1.86 kHz, D = 1.42 km/s.

Three TDWs with decreasing frequencies f = 2.73 → 2.58 kHz and velocities D = 1.39→1.31 km/s were obtained as g_Σ = 234→180 kg/(s·m^2) and ϕ = 1.29 → 1.25. At g_Σ ≈ 180 kg/(s·m^2), a transitional process was observed with reduction of the number o0f TDWs (see Fig. 2). It is seen that the transition n = 3 → 2 occurs extremely rapidly (within one round of the wave). The TDW fronts and attached tails (oblique shock waves in reaction products) move from left to right, burning the mixture fed from the injection system, which is illuminated by products of combustion of a small acetylene-oxygen jet. Two TDWs existed until the end of the experiment in the intervals g_Σ = 180→34 kg/(s·m^2) and ϕ = 1.25 → 1.04; their frequency changed as f = 1.86 → 1.51 kHz and the CSD velocity changed as D = 1.42→1.13 km/s. The TDW front height was h ≈ 13 cm for n = 3 and h ≈ 20 cm for n = 2, whereas the ratio of the TDW height to the distance between the waves was h/l ≈ 1/4. The TDW rotation frequency f obtained in this experiment is shown in Fig. 6 as a function of the specific flow rate of the mixture g_Σ (curve 4).

3.2. CSD in methane/hydrogen – air mixtures

One-wave and two-wave CSD regimes with the TDW rotation frequency f = 0.56 – 1.66 kHz and detonation velocity D = 0.76 ÷ 1.3 km/s were obtained for the first time in poorly detonable gaseous mixtures CH_4+ $8H_2$ + air (m_H = 1/2) and CH_4+ $4H_2$ + air (m_H = 1/3) in the intervals of the specific flow rates of the mixture g_Σ = 64 ÷ 850 kg/(s m^2) and equivalence ratios ϕ = 0.8 – 1.02.

3.2.1. Fuel with the composition CH_4 + $8H_2$ (m_H = 1/2). CSD in this mixture existed in a wide range of the specific flow rates of the mixture g_Σ = 716→64 kg/(s m^2) as ϕ =1.02→ 0.78. In the intervals g_Σ = 160÷ 716 kg/(s m^2) and ϕ = 0.96 ÷ 1.02, regimes with two TDWs and TDW rotation frequency f = 1.32 ÷ 1.66 kHz (D = 1.04 ÷ 1.3 km/s) were observed. At g_Σ = 64 ÷ 160 kg/(s m^2) and ϕ = 0.78 ÷ 0.96, CSD was irregular, and the number of waves alternated between one and two (n = 2↔1). The range of frequencies was f = 1.1 ÷ 1.4 kHz (D = 0.86 ÷ 1.1 km/s) for n = 2 and f = 0.65 ÷ 0.87 kHz (D = 1.02 ÷ 1.3 km/s) for n = 1. A typical fragment of the photographic records of CSD is shown in Fig. 3.

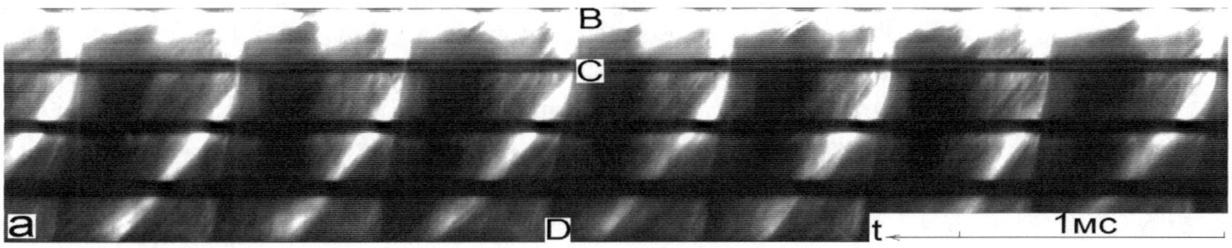

Figure 3. Photographic records (Fragment) of CSD in the CH_4 + $8H_2$ + air mixture: g_Σ = 567 kg/(s m^2), ϕ = 1.02, n = 2, f = 1.63 kHz (D = 1.28 km/s), and p_{c1} = 6.2 10^5 Pa.

The TDW height (BC) was estimated as h ≈ 15 cm, and the ratio of the TDW height to the distance between the waves was h/l ≈ 1/5, where l = 79 cm. The values of the CSD parameters for the FAM with the CH_4 + $8H_2$ fuel are presented in the first row of Table 1.

Table 1. CSD parameters for two compositions of the methane-hydrogen fuel.

Fuel	$p_{m,a}/p_a$	g_Σ, kg/(s·m^2)	ϕ	f, kHz	n	D, km/s
CH_4 +8H_2	24 →4	716→160	1.02→0.96	1.66→1.32	2	1.3→1.04
CH_4 +8H_2	4 →2	160→64	0.96→0.78	1.4→0.65	2↔1	1.1→1.02
CH_4 + 4H_2	25→18	849→645	0.94→0.95	1.36→1.33	2	1.1→1.04
CH_4 + 4H_2	18→2.5	645→ 100	0.95→0.97	0.83→0.56	1	1.28→0.84

3.2.2. Fuel with the composition CH_4 + $4H_2$ (m_H = 1/3). For the mixture with this composition in the range of the specific flow rates of the FAM g_Σ = 849 → 645 kg/(s·m^2) and ϕ = 0.94 → 0.95, we obtained two TDWs (n = 2) with a monotonically decreasing (in time) rotation frequency f = 1.36 → 1.33 kHz and velocity D = 1.07→1.05 km/s. The TDW height was estimated as $h \approx$ 25 cm, and the ratio h/l was $h/l \approx$ 1/6. At $g_\Sigma \approx$ 645 kg/(s·m^2), we observed a transition to the one-wave CSD regime. One TDW existed until the end of the experiment in the intervals g_Σ = 645 → 100 kg/(s·m^2) and ϕ = 0.95 → 0.97; its frequency changed as f = 0.83 → 0.56 kHz and the CSD velocity changed as D = 1.28 → 0.84 km/s. The TDW front height was $h \approx$ 50 cm, and its ratio to the distance between the waves was $h/l \approx$ 1/3. A fragment of the real-scale photographic records of the one-wave CSD is shown in Fig. 4.

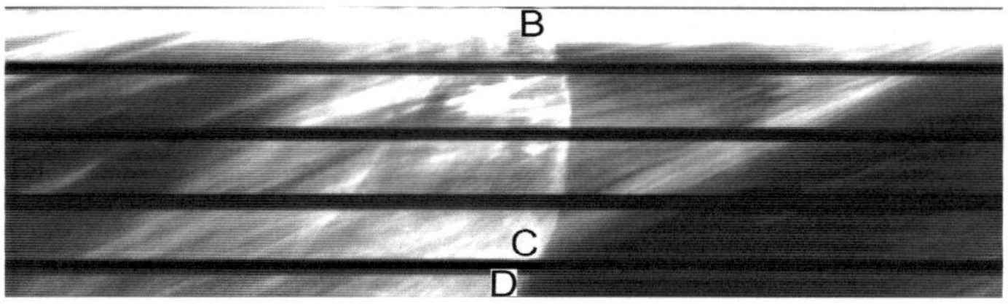

Figure 4. Fragment of the photographic records of CSD in the CH_4 + $4H_2$ + air mixture: a) one-wave regime of CSD (g_Σ = 620 kg/(s·m^2), ϕ = 0.95, f = 0.826 kHz, n = 1, and D = 1.26 km/s).

The experimental results for CSD in DK-500 for the examined CH_4/H_2 compositions are listed in Table 1, and the TDW rotation frequency f as a function of the specific flow rate of the mixture is plotted in Fig. 6 (points *1* and *2*).

3.3. CSD in the heterogeneous kerosene/hydrogen – air mixture

CSD regimes with the number of TDWs from one to five were obtained in the heterogeneous FAM liquid kerosene/H_2 – air in the interval of the specific flow rates of the mixture g_Σ = 230 ÷ 565 kg/(s m^2) with the mass fraction of hydrogen in the two-phase fuel being varied in the interval m_H = 8.4 ÷ 42 %. At g_Σ = 527 kg/(s·m^2), ϕ_f = 0.32, ϕ_H = 0.51, and m_H = 42 %, a five-wave CSD regime with the TDW frequency f = 4.86 kHz, detonation velocity D = 1.48 km/s, and TDW front height h ≈ 60 ÷ 70 mm was obtained. At g_Σ = 485 ÷ 557 kg/(s·m^2), $\phi_f \approx$ 0.7, ϕ_H =0.38 ÷ 0.41, and m_H = 21 %, a three-wave CSD regime with f = 2.51 ÷ 3.04 kHz, D = 1.32 ÷1.6 km/s, and h ≈ 100 ÷ 120 mm was observed (Fig. 5).

At g_Σ = 362 kg/(s·m^2), ϕ_f = 1.29, ϕ_H = 0.27, and m_H = 8.4 %, a one-wave CSD regime with f = 1.02 kHz and D = 1.56 km/s was obtained. The number of TDWs and their rotation frequency decrease with decreasing mass fraction of hydrogen in the two-phase fuel and specific flow rate of air, whereas the TDW front height increases. The dependence of the frequency f on g_Σ is shown in Fig. 6 (points *3*).

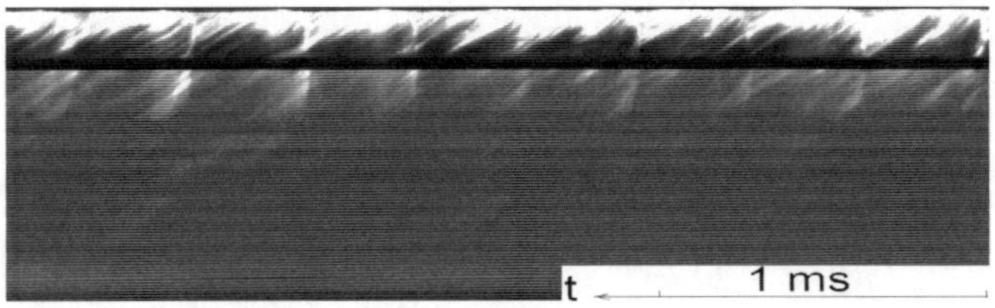

Figure 5. Fragment of the photographic records in the kerosene/H_2 + air mixture; $m_H = 21$ %, $g_\Sigma = 520$ kg/(s·m^2), $\phi = \phi_f + \phi_H = 1.08$, $f = 3.01$ kHz, n = 3, and $D = 1.55$ km/s.

4.Analysis of results

The TDW rotation frequency *f* is plotted in Fig. 6 as a function of the specific flow rate of the mixture g_Σ for three examined FAMs. In particular, based on this figure, the CSD velocity can be easily determined by the formula $D = \pi \cdot d_c \cdot f/n$. It is clearly seen that the TDW rotation frequency and the number of TDWs usually increase with increasing g_Σ for all FAMs. The number of TDWs increases to two (n = 2) first for the FAM with the syngas at $g_\Sigma \approx 34$ kg/(s·m^2), then for the FAM with the fuel CH_4 + $8H_2$ at $g_\Sigma \approx 160$ kg/(s·m^2), then for the FAM with the fuel kerosene/H_2 at $g_\Sigma \approx 420$ kg/(s·m^2), and finally for the FAM with the fuel $CH_4 + 4H_2$ at $g_\Sigma \approx 645$ kg/(s·m^2).

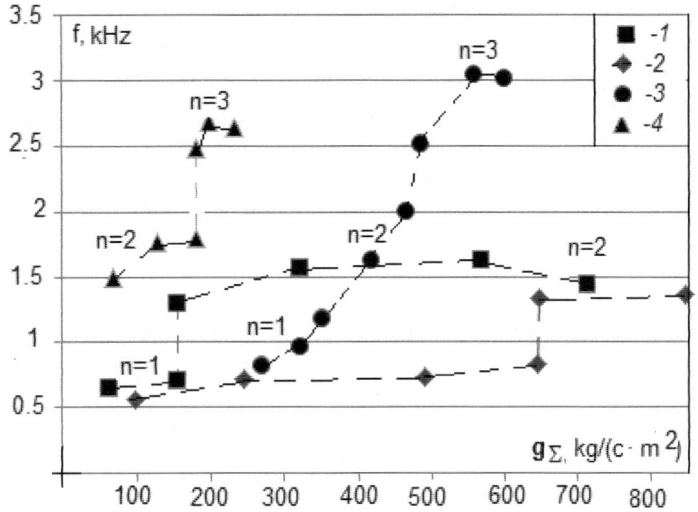

Figure 6. TDW frequency versus the specific flow rate of the mixture in DK-500: *1* – $CH_4 + 8H_2$, *2* - $CH_4 + 4H_2$, *3* – kerosene + H_2 ($m_H = 0.084$), *4* - $CO + 2H_2$.

A comparison of frequencies for the aviation kerosene/H_2 – air mixture ($m_H = 0.084$, points *3*) and CH_4/H_2 – air mixture (points *1, 2*) confirms that the methane/hydrogen – air mixture is the least detonable one. One more feature of CSD for FAMs with the methane-hydrogen fuel observed in Fig. 8 should be noted. In a wide range of the specific flow rates of this mixture (160 kg/(s·m^2) < g_Σ < 645 kg/(s·m^2)), the TDW rotation frequency *f* remains almost constant and the number of waves is identical. Low wave velocities for the CH_4/H_2 – air mixture (points *1, 2*) – $D \approx 1.0$ km/s are worth noting. At $m_H < 0.33$, this mixture cannot compete with the kerosene-hydrogen mixture. Thus, to reach the TDW frequency $f \approx 1.3$ kHz in the CH_4+ $4H_2$ – air mixture, the specific flow rate of the mixture should be twice greater than that in the kerosene/H_2 – air mixture. If preliminary incomplete thermal oxidation of methane to syngas is performed ($CH_4 + 1/2O_2 \rightarrow CO + 2H_2$), then CSD in the syngas-air mixture is obtained in DK-500 at the specific flow rates of the mixture smaller by an order of magnitude than that for the $CH_4 + 4H_2$ fuel (see Fig. 6).

Based on the experimental results [3] and those obtained in the present work, we can conclude that the TDW structures in the case of continuous spin detonation in flow-type annular combustors are enlarged in the following sequence:

$$H_2 \rightarrow CO + 2H_2 \rightarrow CH_4 + 8H_2 \rightarrow \text{kerosene}/H_2 \rightarrow CH_4 + 4H_2.$$

5. Conclusions

Thus, multiwave regimes of continuous spin detonation in a gaseous syngas-air mixture, poorly detonable CH_4/H_2-air mixture, and heterogeneous aviation kerosene/H_2-air mixture were obtained in a DK-500 flow-type annular cylindrical combustor and studied. The limits of CSD existence in terms of the specific flow rate of the fuel-air mixture (minimum values) were determined. The TDW structure in all three examined FAMs is similar to the TDW structure in acetylene-air and hydrogen-air mixtures [3]. It was found that all gaseous FAMs that have been studied up to now, including the least detonable methane-air mixture, with addition of hydrogen can be burned in a flow-type annular cylindrical combustor in the CSD regime. The experimental data obtained in the present study offer a possibility of systematic and targeted investigations of specific features of CSD regimes in heterogeneous fuel-air mixtures in annular combustors of air-breathing engines.

Acknowledgment

This work was supported by the Russian Foundation for Basic Research (Grant No. 16-01-00102a).

6. References

[1] Voitsekhovskii B V 1959 Steady detonation *Dokl. Akad. Nauk SSSR* **129** (6) pp 1254-1256 [in Russian]
[2] Bykovskii F A and Zhdan S A 2013 *Continuous spin detonation* (Novosibirsk: Izd. Sib. Otd. Ross. Akad. Nauk) p 423
[3] Bykovskii F A and Zhdan S A 2015 Current status of research of continuous detonation in fual-air mixtures (review) *Combust. Expl., Shock Waves* **51** (1) pp 21-35
[4] Keim W, ed. 1987 *Catalysis in Cl Chemistry* (Lerningrad: Khimiya) p 296
[5] Bykovskii F A, Zhdan S A and Vedernikov E F 2013 Continuous spin detonation of synthesis gas-air mixtures *Combust. Expl., Shock Waves* **49** (4) pp 435-441
[6] Bykovskii F A, Mitrofanov V V and Vedernikov E F 1997 Continuous detonation combustion of fuel-air mixtures *Combust. Expl., Shock Waves* **33** (3) pp 344-353

On the structure of combustion waves in poorly premixed gaseous mixture

M.F. Ivanov, A.D. Kiverin, N.A. Metelkin, I.S. Yakovenko

Joint Institute for High Temperatures of the Russian Academy of Sciences (JIHT RAS) 125412, Izhorskaya st. 13 Bd.2,Moscow, Russia

E-mail: ivanov_mf@mail.ru

Abstract.
The paper studies numerically the features of combustion wave propagation in poorly premixed hydrogen-oxygen mixture. In present study poorly premixed mixtures are taken to be those filling the combustor volume with random deviations of components concentrations from the average values on the spatial scales of the order of 0.1 – 1.0 mm. The numerical analysis shows that at relatively small spatial scales of concentration non-uniformities (below 0.5 mm) and at relatively low dispersion of concentration the rate of flame acceleration increases that leads to faster transition to detonation compared with the case of ideally premixed mixture. Enhancement of flame acceleration is related with continuous formation of perturbations of the flame front, among which the greatest impact belongs to those with short wavelengths. On the one hand these perturbations intensify the evolution of Darrieus-Landau instability and on the other hand act on the flame front in similar way as the turbulent pulsations do. At larger spatial scales of the non-uniformities or at increased dispersion of concentrations the flame acceleration rate decreases since the greater role in the process evolution now belongs to chemical factor associated with concentration deviation from the stoichiometric compound.

1. Introduction

One of the most hazardous processes arising during the severe accidents on the nuclear power plants or other industrial sectors involving hydrogen technologies is the explosion of hydrogen accidentally released into the atmosphere. Ejected hydrogen fills the technical rooms and stirs with air forming a flammable mixture. Due to the paramount importance for explosive safety the issue of hydrogen release into the air and its subsequent combustion is widely studied in number of works. As a rule these works consider jets of pressurized hydrogen in the air environment and ignition of resultant so-called non-premixed mixtures. In this case the dynamics of hydrogen jet is determined by the pressure inside the hydrogen-containing vessel and by the size of outlet orifice through which hydrogen flows into the air [1, 2, 3]. Volumetric hydrogen release into the atmosphere is less common but still possible situation. When large amounts of hydrogen releases at relatively low overpressure one can expect formation of volumetric clouds of stirring air and hydrogen. After the release mixture components spatial distributions evolve through the several stages from the fast turbulent mixing to the slow diffusive relaxation. On the final stage the large-scale structures containing different components have already dissipated however the mixture is still locally non-uniform. The relaxation to the fully uniform state could last for a relatively long time compared with characteristic times of ignition and flame propagation through the mixture. For example, in [4] nearly uniform mixture inside spherical vessel of 2 m radii was formed only after relaxation of convective and diffusive processes lasting for 30 minutes after turbulent stirring of separately fed hydrogen and air. Even longer life time of small concentration non-uniformities could be observed during accidental accumulation of combustible mixture in volumes (e.g., accumulation of hydrogen-air mixture in the upper part of the reactor room at the nuclear power plant). The origin of such accumulation is related with slow leakage of one of the components and its subsequent relatively slow redistribution in the volume due to convection and diffusion. The combustion initiated at the stage of mixture composition relaxation would propagate like in premixed case but under the action of certain disturbances due to the non-uniform distribution of fuel concentration. Further let call such a situation as combustion of poorly premixed mixture.

Although the applied significance of poorly premixed mixtures combustion is evident it is still less studied compared with combustion of fully premixed mixtures. The main problem is the turbulent nature of the stirring process in most of practically important cases. This determines a high complexity and unpredictability of local spatio-temporal evolution of the concentration field structure defined by convection, viscous dissipation and diffusive smoothing of concentration fluctuations. Nowadays most of works devoted to non-uniform mixtures combustion are mainly associated with initial stage of combustible mixture formation. As a rule both experimental and theoretical research works in this field are carried out in the simplified problem setup where the fuel concentration is distributed via gradient, orthogonal [5, 6] or oriented along the direction of flame propagation [7, 8, 9, 10]. However even the results obtained in such simple cases allow to elucidate the features intrinsic to the process of flame propagation under the conditions of fuel stratification. In this paper let consider numerically the features of flame propagation in poorly premixed mixture which state can be treated as one corresponding to the stage of non-linear stabilization and decay of non-uniformities formed in the process of stirring. By this moment a wide spectrum of concentration perturbations is formed with wavelengths from mixing zone width down to tenths of millimeter. The leading role in the process of stirring and formation of these perturbations in mixture components concentrations belongs to the joint action of Rayleigh-Taylor and Kelvin-Helmholtz instabilities. Mentioned estimation of the range of wavelengths agrees well with experimental data [11]. Obviously, on the stabilization stage the basic mechanism of concentration perturbations decay is the diffusion while the main role in perturbations support belongs to the residual inertia of convective flows and the slow leakages of the mixture components. Taking into account that the stirring represents a turbulent process and therefore on the final stages of stirring the mixture components are distributed chaotically it is reasonable to consider combustion in the mixture with initially given concentrations fields characterized by random deviations of the fuel and oxidizer concentrations from the averaged values.

2. Problem setup and numerical algorithm

To estimate the degree of influence of concentrations perturbations on the processes involved in gaseous combustion let consider as the example the problem of flame propagation through two-dimensional channel filled with stoichiometric hydrogen-oxygen mixture. Chosen problem setup is conventional and close to that considered numerically in [12]. The solution provides all the stages of flame acceleration including transition to detonation that allowed us to analyze in details the basic mechanisms determining both flame acceleration and transition to detonation [12]. Model of chemical kinetics used here is not the same as in [12] and spatio-temporal scales of deflagration-to-detonation transition (DDT) are slightly different but the qualitative features of flame evolution remained unchanged.

Gasdynamics of the gaseous mixture combustion was modeled via solving governing equations system based on Navier-Stokes equations for multicomponent mixture with account for viscous effects, thermal conductivity, multicomponent diffusion and energy release from chemical transformations. Transport coefficients were evaluated from the first principles of gas kinetics theory [13]. Thermodynamic quantities were calculated from standard thermophysical tables using high-order polynomial interpolation [14]. Chemical reactions were modeled with chemical kinetic scheme of hydrogen oxidation from [13]. Numerical solution of the governing system was obtained with euler-lagrange method [15] modified to ensure stability and second-order in space precision of the numerical solution procedure.

To study the influence of mixture concentration spatial non-uniformities on the flame propagation, calculations were performed with initially defined deviations of hydrogen and oxygen concentration from the mean value. Deviations of the hydrogen concentrations C_{H_2} from the mean, mostly stoichiometric condition were calculated as random quantities, normally distributed with defined dispersion equal to Ω. In order to perform detailed analysis of concentration non-uniformities influence on the combustion process, computations were carried out in two setups. In first setup dependence between the flame propagation speed and dispersion of the concentration deviation was analyzed. In this case small rectangular areas with certain size were randomly picked inside computational domain. In each area initial concentrations were defined with certain dispersion varied for each calculation. For every value of dispersion a set of calculations were performed (from 3 to 6 calculations in each set) with independently generated random deviations. For further analysis results acquired in each set were averaged. Obtained results allowed us to plot a relation between characteristics of flame non-stationary evolution and concentrations dispersion. The second setup was aimed to obtain dependence between the flame propagation speed and concentration non-uniformities spatial scales. In this case the size of rectangular areas was varied from one calculations set to another, while the value dispersion of concentrations deviation in these areas was fixed. On the basis of calculations the dependence between combustion process development parameters and spatial scales of concentration non-uniformities was acquired. Presented results were also averaged for each set of calculations.

Combustion process was modeled inside semi-opened channel of 5 mm width and 100 mm length with adiabatically isolated non-slip walls. Non-slip boundary conditions imposed on the side walls allowed to take into account affect of boundary layer. Channel was filled with hydrogen-oxygen mixture at 1 atm pressure and 300 K temperature. In contrast to [12] components concentrations were non-uniform and only average values were correspond to stoichiometric conditions. Ignition took place near left end-wall of the channel in thin heated layer of 1 mm width. Main results were obtained with high resolution on fine computational grid with cell size 0.05 mm x 0.05 mm, which allowed to precisely reproduce deflagration-to-detonation process in hydrogen-oxygen mixtures [16].

3. Results and discussion

Both experimental and theoretical studies [12, 17] show that the flame front in the premixed stoichiometric hydrogen-oxygen mixture propagates through the semi-opened channel with acceleration up to the sonic speed in the fresh mixture and then the mechanism of transition to detonation switches on. Herewith the evolution of combustion wave passes four basic stages. The first stage represents formation of a stable flame front in the vicinity of ignition region that is fully determined by the characteristics of the ignition source. On the second stage flame propagates with acceleration moving with the flow of the fresh mixture pushed by the expanding combustion products. As the flow is dragged in the boundary layer region and as the flame front itself is subjected to hydrodynamic instability front surface grows and becomes curved. As the flame evolution transits to the third stage the flame structure stabilizes under the influence of nonlinear factors and predomination of long wavelength perturbations. Flame evolution achieves fourth stage once the flame front accelerates up to the sonic speed. On this stage the flame propagation is defined by the peculiarities of combustion in the supersonic flow that results in formation of sufficiently strong shock wave on the scales of the flame front that in turn leads to the onset of detonation.

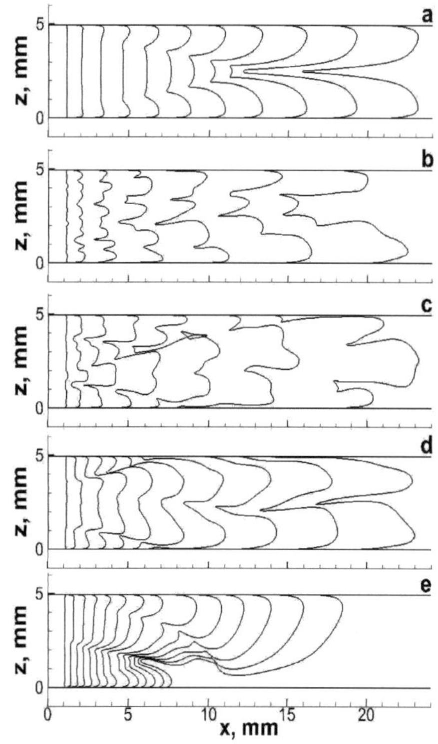

Figure 1. Evolution of the flame front propagating through the channel filled with poorly premixed hydrogen-oxygen mixture. Subsequent positions of the flame front are shown with equal time intervals equal to 10 us. (a) – fully premixed stoichiometric mixture; (b) – δ=0.25 mm; (c) – δ=0.5 mm; (d) – δ=1.25 mm; (e) – δ=2.5 mm; Ω=0.21.

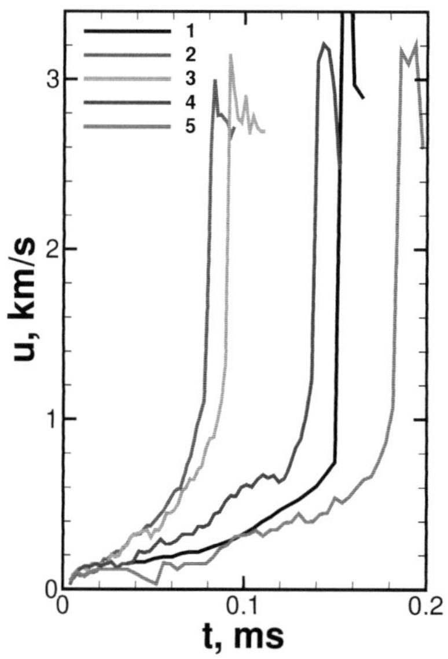

Figure 2. History of the flame speeds in different poorly premixed mixtures. 1 fully premixed stoichiometric mixture; 2 – 0.25 mm; 3 – 0.5 mm; 4 – 1.5 mm; 5 – 2.5 mm; Ω=0.21.

Consider now the influence of spatial non-uniformities in the distribution of combustible mixture components on the flame speed. We will analyze the flame propagation in the mixtures with different degrees of stirring from those almost premixed and characterized by small-scale perturbations to those less premixed and characterized by large-scale perturbations. This should allow us to understand the effect of non-uniformities spatial scales on both acceleration and deceleration of the flame front on different stages of flame evolution. Results of numerical analysis show that the spatial non-uniformity in concentrations of reacting mixture components could affect the combustion on all the stages of its evolution. Thus on the first stage of flame initiation the non-uniformity determines the topology of the forming flame front. In turn the formed spectrum of the perturbations defines the initial conditions for evolution of flame front instability, which plays a great role in further evolution of the flame front and its interaction with the flow. Figure 1 illustrates how the degree of concentration spatial non-uniformity affects

IOP Conf. Series: Journal of Physics: Conf. Series **899** (2017) 042003 doi:10.1088/1742-6596/899/4/042003

the evolution of the flame front in the process of its propagation through the channel filled with poorly premixed mixture. Herewith one can clearly see that the role of small-scale non-uniformities (δ=0.25 mm and δ=0.5 mm in Fig. 1) occurs to be much more significant compared with large-scale ones (δ=1.25 mm and δ=2.5 mm in Fig. 1) that as a result causes faster flame propagation on the second stage of flame evolution in channel. This can be explained by the fact that the initial linear stage of flame front instability is mainly determined by the evolution of short-wavelength perturbations. The presence of small-scale non-uniformities in the flow and their continuous influence on the flame front on the second stage of the process causes saturation of the spectrum of flame front perturbations with short wavelengths. On the following non-linear stage the energy is transferred from these short wavelengths to the longer ones that in turn provides additional acceleration of the flame front on the third stage of the process. This can be clearly seen from figure 2 representing the histories of flame speeds in poorly premixed mixtures characterized by the same dispersion of concentration and different spatial scales of the non-uniformities. Fig. 2 illustrates a notable increase in acceleration rate in case of poorly premixed mixtures with small-scale non-uniformities (δ=0.25 – 1.00 mm) compared with the case of premixed stoichiometric compound. With the increase in spatial scales of non-uniformities the factor of concentration deviation from the stoichiometric one becomes to play a greater role than the factor of flame front perturbations generation. Due to this the overall process of flame acceleration is alike to that in non-stoichiometric hydrogen-oxygen mixtures where the onset of detonation arises on larger spatio-temporal scales.

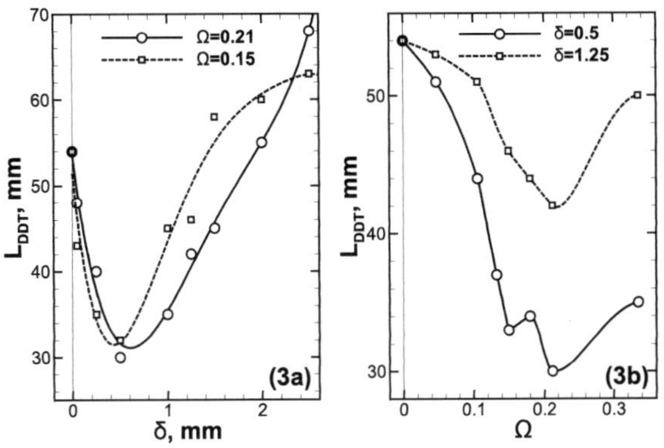

Figure 3. Dependence of pre-detonation length on the non-uniformities spatial scale at fixed dispersion value (a) and on the dispersion value at fixed spatial scale of the non-uniformities (b).

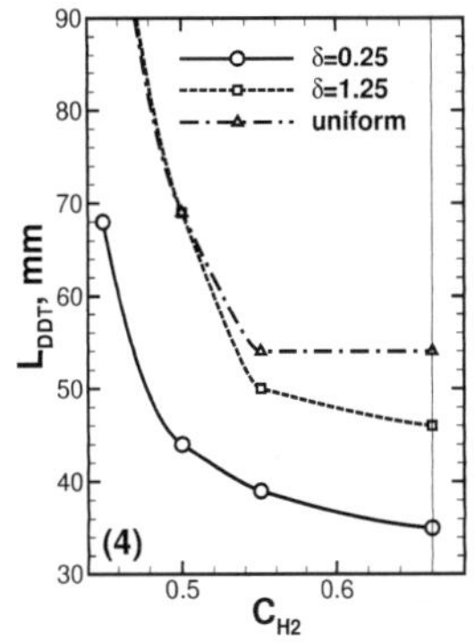

Figure 4. Dependence of pre-detonation length on the fuel concentration at different δ; Ω=0.15.

To get more comprehensive understanding of the role of chemical composition non-uniformities in the processes of hydrogen-oxygen flame acceleration and DDT let consider the dependence of pre-detonation length (the distance passed by the flame front before onset of detonation) on the non-uniformities characteristics (see. Fig. 3). Figure 3a shows that the dependence of pre-detonation length (and therefore of corresponding time scale) on the non-uniformities scales is non-monotonic. The minimum in the dependence corresponds to $\delta \sim 0.25 - 0.50$ mm that by the order of magnitude corresponds to the theoretical estimate for critical wavelength λ_c below which hydrodynamic instability of the flame front (Darrieus-Landau instability) does not evolve. According to the estimate proposed in [18] and taking into account the experimental data, empirical estimations and numerical results for stoichiometric hydrogen-oxygen ($\nu/U_f \approx 10^{-5}$ m, $\theta \approx 7.3$) one can obtain $\lambda_c \approx 0.25$ mm. Perturbations of smaller scales are not directly involved in the evolution of hydrodynamic instability of the flame front. However their interference on the flame surface as well as the expansion of the flame front perturbed in such a way determine excitation of the modes with wavelengths $\sim \lambda_c$. The perturbations generated by the spatial non-uniformities of larger scales have weaker influence on the Darrieus-Landau instability evolution as the rate of their increase is smaller. Furthermore as it was already mentioned above, with the increase in non-uniformity spatial scales the chemical factor becomes to play a greater role that determines local flame deceleration and as a result integral deceleration of the whole process. The longer process the higher influence of convection and diffusion on the flow structure. As a result the influence of spatial non-uniformities on the structure of the flame surface decreases with time. Due to this the effect of dispersion in mixture composition on the flame evolution is limited in time and decreases with the flame acceleration.

One more factor affecting the scales of DDT is the dispersion value, which characterize the deviation of mixture composition from the average one inside the single non-uniformity. At the fixed value of the non-uniformity spatial scale the increase in dispersion value leads to the increase in amplitude of flame surface perturbations that in turn causes increase in the acceleration rate and faster onset of detonation. However further increase in the deviation of the mixture composition from the stoichiometric one leads to local flame deceleration when entering the regions containing non-stoichiometric mixture. As a result the average flame deceleration can be observed (see Fig. 3b).

In case of averagely non-stoichiometric compositions the non-uniformity in fuel concentration distribution affects the flame surface evolution in the similar way as in stoichiometric case discussed above. Small-scaled non-uniformities lead to the increase in detonability. Figure 4 represents the decrease in pre-detonation length related with the non-uniform composition of lean hydrogen-oxygen mixtures. As one can see the higher rate of acceleration of lean flames is achieved due to the small-scale perturbations just as in the case of stoichiometric mixture. At the same time the effect of perturbations of larger scales weakens with the decrease in average hydrogen concentration and already in equimolar mixture the non-uniformities of δ=1.25 mm cause no additional acceleration of DDT. Presented results indicate a possibility of qualitative changes in the detonation onset predictions for mixtures occupying a certain volume since the estimate of pre-detonation length can vary depending on the degree of stirring. In poorly premixed mixture the pre-detonation length can be significantly shorter than in fully premixed mixture with the same average composition. It can be clearly illustrated on the example of mixture with 45% content of hydrogen. In this case we did not observe DDT in the semi-opened channel of 100 mm length filled with fully premixed mixture. However in case of poorly premixed mixture characterized by small-scaled non-uniformities in hydrogen concentration we observed DDT already at the distance of 68 mm.

On the final fourth stage of flame acceleration on which the flow velocity achieves the sonic speed the physicochemical processes evolving on the scales of reaction zone become to play the leading role [16]. Depending on the mixture composition the process could evolve in two paths: (1) the additional acceleration of the flame front takes place due to the additional compression inside the reaction zone by the chocked compression waves; (2) the regime of flame propagation stabilizes near the sonic speed in the combustion products when the compression rate is not enough to accelerate combustion wave. Thereby in case of non-uniform concentration distribution ahead the flame front the subsequent burning of volumes containing mixtures of different compositions could cause changing in the regimes of supersonic combustion wave. Depending on the dispersion value the DDT can be either accelerated or prevented. After the formation of stable detonation wave the non-uniformities could still affect the process, leading to the disturbance of the reaction zone and as a result to the instability of detonation front.

4. Conclusions

Considered here effect of flame acceleration due to the perturbations in mixture concentration formally can be compared with the flame acceleration related with the flow velocity fluctuations in the turbulized flow. Both factors are intrinsic to the scenarios of accidental hydrogen explosion and are of importance for estimations of the hazardous impact of the explosion. At the first sight the character of influence of both factors should be similar however the quantitative effects are different. In case of flow turbulization the increase in perturbations magnitude (up to the certain limit) leads to the more intensive effect on the flame acceleration. In case of mixture composition perturbations there are two main opportunities. In case of small-scale non-uniformities the increase in perturbations magnitude favors the increase in flame acceleration rate and corresponding decrease in pre-detonation length. Herewith the role of active centers (regions with near stoichiometric mixture composition) is greater than the role of neutral centers (with significant deviation from stoichiometry). On the contrary in case of large-scale non-uniformities the effect of local flame deceleration becomes predominant due to the greater role of neutral centers. Therefore deceleration of DDT process takes place and even the prevention of DDT becomes possible. On the base of distinguished peculiarities one can conclude that the specific feature of considered here instability is its maximal development already at relatively small deviations in the perturbed parameter (fuel concentration) from its average value. In practice the possible presence of weak concentration perturbations in the nearly premixed combustible mixture (that corresponds to the case of small-scale non-uniformities with small dispersion value) their affect on the flame acceleration and DDT could lead to significantly different results from those obtained via prognostic calculations considering combustion of fully premixed mixture.

5. Acknowledgements

The work was funded by Russian academy of sciences (Grants No. I.13P Thermal physics of high energy densities and I.24P Combustion and explosion).

References

[1] Kim Y R, Lee H J, Kim S and Jeung I S 2013 *Proc. Combust. Inst.* **34** 2057 – 2064
[2] Golovastov S and Bocharnikov V 2012 *Int. J. Hydrogen Energy* **37** 10956 – 10962
[3] Duan Q L, Xiao H H, Gao W, Wang Q S, Shen X B, Jiang L and Sun J H 2015 *Int. J. Hydrogen Energy* **40** 8281 – 8289
[4] Petukhov V A, Bublik N P, Gusev P A, Gutkin L D and Solntsev O I 2016 *High Temp.* **54** 99 – 104
[5] Ettner F, Vollmer K G and Sattelmayer T 2014 *Journal of Combustion* **2014** 15
[6] Vollmer K G, Ettner F and Sattelmayer T 2012 *Combust. Sci. Technol.* **184** 1903 – 1915
[7] Badr O and Karim G 1984 *J. Fire Sci.* **2** 415 – 426
[8] Karim G A and Tsang P 1975 *J. Fluids Eng.* **97** 615 – 617
[9] Kang T and Kyritsis D C 2005 *Combust. Sci. Technol.* **177** 2191 – 2210
[10] Pires Da Cruz A, Dean A M and Grenda J M 2000 *Proc. Combust. Inst.* **28** 1925 – 1932
[11] Inogamov N A, Demyanov A Y and Son E E 1999 *Hydrodynamics of mixing (in Russian)* (MIPT Publishing)
[12] Ivanov M F, Kiverin A D, Liberman M A and Fortov V E 2010 *Dokl. Phys.* **55** 480 – 484
[13] Warnatz J, Maas U and Dibble R 2006 *Combustion: Physical and Chemical Fundamentals, Modeling and Simulation, Experiments, Pollutant Formation* (Springer-Verlag Berlin Heidelberg)
[14] Heywood J B 1988 *Internal combustion engine fundamentals* (McGraw-Hill Education)
[15] Belotserkovsky O M and Davydov Y M 1982 *Coarse-particle method in hydrodynamics (in Russian)* (Russian Publ. Inc. Nauka, Mir)
[16] Kiverin A D, Yakovenko I S and Ivanov M F 2016 *Int. J. Hydrogen Energy* **41** 22465 – 22478
[17] Kuznetsov M, Alekseev V, Matsukov I and Dorofeev S 2005 *Shock Waves* **14** 205 – 215
[18] Pelce P and Clavin P 1982 *J. Fluid Mech.* **124** 219 – 237

IOP Conf. Series: Journal of Physics: Conf. Series **899** (2017) 042004 doi:10.1088/1742-6596/899/4/042004

Measurements of detonation propagation in the plastic explosive in charges of small diameters using synchrotron radiation

Kashkarov A. O.[1,3], Pruuel E. R.[1,3], Ten K. A.[1], Rubtsov I. A.[1,3], Kremenko S. I.[1,3], Prosvirnin K. M.[2], Kiselev A. N.[2], Kosolapov I. E.[2], Kostitsyn O. V.[2], Loboyko B. G.[2], Muzyrya A. K.[2]

[1] Lavrentyev Institute of Hydrodynamics SB RAS, Novosibirsk, Russia
[2] RF-NC Academician Zababakhin All-Russian Research Institute of Technical Physics, Sneginsk, Russia
[3] Novosibirsk State University, Novosibirsk, Russia

E-mail: kashkarov@hydro.nsc.ru

Abstract. During the detonation of charges of plastic explosives based on PETN and RDX, the spatial structure of the arising flow and the degree of compression of the matter behind the detonation front were obtained using the synchrotron radiation facilities. The experiments were carried out with charges of small diameters, which made it possible to control the symmetry of the undisturbed charge of the explosive and the flow behind the detonation front. In addition, the advanced detector was used to increase the frame rate by 4 times in comparison with the authors' earlier works.

1. Introduction

Registration of fast processes accompanying explosive phenomena by using the synchrotron radiation (SR) as the source of X-ray pulses is the efficient method used to the study of detonation and the shock wave phenomena for more than ten years. The authors of this article use and continuously modernize the experimental station at the Siberian Synchrotron and Terahertz Radiation Centre at the Budker Institute of Nuclear Physics [1, 2]. This station enables to obtain a multi-frame slit X-ray film with a frame exposure time of about 1 nanosecond and a spatial resolution of 0.1 mm. The similar station for recording the SR during high-speed processes was created by colleagues at the Argonne National Laboratory in 2012 [3, 4].

The way of obtaining information about the structure of the flow behind the front of the detonation wave is based on the registration of the SR beam passing through the detonating charge. The identity of the X-ray pulses of the SR by the spectrum and the intensity distribution in the beam during the experiment makes it possible to calibrate the detector and obtain the distribution of the compression of matter along the beam for the process being studied. Also, an important feature is the fact that the registration of the distribution of the compression of substance in a detonating charge of an explosive does not introduce perturbations into the arising flow. This circumstance is especially important for studying the detonation process in charges of small diameters. In this case the use of contact sensors can lead to catastrophic changes in the observed phenomenon.

To register the radiation passing through the expansion area of the detonation products, the new detector was launched (2017) and used in this work. Compared with the previous detector [5], the time resolution was improved by 4 times. Now, the interval between frames is 124 ns. The number of frames during the experiment was also increased up to 100. Thus, it is possible to trace the detonation process starting from the unperturbed state of the material to the late stages of expansion of the detonation products.

2. Experimental setup and reconstructing distirbution of compression of substance

In this work, we measured the detonation parameters in cylindrical charges of plastic explosives based on PETN and RDX with diameters of 5, 10, and 15 mm. The measurement of the flow, which arises behind the front of the detonation wave, was carried out in the cross section of the charge as shown in figure. 1. The intensity distirbution of the SR beam passed through the detonating charge of high explosive is recorded by the one-coordinate multichannel detector. The amount of substance along the SR beam was calculated from the measured intensity distribution using the previous calibration of the detector. For this purpose the attenuation of the intensity for the known amount of the mass of the investigated explosive along the SR beam is determined in the static regime. The integral current in the accelerator ring completely determines the spectrum and the intensity distribution of the synchrotron radiation beam, so the calibration can be carried out regardless of the time of the experiments. However, the detonation products and fragments of experimental assembly that reach the windows of the explosive chamber during an explosion can affect the data recorded, so the calibrating just before each experiment considerably improves results.

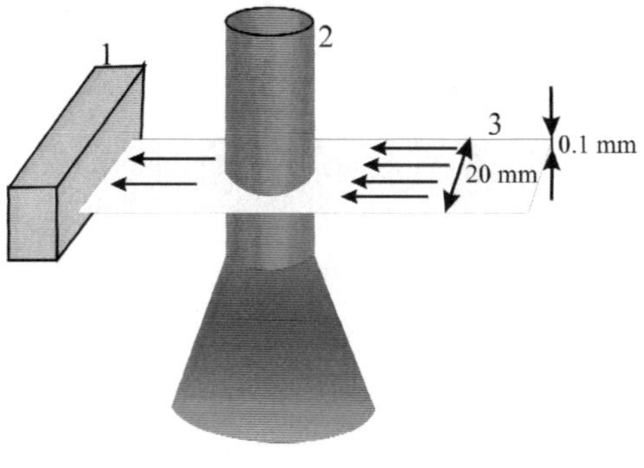

Figure 1. Scheme of the experiment. 1 - detector of the SR beam, 2 - detonating charge of explosive, 3 - SR beam.

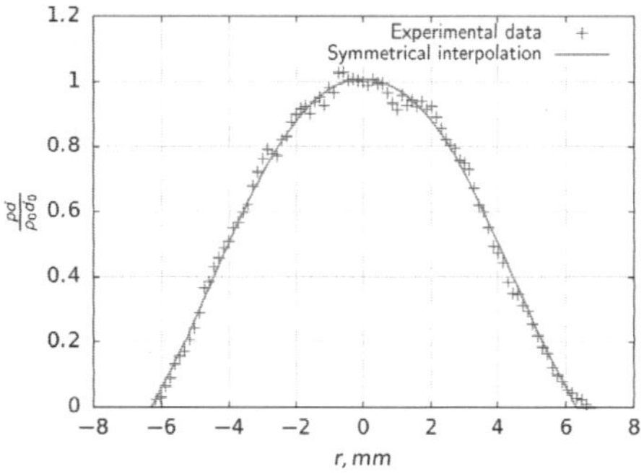

Figure 2. Amount of relative mass behind the detonation front of PETN-based. d=10mm, $t \approx 0.3\mu s$.

The charge was initiated by the generator of plane shock wave. The measurements was carried out at the distance from the face of the initiation that sufficient to establish the stationary detonation regime. The assumption of the stationary detonation front should be made to compare the experimental data for the charges of different diameters and explosive compositions.

Charges of small diameters can be completely placed in the region of registration of the SR beam, which makes it possible to control the symmetry of both the undisturbed charge and the flow arising behind the detonation front. The relative amount of mass for the charge of the explosive based on PETN of 10 mm in diameter is shown in figure. 2. The cylindrical symmetry of the mass distribution is of fundamental importance, since it gives the possibility to set the task of tomography of the compression of substance in the process using only one shot. For the tomography problem, it is necessary to smooth out the noisy experimental data. However, there are natural discontinuities at the detonation front and at the charge edge, typical for the

IOP Conf. Series: Journal of Physics: Conf. Series **899** (2017) 042004 doi:10.1088/1742-6596/899/4/042004

explosion process. To solve the tomography problem is used the method [6] developed earlier by the authors of this work. This method based on the use of a priori information to smooth experimetnal data. The essence of the method is set out below.

Measuring the amount of mass along the SR beam during detonation of a cylindrical charge of explosive, we should obtain a distribution of the typical form shown in figure. 3. First, we register the unperturbed state of the charge of the explosive, then we see that the detonation front passes through the registration cross section, and later the side expansion of the detonation products proceeds. After a while, the detonation products go outside the registration area of the SR beam. This distribution of mass is equivalent to the distribution of the compression of the substance, obtained by setting the corresponding values in the nodes of the grid shown in figure 4. For this purpose, the spline function is calculated through the grid nodes along each line shown in figure 4 starting from the detonation front down. The spline function gives us the values not only at the grid nodes but also between the nodes. In the experiment the mass distribution is measured in the horizontal cross-sections of the grid (one of them is indicated by a dashed line). This mass distirbution can be obtained with use of the another spline function, which is calculated along the dashed line inside the region filled with the explosion products. The geometry of the grid is determined by the curvature of the front and the border of the region filled with the explosion products. Thus, none of the splines intersects the discontinuities. Minimizing the deviations between the experimental and calculated data by varying the grid parameters, we obtain the distribution of the compression of the substance corresponding to the X-ray shadow observed in the experiment.

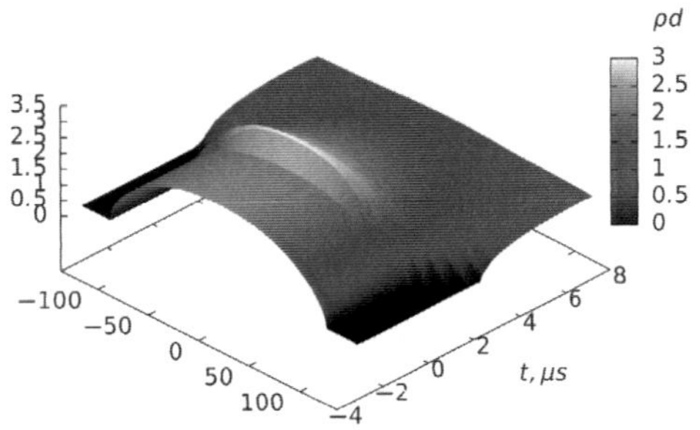

Figure 3. Expected distribution of mass along the SR beam.

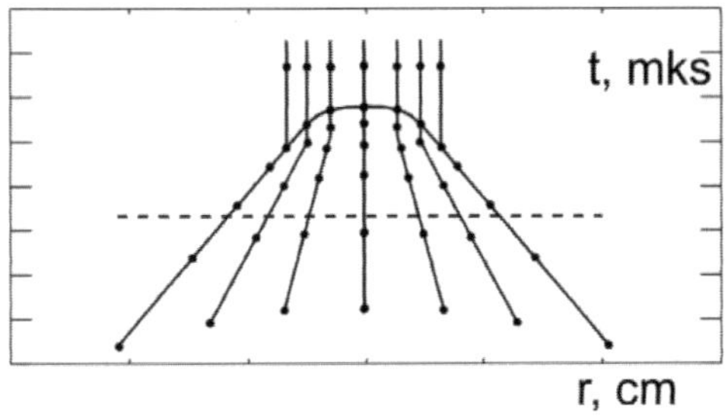

Figure 4. Flow parametrization scheme. The dashed line corresponds to the SR beam in one of the frames.

For the compositions under study, the curvature of the detonation front is very large. For charges of small diameters, the detonation front pass the observation cross-section in only tens of nanoseconds, while the time interval between frames is 124 nanoseconds. This circumstance leads to the impossibility of the reliable determination of the curvature of the front directly in these experiments, and so the curvature was determined in the separate experiments. The similar situation arises with the determination of the state of substance in the von Neumann spike, since the typical width of the chemical reaction zone for the considered explosives is also tens of nanoseconds [7]. Therefore, at best, the first frame behind the detonation front will show the mass distribution close after the Chapman-Jouguet state.

When the detonation front passed through the explosive charge, the detonation products quickly exceed the border of the registration area, due to intensive side expansion. The considered method of reconstructing the flow structure allows one to obtain data in this case due to the requirement of smoothness of the solution, but the error will substantially increase. Previously, it was usually possible to register no more than one frame before the expand of detonation products outside the observation area. The 4-fold decrease in the interval between frames to 124 nanoseconds significantly increases the accuracy of determining the structure of the compression of substance behind the detonation front.

3. Results

The solution of the tomography problem results in the radial function of the dynamics of the compression of substance in the detonating charge for each experiment. This function can be shown as a three-dimensional surface (figure. 5) or a map (figure. 6) that allows us to compare the flow structures behind the detonation front for the explosive compositions investigated.

The non-constant magnitude of the compression of substance in the undisturbed region of the charge is related both to the quality of the production of the charges and to the experimental error in measuring the amount of substance along the SR beam. This error does not exceed 5% for static objects. This value agrees well with the obtained compression of the substance before the detonation front. It is expected that the total error for the dynamic flow region will be of the similar order.

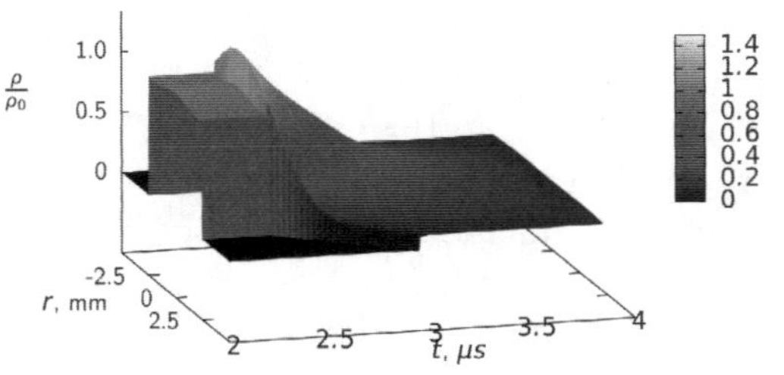

Figure 5. The distirbution of the substance compression of RDX-based charge. Diameter 5 mm.

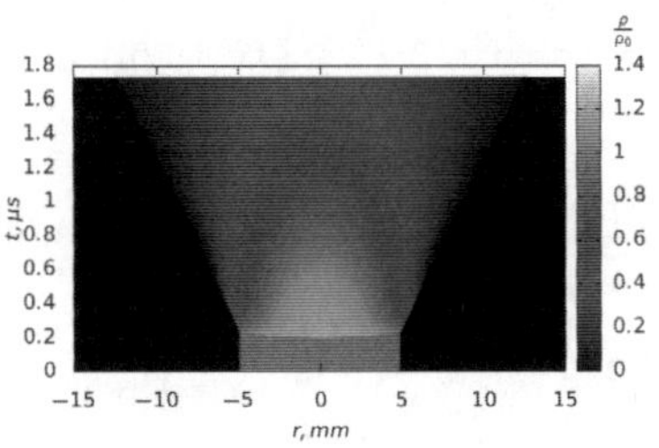

Figure 6. The distirbution of the substance compression of PETN-based charge. Diameter 10 mm.

Figure 7 shows the diagrams of the compression of substance along the charge axis for various explosive compositions and diameters. The quantitative difference in the flows behind the detonation front are obvious. For clarity, the profiles were shifted in time to align detonation fronts. It should be noted that the moment when the detonation front passed through the observed cross section of the charge is determined with an accuracy not exceeding 124 ns. This is due to the frame rate and the almost flat shape of the detonation front.

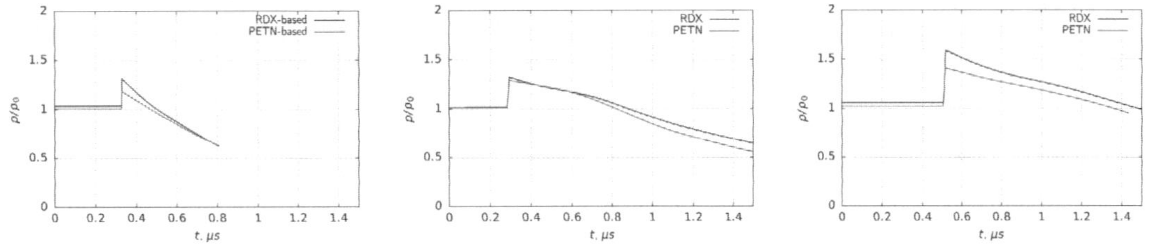

Figure 7. Compression of substance on the charge axis.

Measurements show that the degree of compression of substance behind the detonation front for charges based on PETN is smaller, and the subsequent reduction of compression occurs faster. However, the obtained value of the maximum compression of substance can hardly be referred to the state at the von Neumann spike or at the Chapman-Jouguet point. The values obtained for the maximum compression of the substance are rather referred to later states behind the detonation front.

4. Conclusions

The paper presents quantitative data on the dynamics of compression of substance in the area of expansion of the detonation products. This allows to compare the detonation processes in charges consisting of various explosives. It was necessary to make a number of approximations to obtain the data. First of all, we had to consider detonation stationary, that is valid for charges of sufficient length. Further, we had to assume that the arising flow is axisymmetric. The validity of the last approximation was checked during the experiment for charges of small diameters.

On the other hand, obtaining the compression distribution of the substance entails smoothing out the experimental data, and this can lead to partial loss of the data. Therefore, in order to calibrate numerical models, the comparison of the calculated mass distribution along the SR beam with the experimental one will give better results than the use of the compression distribution of the substance for these purposes.

In [8] authors used the VISAR technique and the laser-geterodyne technique (PDV-technique) to study the plasticized PETN of close compositions. They recordered time profiles having the near-front peak and the plateau behind the detonation front with the follow-on dip. According to the authors, the near-front peak corresponds to the Chapman-Jouguet state. In this paper, we do not observe any specific features on the compression diagrams that could be indicate the presence of this plateau.

Acknowledgments

This work was supported by Russian Foundation for Basic Research (Grant No. 16-29-01050).

References

[1] Pruuel E R, Ten K A, Tolochko B P, Merzhievskii L A, Luk'yanchikov L A, Aulchenko V M, Zhulanov V V, Shekhtman L I and Titov V M 2013 *Doklady Physics* **58**(1) 24–8 doi: 10.1134/S1028335813010035

[2] Titov V M, Pruuel E R, Ten K A, Luk'yanchikov L A, Merzhievskii L A, Tolochko B P, Zhulanov V V and Shekhtman L I 2011 *Combust., Expl., Shock Waves* **47**(6) 3–15 doi: 10.1134/S0010508211060013

[3] Luo S N, Jensen B J, Hooks D E, Fezzaa K, Ramos K J, Yeager J D, Kwiatkowski K and Shimada T 2012 *Review of Scientific Instruments* **83** 073903 URL `http://dx.doi.org/10.1063/1.4733704`

[4] Firestone M A, Dattelbaum D M, Podlesak D W, Gustavsen R L, Huber R C, Ringstrand B S, Watkins E B, Jensen B, Willey T, Lauderbauch L, Hodgin R, Bagge-Hansen M, van Buuren T, Seifert S and Graber T 2017 *AIP Conference Proceedings* **1793** 030010 URL `http://aip.scitation.org/doi/abs/10.1063/1.4971468`

[5] Shekhtman L I, Aulchenko V M, Bondar A E, Dolgov A D, Kudryavtsev V N, Nikolenko D M, Papushev P A, Pruuel E R, Rachek I A, Ten K A, Titov V M, Tolochko B P, Zhilich V N and Zhulanov V V 2012 *Journal of Instrumentation* **7** C03021 URL `http://stacks.iop.org/1748-0221/7/i=03/a=C03021`

[6] Pruuel E R, Ten K A, Titov V M, Lukianchikov L A, Merzhievsky L A, Zubkov P I, Tolochko B P, Zhogin I L, Zhulanov V V, Aulchenko V M and Shekhtman L I 2010 *Proc. of the 14th International Detonation Symposium.* (Office of Naval Research) pp 345–351

[7] Loboiko B G and Lubyatinsky S N 2000 *Combustion, Explosion and Shock Waves* **36** 716–733 URL `http://dx.doi.org/10.1023/A:1002898505288`

[8] Vorobyov A B, Tarzhanov V I, Kozlov E A, Eskov A N, Kuchko D P, Ralnikov M A and Shirokobokov A E 2017 *XIII Zababakhin scientific talks: Abstracts.* (RFNC-VNIITF) p 68

Influence of mixture content on the minimum sparkplug ignition energy of a coal dust suspension in the air

A Yu Krainov and K M Moiseeva

Tomsk State University, Tomsk, 634050 Russia

E-mail: Moiseeva_KM@t-sk.ru

Abstract. This paper provides a numerical simulation of a coal-dust suspension sparkplug ignition in the air. The mathematical model is based on the dual-velocity two-phase model of the reacting gas-dispersion medium. The model consists of the continuity equation, the impulse–conversation equation, the gas and the particle energy-conversation equations, the particle burn-out equation and the component mass-conversation equations which take into account diffusion and burn-out. The energy equations consider the thermal conductivity and chemical reactions on the particle surface. The research investigates the influence of particle size on the minimum sparkplug ignition energy.

1. Introduction

The problem of coal-dust sparkplug ignition is the question of the day in coal industry. The researches [1] – [4] investigate ignition features of coal-dust suspension and flame propagation in the air and in a reactive gas. The work [1] suggests the explosion emerging scenarios in mines. It offers assumption about the influence of coal-dust and reactive gas content on the possibility of their mixture to ignite. The paper [2] provides the data on the explosive risk of coal-dust suspension in the air and in a reactive gas. The coal-dust-reactive gas-air mixture is capable of exploding under the low concentration of coal particles in it.

We have investigated the influence of the fine coal dust powder fraction on the flame propagation rate [3]. The presence of the small coal dust particles in the methane-air mixture increases the flame propagation rate under the low initial concentration of reactive gas. If the mixture is close to the stoichiometric condition the particles decrease the flame propagation rate. The pressure growth rate in the volume increases with the presence of coal dust suspension in comparison to its absence

Papers [5] – [9] are devoted to the problem of sparkplug ignition. The minimum ignition energy of coal-dust suspension in the air depends on coal particle mass concentration [5, 6]. The small particles decrease the minimum ignition energy. The larger particles increase the minimum ignition energy when the mixture is close to the stoichiometric condition, and decrease when the mixture is far from stoichiometric. The large particles do not influence the minimum ignition energy; it is equal to the mixture ignition energy [7].

The research [8] has investigated the radiation transport influence on the sparkplug ignition of the heterogeneously reacting suspension. The study provides the analytic formula to determine the critical activation energy of suspension sparkplug ignition taking into account radiation transport. The motion of suspension during heating is not taken into account in the papers [7, 8]. Later on, the authors have investigated the effect of thermal expansion on the minimum sparkplug ignition energy of the gas [9].

According to the study the thermal-diffusion model of the gas sparkplug ignition understates the minimum activation energy by more than two times.

In the present study we investigate the sparkplug ignition of coal-dust suspension in the air using mathematical models from [7 – 10] and experimental data from [5 – 6]. We have developed the model which takes into account the influence of thermal and dynamic interactions between particles and gas. The aim of the research is to determine the effect of the mixture proportion on the sparkplug ignition energy.

2. Mathematical model

The following assumptions are made: coal dust is uniformly distributed within the volume, the suspension contains $N-$fractions, which mass concentration equals $m_{dust} = \sum_{i=1..N} m_{dust,i}$. The instantaneous filiform ignition source is located in the center of the volume. The outer boundary of the computational domain is supposed to be infinitely far from the ignition source. The interaction between coal dust fractions and the electrode heat waste are neglected. Other assumptions are the same as in [10].

The mathematical model is based on the dual-velocity two-phase model of the reacting gas-dispersion medium [11]. It consists of the continuity equation, the impulse–conversation equation, the gas energy-conversation equation considering thermal conductivity, the particle energy-conversation equation considering chemical reaction on its surface, the component mass-conversation equations taking into account the particle burn-out and diffusion and the particle burn-out equation. Under the made assumptions the model has the following form:

The gas continuity equation:

$$\frac{\partial r\rho_g}{\partial t} + \frac{\partial r\rho_g u_g}{\partial r} = r\sum_{i=1..N} G_i. \tag{1}$$

The gas impulse–conversation equation:

$$\frac{\partial r\left(\rho_g u_g\right)}{\partial t} + \frac{\partial r\left(\rho_g u_g^2 + p\right)}{\partial r} = p - r\sum_{i=1..N} \tau_{tr,i} + r\sum_{i=1..N} G_i u_{k,i}. \tag{2}$$

The gas energy-conversation equation:

$$\frac{\partial r\rho_g\left(\varepsilon_g + \frac{u_g^2}{2}\right)}{\partial t} + \frac{\partial r\left[\rho_g u_g\left(\varepsilon_g + \frac{u_g^2}{2}\right) + pu_g\right]}{\partial r} = \frac{\partial}{\partial r}\left(r\,\lambda_g\left(T_g\right)\frac{\partial T_g}{\partial r}\right) + $$
$$+ r\sum_{i=1..N}\left(Gc_k T_{k,i} - u_{k,i}\tau_{tr,i} + G_i\frac{u_{k,i}^2}{2} + \alpha_{k,i}\, n_{k,i} S_{k,i}\left(T_{k,i} - T_g\right)\right). \tag{3}$$

The oxygen mass balance equation:

$$\frac{\partial r\rho_{O2}}{\partial t} + \frac{\partial r\rho_{O2}u_g}{\partial r} = \frac{\partial}{\partial r}\left(rD_g\left(T_g\right)\frac{\partial\rho_{O2}}{\partial r}\right) - r\alpha_1\sum_{i=1..N} G_i. \tag{4}$$

The particle mass balance equation:

$$\frac{\partial r\rho_{k,i}}{\partial t} + \frac{\partial r\rho_{k,i}u_{k,i}}{\partial r} = -rG_i,\ i = 1..N. \tag{5}$$

The particle impulse–conversation equation:

$$\frac{\partial r\left(\rho_{k,i}u_{k,i}\right)}{\partial t} + \frac{\partial r\rho_{k,i}u_{k,i}^2}{\partial r} = r\tau_{tr,i} - rGu_{k,i},\ i = 1..N. \tag{6}$$

The particle energy-conversation equation:

$$\frac{\partial r\rho_{k,i}\left(\varepsilon_{k,i}+\frac{u_{k,i}^2}{2}\right)}{\partial t}+\frac{\partial r\rho_{k,i}u_{k,i}\left(\varepsilon_{k,i}+\frac{u_{k,i}^2}{2}\right)}{\partial r}=-r\alpha_{k,i}S_{k,i}n_{k,i}\left(T_{k,i}-T_g\right)+ \quad (7)$$
$$+rQG_i-rG_ic_kT_{k,i}-rG_i\frac{u_{k,i}^2}{2}+r\tau_{tr,i}u_{k,i},\ i=1..N.$$

The particle number concentration equation:

$$\frac{\partial rn_{k,i}}{\partial t}+\frac{\partial rn_{k,i}u_{k,i}}{\partial r}=0,\ \ i=1..N. \quad (8)$$

The perfect-gas law:

$$p=\rho_gR_gT_g. \quad (9)$$

The particle radius equation:

$$r_{k,i}=\sqrt[3]{\frac{3\rho_{k,i}}{4\pi\rho_k^0n_{k,i}}},\ \ i=1..N. \quad (10)$$

The initial condition:

$$T_g\left(r,t_z\right)=T_b+\frac{Q_z}{4\pi\lambda_bt_z}\exp\left(-\frac{r^2}{4\chi_bt_z}\right),\ T_{k,i}\left(r,t_z\right)=T_b,\ \rho_{O2}\left(r,t_z\right)=\rho_{O2,b},$$
$$\rho_{k,i}\left(r,t_z\right)=\rho_{kb,i},\ u_g\left(r,t_z\right)=u_{k,i}\left(r,t_z\right)=0,\ \rho_g\left(r,t_z\right)=\rho_b,\ n_{k,i}\left(r,t_z\right)=n_{kb,i},\ \ i=1..N. \quad (11)$$

The boundary conditions:

$$\frac{\partial\rho(0,t)}{\partial r}=\frac{\partial\rho_{O2}(0,t)}{\partial r}=\frac{\partial T_g(0,t)}{\partial r}=0,\ u_g(0,t)=0,$$
$$\frac{\partial\rho_{k,i}(0,t)}{\partial r}=\frac{\partial n_{k,i}(0,t)}{\partial r}=\frac{\partial T_{k,i}(0,t)}{\partial r}=0,\ \ u_{k,i}(0,t)=0,\ i=1..N, \quad (12)$$
$$\frac{\partial\rho_{O2}(\infty,t)}{\partial r}=\frac{\partial T_g(\infty,t)}{\partial r}=0.$$

Where: ρ – density; ρ_{O2} – oxygen partial density; ρ_{CH4} – methane partial density; u – velocity; t – time; r – radial coordinate; r_k – particle radius; p – pressure; Q – reaction heat; Q_z – spark ignition energy; k_0 – chemical reaction-rate constant; T – temperature; E – energy of activation; R_u – the molar gas constant; S_m – cross-sectional area; η – absolute viscosity coefficient of the gas; c_p – the gas heat capacity at constant pressure; c_v – gas heat capacity at constant volume; c_k – gas heat capacity of the coal; $\varepsilon_k=c_kT_k$ – the particle internal energy; $\varepsilon_g=\frac{p}{\rho_g(\gamma-1)}$ – gas internal energy; $\lambda_g=\lambda_{st}\left(\frac{T_g}{T_v}\right)^{0.667}$ – thermal conductivity coefficient; $D_g=D_{st}\left(\frac{T_g}{T_v}\right)^{0.667}$ – diffusion coefficient [12]; $\alpha_k=\frac{Nu_k\lambda_g}{2r_k}$ – gas-particles heat exchange coefficient; $\gamma=\frac{c_p}{c_v}$ – adiabatic exponent; $\alpha_1=\frac{\mu_{O2}\nu_{O2}}{\mu_C\nu_C}$ – oxygen consumption coefficient in reaction with coal dust; $\chi_g=\frac{\lambda_g}{c_v\rho_g}$ – heat diffusivity. Indexes: b - initial conditions of the gas parameters, z –spark characteristics, k - particle parameters, g – gas parameters, $i=1..N$ – fraction number, numeration begins from largest to the smallest size. The rate of particle mass changing is defined by $G=n_kS_kj_1\rho_{O2}$, where $j_1=\frac{\beta_mk_{02}\exp(-E_2/R_uT_k)}{\beta_m+k_{02}\exp(-E_2/R_uT_k)}$ – heterogeneous reaction rate, $\beta_m = \frac{\lambda_g(T)Nu_D}{c_g\rho_g r_k}$ – particles mass-transfer coefficient [13]. The friction force is determined by $\tau_{tr} = n_k F_{tr}$, where $F_{tr} = \frac{C_r S_m \rho_g (u_g - u_k)|u_g - u_k|}{2}$ – interaction force of a single particle with the gas, $C_r = \frac{24(1+0.15\,\mathrm{Re}^{0.682})}{\mathrm{Re}}$ – friction coefficient, $\mathrm{Re} = \frac{2\rho_g r_k |u_g - u_k|}{\eta}$ – Reynolds number, $Nu_k = 2 + (Nu_l^2 + Nu_t^2)^{0.5}$ – Nusselt number, where $Nu_l = 0.664\,\mathrm{Re}^{0.5}$, $Nu_t = 0.037\,\mathrm{Re}^{0.8}$ [10].

We have solved the system of equations (1) – (12) using the method from [10]. The spatial step around the ignition source area (up to coordinate $r = 10^{-3}$ *m*) was set to $\Delta h_{const} = 10^{-6} m$. After the coordinate $r = 10^{-3} m$ the spatial step was defined as geometric progression $\Delta h_{i+1} = 1.005 \cdot \Delta h_i$. The time step was calculated by Courant's stability criterion.

3. Results and discussion

In this study we used thermophysical and kinetics parameters of the coal dust from [19] to solve the problem. The coal particle size r_k varied from 10^{-8} to $3 \cdot 10^{-6}$ *m*. The number of the fractions and fraction composition in percentages was also varied to evaluate the influence of the changing values on the ignition. The aim of the calculations was to determine the minimum spark ignition energy Q_z for each composition when the flame front occurred, propagated by the volume. The dimension of the ignition energy is *J/m*. To define the energy in *J*, the anode-cathode spacing has to be taken into account, which ranges from $4 \cdot 10^{-3}$ to $6 \cdot 10^{-3} m$ [5].

The curve of the minimum sparkplug ignition energy depending on fraction mass concentration $m_{dust,\ 2}$ in percentages is presented in Fig. 1. The total mixture mass is $m_{dust} = 0.4\ kg/m^3$. The value of $m_{dust,\ 2}$ =0% corresponds to the monodisperse mixture with particle size $r_{k,1} = 2 \cdot 10^{-6}\ m$, The value of $m_{dust,\ 2}$ =100% - the monodisperse mixture with $r_{k,2} = 2 \cdot 10^{-7}\ m$. According to the Fig.1, under the constant mass m_{dust}, the mixtures with the smallest fractions have the lesser value of the minimum ignition energy.

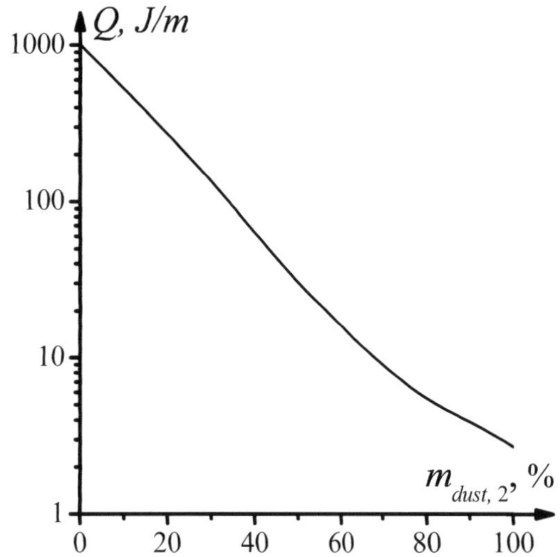

Fig. 1. $m_{dust} = 0.4\ kg/m^3$, $r_{k,1} = 2 \cdot 10^{-6}\ m$, $r_{k,2} = 2 \cdot 10^{-7}\ m$

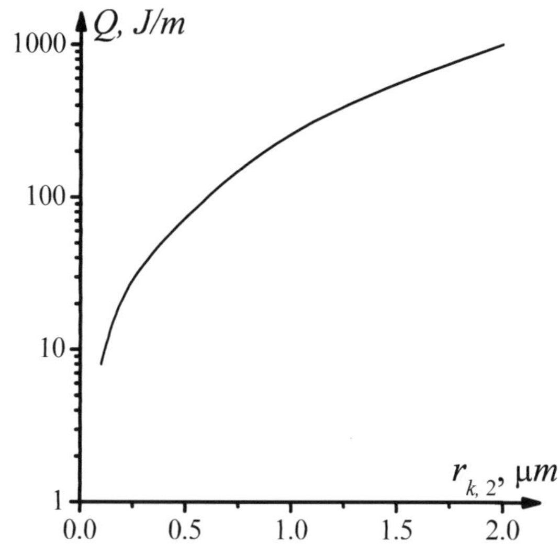

Fig. 2. $m_{dust} = 0.4\ kg/m^3$, $r_{k,1} = 2 \cdot 10^{-6}\ m$, $m_{dust,1} = 0.5\ m_{dust}$, $m_{dust,2} = 0.5\ m_{dust}$

IOP Conf. Series: Journal of Physics: Conf. Series **899** (2017) 042005 doi:10.1088/1742-6596/899/4/042005

The more large particles are in the mixture the more minimum sparkplug ignition energy. We conducted several calculations for the mixture with different values of radius of $r_{k,2}$. The calculation result for the fraction ratio $r_{k,1}/r_{k,2} = 1$ and varied radius $r_{k,2}$ is shown in Fig.2. It should be noted that with the increase of fraction radius $r_{k,2}$ the minimum ignition energy rises.

The obtained results show that for the bi-disperse suspension the presence of finely-divided fraction in it leads to significant decrease of minimum sparkplug ignition energy. To show the influence of dispersity on minimum ignition energy of a polydisperse suspension, the parameters of coal particles to make calculations are taken from [1]. The work [1] provides data about dust laden air in mine working: from 40 to 80 % of coal particles have the diameter less than $1.3 \cdot 10^{-6}$ *m*, 15 ÷ 35 % less than $2.6 \cdot 10^{-6}$ *m*, 5 ÷ 20 % less than $4 \cdot 10^{-6}$ *m* and 3 ÷ 10 % of particles have the diameter more than $4 \cdot 10^{-6}$ *m*. In this research the investigated suspension has the fractions with the following radiuses: $r_{k,1} = 4 \cdot 10^{-6}\ m,\ \ r_{k,2} = 2 \cdot 10^{-6}\ m,\ \ r_{k,3} = 1.3 \cdot 10^{-6}\ m,\ \ r_{k,4} = 0.65 \cdot 10^{-6}\ m$. The fraction composition in percentages varies within mentioned range. The values of minimum ignition energy spread from 18 *J/m* for the suspension with 70% of $r_{k,4}$ to 35 *J/m* for 40 % of $r_{k,4}$.

According to obtained results the presence of the small particles has the significant impact on the minimum sparkplug ignition energy. The number of large particles slightly effect on the ignition energy. The Fig.3,4 show the ignition energy dependence on fraction mass concentration $m_{dust,\ 4}$. The area between curves determines the values of minimum spark ignition energy. The spread of the values for the same $m_{dust,\ 4}$ relates to the different mass value of large particles with the radius $r_{k,3}$. The upper curves on Fig.3,4 correspond to the largest mass of the particles $r_{k,1} = 4 \cdot 10^{-6}\ m$ and $r_{k,2} = 2 \cdot 10^{-6}\ m$. The lower curves correspond to the suspension with high content of $r_{k,3} = 1.3 \cdot 10^{-6}$ *м*. In Fig.4 the anode-cathode spacing is $4 \cdot 10^{-3} m$.

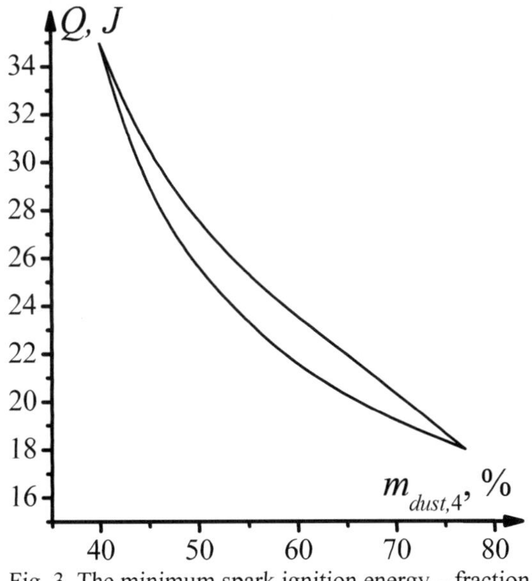

Fig. 3. The minimum spark ignition energy – fraction mass concentration curve, $m_{dust} = 0.4$ kg/m^3.

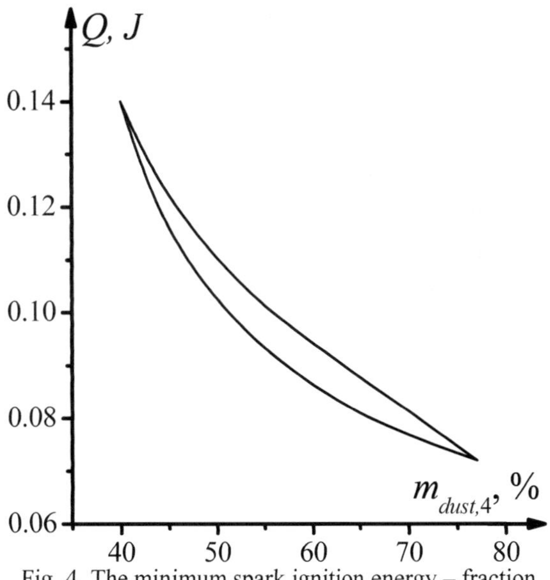

Fig. 4. The minimum spark ignition energy – fraction mass concentration curve, $m_{dust} = 0.4$ *kg/m*^{3}.

The spread of values in Pic.4 is satisfactorily agreed with the values of minimum ignition energy determined in [5]. The work [5] provides data about the minimum ignition energy for long flaming coal with the mass m_{dust} from different coalfields. The minimum ignition energy of coal dust from different coalfields varies from 4 to 80 *mJ*. The differences in the obtained energy values are due to the various coal dust compositions [5]. The obtained result in this research is in a good agreement with those collected in [5] under the fraction concentration of small particles more than 65%.

4. Conclusions

We numerically investigated the sparkplug ignition problem of bi and poly- dispersed coal-dust suspension. The paper provides data about the effect of particle size and fraction composition in percentages on minimum spark ignition energy. The larger the particles the more energy is needed to ignite the coal-dust suspension. Spread of sparkplug ignition values for bi-dispersed suspension depending on the content is wider than for poly-dispersed.

Acknowledgments
This work was supported by the Russian Science Foundation grant No. 16-33-60091 mol_a_dk.

References:

[1] Vasil'ev A. A., Pinaev A. V, Trubitsyn A. A., etc. 2017 *Combust., Expl.,Shock Waves* 1**(53) 8-14.**
[2] Kalyakin S, Kostenko V, Zav'yalova E and Shtrokh L 2012 *Aktualne problemy zwalczania zagrożeń górniczych: II konferencja naukowo-techniczna* (*Brenna*) No 7-9 (Brenna) p 176
[3] Dement'ev A A, Moiseeva K M, Krainov A Yu and Paleev D. Yu 2016 *JOEP* **6 (89)** 1538–1546.
[4] Baklanov A M, Valiulin S V, Dubtsov S N, Zamashchikov V V, Korzhavin A A, Onischuk A A, Purtov P A, Klishin V I, Paleev D Y and Kontorovich A E 2015 *Doklady Phys. Chem.* 461 (1) 57-60
[5] Korolchenko A. Ya. 1986 *Fire and explosion hazard of industrial dust* (Moscow: Himija)
[6] Neychlelyaev MI, Lyubimova AI, Petrukhin PM, etc. 1992 *Combating coal dust explosions in mines* (Moscow: Nedra)
[7] Krainov A. Yu., Baimler V. A. 2002 *Combust., Expl.,Shock Waves* **3(38)** 30-36.
[8] Krainov A. Yu. 2001 *Combust., Expl.,Shock Waves* **3(37)** 16-24.
[9] Krainov A. Yu., Baimler V. A. 2002 *Combust., Expl.,Shock Waves* **4(38)** 9-13.
[10] Krainov A. Yu. Moiseeva K. M. 2016 JPCS. **754**, 052003.
[11] Nigmatulin R I 1990 *Dynamics of Multiphase Media* (CRC Press; Rev Sub edition)
[12] Shkadinskii K G and Barzykin V V 1968 *Combust., Expl.,Shock Waves* **4 (2)** 100-4
[13] Frank-Kamenetskiy D A 1987 *Diffusion and Heat Transfer in Chemical Kinetics* (Moscow: Nauka)

IOP Conf. Series: Journal of Physics: Conf. Series **899** (2017) 042006 doi:10.1088/1742-6596/899/4/042006

PIV/OH PLIF investigation of flow and flame front dynamics of acoustically perturbed conical flame

Z Kravtsov[1,2], D Sharaborin[1,2]

[1]Kutateladze Institute of Thermophysics, 1 Ak. Lavrentyeva Avenue, Novosibirsk, 630090, Russia
[2]Novosibirsk State University, 2 Pirogova Street, Novosibirsk, 630090, Russia

E-mail: zakhar.d.kravtsov@gmail.com

Abstract. The paper reports on the investigation of the flow and flame front dynamics in a premixed conical methane/air flame perturbed by an acoustic field. The investigation is carried out by using stereoscopic particle image velocimetry and planar laser-induced fluorescence of HCHO to measure simultaneously the instantaneous velocity field and flame front location in the axial plane. The optical techniques provided information about local velocity and velocity gradients, stretching and curvature of the flame front. Flow conditions when the flame is covered from the top by a cooled metallic flat plate are also considered. It is found that the flow and flame front lose axial symmetry when the obstacle is present because of disruption of the global instability mechanism of the combustion products plume driven by bouncy force.

1. Introduction

Impinging flames, viz., jet-flames which flow against a solid obstacle, are studied because for some conditions self-induced instabilities may arise [1]. This feature allows to analyze thermoacoustic combustion phenomenon for well controlled conditions [2]. Planar laser-induced fluorescence (PLIF) is widely used for analysis of the local flame front shape, local gas temperature, species concentration (such as OH and NO) and heat release [3]. Such experimental data are very useful for better understanding of combustion processes. Accurate temperature and concentration measurements are also necessary for validation of theoretical models and numerical codes. Particle image velocimetry (PIV) method is now a standard tool for measurements in non-reacting flows [4] and often used for diagnostics of flows with combustion [5]. The advantages of the PIV method for reacting flows are almost negligible intrusiveness and possibility to measure instantaneous velocity field and estimate the velocity gradient. The latter feature allows to detect coherent flow structures, such as large-scale vortices.

In the present study an experimental investigation of flow and flame front dynamics was carried out for laminar impinging methane/air flame by using PIV and HCHO PLIF. To analyse unsteady behaviour of the flow/flame system, an external acoustic perturbations were imposed.

2. Experimental setup

The measurements were carried out for the methane/air flames at atmospheric pressure. The flames were organized in an open combustion rig (see details in [6]) by using a contraction axisymmetric nozzle with the inner diameter at the outlet of $d = 15$ mm. The equivalence ratio ϕ of methane/air mixture issued from the nozzle was 0.7. The jet Reynolds number was fixed as 1700 (bulk velocity of the mixture was $U_0 = 1.7$ m/s). A metalic plate with diameter 80 mm, cooled by a recirculating water flow with temperature of 53°C, was placed normally to the jet axis for organization of the impinging flows. External perturbations of the flow/flame were provided by a loudspeaker, oriented radially to the nozzle rim at the distance 20 cm. The loudspeaker was connected to an amplifier, harmonic function generator, and electric power meter. The electric power supplied the loudspeaker was approximately 3W.

A photograph of the experimental setup is shown in the Figure 1. To provide PIV measurements, the flow issued from the nozzle was seeded by 4 μm TiO_2 particles. The surrounding air was seeded by using a fog generator. A pair of PIV cameras was mounted as shown in Figure 1. The used ImperX IGV-B2020 CCD cameras were oriented horizontally. The cameras were equipped with Sigma AF #50 lenses and band-pass optical filters (60% transmittance at 532 nm and full width of 10 nm at the half maximum). The seeding particles were illuminated by the second harmonic of a Nd: YAG laser with 200 mJ energy per each pulse. The laser beam was converted into the laser sheet with the use of a system of cylindrical and spherical lenses. Each camera captured 4 Mpix images. The time separation between two laser pulses was 40 μs.

For each flow case, 1000 instantaneous three-component velocity fields were measured. The PIV velocity fields were calculated by an in-house ActualFlow software with an iterative cross-correlation algorithm with continuous shift and deformation of the interrogation areas. Final size of the interrogation area for each velocity vector was equal to 32 × 32 pixels. Spatial overlap rate between the neighbour interrogation areas was 50%. The used cross-correlation algorithm accounted for the number of tracer particles in each interrogation area. If the number of particles was less than five, the velocity vector in this spatial domain was not calculated.

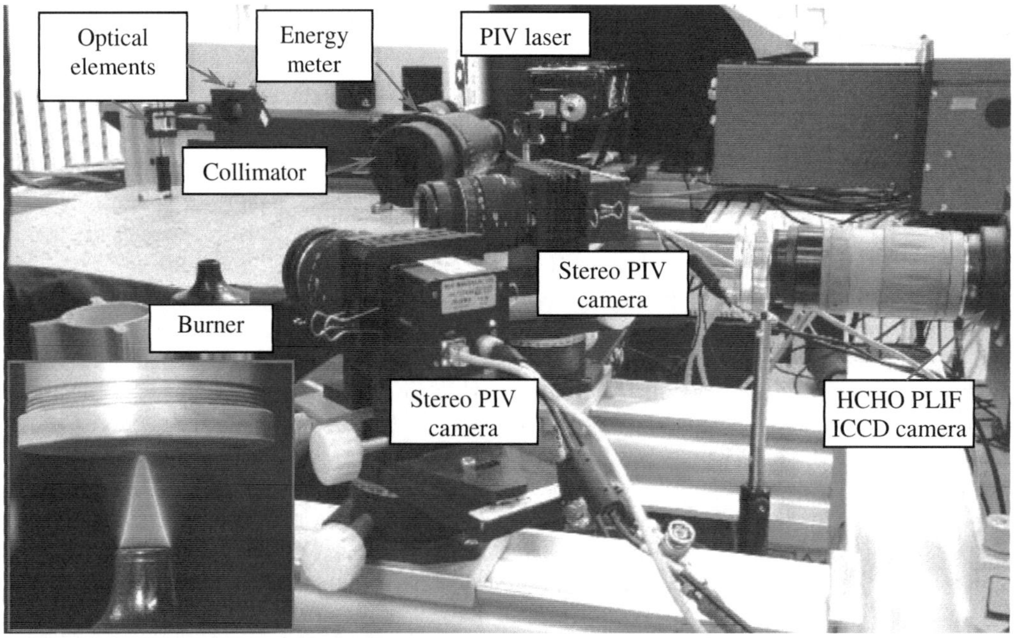

Figure 1. Photograph of the PIV/PLIF setup. Inset shows photograph of impinging flame

For the PLIF measurements radiation of the third harmonic (355nm) of a Nd:YAG laser (Quantel Brilliant B with 45 mJ energy per each pulse) was used to excite HCHO fluorescence. RMS of the pulses energy variation was below 5%. After the collimator optics, the height of the light sheet was 50 mm. The thickness of laser sheet was below 0.8 mm in the measurement region. Laser sheet

IOP Conf. Series: Journal of Physics: Conf. Series **899** (2017) 042006 doi:10.1088/1742-6596/899/4/042006

illuminated the axial plane of the reacting flow. The fluorescence of HCHO radicals was collected by a registration system consisted of an image intensifier (LaVision IRO) and sCMOS camera (LaVision, Imager sCMOS, 16 bit, resolution 2560×2160 pixel), equipped with a Sigma AF #50 lens and appropriate band-pass optical filter. The IRO photocathode (S20 multialkali) provided quantum efficiency about 25% for wavelengths in the range of interest. The exposure time for each image was 200 ns.

The raw PLIF images were processed by LaVision FlameMaster software to account for non-uniform spatial distribution of the laser sheet intensity, non-uniform spatial sensitivity of the registration system. Background signal and dark current were removed to sufficient extent by using PLIF images captured without combustion. In addition, an in-house image processing algorithm was used to determine local curvature of the flame front. The algorithm was based on the intensity gradient of the HCHO signal for determination of the flame front location. Local flame shape was approximated by the using Fourier series.

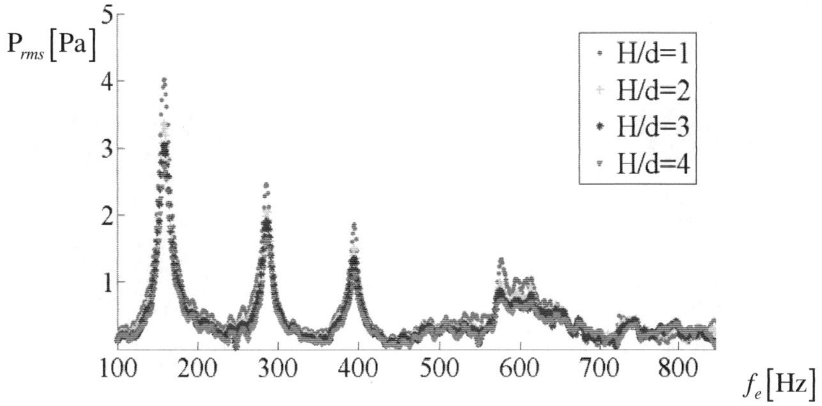

Figure 2. Amplitude of the pressure waves passed through the gas supply path of the combustion rig

To investigate resonance characteristics of the combustion rig for different distances between the nozzle and flat plate (H) a precision microphone Type 2250 by Bruel & Kjaer was used. Intensity of the pressure pulsations at the nozzle outlet was measured under conditions of superimposed external harmonic pressure oscillations (with frequency f_e) in the supply path. The results are shown in Figure 2. The resonance frequencies of the combustion rig (pipelines) are 160, 290, 395 Hz. The frequencies are the same for the configurations with and without the impingement surface, but the amplitude is lower in the latter cases.

3. Results

Figure 3 shows spatial distributions of the time-averaged velocity field for the reacting flows with and without the impingement plate. The contraction nozzle produced nearly "top-hat" velocity profile at the outlet. Thus, the velocity in inside the cone was almost uniform. Velocity of the gas passed through the flame cone increased due to the combustion, local heat release, temperature rise and density decrease. The impingement surface blocked the axial flow and produced flow in the radial direction.

IOP Conf. Series: Journal of Physics: Conf. Series **899** (2017) 042006 doi:10.1088/1742-6596/899/4/042006

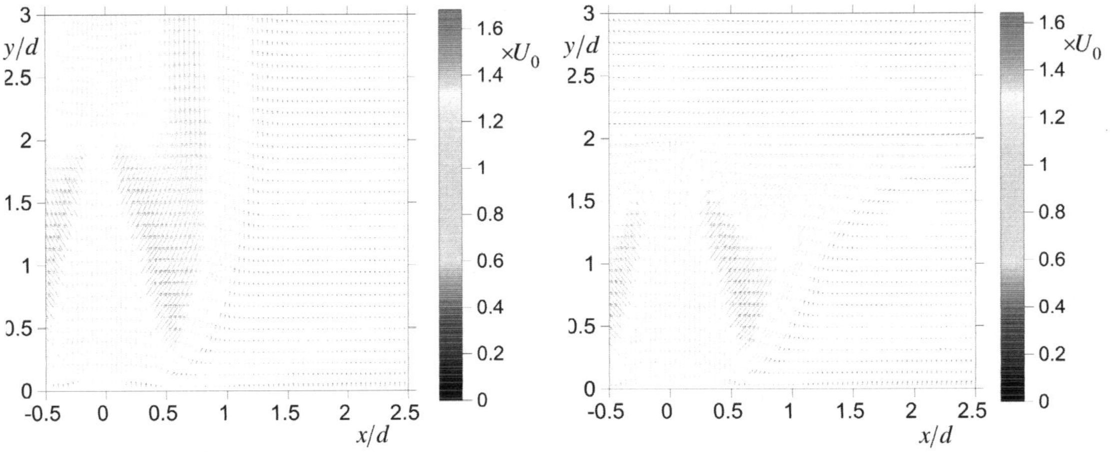

Figure 3. Average velocity field for conical flames (left) without impingement surface and (right) with impingement surface for $H/d = 2$

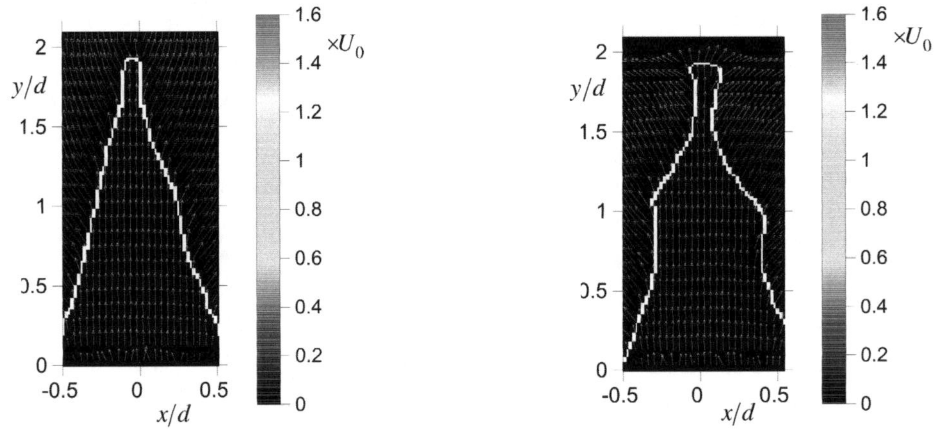

Figure 4. Examples of the instantaneous velocity field for acoustically perturbed conical flames (left) without impingement surface and (right) with impingement surface for $H/d = 2$. White lines correspond flame front location detected from HCHO PLIF data

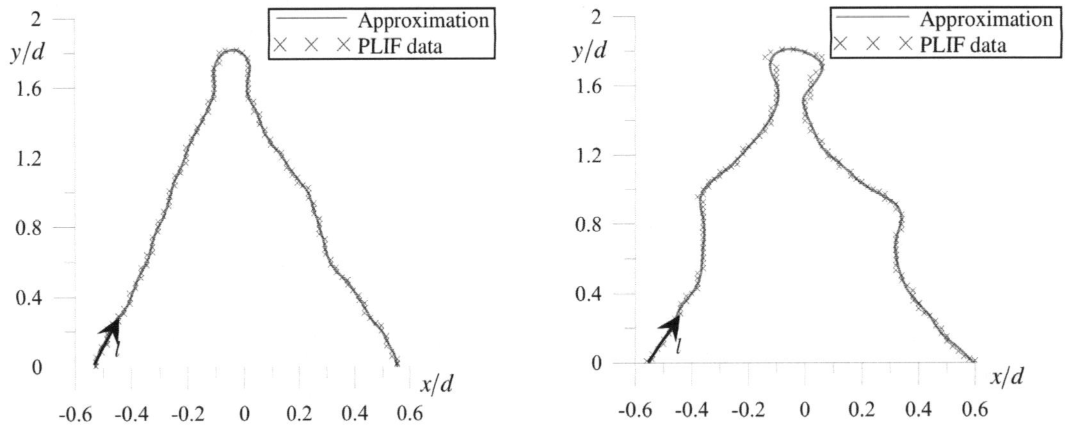

Figure 5. Examples of approximation of the local flame shape. Dark-blue crosses shows coordinates of the pixels from HCHO PLIF data and red lines show result of approximation by smooth curves based on decompositions by Fourier series

Figure 4 shows the instantaneous velocity snapshots for the region of flame front. The flame front cross-cut is visualized by white solid lines, determined from local intensity of HCHO signal, appeared at the preheat zone of the flame front. The external acoustic field resulted in oscillations of the jet flow issued from the nozzle and deformations of the flame front. Moreover, for the case of the impinging flow, the flame front and flow are clearly not symmetric.

To quantify flame front deformations (for the considered 2D data), a local flame front curvature was evaluated. This was done by evaluating coordinates x and y of each n-th pixel at the digitized boundary of the flame front. Figure 5 shows coordinates of the pixel and approximation by smooth curves based on decompositions by Fourier series. The local curvature k was evaluated according to the following equation:

$$k = \frac{x'y'' - x''y'}{\left(x'^2 + y'^2\right)^{3/2}}, \quad (1)$$

where prime indicates derivation with respect to n. The examples of the curvature evaluation are shown in Figure 6. The maximum of the curvature reached maximum at the flame front tip and corresponded to typical curvature radius of 0.1 mm.

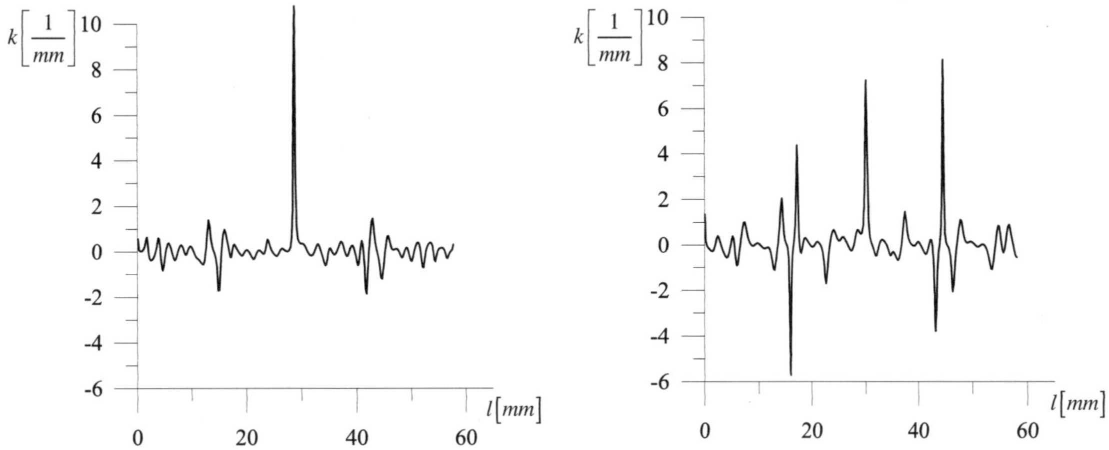

Figure 6. Flame front curvature along the conical flames (left) without impingement surface and (right) with impingement surface for $H/d = 2$

4. Conclusions

Experimental investigation of the flow and flame front dynamics in a premixed conical methane/air flame perturbed by an acoustic field was performed by using the stereoscopic particle image velocimetry and planar laser-induced fluorescence techniques. Two flow/flame conditions are considered, viz., open flame and flame blocked from the top by a flat cooled metallic surface. The instantaneous velocity field was measured in the axial plane simultaneously with detection of the flame front location. The optical techniques provided information about local velocity and velocity gradients, stretching and curvature of the flame front. The external acoustic field was superimposed to promote oscillations of the jet flow issued from the nozzle and deformations of the flame front. The flame from curvature reached maximum (approximately 10 mm^{-1}) at the flame front tip.

Acknowledgments

Financial support has been obtained from Russian Science Foundation (grant № 16-19-10566). A.S. Lobasov and V.M. Dulin are acknowledged for their assistance during the data processing and manuscript editing.

References

[1] Durox D, Schuller T, Gandel S 2002 Self-induced instability of a premixed jet flame impinging on a plate. *Proc. Combust. Inst.* **29** 69-75
[2] Fernandes EC, Leandro RE 2006 Modeling and experimental validation of unsteady impinging flames. *Combust. Flame* **146** 674-686
[3] Hanson RK, Seitzman JM, Paul PH 1990 Planar laser-fluorescence imaging of combustion gases. *Appl. Phys. B.* **50** 441-454
[4] Raffel M, Willert C and Kompenhans J 1998 *Particle Image Velocimetry* (Springer, Berlin Heidelberg, New York)
[5] Stella A, Guj G, Kompenhans J, Raffel M and Richard H 2000 Application of particle image velocimetry to combusting flows: design considerations and uncertainty assessment *Exp. Fluids* **30** 167-80
[6] Alekseenko SV, Dulin VM, Kozorezov YS, Markovich DM, Shtork SI, Tokarev MP 2011 Flow structure of swirling turbulent propane flames. *Flow Turbul. Combust.* **87** 569-595

IOP Conf. Series: Journal of Physics: Conf. Series **899** (2017) 042007 doi:10.1088/1742-6596/899/4/042007

Researching of the reduction of shock waves intensivity in the “pseudo boiling” layer

G I Pavlov[1], D A Telyashov[1], A V Kochergin[1], P V Nakoryakov[1], E A Sukhovaya[1]

[1] Kazan National Research Technical University — KAI named after A. N. Tupolev

E-mail: pavlov16@mail.ru

Abstract. This article applies to the field of acoustics and deals with noise reduction of pulsating combustion chambers, in particular the reduction of the shock waves' intensity with the help of pseudo boiling layer. In the course of work on a test stand that included a pulsator, a compressor with the receiver and a high pressure fan was simulated gas jet flowing from the chamber pulsating combustion and studied the effect of different types of fluidization on effect of reducing the sound pressure levels. Were obtained the experimental dependence of the sound pressure levels from parameters such as: height of the layer of granules; diameter of the used granules; amplitude of the pressure pulsations in the gas stream at the entrance to the camera; frequency of pressure pulsations. Based on the results of the study, it was concluded that the using of a pseudo boiling layer is promising for reducing shock wave noise.

1. Introduction

Pulsed combustion chamber (PCC) has been used in heat power engineering for a long time. They are also used as engines on unmanned aerial vehicles [1]. This is due to the simplicity of the design, the completeness of combustion and high efficiency. On performance characteristics, (PCC) with an aerodynamic valve (AV) is preferable. Such chambers are cheaper to manufacture and have a great service resource. However, their widespread introduction is limited to an increased level of noise. The complexity of solving the noise problem of PCC is related to certain features characteristic of combustion chambers - in (PCC) fuel combustion occurs in a resonant pulsating mode. Moreover, oscillatory processes in (PCC) are low-frequency, than the adoption of this or that design decision aimed at noise reduction is limited. Lots of researches are engaged in noise reduction (PCC). A number of interesting ideas and constructive schemes have been published in the scientific and technical literature [2,3]. Despite this, effective silencers for (PCC) are not manufactured by industry. (PCC) with an aerodynamic valve has two sources of noise: a gas jet flowing out of the aerodynamic valve; High-temperature combustion products emitted from the (PCC) through the resonant tube. The most problematic in terms of noise cancellation is noise (AV), since (AV) at one moment of time should ensure reliable locking of the cavity of the combustion chamber, at another moment of time - to ensure the full flow of the air oxidizer into the combustion zone. (PCC) with an aerodynamic valve, as a device, has been known for a long time. However, there is no work with detailed studies of the

IOP Conf. Series: Journal of Physics: Conf. Series **899** (2017) 042007 doi:10.1088/1742-6596/899/4/042007

mechanism of noise generation (AV) in open printing. Without a clear representation of the physical picture flowing in the aerodynamic valve, it is not possible to solve the problem of noise reduction.

2. Researches

It has been established by experiments that the main noise source (AV) is low-intensity shock waves. The shape of the acoustic signal recorded at the valve cutoff is not sinusoidal, has an impulse character (Figure 1). At the same time, pressure pulsations recorded directly in the combustion chamber vary according to the harmonic law (Figure 2).

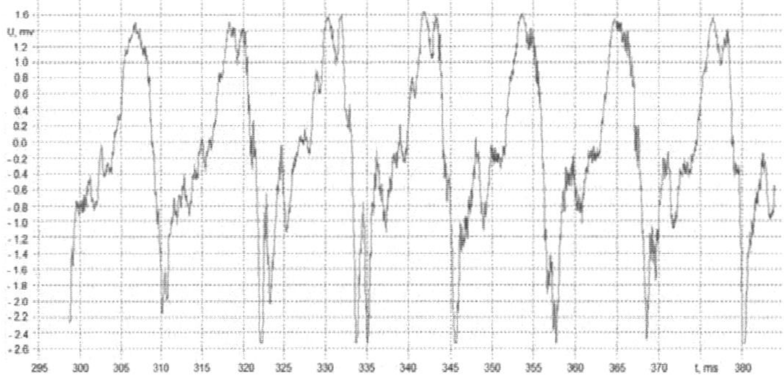

Figure 1.The shape of the electrical signal recorded at the output of the microphone, located near the AV

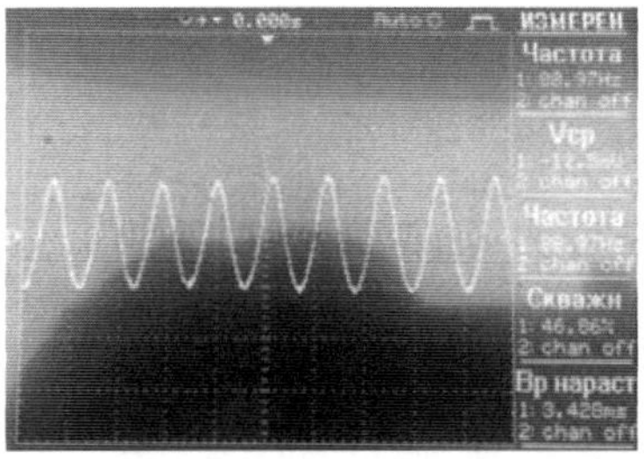

Figure 2. The shape of the electrical signal recorded at the output of the piezoelectric transducer

(The piezoelectric transducer is connected to the PCC)

The revealed fact narrows down the field of research and concretizes the direction of further research. The mechanism of formation of shock waves in the aerodynamic tube is the following. After the combustion products are released from the combustion chamber, the process of filling the chamber with a fresh combustible-air mixture begins. The ignition of this mixture is carried out by high-heated gases, which remain in the wall layer of the resonance tube when the combustion products flow out of it. In the zone of formation of a fresh combustible mixture, heated gases are "thrown in" when the pressure wave moves in the opposite direction (from the open end of the resonance tube towards the combustion chamber) and become fire ignition of the fresh mixture. The fire ignitions represent a set of small vortexes. From these sources, the gas mixture is ignited by micro volumes that form a string of compression waves. These waves rush towards the aerodynamic valve and, overlapping each other, form a compression wave of shock type. This idea of the noise generation mechanism in the aerodynamic valve allows you to choose more specific technical solutions aimed at reducing noise intensity. The authors of the work proposed for the first time the noise suppression effect of gas jets by dispersed particles in a suspended state. An attempt to use the above-stated effect to reduce the intensity of shock waves generated by the (AV) pulsed combustion chamber was the main goal of this work's authors. However, without detailed studies of the influence of the fluidized bed on the intensity of shock waves, it was not possible to achieve these goals. A gas jet flowing from the (AV) pulsed combustion chamber was simulated on the stand. Cold air was used as a working fluid. Schematic diagram of the test bench is shown at Figure 3.

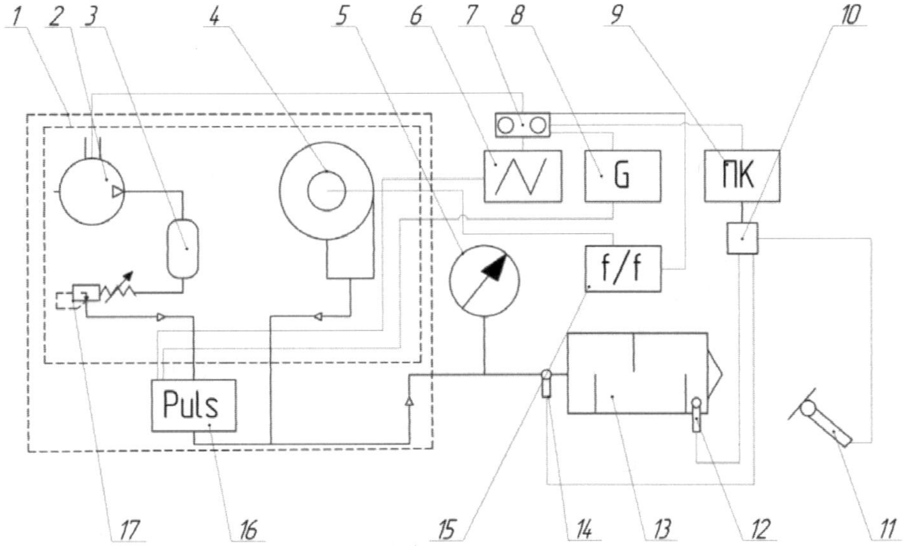

Figure 3. Schematic diagram of the test stand

The test stand included a high-pressure fan, a compressor with a receiver, a pulser, a control panel, a system for measuring the pulsations of pressure in the flow before and after the fluidized bed, and also in the near acoustic field, a manometer. The pulser was driven by a fan. Necessary amplitude of pulsations was established in front of the pulser by changing the pressure of the compressed air. The pressure pulsations in the flow were measured by a piezoelectric sensor and in a near acoustic field by a condenser microphone. The object of the research and the microphone were located in a separate soundproof room. The six-channel "Ecophysics" measuring system with the "Signal +" software was used as the recording equipment.

The hydraulic resistance of the gas-dynamic tract was determined by the manometer. The object of the research was a model of a sound attenuator with a fluidized bed. The general view of the model is shown in Figure 4.

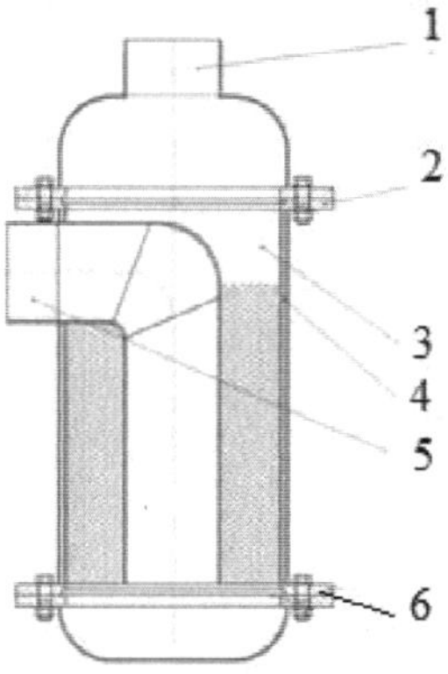

Figure 4. Technical diagram of a sound attenuator based on a fluidized bed

1 – output nozzle; 2 - upper flange connection with the mesh unit; 3-chamber; 4 - cylindrical body of the chamber; 5-input nozzle; 6- lower flange connection with the mesh unit.

3. Experimental activities

The model consisted of the following main parts: a vertically located cylindrical chamber 3 partially filled with granules, an input 5 and an output device 1. For visualization the interaction of the granules with a pulsating gas jet the window was provided on the side surface of the chamber (Figure 5)

Figure 5. General view of the chamber with a window

Prior analysis of the experimental data indicates that in other conditions being equal the effect of reducing the sound pressure level does not depend significantly on the type of fluidization. Further studies were carried out on a model with a homogeneous fluidized bed. The research tasks were to obtain experimental dependences of the sound pressure level on parameters such as: the height of the layer of granules; diameter of used granules; amplitude of pressure pulsations in the gas flow at the entrance to the chamber; frequency of pressure pulsations. The hydraulic resistance of the fluidized bed was also determined.

An effectiveness evaluation of the smoothing efficiency of the pulsating gas flow was carried out by comparing the amplitude of the pressure pulsations obtained at the input and output of the noise-damping device. Plastic balls were used as the dispersed particles. The size of the balls conformed to the following diameters in mm: 4,6,8,10. The experiments were carried out at frequencies equal to 50, 80, 200, 500, and 650 Hz. They corresponded to the values of the frequencies that are energy-carrying for the acoustic signal spectrum (AV) of the PCC. Table 1 shows the sound pressure levels in the near acoustic field of the sound attenuating device. The height of the fluidized bed was 40 mm.

Table 1

	Granule diameter, mm				
Frequency, Hz	4	6	8	10	Without granules
50	77 dBa	72 dBa	74 dBa	74 dBa	84 dBa
80	79 dBa	76 dBa	78 dBa	77 dBa	89 dBa
200	89 dBa	83 dBa	88 dBa	83 dBa	97 dBa
500	96 dBa	91 dBa	94 dBa	93 dBa	104 dBa
650	99 dBa	95 dBa	97 dBa	97 dBa	109 dBa

From the experimental data given in table 1, shows that the height of fluidized bed height of 40 mm significantly affects the reduction of sound pressure level in the gas stream. In this case, granules with a diameter of 6 mm are most suitable. Increasing the height of the fluidized bed affects the hydraulic losses. In all studied cases (with granules of different diameters) backpressure gas path increased in 2 ... 2.5 times compared with the fluidized bed 40 mm in height.

Based on the results of the studies a technical scheme of a noise silencer for (AV) is proposed. Ceramic granules were used as dispersed particles. The general view of the granules is shown in Figure 6.

Figure 6. Ceramic granules (diameter 6 mm)

The working fuel for (PCC) was propane. The experimental model of PCC with noise silencers is shown in Figure 7.

Figure 7. Experimental model of the AV silencer on the test stand

A detailed description of the experimental model is given in research [9]. The efficiency of the silencer and its effect on the work of the PCC were evaluated In the course of the experiments. It was noted that in the range of the thermal power of the PCC equal to 40 kW to 120 kW the sound pressure level in the near acoustic field did not exceed 82 dB. The concentration of harmful substances at the exit from the resonator tube remained at the same level as without a noise-damping device. This indicates that the proposed noise silencer does not interfere with the operation of the aerodynamic valve.

4. Conclusion

On the basis of the done work it is possible to draw an important conclusion about the prospects of using a fluidized bed to reduce shock-wave noise. The resulting effect can be explained by the strong fragmentation and reflection of shock waves in the layer of granules and also by the friction of gases on the surface of dispersed particles. By "tossing" the granules with a pulsating jet the transparency of the layer increases and the hydraulic resistance decreases as a consequence.

5. References

1. Bychenok V.I. Heat power engineering of working process in pulsated combustion apparatus / / Dis. Doc. Tech. Sciences: 05.14.04 / B.C. Severyanin - Protected 2004 - Voronezh, 2004. - 339 p.

2. L.D. Landau, E.M. Lifshitz. Hydrodynamics. Theoretical physics volume 6 // 3rd revised edition. - Moscow: Science. Ch. Ed. Physical mat. Lit., 1986. 736 s

3. NI. Ivanov. Engineering acoustics. Theory and practice of noise control / / Textbook - Moscow: University Book., Logos 2015. 432 c.

4. I.G. Filippov, O.A. Egorychev. Nonstationary Oscillations and Diffraction of Waves in Acoustic and Elastic Media, M., Mashinostroenie, 1977. 304 p.

5. Journal of Applied Mechanics and Technical Physics // Nauka, No. 3 (133), 1982. 176 p.

6. V.N. Alekseev, A.G. Semenov, A.T. Skvortsov. Scattering of sound by a potential flow arising from the motion of a sphere // Acoustic journal. Volume 41. № 6. 1995. 876 - 882.

7. S.S. Zabrodsky. Hydrodynamics and heat transfer in a fluidized bed. M. - L., Gosenergoizdat, 1963. 488 p.

8. Link to drawing No. 5: http://900igr.net/up/datas/174519/003.jpg (circulation date on February 25, 2017)

9. Energy Resource Efficiency and Energy Saving in the Republic of Tatarstan // Proceedings of the XVI International Symposium "Energy and Resource Efficiency and Energy Saving" Kazan, 2016. - 333-337 p.

Features of the gas dynamics and local heat transfer in intake system of piston engine with supercharging

L V Plotnikov[1]

[1]Ural Federal University named after the first President of Russia B.N. Yeltsin, ul. Mira 19, Ekaterinburg 620002, Russia

E-mail: plotnikovlv@mail.ru

Abstract. Comparison of experimental research results of gas dynamics and instantaneous local heat transfer in the intake pipes for piston internal combustion engines (ICE) without and with supercharging are presented in the article. Studies were conducted on full-scale experimental setups in terms of gas dynamic nonstationarity, which is characteristic of piston engines. It has been established that the turbocharger installation in a gas-air system of piston internal combustion engine leads to significant differences in the patterns of change in gas-dynamic and heat transfer characteristics of flows. These data can be used in a modernization of piston engines due to installation of a turbocharger or in a development of gas-air systems for piston ICE with supercharging.

1. Introduction

One of the major trends of development of modern internal combustion engines (ICE) is to increase their power and economical efficiency. A possible solution of this problem is to install the turbocharger (TC) on a piston engine. Usually the turbocharger installation is viewed solely as an effective way of increasing the mass airflow through an engine cylinders and a method of improvement of its technical and economic indicators [1-4]. The question about the influence of TC on gas dynamic and heat transfer characteristics of flow in a gas-air systems of piston engine is hardly considered in the literature. However, it is clear that the installation of TC will lead to a significant change in the thermal and mechanic characteristics of the gas flow in a gas-air system. For example, some studies of the effect of a turbine in the exhaust pipe on gas dynamic and heat transfer characteristics of pulsating flows were carried out earlier [5, 6]. It has been established that maximum speed of air flow in the exhaust pipe with the turbocharger was lower than in the pipe without TC (the differences reach 40 %). The presence of the turbocharger in exhaust system leads to some smoothing of fluctuations of the flow velocity during the whole working cycle of piston ICE. Also, it has been determined that decreasing the intensity of heat transfer was observed in the exhaust system with the turbocharger, it was characteristic of all rotation frequencies of TC and piston ICE. The decrease in a heat transfer intensity in the exhaust system with TC was 10-15 % at an excess pressure of p_b = 1,0 bar and 15-20 % at an excess pressure of p_b = 2,0 bar. Reduction in a heat transfer intensity to the walls of the exhaust pipe in engines with TC should positively affect the working process and the technical and economic parameters of the piston ICE. Because greater amount of heat will work in a turbocharger, and not to be lost through walls of the exhaust pipe in this case.

Comparison of results of experimental research of gas dynamics and instantaneous local heat transfer in the intake pipes for piston internal combustion engines without and with supercharging are presented in this article. The studies were carried out taking into account a gas-dynamic nonstationarity.

2. Experimental setups and measurement equipments

Experimental setups for experimental studies of gas dynamics and local heat transfer in the intake systems have been designed and manufactured. They are full-scale models of single-cylinder engine with dimensions of 8.2/7.1 (piston diameter – 82 mm stroke – 71 mm) with or without turbocharger (Figure 1). The prototype for the created models was a car engine of the Russian VAZ company. The drive of a crankshaft was carried out using an electric motor. The frequency of crankshaft rotation was varied in the range from 600 to 3000 rpm. The turbo was carried out by means of the turbocharger TKR-6. Frequency of TC rotor rotation was changed in the range from 35 000 to 46 000 rpm. A detailed description of experimental setups is presented in [7].

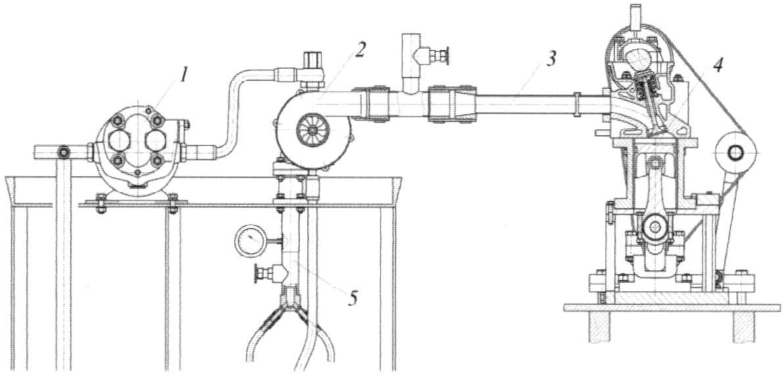

Figure 1. Scheme of the experimental setup: *1* – oil pump; *2* – turbocharger; *3* – intake pipe; *4* – piston engine; *5* – compressed air supply system.

Automated system of collecting and processing experimental data was created for the research. Instantaneous values of air flow velocity w_x and local heat transfer coefficient α_x was determined using the thermal anemometer of the constant temperature. The applicability of hot-wire method for studying gas exchange processes in piston engines is illustrated in the following works [1, 2]. In this work, the sensitive element of the anemometer sensor was nichrome thread which had a diameter of 5 μm and length 5 mm. A sensor with a free thread was used to measure the air flow velocity. The thread was placed perpendicular to the axis of the investigated pipe. The local heat transfer coefficient was determined using the thread, which was placed on the fluoropolymer substrate. The fluoroplastic substrate was mounted flush with the wall of the pipe. Measurements of the frequency of crankshaft rotation and the rotor of the turbocharger was carried out by means of tachometers. The WIKA pressure sensor was used to determine the instantaneous values of the static flow pressure in the pipes.

Traditional straight duct with round cross section was chosen as the starting base for research. This choice is explained by a small number of experimental data on gas-dynamic and heat transfer characteristics of flows in gas-air systems of the piston ICE and a weak knowledge of this subject in the literature. The length of intake pipe was 300 mm. Inner diameter of the intake pipe was 32 mm.

3. Local heat transfer determination method

Thermal anemometry method to study a gas-dynamics and local heat transfer in gas flows were chosen on the basis of literature analysis [8, 9]. It is known that the small dimensions of sensing element of the thermo-anemometer sensor do not introduce significant changes in a nature of gas flow; high sensitivity allows one to record the fluctuations of a variables with small amplitudes and large frequencies;

simplicity of a hardware scheme gives the opportunity to securely record an electrical signal from the output of the thermal anemometer for storage and further processing on a personal computer. It should be noted that high speed of the thread sensors was a key factor in its selection because other types of sensors does not provide the necessary time resolution.

As noted above, the thermal anemometer of constant temperature was used to determine a local heat transfer coefficient α_x. The sensitive element of the thermo-anemometer sensors was a nichrome filament with a diameter of 5 μm and a length of 5 mm. This thread was installed on a fluoroplastic substrate, whose surface was mounted flush with the wall of the pipeline. It should be noted that the filament diameter was less than the average microroughness of the substrate.

Preliminary design evaluation and pilot experiment have shown that it was not possible to provide the implementation of the known techniques for a direct determination of a local heat transfer coefficient using wire sensors in this case. Therefore, in this study the determination of a local heat transfer coefficient was based on indirect calibration based on known empirical relationships, i.e. by the method proposed by S.S. Kutateladze. It is based on the basic indicator of a local heat transfer of a well-studied process. For this case, stationary heat transfer was chosen in a long straight pipe ($l / d \geq 50$) with a circular cross section. The calibration was to relate a calculated heat transfer coefficient α (W/(m^2·K)) for a long straight pipe and a signal value from the thermo-anemometer sensor U (V).

Thus, in this paper the determination of a local heat transfer in a pulsating flow of gases is based on the idea of hydrodynamic analogy of heat transfer (Reynolds analogy). It is based on the assumption of the unity of a transfer processes of momentum and heat in turbulent flow and establishes the quantitative relationship between heat transfer and hydraulic resistance (Stanton criterion).

4. Features of the gas dynamics and local heat transfer in the intake system of the piston engine with supercharging

It has been established that there are significant differences in the patterns of change of air flow rate in piston engines with supercharging and without it. A sharp and rapid increase of the air flow velocity in the intake pipe is observed immediately after the valve opening in an engine with a turbocharger (Figure 2). This is due to the fact that the pressure in the intake pipe is above atmospheric pressure and exceeds the pressure in the engine cylinder at the end of the exhaust.

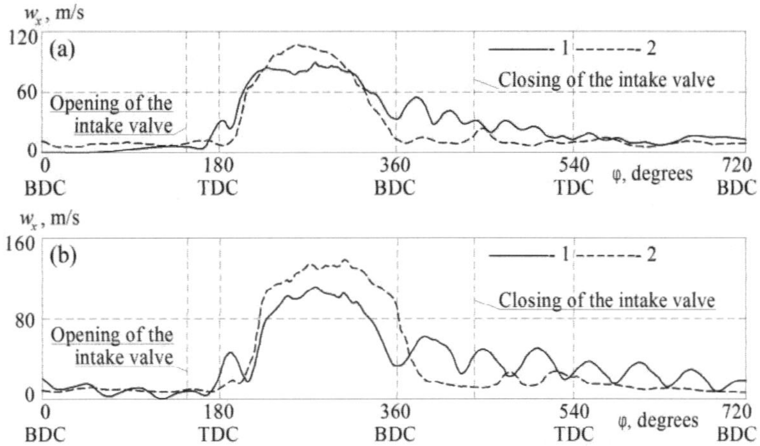

Figure 2. Local (l_x = 140 mm) flow velocity w_x in intake pipe as a function of crankshaft rotation angle φ of piston ICE with (n_{TC}= 35000 rpm) and without turbocharger for the crankshaft rotation speed: (a) n = 1500 rpm; (b) n = 3000 rpm. System configuration: (1) without and (2) with turbocharger.

It has been established that maximum values of air flow rate w_x in the intake pipe in the engine with a supercharger are 15-25 % higher than in the engine without the TC (Figure 2). It has been established that the differences in maximum values of air flow velocity are increased with increasing rotation frequency of the turbocharger rotor. The differences reach 40 % at n_{TC} = 45000 rpm.

It should be noted that the air flow velocity in the intake pipe is not equal to zero after closing the intake valve. Pulsations of the flow velocity with a fairly high amplitude are observed in the pipe. This phenomenon is typical for the intake systems of piston engines with supercharging and without it. See also article [10].

Similar effects are recorded in the graphs of change of the flow pressure in the intake system (Figure 3). Fluctuations in the flow pressure p_x in the intake pipe after closing the valve are observed. The fluctuations are most characteristic of the engine without the TC. Small fluctuations in flow pressure during the entire work cycle are observed in the intake pipe of the engine with supercharging. These fluctuations are the result of the work of the turbocharger blade apparatus [11].

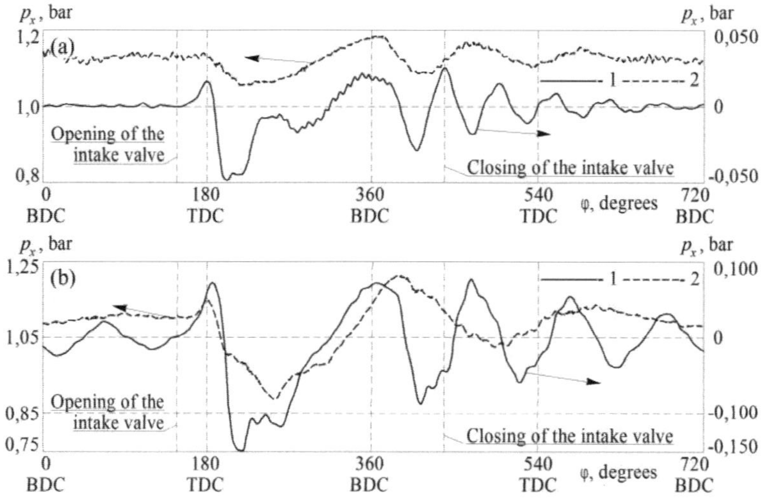

Figure 3. Local (l_x = 140 mm) flow pressure p_x in intake pipe as a function of crankshaft rotation angle φ of piston ICE with (n_{TC}= 35000 rpm) and without turbocharger for the crankshaft rotation speed: (a) n = 1500 rpm; (b) n = 3000 rpm. System configuration: (1) without and (2) with turbocharger.

The maximum and average values of α_x in the intake pipe of a piston internal combustion engine with a turbocharger is significantly increased, compared with values for the engine without TC (Figure 4). This is true for all frequencies of the crankshaft rotation and the turbocharger rotor. For example, the maximum values of local heat transfer coefficient in the intake pipe of the piston ICE with a turbocharger are almost 2 times higher (at engine speed of 600 rpm) compared to α_x in internal combustion engines without TC. The difference is about 75% at n = 3000 rpm. The increase of the local heat transfer coefficient in the intake pipe in the presence of TC can lead to the following:

- heating air from the hot walls of the pipe during the intake process;
- increasing thermal stresses in the parts and components of intake system.

It is shown that the pulsations of the local heat transfer coefficient in the intake pipe of the internal combustion engine with supercharging are observed on all modes of operation. Pulsations of α_x have a large amplitude in the case of installation of the turbocharger than the engine without it. The fluctuations intensity of the local heat transfer coefficient is decreased with increasing rotor speed of the turbocharger n_{TC}. The fluctuations intensity of the α_x is increased with increasing speed of the crankshaft n.

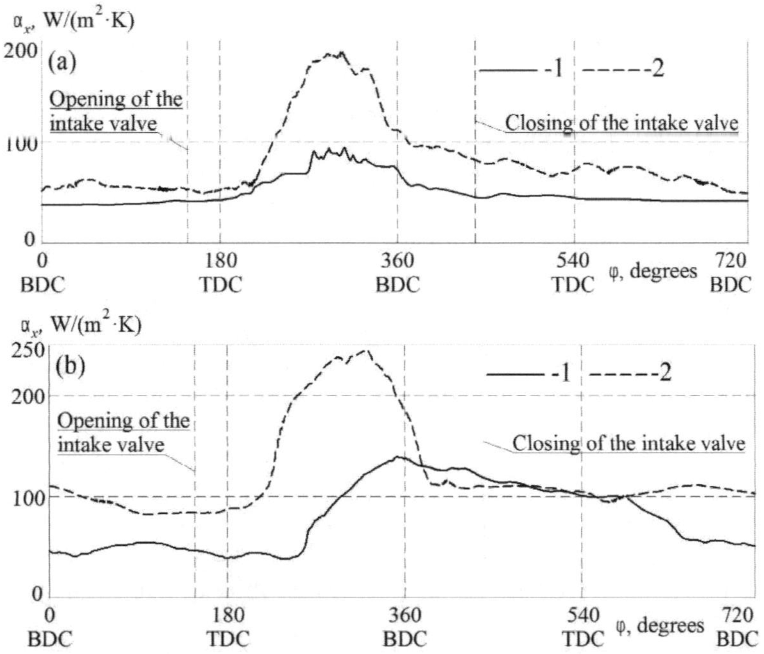

Figure 4. Local (l_x = 150 mm) heat exchange coefficient α_x in intake pipe as a function of crankshaft rotation angle φ of piston ICE with (n_{TC}= 35000 rpm) and without turbocharger for the crankshaft rotation speed: (a) n = 1500 rpm; (b) n = 3000 rpm. System configuration: (1) without and (2) with turbocharger.

Data about the instantaneous local heat transfer in a gas-air systems of piston internal combustion engines with supercharging are necessary for:

- calculations of heating air in the intake process;
- to determine the dynamics of the temperature distribution in the parts and components of intake system;
- calculations of thermal stresses.

5. Conclusions

The main findings of the research are as follows:

1. It has been established that the turbocharger installation in the intake system of a piston internal combustion engine leads to significant differences in the patterns of change in gas-dynamic flow characteristics. Maximum values of air flow velocity in the intake pipe in the engine with a supercharger are 15-50 % higher than in the non-turbo engine; these differences are increased with an increase of the rotor speed of turbocharger;
2. Maximum values of heat transfer coefficient in the intake pipe of a piston engine with a supercharger are increased up to 2 times. The average values of a local heat transfer coefficient in the intake pipe of a piston ICE with supercharging are increased by up to 45 % compared to engines without turbocharger.
3. The obtained data can be used to design intake and exhaust systems when upgrading the existing engine or designing a new internal combustion engine with supercharging.

Acknowledgments

The work was supported by the Russian Foundation for Basic Research (grant No. 16-38-00004).

References

[1] Draganov B H, Kruglov M G and Obuhova V S 1987 *The design of the intake and exhaust ducts of internal combustion engines* (Kiev: Vishcha shkola publ.) p 175 [in Russian]
[2] Vihert M M and Grudskij Yu G 1982 *Construction of the intake systems of high-speed diesel engines* (Moscow: Mashinostroenie publ.) p 151 [in Russian]
[3] Kavtaradze R Z 2008 *The theory of piston engines. A special Chapter* (Moscow: Bauman Press) p 720 [in Russian]
[4] Sharoglazov B A, Farafontov M F and Klement'ev V V 2010 *Internal combustion engines: theory, modeling and calculation processes* (Chelyabinsk: IuUrGU publ.) p 382 [in Russian]
[5] Brodov Yu M, Grigoriev N I, Zhilkin B P, Plotnikov L V and Shestakov D S 2014 *Bul. of the South Ural state Un.* (*Ser. Energy* vol. 14) **2** 5–11 [in Russian]
[6] Plotnikov L V, Zhilkin B P and Brodov Yu M 2015 *Newsletter of Higher Education. Inst.* (*Mechanical Engineering* vol. 669) **12** 35–44 [in Russian]
[7] Zhilkin B P, Lashmanov V V, Plotnikov L V and Shestakov D S 2015 *Process Improvement in gas-air path of piston internal combustion engines* (Ekaterinburg: Ural Press) p 228 [in Russian]
[8] Povh I L 1974 *Aerodynamic experiment in mechanical engineering* (Leningrad: Mashinostroenie publ.) p 480 [in Russian]
[9] Bradshaw P 1971 *An introduction to turbulence and its measurement* (Oxford: Pergamon Press) p 238
[10] Zhilkin B P, Shestakov D S and Plotnikov L V 2012 *Tyazh. Mashinostr.* **2** 48–51 [in Russian]
[11] Shestakov D S, Plotnikov L V, Zhilkin B P and Grigor'ev N I 2013 *Dvigatelestroen.* **1** 24–27 [in Russian]

IOP Conf. Series: Journal of Physics: Conf. Series **899** (2017) 042009 doi:10.1088/1742-6596/899/4/042009

Dynamics of force impulses and bubble oscillations during gas burning on a thrust wall in water

V S Teslenko, A P Drozhzhin and R N Medvedev

Lavrentyev Institute of Hydrodynamics SB RAS, Lavrentyev Prosp. 15, 630090, Novosibirsk, Russia

E-mail: teslenko@hydro.nsc.ru

Abstract. Comparative experiments are formed for measuring the force impulses during combustion of a stoichiometric propane-oxygen mixture on a thrust wall in cylindrical combustors and on an open wall shaped as a flat disk. For identical gas charges, the mean specific impulse generated by pulsed burning of the gas on an open thrust wall is found to be higher than that in the case of gas charge burning in cylindrical barrels.

1. Introduction

A series of experiments with measurements of force impulses and shadowgraphy of bubble oscillations in various submerged combustion chambers was performed for the purpose of developing effective methods of pulsed burning of gases in water for thrust generation [1, 2]. Previous experiments and model calculations with cylindrical combustors showed that the amplitude of the first force impulse depends weakly on the volume of the gas burned in the cylindrical combustor. The amplitudes of the subsequent force impulses arising at the bubble collapse instants depend on the combustor geometry. For cylindrical combustors, the time between the first and second impulses (period of the first oscillation of the bubble in the barrel) depends on the barrel length [3]. As the barrel length increases, the bubble oscillation period also increases. The accompanying edge effects in the form of cavitation vortex rings at the exit of the cylindrical barrel and cavitation inside the barrel [4] lead to dissipative losses of the impulses; therefore, these phenomena should be avoided.

2. Experimental

The arrangement of experiments is illustrated in figure 1. A plastic cylindrical chamber (2) with an inner diameter d, outer diameter D, and height H was placed vertically in a plexiglas cuvette (1) 0.5 m high, which was filled with water. The chamber was filled with a combustible mixture (3) up to a needed level. The mixture was ignited by a spark generated by a high-voltage source (5) with the energy up to 2 J. Burning of predetermined portions of the gas mixture (3) ensured generation of force impulses on a thrust wall (4) with force transfer to a dynamometer (6). The dynamometer was fixed on a horizontal rigid beam (7). The force impulses on the thrust wall were measured by the dynamometer (6), 40 mm in diameter and 15 mm high, which was made of lead zirconate-titanate and was acoustically uncoupled along the axis. The electric signal from the dynamometer passed to an emitter-follower with a time constant $\theta \approx 10$ s and was recorded by a TDS-210 digital oscilloscope (8). Simultaneously with force impulse measurements, the pressure in the combustion chamber (3) was measured by tourmaline sensors (9) with the characteristic resolution of 0.3 μs and the time constant of more than 1 s. The pressure sensor was mounted in the upper end of the chamber on the thrust wall.

All parameters were measured within the error from 5 to 15%. The experiments were performed with gas charges of a stoichiometric propane-oxygen mixture ($C_3H_8 + 5{\cdot}O_2$) with the volume V_g from 1 to 3 ml. Shadowgraphs of the hydrodynamic processes were taken by a MotionXtra HG-LE digital camera (10). The system was controlled by a remote control panel (11), and the measured results were fed to a computer (12).

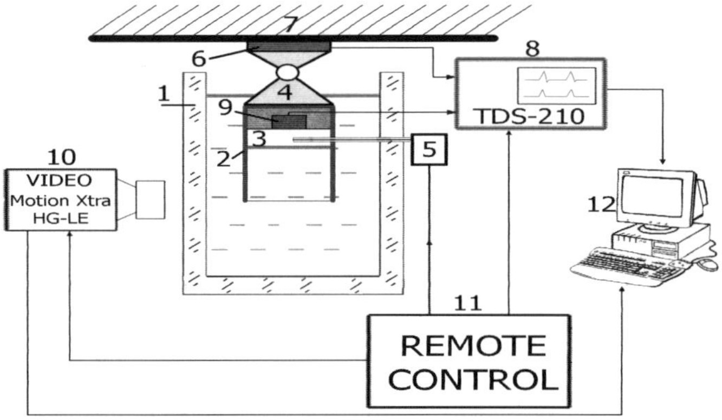

Figure 1. Arrangement of experiments.

3. Experimental results

Figure 2 shows several shadowgraphs, which illustrate the dynamics of the processes in a cylindrical chamber with a diameter d = 29 mm and length H = 91 mm in the course of burning a stoichiometric propane-oxygen mixture with a volume V_g = 3 ml.

Figure 2. Shadowgraphs illustrating the processes in a cylindrical chamber (d = 29 mm and H = 91 mm) for a gas mixture with V_g = 3 ml.

For numerical comparisons of the force impulses (figures 4 and 6), the thrust oscillograms were divided into four intervals (figure 3): 1) the interval (t_0 - t_1) corresponds to generation of the first positive impulse I_1 (first phase) owing to pressure generated by combustion of the gas mixture; 2) the interval ($t_1 - t_2$) corresponds to the negative phase $I_{(-1)}$ in the course of cavity expansion; 3) the interval ($t_2 - t_3$) corresponds to the positive impulse I_2 (second positive phase) due to a collapse of the cavity at the end face of the thrust wall; 4) the interval ($t_3 - t_4$) corresponds to the second negative phase $I_{(-2)}$ due to secondary expansion of the cavity.

Figure 4 shows the force impulse oscillograms (figure 4a) and the relationships of the first four phases (figure 4b) calculated in accordance with the scheme in figure 3 for the case of burning the gas mixture presented in the shadowgraphs (see figure 2).

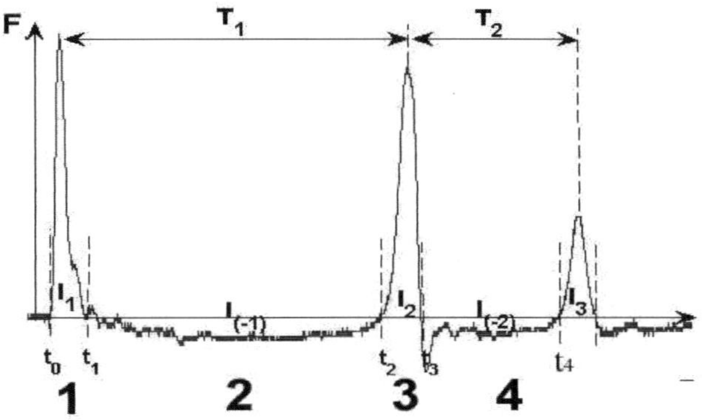

Figure 3. The scheme of dividing the thrust force into time intervals.

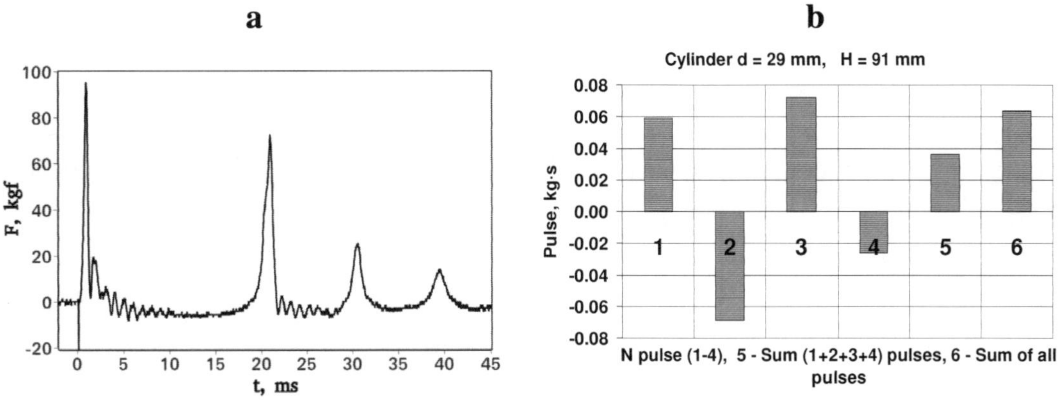

Figure 4. a) the oscillogram of thrust for gas mixture (V_g = 3 ml) combustion in cylindrical chamber (d = 29 mm, H = 91 mm), b) thrust integrals with time for initial 4th phases (1-4), their sum (5) and the full integral (6) over the oscillogram.

The frames illustrating the hydrodynamic processes that occur during combustion of a propane-oxygen mixture with a volume V_g = 2 ml on the open thrust wall shaped as a disk of the diameter D = 98 mm are presented in figure 5. Figure 6 shows the force impulse oscillograms (figure 6a) and relationships of the numerical values of the force impulses (figure 6b) for the case of combustion of a gas charge with a volume V_g = 3 ml.

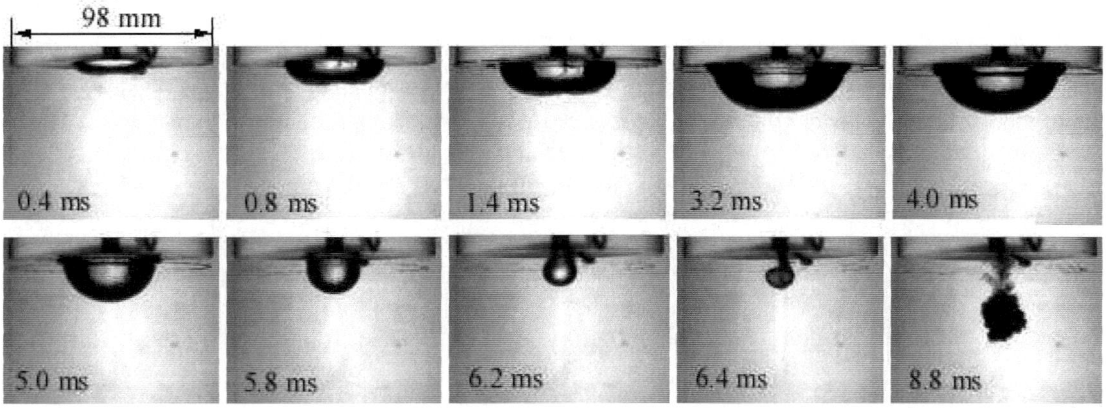

Figure 5. Frames illustrating bubble oscillations on a disk-shaped thrust wall for a gas charge with V_g = 2 ml.

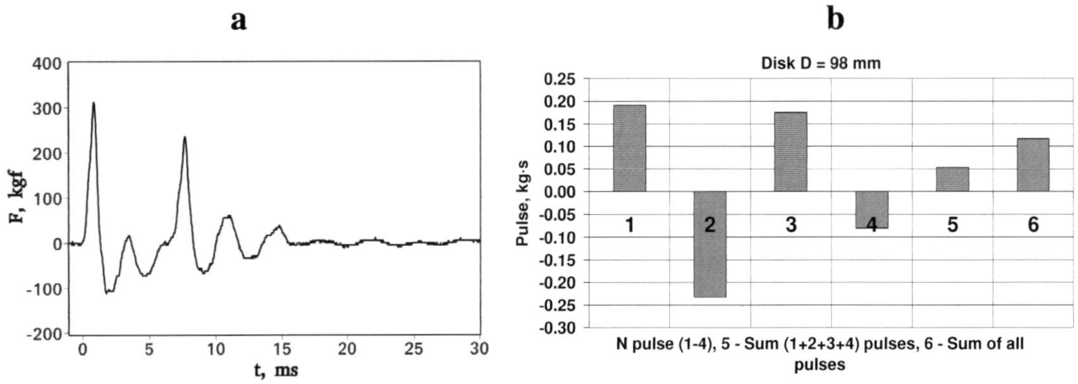

Figure 6. a) the oscillogram of thrust for gas mixture (V_g = 3 ml) combustion at open wall shaped as a disk (D = 98 mm), b) thrust integrals with time for initial 4th phases (1-4), their sum (5) and the full integral (6) over the oscillogram.

Figure 7 shows the experimental dependence of the period of the first oscillation of the bubble (T_1) as a function of the initial volume of the gas charge V_g for the case of gas mixture combustion at flat wall.

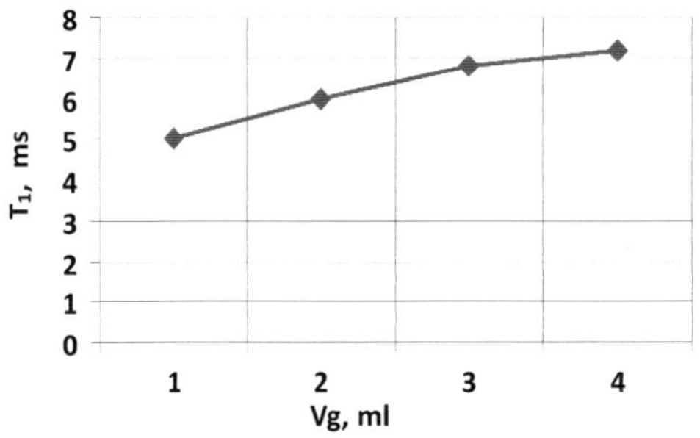

Figure 7. The dependence of the first period of bubble pulsation at flat wall on the volume of initial gas mixture V_g.

4. Analysis of results

It is clearly seen from the presented results that the amplitude values of the thrust (kgf) and force impulses (kg·s) for identical volumes of the gas charge (V_g = 3 ml) are higher in the case of an open thrust wall than in the case of burning the gas charge in a cylindrical barrel.

As the quantitative comparisons of the thrust characteristics were performed for identical gas charges, the results can be compared in terms of the following parameters: ***specific impulse*** and ***averaged specific thrust.***

The specific impulse per one cycle of gas combustion was determined as

$$I_m = \frac{1}{m}\int_0^{T_N} F(t)dt, \tag{1}$$

where $F(t)$ is the force acting on the thrust wall, m is the mass of the charge of the combustible gas mixture, and T_N is the total duration of all phases during one cycle of gas combustion.

As it follows from the oscillograms (see figures 4a and 6a), the amplitudes and duration of the phases depend on the geometry of the gas charge and of the device used for gas charge burning. In our experiments, the calculations were performed for all phases (positive and negative) with summation of the first four values (5 in figures 4b and 6b) and with integration with respect to time of the entire thrust oscillogram during one cycle of gas charge combustion (values 6 in figures 4b and 6b).

As the amplitude values of subsequent force impulses (after the second one) drastically decrease, the thrust parameters were compared by two methods to ensure meaningful comparisons: a) based on the sum of the first four intervals (values 5 in figures 4b and 6b); b) based on complete integration of the entire oscillogram (up to 45 ms, values 6 in figures 4b and 6b). The experiments allowed one the following results to be obtained.

1a) In the case of gas charge combustion in a cylindrical chamber (H = 91 mm), the ***specific impulse*** I_m after summation of the first four intervals was from $7.7 \cdot 10^3$ to $1.1 \cdot 10^4$ s.

1b) In the case of gas charge combustion in a cylindrical chamber (H = 91 mm), the ***specific impulse*** I_m after complete integration was from $1.4 \cdot 10^4$ to $1.9 \cdot 10^4$ s.

2a) In the case of gas charge combustion on a disk, the ***specific impulse*** I_m after summation of the first four intervals was from $1.1 \cdot 10^4$ to $1.4 \cdot 10^4$ s.

2b) In the case of gas charge combustion on a disk, the ***specific impulse*** I_m after complete integration was from $2.5 \cdot 10^4$ to $5.7 \cdot 10^4$ s.

The ***averaged specific thrust*** was determined as

$$\langle I_N \rangle = \frac{1}{mT_N} \int_0^{T_N} F(t)dt, \qquad (2)$$

where $F(t)$ is the total force acting on the thrust wall in the motion axis direction and T_N is the duration of all positive and negative phases within one cycle of gas charge combustion.

Based on this definition, we find the averaged values of the specific thrust for the first four intervals.

1) In the case of combustion of the gas charge with V_g = 3 ml in a cylindrical barrel, the duration of the first four intervals was T_N = 33 ms; therefore, the averaged specific thrust $I_N = I_m / T_N$ was from $0.23 \cdot 10^6$ to $0.33 \cdot 10^6$.

2) In the case of combustion of the gas discharge with V_g = 3 ml on a disk, the duration of the first four intervals was T_N = 10 ms; therefore, the averaged specific thrust $I_N = I_m / T_N$ was from $1.1 \cdot 10^6$ to $1.4 \cdot 10^6$.

The results obtained in this study show that gas combustion on an open thrust wall is more effective than in cylindrical chambers in terms of both the specific impulse and the averaged specific thrust. The parameter of the averaged specific thrust shows the maximum possible frequency of fuel burning without the loss of the specific impulse. It is seen that this parameter is higher in the case of gas charge combustion on a disk owing to significant reduction of the period of bubble oscillations. This is similar to an increase in the averaged power of the internal combustion engine due to an increase in the number of revolutions of the crank shaft.

5. Conclusions

The experimental results show that the specific impulse in the case of gas combustion on a disk is at least not lower and sometimes is even higher than the specific impulse of gas combustion in a cylindrical chamber. Possible reasons are the absence of friction-induced losses and the edge effects at the barrel edges [2, 3].

The present results suggest that it is possible to burn gases in water on a flat wall with frequencies higher than the frequency of gas charge combustion in cylindrical chambers, other conditions being identical. The maximum possible frequency of gas combustion on a flat wall ν_{max} is determined as $\nu_{max}=1/T$, where T is the sum of the period of hemispherical bubble oscillations and the time necessary for the next gas charge to form.

An increase in the combustion frequency with retaining the specific impulse value, which is provided by gas combustion on a flat wall as compared to a cylindrical chamber, makes it possible to

IOP Conf. Series: Journal of Physics: Conf. Series **899** (2017) 042009 doi:10.1088/1742-6596/899/4/042009

increase the averaged thrust of underwater propulsion and to decrease losses due to cavitation and friction.

References

[1] Teslenko V S, Drozhzhin A P, Medvedev R N and Batraev I S 2014 In-water gas combustion in linear and annular gas bubbles *Thermophysics and Aeromechanics* **21**(4) 479–88

[2] Teslenko V S, Drozhzhin A P and Medvedev R N 2017 In-water gas combustion for thrust production *Thermophysics and Aeromechanics* **24**(4) (In press)

[3] Medvedev R N, Drozhzhin A P and Teslenko V S 2015 Threshold of cavitation vortex ring formation during liquid outflow from a submerged tube *Modern Science: Researches, Ideas, Results, Technologies* **16**(1) 86–9

[4] Medvedev R N, Drozhzhin A P and Teslenko V S 2016 Thrust generation by pulse combustion of gas in a submerged chamber *International Journal of Multiphase Flow* **83** 232–8

IOP Conf. Series: Journal of Physics: Conf. Series **899** (2017) 042010 doi:10.1088/1742-6596/899/4/042010

Numerical investigation of flow structures with an oblique detonation wave in a hypersonic annular cylindrical chamber

Anatoliy V Trotsyuk

Lavrentyev Institute of Hydrodynamics, Siberian Branch of the Russian Academy of Science, Novosibirsk 630090, Russia

E-mail: trotsyuk@hydro.nsc.ru

Abstract. A new supersonic flow-type annular detonation combustor is designed in which steady oblique detonation waves in the channel are generated using a compression body in the form of a solid single-wound spiral with a constant pitch angle. A two-dimensional unsteady mathematical model of the reacting flow in this device is formulated. The flow dynamics at the start of the chamber operation and steady supersonic flow structures for a stoichiometric hydrogen-air flow with an inlet Mach number M_0=5 are numerically investigated. Two-dimensional numerical simulation is carried out for different spiral angles and geometrical dimensions of the chamber. A bifurcation of steady flow structures with respect to the initial condition of the problem is observed.

1. Introduction

The detonation combustion of mixtures in supersonic flows is currently considered to be a more effective method of obtaining the release of the chemical energy of fuel in scramjet engines than the widely used method based on turbulent diffusion combustion. At present, there are two different concepts of flow-type detonation chambers (DC). In the first concept, unsteady combustion of the mixture occurs in rotating transverse detonation waves (TDWs) in an annular cylindrical chamber [1, 2]. The second version uses steady oblique detonation waves (ODW) formed over the surface of the compression body [3]. In the simplest case of this concept, the DC is a rectangular channel in which at least one of the walls is a flat wedge. Each of these DC concepts has its own advantages and disadvantages.

In this paper, we propose a new design of a flow-type DC, which, in a sense, is a combination of the DC geometries described above. The chamber is an annular channel located between inner and outer cylindrical walls. The compression surface is an infinitely thin, solid, single-wound spiral that covers the entire cross-section of the annular channel from the inner to the outer wall of the DC, see Figure 1. The spiral has a constant pitch angle α with respect to the axis of the chamber. The width of the annular channel is considered to be much smaller than the radius of the chamber, so that the gas flow can be considered to be uniform along the radius of the DC. This makes it possible to model the flow in the DC in a two-dimensional formulation in the same manner as is usually done for a simulation of rotating DWs in annular chambers [1,2]. At the DC inlet, there is a hypersonic flow (M_0 > M_{CJ}) of a uniformly mixed fuel-air mixture. Here M_0 is the freestream Mach number and M_{CJ} is the Mach number of the Chapman-Jouguet DW.

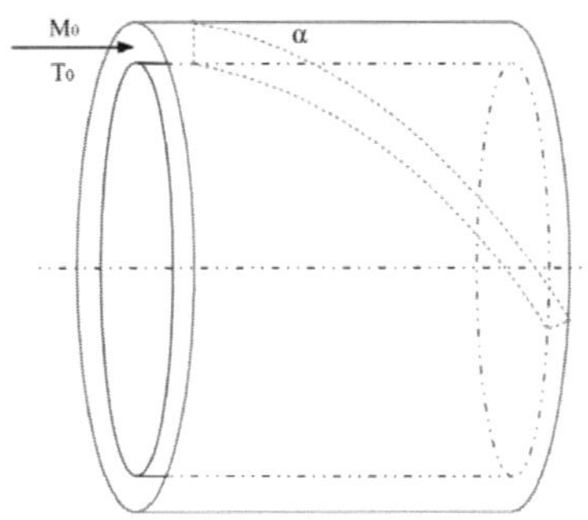

Figure 1. Circuit design of the annular cylindrical chamber with single solid spiral.

This paper presents a numerical study of the possible flow regimes with a steady ODW in the proposed annular DC with varying geometrical dimensions of the DC, spiral pitch angle α, and initial flow parameters. Two-dimensional modeling makes it possible to determine the main structures constituting the reacting flow in the DC considered, the degree of steadiness of the flow, the mechanism of interaction of these structures with each other, and their variations with changes in the above-mentioned parameters of the problem.

2. Governing equations and model of chemical kinetics

The dynamics of the compressible chemically reactive medium is described by the two-dimensional Euler equations.

The chemical reaction in the DW is described according to the two-stage model of the detonation kinetics (induction period and main heat release stage) first proposed in [4]. The duration of the induction period is determined by known empirical formula [5] for hydrogen mixtures. The main heat release take place after the induction period. The changes in internal energy and all thermodynamic parameters of the gas are described at this stage by the equations of the approximate kinetic model [6, 7]. The model is highly accurate and consistent with the second law of thermodynamics. The constants of the model have a clear physical meaning and are calculated from the tabulated thermochemical parameters of the mixture before the two-dimensional numerical calculation. This two-stage kinetic model has been used for numerical simulation of the two-dimensional cellular DW structure in hydrogen–oxygen mixtures in our previous study [8]. A good agreement has been obtained between the numerical results and experimental data on the size a_0 over a wide range of initial pressures and degrees of dilution with argon.

The system of governing equations was closed by the well-known thermal equation of state for an ideal gas.

3. Numerical method

The resultant systems of equations were solved numerically using the code based on the Godynov-type finite-volume scheme [9] with the fourth-order MUSCL TVD reconstruction [10] and the advanced HLLC algorithm [11] for an approximate solution of the Riemann problem. In implementation of HLLC algorithm for the case of a chemically reacting mixture, the "energy relaxation method" [12] was used. This method eliminates the problem of numerical solution of the Riemann problem for a medium with a complicated nonlinear equation of state (including that with a variable ratio of specific heats). Integration in time was performed with second-order accuracy by using additive semi-implicit Runge-Kutta methods [13]. The time step was determined at each time layer of the solution from the stability condition [9]. In the present simulations, the values of the Courant number were CFL=0.3–0.4. A uniform fixed grid was used in both directions.

At the inlet boundary x=0 of the solution domain, incoming unperturbed supersonic flow parameters were specified. In the external peripheral grid cells at the outlet boundary x=L, low counterpressure for the guaranteed critical gas flow from the DC outlet into the surrounding virtual space were specified. At the upper and lower boundaries, periodic boundary conditions were defined at $x \leq L_1$ and the conditions of an impervious solid wall at $x > L_1$.

The code is parallelized with MPI library using the domain decomposition technique.

4. Results and Discussion

Numerical simulation was performed for a stoichiometric hydrogen-air mixture. At the DC inlet (x=0), the flow of the mixture was assumed to be uniform at a static pressure p_0=0.166 atm and a static temperature T_0=850 K, and the values of α and M_0 were varied. For these initial parameters of the mixtures, the Mach number M_{CJ}=2.73. The study was started with a strongly overdriven DW as a simpler type of DW with a smooth front, since the front in this case does not have a cellular structure. We used a sufficiently large Mach number M_0=5 and a large spiral angle α=30°. The length of the DC was chosen to be L=1.5 cm during preliminary calculations, and the height H of the two-dimensional

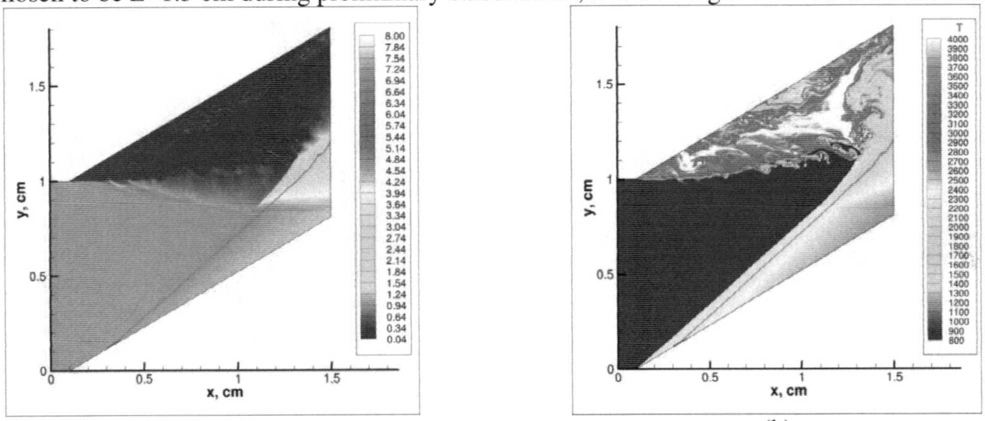

(a) (b)

Figure 2. Stationary structure of flow field in a detonation chamber for H=1 cm and α=30°: (a) normalized density; (b) temperature (K). Start from a uniform supersonic flow.

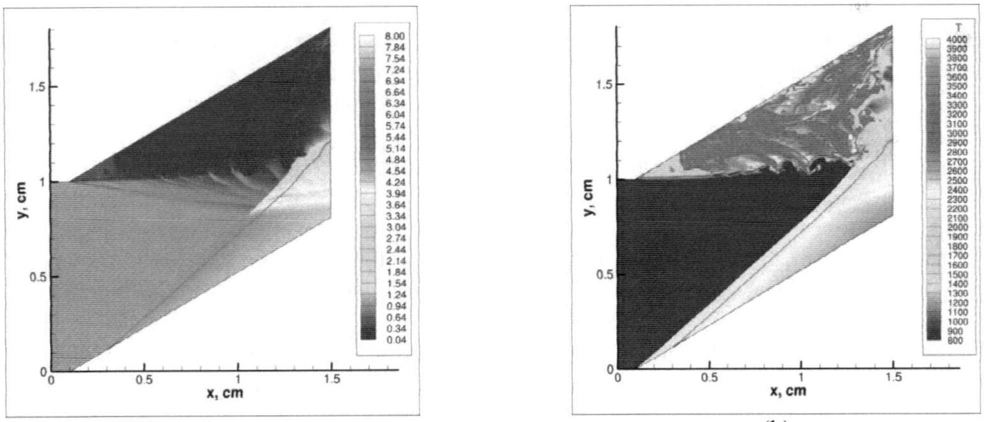

(a) (b)

Figure 3. Stationary structure of flow field in a detonation chamber for H=1 cm and α=30°: (a) normalized density; (b) temperature (K). Start from a quiescent mixture.

solution domain was varied; $H=2\cdot\pi\cdot r_c$, where r_c is the radius of the annular layer of the mixture. The beginning of the spiral was set at $x=L_1$=0.1 cm.

Figures 2 and 3 show the normalized density and temperature fields of the mixture for H=1 cm. As can be seen from the figures, the main flow components in the DC are an overdriven oblique DW, formed over the spiral compression surface, and a rarefaction region formed on the back side of the spiral (i.e., behind the rarefaction surface). The crimson solid line in the figures shows the boundary of the induction zone, behind which a heat release zone begins.

It is seen from Figs. 2 and 3 that for the given value of H, there is a strong interaction between the ODW and the flow region behind the rarefaction surface. Investigation of the flow dynamics shows that the emerging oblique DW interacting with the classical two-dimensional rarefaction wave (RW) paradoxically initiates induction combustion in the latter. That is, detonation combustion is initiated in a flow zone with decreasing gas parameters (temperature, density, etc.). A particularly interesting finding in this study is the existence of two steady-state (on average) solutions depending on the initial conditions of the problem, which occur under the same boundary conditions, i.e., a bifurcation effect with respect to the initial solution vector. Figure 1 shows the obtained steady flow at the start (t=0) from a uniform flow of the mixture $\mathbf{u}=(M_0 \cdot c_0, 0)$, were **u** is the two-dimensional gas particle velocity vector and c_0 is the initial speed of sound. Physically, this start corresponds to the instantaneous placement of the compression body under study in the initial uniform supersonic flow. At subsequent

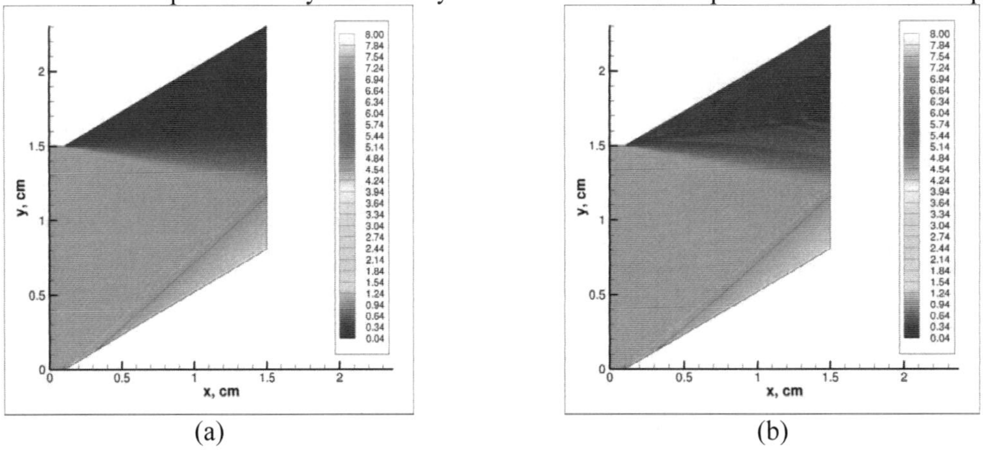

Figure 4. Stationary structure of flow field in a detonation chamber for H=1.5 cm and α=30°, normalized density. Start from: (a) a uniform supersonic flow; (b) a quiescent mixture.

times, an unsteady shock- wave pattern is formed around the investigated body in the solution domain; the final configuration is shown in Figure 2. When starting from a quiescent mixture, $\mathbf{u}=(0,0)$, steady flow occurs shown on Fig. 3. The start from this state corresponds to the removal of the virtual wall at the left boundary x=0 and the propagation of a plane shock wave (SW) from left to right in the solution domain and its interaction with the compression body.

It can be seen that the obtained steady-state solutions differ greatly from each other in the size, shape, and structure of the flow and the values of the RW parameters. In the case of Fig. 2, the first characteristic of the two-dimensional RW is almost immediately followed by an oblique shock wave with a jump in the gas parameters which then decrease toward the DC outlet. Penetration of this jump into the detonation products behind the ODW causes a jump in the parameters in the energy release region $1 \leq x \leq 1.5$ cm (see Fig. 1). When starting from the state of rest, we have a 2D rarefaction wave behind the spiral surface, which is perturbed by compression waves, and the gas parameters in this flow region continuously (though not monotonically) decrease toward the DC outlet. In the detonation products behind the ODW, we have a region of rapidly decreasing flow parameters (see Fig. 3).
There is also a significant difference in gas temperature in the reverse flow region, cf. Figs. 2 and 3 at $y \approx 1.3$. The more uniform and low-temperature region behind the rarefaction surface is obtained in the case of starting from $\mathbf{u}=(0,0)$. Before the interaction with the RW, the flow in the ODW has no differences.

Thus, the flow region behind the rarefaction surface of the spiral has a complex structure with unsteady turbulent shear layers, reverse flow zones, and combustion regions. These flow elements are essentially unsteady, but, on the average over time, this region is fairly stable in terms of its shape and the average values of the gas parameters in it. The differences in the obtained steady-state (on average) solutions are due to the difference between the SW and the RW formed during the transient process, which interact differently with each other at the start from different initial conditions. Investigation of the time stability of the obtained two solutions and the dependence of the 2D modeling results on the difference grid resolutions used shows that these structures are steady (on average) and do not transform into one another or degenerate into any third similar flow configuration.

The flow structure is found to change suddenly as the parameter H is changed. Figure 4 gives the steady field of the normalized flow density for H=1.5 cm and the same values of the other parameters as previously. Figure 4a corresponds to the start from a uniform flow, and Fig. 4b to the start from the state of rest. It is seen that, in both cases, there is no interaction between the RW and the smooth oblique DW. In the first case, we have a classical smooth two-dimensional RW behind the back side of the spiral and combustion is completely absent. In the case of starting from **u**=(0,0), the RW has small perturbations and a small stagnation zone with detonation products remaining from the preceding

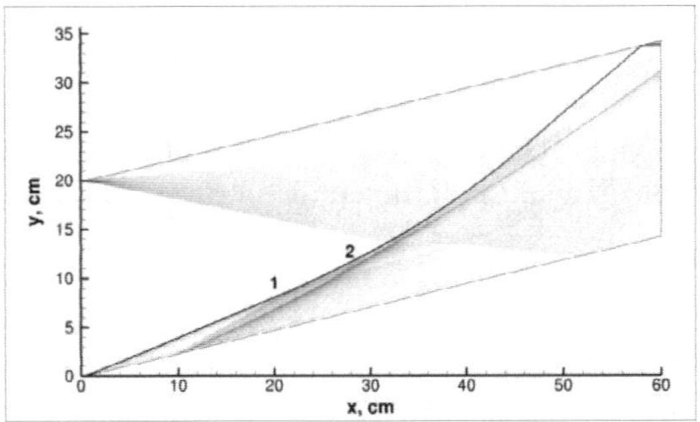

Figure 5. Stationary structure of flow field in a detonation chamber for H=20 cm.

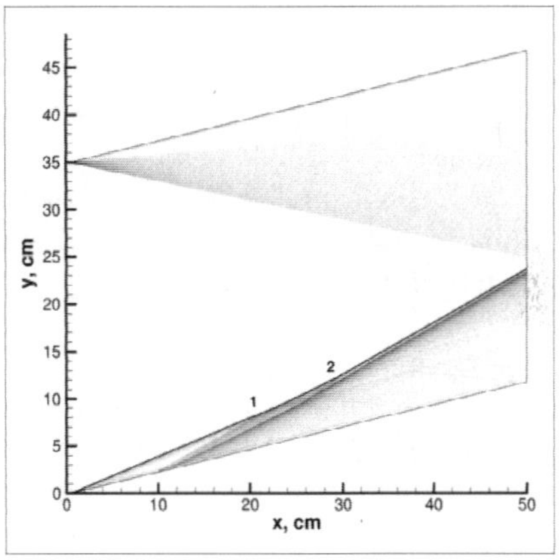

Figure 6. Stationary structure of flow field in a detonation chamber for H=35.

nonstationary transient process. It can be said that with decreasing interaction between the flow components, the bifurcation effect on the initial solution vector is extremely weak.

Next, the flow structure was studied at smaller pitch angles of the spiral. The dimensions of the solution domain were considerably increased due to a decrease in the temperature of the gas in the induction zone with decreasing angle α and due to the exponential dependence of the length of the induction zone on temperature. Figures 5 and 6 show the numerical Schlieren visualization of the obtained steady flow fields for α=13.4° and M_0=5. The calculation for H=20 cm and L=60 cm is presented in Fig. 5, and that for H= 35 cm and L=50 cm in Fig. 6; L_1=0.5 cm in both cases. In these figures, we have the classical λ-structure of the formation of an oblique DW previously observed in the calculation of DWs on a wedge and cone [3]. Digits 1 and 2 denote two jumps in the induction zone, which, sequentially interact with the leading shock front to form a steady oblique DW over the compression surface of the annular spiral. In both cases, we have a classical two-dimensional RW behind the rarefaction surface of the spiral and there are no perturbations or combustion zones in the RW. As can be seen in Fig. 5, the geometry of the solution domain allows a strong interaction between the ODW and the RW. However, careful calculations show that, for this geometry, there is no bifurcation of the steady-state solution for the initial state. Thus, the bifurcation effect can be completely eliminated by decreasing the spiral angle and hence the intensity of the oblique SW and the RW during the transient process. In the case of Fig. 6, the geometry does not allow interaction between the flow components, and we have an isolated ODW and an unperturbed two-dimensional RW. In this case, the bifurcation effect was also not observed.

In the case of Fig. 5, we have an almost ideal situation in which the geometrical parameters of the proposed annular DC and the spiral angle are selected so that for an inlet flow Mach number M_0=5, almost all of the reacting mixture supplied to the DC is burned in the ODW.

5. Concluding Remarks

A two-dimensional unsteady mathematical model was formulated for the flow in a supersonic flow-type annular detonation chamber with a compression body in the form of a solid single-wound spiral with a constant pitch angle α with respect to the axis of the chamber.

The flow dynamics at the start of the chamber was studied numerically, and steady-state solutions were obtained for the supersonic flow of a stoichiometric hydrogen-air mixture with an inlet Mach number M_0=5 in the chamber. The angle of the spiral and the geometrical dimensions of the chamber were varied.

It is found that the main components of the steady flow in the chamber are an oblique overdriven detonation wave and a two-dimensional rarefaction wave. For $\alpha=30^{\circ}$, a bifurcation of the solution with respect to the initial condition problem was observed. The existence of this effect was shown to depend on both the intensity of these two main flow components and the geometry of their interaction with each other.

With decreasing spiral angle, an oblique detonation with the classical λ-structure was obtained. The calculations for $\alpha=13.4^{\circ}$, H=20 cm, and L=60 cm resulted in a quite efficient detonation combustor configuration of the proposed design.

Acknowledgements

The authors gratefully acknowledge the financial contribution from the Russian Foundation for Basic Research (Grant No. 17-03-01351). The computations were performed on an MVS-100K supercomputer at the Joint SuperComputer Center (JSCC) of the Russian Academy of Sciences (Moscow).

References

[1] Zhdan S A Bykovskii F A and Vedernikov E F 2007 Mathematical modeling of a rotating detonation wave in a hydrogen-oxygen mixture *Combust. Explo. Shock Waves* **43** 449–59

[2] Bykovskii F A and Zhdan S A 2013 *Continious Spin Detonation* (Novosibirsk: Lavrentyev Institute of Hydrodynamics Press) 422 p in Russian

[3] Berlyand A T, Vlasenko V V and Svishchev S V 2001 Stationary and Nonstationary Wave Structures that Arise in Stabilization of Detonation Over a Compression Surface *Combust. Explo. Shock Waves* **37** 82–98

[4] Korobeinikov V P, Levin V A, Markov V V and Chernyi G G 1972 Propagation of blast waves in a combustible gas *Acta Astronaut.* **17** 529–37

[5] White D R 1967 Density induction times in very lean mixtures of D_2, H_2, C_2H_2, and C_2H_4 with O_2 *Proc. 11th Symposium (Int.) on Combustion* The Combustion Institute (Pittsburgh: Academic Press) 147–54

[6] Nikolaev Yu A and Fomin P A 1982 Analysis of equilibrium flows of chemically reacting gases *Combust. Explo. Shock Waves* **18** 53–8

[7] Nikolaev Yu A and Zak D V 1988 Agreement of models of chemical reactions in gases with the second law of thermodynamics *Combust. Explo. Shock Waves* **24** 461–4

[8] Trotsyuk A V 1999 Numerical simulation of the structure of two-dimensional gaseous detonation of an H_2–O_2–Ar mixture *Combust. Explo. Shock Waves* **35** 549–58

[9] Godynov S K 1976 *Numerical Solution of the Multi-Dimensional Problems of Gaseous Dynamics* (Moscow: Nauka Press) 400 p in Russian

[10] Yamamoto S and Daiguji H 1993 Higher-order-accurate upwind schemes for solving the compressible Euler and Navier-Stokes equations *Comput. Fluids* **22** 259–70

[11] Batten P, Leschziner M A and Goldberg U C 1997 Average-state Jacobians and implicit

IOP Conf. Series: Journal of Physics: Conf. Series **899** (2017) 042010 doi:10.1088/1742-6596/899/4/042010

methods for compressible viscous and turbulent flows. *J. Comput. Phys.* **137** 38−78
[12] Coquel F and Perthame B. 1998 Relaxation of energy and approximate Riemann solvers for general pressure laws in fluid dynamics *SIAM J. Numer. Anal.* **35** 2223−49
[13] Shen J W and Zhong X 1996 Semi-implicit Runge-Kutta schemes for non-autonomous differential equations in reactive flow computations *AIAA 27th Fluid Dynamics Conf., New Orleans, LA, June 17-20, 1996* AIAA Paper No.96−1969

IOP Conf. Series: Journal of Physics: Conf. Series **899** (2017) 042011 doi:10.1088/1742-6596/899/4/042011

Experimental and theoretical research of the interaction between high-strength supercavitation impactors and monolithic barriers in water

A N Ishchenko, S A Afanas'eva, V V Burkin, A S Diachkovskii, A I Zykova, M V Khabibullin, A V Chupashev , N T Yugov

National Research Tomsk State University, Tomsk, Russia

E-mail: Lex_okha@mail.ru

Abstract. The article describes experimental and theoretical research of the interaction between supercavitating impactors and underwater aluminum alloy and steel barriers. Strong alloys are used for making impactors. An experimental research technique based on a high-velocity hydro-ballistic complex was developed. Mathematical simulation of the collision the impactor and barrier is based on the continuum mechanics inclusive of the deformation and destruction of interacting bodies. Calculated and experimental data on the ultimate penetration thickness of barriers made of aluminum alloy D16T and steel for the developed supercavitating impactor are obtained.

Overcoming of distance in water with high-velocity body penetration and efficient overcoming of underwater barriers at depth is a complex technical task. A solution of this tack requires from researchers to develop new approaches. The aim of this work is to develop an experimental and calculation methodology for predicting the results of the supercavitating impactors (SI) interaction with underwater barriers, in particular their ultimate penetration thickness.

Experiments on the investigation of SI interaction with underwater barriers were carried out on the hydro-ballistic complex [1]. Preliminary studies of SI motion in water at velocities of more than 1000 m/s have shown that steel SI do not withstand loads and are destroyed when moving in water [2,3]. To solve this problem, it is necessary to use more durable materials, for example tungsten alloys, capable of withstanding loads when moving in water in the velocity range 1000 ... 1500 m/s [4].

Figure 1 shows the appearance of the SI sample developed for this purpose from the alloy tungsten + nickel + ferrum (WNiFe).

Figure 1. Appearance of the SI from WNiFe

The plates made of aluminum alloy D16T and steel were used as underwater barriers.

Figures 2 and 3 show the results of SI interaction with the barrier made of aluminum alloy D16T 40 mm thick after passing in water 3.2 m at the water entry velocity $V = 1098$ m/s. The external state of the inlet and outlet barrier bores suggests that the SI interaction with the barrier occurred at a right

IOP Conf. Series: Journal of Physics: Conf. Series **899** (2017) 042011 doi:10.1088/1742-6596/899/4/042011

angle, which characterizes the stable water motion regime. The velocity of SI collision with the barrier is V_0 = 1022 m/s. The residual SI velocity after penetration of the barrier is 913 m/s. The high magnitude of the after-barrier velocity indicates an even greater supply of the impactor kinetic energy.

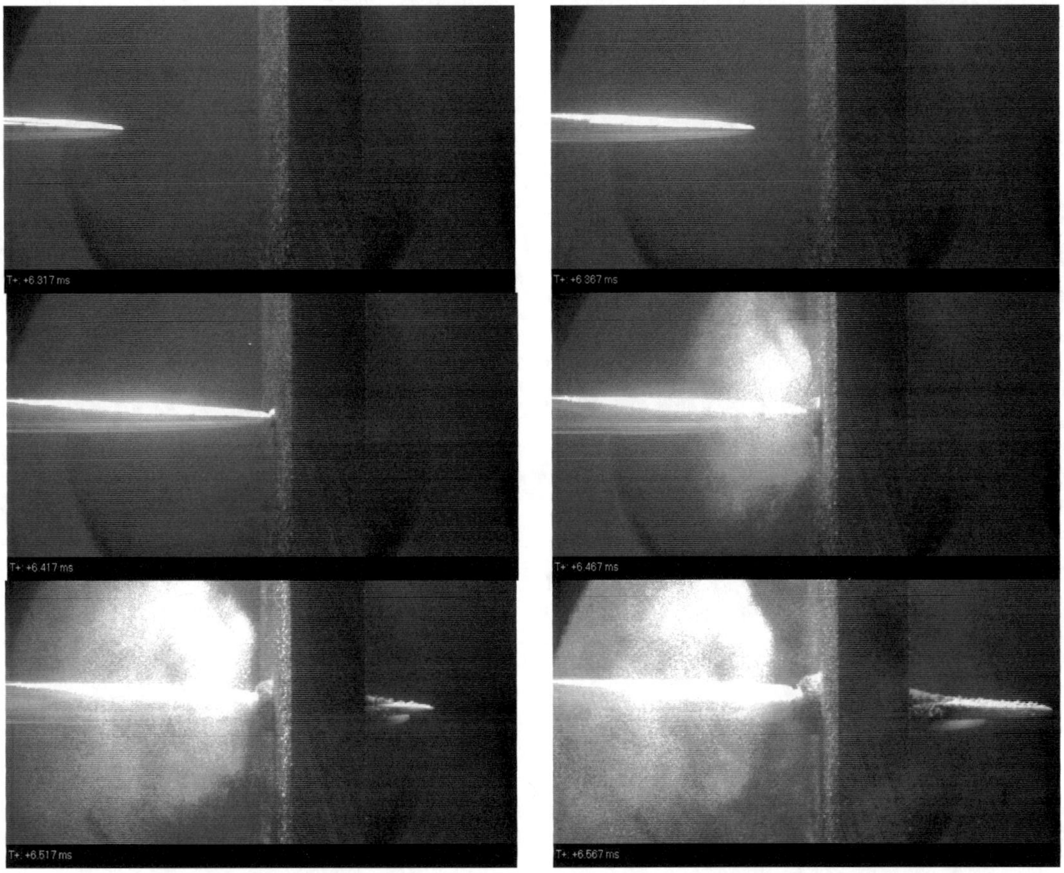

Figure 2. Photographic record of SI interaction with barrier made of alloy D16T thick 40 mm at V_0 = 1022 m/s

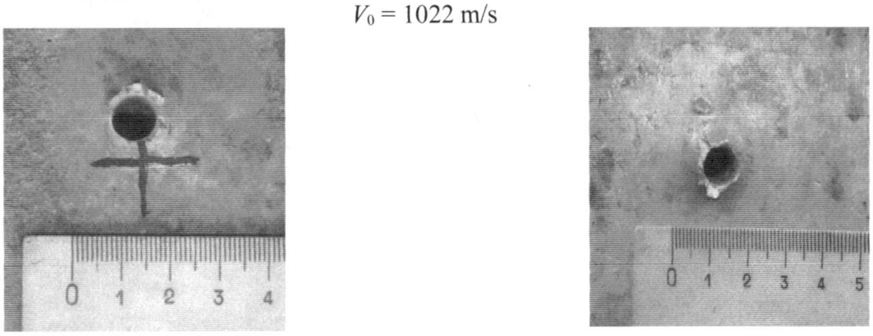

a *b*

Figure 3. Result of SI collision with barrier made of alloy D16T thick 40 mm at V_0 = 1022 m/s: *a* – inlet bores; *b* – outlet bores.

An analysis of the SI penetration interaction with the barrier is carried out using a mathematical model describing the collision process in the continuum mechanics framework [5] in an axisymmetric statement by the large particles numerical method [6] and a complete dimensioned formulation by a numerical finite element method [7, 8].

Figure 4 shows the results of mathematical modeling under conditions that repeat the experiment described above. The interaction process lasts about 150 microseconds. The impactor hole through the barrier. In this case, it deforms at the head, acquiring a shape close to cylindrical. In the penetration process it loses velocity to 858 m/s. The discrepancy of after-barrier velocity with the experiment was 6 %.

Figure 4. Chronogram of SI collision with barrier made of alloy D16T thick 40 mm at V_0 = 1022 m/s

By mathematical modeling the limiting penetration thickness of the barrier from the alloy D16T is determined. This thickness for the SI at the impact velocity V_0 = 1022 m/s is 83 mm (Figure 5). When interacting with the barrier, the SI is triggered and the impactor velocity is total lost.

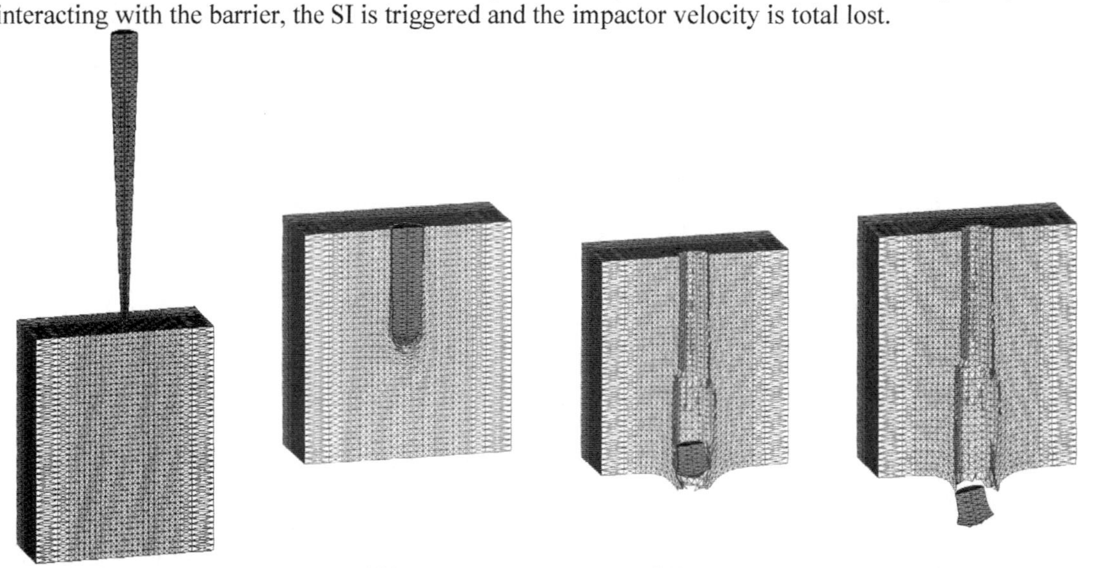

Figure 5. Chronogram of SI collision with barrier made of alloy D16T thick 83 mm at V_0 = 1022 m/s

Figures 6 and 7 show the experiment results where the water entry SI velocity is $V = 1131$ m/s and SI interaction with the compound barrier. The first barrier is a steel plate 45.4 mm thick; the second barrier is plate of D16T alloy 40 mm thick as "marker" at a distance of 195 mm from the first. After passing through the water at 3.2 m, the SI velocity when touching the barrier was $V_0 = 1054$ m/s. The SI breaks the first barrier through and leaves a small depression which is the "marker". The movement of the spall steel barrier fragment is also recorded.

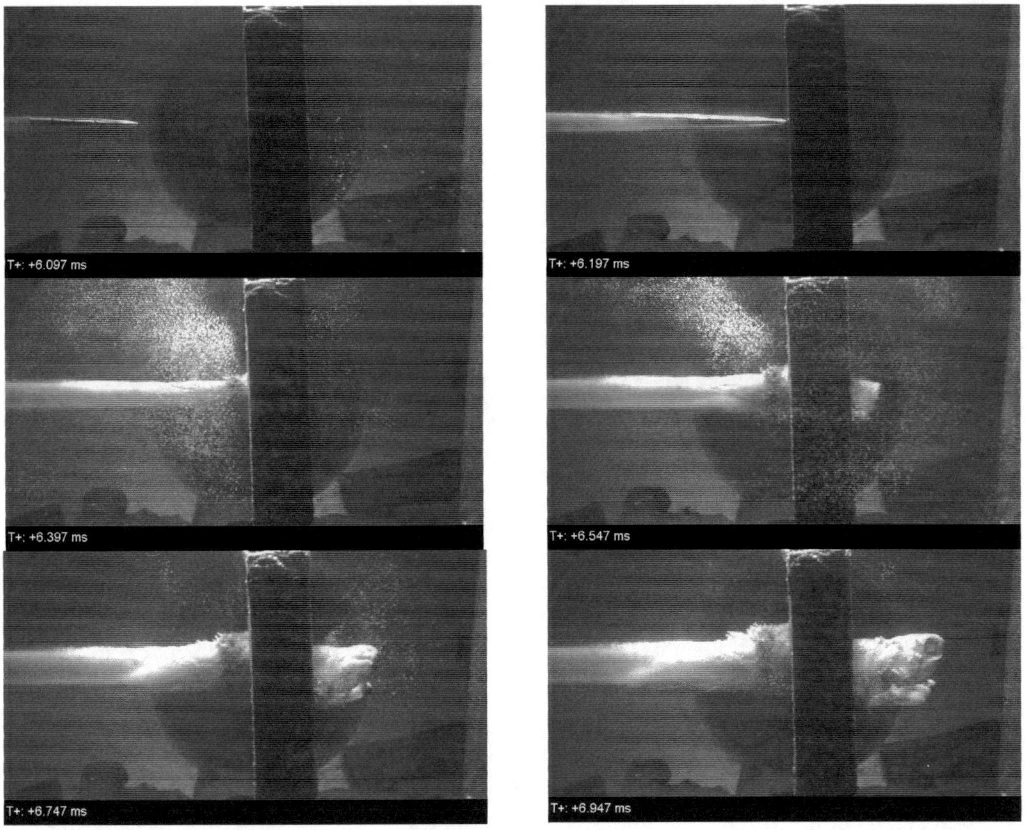

Figure 6. Photographic record of SI interaction with steel barrier 45 mm thick at $V_0 = 1054$ m/s

a *b* *c*

Figure 7. Result of SI collision with steel barrier 45 mm thick at $V_0 = 1054$ m/s: *a* – inlet bores; *b* – outlet bores, c – crater on the barrier «marker» face.

Figures 8 and 9 show the mathematical simulation results of the collision the SI and the compound barrier under repeating experiment conditions in the cylindrical coordinate system z (cm), r (cm) as the distribution of the mass velocity vector related to the mass center velocity of the projectile V_c (in the left half-plane), pressure distribution (GPa) (in the right half-plane) as a function of time t.

IOP Conf. Series: Journal of Physics: Conf. Series **899** (2017) 042011 doi:10.1088/1742-6596/899/4/042011

The steel barrier penetration process lasts about 200 microseconds. Firstly, the impactor pierces the barrier and further deforms at the head, acquiring a compact shape. When moving to the barrier rear side, impactor pushes ahead a steel barrier fragment as called - a "cork". A burn through was formed in the barrier. The impactor rest weighing 57 grams and the steel "cork" interact with the "marker" barrier at a speed of 201 m/s lasts about 150 μs until their full stop in the barrier at insignificant depth.

The analysis results show that the steel plate 45.4 mm thick is broke through by the SI almost at the limit of through penetration. It has been experimentally established that when the impact velocity of the SI is reduced to 894 m / s, penetration of the same barrier is not observed.

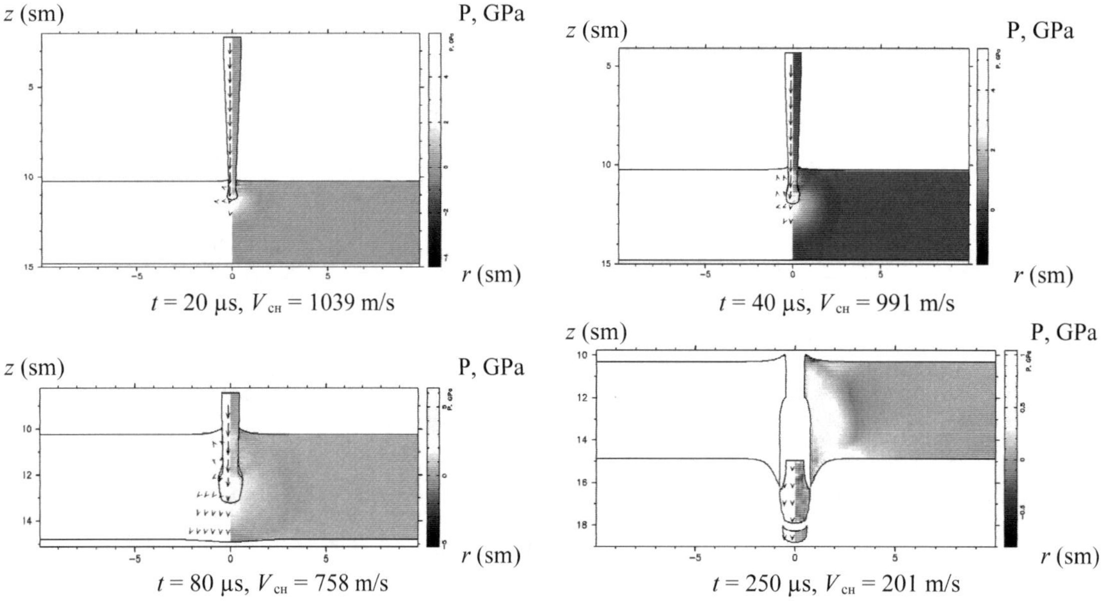

Figure 8. Chronogram of SI interaction with first steel barrier 45.4 mm thick at $V_0 = 1054$ m/s.

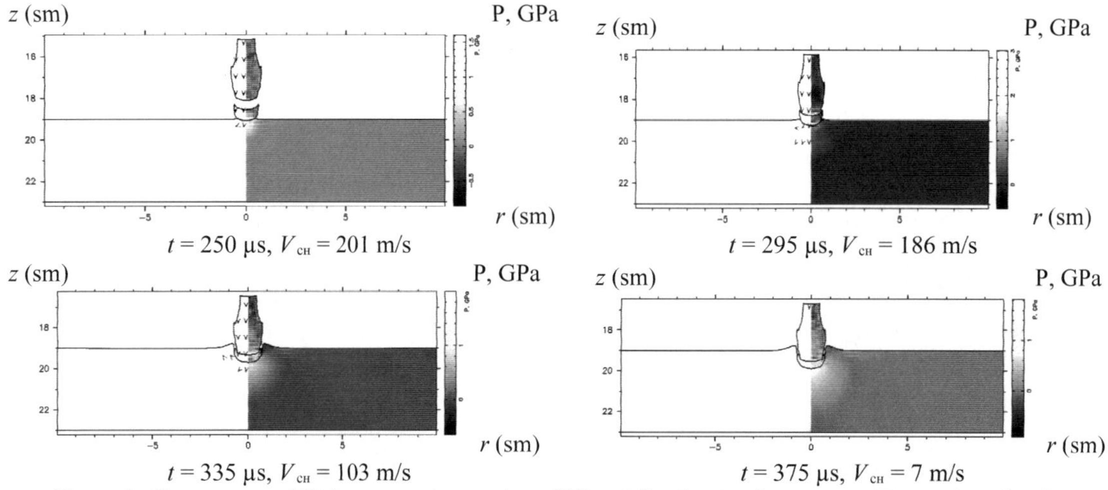

Figure 9. Chronogram of SI fragments interaction of SI and first barrier fragments with «marker» barrier at $V_{\text{сн}} = 201$ m/s

Thus, the calculation and experimental technique for investigation of the interaction of the SI with underwater barriers was developed on the basis of the high-speed hydro-ballistic complex. For the numerical realization of predicting the interaction of the SI with underwater barriers, methods for dynamical interaction of impactors and barriers have been developed in the axisymmetric statement by the large particles numerical method and the complete dimensioned formulation by the numerical finite element method in the framework of continuum mechanics with allowance for the deformation and destruction of interacting various materials bodies.

The calculated and experimental data about the ultimate penetration thickness of barriers from aluminum alloy D16T 83 mm and steel 45 mm at the impact velocity of about 1000 m/s for the developed SI are obtained.

The work was carried out with the financial support of the Russian Federation Ministry of Education and Science in the framework of state task No. 9.9036.2017/БЧ.

References

[1] Burakov V A, Burkin V V, Ishchenko A N, Korol`kov L V, Stepanov E Yu, Chupashev A V, Agafonov S V and Rogaev C S 2015 Experimental ballistic complex // *Patent of invention* № 2591132. Application № 2015113676 from 13.04.2015

[2] Afanas`eva S A, Belov N N, Burkin V V, D`yachkovskii A S, Evtyushkin E V, Ishchenko A N, Monakhov R Yu, Rodionov A A, Khabibulin M V, Yugov N T and Zykov E N 2013 Special features of high-velocity projectile interaction with barriers protected by a water layer *Russian Physics Journal* RUPJ04(2013) **56** pp 370–377

[3] Akinshin R N, Afanas`eva S A, Borisenkov I L, Burkin V V, D`yachkovskii A S, Ishchenko A N, Korol`kov L V, Moiseev D M and Khabibulin M V 2016 Special features of high-velocity interaction of supercavitating solids in water *AIP Conf. Proc.* **1698** 040010

[4] Chernyak G B and Povarova K B 2014 *Vol'fram v boepripasakh [Tungsten in ammunition]* (Moscow: CNIIKhM) p 356

[5] Sedov L I 1973 *Continuum mechanics* (Moscow: Nauka) p 536

[6] Khabibulin M V and Afanas`eva S A 2012 Calculation of the phenomena occurring in condensed medium as a result of intense impulse actions, in the axisymmetric setting *Federal Service for Intellectual Property. Certificate of computer software registration* № 2012617301 (Moscow) p 80

[7] Yugov N T, Belov N N and Yugov A A 2010 Calculation of adiabatic nonstandard flows in a three-dimensional formulation (RANET-3) *Federal Agency for Intellectual Property, Patents and Trademarks. Certificate of computer software registration* № 2010611042 (Moscow)

[8] Afanas`eva S A, Burkin V V, D`yachkovskii A S, Zykov E N, Ishchenko A N, Korol`kov L V, Monakhov R Yu, Rodionov A A, Khabibulin M V and Chupashev A V 2014 Theoretical and Experimental Analysis of the High-Velocity Interaction of Solid Bodies in Water // *Journal of Engineering Physics and Thermophysics* JOEP02(2014) **87** 2 pp 399–408

IOP Conf. Series: Journal of Physics: Conf. Series **899** (2017) 052001 doi:10.1088/1742-6596/899/5/052001

Calculation and experimental researches into the heat exchange surface formed by oppositely directed truncated cones with saddle-shaped connection straps

A Ye Baranov, D N Ilmov, V A Mavrov, Yu N Mamontov, A S Skorokhodov

SSC Keldysh Centre, Moscow, Russia

E-mail: ilmovdn@mail.ru

Abstract: The article has presented experimental data on the investigation of thermal and hydraulic characteristics of heat exchangers made of thin-walled panels with the surface formed by oppositely directed truncated cones with saddle-shaped connection straps. An approach to the mathematical description of surfaces of given class have also been proposed, and numerical modeling of stream and heat transfer have been performed. Results of numerical modeling have been compared with the experimental ones.

1. Introduction

At the present time a concept to the creation of the radically new spacecraft with nuclear power systems of 1 MW power and more is under consideration [1]. Special operating features of such a system imply high temperatures (over 1000 K) and big differential pressures (tens of atmosphere) between the heat-transfer agent circuits. In so doing, there are weight and size restrictions, which eliminate the use of conventional constructions for practical purposes. At SSC Keldysh Centre a new design based on stamped plates has been developed [2]. As the surface structure of heat exchanger plates is unique, the research on its thermal and hydraulic characteristics is urgent. For this purpose the "air-air" research with using heat exchangers with a different quantity of the plates was conducted. Progress of these experimental researches was carefully considered in [3].

2. Experimental

Shown in Fig. 1 is the schematic sketch of the plates and the panel for the creation of experimental cross-flow models of the heat exchangers.

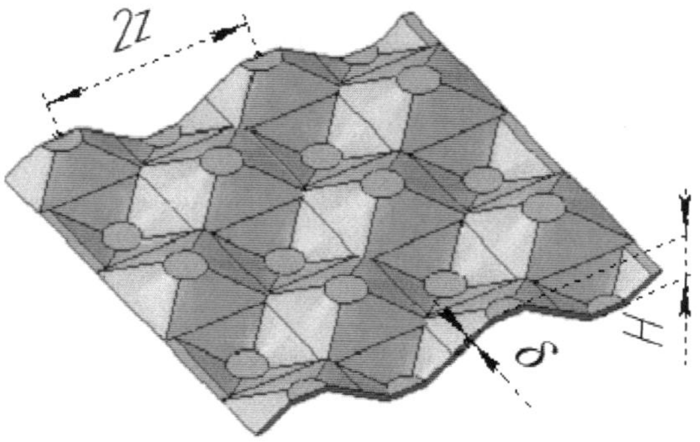

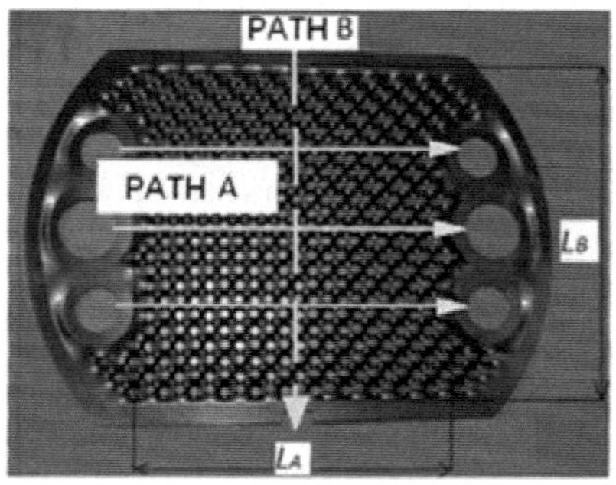

Fig. 1. The schematic relief of the plate (on the left) and the plate for the creation of heat exchanger models (on the right).

The following designations were introduced: 2 z – the spacing between the adjacent bulges on the plate (8 mm); *H* – the height of the plate after stamping (1.5 mm); δ – the thickness of the original blank (0.2 mm); h=H-δ – the maximum height of the flow section. The coefficient of surface development (ribbing) for this plate is calculated by the formula:

$$k_F = \frac{F_1}{L_A L_B} = 1 + \frac{\pi}{4z}\left(\sqrt{h^2 + z^2} - z\right), \quad (1)$$

where L_A, L_B и F_1 are the width, the length and the area of the plate surface after the stamping.

Refined manufacturing process yields the surface development 3 – 6 %, besides; the specific shape of channels makes the flow turbulent, which ensures heat exchange intensification.

The equivalent diameter of channels shaped in matrix is defined by their summary volume V and the total area of the "wetted" surface F_F:

$$d_E = 4V / F_F = 2h / k_F . \quad (2)$$

In the heat exchangers, the heat-transfer agent, being under lesser pressure and greater temperature (hot), is supplied to the cavity A. The cavity A is formed by a space between the plates and the two collectors, formed by the two 12 mm - diam. openings and one 16 mm - diam. opening. Another heat-transfer agent circulates in the external cavity B. It comes to the plates from a segment formed by the outer case and the heat-exchange matrix. The path length of the heat-transfer agents - L_A=88 mm L_B=94 mm (Fig. 1). The coefficient of surface development k_F =1.04; the equivalent diameter d_E=2.5 mm. The effective heat exchange area was derived from the formula $F = k_F L_A L_B N$ (where N is the number of panels). The areas of the flow section of the heat-transfer agents are $S_A = hL_B N_A$, $S_B = hL_A N_B$.

An experimental installation providing the air pumping at the mass flow rate up to 500 g/s and a feasibility of its heating to 600 °C, was made for testing of the heat exchangers. Experiments were carried out in the steady-state mode in which the air mass flow rate G, temperatures and static pressures at the heat-exchanger inlets/outlets were recorded.

On mathematical processing of results it has been suggested that the air is an ideal gas with the gas constant R=287 J/(kg K). Thermal and physical properties were believed to be dependent on the mean temperature in the hot and the cold cavities from the data [4], the metal thermal conductivity λ_{ME}=16 W/(mK). The amount of transferred heat per time unit Q was determined from the thermal balance. The thermal efficiency was further determined:

$$\eta = Q / Q_{ID}, \quad (3)$$

where $Q_{ID} = c_{PC} G\left(T_{H1} - T_{C1}\right)$ is the ideal heat power (c_{PC} is the heat capacity of the cold heat-transfer agent, T_{H1} and T_{C1} are the inlet temperatures of the hot and cold heat-transfer agents). At the known thermal efficiency, the quantity of units of carrying over of warmth Z (NTU) was determined for a single crossed current from solving the following equation [5]:

$$\eta = 1 - \exp\left\{\left[\exp\left(-Z^{0.78}\right) - 1\right] Z^{0.22}\right\}. \quad (4)$$

After determination of the quantity of units of carrying over of warmth, the overall heat transfer coefficient was calculated:

$$K = Z c_{PC} G / F . \quad (5)$$

On the other hand, the common heat transfer coefficient is defined by the convective heat exchange factors α_A and α_B of the heat-transfer agents and the thermal resistance of the plate, δ / λ_{ME}, by the formula:

$$K = \left(1/\alpha_A + 1/\alpha_B + \delta/\lambda_{ME}\right)^{-1}. \quad (6)$$

For the average values of the convective heat exchange factors, it was assumed that $\alpha_A = \alpha_B = \alpha$. This approximation is completely justified, as $s_A \approx s_B$, and the mass flow rates of the heat-transfer agents were equal.

Two closely related dependences in the dimensionless form were obtained: $\mathrm{Nu}_A = \alpha d_E / \lambda_A$ as a function $\mathrm{Re}_A = G d_E / (s_A \mu_A)$ and $\mathrm{Nu}_B = \alpha d_E / \lambda_B$ as a function $\mathrm{Re}_B = G d_E / (s_B \mu_B)$. Through interpolation the final result was represented in the form of the general averaged dependence $\mathrm{Nu}/\mathrm{Pr}^{0.4} = f(\mathrm{Re})$, where $\mathrm{Pr}=\mu c_p / \lambda$ is the Prandtl number.

The drag coefficient was calculated by experimental results separately for the hot and the cold cavities on a basis of the generally accepted dependence:

$$\Delta p = \xi \frac{\rho U^2}{2} \frac{L}{d_E}, \tag{7}$$

where $U = G/(\rho S)$ is the mean – average velocity, ξ is the required drag coefficient, Δp is the pressure losses.

Apart from experimental determination of dimensionless characteristics of surfaces for the heat exchangers, the task of development of mathematical modeling methods was set. Numerical modeling was grounded on solving Navier-Stokes averaged equations, that are also referred to as Reynolds equations [6]. Parameters in the equations (velocity, density and others) are time-averaged. The steady-state continuity equation, the equations of motion and energy for the compressible medium (ideal gas) are solved. Averaging the Navier-Stokes equations and the energy equations leads to the appearance of unknown terms in them, related to turbulent pulsations. To close a set of equations, the Boussinesq hypothesis of supplemental turbulent viscosity and heat conduction is applied. The turbulent viscosity: $\mu_T = \rho \mathrm{v}_T l_T$, where v_T, l_T are the velocity scale and the spatial scale of turbulent pulsations. For determination of the latter, the semi empirical model $k-\varepsilon$ of the second closure order is used. It suggests a solution of the two transfer equations for kinetic turbulent energy «k» and for turbulent dissipation rate «ε». The form of the equations of the $k-\varepsilon$ model and values of semi empirical constants can be found e.g. in [7]. An analysis of the efficiency of the application of various models of turbulence to the assigned task was not within the scope of this article. However, we note that the application of various models, including "large eddy simulation", as well as the direct solution of Reynolds equations without turbulent components, led to very close results.

Solving the set of the Reynolds equations and the equations of the $k-\varepsilon$ model is labour-consuming enough, but a well developed problem. Let us consider the application of this mathematical instrument to the task set.

For numerical modeling the two approaches can be proposed: thorough reproduction of geometry (3D scanning), which demands great computational expenditures or search for simplified analogues. As an equivalent, a columnar surface with the "contracted" section, similar in the area and hydraulic diameter, is here proposed (Fig.2).

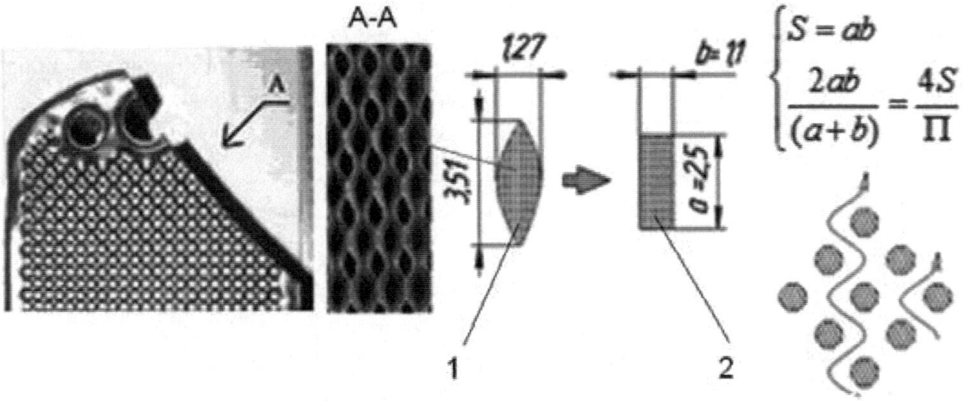

Fig. 2. Construction of the equivalent design domain by the "contracted" sections (1 –the "contracted" section, 2 – the section of the design model).

Fig. 3 shows the velocity and temperature fields, obtained with the aid of the given simplified model. The solution was fulfilled with using the software package FLUENT ANSYS. It is seen that the jet flow pattern, uniform in width, is being formed rather quickly.

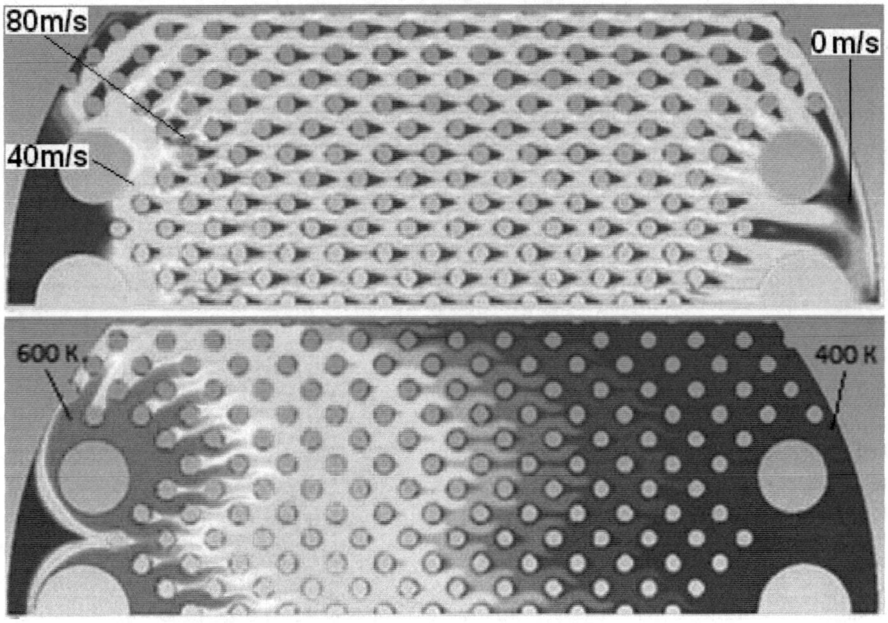

Fig. 3. The velocity and temperature fields as a result of the modeling.

Given in Fig.4 are the values of the drag coefficients for the cavities A and B, experimentally obtained through the numerical modeling on the 20-panel heat exchanger. The drag coefficient of the inner cavity A is higher 1.5 to 1.6 times, than that of the cavity B. The distribution of experimental points needs some explanation. In the process of the research, it became clear, that the drag depends on the pressure differential between the cavities. This differential increases with the flow rate. One series of experimental data was acquired at the start of the installation, when the inner cavity A is contracted and its drag coefficient has higher values, the outer cavity B is extended and its drag coefficient has relatively low values. Another series of experiments was conducted at the start, when the pressure differential between the circuits was opposite. It is seen, that a deformation of the heat exchange matrix changes the drag of the circuits A and B by 15 to 20 %. The approximation lines of experimental points were built for both series of the data. Just these curves should be compared with those obtained by the modeling. Thus, the modeling results accord well with the experimental results. It is precisely the channels "contracted" section, that supposedly governs pressure losses.

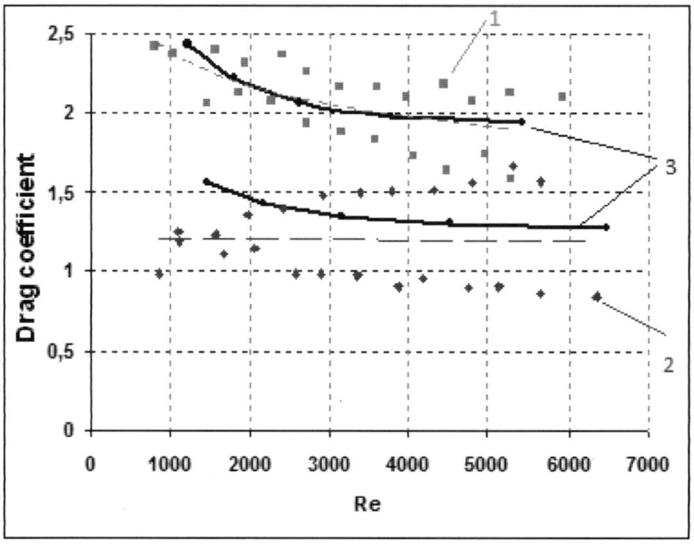

Fig. 4. The dependence of the drag coefficient on the Reynolds number (Re) for the 20-panel heat exchanger: 1 – the circuit A; 2 – the circuit B; 3 – the modeling for the circuits A and B.

Presented in Fig.5 are the values of the dimensionless heat-transfer coefficient $Nu/Pr^{0.4}$ at different Reynolds numbers, obtained in the experiments on the 2-, 4-, 20- and 40 panel models of the heat exchangers. The empiric results agree closely with the approximating function and between each other:

$$Nu=0{,}18\,Pr^{0.4}\,Re^{0{,}69} \qquad (8)$$

The dependence obtained by the modeling accords satisfactorily with the experimental results. At the Re numbers less, than 2500, the calculation yields the overstated Nu number; at the great Re numbers – the understated. Nevertheless, in the range $2000 < Re < 4000$ the modeling error does not exceed 10%.

It may be inferred that the proposed approach to the simplified construction of the design domain of heat exchange surfaces, based on the equality of the flow areas and the equivalent diameters of the contracted section, actual and model channels, has been completely justified in case of calculation of the drag coefficient, and may be as a whole considered applicable to the modeling of heat exchange.

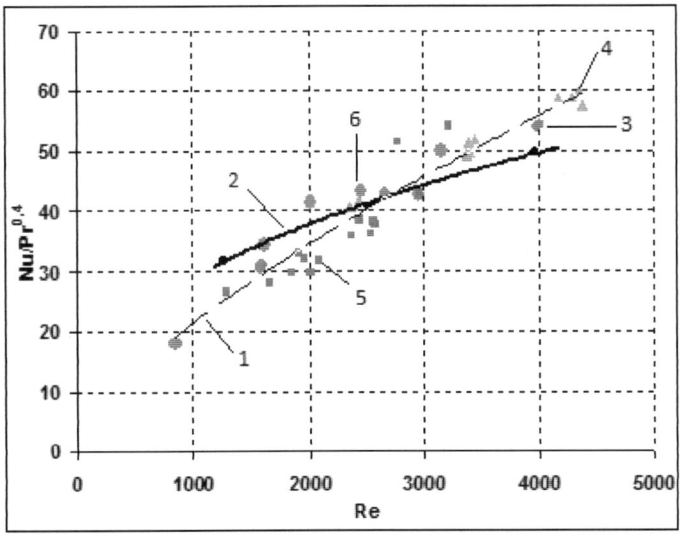

Fig 5. The dimensionless heat-transfer coefficient as a function of the Reynolds number: 1 – the approximation of experiments; 2 –the modeling; 3-6 –testing of 2-, 4-, 20- and 40-panel heat exchangers.

3. Conclusion

Thus, the computational and experimental research into thermal and gas dynamic characteristics of stamped plates with the specific relief of the surface was performed. The dependences of the dimensionless heat-transfer coefficient and the drag coefficient on the Reynolds number were obtained. The mathematical model for research of the working process regularities in energy-intensive heat-exchange devices of new-generation space systems was developed and verified.

References

1. A. S. Koroteev, "New stage of cosmic thermal engineering development," Vestnik Ross. Akad. Nauk 2012, Vol.82, No. 4, P. 317–322.
2. N. I. Filatov, A. E. Baranov, D. A. Lyalin, V. A. Kozhevnikov, and V. A. Mavrov, RF Patent No. 2529288, "Heat exchange unit plate packet," 2014.
3. Investigation of Heat Exchangers for Energy Conversion Systems of Megawatt_Class Space Power Plants. D. N. Ilmov, Yu. N. Mamontov, A. S. Skorohodov, V. A. Smolyarov, and N. I. Filatov. Thermal Engineering, 2016, Vol. 63, No. 1, P. 35–41.
4. N. B. Vargaftic, Handbook of Thermophysical Properties of Gases and Liquids. Nauka, Moscow, 1972. 720 p.
5. P. I. Bazhan, G. E. Kanevets, and V. M. Seliverstov, Handbook on Heat Exchange Units. Mashinostroenie, Moscow, 1989. 369 p.
6. A. I. Leontiev. The theory of a heat and mass transfer. Vysshaya Shkola, Moscow, 1979. 495 p.
7. D. Anderson, J. Tannehill, R. Pletcher. Computation fluid mechanics and heat transfer. Mir, Moscow. 1990. 384 p.

IOP Conf. Series: Journal of Physics: Conf. Series **899** (2017) 052002 doi:10.1088/1742-6596/899/5/052002

3D numerical modelling of the thermal state of deep geological nuclear waste repositories

Butov R. A., Drobyshevsky N. I., Moiseenko E. V., Tokarev Yu. N.

Nuclear Safety Institute of Russian Academy of Sciences (IBRAE RAN),
52, B. Tulskaya str., Moscow, 115191, Russia

bra@ibrae.ac.ru

Abstract. One of the important aspects of the high-level radioactive waste (HLW) disposal in deep geological repositories is ensuring the integrity of the engineered barriers which is, among other phenomena, considerably influenced by the thermal loads. As the HLW produce significant amount of heat, the design of the repository should maintain the balance between the cost-effectiveness of the construction and the sufficiency of the safety margins, including those imposed on the thermal conditions of the barriers. The 3D finite-element computer code FENIA was developed as a tool for simulation of thermal processes in deep geological repositories. Further the models for mechanical phenomena and groundwater hydraulics will be added resulting in a fully coupled thermo-hydro-mechanical (THM) solution. The long-term simulations of the thermal state were performed for two possible layouts of the repository. One was based on the proposed project of Russian repository, and another features larger HLW amount within the same space. The obtained results describe the spatial and temporal evolution of the temperature filed inside the repository and in the surrounding rock for 3500 years. These results show that practically all generated heat was ultimately absorbed by the host rock without any significant temperature increase. Still in the short time span even in case of smaller amount of the HLW the temperature maximum exceeds 100 °C, and for larger amount of the HLW the local temperature remains above 100 °C for considerable time. Thus, the substantiation of the long-term stability of the repository would require an extensive study of the materials properties and behaviour in order to remove the excessive conservatism from the simulations and to reduce the uncertainty of the input data.

1. Introduction

The deposition in deep geological repositories is now considered to be one of the most promising methods of high-level radioactive waste (HLW) and spent nuclear fuel (SF) disposal. The projects of such repositories now exist in several countries: Finland, Sweden, USA, Belgium, France, etc [1]. The progress of these projects varies from the research stage like in Belgium to the construction of the actual repository like in Finland. A project of a deep geological repository for both high- and low-level radioactive waste (LLW) is being developed in Russia as well [2].

The waste is separated from the environment by a number of barriers, natural and engineered. Natural barriers are represented by host medium, which is rock, salt or clay for existing projects. In the Russian projects the granitoid rock, namely gneiss, granite and diorite, is chosen. Engineered barriers include glass that encapsulates the vitrified waste, steel casings, layers of concrete, bentonite or other materials. These engineered barriers should maintain integrity for considerable time – at least several centuries.

The HLW and SF emit a significant amount of heat, which poses a problem of efficient heat removal, since the mechanical strength and corrosion resistance of the barriers decrease at high temperatures. Therefore the assurance of an acceptable heat regime in the repository is one of the primary objectives. The heat regime depends on a number of factors, including the repository geometry, filling schedule, heat generation in the waste, thermal properties of the materials. Since the construction of underground vaults and tunnels is quite expensive, the choice of the repository structure which will be cost-effective while providing sufficient heat removal from the waste is one of the primary objectives.

In this paper we address the problem of numerical modeling of the heat regime evolution for a deep geological HLW repository. A prototypic repository layout based on the proposed project of the Russian repository along with an alternative one with more dense waste storage was considered. Since the repository has a complicated spatial structure, fully 3D approach to the modeling was used. The temporal evolution of the temperature field within the repository and in the surrounding rock was obtained for the time span of 3500 years for both repository layouts.

2. Method for thermal state calculation

A code for simulation of thermal and mechanical state of HLW and SF repositories named FENIA (Finite Element Nonlinear Incremental Analysis) that uses 3D finite-element approach is being developed. Currently it solves thermal problem with energy conservation equation given as:

$$\rho(t,X)c(t,T,X)\frac{\partial T}{\partial t}-\frac{\partial}{\partial x_i}\left(\lambda_{ij}(t,T,X)\frac{\partial T}{\partial x_j}\right)=q(t,X), i,j=1..3, X=(x_1,x_2,x_3), \qquad (1)$$

where ρ is the density, t – the temporal coordinate, X – the vector of spatial coordinates (x_1,x_2,x_3), c – the heat capacity, T – the temperature, λ – the heat conductivity tensor, q – the heat source or sink. The mesh is supposed to be constructed primarily of hexahedral or tetrahedral finite elements, but arbitrarily shaped convex elements can be used as well. The first- and second order shape functions are used for the solution.

The dependence of the coefficients in (1) on temporal and spatial coordinates as well as on the temperature incorporates a number of various phenomena. E.g. heat capacity and conductivity temperature dependence of the materials is considered and can be taken either from the materials database or from the user input, while density change with temperature is supposed to be negligible.

Since the equation (1) is solved in the whole domain that includes the repository and surrounding rock formations, the functions $\rho(X)$, $c(X)$, $\lambda(X)$, $q(X)$ can have discontinuities at the material boundaries. The code is designed to efficiently treat these discontinuities. Another source of the spatial dependence of the thermal properties of the host rock is the effect of the excavation process. The material properties in the excavation damage zone (EDZ) can differ from those in the bulk rock due to the change in fracture structure.

The heat sources or sinks result from the decay heat in the HLW and SF and from chemical reactions in the engineered barriers. The temporal dependence of the parameters is used for interaction with other codes: the processes that influence the material properties, like water inflow or chemical reactions, can be modeled externally, and the results can be time-dependently inputted to the FENIA code. Traditionally codes like COMSOL Multiphysics are used for such simulations, but we are mainly oriented on the use of modules created in our institution, since it will make further tight coupling easier.

Further the equilibrium equation for modeling of mechanical behavior and the simplified continuity equation for groundwater flow simulation will be added. This will most likely require an account for additional dependencies: e.g. the density of some materials, like bentonite, as well as their conductivity tensors, will depend on the amount of accumulated water. Long-term application of the

IOP Conf. Series: Journal of Physics: Conf. Series **899** (2017) 052002 doi:10.1088/1742-6596/899/5/052002

pressure from the host rock which will be modeled in the mechanical module will also alter the temporal dependence of the properties due to the fracture closing. After the implementation of mechanical and hydrological models the FENIA will be a fully THM-coupled solution for modeling of processes in deep geological repositories.

3. Prototypic repository layout

According to the repository concept the vitrified HLW will be placed in vertical boreholes several dozen meters long located at depth of 400 – 500 m. The boreholes are arranged in rows and within each row are connected at the upper and the lower end by the tunnels – so called disposal chambers. After filling the boreholes with HLW these chambers are used for LLW disposal, but since the impact of the LLW in the overall heat generation is negligible, they are not considered in this paper. The ends of the disposal chambers are in turn connected by two parallel tunnels. The sketch of the layout of one level formed by the tunnels with distances between the chambers and the boreholes taken in the simulation is shown in Figure 1. The repository is supposed to consist of two parts with 14 chambers in each part. One part will be used for disposal of the HLW with decay heat about 1 kW/m^3, another – about 1.5 kW/m^3. The boreholes in the first part will be placed more densely – 20 holes per chamber, while in the second part there will be 13 holes in each chamber. It is supposed to be 27 tons of the vitrified HLW in each borehole. Overall area covered by the repository is about 360 by 700 m.

The engineered barriers (canisters and containers) that separate the vitrified HLW from the environment form several layers of steel separated by concrete or bentonite. The space between the container and the borehole wall is filled with thixotropic slurry. The decay heat of the HLW is supposed to decrease by 2.5% per year.

The host rock massif is formed by granitoids: granite, gneiss and diorite. The ambient temperature of the rock at this depth is 5 – 10 °C. The average annual temperature on the ground surface is supposed to be about 0 °C.

An alternative repository layout was considered so that to assess the possibility to store more HLW within the proposed area. In this layout extra tunnels for HLW were added. The tunnels connect the adjacent disposal chambers and are placed at the same distance as the boreholes. Each tunnel holds 4.5 tons of vitrified HLW in the first part of the repository and 6 tons – in the second part.

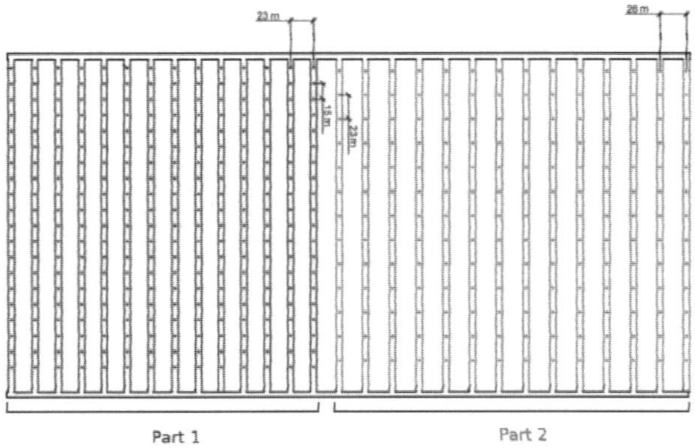

Figure 1. Layout of the repository level with distances used in the simulations.

4. Repository model

A simulation model for the repository and the surrounding host rock was created. The 3D mesh consists of 15 million hexahedral finite elements and covers the repository and 400 m of host rock around it in each direction. Since the repository has two symmetry planes, horizontal at the boreholes middle and vertical at the disposal chambers middle, the mesh covered one quarter of the region. The height of the boreholes was takes as 75 m, thus with account for the symmetry total simulated area is about 1100 m wide, 1450 m long and 900 m high. The size of the area is based on the depth of the repository.

Several simplifications were used in the modeling. Heat generation was taken as uniformly distributed along the borehole or tunnel – the bentonite separators between the containers were not considered. Some materials were grouped into a single one with averaged properties. E. g. the slurry and the containers made of steel, concrete and bentonite were taken as a homogeneous medium. Thermal properties of gneiss at temperatures of 20 – 100 °C [3] were used for modeling the host rock, which was also supposed to be homogeneous – without an account for the EDZ. All tunnels that do not contain the HLW were supposed to be filled with the material having the same properties as the host rock. In an actual repository they will be filled with LLW containers. Since the thermal properties of the granitoids can vary significantly [3], the possible error due to this simplification is of the same order of magnitude as one resulting from the uncertainty of the host rock properties. Thus three media were modeled in the simulations: vitrified HLW, engineered barriers and host rock. Thermal properties for these media used in the simulations are shown in Table 1.

Gradual filling of the repository is modeled – every year the boreholes (and the tunnels for the alternative layout) connected to one disposal chamber are filled starting from the first part. Thus the complete loading of the repository takes 27 years.

The initial temperature in the whole simulation domain was taken as 9 °C. Boundary conditions of the first kind with 9 °C temperature were placed at all boundaries.

Table 1. Thermal properties of the materials used in the simulations.

Material	Heat conductivity, W/(m·K)	Heat capacity, J/(kg·K)	Density, kg/m^3
Vitrified HLW	2.4	800	2500
Host rock	2.91	840	2700
Engineered barriers	8.1	1500	2800

5. Simulation results

The temperature field evolution in the repository and surrounding rock during 3500 years was simulated with the FENIA code for both layouts. The temperature in the first part of the repository reaches maximum value between 50 – 100 years from the beginning of the HLW loading. Figure 2 shows the temperature distribution after 55 years for the original layout across three horizontal cross-sections: at the middle of the boreholes, at the tops of the boreholes and 100 m up from the boreholes middle. Here whole first part of the repository has temperature 100 – 113 °C while in the second part it exceeds 100 °C only in the HLW, the containers and the rock in the vicinity of the containers since the second part is filled later.

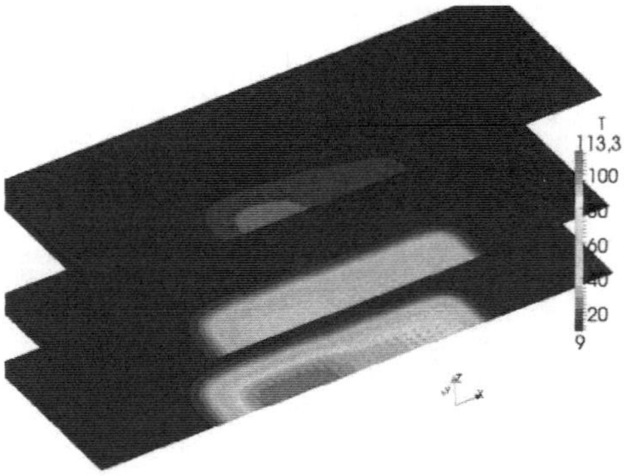

Figure 2. Temperature distribution after 55 years for the original layout across three horizontal cross-sections.

Figure 3a shows the temperature profiles along the symmetry axis of the repository at the middle of the boreholes for several time moments: 100 years, 270 years and 3500 years from the beginning of the HLW loading. The chambers in the first part of the repository contain even number of boreholes (20), while in the second part – odd number (13). Therefore in the second part the symmetry axis crosses the boreholes with the HLW which results in sharp peaks on the graph for 100 years. After 100 years in whole repository, even inside the HLW of the second part, the temperature does not exceed 100 °C. After 275 years it decreases to less than 50 °C, and after 3500 years practically gets back to the initial values. The temperature distribution along the vertical line in the center of the first part of the repository for the same time moments is shown in Figure 3b. Zero coordinate is set to the middle of the boreholes. One can see that the temperature increase near the ground surface – coordinates 400 m and higher – will not be more than several degrees. Very low heat flux values at the simulation domain boundaries were obtained.

a) b)

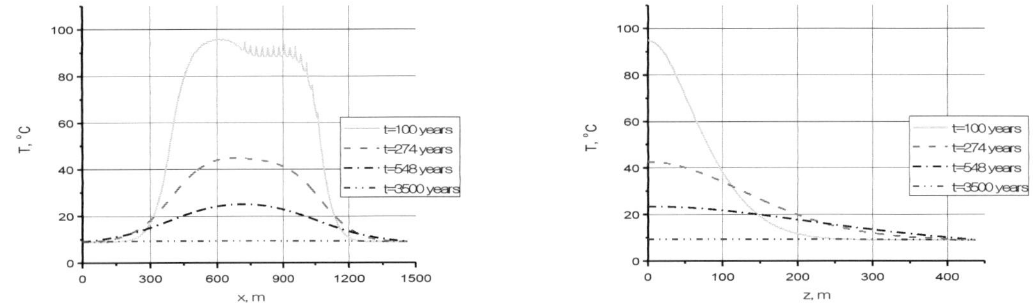

Figure 3. Spatial temperature profiles for different time moments for the original layout. a) along the symmetry axis of the repository at the middle of the boreholes b) along the vertical line in the center of the first part of the repository.

The results for the alternative layout show considerably higher temperature maxima with slower decrease. Figure 4 shows the temperature distribution after 60 years for the alternative layout across the same horizontal cross-sections as in Figure 2. Here the maximum temperature reaches 135 °C. As it is shown in Figure 5a the temperature remains above 100 °C in the whole repository after 125 years, while in the original layout it dropped below 100 °C after 100 years. However, after 3500 years the repository cools down to the initial temperature as well. The vertical temperature distribution presented in Figure 5b shows very small temperature increase near the ground surface with low heat fluxes, like in the original layout.

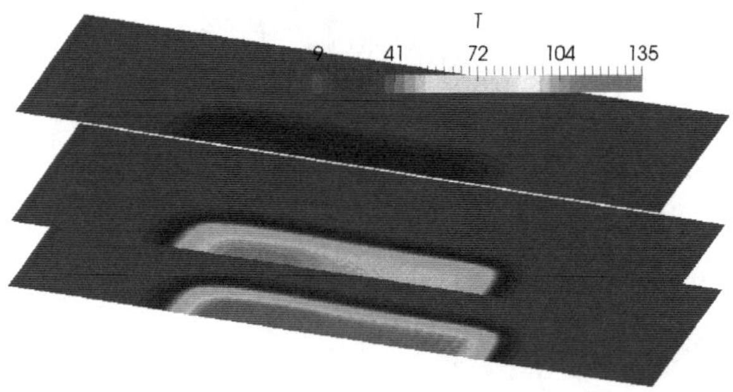

Figure 4. Temperature distribution after 60 years for the alternative layout across three horizontal cross-sections.

IOP Conf. Series: Journal of Physics: Conf. Series **899** (2017) 052002 doi:10.1088/1742-6596/899/5/052002

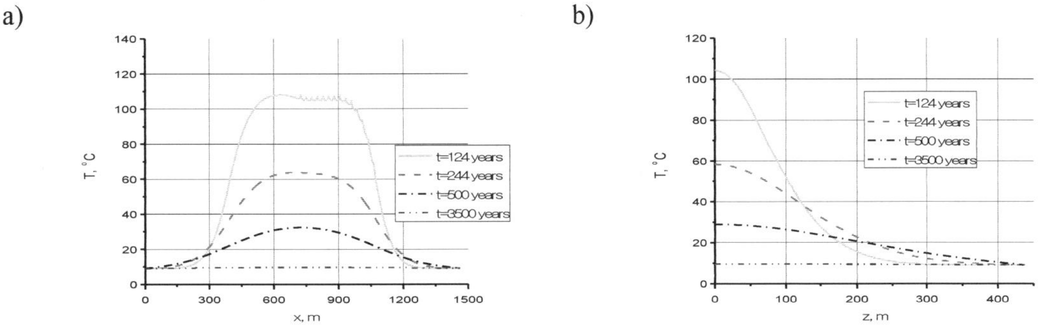

Figure 5. Spatial temperature profiles for different time moments for the alternative layout. a) along the symmetry axis of the repository at the middle of the boreholes b) along the vertical line in the center of the first part of the repository.

6. Conclusions

The detailed 3D temperature field evolution during 3500 years was obtained with the FENIA code for two prototypic deep geological HLW repository layouts: based on the proposed project and the alternative one with more waste stored. The resulting heat fluxes from the simulation area boundary were close to zero, so it can be concluded that practically all generated heat was ultimately absorbed by the host rock without any significant temperature increase. Virtually no heat reached the ground surface in both layouts.

However, even in the original layout the maximum temperature exceeded 100 °C, and in the alternative layout high temperature values kept for about 100 years. This means that the increase of the overall power density in the repository would pose significant thermal stress on the engineered barriers. One should note that the presented simulations are quite conservative as they do not consider heat transfer by water and steam as well as the impact of ventilation of not yet filled tunnels and deposition chambers. Still, if the need for disposal of larger amount of the HLW or of the waste with higher decay heat arises, it would require thorough modelling of the heat transfer in the repository to ensure the acceptable thermal regime. Anyway, an extensive study of the materials properties and behavior under long-term thermal stress will be necessary including the large-scale experiments in the underground research laboratory under the realistic conditions to lower the conservatism and to decrease the uncertainty of simulation parameters.

The present study was primarily focused on obtaining the estimates for the global temperature field in whole repository area, and the engineered barriers were modeled quite roughly. The local simulations of the containers and nearby host rock with high mesh resolution will be needed to obtain a detailed picture of the heat state of the engineered barriers. This will also require more elaborate input data, primarily the thermal properties of the materials, including the host rock. This data will be available when the underground research laboratory at the proposed repository location will be established, and the final choice of the container materials will be made.

References

[1] 2003 Scientific and technical basis for geological disposal of radioactive wastes *Technical Report Series* **413** (Vienna : International Atomic Energy Agency).

[2] Kudryavtsev E G, Gusakov-Stanyukovich I V, Kamnev E N, Lobanov N F and Beygul V P 2009 Construction of a Deep Geological Disposal Facility for Final Isolation of High-Level Waste in Nizhnekansky Rock Massif (Krasnoyarsk region). *CEG Workshop on Disposal of Radioactive Waste and Spent Nuclear Fuel – Experience and Plans. Bommersvik, Sweden, 24-26 February 2009*

[3] Anderson E B, Belov S V, Kamnev E N, et al 2011 *Underground Disposal of Radioactive Wastes* (Gornaya Kniga: Moscow)

Calculations of hydrophysical fields in the coastal regions of the Black Sea with high spatial resolution

N Evstigneeva

Marine Hydrophysical Institute RAS, Sevastopol, Russian Federation

E-mail: naevstigneeva@yandex.ru

Abstract. Numerical experiments have been carried out using a hydrodynamical model with nonlinear equations of motion and heat and salt advection to reconstruct the fields of hydrophysical parameters in the coastal regions of the Black Sea taking into account the real atmospheric forcing and river discharges for the winter and summer seasons of 2006. A higher spatial resolution allowed to get a detailed meso- and submesoscale structure of hydrophysical fields in the upper and deep layers of the Southern Coast of Crimea and the north-western shelf and to obtain quantitative and qualitative characteristics of the eddies and jets more accurately in comparison with previous calculations.

1. Introduction

The tasks of construction of coastal and port facilities, tourism, fishing and fish farming, mining and environmental monitoring are inseparably connected with a study of hydrodynamics of the waters of the shelf zones, their mixing and exchange with open parts of the sea. Nowadays a reconstruction and study of spatio-temporal variability of hydrophysical fields in the coastal areas on a scale of a few kilometers and days is one of the urgent problems of oceanography. New results on meso- and submesoscale features of circulation in the various regions of the World Ocean were obtained in the following investigations on modeling, forecasting and observation of hydrophysical fields with high spatial resolution [1–10].

Information about small-scale eddies in the coastal zone of the Black Sea according to satellite monitoring data was presented in [1, 2]. Mechanisms of their possible formation were suggested. In [3] areas of eddy activity in the White Sea were found with the help of remote and contact observations, characteristics of submesoscale eddies were mapped. In [4] according to satellite radar data it was found that small-scale eddies of the Baltic, Black and Caspian Sea were associated with either hydrologic fronts or peripheral areas of mesoscale eddies. In [5] water circulation of the Adler-Sochi region was reconstructed with an improved spatial resolution of the coastal zone due to the changing grid step (with uniform spatial resolution ~4 km and uniform, with a step reduction to 50 m). It was shown that an eddy circulation was reproduced more adequately by the model with a higher spatial resolution. In [6] mesoscale eddies, generated by the instability of alongshore wind currents, were described with the help of an idealized system of currents similar to the California Current. In [7] mesoscale variability of the World Ocean, based on altimetric data, was studied using an automated procedure of identification and tracking of mesoscale structures. 35 eddies with a lifetime of more than 16 weeks were described. In [8] a seasonal cycle of submesoscale flows in the upper ocean layers was investigated in an idealized model domain analogous to mid-latitude open ocean regions. It was shown that submesoscale processes became much stronger when a resolution was increased. In [9] mechanisms of generation and instability of submesoscale flows (density fronts, filaments, topographic traces and coherent vortices) in the ocean were considered. The paper [10] was devoted to numerical modeling of submesoscale structures in the Gulf of Finland of the Baltic Sea during coastal upwelling events in July–September 2006 using POM-model.

The aim of this investigation is a reconstruction of hydrophysical fields in the winter and summer seasons with high spatial resolution on the basis of the hydrodynamic model [11, 12] in the coastal regions of the Black Sea, a determination of possible mechanisms of the formation of meso- and submesoscale features of circulation and their dependence of the season. In addition, to show an importance of a use of high spatial resolution on reconstruction of meso- and submesoscale structure of fields on the basis of a comparison with the results of the calculation with a coarser grid size (~1.6 km). The work, presented in this paper, is a continuation of our previous investigation [13].

We consider these areas of the Black sea (a coastal zone of the Southern Coast of the Crimea and the north-western shelf) because they contribute significantly to the dynamics and energy of the circulation.

2. Statement of the problem and description of numerical experiments

The main tool, used in this investigation, is a z-coordinate three-dimensional nonlinear energy-balanced prognostic model of baroclinic currents of the Black Sea, which was developed at the Marine Hydrophysical Institute under the guidance of Dr. Sc. Demyshev S.G. and has been successfully used for solving theoretical and applied problems. The system of model equations using the Boussinesq approximation, hydrostatic approximation and incompressibility of seawater in the Gromeko–Lamb form, the boundary conditions on the surface, at the bottom, on the solid lateral walls were written as follows [11]. System of equations, including equations of motion, heat and salt advection, continuity equation, equation of state and equation for the pressure with the boundary and initial conditions was solved numerically, using a C-grid finite-difference approximation. It should be noted that the sea level was calculated from the discretized analogue of the continuity equation. The following steps were made in [12] in order to adapt the model [11] to the conditions of the coastal zone of the Black Sea with open boundaries. New data arrays of the region bathymetry were processed, boundary conditions on the open boundaries were selected and implemented, model parameters were chosen on the basis of preliminary experiments, river inflow locations and depths of estuaries were assigned, initial fields as well as fields of wind stress, heat flows, short-wave radiation, precipitation and evaporation were processed in order to be used in the model.

Experiments 1 and 2 were conducted for the coastal zone of the Southern Coast of Crimea (SCC), located between parallels 44.25 and 44.72°N and the meridians 33.95 and 34.55°E, from 1 to 31 of January, 2006 and from 1 to 31 of July, 2006 respectively. The region had three open boundaries: western, southern and eastern, a maximum depth was 2 km. We used high spatial resolution (horizontal grid 350×350 m and 38 vertical layers from 1 m to 2 km) and took into account a detailed representation of the bottom relief with a resolution of ~500 m. The time step was equal to 10 s. Horizontal coefficients of turbulent viscosity and diffusion were equal to $\nu_H = 35\cdot10^4 cm^2/s$, $\kappa_H = 35\cdot10^4 cm^2/s$.

Experiments 3 and 4 were conducted for the north-western shelf (NWS) of the Black Sea from 1 to 31 of January, 2006 and from 1 to 31 of July, 2006 respectively. We considered a region of the Black Sea, limited by latitude 45.5°N and located between meridians 29.5° and 33.5° E, a maximum depth was 50 m. We used high spatial resolution (horizontal grid 500×500 m and 44 vertical layers from 1 to 49 m). The time step was equal to 10 s. We used more detailed presentation of bottom topography (with a resolution of ~1.6 km), interpolated to 500×500 m. It has been obtained by digitization of navigation maps by staff of Shelf hydrophysics and Waves Theory departments of Marine Hydrophysical Institute. Horizontal coefficients of turbulent viscosity and diffusion were equal to $\nu_H = 3\cdot10^5 cm^2/s$, $\kappa_H = 3\cdot10^5 cm^2/s$. We took into account the discharges of three rivers: Dnieper, Dniester, and South Bug.

Fields of currents, temperature and salinity, obtained from the model for the entire sea on a 1,64 × 1,64 km horizontal grid [14], were used to specify initial and boundary conditions at the open boundaries of the domain for experiments 1–4. The vertical coefficients of turbulent exchange of momentum and diffusion were calculated according to the Philander–Pacanowski approximation [15]. The fields of tangential wind stress, heat fluxes, as well as precipitation and evaporation, obtained from data of the regional atmospheric model ALADIN, were linearly interpolated to the selected grids for each day.

The following boundary conditions were specified on the open boundaries. The components of the current velocity, temperature and salinity (the Dirichlet conditions) were assigned in the boundary regions where water flowed into the domain; conditions $\partial u/\partial n = 0$, $\partial v/\partial n = 0$ for u, v and radiation condition [16] for T, S were specified in the boundary regions where water flowed out of the domain.

For the classification of obtained eddies we estimated a value of the local baroclinic deformation radius (R_d) for the selected coastal areas of the Black Sea. It was equal to ~6–7 km. We assumed that submesoscale eddies had a radius bigger than local baroclinic Rossby deformation radius and Rossby number was less than unity ($R>R_d$, $R_0<1$). We assumed that submesoscale eddies had a radius smaller than R_d and Rossby number was of the order of unity ($R<R_d$, $R_0 \approx 1$).

3. The results of experiments

3.1. SCC, winter

The Rim Current (RC) is the main element that determines a circulation of the Black Sea in general and the selected region specifically. The winter period was characterized by a dominance of the alongshore jet of the RC, it was directed to the west and southwest during the calculating period from 1 to 31 January, the maximum velocity in the upper horizons was 140 cm/s. Submesoscale eddy formations of various signs of rotation were obtained in the upper 30-m water layer between the coastline and the RC. A possible mechanism of their formation was the flow of the RC over inhomogeneities of the bottom relief and coastal orography. Those small-scale structures had a short lifetime (1–2 days) and were generated near the cape Ayu-Dag, in the eastern part of the region and in the Gulf of Yalta. Figure 1 shows current fields at the depth of 5 m on January 12 and 20, in which the following eddies were reconstructed: cyclonic in the vicinity of the cape Ayu-Dag ($R \approx 5$ km$<R_d$, $R_0 \approx 1.1$, figure 1a) and anticyclonic in the eastern part of the region $R \approx 5$ km $<R_d$, $R_0 \approx 0.8$, figure 1b).

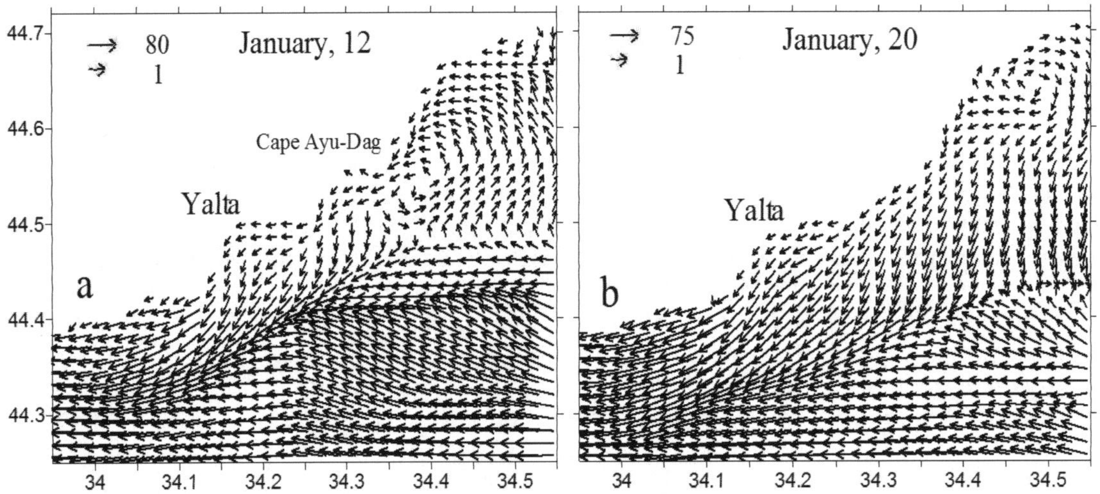

Figure 1. Current fields (cm/s), calculated at the depth of 5 m (every fourth arrow was drawn): a – January 12, b – January 20.

We compared current fields with the results with a more coarse resolution [14] and found a correlation between them during the calculating period, however a number of eddies was absent and a field structure was smoother in the experiment with a lower resolution.

The following structure was typical for the temperature and salinity fields: warmer and less saline waters were located along the coast (10–11°C, less than 18‰), colder and more saline waters – in the deep-water part of the region (8–9°C, more than 18‰).

3.2. SCC, summer

A circulation was characterized by a certain weakening of the jet of RC (for example, figure 2a) and meso- and submesoscale eddy structures were generated during the calculation period from 1 to 31 of July. Anticyclonic eddy formations with a radius of about 20 km and a lifetime of 3–4 days were obtained in the upper 100-m layer between the coastline and the RC from 10 to 12, from 18 to 21, from 24 to 26 of July, moving along the direction of the RC (for example, figure 2b). Analysis of the equations of energetics showed that processes of baroclinic instability of the RC were observed in the coastal zone. It was a possible cause of the formation of mesoscale eddies. Eddies were generated in the regions, which were characterized by negative values of the buoyancy force and, as a consequence, a transition from kinetic energy to potential energy took place. A correspondence was found between obtained features of the circulation and a number of satellite images.

Submesoscale features of circulation with a radius less than 5 km were reconstructed in the upper water layer, for example, anticyclonic eddy formations in the eastern part of the region ($R \approx 4$ km$<R_d$, $R_0 \approx 0.9$, figure 2a). The analysis showed that they were formed as a result of the flow of the RC along the coastline of the SCC and inhomogeneities of the bottom relief.

We compared current fields with the results with a more coarse resolution [14] and found a correlation between them during the calculating period, however a use of higher spatial resolution allowed to reconstruct a field structure more accurately.

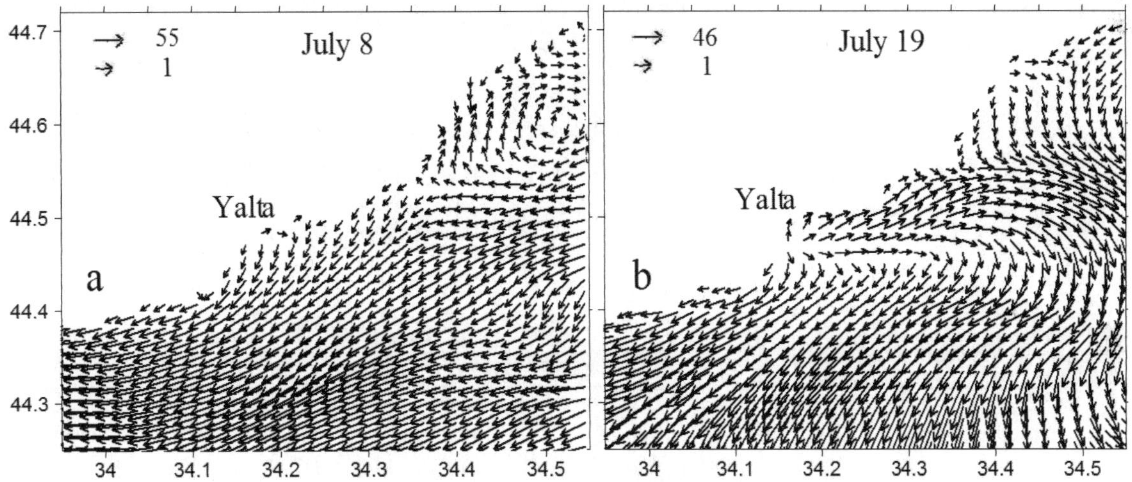

Figure 2. Current fields (cm/s), calculated in the upper water layer (every fourth arrow was drawn): a – July 8, b – July 19.

Analyzing thermohaline fields, we noted thickening of isotherms and isohalines along the coast, separating warmer and less saline waters (along the coastline) and colder and more saline waters (in the deep-water part).

Vertical structure of the velocity, temperature, salinity was also investigated in order to estimate the influence of horizontal resolution on vertical profiles. We found a correlation between the results of experiments with a fine and coarse resolution during the calculating period, however some features of the circulation, associated with eddy formation, were absent in the experiment with a coarse resolution. In particular, anticyclonic eddies were distinguished as areas of deepening of the isopycnic surfaces, as well as reduced salt reserve and maximum heat reserve.

3.3. NWS, winter

Surface shelf currents were intense (maximum velocity was 100 cm/s), cyclonic vorticity was observed in the fields (the prevailing direction of currents was west and south-west) during the calculating period from 1 to 31 of January.

A field of currents had a complex mesoscale structure, characterized by eddies and jets. Eddies of various sizes and rotating signs were reproduced in the upper layer of water near estuary of the river Dniester, near Odessa, in central part of the region, in areas of Tendrovsky and Karkinit bays, as well as near the open boundary during the calculating period.

Eddy structures, associated with to the topography of the bottom, were revealed in the field of currents. There was a correspondence between the zones of eddy generation and topographic inhomogeneities located between 30.8 and 31.2°E and 31.8 and 32° E from 14 to 19 of January. A possible mechanism of formation of those eddies – influence of inhomogeneity in the bottom topography on the fluid flow.

Figure 3 shows mesoscale cyclonic eddies, reconstructed on January 19 at the depth of 10 m. Bottom relief played a crucial role in the formation of jets in the western part of the region (for example, intense jet about 15 km wide, directed to the north, was limited by 20–28 m isobaths).

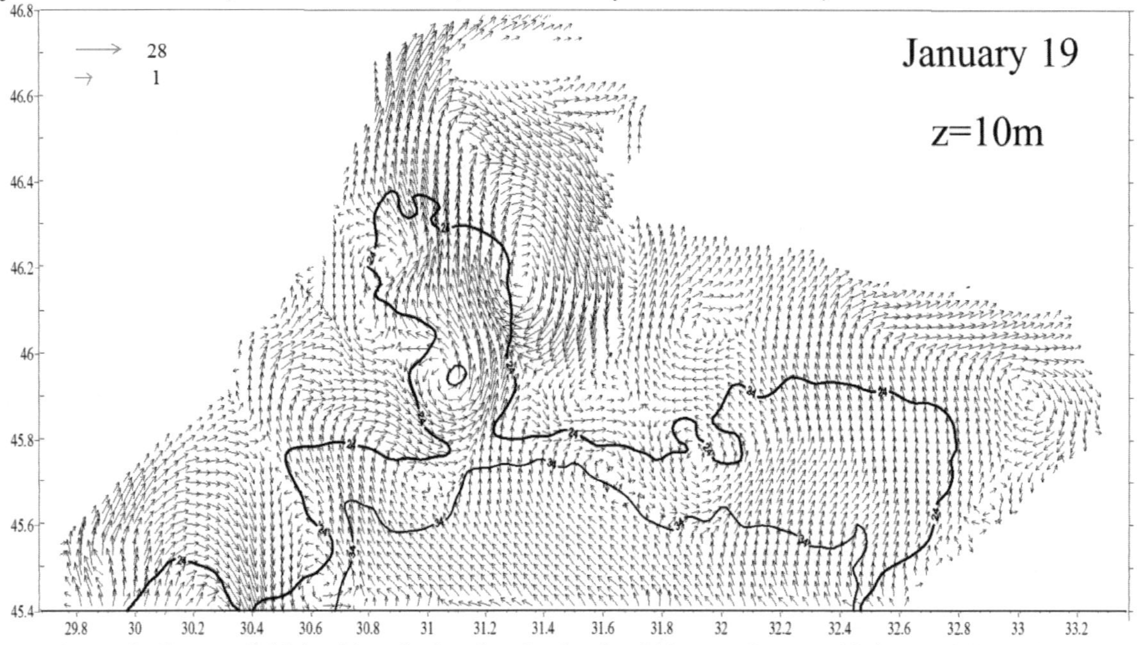

Figure 3. Current field (cm/s), calculated at the depth of 10 m on January 19 (every fifth arrow was drawn, isobaths 24 and 32 m were marked).

Intense jets, which trajectory coincided with the isobaths of 20–28 m, were directed to the south. They were observed below the depth of 10 m from 1 to 12 of January. Meandering of those currents could lead to the formation of eddies of different signs of rotation in the western part of the north-western shelf.

We compared current fields with the results with a more coarse resolution [14] and found a correlation between them during the calculating period, however a use of higher spatial resolution and more detailed bathymetry allowed to reconstruct the field structure more accurately. In particular, a number of eddies was absent and a field structure was smoother in the experiment with a lower resolution.

Thermohaline fields were also analyzed. Under the influence of the northern winds and heat fluxes with negative values, there was an intensive cooling of the shallow areas. The values of temperature varied from 0–1.0°C along the coastline to +7–8°C in the open part of the region. The salinity of the sea surface layer in the north-western part was 17–18‰, it decreased to 16‰ near the shores, salinity was about 0 – 1‰ in the region of river estuaries. A zone of relatively warm water, limited by an isotherm of 3°C and corresponding to an eddy formation in the field of currents, was formed in the center of the region.

3.4. NWS, summer

The main direction of surface shelf currents was western, southern and south-western (maximum velocity was 50 cm/s) during the calculating period from 1 to 31 of July. Eddy formations were reproduced in the upper 30-meter layer near Odessa, near estuary of the river Dniester, in areas of Tendrovsky and Karkinit bays, as well as in central part of the region and near the open boundary during the calculating period.

Mesoscale cyclonic eddy, which was repeatedly registered in satellite observations, was generated at a depth of 1–24 m between the meridians 30.8 and 31.2 E. A possible mechanism of formation of that eddy – influence of inhomogeneity in the bottom topography on the fluid flow. According to the calculation results, it was obtained from 5 to 8 and from 14 to 20 of July 2006. The lifetime of those eddies was 1–3 days.

Intense jets were generated and developed in the upper horizons, as well as deep horizons. Figure 4 shows current field at the depth of 10 m on January 16. We compared current fields with the results with a more coarse resolution [14], and found a correlation between them during the calculating period, however a use of higher spatial resolution and more detailed bathymetry allowed to reconstruct the field structure more accurately. In particular, a number of eddies was absent and a field structure was smoother in the experiment with a lower resolution (figure 4b).

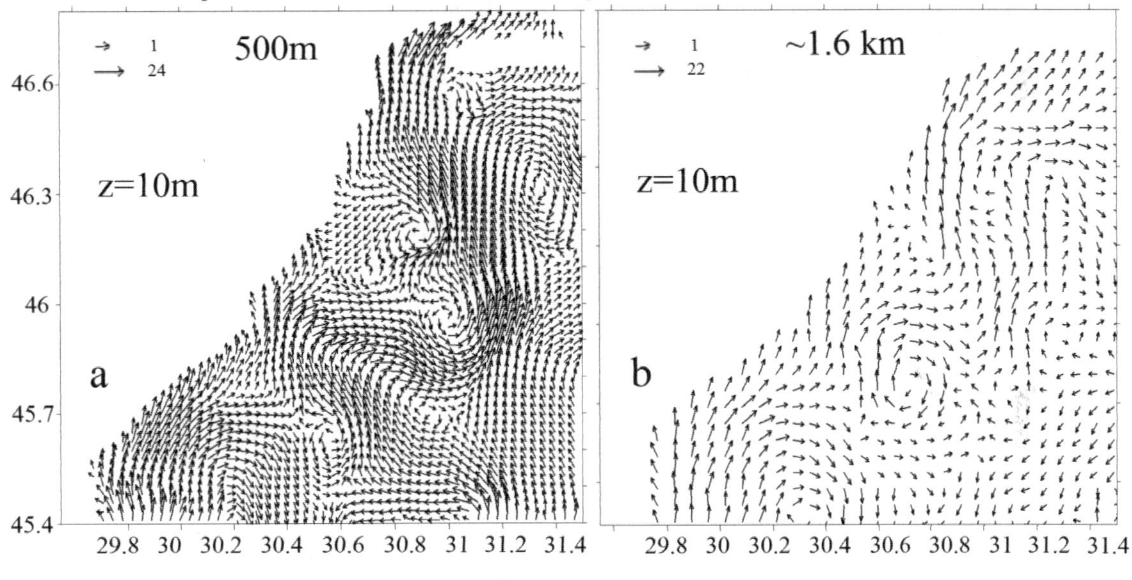

Figure 4. Current fields (cm/s) for the western part of NWS, calculated on July 16 at the depth of 10 m: a – with a resolution of 500 m (every fifth arrow was drawn), b – with a resolution of ~1.6 km (every second arrow was drawn).

4. Conclusions

Prognostic calculations of hydrophysical fields during the winter and summer periods with a spatial resolution of 350 m in the region of the SCC and 500 m in the area of the NWS have been carried out on the basis of the hydrodynamic model, adapted to the conditions of the coastal zone of the Black Sea with open boundaries. Possible reasons of the formation of meso- and submesoscale features of the coastal circulation were described.

A dominance of RC was typical for the region of SCC in the winter season, submesoscale eddy formations of various signs of rotation were obtained in the upper layer between the coastline and the RC (near the cape Ayu-Dag, in the eastern part of the region, in the Gulf of Yalta) by flowing the RC along coastal orography and the bottom relief.

A certain weakening of the jet of RC was noted in the summer season for the region of SCC, anticyclonic mesoscale eddy formations were generated between the coastline and the RC, moving along the direction of the RC. A possible mechanism of their formation was a baroclinic instability of the RC.

A field of currents of NWS in the winter and summer seasons had a complex meso- and submesoscale structure, characterized by eddies and jets. Eddies of various sizes and rotating signs were reproduced in the upper layer of water near estuary of the river Dniester, near Odessa, in central part of the region, in areas of Tendrovsky and Karkinit bays, as well as near the open boundary during the calculating period. A possible mechanism of formation of eddies in the central part – influence of inhomogeneity in the bottom topography on the fluid flow.

Bottom relief played a crucial role in the formation of jets in the western part of the shelf. Meandering of those currents could lead to the formation of eddies of various signs of rotation.

In comparison with previous calculations, new meso- and submesoscale eddies and jets have been reconstructed due to higher spatial resolution and more detailed bathymetry and coastline, some of them were confirmed by observational data.

This work was supported by the Russian Foundation for Basic Research (project no. 15–05–05423 A).

References

[1] Kostyanoy A, Ginzburg A, Sheremet N et al. 2010 Small-scale vortices of the Black Sea *Modern problems of remote sensing of the Earth from space* **7** 1 248–259.
[2] Zatsepin A, Baranov V, Kondrashov A et al. 2011 Submesoscale eddies at the Caucasus Black Sea shelf and the mechanisms of their generation *Oceanology* **51** 4 554–567.
[3] Rodionov A, Romanenko D, Zimin A et al. 2014 Submezoscale structures of the White Sea waters and their dynamics. State and direction of the investigations *Fundamental and Applied hydrophysics* **7** 3 29–41
[4] Karimova S. 2012 Statistical analysis of the submesoscale eddies of the Baltic, Black and Caspian seas by satellite radar data *Earth Research from Space* **3** 31–47.
[5] Diansky N, Fomin V, Zhokhova N et al. 2013 Simulations of currents and pollution transport in the coastal waters of Big Sochi *Izvestiya – Atmospheric and Ocean Physics* **49** 6 611–621.
[6] Capet X, McWilliams J, Molemaker M et al. 2008. Mesoscale to sub-mesoscale transition in the California current system. Part I: flow structure, eddy flux, and observational tests *J. Phys. Oceanogr.* **38** 29–43.
[7] Chelton D, Schlax M, Samelson R 2011 Global observations of nonlinear mesoscale eddies *Progress in Oceanography* **91** 167–216.
[8] Brannigan L., Marshall D., Naveira-Garabato A., GeorgeNurser A. 2015 The seasonal cycle of submesoscale flows *Ocean Modelling* **92** 69–84.
[9] McWilliams James C 2016 Submesoscale currents in the ocean *Proc. R. Soc. A.* **472**, 2189, http://dx.doi.org/10.1098/rspa.2016.0117.
[10] Vali G, Zhurbas V, Lips U, Laanemets J 2016 Submesoscale structures related to upwelling events in the Gulf of Finland, Baltic Sea (numerical experiments) *J. Mar.Syst.* http://dx.doi.org/10.1016/j.jmarsys.2016.06.010.
[11] Demyshev S, Korotaev G 1992 Numerical energy-balanced model of baroclinic currents of the ocean on the C-grid *Numerical models and results of calibration calculations of currents in the Atlantic Ocean* 163–231.
[12] Demyshev S, Evstigneeva N 2013 Numerical analysis of hydrophysical fields on the northwestern shelf of the Black Sea *Oceanology* **53** 5 511–524.
[13] Demyshev S, Evstigneeva N 2016 Modeling meso- and sub-mesoscale circulation along the eastern Crimean coast using numerical calculations *Izvestiya – Atmospheric and Ocean Physics 52* 5 560–569.
[14] Demyshev S, Dymova O Numerical 2013 Analysis of the Mesoscale Features of Circulation in the Black Sea Coastal Zone *Izvestiya – Atmospheric and Ocean Physics* **49** 6 655–663.
[15] Pacanowski R, Philander S 1981 Parameterization of vertical mixing in numerical models of tropical oceans *J. Phys. Oceanogr* **11** 11 1443–1451.
[16] Orlanski I 1976 A simple boundary condition for unbounded hyperbolic flows *J. Comput. Phys.* **21** 251–269.

IOP Conf. Series: Journal of Physics: Conf. Series **899** (2017) 052004 doi:10.1088/1742-6596/899/5/052004

Hybrid methods for simulating hydrodynamics and heat transfer in multiscale (1D-3D) models

S A Filimonov[1], E I Mikhienkova[2], A A Dekterev[1,2], D V Boykov[1]

[1] Kutateladze Institute of Thermophysics SB RAS, Lavrenteva av. 1, 630090, Novosibirsk, Russia
[2] Siberian Federal University, Svobodniy 79/10, 660041, Krasnoyarsk, Russia

E-mail: dekterev@mail.ru

Abstract. The work is devoted to application of different-scale models in the simulation of hydrodynamics and heat transfer of large and/or complex systems, which can be considered as a combination of extended and "compact" elements. The model consisting of simultaneously existing three-dimensional and network (one-dimensional) elements is called multiscale. The paper examines the relevance of building such models and considers three main options for their implementation: the spatial and the network parts of the model are calculated separately; spatial and network parts are calculated simultaneously (hydraulically unified model); network elements "penetrate" the spatial part and are connected through the integral characteristics at the tube/channel walls (hydraulically disconnected model). Each proposed method is analyzed in terms of advantages and disadvantages. The paper presents a number of practical examples demonstrating the application of multiscale models.

1. Introduction

In the contemporary world, the numerical simulation is practically indispensable when studying natural systems and designing technical objects. At solving problems of hydrodynamics, different modeling approaches are used depending on the type of the object (system, device, etc.). If the object can be represented in the form of a set of extended elements, in which the ratio of the characteristic dimensions differs in 10-1000 times, such as piping, channels, ducts, etc., then the modeling of such objects typically is carried out based on network (hydraulic) simulation [1, 2]. At that, the calculation of the desired characteristics of such a system (pressure, velocity, temperature, etc.) is carried out by methods based on the theory of hydraulic circuits (THC) [1]. The hydraulic circuit consists of a set of nodes and edges (a simple directed graph [3]). The laws of mass and energy conservation (Kirchhoff's laws) are ensured in the nodes, while the edges provide for the law of momentum, which describes the dependence of the pressure drop along the length of the edges depending on the flow rate.

Use of the computational fluid dynamics (CFD) methods allows obtaining the spatial distribution of flow characteristics and heat transfer in "compact" objects, that is the objects, whose ratio of the characteristic dimensions differs in 1-10 times [4]. These methods are based on the numerical solution of the Navier-Stokes equations. As a consequence, we obtain fields describing concerned object characteristics distributed over the volume, for example: velocity, pressure, temperature, concentration, turbulence characteristics, etc.

Each of these simulation approaches has its advantages and disadvantages. The undoubted advantage of network simulation is the relative simplicity of constructing a topology of a model and a low computational cost. The disadvantages may include the strong dependence of the model on empirical data (e.g., hydraulic resistance), characterizing the properties of the element (node or edge) as well as the possibility of obtaining only system characteristics averaged over the network element. Strong point of spatial CFD simulation is the possibility of modeling an object with complex geometry and describing the detailed flow structure, as well as less dependence of the model on empirical data, however, such simulation requires significant computational costs. Both simulation approaches have found wide application in solving various practical problems. Nevertheless, there are cases where simulation of the studied object within a single approach is quite difficult. In this case it is necessary to build multiscale model, which allows describing objects consisting of "compact" and extended elements. Examples of such simulation objects can be found in various areas of life:

- blood-vascular system consisting of cardiac muscle, aorta, major arteries and veins, on the one hand, and networks of fine blood vessels and capillaries, on the other hand;
- thermal and water supply system containing long pipeline sections and complex regulating, distributing and collecting devices;
- natural waterway systems, which include river channels with canals and delta, as well as hydro-technical utilities.

2. Basic methods of constructing multiscale models

There are several ways of implementing multiscale models, among which the main are the follows:

1. *fully-separated model*: the spatial and the network parts of the task are calculated separately;
2. *hydrodynamically unified model*: the spatial and the network parts are calculated simultaneously;
3. *hydrodynamically disconnected model*: network elements "penetrate" the spatial part, and the connection is implemented by using the integral characteristics at the walls of the tube/channel.

2.1 Fully separated model

When applying this model, spatial and network parts of the task are calculated separately and usually by means of different software products, at that, the calculation data obtained from one of the model parts determine the input parameters of the another part. The most striking example of the application of this method is the determination of the hydraulic resistance of non-standard element of the network model by means of spatial methods, or the definition of the input parameters of a complex device, after having calculated the lead-in pipeline fittings by network methods. Undoubted advantage of this method is the possibility to use ready-made software products to calculate both parts of the task without any modifications, because the transmission of information between the parts of the model is carried out through the peripheral data transfer. Though use of this modeling technique requires fulfillment of one of the following conditions that severely restricts its applicability:

- for the element, which is simulated by a three-dimensional calculations, there must be opportunity to build the relationship between flow rates in various inlets of the spatial element and pressure drops between them that may require a very large number of 3D calculations;
- pressure loss in the 3D element must have a negligibly small network effects on the network at large.

However, if one of the conditions is satisfied, then for further wide-sweeping calculations we can use only the network model, and hence reduce the total time required for the computational research.

Let`s demonstrate the application of the completely separated simulation method on the example of calculation of the exhaust ducts of the hot air supply system for the boiler at Yavinsky state district power plant (GRES). To determine the boundary conditions of the spatial model of the combustion chamber (figure1a) we have constructed a network model of the hot air supply system (figure 1b). The comparison of calculation results of the network model with the experimentally measured data on the hot air flow rate revealed a large discrepancy between the experimental data and calculation. Analysis of the results has shown that the problem lies in the duct, distributing out the hot air to the burner tiers (figure 1c). It should be noted that the duct design has one inlet and four outlets that in general case would require a large number of 3D calculations to determine the parameters of the network model. However, in this task the pressures at all four duct outlets are similar (equal to the pressure in the combustion chamber), and thus all the settings required for the network calculation can be obtained from a single 3D calculation.

The analysis of the spatial calculation of the duct revealed the presence of a large vortex in the delivery duct, which locks the flow and causes the non-uniformity of flow rates at the four outlets. Not considering the effect of resistance generated by the vortex leads to the discrepancy between the experimental data and results of the network simulation.

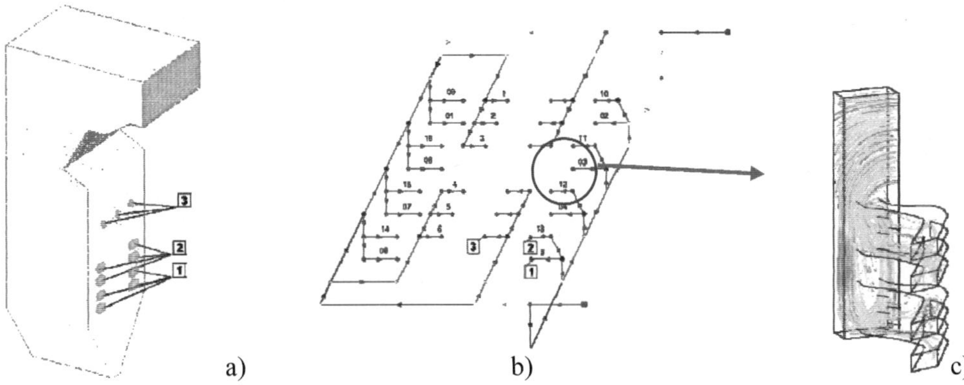
a) b) c)

Figure 1. The model of the hot air supply system for the boiler at Yavinsky GRES.
a) spatial model of the combustion semi-chamber of the boiler;
b) a network model of the hot air supply system;
c) delivery duct of the secondary blast; 1, 2, 3– burner tiers

2.2 The hydrodynamically unified model

If building a separated model for a certain task is impossible, i.e. it is impossible to obtain a definite dependence of the parameters of one part of the model on the calculation result of the other part, then it is necessary to use a hydrodynamically unified multi-scale model. In this approach, both model parts are calculated simultaneously, and the data exchange between the parts occurs iteratively. In the vast majority of applications of this method, different parts of the model are calculated by means of different software products. Therefore, there is a need to transfer data from one program to another using either third-party libraries, for example, MpCCi library is used to link ANSYS and Flowmaster [5], or via additional exchange files [6]. Even if the software package allows simulating both model parts, even then the problem of boundary conditions transfer remains relevant. For example, in [7] dealing with the simulation of blood flow, the values of flow rate are transferred downstream, while pressure values are transferred upstream.

The SigmaFlow software package formulates a single equation for the pressure correction for the entire computational domain, including the network and the spatial parts. This allows calculating the associated pressure field, while flow rates at the boundaries are calculated by interpolation (see [8]). This approach provides a faster convergence compared to traditional methods, as well as leads to unambiguous determination of the flow direction.

The example of practical use of such approach is the problem on simulation of the gas removal system at the aluminum plant facility . The gas removal system is designed to collect from electrolysis baths harmful gases and transfer them to the gas cleaning devices. The problem consists in the equalizing of flow rates from each bath in four spreads (1st and 2nd spreads contain 24 baths, 3rd spread contains 26 baths, and 4th spread contains 20 bath). The equalizing of the gas flow rates removed from the baths is achieved by means of the regulating gate valves. The main feature of the gas removal system consists in presence of the central collecting receiver, where pressure losses reach half of the total pressure loss in the investigated part of the system. In the general case, the simulation of this problem within the framework of one of the main approaches (fully 3D or fully THC) is impossible. The size of the computational grid to build a fully three-dimensional model of the whole facility will be too large, since the lengths of certain linear sections are more than 100 m. When building a fully-netted model it is necessary to know the hydraulic resistance of the collecting drain for different operating modes. The use of separated model with the definition of hydraulic resistance of the collecting drain and each branch tube based on preliminary spatial calculations is also not suitable, since the flow distribution along the ducts varies significantly in the regimes before and after flow rates equalization. Building a multi-scale model allows removing this obstacle. Extended elements are arranged in the form of a network, a collecting drain is presented in the form of spatial element (see figure 2), and the resistance of the collecting drain is calculated for specific flow rates.

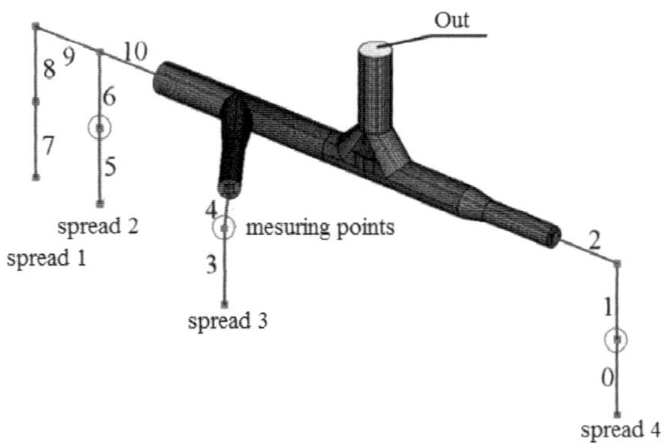

Figure 2. Multi-scale model of the gas removal system at the facility of aluminum plant. spread 1-4 – are the numbers of the spreades, 0-10 – are the numbers of calculation edges, circles show measuring points.

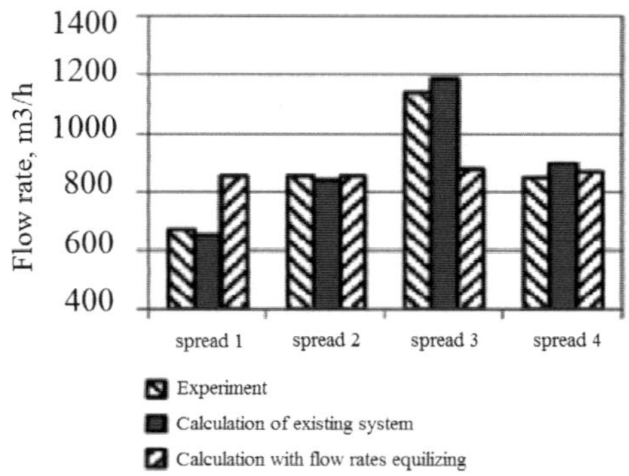

Figure 3. The calculation results of the gas removal system.

Exhaust duct network and the geometry of the collecting drain were constructed according to the design drawings.

Network elements of the hybrid model are divided into three groups:

1. Edges, whose parameters are defined based on the geometric characteristics (edges 2, 9, and 10).
2. Edges, which model the area for installation of the regulating gate valve (edges 1, 4, 6, and 8). Minimal length and resistance correspond to the opened gate valve.
3. Edges describing the spread of below building exhaust duct (edges 0, 3, 5, and 7).

The parameters of the third group edges were determined based on the experimental data.

The results of the hybrid calculation were compared with experimental data. The rarefaction at the outlet of the collecting drain amounted to 2416 Pa. After the equalization procedure of the flow rates from electrolysis baths, that is, identification of the additional resistance, a re-calculation was carried out, whose results are presented in figure 3. The rarefaction in the system required for obtaining the desired gas flow rates from the baths, rose up to 2868 Pa.

2.3 A hydrodynamically disconnected model

When modeling objects characterized by hydrodynamically disconnected or loosely connected flow of several fluids, we can use the third method to build multi-scale models. In this kind of models, the network one-dimensional elements "penetrate" spatial area. Tube bundles in heat exchangers are an example of such object. The connection between the network and the spatial parts of the model is

IOP Conf. Series: Journal of Physics: Conf. Series **899** (2017) 052004 doi:10.1088/1742-6596/899/5/052004

described through the source terms of the transport equation that are adjusted during each global iteration. Experience of the performed calculations has shown that additional inner iterations for adjustment of the source terms are not required.

When calculating tube-and-shell heat exchangers by empirical formulas, a number of problems arise due to the neglect of the uneven distribution over the heat exchanger volume of not only temperature, but physical properties of the coolant. This may lead to incorrect determination of thermal head and the hydraulic resistance of the tube bundle and consequently result in incorrect calculation of heat transfer and pressure drop.

When modeling tube-and-shell heat exchanger, it was necessary to carry out calculation of tube bundle consisting of six hundred tubes that would require the use of a computational grid consisting of about hundreds of millions of cells, and therefore the use of large computing power. When modeling the heat exchanger, employing the hybrid technique, each passage of the pipe from one tube board to the other is represented as a separate network element. A hybrid model of interaction between network and spatial elements allows calculating the heat fluxes and resistance coefficients, determining the temperature fields (figures 4, 5) and the substance properties, as well as the pressure drop in the system.

Figure 4. Temperature field in the central cross-section of the multi-scale heat exchanger model

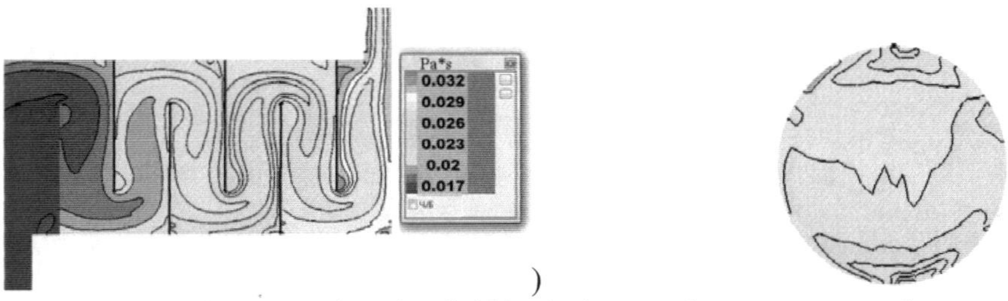

) b)

Figure 5. Viscosity field in the heat exchanger cross-section:
a) longitudinal cross-section; b) transverse cross-section.

At that, for the simulation we used the spatial grid consisting of 125,000 cells and a network model consisting of 25,000 edges, describing the pipelines that allowed performing calculations by means of a personal computer.

2.4 Combined model

The above described models can be combined. An example of a combination of hydrodynamically unified model and disconnected model is demonstrated on the example of heat transfer calculation in the cooling unit of the electronic chip with branched microchannel flow system. This is the conjugated heat transfer problem in the system with different spatial scales. The combination of the two methods consists in the fact that on the one hand the hydraulic part of the task consists of both spatial and network elements (channels), which requires the associated description, while on the other hand, thermal part of the problem describes the interaction between the cooling fluid and a chip through the energy exchange using hydrodynamically disconnected model. Comparison of the results was carried out based on the experimentally obtained temperature field and fully-spatial and hybrid calculations (figure 6). Quantitative verification of temperature distribution along the two given lines is shown in figure 7. The result of the comparison shows the adequacy of the application of the hybrid approach for solving this class of problems.

IOP Conf. Series: Journal of Physics: Conf. Series **899** (2017) 052004 doi:10.1088/1742-6596/899/5/052004

Conclusions

Despite the significant increase in computational power, many complicated problems of hydrodynamics and heat transfer cannot be resolved based on solving spatial task only at the expense of detail grid partitioning, even with wide-sweeping parallelization that is just extensive approach. One type of intense approach is the application of hybrid models. In this case, depending on the level of necessary information, different task areas are calculated using various models with different spatial resolution. The paper discusses ways of implementing hybrid models.

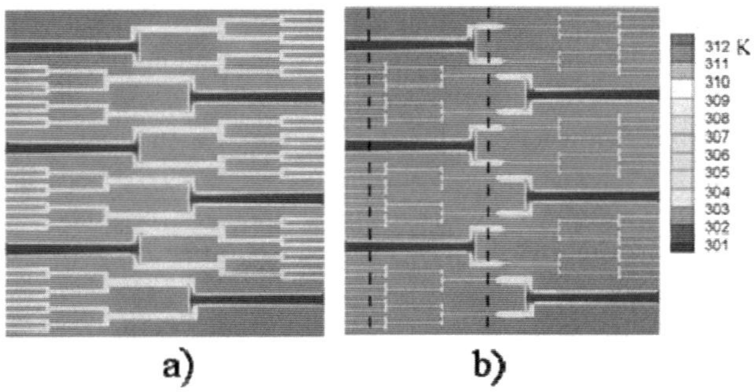

Figure 6. Comparison of temperature fields in the central cross-section:
a) spatial calculation; b) hybrid calculation.

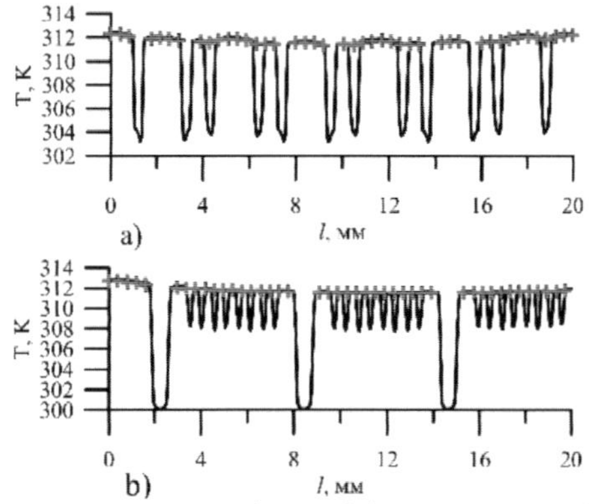

Figure 7. Comparison of the temperature in two cross-sections highlighted by the dotted line in figure 6b:
a) left cross-section, b) central cross-section; solid line shows spatial calculation, while markers correspond to hybrid calculation.

References

[1] Merenkov AP and Khasilev V.Ya. 1985 *Theory of hydraulic circuits*. (Moscow: Nauka). .

[2] Atavin AA. 1975 *Calculation of transient flow of water in ramified systems of riverbeds or channels*. Collected papers: Dynamics of a continuous medium **22**. pp. 25-36.

[3] Wilson R. 1996 Introduction to graph theory (Edinburgh Gate, Harlow, Essex CM20 2JE, England) p. 180

[4] Patankar S.V. 1980 Numerical heat transfer and fluid flow. (New York: Hemisphere Publishing Corporation)

[5] MpCCI 4.4.2-1 Documentation. 2015.

[6] Wedding WC. 2014 *Multiscale modeling of the mine ventilation system and flow through the gob. Ph.D. thesis in Mining Engineering*. (University of Kentucky).

[7] Dobroserdova, T.K. 2013 *Numerical simulation of blood flow in the presence of vascular implants or pathologies. Ph.D. thesis in physics and mathematics* (Moscow)

[8] Mikhienkova E.I., Filimonov S.A., and Dekterev A.A. 2016 *A hybrid algorithm for the joint calculation of multi-scale hydraulic problems taking into account thermal processes*. Computational methods and programming: new computing technology, **17** 4, pp. 380-392.

The reported study was funded by Russian Foundation for Basic Research, Government of Krasnoyarsk Territory, Krasnoyarsk Region Science and Technology Support Fund to the research project №16-48-242085

IOP Conf. Series: Journal of Physics: Conf. Series **899** (2017) 052005 doi:10.1088/1742-6596/899/5/052005

Numerical modeling of wind waves in the Black Sea generated by atmospheric cyclones

V V Fomin

Department of Numerical Technologies and Mathematical Modeling, Marine Hydrophysical Institute RAS, 2 Kapitanskaya Str., Sevastopol, 299011, Russia

E-mail: fomin.dntmm@gmaile.com

Abstract. The influence of the translation speed and intensity of atmospheric cyclones on surface wind waves in the Black Sea is investigated by using tightly-coupled model SWAN+ADCIRC. It is shown that the wave field has a spatial asymmetry, which depends on the velocity and intensity of the cyclone. The region of maximum waves is formed to the right of the direction of the cyclone motion. Speedier cyclones generate wind waves of lower height. The largest waves are generated at cyclonic translation speed of 7–9 m/s. This effect is due to the coincidence of the characteristic values of the group velocity of the dominant wind waves in the deep-water part of the Black Sea with the cyclone translation speed.

1. Introduction

Wind waves are among the main factors determining the safety of navigation, coastal zone dynamics that influence the coastal infrastructure, ecological processes and recreation potential. Severe storms are dangerous to vessels in open sea and coastal waters, storm waves can also cause severe damage along and on the coast. Therefore, the ability to predict wind waves is an important challenge and is of great value to many user communities. Recently, a large number of works dedicated to various aspects of mathematical modeling of the Black Sea wind waves has been published. The analysis of the storm climate in the western Black Sea on a continuous set of retrospective data covering the historical period of 63 years was performed in [1]. Numerical modeling of wind waves in the bays of the Crimean Peninsula was carried out in [2]. Assessments of interannual and seasonal wind waves parameter variability in the Black Sea were published in [3]. The performance of a wind-wave modeling system applied to the Black Sea basin was evaluated in [4]. In [5, 6], the results of numerical simulation of storm waves in the Black Sea using different wind forcing were presented. Spatiotemporal variability of the Black Sea wave climate for the last 37 years was investigated in [7].

The aim of this work is to study the sensitivity of wind wave fields in the Black Sea to changes in the characteristics of atmospheric cyclones. For the Black Sea, this problem has been studied very little. As is known [8], that atmospheric cyclones are capable of generating extreme wind waves. For specific basin, the intensity of wind waves is determined by the trajectory, intensity, and translation speed of atmospheric cyclones [9, 10].

2. Model description

The tightly-coupled model SWAN+ADCIRC [11] integrates two numerical models – Simulation Waves Nearshore (SWAN) [12, 13] and Advanced Circulation Model for Shelves Coasts and

Estuaries (ADCIRC) [14]. Both models are used to simulate the wind waves and storm surge in marine basins and lakes.

SWAN predicts the evolution in geographical space (x, y) and time t of the wave action spectral density $N(x, y, t, \theta, \sigma)$, where θ is the wave direction and σ is the relative frequency, by means of the action balance equation

$$\frac{\partial N}{\partial t}+\frac{\partial}{\partial x}\left[(c_x+U)N\right]+\frac{\partial}{\partial y}\left[(c_y+V)N\right]+\frac{\partial}{\partial \theta}(c_\theta N)+\frac{\partial}{\partial \sigma}(c_\sigma N)=\frac{S}{\sigma}, \tag{1}$$

where (c_x, c_y) are group velocity components; (U, V) are ambient current velocity components; c_θ and c_σ are propagation velocities in the θ and σ spaces. In (1) S is the source which represents the effects of wind generation, dissipation and nonlinear wave-wave interactions. The choice of SWAN parameters are given by Booij et. al. [12].

The ADCIRC basic equations are defined as follows

$$\frac{\partial U}{\partial t}+U\frac{\partial U}{\partial x}+V\frac{\partial U}{\partial y}-fV=-g\frac{\partial}{\partial x}\left[\eta+\frac{P_\mathrm{A}}{g\rho_0}\right]+\frac{\tau_{sx}-\tau_{bx}}{\rho_0 H}+\frac{M_x}{H}, \tag{2}$$

$$\frac{\partial V}{\partial t}+U\frac{\partial V}{\partial x}+V\frac{\partial V}{\partial y}+fU=-g\frac{\partial}{\partial y}\left[\eta+\frac{P_\mathrm{A}}{g\rho_0}\right]+\frac{\tau_{sy}-\tau_{by}}{\rho_0 H}+\frac{M_y}{H}, \tag{3}$$

$$\frac{\partial \eta}{\partial t}+\frac{\partial HU}{\partial x}+\frac{\partial HV}{\partial y}=0, \tag{4}$$

where (U, V) are current velocity components; η is the sea level; f is the Coriolis parameter; g is the acceleration of free fall; P_A is the atmospheric pressure; ρ_0 is the water reference density; $H=h+\eta$ is the total water depth; h is the unperturbed depth of water; M_x, M_y are the horizontal eddy-viscosity terms; (τ_{sx}, τ_{sy}) are surface stresses; (τ_{bx}, τ_{by}) are bottom stresses components. The following notations are used in (2)–(4):

$$\tau_{sx}=\rho_a C_a W_x\sqrt{W_x^2+W_y^2}+\tau_{wx},\ \tau_{sy}=\rho_a C_a W_y\sqrt{W_x^2+W_y^2}+\tau_{wy}, \tag{5}$$

$$\tau_{bx}=\rho_0 C_d U\sqrt{U^2+V^2},\ \tau_{by}=\rho_0 C_d V\sqrt{U^2+V^2}, \tag{6}$$

where (τ_{wx}, τ_{wy}) are wave stresses components; ρ_a is the air density; (W_x, W_y) are wind velocity components; C_a is the surface stress coefficient; C_d is the bottom stress coefficient. In (5), (6) the stress coefficients are given by: $C_a=(0.75+0.067\sqrt{W_x^2+W_y^2})10^{-3}$, $C_d=gn^2/H^{1/3}$, where $n=0.025$ is Manning's roughness.

SWAN computes wave fields by using of wind speed from an atmospheric model, as well as sea level and currents from ADCIRC. In its turn, in ADCIRC wave stresses obtained from SWAN is used. SWAN and ADCIRC are executed sequentially on the same unstructured grid and use the same set of computational cores. Both models use Message Passing Interface (MPI), which allows the use of parallel computing.

3. Results and discussion

In numerical experiments the finite element mesh consisted of 157,868 nodes and included the Black Sea and Sea of Azov. The size of triangular elements edges varies from 60 to 4000 m. Integration time steps in ADCIRC and SWAN were 3 s and 600 s respectively. The angular resolution in SWAN was set to 10°. A grid with 40 nodes was used for frequency coordinate in the range of 0.03–1 Hz.

The atmospheric pressure and wind speed components in the cyclone were set by the analytical model of a symmetric vortex proposed by Holland [15]:

$$P_A = P_C + \Delta P \exp[-(R/r)^B],\ W_g = \left[\frac{r}{\rho_a}\frac{dP_A}{dr} + \left(\frac{fr}{2}\right)^2\right]^{1/2} - \frac{fr}{2},$$

$$W_x = -\frac{\mu W_g}{r}(r_x \sin\alpha + r_y \cos\alpha),\ W_y = \frac{\mu W_g}{r}(r_x \cos\alpha - r_y \sin\alpha),$$

where $r = \sqrt{r_x^2 + r_y^2}$ is the distance from the center of the cyclone (x_c, y_c); $r_x = x - x_c$; $r_y = y - y_c$; P_C is the central low atmospheric pressure; $\Delta P = P_N - P_C$ is the atmospheric pressure anomaly; $P_N =$ 1000 mb is the ambient atmospheric pressure; W_g is the gradient wind speed; $R = 75$ km is the radius of the maximum wind speed; $B = 2.5$ is the scaling parameter; $\mu = 0.8$ is the reduced factor; $\alpha = 110°$ is the included angle of the wind speed.

Trajectory of motion of the cyclone center was determined as follows: $x_c = x_0 + Ct$; $y_c = y_0$, where $x_0 = 25°$E and $y_0 = 43.5°$N are the coordinates of the cyclone center for $t = 0$; C is the cyclone translation speed. The anomaly of atmospheric pressure ΔP, which determines the intensity of wind waves, varied within 10–20 mb. The values of C varied from 2 to 14 m/s.

Let's consider the effect of the cyclone translation speed on the spatial structure of the wave fields in the Black Sea. The distributions of the significant wave height (SWH) and the mean wave direction for the three characteristic values of C are shown in figures 1–3. Here the moments of time are chosen so that the geographical location of the cyclones is the same. It follows that a cyclone with a symmetrical wind distribution generates a wave field with a radial asymmetry caused by the cyclone's translational motion. The region of maximum wave heights is formed in the southern part of the cyclone (to the right in relation to the direction of the cyclone) and its configuration is determined by the value of C.

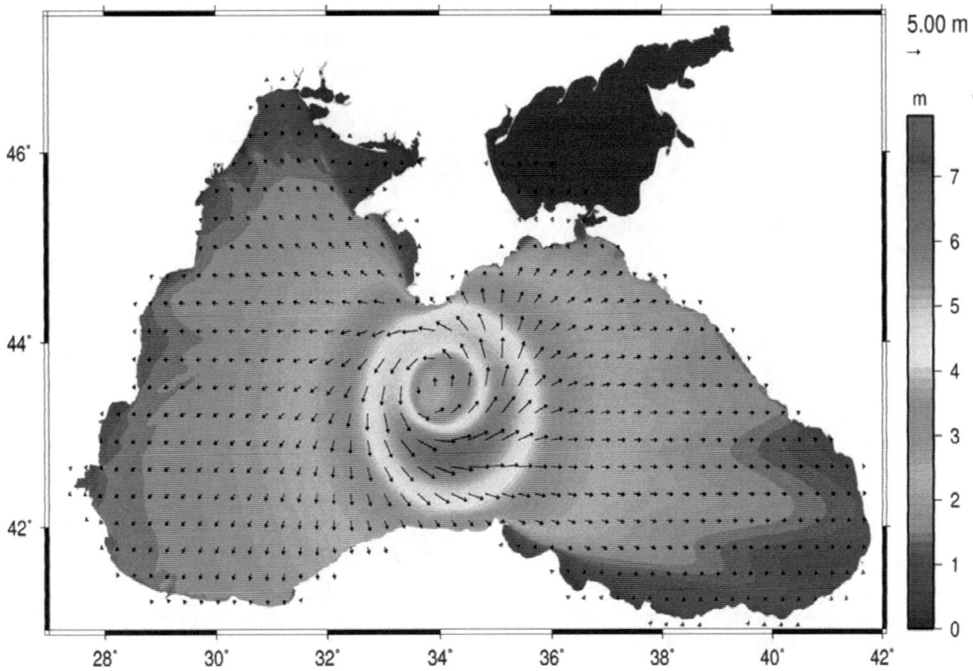

Figure 1. The distribution of significant wave height and mean wave direction for $C = 3.5$ m/s, ΔP = 15 mb and $t = 60$ h.

IOP Conf. Series: Journal of Physics: Conf. Series **899** (2017) 052005 doi:10.1088/1742-6596/899/5/052005

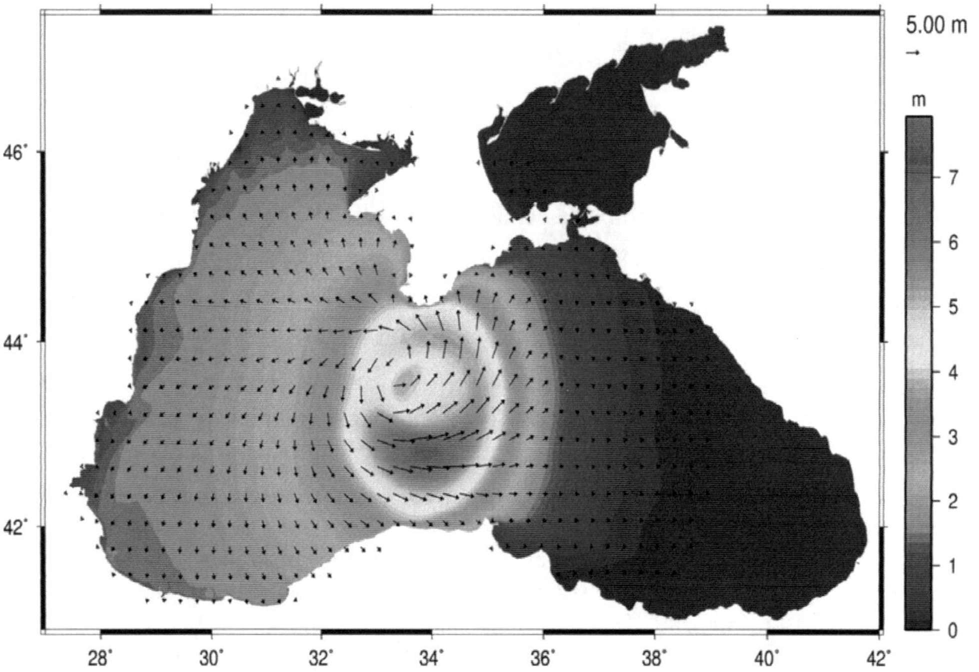

Figure 2. The distribution of significant wave height and mean wave direction for C = 7 m/s, ΔP = 15 mb and t = 30 h.

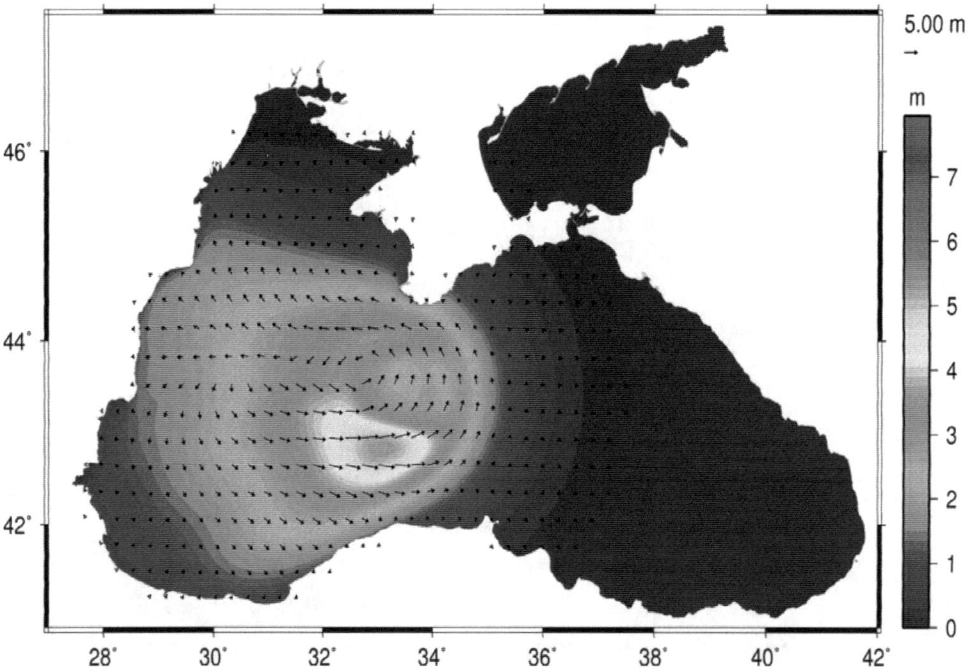

Figure 3. The distribution of significant wave height and mean wave direction for C = 14 m/s, ΔP = 15 mb and t = 15 h.

At C = 3.5 m/s, the asymmetry of the wave field is small (figure 1). In the eye region of the cyclone, the area of the wave height minimums with values of SWH ~ 2 m is well traced. In turn, at C = 7 m/s, such a region of minimums does not arise (figure 2), and the region of maximums in the southern part of the cyclone becomes more localized and the wave height in it increases by ~ 10%. At C = 14 m/s, the intensity of the wave field decreases noticeably, and its asymmetry increases (figure 3). This finds itself in the shift of the region of SWH maximums in the direction opposite to the direction of the cyclone. In all three cases, the mean wave direction does not coincide with the direction of the wind in the cyclone, which is especially noticeable at C = 14 m/s.

Calculations show that with an increase in the cyclone translation speed from 2 to 9 m/s, the wave heights also increase monotonically. On the contrary, when the cyclone translation speed exceeds 9 m/s, the wave heights begin to decrease monotonically. To illustrate this effect, figure 4 shows the dependence of the maximum SWH on the cyclone translation speed. Hence, it is clear that at a cyclone velocity of 7–9 m/s, the height of wind waves in the Black Sea will be the highest. This result is valid for the case when the cyclone moves over the deep sea part of the Black Sea.

The presence of a maximum of SWH can be explained by the resonant mechanism of wind wave generation by a moving cyclone [8]. The wind speed in the cyclone affects both the height of the generated waves and their period. In turn, the wave period determines the group velocity of the waves C_g (the velocity with which the energy of the dominant waves propagates). The higher the wind speed, the higher the C_g value. At $C_g \sim C$, the waves propagate at the same velocity as the cyclone. Remaining under the influence of strong wind for a long period of time, the waves will receive maximum energy from the wind. If $C < C_g$, then the waves will propagate ahead of the cyclone. Otherwise, if $C > C_g$, then the waves will lag behind the cyclone.

Let's consider the effect of atmospheric pressure anomaly (cyclone intensity) on wind waves. In figure 5 for C = 8 m/s and three values of ΔP, the time dependence of the dimensionless quantity S/S_0 is shown, where S is the part of the sea surface area for which the condition SWH $\geq$ 3.5 m is satisfied; $S_0 = 2\pi R^2$ = 35.3 km^2 is the characteristic scale of the area. Curve 1 corresponds to ΔP = 10 mb (maximum wind velocity, MWV = 18.7 m/s); curve 2 corresponds to ΔP = 15 mb (MWV = 23.4 m/s) and curve 3 corresponds to ΔP = 20 mb (MWV = 27.5 m/s). As can be seen, with increasing atmospheric pressure drop, the intensity of the storm increases. The maximum values of S/S_0 and SWH are respectively equal to: 0.54 (4.75 m); 3.57 (7.32 m); 5.73 (9.64 m).

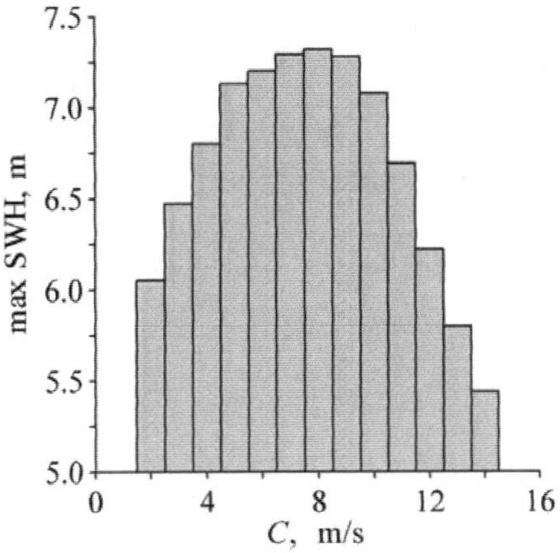

Figure 4. Maximum SWH values of waves generated by cyclones moving with different cyclone translation speeds for $0 < t < 4$ days and ΔP = 15 mb.

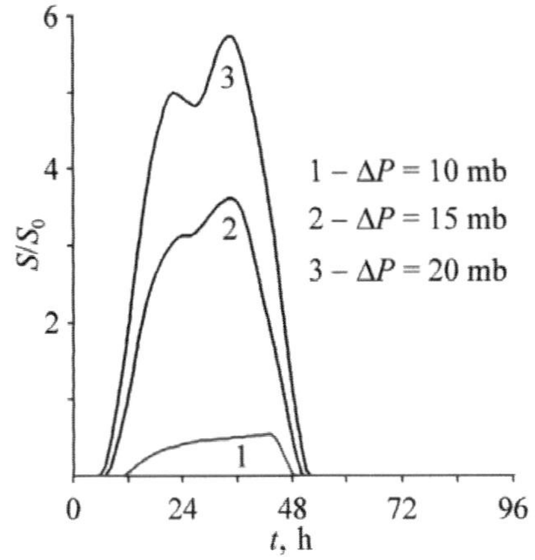

Figure 5. The area of intense waves S/S_0 as a function of time for different pressure anomalies ΔP and C = 8 m/s.

4. Conclusion

Mathematical modeling of wind waves in the Black Sea, generated by atmospheric cyclones with an asymmetric structure, is performed. The calculations were performed on the basis of the numerical model SWAN+ADCIRC using an unstructured calculation grid. An analysis of the results of numerical experiments showed that the wave field has a spatial asymmetry, the nature of which is determined by the translation speed and the intensity of the cyclone. The region of maximum wave heights is formed in the southern part of the cyclone (to the right in relation to the direction of the cyclone). Speedier cyclones generate wind waves of lower height. The highest values of wave heights are achieved at cyclonic velocities of 7–9 m/s. This effect is due to the coincidence of the characteristic values of the group velocity of the dominant wind waves in the deep-water part of the Black Sea with the cyclone translation speed.

5. Acknowledgment

This work was supported by Marine Hydrophysical Institute, research project No. 0827-2014-0010. The author thanks Crystal Fulcher from UNC-Institute of Marine Sciences for providing the source codes of SWAN+ADCIRC. Calculations were carried out on a computing cluster of Marine Hydrophysical Institute. Some ADCIRC visualizations were produced with FigureGen (Dietrich et. al., 2013).

References

[1] Valchev N N, Trifonova E V and Andreeva N K 2012 *Nat. Hazards Earth Syst. Sci.* **12** 961–77
[2] Fomin V V, Alekseev D V and Ivancha E V 2012 *Turkish Journal of Fisheries and Aquatic Sciences* **12** 363–69
[3] Arkhipkin V S, Gippius F N, Koltermann K P and Surkova G V 2014 *Nat. Hazards Earth Syst. Sci.* **14** 2883–97
[4] Rusu L, Bernardino M and Guedes Soares C 2014 *Journal of Operational Oceanography* **7** 5–20
[5] Van Vledder G and Akpınar A 2015 *Applied Ocean Research* **53** 161–78
[6] Myslenkov S A, Shestakova A A and Toropov P A 2016 *Russian Meteorology and Hydrology* **41** 706–13
[7] Divinsky B V and Kosyan R D 2017 *Continental Shelf Research* **136** 1–19
[8] Young I R 2003 *Marine Structures* **16** 201–18
[9] Moon I J, Ginis I and Hara T 2004 *J. Atm. Sci.* **61** 2334–48
[10] Liua H, Xiea L, Pietrafesab J and Baoa S 2007 *Ocean Modelling* **18** 37–52
[11] Dietrich J C, Zijlema M, Westerink J J, Holthuijsen L H, Dawson C N, Luettich R A, Jensen R E, Smith J M, Stelling G S and Stone G W 2011 *Coastal Engineering* **58** 45–65
[12] Booij N, Ris R C and Holthuijsen L H 1999 *J. Geophys. Res.* **104** 7649–66
[13] Zijlema M 2010 *Coastal Engineering* **57** 267–77
[14] Luettich R A and Westerink J J Formulation and Numerical Implementation of the 2D/3D ADCIRC Finite Element Model Version 44.XX 2004 available from http://www.unc.edu/ims/adcirc/adcirc_theory_2004_12_08.pdf (accessed 15.05.17)
[15] Holland G J 1980 *Mon. Wea. Rev.* **108** 1212–18

Numerical simulation of the impact of high-speed metallic plates using two approaches

S V Fortova[1], P S Utkin[1], T S Narkunas[2] and V V Shepelev[1]

[1] Institute for Computer Aided Design of the Russian Academy of Sciences, 19/18 2nd Brestskaya str., Moscow, Russia
[2] Moscow Institute of Physics and Technology, 9 Institutskiy per., Dolgoprudny, Moscow Region, Russia

E-mail: sfortova@mail.ru

Abstract. The paper is devoted to the mathematical modeling of the problem of two metal plates impact using two approaches. In the first approach the problem is solved using three-dimensional Euler equations and the stiffened-gas equation of state for the media. The parameters of the equation of state are calibrated using wide-range equations of state computations of the parameters of shock waves which form after the impact. The second approach is based on one-dimensional two-fluid seven-equation model. In simulations of metal plates impact we get two shocks after the initial impact that propagate to the free surfaces of the samples. The characteristics of shock waves are close (maximum relative error in characteristics of shocks is not greater than 7%) to the data from the wide-range equations of states computations.

1. Introduction

The study of the mechanisms and the laws of hydrodynamic instability development and turbulent flows in different fundamental and practical problems is among the most complicated modern issues of the continuum media mechanics. The best practices in natural and numerical experiments show that the modeling of the hydrodynamic instability development should be based on three-dimensional (3D) models the program realization of which demands huge computational costs.

The authors of the paper have developed for several years expandable software for the numerical solution of 3D Euler equations using different numerical schemes including high-order methods [1, 2]. The software provides the solution of some problems of the hydrodynamic instability development theory. The obtained results however relates to the flows of ideal gas. At the same time the effects of hydrodynamic instability development are also take place in the media with much more complex equations of state (EOS). In natural experiments [3] on high-speed impact of metal plates, the 3D structures were found on the contact surface of the plates and one of the possible reasons for the effect is the development of Rayleigh–Taylor instability. In our previous works [4, 5] we got the principal effect of 3D structures formation on the contact surface between colliding plates in the statement qualitatively corresponded to the experiment in [3]. The 3D Euler equations simulations for the media with the ideal gas equation of state were used. The quantitative correspondence however was not good. So the goal of the current work is the development of the mathematical model and the computational algorithm from [4] to obtain better agreement between the parameters of the process of high-speed metal plates impact.

The problem has a long history of investigations and for the theoretical studies different numerical approaches were used each with its own pros and cons. In the early studies of S.K. Godunov with co-authors, the problem was considered in two-dimensional Lagrangian statement [6, 7]. In [4, 8] as a result of two- and tree-dimensional gas-dynamics modeling the multidimensional structures at the contact surface were obtained and the assumptions about the mechanisms of the structures formation were done. In [9], the numerical simulations of waves formation under an oblique impact of metal plates during explosion welding on the basis of Maxwell relaxation model and molecular dynamics method were carried out. It was shown that the numerical simulation correctly reproduces the formation and evolution of waves on the contact boundary. The problem was also considered in [10] using powerful Euler–Lagrangian 3D computer code taking into account wide-range EOS of the metals.

The problem of high-speed impact of two plates is actually two-fluid problem at least at the initial stages of the process when the metals behave as non-mixing compressible fluids. At the same time, there is a lack of the works with consideration of the problem using the models of heterogeneous media mechanics. Another goal of the paper is the formulation of the one-dimensional (1D) two-fluid mathematical model, computational algorithm and fitting of the parameters of the numerical technology for both qualitative and quantitative description of the initial stage of two metal plates impact in the statement of [3].

2. Statement of the problem

Consider the interaction of the lead plate with the thickness 2 mm (parallelepiped 5 mm × 5 mm × 2 mm in 3D, a segment [3 mm; 5 mm] in 1D statement) and the initial density 11 300 kg/m^3 with the steel plate with the thickness 3 mm (parallelepiped 5 mm × 5 mm × 3 mm in 3D, a segment [0 mm; 3 mm] in 1D statement) and the initial density 7 900 kg/m^3 (see figure 1). The lead plate is thrown to the direction of steel one with the velocity 500 m/s. It is assumed that at the initial stage of the impact during first 10 μs the metals behave as pseudo-fluids [3] so the gas dynamics approach is valid. The initial pressure is 10^5 Pa everywhere.

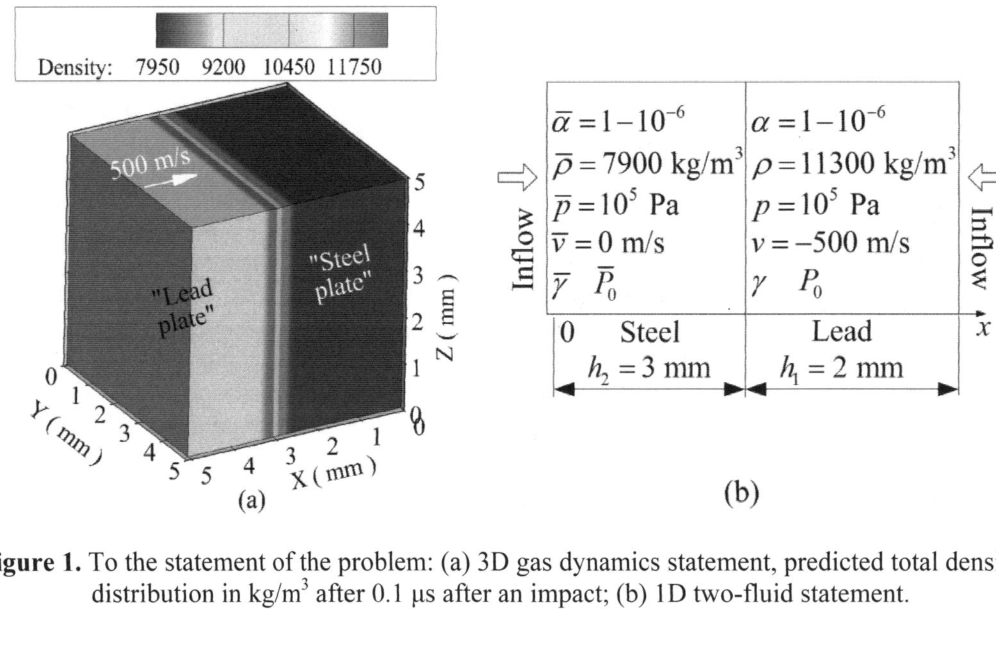

Figure 1. To the statement of the problem: (a) 3D gas dynamics statement, predicted total density distribution in kg/m^3 after 0.1 μs after an impact; (b) 1D two-fluid statement.

We set the inflow conditions at the boundaries $x = 0$ mm and $x = 5$ mm, periodic conditions at the others in 3D statement. The inflow values at $x = 0$ mm corresponds to the initial conditions for the steel plate and at $x = 5$ mm to the lead one. As a result of metal plates interaction, two shock waves (SW) are formed. The computation lasts up to the moment of SW arrival to the boundaries $x = 0$ mm or $x = 5$ mm. Computational grid is uniform with the cell size equal to 10^{-3} mm in 1D statement and contains 5 000 cells. Computational grid is uniform and contains $100 \times 100 \times 100 = 10^6$ cells in 3D statement.

3. Mathematical models and numerical algorithms

For the 3D statement the mathematical model is based on 3D non-stationary two-component Euler equations [1, 2] supplemented by stiffened gas EOS [7] with the parameters γ and P_0. Numerical algorithm is based on the spatial directions splitting technique and hybrid grid-characteristics method [1, 2]. The Roe averaging procedure is used for all matrixes (Jacobian, the matrixes of left and right eigenvectors for the determinative system of equations) coefficients computations. The procedure takes into account the stiffened gas EOS [11]. The numerical approach in use is valid for the general EOS if the partial derivatives $(\partial p/\partial \varepsilon)_\rho$ and $(\partial p/\partial \rho)_\varepsilon$ are known (here and further the notations are standard).

The second approach is based on the system of equations from [12] which describes two-fluid compressible flows. Governing system of equation comprises the mass, momentum and energy conservation laws for both fluids. The specific features of the model are compressibility of fluids, velocities and pressures non-equilibrium, the absence of conservative form of notation. Conservation laws are supplemented by transfer equation for the volume fraction for one of the fluids. The stiffened gas equations of state are used for each fluid with different values of γ and P_0 parameters. Computational algorithm for one time step integration consists of three stages. At first, the hyperbolic step is carried out using HLL numerical scheme and then the velocities relaxation procedure and the pressures relaxation procedure [13].

4. Results of the numerical experiments

4.1. 3D gas dynamics simulations

To reproduce quantitatively reasonable characteristics of metal plates impact, the parameters of stiffened gas EOS should be fitted using either experimental data or the results of simulations taking into account the wide-range EOS for metals. We use the results of simulations [14] obtained using wide-range EOS [15] and the numerical approach [10]. The results from [14] will be referred as reference values further, see table 1. Here and elsewhere SW in steel plate will be called the "left"" one and in lead plate – the "right" one in accordance with the statement in figure 1.

Table 1. Reference parameters behind the shock waves in lead and steel.

SW	Pressure, GPa	Density, kg/m^3	Gas velocity, m/s	SW speed, km/s
In steel	7.99	8 246	211	4.72
In lead	7.99	12 830	211	1.97

Table 2. Predicted characteristics of SWs in calculation in 3D statement.

SW	Pressure, GPa	Density, kg/m^3	Gas velocity, m/s	SW speed, km/s
In steel	7.24 (–7%)	8 595 (+4%)	272 (+29%)	3.40 (–28%)
In lead	7.24 (–7%)	12 294 (–5%)	272 (+29%)	2.38 (+28%)

Figure 2 illustrates the predicted fields of density and pressure at the time moment 0.5 μs after the impact. The density distribution in figure 1a at the earlier time moment shows both shock waves and contact surface which moves in the direction of the steel plate. The figures 1a and 2 correspond to the parameters of stiffened gas EOS γ = 3.0 and P_0 = 25 GPa. The parameters were found as a result of parametric calculations in which we compared the calculated characteristics of both SWs with the reference values. Table 2 demonstrates the finally achieved agreement with the relative errors.

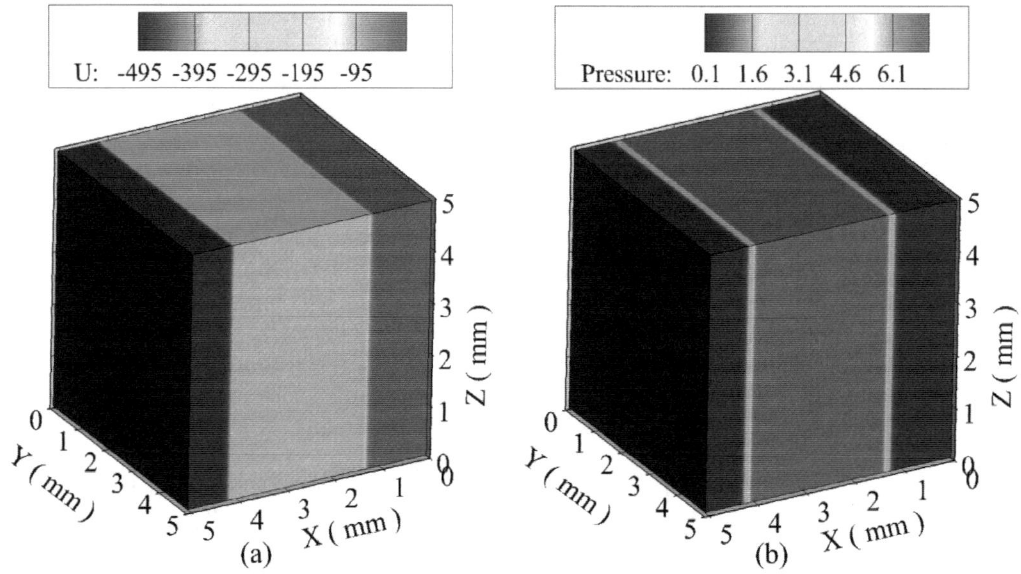

Figure 2. Predicted distributions of (a) velocity in m/s and (b) pressure in GPa after 0.5 μs after an impact in 3D statement.

4.2. 1D two-fluid model simulations

To satisfy the hyperbolicity conditions for the determinative system of equations, all computational cells should contain both fluids so on the interval [0 mm; 3 mm] we set the volume fraction of lead equal to 10^{-6} and on the interval [3 mm; 5 mm] we set the volume fraction of steel equal to 10^{-6}. A series of numerical experiments was carried out in order to fit the results of the calculations with stiffened gas EOS to the results of modeling using wide-range EOS. The parameters γ and P_0 for lead and steel plates were varied and the shock speeds D_L and D_R, contact velocity v_{cont} and pressure p_{cont}, post-shock densities (see table 3) were compared with the reference values (see table 1). For each parameter, the relative error in comparison with the reference value was measured.

Table 3. Predicted characteristics of SWs in calculation in 1D two-fluid statement.

No	Lead		Steel		p_{cont}, GPa	v_{cont}, m/s	D_L, km/s	D_R, km/s
	γ	P_0, GPa	γ	P_0, GPa				
1	3.9	65.5	3.0	30.0	10.1 (+26%)	215 (+2%)	6.0 (+27%)	2.7 (+38%)
2	3.9	65.5	3.0	60.0	11.9 (+49%)	251 (+19%)	6.0 (+27%)	3.8 (+94%)
3	3.9	65.5	4.0	30.0	11.0 (+38%)	232 (+10%)	6.1 (+29%)	3.2 (+64%)
4	3.9	65.5	3.0	15.0	8.5 (+6%)	182 (–14%)	6.0 (+27%)	1.9 (–4%)
5	3.0	65.5	3.0	15.0	8.0 (+0%)	196 (–7%)	5.2 (+11%)	1.8 (–7%)
6	2.9	65.5	3.0	15.0	8.0 (+0%)	198 (–6%)	5.1 (+8%)	1.8 (–7%)
7	2.7	65.5	3.0	15.0	7.8 (–2%)	202 (–4%)	5.05 (+7%)	1.9 (–4%)

We started with the EOS parameters for the steel plate from [7] and some default parameters for the lead one. The largest error was obtained for the contact surface pressure and SW speeds. In calculations No. 2 and 3 the values of γ and P_0 for steel plate were increased. The errors for mentioned parameters became larger. The post-shock densities remained almost the same and it is the general tendency for the following numerical experiments (the maximum relative error is 7% for all numerical experiments). The positive dynamics for the contact pressure and right SW speed was obtained with the decrease of P_0 for the steel plate in calculation No. 4. At the same time, the left SW speed was insensitive to the variation of the EOS parameters in calculations No. 1 – 4. In calculation No. 5 – 7 we successively decreased the value of γ for the lead plate and obtained the error 7% for the left SW speed. Figure 3 illustrates the predicted distributions of steel volume fraction and density, lead density and contact pressure for the EOS parameters No. 7 in table 3. Note that the numerical scheme is quite diffusive and should be improved (see for example [16]) but at the same time provides the robust calculation for the huge pressure gradients up to the 5 orders of magnitude.

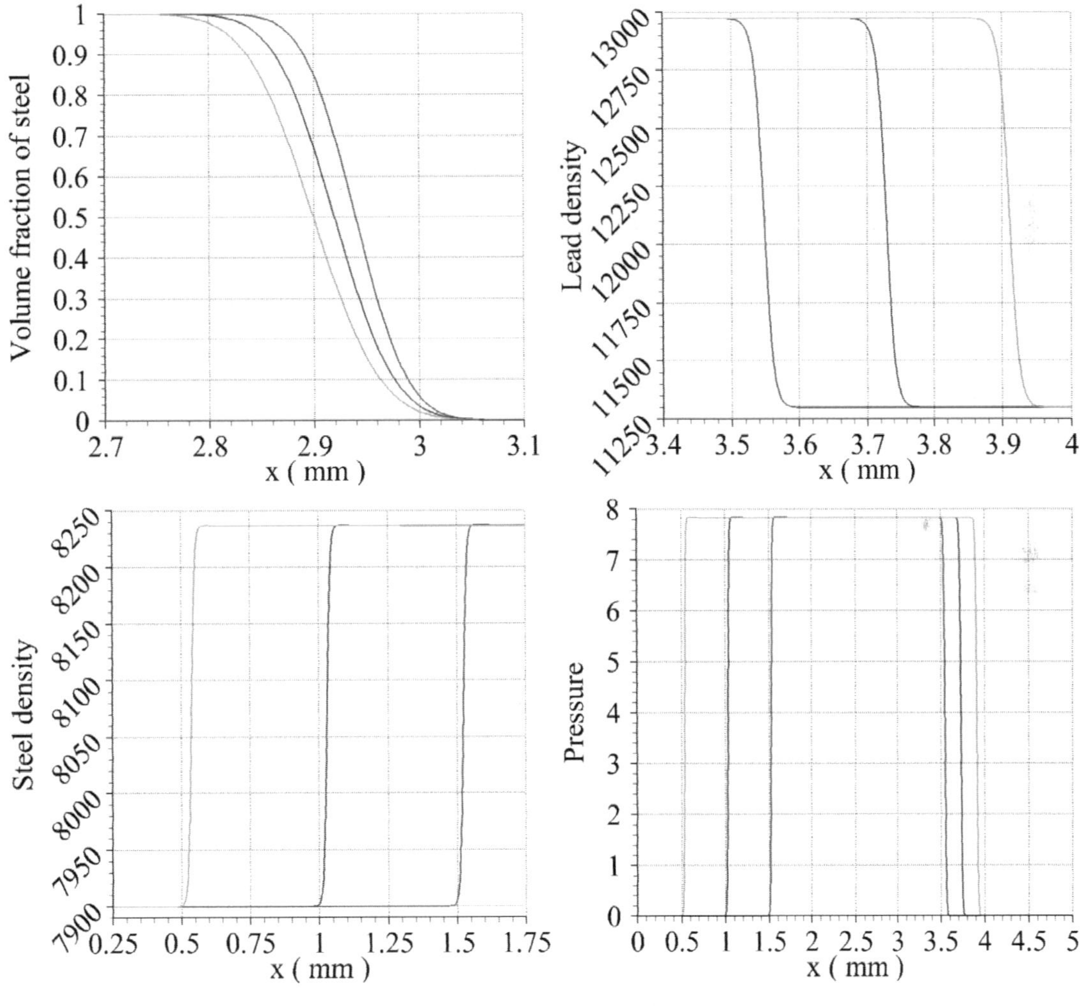

Figure 3. Predicted distributions at the successive time moments in 1D two-fluid simulation. Red lines correspond to 0.3 mcs time moment, blue lines – 0.4 mcs, green lines – 0.5 mcs. Density scales are in kg/m^3, pressure scale is in GPa.

5. Conclusions

The 3D mathematical model based on the multicomponent Euler equations supplemented with the stiffened gas equation of state for the numerical investigation of the metal plates impact is described.

The parameters of the stiffened gas equation of state are calibrated on the basis of the computations using wide-range equation of state for the metals. For the characteristics of the shock waves which are formed after the impact of two metal plates, the maximum error is in the range of 30%. The proposed numerical technique is suitable for the 3D numerical investigations of intensive directed energies flows on the substances taking into account the instabilities development on the contact surface.

The parametric numerical study of two metal plates impact using two-fluid approach in comparison with the simulations taking into account wide-range EOS for the metals is carried out. In calculations, the formation of two SW is obtained with the characteristics with maximum relative error 7% in comparison with the reference values. The determinative system of equations [12] is used. Both metals are considered to be weakly compressible media with stiffened gas equations of state. The computational algorithm is based on the operator splitting approach and HLL method for the approximation of the numerical flux and the non-conservative right-hand side terms.

The work of S.V. Fortova, P.S. Utkin and V.V. Shepelev (sections 1 – 3 and 4.2) is supported by the Russian Science Foundation under grant 17-11-01293 and performed at the Institute for Computer Aided Design of RAS.

References

[1] Fortova S, Kraginskii L, Chikitkin A and Oparina E 2013 *Math. Mod. Comp. Simul.* **5** 607
[2] Belotserkovskaya M, Pronina A, Fortova S and Shepelev V 2016 *Comp. Math. Math. Phys.* **56** 1162
[3] Yakovlev I 1973 *Fizika goreniiy i vzryva* **9** 447
[4] Belotserkovsky O, Fortova S, Troshkin O, Pronina A, Eriklintsev I and Kozlov S 2016 *Math. Mod. Comp. Simul.* **8** 497
[5] Fortova S, Shepelev V, Pronina A and Utkin P 2016 *J. Phys.: Conf. Ser.* **774** Paper No. 012028.
[6] Godunov S, Deribas A, Zabrodin A and Kozin N 1970 *J. Comp. Phys.* **5** 517
[7] Godunov S, Zabrodin A, Ivanov M, Krayko A and Prokopov G 1976 *Numerical Solution of the Multidimensional Gas Dynamics Problems* (Moscow: Nauka)
[8] Demchenko V and Sergeev M 2003 *Fluid Dynamics* **38** 923
[9] Godunov S, Kiselev S, Kulikov I and Mali V 2013 *Proc. Steklov Inst. Math.* **281** 12
[10] Fortov V, Kim V, Lomonosov V, Matveichev A and Ostrik A 2006 *Int. J. Impact Eng.* **33** 244.
[11] Kulikovskii A, Pogorelov N and Semenov A 2000 *Mathematical Aspects of Numerical Solution of Hyperbolic Systems* (Chapman and Hall/CRC)
[12] Saurel R, Abgrall R 1999 *J. Comp. Phys.* **150** 425
[13] Saurel R, Lemetayer O 2001 *J. Fluid Mech.* **431**, 239
[14] *Shock Wave Database*, http://www.ihed.ras.ru/rusbank/
[15] Fortov V, Khishchenko K, Levashov P and Lomonosov I 1998 *Nucl. Instr. Meth. Phys. Res. A.* **415** 604
[16] Liang S, Liu W, Yuan L 2014 *Comp. Fluids* **99** 156

Numerical simulation of the hydrodynamic instabilities of Richtmyer-Meshkov and Rayleigh-Taylor

S.V. Fortova[1], V.V. Shepelev[1], O.V. Troshkin[1], S.A. Kozlov[1]

[1] Institute for Computer Aided Design of the Russian Academy of Sciences (ICAD RAS), 123056, 19/18, 2nd Brestskaya str., Moscow, Russia,

sfortova@mail.ru

Abstract. The paper presents the results of numerical simulation of the development of hydrodynamic instabilities of Richtmyer-Meshkov and Rayleigh-Taylor encountered in experiments [1-3]. For the numerical solution used the TPS software package (Turbulence Problem Solver) that implements a generalized approach to constructing computer programs for a wide range of problems of hydrodynamics, described by the system of equations of hyperbolic type. As numerical methods are used the method of large particles and ENO-scheme of the second order with Roe solver for the approximate solution of the Riemann problem.

1. Introduction.

An important section of modern computational fluid dynamics is the numerical simulation of the growth and development of hydrodynamic instabilities of the continuum. The complexity and specificity of the task and the possibility of comparison with experimental data allows not only to analyze the phenomenon, but also to use such tasks for the study of differential methods and verification of hydrodynamic codes. Examples of phenomena under investigation are wakes in the atmosphere and the ocean, the interaction of a blowing jet with the basic flow, the interaction of intense laser light with matter, many atmospheric phenomena, etc. A characteristic feature of these processes is the presence of high gradients of flow variables and high Reynolds numbers [4]. According to the Kolmogorov–Obukhov theory [5, 6], turbulent motion at high Re involves several ranges, namely, the energy, inertial, and dissipation ranges. For a broad class of phenomena at high Reynolds numbers, the effect of molecular viscosity on general flow characteristics in the energy and inertial ranges is overall insignificant. Accordingly, such phenomena are numerically studied as based on perfect fluid models, namely, on the Euler equations [4]. The work is devoted to numerical modeling of various natural experiments of hydrodynamic instabilities of Richtmyer-Meshkov [1,2] and Rayleigh-Taylor [3] using techniques of the TPS (Turbulence Problem Solver) software complex [7].

2. Experiments.

2.1. Richtmyer-Meshkov instability.

In [1] the nonlinear stage of growth of the RMI is experimentally investigated, immediately following the linear stage described by Richtmyer [2]. The experiments were carried out in a square tube with height of h_0 (figure 1), containing Zone 0 filled with xenon and Zones 1 and 2 filled with argon separated by a lavsan film. In the Zones 0 and 1 both gases are at rest, and the system is in equilibrium. Initial data was set as follows:

$\rho_0 = 5.894\ kg/m^3,\ \rho_1 = 1.784\ kg/m^3,$

$u_{0,1} = \upsilon_{0,1} = 0,$

$p_{0,1} = 50000\ Pa.$

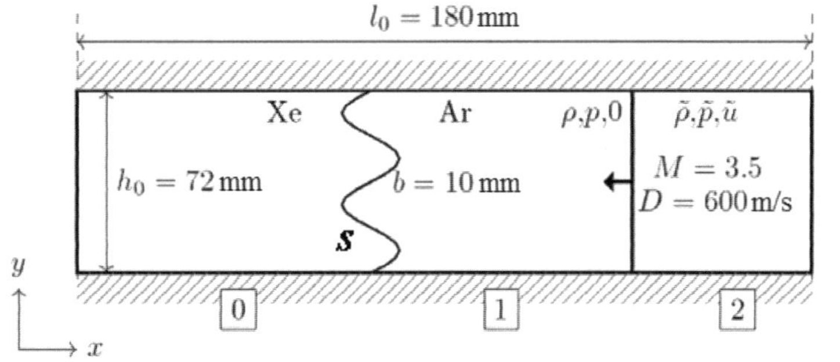

Figure 1. The scheme of the experiment on the development of Richtmyer-Meshkov instability.

To the Zone 1 at speed of 600 m/s (Mach number equals 3.5) from the right side moves the front of Zone 2 which contains compressed argon:$\rho_2 = \tilde{\rho}, u_2 = \tilde{u}, v_2 = \tilde{v}, p_2 = \tilde{p}$. For initial conditions to be correctly specified we need to calculate the state of the medium behind the shock wave. This state should specify the correct front speed. As it is known, the jumps of the values on the discontinuity are connected with the speed of the front by the Rankine-Hugoniot relations [8-10]. Thus we can obtain the initial conditions in the Zone 2 in figure 1:

$$\tilde{u} - D = \frac{\sigma(-D)}{2+\sigma} + \frac{2(1+\sigma)p}{\rho(2+\sigma)(-D)} > 0,$$

$$\tilde{p} = p + \rho(D-u)(\tilde{u}-u),$$

$$\tilde{\rho} = \rho\frac{D-u}{D-\tilde{u}}.$$

Several experiments were carried out at different initial perturbation of the boundary. The length of the initial perturbation varied at a constant amplitude: 72mm (one-mode perturbation), 36 mm (two modes), 24 mm (three modes), 12 mm (6 modes) and 8 mm (9 modes). After 140 μs after the beginning of the interaction between the shock wave and films in the pipe the registration of the flow pattern on a photographic film was made. By this moment in the latter two cases the stage of fully developed turbulence practically occurred. For the first three variants of the experiment in the section 5.1 the results of the numerical calculation are presented.

2.2. Rayleigh-Taylor instability.

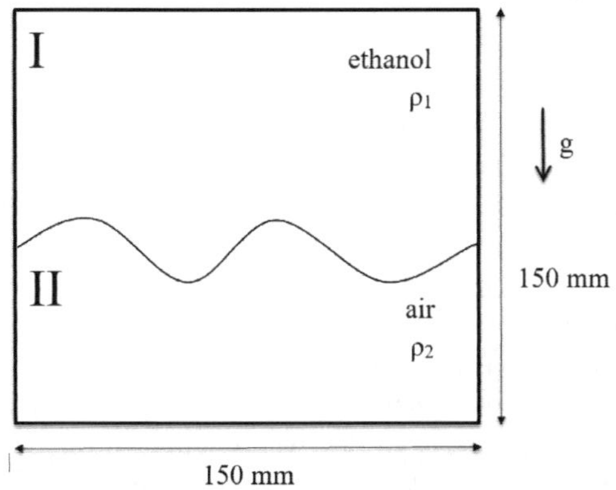

Figure 2. The scheme of the experiment on the development of Rayleigh-Taylor instability. Zone I is filled with ethanol, zone II is filled with air.

Consider one of the well-known experiments [3] to investigate the growth and development of this type of instability. The experimental installation is shown in figure 2. It represents a square-shaped tank which thickness is substantially less than width and height, i.e. the experiment can be considered as two-dimensional. Initially, by a thin partition the tank is separated into two parts, the upper being filled with ethanol, and the lower being filled with the air. The ratio of densities ρ_1/ρ_2 =600. The bending of the partition specifies the initial disturbance of the boundary between the liquids is $\zeta = a_0 \sin(2\pi x/\lambda)$, where $a_0 = 0.5$ mm, $\lambda = 65$ mm. Due to the special rocket engines, the tank is accelerated to significant accelerations in the range of $25 - 70g_0$, where g_0 is the acceleration of gravity. The process is recorded on photographic film.

3. Mathematical formulation.

Traditionally, the classical laws of conservation of mass, momentum and energy of a continuous medium describe the considered problems. With their help, the emergence of vortices can be naturally observed which is discovered in the experiments [1-3]. The basis of the approach is the system of these laws in the approximation of the ideal medium of Euler, describing two-dimensional inviscid flow of compressible continuous medium:

$$\frac{\partial \mathbf{U}}{\partial t}+\frac{\partial \mathbf{F}}{\partial x}+\frac{\partial \mathbf{G}}{\partial y}=\mathbf{R}.$$

Here $\boldsymbol{U}$ is the vector of conservative variables, $\boldsymbol{F}, \boldsymbol{G}$ are flux vectors:

$$\mathbf{U}=\begin{pmatrix}\rho\\ \rho u\\ \rho \upsilon\\ \rho E\end{pmatrix},\ \mathbf{F}=\begin{pmatrix}\rho u\\ \rho u^2+p\\ \rho u\upsilon\\ u(\rho E+p)\end{pmatrix},\ \mathbf{G}=\begin{pmatrix}\rho\upsilon\\ \rho u\upsilon\\ \rho\upsilon^2+p\\ \upsilon(\rho E+p)\end{pmatrix}.$$

The right-hand term of our system is the vector of external forces, which is equal to the zero vector in the case of modeling Richtmyer-Meshkov instability. In the case of Rayleigh-Taylor instability, a flow occurs under a constant external acceleration g, and the vector of external forces becomes:

$$\mathbf{R}=\begin{pmatrix}0\\ 0\\ -\rho g\\ -\rho g\upsilon\end{pmatrix}.$$

Ideal gas equation of state $p = (\gamma - 1)\rho e$ was used.

4. Numerical simulations. Turbulence Problem Solver (TPS) software complex.

Numerical simulations were performed using the Turbulence Problem Solver (TPS) software package [6] developed by the authors. The package is implemented in C++ and provides a unified, object-oriented program interface for implementation of different numerical algorithms for solving hyperbolic problems. Using the software various problems of continuum mechanics has been numerically investigated: Rayleigh – Taylor instability problem, the shear layer problem, Kolmogorov problem of the formation of currents of the "parquet" type for an ideal medium, and others.

5. The results of numerical simulations.

5.1. Richtmyer-Meshkov instability.

For the numerical solution of the system of the Euler equations explicit finite volume method was used. To calculate intercell fluxes of conservative variables, ENO reconstruction of the second order was used. Primitive variables were reconstructed: density, velocity and pressure. Riemann problem was solved approximately using the Roe method. To avoid rarefaction shockwaves, entropy correction procedure was implemented. No artificial viscosity terms required because of sufficient scheme viscosity of the method itself, which allows to model flows even in the zones of large density and pressure gradients. The robustness of the method was verified on standard tests [11].

The system of equations was solved in the rectangular region covered with a uniform rectangular grid of 500x200 cells. The initial conditions were specified according to the experimental setting. The region is divided into three zones, the first two of which are resting the system xenon-argon with a curved boundary (see figure 1). In the third zone the state of argon behind the shock wave is calculated using the Rankine-Hugoniot relations.

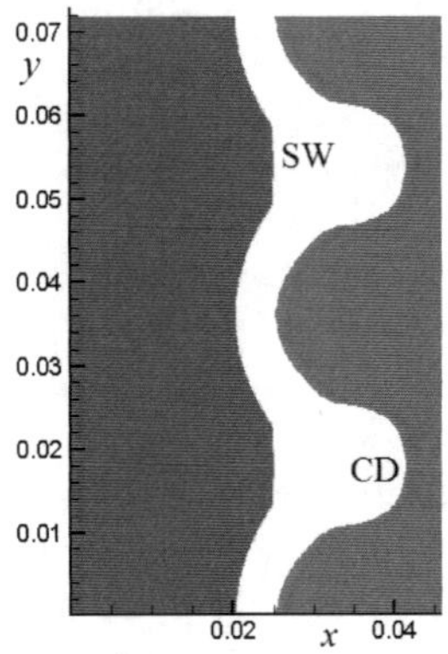

Figure 3. The front of passed shock wave (SW), and the curved boundary of the contact discontinuity (CD) at the time moment of 80 μs.

To calculate flows through the border a single row of ghost cells is used. On the top and the bottom borders reflective boundary condition is used. In this case, the values of the density, pressure and tangential velocity components in the ghost cells are transferred from the inner region, and the values of normal velocity components change its signs to the opposite. On the left and right borders transmissive boundary condition is used, i.e. the values in fictitious cells are completely transferred from the inner region.

IOP Conf. Series: Journal of Physics: Conf. Series **899** (2017) 052007 doi:10.1088/1742-6596/899/5/052007

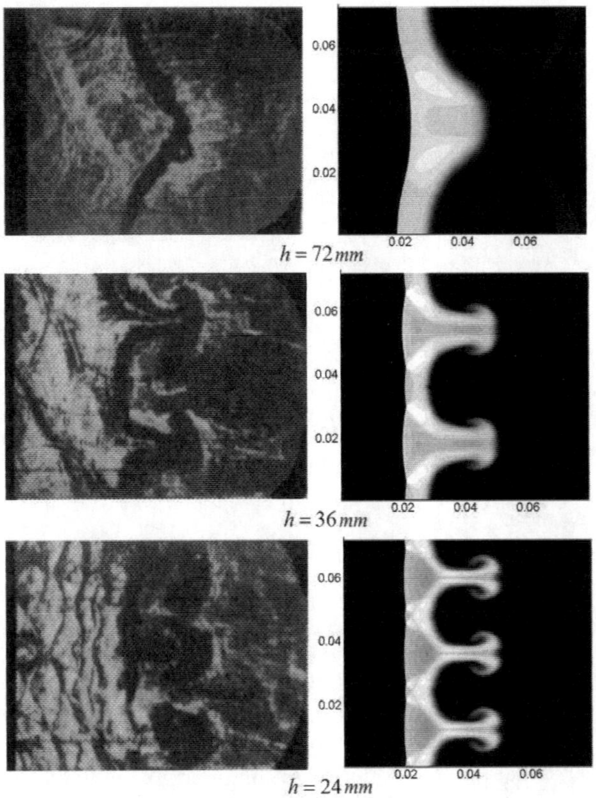

Figure 4. The development of Richtmyer-Meshkov instability at t =140 μs: photographs [1] on the left, the numerical simulation results on the right.

For the three values of the wavelengths of the initial perturbation, the process of passing a shock wave through the contact border was simulated, accompanied by the development of instability. After reaching the film, the wave front begins to propagate into xenon, which leads to the simultaneous wave propagation in both media. Front velocities in these media differ: in xenon wave moves slower and it lags from the wave propagating through the argon. Figure 3 shows the shape of a shock wave front (left boundary) which is fully transmitted through the contact boundary for the case of $h = 36\,mm$. The perturbed contact boundary follows. By the time of 80 μs, nonlinear stage of RMI is evolving.

Figure 4 shows a density distribution in the case of a two-mode perturbation of the boundary at the time moment of 140 μs. The bright part corresponds to the middle region between the shock wave and the contact boundary (see figure 3). In the case of single-mode perturbations, the development of the instability occurs slower. The simulation of the nonlinear stage was turned into a transition stage, which lead to mushroom-shaped structure formation. Thus, due to reducing the length of the initial perturbation, we reduced the duration of each stage. Therefore, by the time moment of 140 μs in the case of $h = 9\,mm$, the pattern of developed turbulence in the mixing layer was recorded on the photographic film.

5.2. Rayleigh-Taylor instability.

Figure 5. The development of the of Rayleigh-Taylor instability: experimental photographs are on the left, and numerical simulation results are on the right. Moments of time 0 ms (perturbed boundary), 8 ms and 16 ms are taken.

The system of the Euler equations in this numerical experiment was solved by the method of large particles of the second order accuracy on the grid of 600x520 cells. To avoid spurious oscillations artificial viscosity was used. The ability of the method of large particles to allow a sufficiently large ratio of the densities with high-quality coincidence pattern of instability tells about its high reliability, which also was tested on standard tests [11]. A growing perturbation on the boundary has a number of minor peculiarities: because of the smallness of the amplitude of perturbation of the perturbed boundary is sort of broken line. Every kink of the line develops into the individual bubble-like structure.

Figure 6 shows the trajectories of the mixing zone boundaries (left) and the evolution of the upper boundary velocity (right). The following procedure was used for determining the positions of boundaries. The density mesh function is averaged for each row of cells with the same vertical coordinate y. Thus, we have a one-dimensional profile of the average density depending on the y coordinate. Further, by this profile, the upper (lower) boundary of the mixing zone are calculated.

They are the values of y coordinates in which the difference between average density and the density of alcohol (air) exceeds 5% of the densities difference. Boundary velocities are obtained by numerical differentiation of the coordinates by the time.

6. Conclusion.

Numerical experiments to investigate hydrodynamic instabilities of Richtmyer-Meshkov and Rayleigh-Taylor were made, using, respectively, the finite volume method based on the approximate solver of ROE and ENO-reconstruction and the method of large particles. For numerical simulation was used Turbulence Problem Solver (TPS) software package. The resulting patterns were obtained of growth and development of hydrodynamic instabilities, showing qualitative agreement with experimental data.

Acknowledgements

The theoretical part of the work is carried out with the support of the Russian science Foundation (Agreement No 17-01-00812\17), the computing part was supported by the RSF (Agreement No. 17-11-01293) and Program of Presidium of Russian Academy of Sciences "Fundamental problems of mathematical modeling".

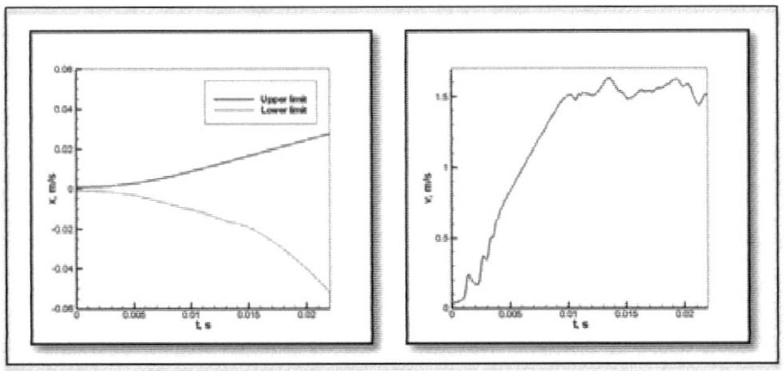

Figure 6. Left: the trajectories of the upper and lower boundaries of the mixing zone; right: the evolution of the velocity of the upper boundary of the mixing zone at the development of Rayleigh-Taylor instability.

References

[1] Aleshin A.N., Gamaliy E.G., Zaytsev S.G., Lazareva E.V., Lebo I.G., Rosanov V.B. Investigation of non-linear and transitive stage of the development of Richmyer-Meshkov instability, JTP Lett., 10 (1988).

[2] Richtmyer, R. D. Taylor Instability in Shock Acceleration of Compressible Fluids. Commun. Pure Appl. Math. 13, 297-319 (1960).

[3] Read K.I., Experimental investigation of turbulent mixing by Rayleigh–Taylor instability, Physica D 12, 45–58 (1984).

[4] L.D. Landau and E.M. Lifshitz, Electrodynamics of Continuous Media (Gostekhteorizdat, Moscow, 1953; Butterworth-Heinemann, Oxford, 1984).

[5] A.N. Kolmogorov, "The local structure of turbulence in incompressible viscous fluid for very large Reynolds numbers," Dokl. Akad. Nauk SSSR 30 (4), 299–303 (1941).

[6] A.M. Obukhov, "Spectral energy distribution in turbulent flow," Izv. Akad. Nauk SSSR, Ser. Geogr. Geofiz. 5(4), 453–466 (1941).

[7] S.V. Fortova, L.M. Kraginskii, A.V. Chikitkin, E.I. Oparina, Software package for solving hyperbolic-type equations, Mathematical Models and Computer Simulations, 2013, 5:6, 607–616.

[8] Belotserkovskii O.M., Oparin A.M., Chechetkin V.M., Turbulence: new approaches, Science, Moscow (2002).

[9] Zel'dovich Ya.B., Raizer Y.P. Physics of Shock Waves and High-Temperature Hydrodynamic Phenomena, Science, Fizmatlit, 2-nd edition, 1966.

[10] S. K. Godunov, Difference method of computation of shock waves, Uspehi Mat. Nauk, (N.S.), 12(1(73)):176–177, 1957.

[11] Toro E. F. Riemann solvers and numerical methods for fluid dynamics: a practical introduction. – Springer Science & Business Media, 2013.

IOP Conf. Series: Journal of Physics: Conf. Series **899** (2017) 052008 doi:10.1088/1742-6596/899/5/052008

An investigation of tritium transfer in reactor loops

O H Ilyasova[1], **N A Mosunova**[1]

[1] Nuclear Safety Institute (IBRAE), Russian Academy of Sciences, ul. Bolshaya Tulskaya 52, Moscow, 115191

e-mail: ilyasova@ibrae.ac.ru

Abstract. The work is devoted to the important task of the numerical simulation and analysis of the tritium behaviour in the reactor loops. The simulation was carried out by HYDRA-IBRAE/LM code, which is being developed in Nuclear safety institute of the Russian Academy of Sciences. The code is intended for modeling of the liquid metal flow (sodium, lead and lead-bismuth) on the base of non-homogeneous and non-equilibrium two-fluid model. In order to simulate tritium transfer in the code, the special module has been developed. Module includes the models describing the main phenomena of tritium behaviour in reactor loops: transfer, permeation, leakage, etc. Because of shortage of the experimental data, a lot of analytical tests and comparative calculations were considered. Some of them are presented in this work. The comparison of estimation results and experimental and analytical data demonstrate not only qualitative but also good quantitative agreement. It is possible to confirm that HYDRA-IBRAE/LM code allows modeling tritium transfer in reactor loops.

1. Introduction

Nowadays the fast reactor technology is one of the most promising ways for liquid-metal reactor development. In a view of great importance to provide radiation safety in these reactors, it is necessary to simulate the main physical processes of radionuclide behavior during normal operation. To carry out such simulations it should be indispensable to develop computer codes for modeling the main physical and chemical processes.

Tritium is a radioactive isotope of hydrogen and its half-life is 12,32 years. The nucleus of a tritium atom consists of a proton and two neutrons. Helium, beta-particle and antineutrino are generating during a radioactive decay of tritium:

$$^3_1H \rightarrow {}^3_2He + \beta + \nu^-$$

Tritium mostly comes from the reactor core by ternary fission and its production rate depends on the reactor power. Due to its good permeability, tritium can leave the reactor core, transfer in channels and permeate through pipe walls even during normal operation. These properties of tritium (good permeability and radioactivity) make computational simulation of tritium transfer very important problem.

In this paper we consider the problem of modeling of the tritium transfer behaviors, such as tritium transport, permeation, leakage, etc. Simulation was carried out by HYDRA-IBRAE/LM code, which is being developed in Nuclear Safety Institute of the Russian Academy of Sciences (IBRAE RAN). The code is based on thermal-hydraulic two-fluid model, containing three one-dimensional conservation equations for two phases (gas and fluid). There are some modules in HYDRA-

IBRAE/LM code for special computational problems, for tritium transfer simulation a module named TRITIUM was developed. TRITIUM module includes following models:

- Tritium source in the primary loop;
- Tritium and hydrogen transfer in the primary loop;
- Tritium and hydrogen source in the secondary loop;
- Tritium and hydrogen source in the tertiary circuit;
- Tritium and hydrogen permeation to the outside through the walls of components and piping;
- Tritium and hydrogen leakage;
- Tritium and hydrogen removal by the purification system installed in the primary and secondary loops.

2. Mathematical model

In HYDRA-IBRAE/LM code, tritium mass transfer processes are considered in channel and basic mass-balance equation is

$$\frac{dM_{ij}}{dt} = \sum_k \frac{M_{ik}}{\tau_{k\to j}} - \frac{M_{ij}}{\tau_j} + S_{ij} - R_{ij} \tag{1}$$

Where M_{ij} – the mass of chemical i in node j, t – time, terms $\sum_k \frac{M_{ik}}{\tau_{k\to j}}$ and $\frac{M_{ik}}{\tau_j}$ are source and sink in the node, S_{ij} and R_{ij} – mass source of chemical and mass sink rate of component in the node. Source and sink terms are determined by the phenomena taking place in the node, e.g. tritium permeation, leakage or tritium removal by purification system. Terms S_{ij} and R_{ij} depend on coolant properties, pipe material and geometrical properties. For the same reactor items a type of the equations (1) is similar.

HYDRA-IBRAE/LM code computes equation coefficients dynamically; it means that the coefficients and values depend on system characteristics changing over time, e.g. temperature, pressure, etc.

3. Analytical tests

3.1. Test 1 (Initial tritium concentration in a channel)

Analytical test on tritium transport in a channel was computed by HYDRA-IBRAE/LM code. There was initial tritium concentration in a horizontal sodium-filled circle tube. The inlet sodium flow rate was constant. Initial and boundary conditions for simulation are listed in Table 1.

Table 1. HYDRA-IBRAE/LM initial and boundary conditions.

Parameter	Value	Unit
Channel length	0,1	m
Diameter	0,03	m
Inlet flow rate	$2\cdot10^{-4}$	m^3/s
Sodium temperature	835	K
Sodium pressure	$1{,}52\cdot10^{5}$	Pa
Initial tritium mass	$2\cdot10^{-12}$	kg

Assuming that there is one node in the channel, the equation (1) is simplified in following way:

$$\begin{cases} \frac{dM_1}{dt} = -\frac{M_1}{\tau} \\ t = 0:\ M_1 = M_0 \end{cases}$$

Where τ is the characteristic time, which is computed by equation:

$$\tau = \frac{V_{ch}}{G_V} = \frac{L}{U} \text{ (in the cylindrical channel),}$$

Where

V_{ch} – channel volume (m^3);

G_V – coolant flow rate (m^3/s) ;

L – channel length (m);

U – coolant velocity (m/s).

Analytical solution of this equation is expressed by

$$M_1(t) = M_0 \exp(-t/\tau)$$

This test was computed using HYDRA-IBRAE/LM code. The computational results are shown in the Fig. 1. The red line is related to HYDRA-IBRAE/LM results, the black one is related to analytical solution.

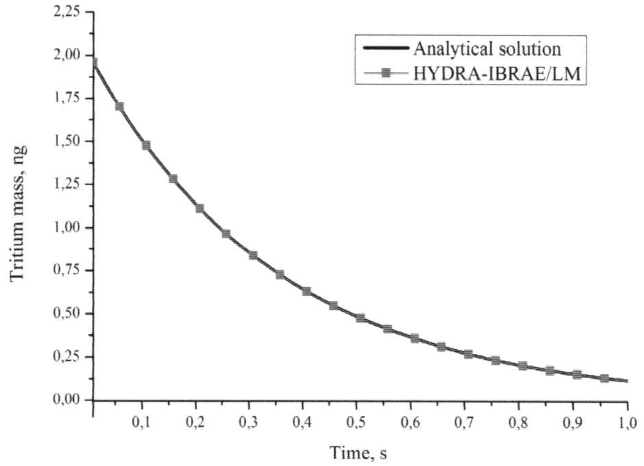

Figure 1. HYDRA-IBRAE/LM results in comparison with analytical solution.

Figure 1 demonstrates very good agreement between analytical solution and HYDRA-IBRAE/LM computations. An average relative error for the analytical test was smaller than 0,01%,

3.2. Test 2 (Tritium source in a channel)

The second analytical test is devoted to the tritium transfer in a channel, which has an initial tritium source. The inlet sodium flow rate was constant. In the first part of the channel there is a tritium source. Initial and boundary conditions for simulation are listed in Table 2. During the estimation, tritium mass in the channel was calculated.

Table 2. HYDRA-IBRAE/LM initial and boundary conditions.

Parameter	Value	Unit
Channel length	0,2	m
Diameter	0,45	m

Inlet flow rate	$2\cdot10^{-2}$	m^3/s
Sodium temperature	835	K
Sodium pressure	$1{,}52\cdot10^{5}$	Pa
Initial tritium source	$4\cdot10^{-12}$	kg/s

Assume that there are two equal parts in the channel. The main equation is simplified:

$$\begin{cases} \dfrac{dM_1}{dt} = Q - \dfrac{M_1}{\tau} \\ \dfrac{dM_2}{dt} = \dfrac{M_1}{\tau} - \dfrac{M_2}{\tau} \\ M_1(0) = M_2(0) = 0 \\ Q = const \end{cases},$$

where M_i – tritium mass in node *i* and τ is the characteristic time, which is computed by equation:

$$\tau = \frac{V_{ch}}{G_V} = \frac{L}{U} \text{ (in the cylindrical channel),}$$

Where

V_{ch} – channel volume (m^3);

G_V – coolant flow rate (m^3/s) ;

L – channel length (m);

U – coolant velocity (m/s).

Analytical solution of this task is expressed by

$$\begin{cases} M_1(t) = Q\tau\left(1 - \exp\left(-t/\tau\right)\right) \\ M_2(t) = Q\tau\left[1 - \left(1 + t/\tau\right)\exp\left(-t/\tau\right)\right] \end{cases}.$$

Figures 2-3 show the comparison between analytical solution and HYDRA-IBRAE/LM results. The red line is related to HYDRA-IBRAE/LM results, the black line is related to the analytical solution.

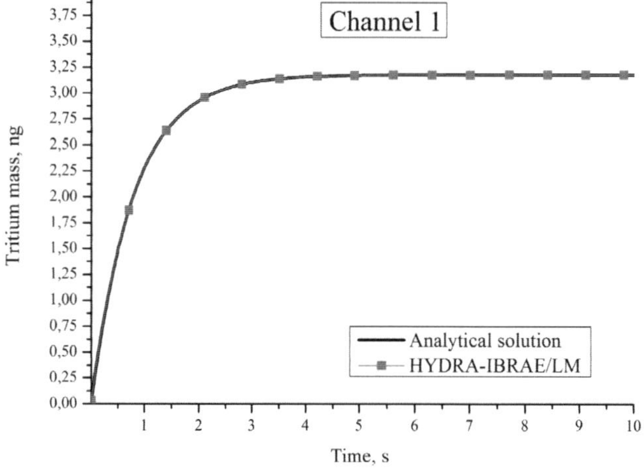

Figure 2. HYDRA-IBRAE/LM results in comparison with analytical solution (node 1).

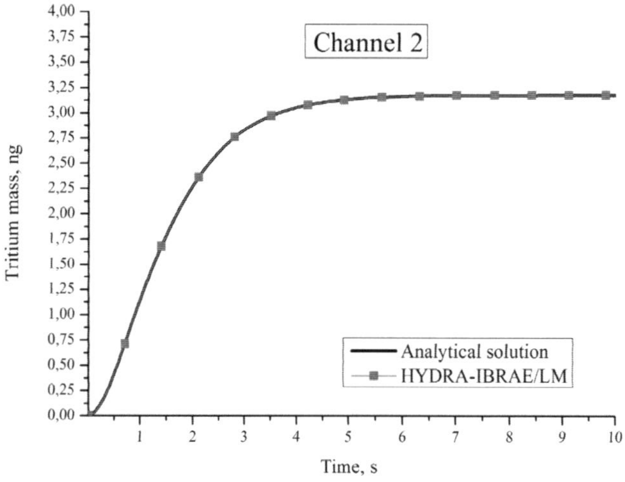

Figure 3. HYDRA-IBRAE/LM results in comparison with analytical solution (node 2).

Average relative error is no more than 0,01%. This result demonstrates good qualitative and quantitative agreement.

4. Comparative computations with THYTAN code

4.1. Modeling of purification system

Unfortunately, suitable experimental data for verification of purification system does not exist, so the purification model was verified by analytical tests and comparative computations with another computer codes. The numerical analysis code, THYTAN (Tritium and HYdrogen Transportation

ANalysis) [1], was developed by the Japan Atomic Energy Research Institute (a former organization of the Japan Atomic Energy Agency). THYTAN was developed in the late 1990s as a tool for estimating tritium transport in the HTGR (high-temperature gas-cooled reactor) hydrogen production system.

We consider a helium-filled channel with a flow into a cold trap. In THYTAN code there is a similar mass balance equation [1]:

$$V_j \frac{dC_{i,j}}{dt} = \sum (F_{total,j-1} \cdot C_{i,j-1}) - F_{total,j} \cdot C_{i,j} - S_{i,j},$$

where

V_j – the volume of node (m^3);

$F_{total,j}$ – the volumetric flow rate of all chemicals in node j (m^3/s);

$C_{i,j}$ – the volume fraction of chemical i in node j (m^3/ m^3);

$S_{i,j}$ – the total amount of volume change rate of chemical i in node j by considering generation, release, permeation, removal, leakage, and isotope exchange reactions (m^3/s).

Tritium behavior in the purification system is expressed by the following equation:

$$\frac{dC}{dt} = -k_{ct} C \frac{F_{PF,He}}{V_j},$$

where

C – tritium concentration in node (m^3/ m^3);

k_{ct} – fractional efficiency of purification system;

$F_{PF,He}$ – helium flow rate at purification system (m^3/s);

V_j – the volume of node (m^3).

THYTAN boundary conditions for this test are listed in Table 3.

Table 3. THYTAN boundary conditions.

Parameter	Symbol	Value	Unit
The volume of node	V_j	1	m^3
Helium flow rate at purification system	$F_{PF,He}$	$1{,}0 \cdot 10^{-3}$	m^3/s
Fractional efficiency of purification system	k_{ct}	0,9	–
Initial tritium concentration	$C_{0,j}$	$1{,}0 \cdot 10^{-9}$	m^3/ m^3

HYDRA-IBRAE/LM input parameters for modeling this test are listed in Table 4.

Table 4. HYDRA-IBRAE/LM input parameters for simulation.

Parameter	Value	Unit
Channel length	10	m
Diameter	0,799	m
Number of cells	5	–

Cold trap coordinate	4,47	m
Helium flow rate at purification system	$1{,}336\cdot 10^{-4}$	kg/s
Helium temperature in channel	723	K
Helium temperature in cold trap	303	K
Initial tritium mass in channel	$5{,}0\cdot 10^{-12}$	kg
Fractional efficiency of purification system	0,9	–

Comparison between results of HYDRA-IBRAE/LM and THYTAN are shown in Figure 4. Red and black lines are related to THYTAN and HYDRA-IBRAE/LM, respectively.

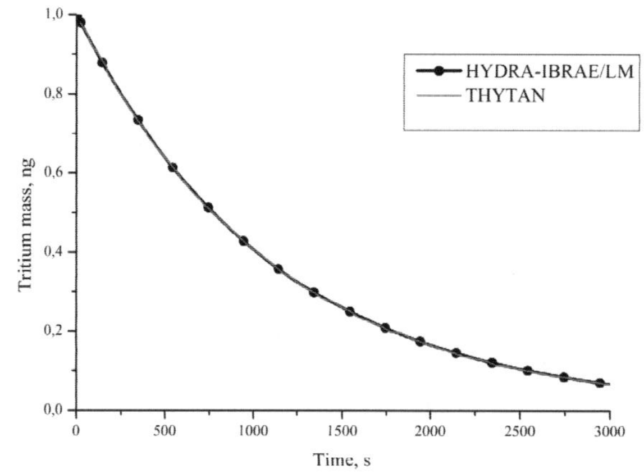

Figure 4. Comparison between THYTAN and HYDRA-IBRAE/LM results.

The results are nearly identical. Both codes uses the similar model and this agreement is predictable. According to this comparison it is possible to say that purification model of HYDRA-IBRAE/LM is used correctly.

4.2. Modeling of tritium leakage

The second comparative test with THYTAN code is related to tritium leakage to the atmosphere. A problem formulates in the following way: in the primary loop with helium inside, there is a leak to the containment vessel. The containment vessel has a leak to the atmosphere. During the simulation the tritium mass in the primary loop and in the containment vessel were computed.

This task was considered by authors of [1]. Figure 5 shows a nodalization scheme of THYTAN for calculation of leakage.

IOP Conf. Series: Journal of Physics: Conf. Series **899** (2017) 052008 doi:10.1088/1742-6596/899/5/052008

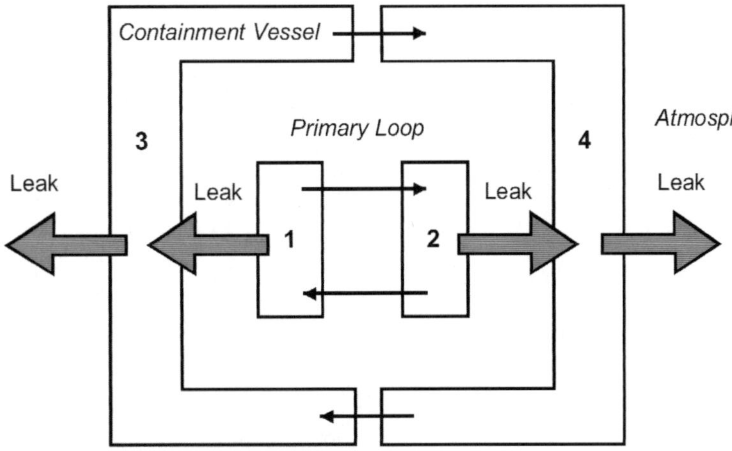

Figure 5. Nodalization scheme (THYTAN) for verification of the leak model.

The same task was simulated by HYDRA-IBRAE/LM code. Parameters for simulation are listed in Table 5.

Table 5. Parameters for simulation (HYDRA-IBRAE/LM and THYTAN).

Parameter	Value	Unit
Initial tritium mass in primary loop	$9{,}6\cdot10^{-10}$	kg
Tritium source in primary loop	$11{,}08\cdot10^{-17}$	kg/s
Primary loop coolant	Helium	
Primary loop volume	232	m^3
Primary loop temperature	809	K
Primary loop pressure	$2{,}33\cdot10^{6}$	Pa
Leak rate from primary to containment vessel	$3{,}7\cdot10^{-5}$	kg/s
Containment vessel coolant	Nitrogen	
Containment vessel temperature	809	K
Containment vessel pressure	$3{,}93\cdot10^{5}$	Pa
Containment vessel volume	$1{,}56\cdot10^{4}$	m^3
Leak rate from containment vessel to atmosphere	$2{,}3\cdot10^{-3}$	kg/s

Figures 6–7 show a comparison between estimation results.

IOP Conf. Series: Journal of Physics: Conf. Series **899** (2017) 052008 doi:10.1088/1742-6596/899/5/052008

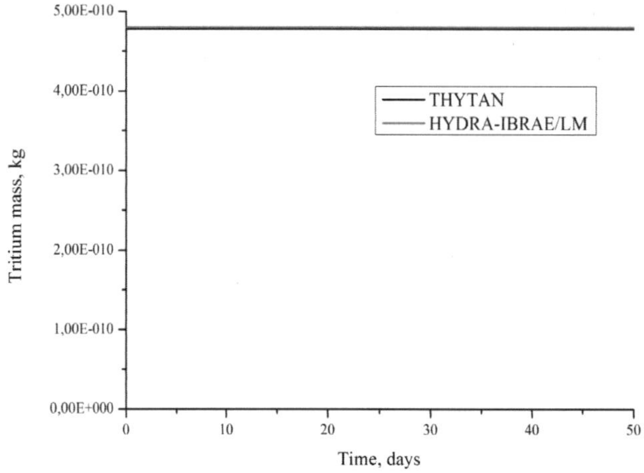

Figure 6. Comparison between HYDRA-IBRAE/LM and THYTAN results (primary loop).

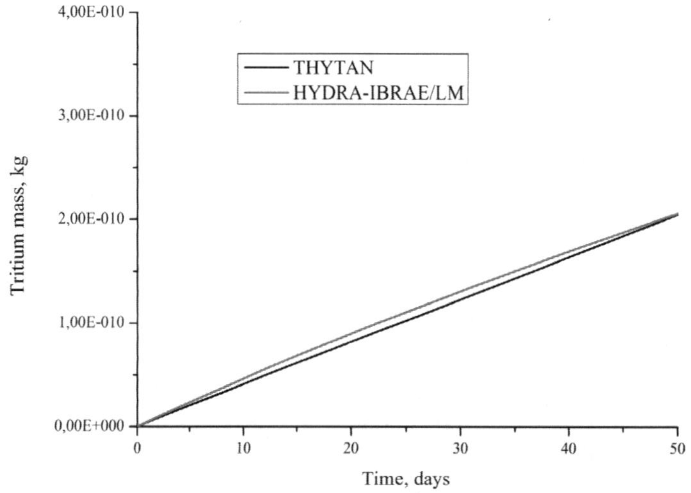

Figure 7. Comparison between HYDRA-IBRAE/LM code and THYTAN results (containment vessel).

According to Figures 6-7, the computational results demonstrate not only qualitative but also good quantitative agreement. The standard deviation of computational results is no more than 1% and 6% for primary loop and containment vessel, respectively. Thus, HYDRA-IBRAE/LM code allows simulating tritium leakage to atmosphere.

5. Experiment simulation

Peach Bottom [2] was the first installation of a High-Temperature Gas-Cooled Reactor (HTGR) in the United States. Its power operation started in 1967 and finished in 1974, when it was shut down for decommissioning. During the operation period the some tritium properties in the steam generator was periodically monitored.

IOP Conf. Series: Journal of Physics: Conf. Series **899** (2017) 052008 doi:10.1088/1742-6596/899/5/052008

The Peach Bottom End-of-Life (PBEOL) program started in March 1975. The main object of this program was to give data for validation and verification of specific codes for predicting physics, thermal, fission product. After shutting down of the Peach Bottom reactor in October 1974, there was an interest in carrying out measurements of the tritium permeation rates in the steam generator tubes. Measurements were carried out in the superheater, the evaporator, and the economizer sections of the Peach Bottom steam generator. Measurements of the tritium permeation rates were made in the operating temperature ranges of these sections:

- 673 – 973 K in the superheater;
- 573 – 694 K in the evaporator;
- 494 – 623 K in the economizer.

The experimental section is shown in Figure 8 in the simplified form. There were two coaxial channels with wall between them, the initial tritium concentration was in the inner channel and there was no tritium in the outer channel.

Figure 8. Experimental section.

In the experiments the permeation rates were measured in the different sections. The superheater walls were made from Incoloy-800, evaporator and economizer tubes were made from low carbon steel.

Simulation of this experiment was carried out by HYDRA-IBRAE/LM code. In this paper we will present results for the superheater section. The permeation rate J was computed by

$$J = \frac{M_2^{HT}}{S \cdot t},$$

Where

M_2^{HT} – tritium mass in the outer channel (kg);

S – permeation area (taken from experimental data [2], m^2);

t – time.

IOP Conf. Series: Journal of Physics: Conf. Series **899** (2017) 052008 doi:10.1088/1742-6596/899/5/052008

Initial parameters are shown in Table 6.

Table 6. HYDRA-IBRAE/LM input parameters for simulation.

Parameter	Value	Unit
Inner channel radius	$6{,}35\cdot10^{-3}$	m
Wall radius	$9{,}525\cdot10^{-3}$	m
Outer channel radius	$30\cdot10^{-3}$	m
Permeation area	$90{,}16\cdot10^{-4}$	m^2
Pressure	$0{,}1013\cdot10^{5}$	Pa
Initial area concentration in the inner channel	0,46	ppbv

HYDRA-IBRAE/LM calculation results of the permeation rates in comparison with experimental data and analytical solution are shown in Table 7. Results are presented for three different temperatures: 673 K, 823 K and 973 K.

Table 7. Comparison of calculation results with experimental data and analytical solution.

Temperature, K	Calculation result, µCi/m2/h	Analytical solution, µCi/m2/h [1]	Experimental data, µCi/m2/h [2]
673	0,1138	0,1202	0,119±0,0595
823	0,644	0,683	–
973	2,158	2,298	1,29±0,645

As shown in Table 7, there is good agreement between calculation results and analytical solution. For lower temperatures a HYDRA-IBRAE/LM result is in the experimental accuracy limits, but for higher temperatures the results are dramatically different from experimental data. It can be explained by existence of the oxide film in the tubes in case of higher temperatures, but there is no such information in the experimental materials.

6. Conclusion

In the paper verification results of modelling of tritium transfer are presented. For this modelling the special module TRITIUM was developed in HYDRA-IBRAE/LM code. This module contains models describing the main physical processes connecting with tritium behavior: tritium transfer in pipes, tritium permeation through tube walls, tritium leakage into atmosphere and tritium accumulation in cold trap. All equation coefficients are computed in dependence of current status of a system, which makes the module useful and flexible tool for tritium transfer simulation.

The results of experiment simulation, comparative calculations and analytical tests are shown in the article. A good agreement between the experimental data and HYDRA-IBRAE/LM results has been obtained for temperatures about 673 K (there is no sufficient experimental data for other temperatures). An average relative error was no more 0,01% for analytical tests. These results

IOP Conf. Series: Journal of Physics: Conf. Series **899** (2017) 052008 doi:10.1088/1742-6596/899/5/052008

demonstrate that HYDRA-IBRAE/LM allows modeling physical processes, connecting with tritium transfer in reactor loops.

References

[1] Ohashi H and Sherman S R *Tritium movement and accumulation in the NGNP system interface and hydrogen plant*, INL/EXT-07-12647.
[2] Yang L, Baugh A and Baldwin N L *Study of Tritium permeation through peach bottom steam generator tubes*, 1977.

IOP Conf. Series: Journal of Physics: Conf. Series **899** (2017) 052009 doi:10.1088/1742-6596/899/5/052009

The molecular dynamics simulation of self-diffusion in gases and liquids

G V Kharlamov[1] **and S V Zhilkin**[2]

[1] Physics Department, Siberian State University of Telecommunications and Information Sciences, 86 Kirov str., Novosibirsk, 630102, Russia
[2] Odin, 4 Oktyabr'skaya magistral', Novosibirsk, 630007, Russia

E-mail: g_kharlamov@ngs.ru

Abstract. The molecular dynamics calculations of self-diffusion coefficients in Lennard-Jones liquids have been carried out. The change of diffusion mechanism has been observed with transition from dense gas to liquid. The diffusion mechanism in liquids isn't connected with the jumping motion of molecules. A good agreement of our results and the experimental data for liquid argon has been obtained.

1. Introduction

The study of transport processes in gases, liquids and solids always attracted attention of scientists. The strict transport theory, based on the solution of Boltzmann's equation, exists for rarefied gases [1]. In this case, the mass, momentum and energy transport occurs via pair collisions of atoms or molecules. In particular according to the kinetic Boltzmann – Chapman – Enskog theory, the self-diffusion coefficient in rarefied gas is related by [1]

$$D_0 = \frac{3\sqrt{\pi m_0 kT}}{8 m_0 \rho \pi \sigma_0^2 \Omega^{(1,1)*}}, \tag{1}$$

where $\Omega^{(1,1)*}$ is the reduced collision integral.

The self-diffusion in solids occurs via jumping motion of atoms in crystal lattice (see for example [2]). Such diffusion mechanism results in the expression

$$D = D_m e^{-\frac{\Delta W}{kT}}, \tag{2}$$

where ΔW is the activation energy of the jumping motion and D_m is the weak temperature dependent coefficient.

The diffusion mechanism in liquids is more complicated and not investigated completely. There are specific models of diffusion in liquids, although less strict [3, 4]. Most of these theories invoke a combination of vibrational and jumping molecules motions and result in an expression like (2) for self-diffusion coefficient. It is doubtful for simple liquids at any rate because the molecules jumps are not found for liquids. Thus, there is the problem of investigating of the diffusion mechanisms in dense gases and in simple liquids.

In the present paper, the results of molecular dynamics calculations of the self-diffusion coefficients in the Lennard-Jones liquids over a wide range of temperature are presented. The molecular dynamics simulation method is very good for this purpose. It has long been used for the calculations of the diffusion coefficients in dense gases and liquids. Many studies are carried out with the use of the Lennard-Jones 12-6 potential [5 - 14]. In our papers [15, 16], the universal dependence of the relation of diffusion coefficient to theoretical Chapman–Enskog values (1) on density for the homogeneous systems is found.

$$f_d = \frac{D_{md}}{D_0} = 1 - 0.547\rho + 0.3\rho^2 - 0.574\rho^3. \quad (3)$$

Large deviations of the diffusion coefficient for systems in the phase transition region from the universal function are observed. However, the ratio D_{md}/D_0 for the liquid phase ($\rho \geq 0.78$) depends on temperature. Obviously, the equation (3) is not correct for very dense systems.

2. The calculation procedure

A calculation cell in the shape of a parallelepiped or a cube with periodic boundary conditions is used. The sizes of the cell and the number of the particles placed in it are chosen depending on the density. The interaction potential is

$$\begin{aligned} &U(r) = 4\varepsilon\left[\left(\frac{\sigma_0}{r}\right)^{12} - \left(\frac{\sigma_0}{r}\right)^{6}\right], \quad r \leq 4.5\sigma_0, \\ &U(r) = a(r-5\sigma_0)^2 + b(r-5\sigma_0)^3, \quad 4.5\sigma_0 \leq r \leq 5\sigma_0, \\ &U(r) = 0, \quad r \geq 5\sigma_0, \end{aligned} \quad (4)$$

where σ_0 and ε are the diameter of atom or a molecule and the depth of the potential respectively. For optimization of the calculation algorithm, the potential has been cut off by the cubic spline on distance r = 4.5 σ_0. It is obvious that the particles do not interact, if the distance between them exceeds 5 σ_0. For this reason, the calculation cell is divided into zones which are identical cubic cells of the size 5 σ_0. The interaction is considered only between particles of adjacent zones. Such procedure considerably reduces an operating time of the computer program.

Reduced units are used in the calculations: the distance is $r = r^*/\sigma_0$, the temperature is $T = kT^*/\varepsilon$, the energy is $U = U^*/\varepsilon$, the density is $\rho = \rho^* \sigma_0^3$ and time is $t = t^*/\sigma_0 (\varepsilon/m_0)^{1/2}$. Here m_0 is the mass of a molecule. The variables labeled by an asterisk are dimensional. The system volume, the temperature and the quantity of particles are constant during the calculation (NVT ensemble). The standard method of molecular dynamics is used. The system of *n* Newton's equations are solved. For integration of the movement equations, the Verlet method of numerical integration with step t = 0.001 is used. The average temperature of the system is calculated by formula $T_{sr} = 2E_k/(3n)$, where n is the quantity of particles, E_k is the kinetic energy of all particles $E_k = \sum_{i=1}^{n} m_0 v_i^2/2$. After several steps of the program, the velocity components of each particle v_x, v_y and v_z are multiplied by factor $\sqrt{T/T_{sr}}$, where *T* is the set temperature of the system.

The velocity autocorrelation function (VACF) of particles *F*(*t*) is calculated by formula

$$F(t) = \langle \vec{v}(0) \cdot \vec{v}(t) \rangle = \frac{1}{n}\sum_{i=1}^{n} \vec{v}_i(0) \cdot \vec{v}_i(t). \quad (5)$$

Here $\vec{v}_i(t)$ is the velocity of a particle at the time moment t. The procedure of the calculation of $F(t)$ is repeated many times. Then the received results are averaged. The number of averaging for one point $F(t)$ reaches the values from 1350000 to 13500000.

The self-diffusion coefficient D is calculated by Green–Kubo formula:

$$D = \frac{1}{3}\int_0^\infty F(t)dt . \quad (6)$$

3. Results and discussion

The self-diffusion coefficients of all investigated systems are given in table 1. The simulations show that the ratio D_{md}/D_0 increases with increasing temperature. In figure 1 these results together with the data for less density systems from [16] are presented. One can see that the ratio D_{md}/D_0 doesn't depend on temperature up to $\rho = 0.61$ at high temperature (T > 1). However, the uniform temperature dependence is observed for the systems of the present paper.

Table 1. Data of the self-diffusion coefficient calculation.

Density ρ	Temperature T	Number of particles	Self-diffusion coefficient D_{md}	D_{md}/D_0
0.7816	4	6347	0.318	0.516
	3		0.242	0.49
	2		0.162	0.45
	1		0.077	0.41
	0.75		0.053	0.376
0.8448	6	6861	0.391	0.518
	5		0.332	0.499
	4		0.266	0.469
	3		0.201	0.44
	2		0.129	0.392
	1.792		0.113	0.377
	1.582		0.096	0.357
	1.372		0.082	0.348
	1.162		0.066	0.328
	1.07		0.059	0.318
	0.983		0.054	0.316
	0.905		0.047	0.299
	0.831		0.041	0.284
	0.77		0.037	0.277
	0.72		0.033	0.265
0.888	4	3000	0.219	0.407
	3		0.157	0.363
	2		0.101	0.321
	1.5		0.067	0.274
	1.1		0.046	0.253
	1		0.041	0.247
	0.9		0.034	0.225
	0.75		0.024	0.196
	0.6		0.016	0.165
	0.5		0.011	0.131
	0.4		0.00614	0.094

In figure 2 the temperature dependences of self-diffusion coefficients are shown. They are linear dependences with a high accuracy

$$D_{md} = \alpha + \beta T. \tag{7}$$

The analytic fit parameters α and β depend on system density: $\alpha = -4.76\ 10^{-3}$ and $\beta = 0.081$ for $\rho = 0.7816$; $\alpha = -0.013$ and $\beta = 0.069$ for $\rho = 0.8448$; $\alpha = -0.019$ and $\beta = 0.059$ for $\rho = 0.888$.

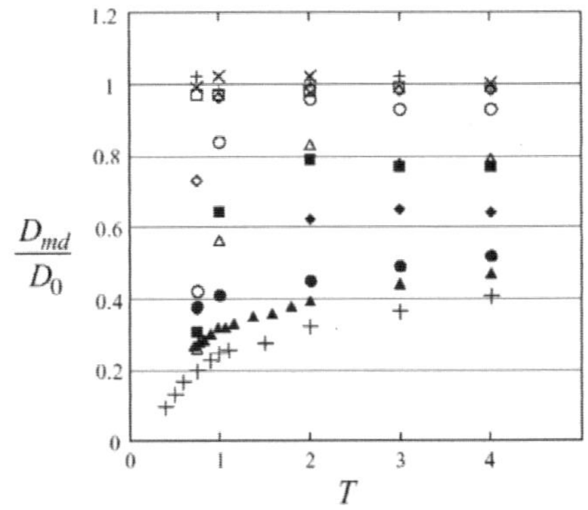

Figure 1. Temperature dependences of the reduced self-diffusion coefficients D_{md}/D_0 at different system densities: $\rho = 3.42\ 10^{-5}$(×), $\rho = 3.43\ 10^{-4}$(+), $\rho = 3.47\ 10^{-3}$(□), $\rho = 3.48\ 10^{-2}$(◊), $\rho = 0.1$(○), $\rho = 0.4$(Δ), $\rho = 0.437$(■), $\rho = 0.61$(♦), $\rho = 0.7816$(●), $\rho = 0.8448$(▲), $\rho = 0.888$(+).

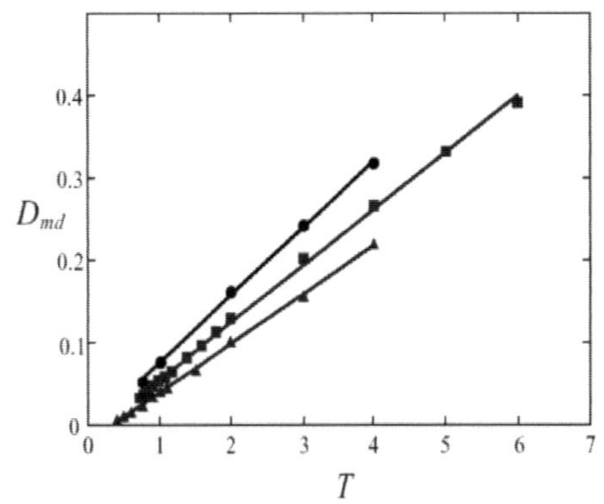

Figure 2. Temperature dependences of the self-diffusion coefficients D_{md} for liquid systems at different densities: $\rho = 0.7816$(●), $\rho = 0.8448$(■), $\rho = 0.888$(▲).

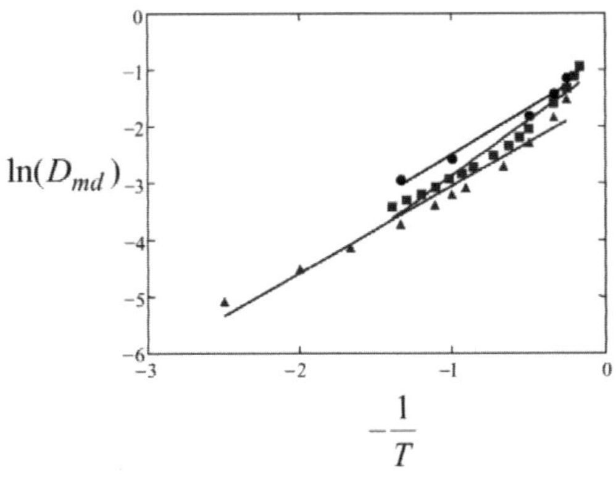

Figure 3. Dependence of $\ln(D_{md})$ on $1/T$ for the liquid systems at different densities: $\rho = 0.7816$(●), $\rho = 0.8448$(■), $\rho = 0.888$(▲).

Thus, we can understand that the diffusion mechanism is changed with transition from dense gas to liquid. In dense gases, the diffusion mechanism is the same as in rarefied gases and the temperature dependences of self-diffusion coefficient are alike (see (1)). A new diffusion mechanism appears in the liquids. One can expect that the new mechanism is connected with jumping motion of molecules in liquids. Then the temperature dependence of self-diffusion coefficient must follow the formula (2). To test this assumption, the dependence of $\ln(D_{md})$ on $1/T$ is presented in figure 3. One can see that the molecular dynamics calculations results don't agree with the dependence (2). Consequently, this diffusion mechanism isn't connected with jumping motion of molecules. Probably, it is connected with motion of molecules group surrounding a molecule. Then the velocity of the molecule is relaxed with increasing the average velocity of the molecules group. In other words, this diffusion mechanism is connected with relaxation of momentum fluctuations in liquids.

Our molecular dynamics simulations allow us to compare the results with experimental data in liquid argon [17]. The parameters $\varepsilon/k = 124$ K and $\sigma_0 = 3.418$ Å are used to scale the experimental data. These parameters are obtained from experimental data on viscosity of gaseous argon [18].The experimental self-diffusion coefficient at 84.56 K is (1.53 ± 0.03) 10^{-5} cm^2/sec. We have obtained the value 1.75 10^{-5} cm^2/sec.

4. Conclusions

In this paper, the molecular dynamics calculations of self-diffusion coefficients in Lennard-Jones liquids have been carried out. The diffusion mechanism change has been observed with transition from dense gas to liquid. The diffusion mechanism in liquids isn't connected with the jumping motion of molecules. A good agreement of our results and the experimental data for liquid argon has been obtained.

To understand the nature of the diffusion mechanism, it is necessary to study fluctuation of density and momentum in liquids. Clusters and nanodrops can appear in dense gases [15, 16]. These structures have unusual properties [19, 20]. They strongly affect self-diffusion coefficient in dense gases. Perhaps, similar structures can appear in liquids. They must influence the diffusion greatly. The molecular dynamics study of such fluctuations is a very important and interesting problem.

References

[1] Hirschfelder J O, Curtiss C E and Bird R B 1964 *Molecular Theory of Gases and Liquids* (New York: Wiley)
[2] Boltaks B J 1963 *Diffusion in Semiconductors* (New York: Academic Press)
[3] Frenkel J 1955 *Kinetic Theory of Liquids* (New York: Dover)
[4] March N H and Tosi M P 2002 *Introduction to Liquid State Physics* (Singapore:Word Scientific)
[5] Meier K, Laesecke A and Kabelac S 2004 *J. Chem. Phys.* **121** 9526
[6] Rowley R I and Painter M M 1997 *Int. J. Thermophys.* **18** 1109
[7] Straub J E 1992 *Mol. Phys.* **76** 373
[8] Heyes D M 1988 *Phys. Rev.* B **37** 5677
[9] Liu H, Silva C M and Macedo E A 1998 *Chemical Engineering Science* **53** 2403
[10] Laghaei R, Nasrabad A E and Eu B C 2006 *J. Chem. Phys.* **124** 154502
[11] Dyer K M, Pettitt B M and Stell G 2007 *J. Chem. Phys.***126** 034502
[12] Yamaguchi T, Kimura Y and Hirota N 1998 *Mol. Phys.* **94** 527
[13] Hammonds K D and Heyes D M 1988 *J. Chem. Soc., Faraday Trans.* **84** 705
[14] Speedy R J, Prielmeier F X, Vardag T, Lang E W and Lüdemann H-D 1989 *Mol. Phys.* **66** 577
[15] Zhilkin S V and Kharlamov G V 2015 *Optika Atmosfery i Okeana* **28** 138 (in Russian)
[16] Kharlamov G V and Zhilkin S V 2016 *Lecture Notes in Engineering and Computer Science: Proc. of The World Congress on Engineering and Computer Science* (San Francisco: USA) p 554
[17] Cini-Castagnoli G and Ricci F P 1960 *J. Chem. Phys.* **32** 19
[18] *Reference-book* http://www.chemway.ru/bd chem/tbl mol /tbl mol ld 1.php
[19] Kharlamov G V, Onischuk A A, Vosel S V and Purtov P A 2010 *Colloids and Surfaces A: Physicochem. Eng. Aspects* **379** 10
[20] Kharlamov G V, Onischuk A A, Vosel S V and Purtov P A 2012 *J. Phys.: Conf. Ser.* **393** 012006

High-accuracy polynomial solutions of the classical Stefan problem

V A Kot[1]

[1] Heat and Mass Transfer Institute, NASci of Belarus, 15 Brovka Str., Minsk, 220072, Belarus

valery.kot@hmti.ac.by

Abstract. High-accuracy polynomial solutions of the Stefan problem for a semi-infinite medium with Dirichlet and Neumann boundary conditions and boundary conditions of general form are presented. The initial temperature of the medium was assumed to be equal to its phase-transition temperature. With the use of the integral method of boundary characteristics, based on the multiple integration of the heat-conduction equation, sequences of identical equalities with different boundary conditions were obtained. On the basis of these equalities, polynomial solutions of different degrees were constructed. High efficiency of the approach proposed was demonstrated by different examples. The polynomial solutions of the second and third degrees surpass in approximation accuracy the analogous known solutions. The accuracy of the calculations of the interphase boundary with the use of the fourth- and fifth-degree polynomials is higher by several orders of magnitude then that of numerical methods.

1. Introduction

The solution of the Stefan problem includes the calculation of the temperature (concentration) profile in a body with determination of the law of movement of the interphase boundary in it [1]. In the overwhelming majority of cases, Stefan problems are solved by numerical methods [2, 3]. Problems with phase transitions, with rare exception [1, 4–6], have no exact solutions. At present, approximate analytical methods are used widely for solving such problems [7–13]. We consider the classical Stefan problem in the following mathematical formulation. It is necessary to find the position of the interphase boundary $s(t)$ and the temperature function $T(x,t)$, satisfying the equation

$$\frac{\partial T}{\partial t} = \frac{\partial^2 T}{\partial x^2}, \qquad 0 < x < s(t), \qquad t > 0 \tag{1}$$

with one of the boundary conditions

$$\text{(i)}: \ T = h(t); \ \ \text{(ii)}: \ -\frac{\partial T}{\partial x} = q(t); \ \ \text{(iii)}: \ \alpha\frac{\partial T}{\partial x} + \beta T = \gamma(t), \quad x = 0, \quad t > 0 \tag{2}$$

the conditions at the moving interphase boundary

$$T(s,t) = 0, \quad -\left.\frac{\partial T}{\partial x}\right|_{x=s} = \frac{1}{\text{Ste}}\frac{\mathrm{d}s}{\mathrm{d}t}, \quad t > 0 \tag{3}$$

IOP Conf. Series: Journal of Physics: Conf. Series **899** (2017) 052010 doi:10.1088/1742-6596/899/5/052010

and the initial condition $s(0)=0$. Here, Ste is the Stefan number [1, 14].

2. Identical equalities and their sequences

By analogy with [11, 12] we construct sequences of identical equalities.

2.1. Dirichlet boundary condition

Using the operators $D_t \equiv \partial / \partial t$ and $L \equiv \partial^2 / \partial x^2$, we write equation (1) in the form

$$D_t T = LT . \tag{4}$$

Let us introduce the integral operators $\mathcal{L}_n^x \equiv \overbrace{\mathcal{L}_x \ldots \mathcal{L}_x \mathcal{L}_x}^{n}$, $\mathcal{L}_n \equiv \overbrace{\mathcal{L}_1 \ldots \mathcal{L}_x \mathcal{L}_x}^{n}$, where $\mathcal{L}_x = \mathcal{L}_1^x \equiv \int_s^x \int_s^x (\cdot) dx^{(2)}$, $\mathcal{L}_1 \equiv \int_s^0 \int_s^x (\cdot) dx^{(2)}$. Applying the operators $\mathcal{L}_n^x$ and $\mathcal{L}_n$ to equation (4), we obtain the following sequence of identical equalities [13]:

$$\left\{ \mathcal{L}_n T + \frac{1}{\text{Ste}} \sum_{k=1}^{n} \underbrace{\int_0^t dt \ldots \int_0^t}_{n-k} \frac{s^{2k}}{(2k)!} dt^{(n-k)} \equiv \underbrace{\int_0^t dt \ldots \int_0^t}_{n} h \, dt^{(n)} = H_n \right\}_n , \quad \forall n \in \mathbb{Z}_+ . \tag{5}$$

2.2. Neumann boundary condition

Let us integrate equation (4) over the region $x \in [0, s]$:

$$\int_0^s \frac{\partial T}{\partial t} \mathrm{d}x = \frac{\mathrm{d}}{\mathrm{d}t} \int_0^s T \, \mathrm{d}x - T(s,t) s' = \int_0^s \frac{\partial^2 T}{\partial x^2} \mathrm{d}x = \frac{\partial T}{\partial y} \bigg|_0^s = \frac{\partial T(s,t)}{\partial x} - \frac{\partial T(0,t)}{\partial x} . \tag{6}$$

In view of the boundary conditions (2)-(ii), (3), from equation (6) we obtain the relation

$$\frac{\mathrm{d}}{\mathrm{d}t} \int_0^s T \, \mathrm{d}x = q - \frac{s'}{\text{Ste}} . \tag{7}$$

Integration of (7) with conditions (3) gives the zero-order equality

$$\hat{\mathcal{L}}_x T + \frac{s}{\text{Ste}} \equiv \int_0^t q \, \mathrm{d}t = Q_1 , \tag{8}$$

where $\hat{\mathcal{L}}_1 \equiv \int_0^s (\cdot) \mathrm{d}x$ is an operator. Integration of equation (4) over the region $x \in [x, s]$ gives

$$D_t \left(\hat{\mathcal{L}}_x T \right) + \beta s' = T_x . \tag{9}$$

Here, $\hat{\mathcal{L}}_x \equiv \int_s^x (\cdot) \mathrm{d}x$ is an additional integral operator and $T_x \equiv \partial T / \partial x$. Let us apply the operator $\mathcal{L}_x$ to equation (9). For its left side we write

$$\mathcal{L}_x \left(D_t \left(\hat{\mathcal{L}}_x T \right) \right) + \mathcal{L}_x \left(s' / \text{Ste} \right) = D_t \left(\mathcal{L}_x \hat{\mathcal{L}}_x T \right) + s' \left(x - s \right)^2 / (2\text{Ste}) . \tag{10}$$

For the right side of (9) we have

$$\mathcal{L}_x T_x \equiv \int_s^x \mathrm{d}x \int_s^x T_x \, \mathrm{d}x = \int_s^x \left(T - T(s,t) \right) \mathrm{d}x = \int_s^x T \, \mathrm{d}x . \tag{11}$$

Using (9)–(11), we arrive at the equation

$$D_t\left(\mathcal{L}_x\hat{\mathcal{L}}_x T\right)+s'\left(x-s\right)^2/(2\mathrm{Ste})=\int_s^x T\,\mathrm{d}x=\hat{\mathcal{L}}_x T\,. \tag{12}$$

Equation (12) at the point $x=0$ has the form

$$D_t\left(\mathcal{L}_1\left(\hat{\mathcal{L}}_x T\right)\right)+s's^2/(2\mathrm{Ste})=\hat{\mathcal{L}}_x T\Big|_{x=0} \tag{13}$$

or, in view of equality (8), the form

$$D_t\left(\mathcal{L}_1\left(\hat{\mathcal{L}}_x T\right)\right)+s(s's+2)/(2\mathrm{Ste})=Q_1\,. \tag{14}$$

Integration of equation (14) gives the first-order identical equality

$$\mathcal{L}_1\left(\hat{\mathcal{L}}_x T\right)+\mathrm{Ste}^{-1}\left(s^3/6+\int_0^t s\,\mathrm{d}t\right)\equiv Q_2\,. \tag{15}$$

Identical equalities of higher orders are obtained in the same way, and we arrive at the sequence

$$\left\{\mathcal{L}_n\left(\hat{\mathcal{L}}_x T\right)+\mathrm{Ste}^{-1}\sum_{k=1}^{n}\underbrace{\int_0^t\ldots\int_0^t}_{n-k}\frac{s^{2k-1}}{(2k-1)!}\mathrm{d}t^{(n-1)}\equiv\underbrace{\int_0^t\mathrm{d}t\ldots\int_0^t}_{n} q\,\mathrm{d}t^{(n)}=Q_n\right\}_n,\quad\forall n\in\mathbb{Z}_{\geq}\,. \tag{16}$$

2.3. Robin boundary condition
On multiple integration of the boundary condition (2)-(iii), we obtain

$$\alpha\underbrace{\int_0^t\mathrm{d}t\ldots\int_0^t}_{n}\frac{\partial T(0,t)}{\partial y}\mathrm{d}t+\beta\underbrace{\int_0^t\mathrm{d}t\ldots\int_0^t}_{n}T(0,t)\mathrm{d}t=\underbrace{\int_0^t\mathrm{d}t\ldots\int_0^t}_{n}\gamma\,\mathrm{d}t=\Upsilon_n,\quad n\in Z_+\,. \tag{17}$$

Performing the formal changes $-\partial T(0,t)/\partial x\rightarrow q$, $T(0,t)\rightarrow h$ in (17), we arrive at the relation

$$\beta\underbrace{\int_0^t\mathrm{d}t\ldots\int_0^t}_{n}h\,\mathrm{d}t-\alpha\underbrace{\int_0^t\mathrm{d}t\ldots\int_0^t}_{n}q\mathrm{d}t=\underbrace{\int_0^t\mathrm{d}t\ldots\int_0^t}_{n}\gamma\,\mathrm{d}t=Y_n,\quad n\in Z_+\,. \tag{18}$$

On the basis of the identical equalities (5) and (16), from (2.27) we obtain the sequence

$$\left\{\mathcal{L}_n\left(\beta T-\alpha\hat{\mathcal{L}}_x T\right)+\frac{1}{\mathrm{Ste}}\sum_{k=1}^{n}\underbrace{\int_0^t\mathrm{d}t\ldots\int_0^t}_{n-k}\left(\beta\frac{s^{2k}}{(2k)!}-\alpha\frac{s^{2k-1}}{(2k-1)!}\right)\mathrm{d}t\equiv Y_n\right\}_n,\quad\forall n\in\mathbb{Z}_+\,. \tag{19}$$

3. Polynomial solutions of the Stefan problem

We represent the solutions of the Stefan problems (1)–(3) in the form of the polynomial $T=\left(1-x/s\right)\sum_{j=0}^{N-1}a_j(t)(x/s)^j$ providing the fulfillment of the boundary condition (2). In this case, the problem is reduced to the determination of the coefficients $a_j(t)$ and the function $s(t)$.

3.1. Dirichlet boundary condition
Let us define the temperature profile by a quadratic parabola ($N=2$). From the first-order equality of sequence (5) we determine the coefficient $a_1(t)$ and, as a result, obtain

$$T=\left(1-\frac{x}{s}\right)\left(h+\frac{2x}{s}\left(\frac{6H_1}{s^2}-h-\frac{3}{\text{Ste}}\right)\right). \tag{20}$$

For determining the function $s(t)$ we will use the second-order identical equality of sequence (5), from which we obtain the constitutive equation

$$\left(\frac{30}{\text{Ste}}+4h\right)s^2-\frac{h'}{6}s^4+\left(\left(\frac{1}{\text{Ste}}-\frac{h}{3}\right)s^2+4H_1\right)(s^2)'=60H_1. \tag{21}$$

Let $h(t)=1$. The exact solution for the semi-space being considered has the form [1, 2]

$$T=1-\text{erf}\left(x/(2\sqrt{t})\right)/\,\text{erf}(\alpha),\quad s(t)=2\alpha\sqrt{t}, \tag{22}$$

where erf is the error function and the constant α is determined from the transcendental equation

$$\sqrt{\pi}\,\alpha\exp(\alpha^2)\,\text{erf}(\alpha)=\text{Ste}. \tag{23}$$

Substitution of the function $s(t)$ in the form $s(t)=2\alpha\sqrt{t}$ into (21) gives the quadratic equation

$$4\beta^2(\text{Ste}-3)-6\beta(15+4\text{Ste})+45\,\text{Ste}=0, \tag{24}$$

where $\beta=\alpha^2$. From (24) we determine the real positive root

$$\alpha=\frac{1}{2}\sqrt{\frac{15+4\text{Ste}-\sqrt{(15+6\,\text{Ste})^2-40\,\text{Ste}^2}}{\text{Ste}/3-1}}. \tag{25}$$

We will estimate the accuracy of formula (25) by the $\varepsilon_\alpha=\left|\alpha-\alpha_{ex}\right|/\alpha_{ex}\cdot 100\%$, where α_{ex} is determined from equation (23). At $\text{Ste}=1$ formula (25) gives $\alpha=0.619975$ to which corresponds ε_α=0.014%. The calculation by the CIM method [8] gives $\alpha=0.618497$ with an error $\varepsilon_\alpha=0.252\%$.

3.2. Neumann boundary condition

At $N=4$ we will determine the polynomial coefficients of the temperature profile at condition (2)-(ii) with the use of the first three equality of (16) and, as a result, obtain the system of equations

$$\begin{pmatrix} 1 & -1 & 0 & 0 \\ 1/2 & 1/6 & 1/12 & 1/20 \\ 1/24 & 1/40 & 1/60 & 1/84 \\ 1/720 & 1/1008 & 1/1344 & 1/1728 \end{pmatrix}\begin{pmatrix} a_0 \\ a_1 \\ a_2 \\ a_3 \end{pmatrix}=\begin{pmatrix} sq \\ Q_1/s-1/\text{Ste} \\ Q_2/s^3-1/\text{Ste}(1/6+\text{S}_1^1/s^3) \\ Q_3/s^5-1/\text{Ste}(1/120+(\text{S}_1^3+\text{S}_2^1)/s^5) \end{pmatrix}. \tag{26}$$

From here we obtain the temperature function

$$T=\left(1-\frac{x}{s}\right)\left[\frac{5s}{64}\left(1-\frac{59}{5}\frac{x}{s}+\frac{151}{5}\frac{x^2}{s^2}-21\frac{x^3}{s^3}\right)q-\frac{13}{8\text{Ste}}\left(1+\frac{x}{s}-\frac{197}{13}\frac{x^2}{s^2}+\frac{315}{13}\frac{x^3}{s^3}\right)+\frac{75}{16s}\left(1+\frac{x}{s}-13\frac{x^2}{s^2}\right.\right.$$
$$\left.\left.+\frac{63}{5}\frac{x^3}{s^3}\right)Q_1+\frac{525}{8s^3}\left(\frac{S_1^1}{\text{Ste}}-Q_2\right)\left(1+\frac{x}{s}-\frac{121}{5}\frac{x^2}{s^2}+27\frac{x^3}{s^3}\right)+\frac{945}{s^5}\left(Q_3-\frac{S_1^3+S_2^1}{\text{Ste}}\right)\left(1+\frac{x}{s}-29\frac{x^2}{s^2}+35\frac{x^3}{s^3}\right)\right],$$

where $S_n^k(t)=\int_0^t\ldots\int_0^t s^{2k-1}/(2k-1)!\,dt^{(n-k)}$. Substitution of this temperature profile into the third-order equality (16) gives the equation

$$
\begin{aligned}
& s^8 q' - 8s^6(10 - s s')q + 240 s^4 (21 - 2 s s')Q_1 + 20160 s^2 \left(\frac{S_1^1}{\text{Ste}} - Q_2 \right)(12 - s s') \\
& -120\,960 \left(\frac{S_1^3 + S_2^1}{\text{Ste}} - Q_3 \right)(55 - 4 s s') - \frac{448 s^5}{\text{Ste}}(45 + s s') = 0.
\end{aligned} \tag{27}
$$

The solution obtained will be analyzed with the use of the test problem [17]: $\{-T_x(0,t) = \mathrm{e}^t, \text{Ste} = 1\}$, $\{ T = \mathrm{e}^{t-x} - 1, \quad s(t) = t \}$. Substitution of the function $s(t)$ in the form $s(t) = at + bt^m$ into (27) gives $s(t) \simeq t + t^9 / 1814\,400 + O(t^{10})$. It is seen (figure 1) that at times as large as $t \approx 3$ the approximate and exact profiles are practically completely coincident. The calculated error norm $\|\varepsilon_T(0.5)\|_1 = 6.27 \cdot 10^{-6}$ is substantially smaller as compared to that of the numerical solutions [15, 16].

Using the function $s(t) \simeq t + t^9 / 1814\,400$, we expand the temperature function $T(0,t)$ into a Taylor series and apply the Pade approximation to it. In the long run we obtain

$$
T(0,t) \approx \frac{t + 0.49210t^2 + 0.02328t^3 + 0.008123t^4}{1 - 0.007905t - 0.139435t^2 + 0.037506t^3 - 0.003511t^4}. \tag{28}
$$

Calculation of the function $T(0,t)$ (figure 2) gives the error $\varepsilon_0 = (T(0,t) / T^*(0,t) - 1) \cdot 100\%$ equal to $\varepsilon_0 = 1.4 \cdot 10^{-5}\%$ and $\varepsilon_0 = 2.6 \cdot 10^{-5}\%$ at $t = 0.5$ and $t = 1$ respectively. In comparison with the numerical schemes [16, 17], these values of $\varepsilon_0(0.5)$ are smaller by two orders of magnitude. Note that formula (28) gives a negligibly small error $(t = 0.5)$ $\varepsilon_0 = 1.06 \cdot 10^{-7}\%$.

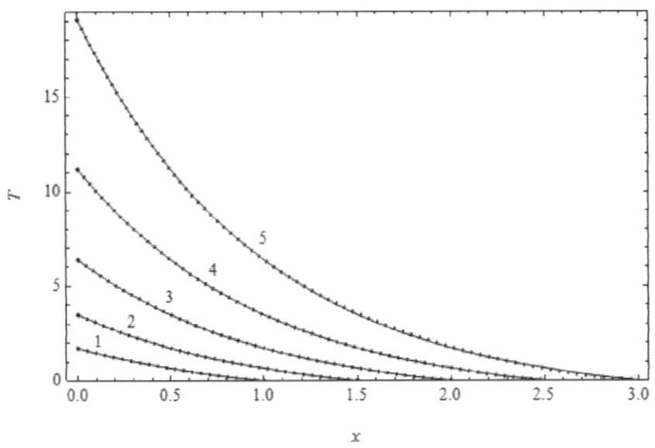

Figure 1. Temperatures given by the exact (full lines) and approximate (N = 4) (dotted lines) solutions at t = 1 (*1*), 1.5 (*2*), 2 (*3*), 2.5 (*4*), 3 (*5*).

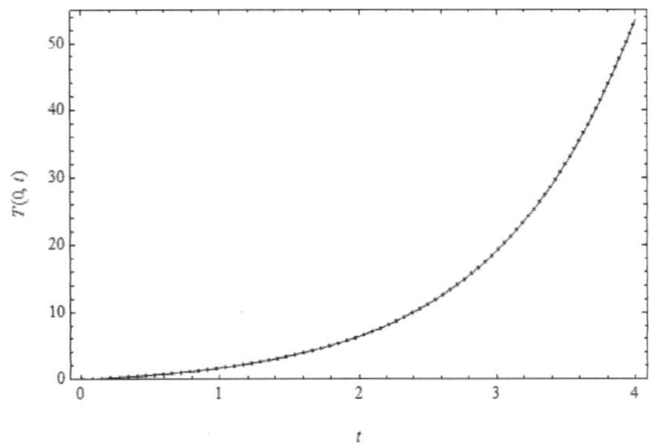

Figure 2. Time change in the surface temperature given by the exact (full line) and the approximate (N = 4) (dotted line) solutions.

3.3. Robin boundary condition

The coefficients of the temperature profile ($N = 3$) will be determined at the boundary condition (2)-(ii) with the use of the first two identical equalities of sequence (19):

$$
\int_0^s (1+x)\,T\,\mathrm{d}x + \frac{s}{2\text{Ste}}(2+s) = \Upsilon_1, \quad \int_0^s T \frac{x^2}{2}\left(1 + \frac{x}{3}\right)\mathrm{d}x + \frac{1}{2\text{Ste}}\left(\int_0^t (2s + s^2)\,\mathrm{d}t + \frac{s^3}{12}(4+s) \right) = \Upsilon_2. \tag{29}
$$

A constitutive equation for the function $s(t)$ can be obtained from the Stefan boundary condition (3), the third-order identical equality of sequence (19), the heat-balance integral (7), or the first-order integral relation (16). We dwell on one of the above-listed variants, namely, the third-order identical equality differentiated with respect to the time t of sequence (16):

$$\frac{\mathrm{d}}{\mathrm{d}t}\int_0^s T\frac{x^4}{24}\left(1+\frac{x}{5}\right)\mathrm{d}x+\frac{1}{\mathrm{Ste}}\left(\int_0^t\left(s+\frac{s^2}{2}\right)\mathrm{d}t+\frac{s^3}{6}+\frac{s^5 s'}{120}+\frac{s^4}{24}(1+s')\right)=\Upsilon_2 . \quad (30)$$

Assuming that α = 1, β = – 1, Ste = 1, и $\gamma(t)=2\mathrm{e}^t-1$ in (2)-(iii), we have the known test problem [16–18]. Expanding the function e^t into a Taylor series and representing the function $s(t)$ in the form $s(t)=at+bt^m$, from equation (30) we find $s(t)=t-t^7/11025+O(t^8)$. The mean square error given by the approach proposed at $t=1$ comprises $\|E_T\|_2=8.545\cdot10^{-4}$.

We also obtained analytical solutions in the form of the fourth- and fifth-order polynomials. At $N=5$ the polynomial coefficients of the temperature profile were determined at the Robin boundary condition (2)-(iii) and the Stefan condition (3) with the use of three identical equalities of sequence (19). Equation (7) was used as a constitutive equation. In this case, the error norm ($t=1$) was $\|E_T\|_2=2.747\cdot10^{-6}$. The indicated problem was solved numerically in the work [17] where the errors at $t = 0.9$ comprised $\|E_T\|_1=3.951\cdot10^{-5}$ (n = 11, Δt = 0.10) and $\|E_T\|_1=1.212\cdot10^{-6}$ (n = 51, Δt = 0.01). As is seen (table 1), the new approach proposed for solving the Stefan problem allows one to calculate the temperature profile with an accuracy close to that of numerical schemes, and these solutions substantially surpass the numerical solutions in the accuracy of calculating the interphase boundary.

4. Conclusion

On the basis of integral equations, high-accuracy solutions of the classical Stefan problem have been obtained. The approach proposed can be used to advantage in different applied problems.

Table 1. Calculation of the temperature front $s(t)$ and the surface temperature $T(0, t)$.

Parameters (t=1)	Accurate values	New scheme (N=5)	Absolute error (E_s / E_T)	Relative error (%) (ε_s / ε_T)
s	1	1.000000056362	5.63621×10^{-8}	5.63621×10^{-6}
$T(0,t)$	1.718281828459	1.718282176243	3.47784×10^{-7}	2.02402×10^{-5}

References

[1] Hill J M 1989 *One-Dimensional Stefan Problems: An Introduction* (London: Chapman & Hall)
[2] Caldwell J 2009 *Therm. Sci.* vol 13 pp 61–72
[3] Mitchell S L and Vynnycky M 2009 *Appl. Math. Comput.* vol 215 pp 1609–21
[4] Lunardini V J 1991 *Heat Transfer with Freezing and Thawing* (London: Elsevier)
[5] Wituła R 2010 *Archives of Foundry Engineering* vol 10 pp 83–8
[6] Tao L N 1979 *Quart. J. Appl. Math. Phys.* vol 30 pp 416–26
[7] Mitchell S L and Myers T G 2012 *Int. J. Diff. Eq.* vol 2012 p 22
[8] Myers T G 2010 *Int. J. Heat Mass Transf.* vol 53 pp 1119–27
[9] Mitchell S L and Myers T G 2010 *SIAM Rev.* vol 52 pp 57– 86
[10] Sadoun N 2006 *Appl. Math. Model.* vol 30 pp 531–44
[11] Кот V A 2016 *Heat Transf. Res.* vol 47 pp 1035–55
[12] Кот V A 2016 *Heat Transf. Res.* vol 47 pp 927–44
[13] Кот V A 2017 *J. Eng. Phys. Thermophys.* vol 89 pp 1289–314
[14] Mosally F A, Wood A S and Al-Fhaid A 2002 *Appl. Math. Comput.* vol 130 pp 87–100
[15] Kutluay S, Bahadir A R and Ozdes A 1997 *J. Comput. Appl. Math.* vol 81 pp 134–144
[16] Kutluay S and Elsen A 2004 *Appl. Math. Comput.* vol 150 pp 59–67
[17] Whue-Teong Ang 2008 *Num. Meth. Part. Diff. Eq.* vol 24 pp 939–49

On the Black Sea deepwater circulation features derived from numerical modeling and measurement data

N V Markova

Marine Hydrophysical Institute of the Russian Academy of Sciences, Kapitanskaya 2, Sevastopol 299011, Russia

E-mail: n.v.markova@mail.ru

Abstract. The Black Sea deepwater currents below the main pycnocline at the depths more than 350 m are considered. Current velocities derived from different datasets are compared. Results of two numerical experiments with MHI non-linear z-coordinate ocean model, estimates of deep Argo float velocities, and in-situ data accumulated in MHI Database are analyzed. It is shown that sub-pycnocline velocities obtained by simulation with MHI model and SKIRON atmospheric forcing generally correlate with Lagrangian deep Argo velocity estimations and MHI Database mean velocities. Existence of deepwater cyclonic and anticyclonic vortexes is drawn from the datasets under consideration. The qualitative difference between modeling results and measurements is the narrow quasi-periodic currents along the northeast continental slope which are not succeeded to verify by available data of Argo and MHI Database collected so far.

1. Introduction

Main structure peculiarities, variability, and dynamic characteristics of the Black Sea surface circulation are generally quite well investigated. A number of large-scale and mesoscale circulation features in the top layers of the sea, such as Rim Current propagating above the continental slope, synoptic-scale cyclonic gyres in the western and east parts of the sea, mesoscale eddies between Rim Current and the coastline, quasi-stationary vortexes such us Sevastopol and Batumi anticyclones, etc are found out. This result was reached as by lots of measurements including surveys and satellite observations, as by hydrodynamical modeling. At the same time, there is a lack of works focused on studying currents in deep Black Sea including numerical simulations. Often the deepwater circulation study is a rather small part of more general research.

Some early publications on the Black Sea dynamics concerned the availability of anticyclonic circulation opposite to Rim Current ("counter-currents", “reverse circulation”) in deepwater layers. A number of theoretical researches and calculations using field data proved the possibility of the existence of reverse circulation. For example, arguments for the counter-currents presence inside the main pycnocline are given in [1, 2]. At the same time, in research [3, 4] based on trajectories of 3 and 7 profiling floats no counter-currents were detected.

Although the discussion about the existence of such currents has a long history, opinions differ widely. Recently in publications concerning the Black Sea dynamics simulations with different numerical models some mentions about anticyclonic currents in deep layers have appeared. Existence of such counter-currents was established, for example, in modern modeling results [5-7]. So it seems to important to revise and compare new modeling results and other available datasets to make more clear the character of the deep Black Sea circulation. Our study starts with four available datasets, we use the results of two different numerical experiments, Argo floats data, and MHI Database current velocity measurements.

2. Experiment 1 (modeling climatic currents)

In the first numerical experiment, the Black Sea climatic fields were simulated with three-dimensional nonlinear z-coordinate MHI model [8] which is one of the most well-established models of the Black Sea dynamics. It was included into the number of international projects and used as a part of operating marine prognosis system till now [9, 10]. The model allows us to reconstruct the full structure of main hydrophysical fields as temperature, salinity, horizontal currents, vertical velocity, and sea level in the basin. The climatic experiment successfully was carried out as part of our mostly methodological study [11], but the results of these computations were not compared with the data of numerical modeling and field measurements in such way we are going to use now. Moreover, it is climatic experiment that allows us to obtain the most general characteristics of hydrophysical fields. So we decided to look at the data calculated before from another side and to use the data for nowadays purposes.

We give a brief description of some particularities of the Experiment 1 running. The procedure of assimilation of mean in-situ temperature and salinity [12] interpolated to the model grid was applied. That technique was developed in [13]. Climatic boundary condition including atmospheric ones were used on the sea surface, and in the straits and river mouths. An annual cycle of the climatic hydrophysical fields on 45 levels vertically (2.5-2100 m) was obtained; the step by both horizontal axes was equal 5 km. The calculated fields were recorded daily.

To analyze the deepwater circulation, we select several standard hydrological horizons below 350 m. Generally the cyclonic type of basin-size deep circulation is presented but there are some structure features found out. There is a number of deepwater mesoscale eddies of different sign in which the maximum velocities at the horizons (5-8 $cm \cdot s^{-1}$) are established.

In Sevastopol anticyclone area, it is found that the sign of rotation of this vortex at the depth of about 400-450 m is changeable. Such behavior (change of the rotation direction) of deep vortexes was described, for example, in [14]. The complicated vertical structure of this vortex has been fixed repeatedly during expeditions [15, 16]. It was supposed that the rotation change was caused by the change of the direction of a barotropic component of a radial pressure gradient [14].

Along the northeastern continental slope at depths below the permanent pycnocline the circulation reverse Rim Current is shown. This feature is presented in the layer of 700-1500 m in spring and summer. The counter-current originates in the region of the Bulgarian coast and is directed anticyclonic towards the north-Caucasian coast. The absolute values of the mean climatic velocities are rather low (see table 1).

Table 1. The Black Sea mean velocities derived from different datasets at some standard deep horizons (z).

Dataset	Mean velocity ($cm\ s^{-1}$)			
	z=500 m	z=750 m	z=1000 m	z=1500 m
Experiment 1 (climatic)	1.1	0.9	0.8	0.8
Experiment 2 (2011)	3.7	3.2	3.1	3.4
Argo	3.6	4.0	5.7	3.5
MHI Database	4.6	2.8	6.2	–

The maximum velocities of the counter-current do not exceeds 3-4 $cm \cdot s^{-1}$, its average velocity is about 1.5-2 $cm \cdot s^{-1}$. The background mean velocities hardly reach 0.5-1 $cm \cdot s^{-1}$. In the area of the north-Caucasian coast which characterized by narrow continental slope, this current is found to start at higher horizons of about 350 m. So, the upper boundary of the climatic counter-current approximately corresponds to the main pycnocline bottom boundary in this region.

The comparison of the modeling results with the data of some non-assimilated hydrological surveys showed that the most differences between the model and in-situ temperature and salinity are in the shallow coastal zone. In this area the variability of the hydrophysical fields is typically most high due to a number of external factors. But the climatic fields are reconstructed under more smooth boundary conditions and do not assume such high-frequently response. In other circumstances, on more large temporal and spatial scales, the behavior of the climatic temperature and salinity is quite close to the measurement data. Corresponding climatic currents picture indicates well known hydrological features of the upper layers but characterized by low velocities. So, we can use climatic calculation data for the purpose of comparison with measurement data preferably averaged, but mostly qualitatively and offshore.

3. Experiment 2 (modeling for 2011)

Recently the second experiment was carried out with the same MHI model, but with some differ model parameters and atmosphere forcing (SKIRON reanalysis data for 2011 is used). The horizontal resolution of the model is improved up to 1.64 km. The experiment is conducted without assimilation of archive temperature and salinity.

It is shown that some mesoscale and sub-mesoscale structures of hydrophysical fields obtained for 2011, are absent in the previous experiment results. It is not unexpected, since the climatic experiment assumes the use of smooth atmospheric fields (wind field, heat and moisture flows) and its spatial resolution is three times worse. Actually, to reconstruct spatial sub-mesoscale features, a high model resolution are strongly required. And to get high-frequent temporal variability, the real forcing is needed.

In 2011 in deep layers the average velocities in relation to climatic ones are 3-4 times higher and equal 3.1-3.7 $cm \cdot s^{-1}$ (see table 1). The maximum values of the velocity are 14-18 $cm \cdot s^{-1}$ and achieved in mesoscale eddies.

Figure 1 shows the currents field at 1100 m horizon in spring, obtained in the climatic simulation (a) and in the modeling hydrophysical fields in 2011 (b). The climatic field is smoother and the velocities are lower. Apparently, the southeastern current along the continental slope of the Caucasus region (counter-current) is present on the both experiment pictures. The number of mesoscale eddies is also shown including ones in the regular area of Batumi anticyclone.

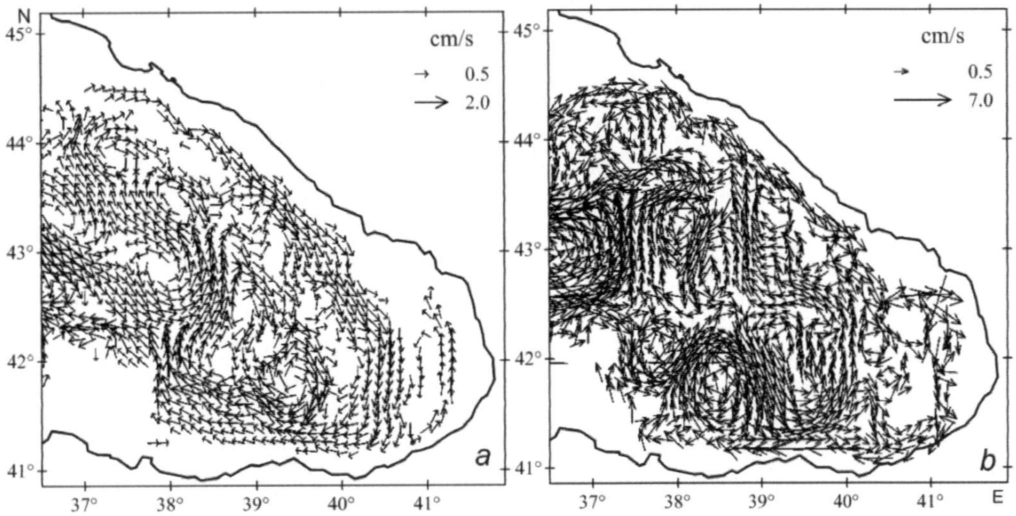

Figure 1. Deep currents in the Eastern Black Sea (April, 11; Depth 1100 m): (*a*) – Experiment 1 (climatic velocities), (*b*) – Experiment 2 (velocities in 2011).

Results of the Experiment 2 allow us to detail some characteristics of counter-current of anticyclonic orientation. It is about 5-9 km in the transverse direction, and propagates at the depth of about 1000 m along the steep and narrow continental slope. Unlike the climatic experiment, in 2011 the intensity of this counter-current is not seasonally depends, but quasi-periodically fluctuations of 1-2-3 weeks are present. It suggests an idea that such currents might be forced by low frequency waves.

4. Calculations with Argo floats data

Any simulation results require verification by in-situ measurement data. For this purpose, the data of the Argo floats, which are freely available (http://www.usgodae.org), can be used. However, the floats are not able to measure currents directly.

The average velocities of the floats underwater movement can be estimated only from the data of their positioning at the surface by the satellites. It seems that the average float velocity at the parking depth is easy calculated, but there is a problem on the assessment of a buoy velocity shift during its ascent and descent stage. The float velocity is ordinary calculated as the ratio of the drifted distance underwater to the time. This method does not take into account the peculiarities of the buoy duty cycle. Therefore, there are errors caused by effect of faster currents in upper layers, during the float ascent for communication with the satellite and its descent to the parking depth. This problem is discussed for example in [17]. Some sophisticated algorithms offered to solve the problem. But as far as we know, there are no published works on successful filter of such errors for the Black Sea floats to date. Further we plan to specify the calculations of Lagrangian velocities of Black sea deep Argo floats as an additional research. But now, as the first iteration, we use the same technique as in earlier works [3, 4] and recent [18] and so confined with a rather rough estimate the value of mean velocity derived from Argo.

Although we use the same calculation method, the most complete decadal Argo data array 2005-2015 is processed. The data of 16 Argo deepwater floats with parking depths of 450 m and lower are considered. Preliminary discussion and data statistical processing was reported in [19]. The selected floats drifted at the parking depth of 450, 500, 750, 1000, 1300 and 1500 dbar. Due to the peculiarities of the bottom relief the buoys sometimes moved above the programmed levels. Therefore, we have to check the information for each float and consider its possible grounding. Taking into account the actual depth of the drift, the found velocities are tied to layers of 350-600, 600-800, 800-1200, and 1200-1600 m. The highest values of the velocities are reached in January-March, and the lowest ones from June till October. The values of the obtained float mean velocities are given in the table 1 (in the rows of standard horizons included into the corresponding layers).

The presence of eddy structures is established on the background of general cyclonic circulation. They can be traced by the trajectories of the Argo floats (figure 2). At the figure we coupled the trajectories into layers of 350-800 and 800-1600 m for easy perusal. Vortexes of predominantly synoptic scale are found in the central and western parts of the sea, and smaller ones in the eastern part. The presence of mesoscale structures in the Batumi, Sevastopol and Sakarya anticyclone zones is apparent. Along the Anatolian coast, a significant part of the trajectories of deepwater floats follow along the continental slope to the east. This behavior is consistent with the mechanism of the spread of heavy Mediterranean waters propagating from the Bosporus Strait.

5. In-situ deep currents measurements

We considered data of direct in-situ measurements of currents velocity in the Black Sea of different years. This massive is collected in Database of currents, which is the part of the MHI Databank [20]. The data of 4000 soundings at 25 oceanographic deepwater stations (locations see at figure 2) are suitable for statistical estimation. We selected data at three standard horizons: 500, 750 and 1000 m. The 1500 m horizon is not taken into account due to lack of data. Statistical processing of the sample is performed. The average deep currents velocity is given in table 1. The mean value is of the order of some centimeters per second. The sample contains a number of extremely values as high as 80 $cm \cdot s^{-1}$, which need to be filtered out. To exclude such non-realistically data, the third quartile of the velocity modules at each horizon is considered. Thus, 25% of the highest values are discarded. After the filtration, the maximums of instantaneous velocities are equal: at a depth of 500 m – 45 $cm \cdot s^{-1}$, at

750 m – 13 cm·s^{-1}, and at 1000 m – 15 cm·s^{-1}. The median values are accordingly equal to 24 cm·s^{-1} at a depth of 500 m, 12 cm·s^{-1} – at 750 m, and 15 cm·s^{-1} – at 1000 m. But the contribution of small near-zero velocities is considerable and the mean values are got to be quite moderate.

At certain stations, a significant increase (2 or more times) in currents velocity from the surface to the maximum depth of soundings (500-1000 m) is obtained. The direction of circulation at some stations radically changes with depth, but at some it is practically constant. At third ones, it was not possible to allocate the prevailing direction of currents, because the probe has perhaps got to a vortex zone. We can assume then the existence of relatively powered dynamical structures on the weak background in the deep Black Sea.

Since the in-situ data belonged of different periods, and the distance between the stations is quite significant, we are not able to unambiguously reconstruct a map of deepwater currents only on their basis. However, the data of the MHI Database of currents allow us to evaluate the variability of the currents vertical structure.

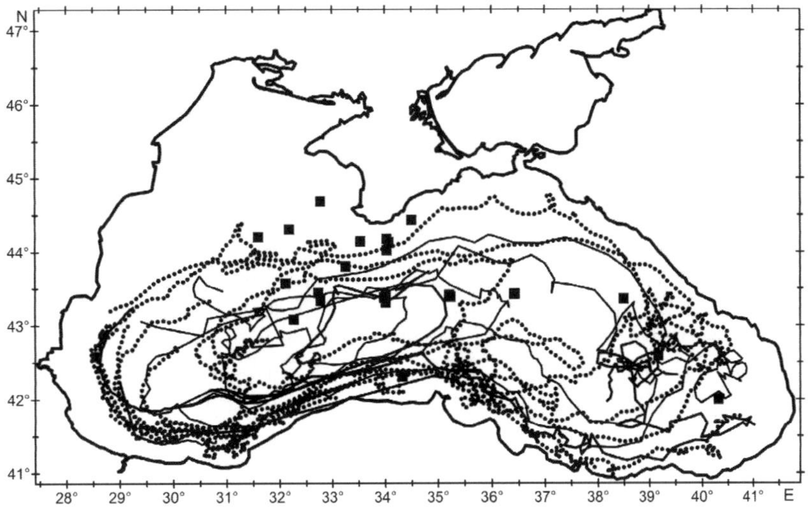

Figure 2. Argo profiling floats trajectories (at 350-800 m – dash lines, at 800-1600 m – solid lines) and MHI Database deepwater stations locations (black squares).

6. Discussion

Numerical modeling seems to be the most informative method for studying the structure of the Black Sea currents. It provides full coverage of the entire basin with data. The use of the MHI climatic model allows us to obtain a four-dimensional structure of the current field, analyze its annual cycle and specify the deep currents features not identical with the surface ones.

As a result of the modeling, elements of reverse circulation are found in the form of anticyclonic counter-currents along the northeastern continental slope at depths of about 1000 m. They are present quasi-periodically in spring and summer as drawn from the climatic experiment, and in all seasons according by the results of experiment with real atmospheric forcing. The counter-current passes to the south of the Crimean Peninsula toward the North Caucasian coast and further to the southeast. In the climatic experiment the counter-current is obtained quite wide (8-20 km in transverse direction), and in Experiment 2 its width is 5-9 km.

By modeling, it is found a number of mesoscale (of the order of several tens kilometers) cyclonic and anticyclonic deepwater vortex structures. It is in these eddies that the highest velocities are attained. The trajectories of deep Argo floats confirm the presence of such features. The high current velocity and direction variability at some in-situ stations could be due to the probe getting into a vortex zone. Thus, in-situ measurements indirectly confirm the presence of vortex structures in deep layers. That conclusion is agreed with results [21] inferred from the processing of survey temperature and salinity data.

The absolute values of climatic currents are low first of all due to the use of averaged atmospheric conditions. This values differ significantly both the results of Experiment 2 and the estimates of the average velocity with measurement data. For comparison, the mean values of the current velocities derived from different datasets are summarized in table 1. It is obvious that the most agreement between the mean velocities from all datasets excluding climatic is achieved at horizons of 500 m (velocity range 3.6-4.6 $cm{\cdot}s^{-1}$), 750 m (2.8-4.0 $cm{\cdot}s^{-1}$) and 1500 m (3.4-3.5 $cm{\cdot}s^{-1}$). The mean velocities from Experiment 2 are most consistent with the estimates from Argo data. The most difference in the mean values is at the horizon of 1000 m, where the counter-currents are modeled and the highest Argo velocities and in-situ ones are reached.

The obtained results generalize the information from the considered datasets. It is clear that the basin circulation is generally cyclonic, but in deep layers it is not as strong and emphasized as in upper ones. There are some dynamic features (vortices, currents) with relatively high velocities that require an additional study to describe the Black sea deepwater circulation more definitely. At date there are some disagreements in important details. In particular, the presence of reverse circulation elements in model fields is not confirmed by data of Argo and in-situ currents measurements under consideration. Therefore, this research will be continued. For a better understanding of the deep Black Sea processes, in future research we assume to use measurement data from alternative sources, carry out additional simulations, improve calculation methods, etc.

References

[1] Bulgakov S N, Korotaev G K and Whitehead J A 1996 The role of buoyancy fluxes in the formation of a large-scale circulation and stratification of sea water: 1. Theory *Izv. Atmos.Ocean. Phys.* **32**(4) 548-556.

[2] Bulgakov S N, Korotaev G K and Whitehead J A 1996 The role of buoyancy fluxes in the formation of a large-scale circulation and stratification of sea water: 2. Laboratory experiments *Izv. Atmos.Ocean. Phys.* **32**(4) 557-564.

[3] Korotaev G, Oguz T and Riser S 2006 Intermediate and deep currents of the Black Sea obtained from autonomous profiling floats *Deep-Sea Res. II* **53**(17-19) 1901-1910.

[4] Gerasimova S V and Lemeshko E E 2011 Evaluation of deep current velocities using ARGO data *Sistemy kontr. okruzh. sredy* (Sevastopol, MHI NASU, in Russian) **15** 187-196.

[5] Arkhipkin V S, Kosarev A N, Gippius F N and Migali D I 2013 Seasonal variations of climatic fields of temperature, salinity and water circulation in the Black and Caspian seas *Moscow Univercity Bulletin. Series 5. Geography* **5** 33-44.

[6] Lukyanova A N et al 2016 The Black Sea deep-water circulation research by results of numerical modeling and in-situ data: INMRAS model numerical experiment *Ecological safety of coastal and shelf zones of the sea* (Sevastopol, Marine Hydrophysical Institute RAS, in Russian) **3** 9-14.

[7] Demyshev S G et al 2016 Numerical experiments on modeling of the Black Sea deep currents *Phys. Oceanogr.* **2** 38-50.

[8] Demyshev S G, Korotaev G K 1992 Numerical energy-balanced model of the baroclinic ocean currents on a C-grid *Chislennye modeli i rezul'taty kalibrovochnyh raschetov techeniy v Atlanticheskom okeane* (Moscow, INM RAS, in Russian) 163-231.

[9] Korotaev G K et al 2011 Development of Black Sea nowcasting and forecasting system *Ocean Science* **7**(5) 629-649.

[10] Korotaev G K et al 2011 Contribution of the Black Sea observing system to ECOOP *Ocean Science Discussions* **8**(4) 1695-1722.

[11] Demyshev S G, Ivanov V A, Markova N V and Cherkesov L V 2007 Construction of the Black Sea currents on the basis of eddy resolving model with assimilation of the climatic temperature and salinity *Ecological safety of coastal and shelf zones and complex use of*

shelf resources (Sevastopol, MHI NASU, in Russian) **15** 215-226.
[12] Belokopytov V N 2004 Thermohaline and hydrologic-acoustical structure of the Black Sea waters PhD thesis (Sevastopol, MHI NASU, in Russian) 160 pp.
[13] Knysh V V, Demyshev S G and Korotaev G K 2002 A procedure of reconstruction of the climatic seasonal circulation in the Black Sea based on the assimilation of hydrological data in the model *Phys. Oceanogr.* **12**(2) 88-103.
[14] Latun V S 1992 Density field structure and water circulation in deep Black Sea anticyclone. *Hydrophysical and hydrochemical studies of the Black sea* (Sevastopol, MHI NASU, in Russian) 92-94.
[15] Morozov A N and Lemeshko E M 2006 Experience of the acoustic Doppler measuring instrument of currents (ADCP) use in the Black Sea *Phys. Oceanogr.* **16**(4) 216-233.
[16] Bulgakov N P and Bulgakov S N 1995 Manifestation of the Black Sea counter-current in the seawater density and hydrostatic pressure *Morskoy gidrofizicheskiy zhurnal* (Sevastopol, MHI NASU, in Russian) **4** 63-76.
[17] Rosell-Fieschi M 2014 Ocean Velocities as Inferred from ARGO Floats: Methodology and Applications PhD Thesis (Institut de Ci`encies del Mar, CSIC) 121 pp.
[18] Milanova M and Peneva E 2016 Deep Black Sea circulation described by Argo profiling floats *Annual of Sofia University St. Kliment Ohridski, Faculty of Physics* **109** 12 p.
[19] Markova N V and Bagaev A V 2016 Velocities of the Black Sea deep currents estimated from the profiling drifters Argo data *Phys. Oceanogr.* **3** 23-35.
[20] Plastun T V 2005 Database of currents from databank MHI NASU *Proc. Int. Sci. Conf. Recent Changes in the Ecosystems of the Black Sea and Sea of Azov. Crimea, Donuzlav, 13-16 Sept. 2005* vol 1, ed E E Sovga et al (Sevastopol: NPC ECOSY-Hydrophysics, in Russian) 153-154.
[21] Oguz T et al 1993 Circulation in the surface and intermediate layers of the Black Sea *Deep-Sea Res. I* **40(**8) 1597-1612.

Acknowledgments
This research was supported by RFBR (project № 15-05-05423 A).

IOP Conf. Series: Journal of Physics: Conf. Series **899** (2017) 052012 doi:10.1088/1742-6596/899/5/052012

Vortex shedding behind an obstacle in a channel under transition to turbulence in steady and pulsating flows

V M Molochnikov[1,3], E I Kalinin[2], A B Mazo[2], A V Malyukov[1], D I Okhotnikov[2] and O A Dushina[1]

[1] Kazan Scientific Center of RAS, Kazan, 420111, Russia

[2] Kazan Federal University, Kazan, 420008, Russia

[3] Kazan National Research Technical University, Kazan, 420111, Russia

e-mail: vmolochnikov@mail.ru

Abstract. Experimental research and DNS of separated flow over a spanwise semicylindrical rib mounted in a rectangular channel have been performed for the cases of steady and pulsating flow with transition to turbulence. Spiral motion of fluid behind the rib directed from the channel walls towards the channel symmetry plane has been shown to govern the formation of large vortices in the mixing layer starting from some critical Reynolds number. The mechanism of origination of such motion has been hypothesized. The effect of frequency and amplitude of forced freestream pulsations on the vortex pattern behind the rib has been described.

1. Introduction

Large vortices play a crucial role in heat transfer enhancement caused by near-wall turbulence promoters. However, there are only a few attempts at studies of the formation of these vortices in conditions of laminar-turbulent transition and they are mainly focused on evolution of instabilities in a separated laminar flow with small perturbations [1].

The majority of researchers attribute the formation of large quasiperiodic vortices in an obstacle wake in unbounded freestream to the global instability of flow on the scale of entire separation region (instability to periodic vortex shedding) [2, 3]. The mechanism of large vortex generation in wall-bounded flows at transitional flow regimes is different. In a rectangular channel, the sidewalls significantly influence the separated flow pattern downstream of the obstacle. This issue has been recognized for the first time ever when the data on the reattachment length obtained by 2D simulation had been observed to deviate from the experiment [4]. The reason for such deviation was shown to be the secondary spiral motion of fluid from the channel sidewalls towards its symmetry plane. With the increase in the Reynolds number, the intensity of this flow grows in the laminar region and falls at transitional and turbulent flow regimes [5]. DNS of a backward-facing step flow [6] showed that when the channel width increases, the Reynolds number corresponding to the onset of instability in the separated shear layer decreases. The loss in stability is accompanied by the transition to essentially three-dimensional behavior of velocity and temperature fields. Some regularities in the spiral motion from the channel sidewalls towards its symmetry plane for the case of backward-facing step flow were shown in [7]. Only the results for Re = 400 are given, although the authors state that they can be extended to the Reynolds number range Re = 100–400. The already cited paper [5] showed that the discrepancy between the reattachment lengths obtained experimentally and computationally (2D model) was initiated when the spiral motion from the channel sidewalls reached the channel symmetry plane. Despite a large number of research papers studying 3D effects in transitional separated flows in rectangular channels, the mechanism of formation of large 3D vortices in such flows has remained beyond the scope of these studies.

Understanding of physical mechanisms of quasiperiodic vortex formation behind the obstacles makes it possible to control the processes in heat exchangers and cooling systems. Forced pulsations of the oncoming flow are a promising control method. However, there are almost no published data on the effect of pulsations on the formation of vortices behind the obstacles in the channels under transitions to turbulence.

In the present paper, the flow over a spanwise rib mounted on the wall of a rectangular channel has been studied experimentally and by direct numerical simulation (DNS) for the case of low Reynolds numbers in the steady and pulsating oncoming flow. The mechanism of laminar-turbulent transition and large vortex formation in the shear layer behind the rib has been studied, the vortex pattern control using the forced periodic freestream pulsations has been considered.

2. Experimental setup and procedure

The experiments were conducted in a rectangular test section (height $H = 20$ mm, width $B = 50$ mm) with a smooth inlet attached to it (figure 1). The spanwise rib (height $h = 3$ mm, radius $R = h$) was mounted at the distance $l = 106$ mm from the test section inlet. The air flow rate through the test section was provided by a vacuum pump. The experiments were carried out in the Reynolds number range $Re = Uh/\nu = 58.5$–439.5 based on the average velocity in the test section, U, the rib height, h, and kinematic viscosity of air, ν. Here, the average velocity, U, varied between 0.28 and 2.12 m/s. Hot-wire measurements revealed no turbulent fluctuations of flow velocity at the channel inlet. The boundary layer thickness exceeded the obstacle height in all the considered regimes. Smoke-wire visualization of flow was performed. The flow pattern was recorded in a light sheet in the channel symmetry plane ($z = 0$) and in the planes parallel to the bottom wall at the distance $y = 1$ and 3 mm from the latter. The frame rate was 800 frame/s. In addition to flow visualization, PIV measurements of instantaneous velocity vector fields were obtained downstream of the rib in the channel symmetry plane. PIV measurements were processed using an iterative algorithm for the calculation of velocity fields with particle displacement compensation. 32×32 pixel interrogation windows with 50% overlap were considered.

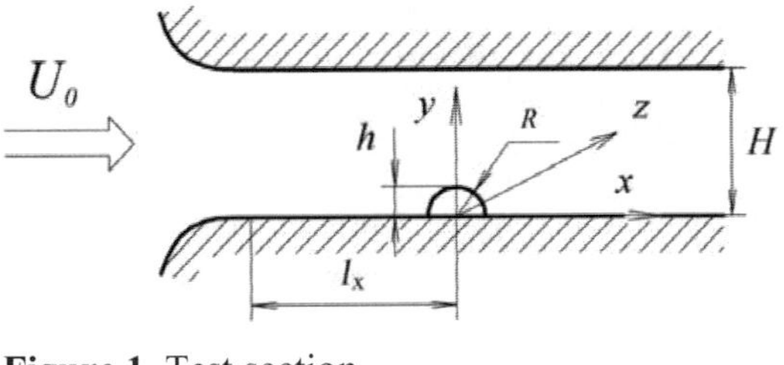

Figure 1. Test section

Direct numerical simulation (DNS) of flow was performed using ANSYS Fluent 14.5 software. The computational domain matched the test section configuration with the semicylindrical rib. 3D unsteady Navier-Stokes equations were solved in dimensionless physical variables for incompressible fluid in the Reynolds numbers range $165 < Re < 480$. All dimensions of the computational domain were normalized by the rib height, h, the time was normalized by h/U, and the pressure was normalized by ρU^2, where ρ is the air density. No-slip condition was set at solid walls; uniform streamwise unit velocity profile and zero excess pressure were set at the inlet; "soft" boundary conditions were set at the channel outlet. The computational grid was built using a locally-structured approach [8]: the domain was filled with a set of structured grid segments, which were connected by non-structured insertions. The grid was refined near the solid walls and in the obstacle wake. In the case of steady flow past the rib, the grid contained 1.5×10^6 cells, the dimensionless spacing of the grid in refinement areas was 0.01, in the middle wake it was 0.3, the time step was

0.02. The total number of grid cells in pulsating flow was 10^6, minimal dimensionless grid spacing in the refinement areas was 0.02, in the middle wake it was 0.15, in the boundary layer on the channel walls it was 0.01, and the time step was 0.05. Calculations on twice finer grid did not reveal any noticeable changes in computational result. The computational results were visualized by the simulation of passive particles released from virtual "smoke wires" located near the rib.

3. Results and discussion

3.1. Steady flow over the rib

Flow visualization downstream of the rib demonstrated a laminar flow with a stable closed recirculation region behind the rib at $Re < 165$ (figure 2 a). The flow pattern in horizontal plane includes a circular motion of fluid from the channel sidewalls towards its center, i.e. corner vortex pairs (figure 2 b). As soon as a critical Reynolds number, $Re = Re_* \approx 196$, is attained, large 3D vortices (vortex clots) are formed near the channel symmetry plane. In $x0y$ plane they look like spanwise vortices (figure 3 a). Corner vortices persist at this regime, they are clearly visible at the distance $y = 3$ mm from the ribbed wall as well (figure 3 b).

When the Reynolds number increases, the region of vortex clot generation onset shifts towards the rib, and the area occupied by these clots grows in transversal direction. The process is accompanied by fragmentation of large vortices into smaller ones. The wake flow downstream of the reattachment line takes the features of turbulent flow.

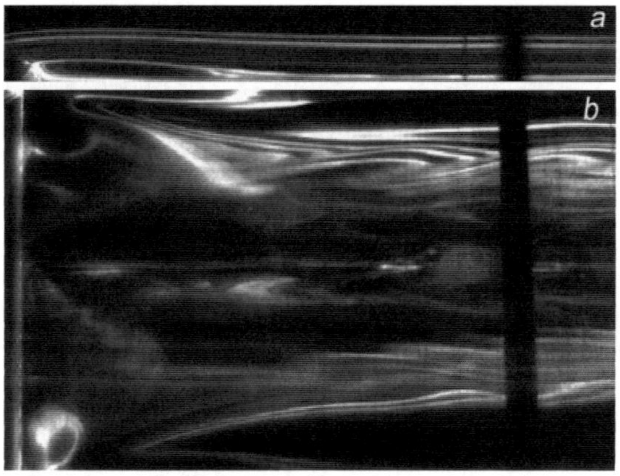

Figure 2. Flow visualization behind the rib at $Re = 135$: (a) – channel symmetry plane; (b) – horizontal plane at $y = 1$ mm

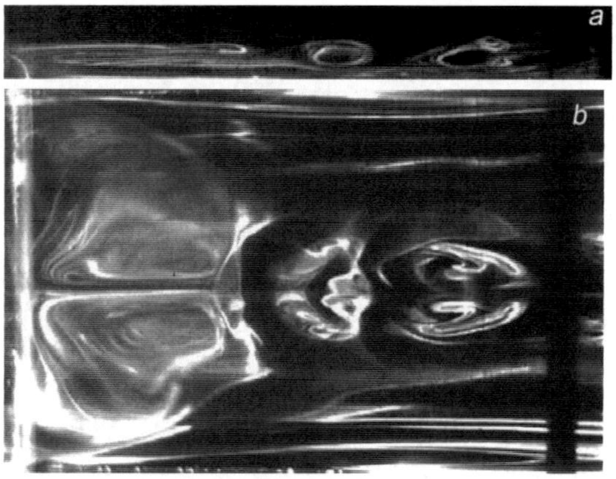

Figure 3. Flow visualization behind the rib at $Re = 233$: (a) – channel symmetry plane; (b) – horizontal plane at $y = 3$ mm

More detailed 3D flow pattern and its quantitative estimation were obtained by DNS. The latter was tested by the comparison of flow patterns, average velocity profiles and fluctuations obtained numerically and experimentally. Good agreement between DNS results and flow visualization together with PIV measurements (figure 4 and 5) was obtained.

Formation of large vortices in the mixing layer at $Re \geq Re_*$ is accompanied by essentially 3D flow in the boundary layer downstream of the obstacle in the channel. Corner vortex pairs shown in figures 2 and 3 appear to represent complex spiral flow directed from the channel sidewalls towards the channel center. The entrained fluid is shown in figure 6. This figure demonstrates the formation of 3D vortex clots near the channel symmetry plane and their shedding.

Thus, corner vortex pairs observed in the experiments include the spiraling fluid visible in the light sheet and stagnation regions not involved into the ascending spiral motion. The part of fluid involved in the spiral motion enters the recirculation region and is periodically shed to the free stream in the form of vortex clots at $Re \geq Re_*$.

In calculations, the increase in dimensionless (normalized by the velocity head) static pressure, *p(x)*, in the channel symmetry plane started upstream of the reattachment point. For example, at Re = 270 the pressure growth started at $x/2h \approx 4.4$ (figure 7), and flow reattachment corresponded to $x/2h \approx 4.75$ [9]. Considerable increase in pressure pulsation amplitude was observed in the region of the adverse pressure gradient ($5 < x/2h < 7.5$ at Re = 270). This is the region of large vortex (vortex clots) generation onset.

The obtained experimental and numerical results revealed the main features of the mechanism of large vortex formation in the rib wake in the channel under transition to turbulence. The core flow above the recirculation region behind the rib develops in conditions of adverse pressure gradient. The reverse flow in the recirculation region is the strongest near the channel symmetry plane due to spiral

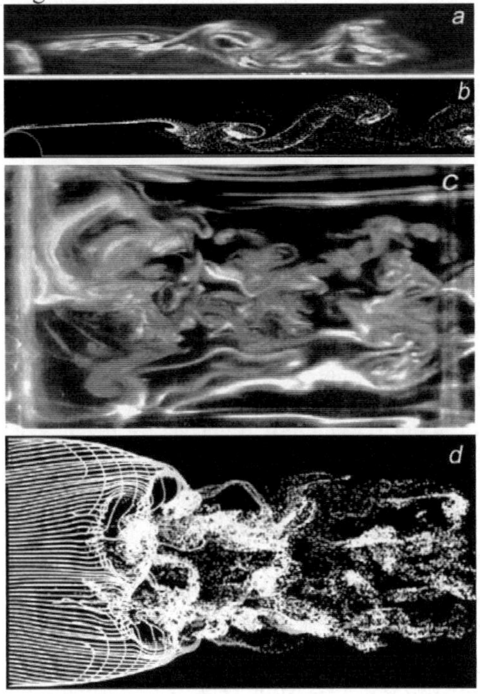

Figure 4. Experimental (a), (c) and computational (b), (d) flow patterns at Re = 270: (a), (b) – channel symmetry plane; (c), (d) – horizont. plane at y = 3 mm

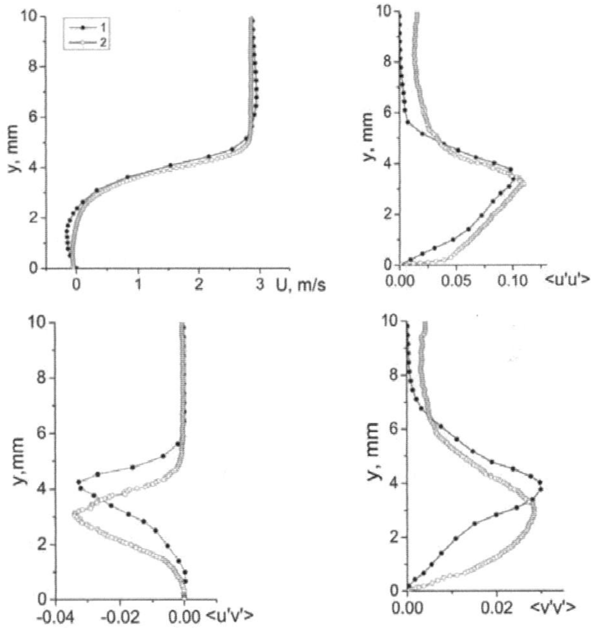

Figure 5. Comparison of experimental and computational profiles of statistical characteristics of flow behind the rib at Re = 439, x/h = 4: 1 – experiment; 2 – calculation

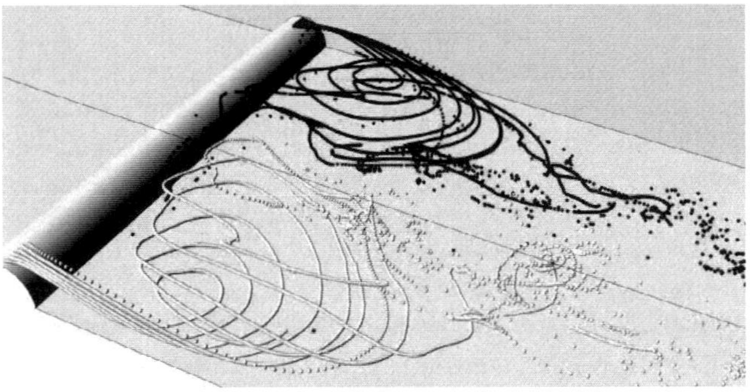

Figure 6. Spiral motion of fluid from the channel sidewalls to its center at Re=439

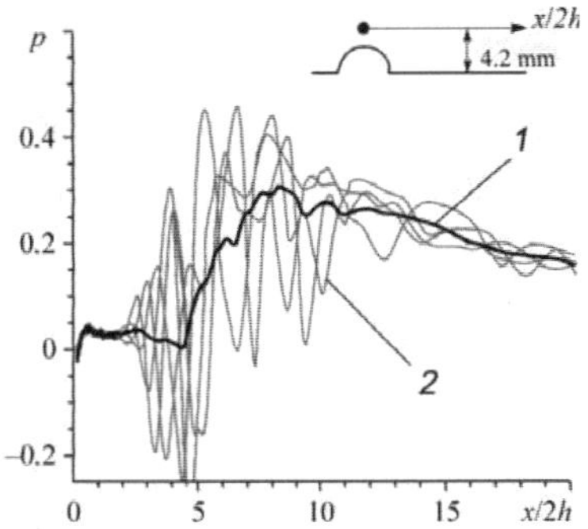

Figure 7. Averaged (1) and instantaneous (2) pressure behind the rib at Re = 270

flow from the sidewalls to the channel center. At low Reynolds numbers the circulation energy of fluid in the closed region behind the rib dissipates to the free stream due to viscous forces at the external boundary of this region and friction forces on the wall; the flow remains stable in this region, separation region is closed, and no vortices are observed. At $Re \geq Re_*$ viscous forces are unable to remove fluid circulation energy accumulated in the separation region, the flow in the adverse pressure gradient region becomes unstable, 3D coherent vortices form inside the separation region near the channel symmetry plane, and they are periodically shed to the main flow. The increase in Reynolds number leads to reduction of the scale of corner vortices and displacement of vortex generation onset in the upstream direction. This process is accompanied by fragmentation of large vortices into smaller ones and augmentation of velocity fluctuations; the obstacle wake downstream of the reattachment line acquires the features of turbulent flow.

The onset of spiral motion (pairs of corner vortices) in the obstacle near wake obviously results from the development of viscous boundary layer in dihedral angles between the bottom horizontal and vertical walls of the channel and from the interaction of this layer with the spanwise rib. According to [10], for the case of flow inside dihedral angles under the adverse pressure gradient, considerable reduction of skin friction is observed when approaching the dihedral angle edge in transversal direction. And it is the region of the pressure growth onset (figure 7) where the most intense displacement of flow from the channel walls is observed (figure 4 and 6).

3.2. Pulsating flow over the rib

In the present work we applied periodic forced flow pulsations to the flow over the rib in order to control the vortex pattern of the flow. The experiments were conducted in the same test section (figure 1), and forced pulsations were generated in a receiver tank *3* mounted at the channel outlet (one wall of the receiver was replaced by a speaker *4*) (figure 8). The speaker was controlled by a pulse generator through a signal amplifier to generate the required frequency and amplitude of pulsations. Forced flow pulsations frequency, f, and relative amplitude $\beta = A/U$ (here A is the amplitude of pulsations) varied in the experiments. The dimensionless frequency and amplitude changed in the ranges $Sh = fh/U_0 = 0.07 - 0.51$, $\beta = 0.07 - 0.3$ for each Reynolds number $Re = 58.5 - 439$. In steady flow $Sh = 0.06$ at $Re = Re_*$ and grows up to 0.13-0.16 at $Re = 228 - 439.5$. For the considered Re, Sh and β, the flow pulsations were shown to significantly alter the flow pattern typical of the steady flow at the same Reynolds numbers. "Starting vortices" are formed behind the rib in the flow acceleration phase. The origination frequency of these vortices is determined by the period of forced flow rate pulsations. The spatial scale of starting vortices depends on the Reynolds number and the parameters of forced unsteadiness: at low frequency of pulsations the linear scale of these vortices is approximately twice the rib height (figure 9) and decreases with the pulsation frequency growth. New pattern behind the rib was revealed in the pulsating flow at low Reynolds numbers and relatively high frequency of pulsations: a closed recirculation region with small starting vortices at its outer edge is formed behind the rib (figure 11).

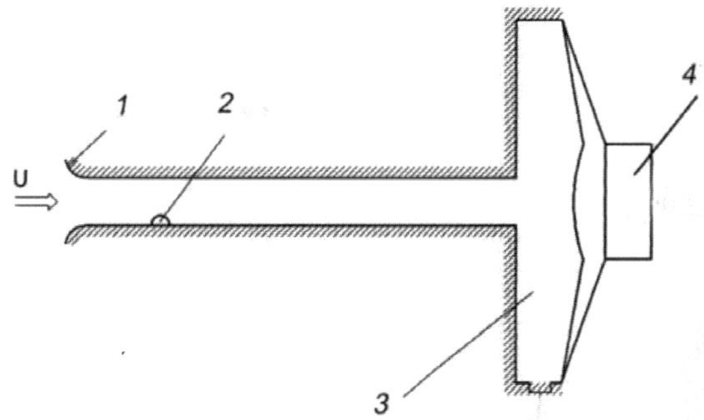

Figure 8. Test section intended for the study of pulsating flow over the rib: *1* – smooth inlet; *2* – rib; *3* – receiver tank; *4* – speaker

The spiral motion of fluid from the sidewalls towards the channel center, while governing the formation of large vortices behind the rib in the steady flow, does not usually reach the channel symmetry plane in the pulsating flow. For this reason, initially, the starting vortices are cylinder-shaped and their transversal dimension depends on the Reynolds number, frequency and

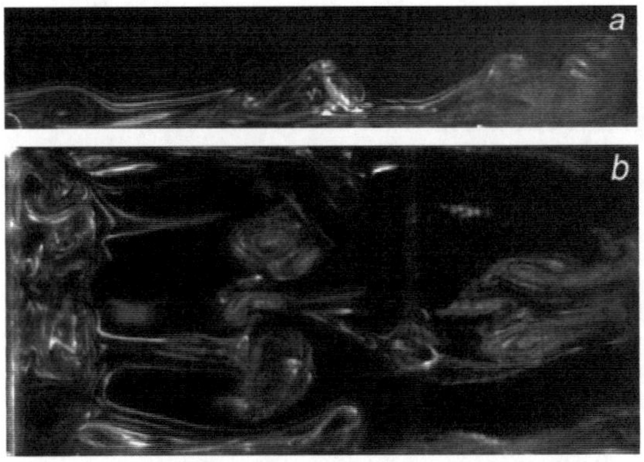

Figure 9. Pulsating flow over the rib at Re = 194, Sh=0.07; β =0.15: (a) – channel symmetry plane; (b) – horizontal plane at y = 1 mm

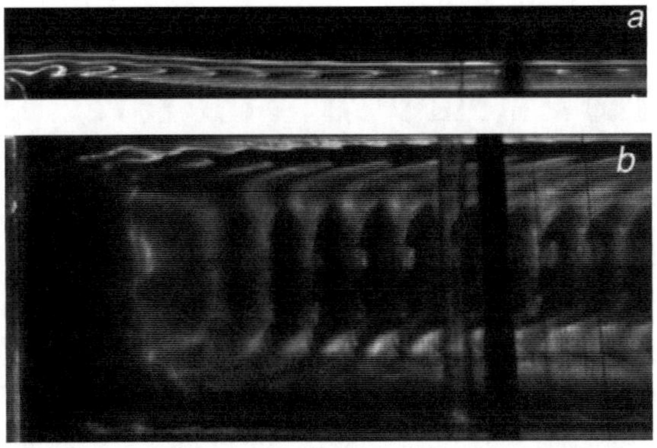

Figure 10. Pulsating flow over the rib at Re = 58.5, Sh=0.51; β =0.15: (a) – channel symmetry plane; (b) – horizont. plane at y = 1 mm

amplitude of forced pulsations.

4. Conclusions

With the growing Re, the "span" of these vortices grows, and the region of spiral motion reduces. The increase in forced pulsation frequency has similar effect on the pattern of starting vortices. Having been shed, each starting vortex disintegrates into several 3D vortex clots, the number of which grows with Re and Sh. The revealed regularities can lay the foundation for active control of the intensity of large vortices formed behind the rib. However, this requires further theoretical studies and laboratory experiments.

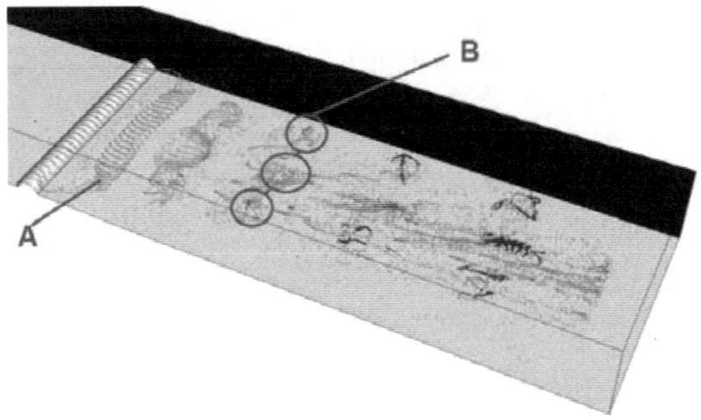

Figure 11. Formation and shedding of a starting vortex behind the rib at Re = 274, Sh = 0.13 and β = 0.15: A – spiral motion; B – vortex clots.

Acknowledgments

This work was supported by the Russian Foundation for Basic Research (grant no.15-01-06172).

References

[1] Yanase S, Kawahara G and Kiyama H 2001 *J. Phys. Soc. Jap.* **70** 3550–55
[2] Dikovskaya I D, Dovgal A V and Sorokin A M 1999 *Thermophysics and Aeromechanics* **6** 25–33
[3] Dovgal A V and Sorokin A M 2002 *Thermophysics and Aeromechanics* **9** 183–90
[4] Williams P T and Baker A 1997 *J. Int. Numerical Method in Fluids* **24** 1159–83
[5] Tylli N, Kaiktsis L and Ineichen B 2002 *Phys. Fluids* **14** 3835–45
[6] Kitoh A, Sugawara K, Yoshikawa H and Ota T 2007 *J. Heat Transfer* **129** 1141–55
[7] Nie J H and Armaly B F 2003 *Trans. ASME* **125** 422–28
[8] Kalinin E I, Mazo A B and Isaev S A 2016 *IOP Conference Series: Materials Science and Engineering* **158** 012047
[9] Dushina O A, Molochnikov V M, Mikheev N I and Paerely A A 2012 *Thermophysics and Aeromechanics* **19** 259–66
[10] Kornilov V I 2000 *Three-Dimensional Turbulent Near-Wall Flows in Streamwise Corners.* (Novosibirsk: Nauka. Siberian Publishing Firm of RAS) p 399

IOP Conf. Series: Journal of Physics: Conf. Series **899** (2017) 062001 doi:10.1088/1742-6596/899/6/062001

Flow structure in the three dimensional waves on the surface of the vertically falling liquid films

A Z Kvon[1,2], A V Bobylev[1,2], V V Guzanov[1,2], D M Markovich[1,2], S M Kharlamov[1,2]

[1]Institute of Thermophysics, Siberian Branch of RAS, Novosibirsk 630090, Russia
[2]Novosibirsk State University, Novosibirsk 630090, Russia

Abstract. Results of simultaneous liquid velocity and film thickness fields measurements in the three dimensional waves are presented. The three dimensional waves formed as a result of decay of the forced two dimensional waves were studied. Liquid velocity fields were reconstructed at different distance from the film surface with reference to the specific area of the wave. It was shown that flow in the three dimensional wave has complicated structure with existence of areas with backflow and transverse flows of opposite directions. Main areas of transverse flows existence are lying under main humps of three dimensional waves.

1. Introduction

Three dimensional (3D) waves on the surface of the vertically falling liquid films are formed as a result of the decay of the two dimensional (2D) waves (fig.1a) and are the last stage of the wave evolution at moderate Reynolds (Re) numbers. Transition to the 3D wave regimes is accompanied by the rivulets formation [1, 2]. In this case distribution of the liquid is not uniform across the plate and results of [1, 2] indicate that transverse fluid flow is directed to the central areas of the growing 3D waves. Despite the fact that characteristics of the rivulet flow were described at [3] it is not clear in which areas of the 3D wave liquid inflow takes place. However this information are needed to understand the mechanisms of the 3D waves evolution and formation of the rivulets. Therefore velocity vector field measurements in the 3D wave are required. Film flows are actively studied, but at the present time results on velocity field measurements in liquid films are presented only in a few papers because of difficulty of carrying out such measurements. For example in [4] authors obtained longitudinal components of velocity V_x (directed along the main stream) at different distances y from the inclined wall for wavy film flow using a photochromic dye activation technique. In [5, 6] modern and most efficient optical method for velocity vector field measurements known as the particle image velocimetry was used. However, as in [4] in these works only $V_x(y)$ was obtained. It should be mentioned that when using PIV methods the stream is pre-seeded by particles, and velocity vectors are determined from the displacement of the particle images on two consecutive frames obtained with known delay in time. Locating the tracers is usually done using special ways of illuminating, as for example in [5], where the stream was illuminated with laser sheet, directed perpendicularly to the wall of the channel along its axis. Thus, two dimensional velocity vector fields of the flow were measured in the plane, illuminated with laser sheet. In the case of the measurements in the three dimensional wave it is necessary to determine the displacement of the tracers at least in two mutually perpendicular directions parallel to the plate and in the different parts of the wave, which is more complicated problem. For this reason there exists lack of experimental data about flow structure in the 3D waves, in particular data about transverse and parallel to the plate components V_z of the velocity vectors, which are connected with cross flow of the liquid. In order to determine areas of the 3D wave in which transverse fluid flows takes place, particle tracking velocimetry (PTV) technique based on using light-field camera was applied. This camera allowed determining position of the tracers in space [7], thus for the first time V_x and V_z components were measured in the whole explored volume of the liquid with binding to the certain areas of the 3D wave.

2. Experimental techniques and operating conditions.

Experiments were performed for the flow regimes described in detail in [1, 2]. Measurements were carried out in the part of the flow in which fast growth of the rivulets due to the 2D-3D wave transition have been observed. The liquid film was formed on the transparent glass plate. Methods of PTV for velocity measurements and laser induced fluorescence for film thickness measurements were applied simultaneously. Velocity vectors and film thickness field were built from the same images, recorded by the light field camera Raytrix 11m that allows recording in double frame mode. In that case two consecutive images are recorded with a preset time delay. The depth resolution that can be achieved by using this camera is 1/40 of the total depth of field [8]. In our case the depth of field of the optical system was 2 mm, thus the depth position of the fluorescent 5 μm polyamide particles with which the stream was seeded was determined with accuracy ± 25 μm. During the LIF measurements fluorescent dye Rhodamin 6G was dissolved with concentration 1.25 mg/L in the working fluid. Intensity of the light emitted by the working solution was proportional to the thickness of the layer and in this way instantaneous film thickness distribution was reconstructed. Recording on the camera was carried out

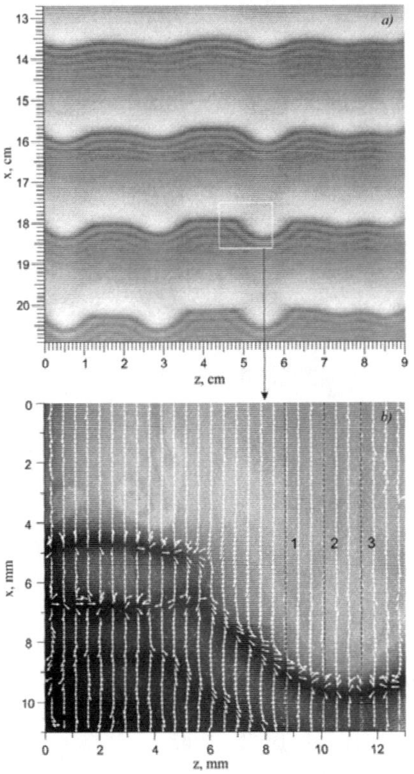

Fig. 1. Measurement area and typical velocity vector field.
a) Part of the LIF image from the fig. 6 in paper [1] illustrating 3D wave formation on the surface of the water film at Re=40, F=19 Hz. Region marked with white square is the measuring area of this work;
b) Flow direction in different parts of the wave at distance 108 μm from the plate. Trajectories for the cross sections 1, 2, 3 are shown in fig.2b.

through the glass from the dry side of the plate in the flow area shown in fig. 1a. Fluorescence of the solution and polyamide particles in the measurement area were excited with pulsed Nd:YAG laser

IOP Conf. Series: Journal of Physics: Conf. Series **899** (2017) 062001 doi:10.1088/1742-6596/899/6/062001

with wavelength 532 nm. More detailed description of the LIF method and its error analysis are described in [2]. Error in measurements of the film thickness did not exceed ±10 μm. Spatial resolution of the recorded images was 6.7 μm/pix. Obtained results which are illustrated below related to the flow regime characterized with Reynolds Re = 40, where Re = q/ν, q – specific volumetric flow rate, ν – kinematic viscosity. The measurements were carried out for the case of the regular decay of the periodic 2D waves, generated by the flow rate modulation with frequency F = 19 Hz. Formations of the 3D waves under these conditions occur at the same position and with a frequency equal to the frequency of the generated 2D waves. Recording frequency was chosen so that forming 3D waves were registered at the same place. This allowed us to collect a volume of the experimental data sufficient for the velocity vector field reconstruction.

2.1. Data processing

The processing of the obtained experimental data included stage of the image filtration, after which the film thickness fields were reconstructed and the coordinates of tracers were determined. At final stage images of the particles were sorted by layers depending on the depth of their immersion. These layers were parallel to the surface of the plate and centers of the neighboring layers where located with spacing of 36 μm from the wall (along the axis y). Velocity vector fields were calculated for the each layer using PTV processing algorithms of the standard software ActualFlow. All tracers laying at distances ± 36 μm from the center of the layer were considered belonging to that layer. Inaccuracy of the velocity measurements at $y \geq 72$ μm was 30% due to insufficient depth positioning of the particles in the shear flow. Nevertheless obtained results allowed us to detect areas of the wave with typical liquid flow direction.

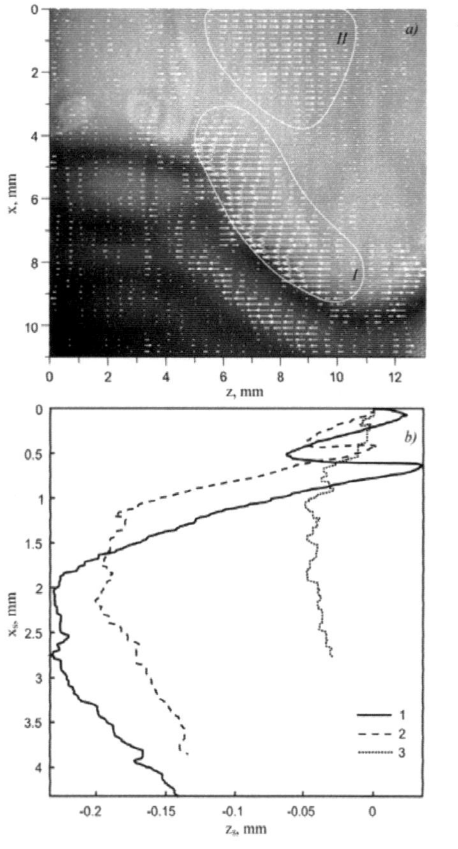

Fig. 2. Transverse displacement of the liquid in different parts of the wave.
a) Transverse velocity V_z at distance 180 μm from the plate. *I* – region of the liquid outflow from the wave, *II* – region of the liquid inflow to the wave; *b)* trajectories of the movement of elementary volumes of liquid along the cross sections shown in fig. 1b at the distance 108 μm from the plate.

3. Results and discussion

As an example velocity vector field in layer with centre at y = 108 μm is shown in fig. 1b. Intensity of the background in this picture is related to the film thickness. Film thicknesses were 90 μm and 400 μm in minimum of the capillary precursor and in maximum of the main hump correspondingly. For illustrative purposes length of each velocity vector was normalized with their magnitude and shows only direction of the flow. As seen, liquid flow in the wave has complicated structure. Liquid inflow can be seen in direction to the central part of the wave in the area of the capillary precursor along its minimum. Areas of the liquid outflow in the opposite direction from the center of the wave were observed in the regions along the capillary precursor maximum. Moreover backflow against gravity was observed in some areas of the capillary precursor minimum. These results show qualitative agreement with results [9], where backflow in the area of the capillary precursor was studied experimentally and numerically and with results of the computational simulation [10], where it was shown that significant cross flows of the liquid take place along the areas of the minimum and maximum of the capillary precursor. Under the main hump of the wave two spatially separated regions with transverse liquid flows in opposite directions can be singled out, as shown in fig. 2a. The flow in the region I, neighboring the leading edge of the wave is directed from its central part. In contrast, in the region II under the trailing edge of the wave, the flow is directed towards wave's central part. Typical velocity magnitude in the marked areas in the case shown in fig. 2a at y = 180 μm was V_x = 0.25 m/s, V_z = 0.06 m/s for the region I and V_x = 0.3 m/s, V_z =0.03m/s for the region II. With moving from the plate wall the increasing of the V_x up to 0.4 m/s and V_z up to 0.08 m/s in both regions are observed while the wave velocity $V \approx 0.42$ m/s. In the figure 2b trajectories of the elementary volumes of liquid are shown when the 3D wave passes by. Trajectories were built based on assumption that the liquid doesn't leave the boundaries of the layer and V_z doesn't change in z - direction within transverse displacements of elementary volume. Trajectories were built in sections marked with dashed lines in fig.1b for the layer with center lying at y = 108 μm. The beginning of the trajectories corresponds to the position of the elementary volumes in the bottom part of the fig 1.b (in area of the capillary precursor), and the end – their position in the tail of the 3D wave, after passing of the main hump. It can be seen from fig. 2b, that transverse displacements of the liquid in the area of the capillary precursor are small in comparison with transverse displacements in the regions I (0.6 mm $< x_s <$ 1.2 mm) and II (2 mm $< x_s <$ 2.6 mm) under the main hump of the wave. Since the region I is decreasing and the region II is increasing in size with growth distance y from the wall it can be supposed that major liquid inflow to the wave during the rivulets formation takes place in the tails part of 3D waves. Thus it was experimentally shown that the liquid flow in the 3D wave has complicated structure with existence of the backflow areas and transverse flows, and the main regions of transverse flows are lying under the main humps of 3D waves.

4. Acknowledgements

This work was supported by the Russian Foundation for Basic Research (project n.15-01-06702-a).

5. References

[1] Alekseenko S, Guzanov V, Markovich D, Kharlamov S. *Technical Physics Letters*. 2012. V. 38 N. 8. P. 739.
[2] Kharlamov S, Guzanov V, Bobylev A et al. Physics of Fluids. 2015. V. 27. N. 11. P. 114106.
[3] Guzanov V, Bobylev A, Kvon A et al. Technical Physics Letters. 2016. V. 42 N. 3. P. 234.
[4] Moran K, Inumaru J, Kawaji M. International Journal of Multiphase Flow. 2002. V. 28. N. 5. P. 731.
[5] Adomeit P, Renz U. International Journal of Multiphase Flow. 2000. V. 26. P. 1183.
[6] Zadrazil I, Markides C. International Journal of Multiphase Flow. 2014. V. 67. P. 42.
[7] Cenedese A, Cenedese C, Furia F et al. 16th Int Symp on Applications of Laser Techniques to Fluid Mechanics. 2012. Lisbon. Portugal.
[8] Seredkin A, Tokarev M. Vychislitel'nye Metody i Programmirovanie. 2016. V. 17. N. 3. P. 224.
[9] Dietze G, Leefken A, Kneer R. Journal of Fluid Mechanics. 2008. V. 595. P. 435.
[10] Dietze G, Rohlfs W, Nährich K et al. Journal of Fluid Mechanics. 2014. V. 743. P. 75.

Comparison between premixed and partially premixed combustion in swirling jet from PIV, OH PLIF and HCHO PLIF measurements

A.S. Lobasov[1,2], L.M. Chikishev[1,2], V.M. Dulin[1,2]

[1]Kutateladze Institute of Thermophysics, 1 Ak. Lavrentyeva Avenue, Novosibirsk, 630090, Russia
[2]Novosibirsk State University, 2 Pirogova Street, Novosibirsk, 630090, Russia

E-mail: Alexey.Lobasov@gmail.com

Abstract. The present paper reports on the investigation of fuel-rich and fuel-lean turbulent combustion in a high-swirl jet. The jet flow was featured by a breakdown of the vortex core, presence of the central recirculation zone and intensive precession of the flow. The measurements were performed by the stereo PIV, OH PLIF and HCHO PLIF techniques, simultaneously. Fluorescence of OH* in the flame and combustion products was excited via transition in the (1,0) vibrational band of the $A^2\Sigma^+ - X^2\Pi$ electronic system. The fluorescence was detected in the spectral range of 305-320 nm. In the case of HCHO PLIF measurements the A-X 4_0^1 transition was excited. The jet Reynolds number was fixed as 5 000 (the bulk velocity was $U_0 = 5$ m/s). Three cases of the equivalence ratio ϕ of methane/air mixture issued from the nozzle were considered 0.7, 1.4 and 2.5. In all cases the flame front was subjected to deformations due to large-scale vortices, which rolled-up in the inner (around the central recirculation zone) and outer (between the annular jet core and surrounding air) mixing layers.

1. Introduction

Currently, swirling jet flows are often used to stabilize flames in gas turbines and furnaces, since the swirl provides efficient and compact combustion with good ignition and stability characteristics for a wide range of operating parameters [1,2]. Structure of turbulent jet flows with swirl and combustion has been investigated in a number of works [3-5]. Structure and dynamics of swirling jet flows is complex due to such effects as formation of spiral/helical vortices, precession of the swirling jet's vortex core, breakdown of the vortex core and formation of the central recirculation zone [1-6].

Planar laser-induced fluorescence (PLIF) is widely used for diagnostics of flows with combustion. It provides information on local flame front deformation and extinctions and local temperature and heat release [7, 8]. Experimental data on the temperature fields and concentration of reactants (such as OH, CH, and NO) is necessary for better understanding of combustion processes, development of theoretical models and verification of numerical codes.

Particle image velocimetry (PIV) method has become a standard tool for velocity measurements in non-reacting flows [9] and often used for diagnostics of flows with combustion [10]. The advantages of PIV method for reacting flows diagnostics are that it is almost non-intrusive and provides estimation of planar velocity fields. The latter feature allows to detect coherent structures in flows, such as large-scale vortices.

The present paper reports on the experimental study of flame stabilisation in a high-swirl jet of methane/air mixture for different equivalence ratios. Three cases of combustion regimes are studied, namely, fuel-lean and fuel-rich premixed flames and fuel-rich partially premixed lifted flame.

2. Experimental setup

Measurements were carried out for the swirling flames of the methane/air mixtures at atmospheric pressure. The flames were organized in an open combustion rig (see details in [11]) by using a contraction axisymmetric nozzle with the exit diameter of $d = 15$ mm. A vane swirl was installed inside the nozzle (more details in [6]) to generate high-swirl flows. The swirl rate based on definition in [1] was 1.0, which is well above the critical value of 0.6 for the vortex breakdown in jet flows. Three cases of the equivalence ratio ϕ of the methane/air mixture issued from the nozzle were studied, namely, $\phi = 0.7$, $\phi = 1.4$ and $\phi = 2.5$. The Reynolds number of the air jet without methane was fixed as 5 000 (bulk velocity $U_0 = 5$ m/s).

A photo of the PIV/PLIF experimental setup is shown in Figure 1. To provide PIV measurements, the flow issued from the nozzle was seeded by 4 μm TiO_2 particles. The surrounding air was seeded by using a fog generator. A system of two CCD PIV cameras (ImperX IGV-B2020) were oriented horizontally as shown in Figure 1. Each camera captured 4 Mpix images. The cameras were equipped with Sigma AF #50 lenses and band-pass optical filters (60% transmittance at 532 nm and with FWHM of 10 nm). The seeding particles were illuminated by the second harmonic of the radiation of a double-head pulsed Nd:YAG laser with 200 mJ energy per each pulse. A beam was converted into a laser sheet by using a system of cylindrical and spherical lenses. Duration of each laser pulse was approximately 10 ns. Time separation between two PIV laser pulses was 35 μs.

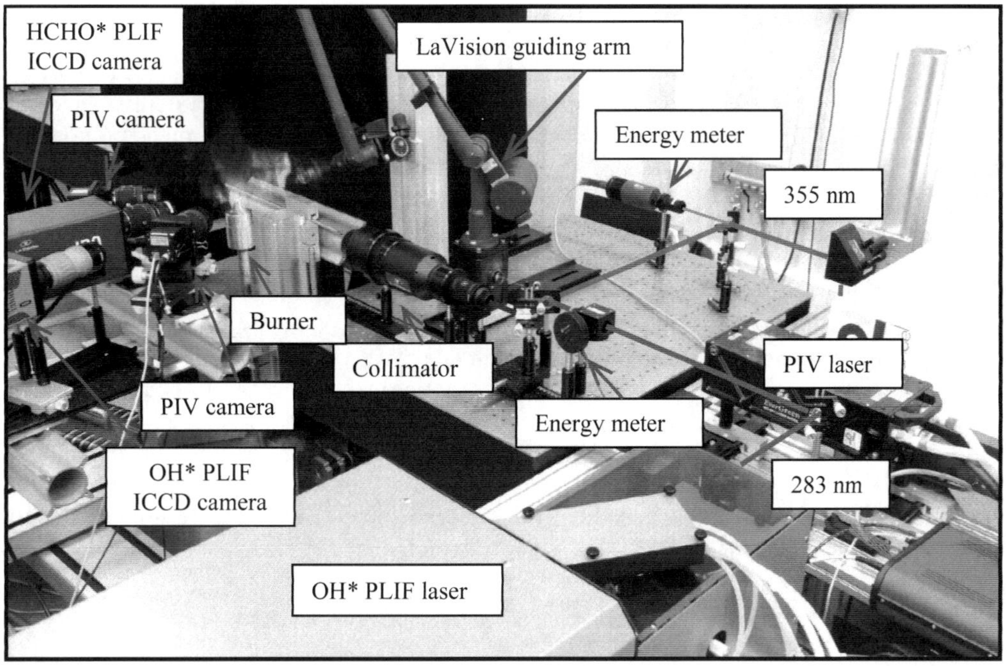

Figure 1. Photo of the PIV/PLIF experimental setup

IOP Conf. Series: Journal of Physics: Conf. Series **899** (2017) 062002 doi:10.1088/1742-6596/899/6/062002

To have a sufficient amount of statistical data for each flow case, a set of 1 000 instantaneous three-component velocity fields were measured via two independent runs. The velocity fields were evaluated by an iterative cross-correlation algorithm with image deformation. The final size of the interrogation are was equal to 32 × 32 pixels. Spatial overlap rate between neighbour interrogation areas was 50%. To take into account possible non-uniformity of flow seeding, the used cross-correlation algorithm accounted for the number of particles in each computational domain. If the number of particles was less than five, the velocity vector in the domain was not calculated. The processing algorithms were described in previous studies [6, 12].

In the case of the OH PLIF measurements a tunable dye laser (Sirah) was used. The tunable laser was pumped by the second harmonic (532 nm) of the radiation of a high-energy pulsed Nd:YAG laser (QuantaRay, 1 J per pulse for 532 nm). The output energy of the dye laser pulses was monitored. The average laser pulse energy in the range of 282-284 nm was approximately 5 mJ. The RMS of the energy variation was below 5%. The third harmonic (355nm) radiation of a pulsed Nd:YAG laser (Quantel Brilliant B, 90 mJ per each pulse for 355 nm) was used for excitation of the HCHO fluorescence. The RMS of the energy variation was also below 5%. Laser beams for OH and HCHO fluorescence were combined by using a dichroic mirror. After a collimator optics, the height of PLIF light sheet was 50 mm. The laser sheet thickness was below 0.8 mm in the measurement region. The sheet illuminated the axial (vertical) plane of the reacting flows.

The fluorescence of OH* was collected by a registration system consisted of a UV-sensitive image intensifier (LaVision IRO) and sCMOS camera (LaVision, Imager sCMOS, 16 bit images with resolution of 2560×2160 pixels) and equipped with a UV-lens (100 mm, f# = 2.8). Photocathode (S20 multialkali) of the IRO provided quantum efficiency about 25% for wavelengths in the range of interest (300-320 nm). The fluorescence of HCHO was collected by a 16-bit ICCD camera (Princeton instruments PI-MAX-4 with GEN II photocathode with quantum efficiency 25% in UV spectral region), equipped with a Sigma AF #50 lens. Appropriate band-pass optical filters were mounted on the lenses to detect OH and HCHO fluorescence. In both cases the exposure time for each PLIF image was 200 ns.

For calibration of the dye laser wavelength, excitation spectrum of OH* fluorescence was measured. The tunable laser wavelength was varied for the spectral range of 282-284 nm with the step of 2 pm. The excitation spectrum of OH radicals was compared to the spectrum calculated by using the LifBase software [13]. The raw PLIF images contain different types of systematic and random error. The systematic errors were caused by spatially non-uniform laser sheet intensity, non-uniform spatial sensitivity of the photocathode and CCD, background image and dark current. The sequence employed for correcting systematic errors is as follows: background subtraction, laser sheet correction, white sheet correction and laser shot-to-shot fluctuation correction.

To make sure that PIV and PLIF measurement planes coincided, prior to each run a photo-sensitive paper was placed into the measurement volume and exposed to a single shot of each laser. This test was performed to ensure that laser sheet planes were well aligned. Synchronization between PIV, OH and HCHO PLIF systems (see scheme in Fig. 2) was provided by combination of an 8-channel synchronizer (BNC, model 575), LaVision processing timing unit and by an in-house programmable synchronizing processor. The entire measuring system was operated at a repetition rate of three acquisitions per two seconds. The OH and HCHO images were registered between the first and second pulses of the PIV laser, with a delay after the first pulse of 10 μs and 20 μs, respectively. The time delay between laser pulses of the PIV and PLIF lasers and short PLIF exposure provided negligible cross-talk between the different systems.

IOP Conf. Series: Journal of Physics: Conf. Series **899** (2017) 062002 doi:10.1088/1742-6596/899/6/062002

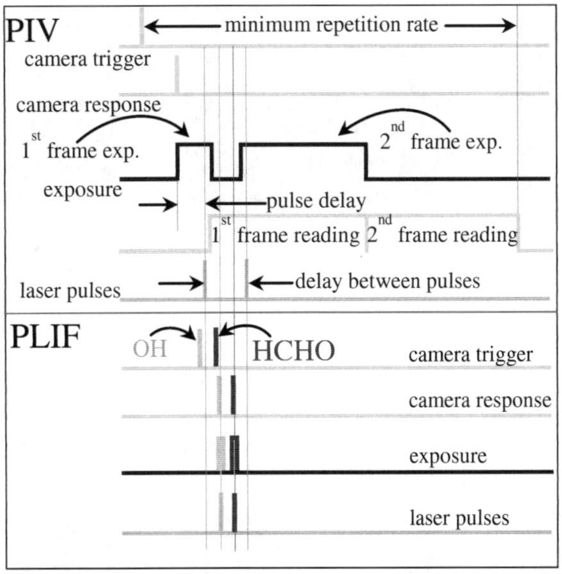

Figure 2. Schematics of temporal synchronization between PIV, OH PLIF and HCHO PLIF systems

3. Results

Based on 2D stereo PIV and 2D PLIF measurements, flame front stabilization in the high-swirl flow is analyzed for three considered combustion regimes. The focus is placed on the flow pattern, large-scale flow features (e.g., recirculation zone, flow precession) and shape of the reaction zones. Figure 3 shows flame photographs and examples of the stereo PIV, OH PLIF and HCHO PLIF snapshots which are captured simultaneously. Intensity of HCHO fluorescence provides information about preheat zones, whereas OH* fluorescence is in the flame front and hot combustion products.

For the premixed combustion of fuel-lean ($\phi = 0.7$) and fuel-rich ($\phi = 1.4$) methane/air swirling jets, the flame front has a shape of inverted cone, penetrating inside the nozzle. According to the PIV data, these combustion regimes corresponds to an annular swirling jet surrounding the central recirculation zone with weakly turbulent slow reverse flow. The HCHO snapshots show that the methane combustion took place in the inner mixing layer between the central recirculation zone and the annular swirling jet. The OH PLIF data shows that the hot combustion products concentrated inside the recirculation zone. The main difference between OH PLIF intensity for the fuel-lean ($\phi = 0.7$) and fuel-rich ($\phi = 1.4$) is that in the latter case OH presents downstream the recirculation zone, during unburnt hydrocarbons mixing and reacting with the surrounding air.

Combustion of the fuel-rich mixture with $\phi = 2.5$, viz., when the fuel concentration is above the flammability limit for the homogenous mixture, occurs after the jet mixing with surrounding air. Thus, the flame burns at a certain distance downstream the nozzle exit. The PIV data shows, that the annular swirling jet surrounds the central recirculation zone with unsteady reverse flow. The OH PLIF data shows that there is no OH* in the swirling jet, issuing from the nozzle and entraining the surrounding air, near the nozzle exit. Hot combustion products with OH* are concentrated inside the recirculation zone and also observed in the outer mixing layer. According to HCHO PLIF, the hydrocarbon combustion occurs both in the inner and outer mixing layers. Fuel preheat takes place during mixing of the hot gases inside the recirculation zone with the jet issuing from the nozzle and it burns in the low-speed regions of the mixing layer after mixing with the surrounding air.

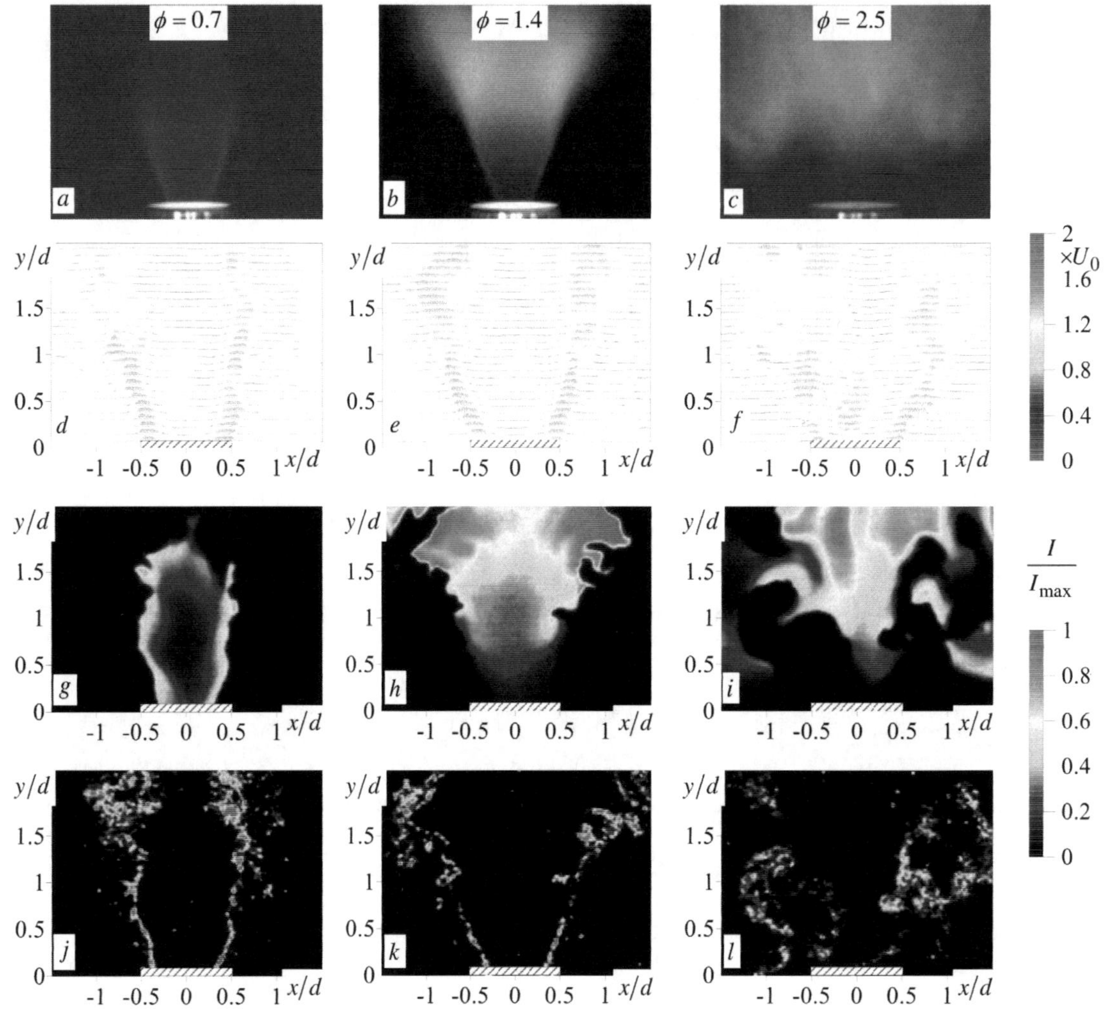

Figure 3. Photographs (a-c) and examples of the simultaneously captured stereo PIV (d-f), OH PLIF(g-i) and HCHO PLIF (j-l) snapshots for the axial plane of the methane/air swirling flames for (a,d,g,j) $\phi = 0.7$, (b,e,h,k) $\phi = 1.4$, (c,f,i,l) $\phi = 2.5$

4. Conclusions

The present work reports on the combined PLIF/PIV measurements for investigation of premixed and partially premixed combustion of methane in a swirling jet with vortex breakdown and central recirculation zone. The stereo PIV measurements were carried out simultaneously with OH PLIF and HCHO PLIF. Analysis of the reacting zones shapes in the high-swirl flow was carried out for three different conditions, viz., fuel-lean and fuel-rich flames with the flame shape like an inverted cone and lifted flame fuel-rich jet mixing with the surrounding air. For the first two cases, the flame front was located in the inner mixing layer between low-speed recirculation zone and annular swirling jet. For the fuel-rich flame the fuel burned in the outer mixing layer after being preheated by the recirculating at the jet axis hot gases and mixed with the surrounding air.

Acknowledgments

The research is supported by Russian Science Foundation (grant № 16-19-10566). The authors are grateful to Professor Kemal Hanjalic and Dmitry Markovich for fruitful discussions and recommendations.

References

[1] Gupta A, Lilley D and Syred N 1984 *Swirl Flows* (Tunbridge Wells, Kent: Abacus Press)
[2] Weber R and Dugué J 1992 Combustion accelerated swirling flows in high confinements. *Progr. Energy Combust. Sci.* **18** 349-367
[3] Syred N, Beer J 1974 Combustion in Swirling Flows: A Review *Combust. Flame* **23** 143-201
[4] Syred N 2006 A review of oscillation mechanisms and the role of the precessing vortex core (PVC) in swirl combustion systems *Prog. Energy Combust. Sci.* **32** 93-161
[5] Candel S, Durox D, Schuller T, Bourgouin J and Moeck J 2014 Dynamics of swirling flames *Ann. Rev. Fluid Mech.* **46** 147-173
[6] Alekseenko S, Dulin V, Kozorezov Y, Markovich D, Shtork S and Tokarev M 2011 Flow Structure of Swirling Turbulent Propane Flames *Flow Turbulence Combust.* **87** 569-595
[7] Eckbreth A 1996 *Laser Diagnostic for Combustion Temperature and Species* (Gordon and Breach, Netherlands)
[8] Koshe-Hoinghaus K and Jeffries (Eds.) J 2002 *Applied Combustion Diagnostics* (Taylor & Francis, New York)
[9] Raffel M, Willert C and Kompenhans J 1998 *Particle Image Velocimetry* (Springer, Berlin Heidelberg, New York)
[10] Stella A, Guj G, Kompenhans J, Raffel M and Richard H 2000 Application of particle image velocimetry to combusting flows: design considerations and uncertainty assessment *Exp. Fluids* **30** 167-180
[11] Alekseenko S, Dulin V, Kozorezov Yu and Markovich D 2012 Effect of high-amplitude forcing on turbulent combustion intensity and vortex core precession in a strongly swirling lifted propane/air flame *Combust. Sci. Technol.* **184** 1862-190
[12] Korobeinichev O, Shmakov A, Chernov A, Markovich D, Dulin V and Sharaborin D 2014 Spatial and temporal resolution of the particle image velocimetry technique in flame speed measurements. *Combust., Expl. Shock Waves* **50** 510-517
[13] Luque J, Crosley D 1999 Lifbase: Database and spectral simulation (version 1.5). *SRI International Report*, MP. 99-009

Technique for velocity vector field dynamics measurement on the basis of smoke visualization of flow

N I Mikheev[1, 2], N S Dushin[1], I I Saushin[1]

[1] Kazan Scientific Center of RAS, Kazan, 420111, Russia
[2] Kazan National Research Technical University named after A.N.Tupolev, Kazan, 420111, Russia

E-mail: n.miheev@mail.ru

Abstract. The main difference between Smoke Image Velocimetry (SIV) technique and traditional PIV is that the smoke with continuous intensity in the image is seeded into the flow instead of separate particles. Owing to better smoke reflectivity, relatively primitive equipment is enough to measure the dynamics of velocity vector fields with the frequency of 25 kHz and higher. The image processing algorithm is adapted to high tracer concentration and relatively large displacement of smoke patches between two consecutive frames. The results of SIV testing are presented, including the estimations of the most measurement noise sensitive characteristics of turbulence calculated from spatial derivatives of fluctuations of small-scale turbulence. The measurement results have been shown to agree well with the data obtained by other methods. Application of SIV technique opens new possibilities in the research of flow pattern and turbulence in unsteady and fast processes.

1. Introduction

PIV has become the most widespread optical measurement method so far [1-3]. Traditional PIV has shown good results in the investigation of steady flows. However, when the dynamics of velocity vector fields is measured in unsteady and fast processes, this method has significant limitations in time resolution due to the characteristics of cameras and lasers involved in the measurements. Frame rate is mainly limited by the pumping time of high-power dual pulsed lasers [4, 5]. Efforts to overcome these limitations have been made recently [6, 7]. The main point here is replacement of dual pulsed lasers with a continuous light source, e.g. DPSS laser [6] or LED [8], and employment of high-speed cameras for flow pattern recording. Moreover, DPSS lasers are approximately 100 times cheaper than pulse lasers, and LEDs are even more affordable. Despite obvious advantages of such an approach, there are some major shortcomings associated with short exposure time typical of high-speed cameras and relatively low power of continuous light sources. This reduces the intensity of light scattered by particles and leads to poor illumination of the camera's sensor. Authors [6] note that further improvement of the mentioned time resolved PIV is associated with employment of more powerful continuous light sources and cameras equipped with sensors of increased sensitivity to light.

Another approach to spatial and temporal resolution refinement has been implemented in Smoke Image Velocimetry (SIV) technique [9]. The present paper briefly describes the technique and provides new test results.

IOP Conf. Series: Journal of Physics: Conf. Series **899** (2017) 062003 doi:10.1088/1742-6596/899/6/062003

2. SIV foundations

The main difference between Smoke Image Velocimetry (SIV) technique and traditional PIV is that the former employs many times higher concentration of tracer particles. This makes the tracer images look like not separate particles but smoke with continuous intensity (figure 1). Owing to better reflectivity of smoke compared to separate particles, it is possible to obtain a processable image intensity even in a light sheet generated by a low-power continuous laser at the camera exposure time of 40 μs. Therefore, SIV allows employment of relatively primitive equipment for measurements of dynamics of two-component velocity vector fields with the frequency of about 25 kHz. This opens new possibilities for the research of unsteady processes. Besides, high seeding density contributes to better spatial resolution and measurement noise reduction.

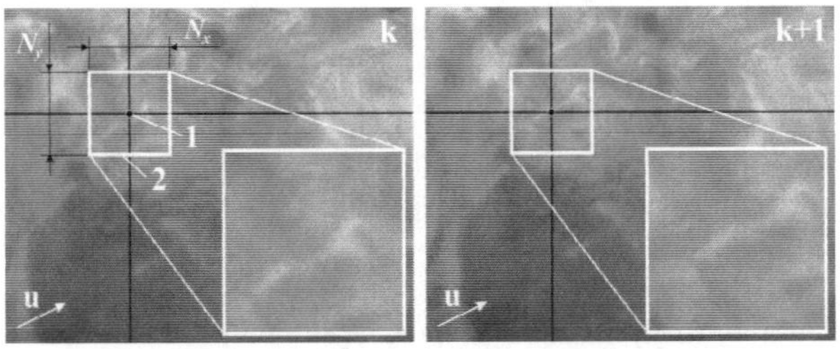

Figure 1. Estimation of interrogation window displacement: 1 – grid node; 2 – reference interrogation window.

The algorithm for estimation of velocity vector field dynamics from smoke visualization is in many ways similar to those used in PIV. However, large dimensions of smoke patches, if compared with separate particles, enable SIV to allow significantly larger displacement of particles between two frames. Instead of searching for the cross-correlation function maximum as is the case with PIV algorithm, SIV searches for similarity between two windows based on minimization of absolute differences of pixel intensities (Sum of Absolute Differences – SAD). This algorithm appeared to be more stable to large displacements. A reference interrogation window is chosen in the vicinity of grid nodes (figure 1). Its dimensions are $N_x \times N_y$ pixel^2. Then each window in the frame k is compared to windows of the same shape and size in the frame k+1 around the same grid node. Similarity between the windows at the displacements of Δi and Δj is calculated from the functional:

$$\Phi_{\Delta i,\Delta j} = \frac{1}{N_x N_y} \sum_{i=1}^{N_x} \sum_{j=1}^{N_y} \left| I_{k,i,j} - I_{k+1,i+\Delta i,j+\Delta j} \right|, \tag{1}$$

where $I_{k,i\,j}$ are grayscale intensities of the pixel with coordinates (i, j) in the frame k; $I_{k+1,i+\Delta i,\,j+\Delta j}$ is the grayscale intensity of the pixel with coordinates $(i+\Delta i,\ j+\Delta j)$ in the frame k+1; Δi, Δj are the displacements of interrogation window coordinates in the frame k+1 relative to the frame k. Coordinates (i, j) are counted from the origin of the considered window.

To speed up the procedure of image processing, a search area is specified around each grid node, which restricts the range of the window displacements, Δi and Δj, in the frame k+1. The search area is specified arbitrarily and can allow for dominating displacement direction at each grid node. The displacements, Δi and Δj, at which the functional Φ is minimal, are determined in the search area. These integer displacements define the position of the window, which is the most similar to the reference one with accuracy up to one pixel.

Further search for the minimum is performed using the approximation of the functional in the vicinity of its minimum. Authors [9] showed that the functional has a sharp minimum, and the shape of $\Phi(\Delta i,\ \Delta j)$ surface near the minimum is nearly conical, i.e. it is nearly a second-order surface. Therefore, to refine the minimum location, values of Φ at nine points closest to the integer approximation are used. These points are located according to 3×3-point arrangement and are displaced from the center by ±1 pixel along each coordinate. The approximating relationship is the following:

$$a_0 + a_1\Delta i + a_2\Delta j + a_{12}\Delta i\Delta j + a_{11}\Delta i^2 + a_{22}\Delta j^2 = \Phi^2 \tag{2}$$

Coefficients a_0, a_1, a_2, a_{12}, a_{11}, a_{22} are estimated by the least-squares method from values of the functional around the minimum. The relationship (2) describes variation of Φ between the grid nodes as well, i.e. Δi, Δj values can be non-integer. The minimum value of the functional and coordinates of the reference interrogation window displacement with subpixel accuracy are estimated from (2) with the obtained coefficients.

3. Testing of SIV

Some experimental testing of SIV measurements of velocity vectors was performed in [9]. It showed that SIV allows large (up to the reference window size) displacements of tracer particles between two consecutive frames. Comparison of SIV results to hot-wire measurements in the turbulent flow with forced velocity pulsations demonstrated almost identical velocity pulsation spectra up to the frequency of 1 kHz, which confirms the reliability of SIV results including the measurements of high-frequency velocity pulsations.

Synthetic images are often used for quantitative estimation of PIV accuracy [1]. The accuracy of submitted SIV technique was also estimated using this approach [5]. To simulate smoke visualization images, images of particles with the diameter of 10 pixels and concentration of C = 0.01014, 0.02027 and 0.04054 particle/pixel^2 were used in [5]. It was shown that employment of SIV for such particle concentrations reduces the random error 44, 31 and 12 times compared with the corresponding PIV results, while the bias error of SIV-SAD algorithm is reduced 9, 6 and 7 times, respectively.

Estimation of turbulence characteristics in the turbulent boundary layer is one of the most representative ways to test an optical method. The development of 2D methods for instantaneous velocity measurements enabled to allow for the turbulence anisotropy when estimating small-scale flow parameters. However, a problem with resolved spatial scales arose at the same time. Unlike the Reynolds stress tensor components, which are defined by large-scale vortex motion [10-12], turbulent energy dissipation is governed by the scales of the order of two Kolmorogov scales [13]. There are also difficulties in estimation of spatial derivatives of flow velocity pulsations that result from their sensitivity to the noise of the measurement method.

Nevertheless, despite constant improvement of 2D and 3D optical methods during more than 20 years, the experimental estimation of dissipation is still complicated by limitations in resolution of PIV technique [10, 11, 13-17]. Measurement volume that is significantly larger than the Kolmogorov scale obviously leads to underestimation of dissipation as the energy of small-scale turbulent pulsations is neglected [14-16]. Spatial resolution of PIV technique has limitations due to maximum allowed seeding density [13, 18]. Refinement of resolved scales in PIV down to less than allowable values leads to abrupt increase in random errors [10, 16], which are observed as the white noise in the instantaneous velocity oscillograms [19].

Major progress in small-scale measurements of velocity gradient tensor components was achieved by combination of PIV with other methods. PTV allows smaller interrogation windows if compared with PIV, but it has limitations on maximum seeding density, which prevents it from being used for experimental estimation of components of the above mentioned tensor [12, 20]. The combination of these two methods (Tomo-3D-PTV [20]) allowed spatially resolved low-error 3D measurements of not only velocity gradient tensor components but also a 3D pressure field. Authors [12] obtained more accurate estimation of dissipation in the range of $25 < y^+ < 140$ in the case of Tomo-3D-PTV with VIC + if compared with Tomo-PIV alone. Nevertheless, Tomo-3D-PTV with VIC + overestimated near-wall dissipation just as Tomo-PIV.

Air flow in a smooth plane channel was considered in test experiments. To study the flow pattern using SIV, the experimental setup schematically shown in figure 2 was employed. The test section *1* was a rectangular 1-meter long channel with a cross section of 75×150 mm^2 and a smooth inlet *10* attached to it. The latter had a contraction ratio of 6:1. A turbulence generating grid *9* was installed downstream of the smooth inlet. The grid had the cell size of 5 mm, steel wire diameter of 1.2 mm and 36% solidity. Flow rate was measured by an ultrasonic flowmeter 3 IRVIS RS4-Ultra mounted downstream of the receiver tank. The relative error in flow rate did not exceed 1%.

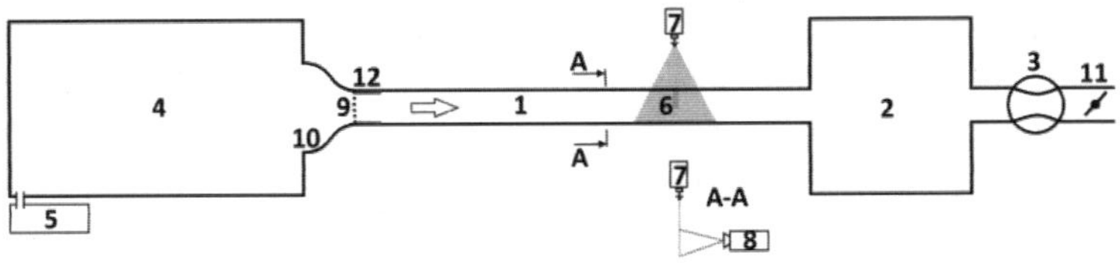

Figure 2. Experimental setup: 1 – test section, 2 – receiver tank, 3 - flowmeter, 4 – smoke conditioning chamber, 5 – smoke generator, 6 - measurement area , 7 – DPSS laser, 8 – high-speed camera, 9 – turbulence generating grid, 10 – smooth inlet, 11 – regulating gate, 12 – abrasive.

To visualize the flow pattern, the air-aerosol mixture (MT-Gravity fluid with medium fog density; Safex aerosol generator) was supplied from the smoke conditioning chamber *4* to the channel inlet. The measurement area *6* was illuminated by a DPSS laser KLM-532/5000-h *7*. The flow pattern in the channel symmetry plane at the distance of L = 0.7 m from the channel inlet was recorded by a monochrome high-speed camera Fastec HiSpec *8* with the frame resolution of 665×110 pixel (scaling factor of 0.0625 mm/pixel), frame rate f = 7083 1/s. The light sheet thickness was 1 mm (16 pixel)

Velocity vector fields were estimated using 16×6 pixel^2 interrogation windows. The window size was chosen according to recommendations of [9]. The spacing between grid nodes at which velocity vectors were obtained was two pixels along both coordinates. When estimating dissipation, to reduce the random errors, the interrogation window size was increased up to 16×16 pixel. Maximum displacement of smoke between two consecutive frames was 10 pixels. Image resolution in Y+ coordinates was 1 pixel = 0.8 Y^+.

Profiles of velocity, turbulent fluctuation intensity, vorticity fluctuations and turbulent energy dissipation were written in wall coordinates (3)

$$y^+ = \frac{yu_\tau}{\nu},\quad U^+ = \frac{U}{u_\tau},\quad \overline{u_i u_j}^{\,+} = \frac{\overline{u_i u_j}}{u_\tau^2},\quad \overline{\omega'\omega'}^{\,+} = \frac{\overline{\omega'_z\omega'_z}}{u_\tau^2\nu},\quad \varepsilon^+ = \frac{\varepsilon\nu}{u_\tau^4}. \tag{3}$$

Vorticity of the vector field, ω_z, was estimated by Stokes' theorem with approximation of circulation, Γ, by a first-order central difference scheme [1]. Friction velocity, u_τ, was estimated as u_τ=0.204 m/s from velocity measurements at a point in viscous sublayer using the relation $u_\tau = \sqrt{\nu\, \partial U_x / \partial y\big|_{y=0}}$. Turbulent energy dissipation was calculated from [16]:

$$\varepsilon = 3\nu\left[\overline{\left(\frac{\partial u_1'}{\partial x_1}\right)^2} + \overline{\left(\frac{\partial u_2'}{\partial x_2}\right)^2} + \overline{\left(\frac{\partial u_1'}{\partial x_2}\right)^2} + \overline{\left(\frac{\partial u_2'}{\partial x_1}\right)^2}\right] + 6\nu\overline{\frac{\partial u_1'}{\partial x_2}\frac{\partial u_2'}{\partial x_1}}. \tag{4}$$

Figures 3-5 compare the profiles of velocity, Reynolds stress and RMS vorticity fluctuations measured by SIV to the ones obtained at the same Re_θ in [8, 21, 22]. The velocity profile agrees well with the theory and other measurement methods across the whole turbulent boundary layer. Larger inclination in the logarithmic region of the velocity profiles obtained by DNS, PIV and SIV is attributed to the streamwise pressure gradient typical of channel flows, while theoretical profile (line) and hot-wire measurements were obtained in the case of free flow over a plate.

Components of Reynolds stress tensor, $\langle u'_i,u'_j\rangle^+$, (figure 4) also agree well with the results obtained by other measurement techniques. Subtle discrepancy in $\langle u'_i,u'_j\rangle^+$ profiles outside the viscous sublayer apparently results from different values of low Reynolds numbers, free-stream turbulence intensity and the channel cross-section shape [23, 24].

The root-mean-square distribution of the wall-parallel vorticity component, figure 5, obtained by SIV is in fairly good agreement with DNS and PIV profiles. The dissipation profile measured by SIV (figure 6) agrees well with the one obtained by an advanced method of VIC+ [12].

SIV allows estimation of the dynamics of flow velocity fields with time resolution that depends on the frame rate of videos. Figure 7 presents the spectra of turbulent fluctuations of two components of flow velocity for the coordinate $y^+=20$. It should be noted that the spectrum was obtained from relatively short video with the duration of 3 seconds. It caused substantial oscillations of amplitude in the low-frequency part of the spectrum. Apart from that, the spectrum was typical of the developed turbulence excluding its high-frequency part with the frequencies exceeding 1000 Hz. In the latter range the measurement noise was pronounced, which in our case had the amplitude of no more than 0.001 m/s. At the friction velocity u_τ =0.204 m/s adopted as a velocity scale in the boundary layer, the linear scale of 0.2 mm corresponded to this frequency. It is comparable to the estimate of Kolmogorov scale λ_K=0.15 mm obtained from the measurement of $<\varepsilon>^+$in the region of $y^+=20$. In other words, the white noise of measurement became visible only outside the frequency range of the maximum energy of turbulent fluctuations. Among the main reasons for such noise was the camera's limited (8-bit) resolution of image intensity [26]. The white noise in PIV measurements of velocity was observed in [19]. However, in the case of time-resolved SIV measurements the noise can be filtered out of the high-frequency spectral range. It was implemented by filtering high-frequency white noise out of velocity component oscillograms. The cutoff frequency was defined from the oscillogram of streamwise component; specifically, it corresponded to the indicative transition from the typical of turbulence pulsation amplitude (decreasing with frequency growth) to approximately constant amplitude (white noise) in high-frequency spectrum range.

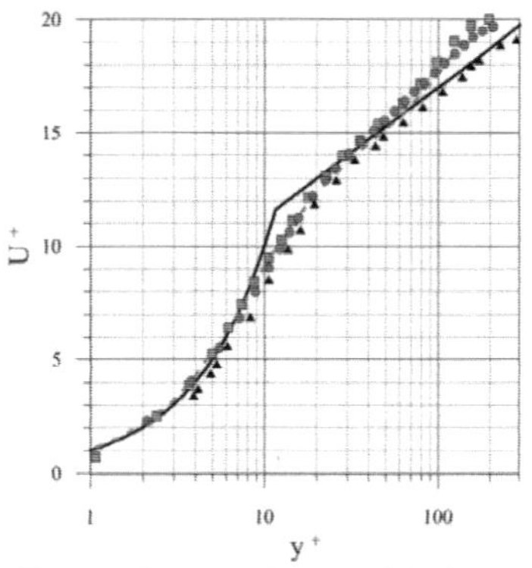

Figure 3. Velocity profile in the boundary layer of turbulent flow: —— logarithmic law; - - - DNS [22]; ▲ hot wire [21]; ■ PIV [8]; ● SIV.

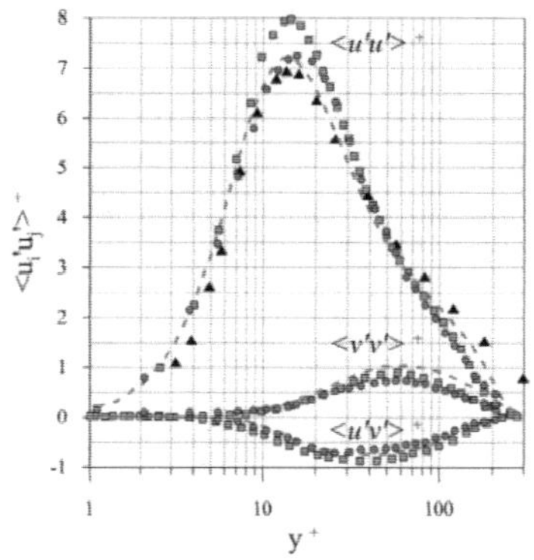

Figure 4. Profiles of turbulent fluctuations: - - - DNS [22]; ▲ hot wire [21]; ■ PIV [8]; ● SIV.

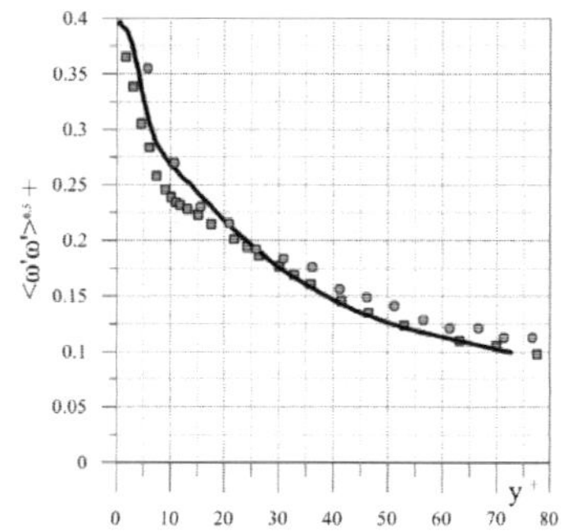

Figure 5. Root-mean-square distribution of the wall-parallel vorticity component: —— DNS [22]; ■ PIV [8]; ● SIV.

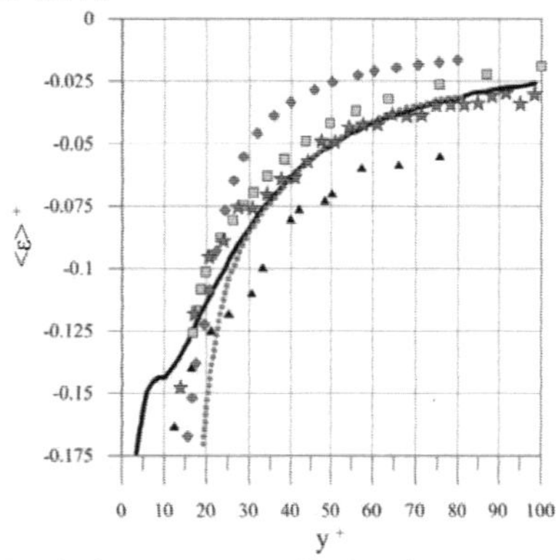

Figure 6. Turbulent energy dissipation: ——DNS [22]; ▴▴▴▴ VIC+ [12]; □ PIV [12]); ◆ PIV [12]; ▲ nine-sensor hot wire [25]; ☆ SIV.

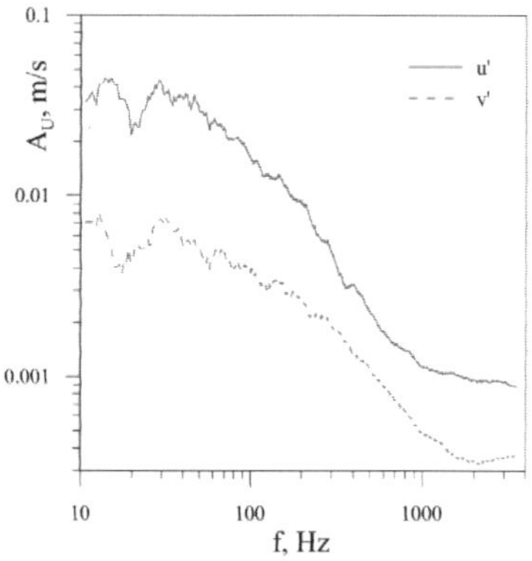

Figure 7. Velocity fluctuation spectra at $y^+=20$.

Conclusions

SIV technique allows employment of relatively primitive equipment for measurements of dynamics of two-component velocity vector fields with the frequency of about 25 kHz. This opens new possibilities for the research of unsteady processes.

Image processing algorithm embedded in SIV is adapted to high tracer concentration and relatively large displacement of smoke between two consecutive video frames.

The profiles of velocity and turbulent fluctuations as well as the profiles of vorticity fluctuations and turbulent energy dissipation estimated from SIV measurements agree well with experimental data obtained by multi-sensor hot-wire anemometers, stereo PIV, 3D-PTV with VIC+ and DNS results.

Acknowledgments

This work was supported by the Russian Science Foundation (Project no.16-19-10336).

References

[1] Raffel M, Willert C and Kompenhans J 1998 *Particle Image Velocimetry. A Practical Guide* (Germany: Springer)
[2] Tokarev M P, Markovich D M and Bilskii A V 2007 *Comput. Technologies* **12** 109-131
[3] Westerweel J, Elsinga G E and Adrian R J 2013 *Annual Review of Fluid Mechanics* **45** 409-436
[4] Mao X, Yu Z, Jaworski A J and Marx D 2008 *Exp. Fluids* **45** 833–846
[5] Zaripov D I, Aslaev A K , Mikheev N I and Dushin N S 2016 *Trudy Academenergo* **1** 42-52
[6] Neal D R, Sciacchitano A, Smith B L and Scarano F 2015 *Meas. Sci. Technol* **26**:074003
[7] Willert C, Stasicki B, Klinner J and Moessner S 2010 *Meas. Sci. Technol* **21**:075402
[8] Willert C E 2015 *Exp. Fluids* **56**:17
[9] Mikheev N I and Dushin N S 2016 *Instrum. Exp. Tech* **59** 880-87
[10] Saarenrinne P and Piirto M 2000 *Exp Fluids* **29**:300–307
[11] Foucaut J M, Cuvier C, Stanislas M and George W K 2016 *Progress in Wall Turbulence 2* 429-39
[12] Schneiders J F G, Scarano F and Elsinga G 2016 *18th Int Symp on the Application of Laser and Imaging Techniques to Fluid Mechanics*
[13] Tokgoz S, Elsinga G, Delfos R and Westerweel J 2012 *Exp Fluids* **53(3)** 561–583
[14] Sheng J, Meng H and Fox R O 2000 *Chem Eng Sci* **55(20)** 4423–4434
[15] Racina A and Kind M 2006 *Exp Fluids* **41(3)** 513–522
[16] Tanaka T and Eaton J K 2007 *Exp Fluids* **42(6)** 893–902
[17] Charonko J and Prestridge K 2016 *18th Int Symp on the Application of Laser and Imaging Techniques to Fluid Mechanics*
[18] Adrian R J and Westerweel J 2011 *Particle image velocimetry* (Cambridge: Cambridge University Press)
[19] Foucaut J M, Carlier J and Stanislas M 2004 *Meas. Sci. Technol* **15**:1046
[20] Novara M and Scarano F 2013 *Exp Fluids* **54** 1-12
[21] Roach P E and Brierley D H 1989 *The Influence of a Turbulent Freestream on Zero Pressure Gradient Transitional Boundary Layer Development Including the Condition Test Cases T3A and T3B, in Numerical Simulation of Unsteady Flows and Transition to Turbulence* eds O. Pironneau et al. (Cambridge: Cambridge University Press)
[22] Schlatter P, Örlü R, Li Q, Brethouwer G, Fransson J H M, Johansson A V, Alfredsson P H and Henningson D S 2009 *Phys Fluids* **21(5):**051702
[23] Huser A and Biringen S 1993 *J. Fluid Mech* **257**:65-95
[24] Balint J L, Wallace J M and Vukolavcevic P 1991 *J. Fluid Mech* **228** 53–86
[25] Honkan A and Andreopoulos Y 1997 *J. Fluid Mech* **350** 29–96
[26] Dushin N S, Mikheev N I, Dushina O A, Zaripov D I, Paereliy A A and Gazizov I M 2016 *Trudy Akademenergo* **4** 23-31

Research of the boundary layer with an adverse pressure gradient by the Smoke Image Velocimetry method

N I Mikheev[1,2], I I Saushin[1], A E Goltsman[1]

[1] Kazan Scientific Center of the Russian Academy of Sciences, Lobachevsky str. 2/31, Kazan, 420111, Russia
[2] Kazan National Research Technical University named after AN Tupolev - KAI, 10, K.Marx St., Kazan, 420111, Russia

E-mail: an116ya@mail.ru

Abstract. The results of an experimental evaluation of velocity profiles, turbulent pulsations, generation and dissipation of turbulent energy in a nonequilibrium boundary layer under the adverse pressure gradient are presented. The profiles of characteristics are estimated by means of the field dynamics of the two-component instantaneous velocity vectors measured by the optical method Smoke Image Velocimetry. The opportunities of using the field measurement method SIV to study the spatial evolution of small-scale characteristics in a boundary layer with a pressure gradient have been showed.

1. Introduction

Flows with an adverse pressure gradient are often found in the equipment, for example, in the diffuser channel or when the airfoil flow. The adverse pressure gradient (APG) as well as the change of the kinematics and the dynamics of a turbulent boundary layer under certain conditions is an initiator of its separation.

Experimental studies of the turbulent boundary layer with APG perfomed by measuring the instantaneous velocity by the point contact methods by [1-13] have sufficiently fully explained the physical basis of such flows. By results of the measurement by the point contact methods it is shown that in the boundary layer with the adverse pressure gradient the velocity profile in a viscous sublayer, in the transition and in the part of logarithmic region corresponds to the velocity profile at zero-pressure-gradient (ZPG) flow. However, related to the outer edge of the boundary layer the adverse pressure gradient leads to the growth of a deviation of the profile shape from the law of the wall. Accordingly, the determination of a universal relation for the describing the influence of the pressure gradient on the velocity profile in the coordinates of the wall law was one of the first actual problem. [14] on the basis of empirical data has proposed to consider the function describing the velocity profile in the boundary layer of the divergent channel as a superposition (the linear combination of two universal functions) of the logarithmic law of the wall and the law of the wake

$$U^+ = 5.5 + 5.5\log_{10} y^+ + \Pi \sin^2\left(\frac{\pi y}{2\delta}\right), \tag{1}$$

where Π according [8, 15] is a linear function of $C_f^{-0.5}$.

It is known the expression proposed by [2] in which the deformation value of a profile is independent of the wall shear but related instead to the local maximum in the Reynolds stress profile. In [2] is also noted the inconformity of the obtained experimental data to some aspects of the mixing length theory valid for the zero-pressure-gradient boundary layer.

In [5] this discrepancy was investigated and described in more detail: the increase of the gradient of the mixing length from κ=0.41 at the beginning of the logarithmic layer to κ=0.78 on the boundary layer edge is found. However, this did not influence the mean velocity profile near the wall which satisfied the law of the wall with the conventional von Karman constant of κ=0.41.

Research [5] has shown that despite the significant differences in the magnitude and distribution of Reynolds stress and triple correlations of velocity fluctuations in comparison with a boundary layer without the pressure gradient, the ratio between various components of stress remained unchanged. This is an indication that the adverse pressure gradient does not fundamentally affect the mechanism of the turbulent energy exchange.

In the works [4, 5, 8] there is noted the appearance of the second peak on the shear stress profiles in the central part over the thickness of the boundary layer in the region of $y/\delta \approx 0.45$ which is almost sixteen times higher than the value of the wall shear stress [5]. It is demonstrates the appearance of a new energy generation zone and its diffusion towards the wall. The turbulence production in the area of $y/\delta \approx 0.45$ is not inferior in intensity to the near-wall zone and depends linearly on the dimensionless parameter of the pressure gradient β.

These facts made it necessary to pay close attention to the equation of turbulent energy balance

$$\underbrace{U\frac{\partial k}{\partial x}+V\frac{\partial k}{\partial y}}_{Advection}=\underbrace{-\overline{uv}\frac{\partial U}{\partial y}+\left(\overline{v^2}-\overline{u^2}\right)\frac{\partial U}{\partial x}}_{Production}\underbrace{-\frac{\partial(\overline{uk})}{\partial x}-\frac{\partial(\overline{vk})}{\partial y}}_{Diffusion}\underbrace{-\frac{1}{\rho}\frac{\partial(\overline{vp})}{\partial y}}_{pv-diffusion}\underbrace{-\frac{\mu}{\rho}\overline{\frac{\partial u_i}{\partial x_k}\frac{\partial u_i}{\partial x_k}}}_{Dissipation}, \quad (2)$$

but everything was complicated by an adjective "point" in the name of measurement method as all terms of the energy balance equation are expressed in spatial gradients of the turbulent fluctuations velocity. Interest in the terms of the energy conservation equation under the conditions of the adverse pressure gradient is also caused by the desire to improve the accuracy of numerical modeling of RANS with the semi-empirical models of turbulence.

Nevertheless, the use of the various unchecked hypotheses of the isotropic and frozen turbulence for flows with the adverse pressure gradient made it possible to reveal the number of interesting differences from the non-gradient case. Dissipation term has been evaluated in [5] in three ways: using the Taylor's hypothesis of frozen turbulence by the isotropic turbulence model [17, 18], using the onedimensional spectral density and by the full ratio. Although the fact that three approaches led to three different dissipation profiles, there is no doubt that the adverse pressure gradient influenced the appearance of a local increase zone in dissipation near the coordinate $y/\delta \approx 0.4$. Similar results were obtained in earlier studies [16, 19]. In [16] there was a twofold increase in dissipation in comparison with the results [19]. However, in [8] the dissipation profile also estimated on the basis of the hypothesis of Taylor's frozen turbulence and isotropy does not differ in shape from the ZPG boundary layer profile. In these studies, the similar influence of the adverse pressure gradient has been found and described for profiles of the generation terms, diffusion and advection, but as in the case of dissipation it is difficult to find the equivalent profiles among the published papers.

Thus, point measurement methods don't provide complete information about the coherent structures and terms of the turbulence energy balance equation which is extremely necessary for understanding the physical processes and development of semi-empirical turbulence models for the numerical solution of the RANS system.

Using a field optical method (PIV or PTV) as an instrument for measuring [20-26] has quite recently allowed to obtain the information about the spatial dynamics of coherent flow structures in channels with an adverse pressure gradient. However, the estimation of the equation terms of the turbulence energy balance isn't presented in these papers. It can be explaining by an actual problem of the PIV method in the balance between the spatial resolution and accuracy of small-scale turbulence structures measurement [27-39].

It is worth mentioning the potential use of the 3D-methods such as Shake-The-Box (STB) [40] and VIC+ [41] for estimating the values and dynamics of the small-scale turbulence characteristics in a boundary layer with a pressure gradient. The STB method is a Lagrangian particle tracking method (LPT) based on iterative particle reconstruction [42] which allows to reconstruct a three-dimensional trajectory of particles with their dense seeding (as in PIV). In the STB method the typical for Tomo-PIV reconstruction of the pixel volumes the cross-correlation method are not used. The results of [43] obtained by means of the Shake-The-Box method showed the potential of a 3D LPT method for reconstructing particle trajectories in flows with a high density of a seeding which would allow to carry out a direct evaluation the strain rate tensor components and dissipation, respectively. The VIC+ method [41] though showed high accuracy in measuring the small-scale characteristics in non-gradient boundary layer, unfortunately it isn't an autonomous method and uses the results of particle velocity measurements (VIC+PIV) or acceleration (VIC+PTV) as the basic data. As a consequence, the accuracy of VIC + depends strongly on the quality of the original data.

The Smoke Image Velocimetry measurement technique (SIV) [44] has been appeared not so long ago. The main difference between the SIV technique and the classical PIV method is the use of a much higher concentration of tracers. For particle seeding the same aerosol generators as PIV are used. Particles also well track the movement of the carrier gas phase, but due to the higher concentration the aerosol on the video image does not look like the separate illuminated particles whereas smoke with a continuous distribution of brightness in the image. In response to the better reflecting ability of smoke in comparison with separate particles it is possible to obtain sufficient brightness for image processing while recording with the high speed camera with normal sensitivity even at a high frame rate in the light sheet created by a low-power continuous laser. Therefore, the SIV technique allows to measure the dynamics of two-component vector fields of the flow velocity with a frequency of 10 kHz using rather simple equipment. It opens up new possibilities for investigating nonstationary processes. With such a sufficiently good temporal resolution, the use of a high tracer concentration also promotes to improve spatial resolution and reduce measurement noise.

In this paper an attempt to evaluate the generation and dissipation profiles of the turbulent energy in the boundary layer of the channel flow with the adverse pressure gradient based on the SIV flow velocity vector field measurements has been made.

2. Experimental setup

The air flow in a smooth plane expanding channel has been studied. Experimental setup for study the flow structure using the optical method of Smoke Image Velocimetry (SIV) is shown in figure 1. The test section *1* was an asymmetric diffuser channel of rectangular cross-section with 5 degrees angle of expansion, a 150 mm width, a 410 mm length, 27 mm entrance section, 60 mm output. Before the diffuser there was a prewired section *13* which was a rectangular 27×150 mm^2 channel with the length of 210 mm and a smooth inlet *10* with 12:1 contraction. A turbulence generating grid *9* with 5 mm cell size, 1.6 mm steel wire diameter was mounted downstream of the smooth inlet *10*. Abrasive P24 (ISO 6344-2) *12* was glued onto the channel perimeter along a 50 mm section. This provided fully developed turbulent boundary layer in the measurement area during the experiments. Channel walls were made of transparent materials (glass and polycarbonate). Stable air flow rate downstream of the test section was provided by a regulating gate *11* and a 1.3 m^3 receiver tank *2* mounted downstream of the latter. Flow rate was measured by an ultrasonic flowmeter *3* IRVIS RS4-Ultra mounted downstream of the receiver tank. The relative error in flow rate did not exceed 1%.

To visualize the flow pattern, the air-aerosol mixture (MT-Gravity fluid with medium fog density and average particle size of 0.1…5 μm; Safex aerosol generator *5*) was supplied from the preparation chamber *4* to the channel inlet. The measurement area *6* was illuminated by a continuous diode-pumped solid-state laser KLM-532/5000-h *7*. The flow pattern in the channel symmetry plane at the distance of L=100 mm from the inlet was recorded by a monochrome high-speed camera Fastec HiSpec *8* with the frame resolution of 665×110 pixel (scaling factor of 0.0615 mm/pixel), frame rate $f = 9578$ 1/s. The camera was equipped with a Navitar 1''F/0,95 lens (focal length 25 mm, manual focus).

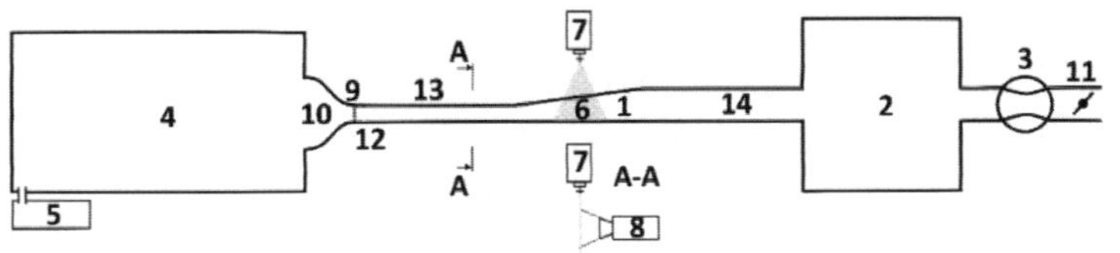

Figure 1. Experimental setup

Flow velocity fields were measured by optical SIV technique based on digital processing of flow pattern video recordings. Velocity vector fields here were estimated by the analysis of turbulent structure displacements visualized by smoke. Interrogation window size was 16×6 and 16x16 pixel. Velocity vectors were calculated from the digital images at the nodes of Cartesian grid with grid spacing of 2 pixels. Maximum displacement of turbulent structures between two consecutive frames was 12 pixels (at the boundary layer edge). Image resolution in Y+ coordinates was 1 pixel = 1.06 Y^+.

Profiles of velocity, turbulent fluctuation intensity and turbulent energy dissipation are written in wall coordinates y+, where shear velocity u_τ was estimated by the Clauser method.Turbulent energy dissipation was estimated by the hypothesis of local axisymmetry [45], which uses much less restrictive assumption to the full dissipation estimating compared to other hypotheses used to estimate the dissipation from 2D measurements of the velocity vector fields. A three-point central difference scheme was used to approximate the derivatives.

The presence of the adverse pressure gradient obviously increases the parametric dimension to describe the similarity laws. In addition to the classical the Re number for ZPG turbulent boundary layers, in the case of APG the dimensionless pressure gradient parameter β proposed by Francis Clauser in 1954 is usually considered. This parameter describes the ratio of the pressure force to the friction force $\beta = (dp/dx)\delta^*/\tau_W$, where dp/dx is the static pressure gradient. During the experiment pressure measurement was not performed, therefore the static pressure gradient was estimated according to the Bernoulli's equation for the adverse gradient of the dynamic pressure calculated from the average velocity in the neighboring sections.

3. Results and discussion

Table 1 presents the main characteristics of the investigated turbulent boundary layer with the adverse pressure gradient calculated from the results of measuring the instantaneous velocity field by the SIV method. Also in Table 1 similar characteristics are presented for the non-gradient boundary layer studied by the authors earlier by means of SIV-measurements, which results were used to analyze the influence the adverse pressure gradient on the flow pattern.

Table 1. Results of the main measured characteristics of turbulent boundary layer

Value	APG	ZPG
Freestream velocity, U_∞ (m/s)	6.14	4.06
Thickness TBL, $\delta_{99\%}$ (mm)	13.44	15.88
Displacement thickness TBL, δ* (mm)	2.81	2.44
Momentum thickness TBL, θ (mm)	1.76	1.58
Dynamic velocity, u_τ (m/s)	0.26	0.204
Dimensionless pressure gradient parameter, β	2.68	0

The velocity profiles obtained by the SIV measurements for ZPG and APG in the coordinates of the wall law y + correspond to the coordinate of y + ≈90 (Figure 2). Starting from y +> 90 the deviation from the logarithmic distribution is clearly traced on the velocity profile with APG: the region appears with a shape similar to the influence of the wake law. In the coordinates of the wall law the velocity profile measured under APG conditions is well described by the Coales ratio (1) with $\Pi = 4.7$.

To date, there is no clear description of the effect of APG on the turbulent pulsation profiles. Moreover, the results of the work (8, 26, 45) have changed the established ideas about this influence. In particular, it was found that under the influence of APG, the peak of the turbulent pulsation profiles shifts from the coordinate $y+ \approx 10$-20 to the region of the middle of the boundary layer thickness and this process is evolutionary along the channel length.

The small adverse pressure gradient realized in this experimental study at a relatively small distance from the measurement section to the entrance into the expanding channel made it possible to fix only the initial stage of the evolution of the boundary layer under the APG conditions. The turbulent pulsation profiles in the region $30 < y+ < 300$ compared to the TBL at ZPG increased almost threefold (Figure 3), which agrees well with the increase in the ratio of the generation level to the dissipation of the turbulent energy $P+/\varepsilon+$ in this section (Figure 4).

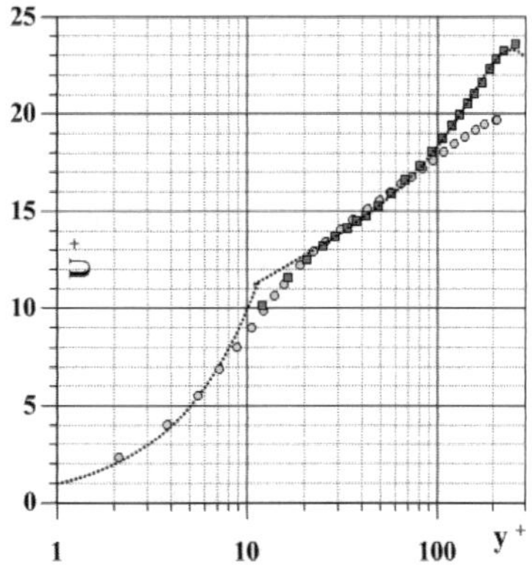

Figure 2. Velocity profiles;
·········- by (1) at Π=4.7;
○ - SIV, ZPG, 16x6;
■ - SIV, APG, 16x16

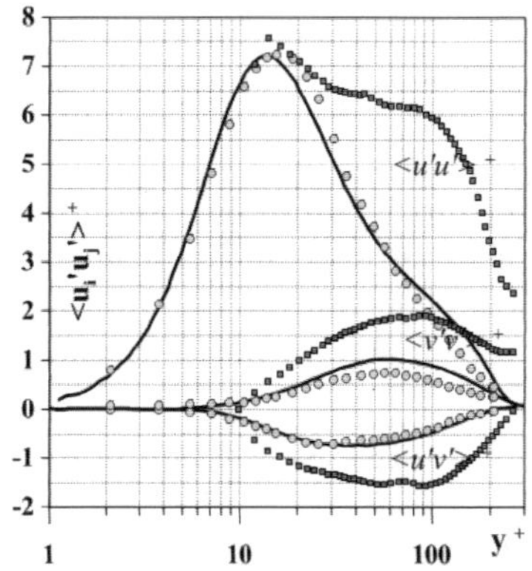

Figure 3. Turbulent pulsation profiles;
——-DNS [46];
○ - SIV, ZPG, 16x6;
■ - SIV, APG, 16x16

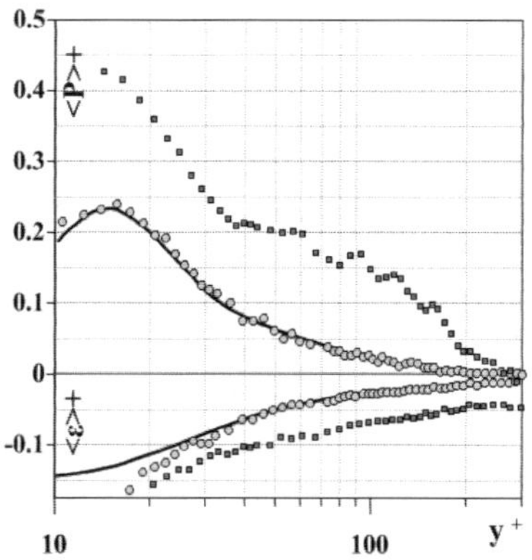

Figure 4. Profiles of generation P + and dissipation ε + of turbulent energy;
——-DNS [47];
○ - SIV, ZPG, 16x16;
■ - SIV, APG, 16x16

4. Conclusions

The optical method SIV with a sufficient accuracy allows to measure the instantaneous velocity vector fields. The results of 2D measurements make it possible to apply more exact hypotheses about the turbulence isotropy in estimating the spatial large-scale and small-scale turbulent characteristics in comparison with point methods. Based on the results of measuring instantaneous velocity vector fields in the coordinates of the wall law, velocity profiles, turbulent pulsations, generation and dissipation of turbulent energy were estimated and well correlated with the theory of the boundary layer with the adverse pressure gradient. The presented in this paper approach is relevant for studying the influence of the adverse gradient on the evolution of the boundary layer along the channel length from the point of view of turbulent energy dynamics, which is one of the most demanded problems of modern mechanics.

Acknowledgements

The work was supported by the Russian Science Foundation (project no.16-19-10336).

References

[1] Ludwieg H, Tillmann W 1950 Investigations of the wall-shearing stress in turb. boundary layers
[2] Perry AE and Schofield WH 1973 *The Physics of Fluids* **16(12)** 2068-2074
[3] Samuel AE and Joubert PN 1974 *Journal of Fluid Mechanics* **66(03)** 481-505
[4] Cutler AD and Johnston J P 1989 *Journal of Fluid Mechanics* **200** 367-387

[5] Skaare PE and Krogstad PA 1994 *Journal of Fluid Mechanics* **272** 319-348
[6] Perry AE and Marušic I 1995 *Journal of Fluid Mechanics* **298** 361-388
[7] Nagano Y, Tsuji T, Houra T 1998 *International Journal of Heat and Fluid Flow* **19(5)** 563-572
[8] Elsberry K, Loeffler J, Zhou MD, Wygnanski I 2000 *Journal of Fluid Mechanics* **423** 227-261
[9] Perry AE, Marusic I, Jones MB 2002 *Journal of Fluid Mechanics* **461** 61-91
[10] Nagib H, Gravante S, Pelivan I 2003 *APS Division of Fluid Dynamics Meeting Abstracts*
[11] Aubertine CD, Eaton JK 2005 *TSFP digital library online*
[12] Maciel Y, Rossignol KS, Lemay J 2006 *AIAA journal* **44(11)** 2450-2464
[13] Vinuesa R, Rozier PH, Schlatter P, Nagib HM 2014 *AIAA journal* **52(2)** 368-384
[14] Coles D 1956 *Journal of Fluid Mechanics* **1(02)** 191-226
[15] Nishri P 1996 *PhD thesis* Tel-Aviv University
[16] Cutler A D, Johnston J P 1989 *Journal of Fluid Mechanics* **200** 367-387
[17] Hinze JO 1975 Turbulence
[18] Townsend AA 1976 The structure of turbulent shear flows
[19] Bradshaw P 1967 *Journal of Fluid Mechanics* **29(04)** 625-645
[20] Knopp T, Schanz D, Schröder A, et al. 2014 *Flow, turbulence and combustion* **92(1-2)** 451-471
[21] Knopp T, Buchmann NA, Schanz D, Eisfeld B et al. 2015 *Journal of Turbulence* **16(3)** 250-272
[22] Reuther NJ, Schanz D, Scharnowski S, et al. 2015 Experimental investigation of adverse pressure gradient turbulent boundary layers by means of large-scale PIV
[23] Fuchs T, Hain R, Kähler CJ 2016 18th International Symposium on Applications of Laser Techniques to Fluid Mechanics
[24] Bross M, Kahler C 2016 18th International Symposium on Applications of Laser Techniques to Fluid Mechanics
[25] Atkinson C, Buchner AJ, Eisfelder M, Kitsios V, Soria J 2016 18th International Symposium on Applications of Laser Techniques to Fluid Mechanics
[26] Hain R, Scharnowski S, Reuther N, Kähler CJ, Schröder A, Geisler R, Cuvier C 2016 18th International Symposium on Applications of Laser Techniques to Fluid Mechanics
[27] Etebari A, Vlachos P 2005 *Experiments in Fluids* **39** 1040-1050
[28] Sheng J, Meng H, Fox RO 2000 *Chem Eng Sci* **55(20)** 4423–4434
[29] Saarenrinne P, Piirto M 2000 *Experiments in Fluids* **29** 300–307
[30] Sharp KV, Adrian RJ 2001 *AIChE J* **47(4)** 766–778
[31] Baldi S, Yianneskis M 2003 *Ind Eng Chem Res* **42(26)** 7006–7016
[32] Racina A, Kind M 2006 *Experiments in Fluids* **41(3)** 513–522
[33] Tanaka T, Eaton JK 2007 *Experiments in Fluids* **42(6)** 893–902
[34] Lavoie P, Avallone G, De Gregorio F et al. 2007 *Experiments in Fluids* **43(1)** 39–51
[35] Atkinson C, Coudert S, Foucaut JM, et al. 2011 *Experiments in fluids* **52(6)** 1641-1656
[36] Kähler CJ, Scharnowski S, Cierpka C 2012 *Experiments in fluids* **50(4)** 1641-1656
[37] Tokgoz S, Elsinga G, Delfos R, Westerweel J 2012 *Experiments in fluids* **53(3)** 561–583
[38] Charonko J, Prestridge K 2016 18th International Symposium on Applications of Laser Techniques to Fluid Mechanics
[39] Foucaut J-M, Cuvier C, Stanislas M, George WK 2016 Progress in Wall Turbulence 2, Series 23
[40] Schanz D, Gesemann S, Schröder A 2016 *Experiments in fluids* **57(70)**
[41] Schneiders JFG, Scarano F 2016 *Experiments in fluids* **57(139)**
[42] Wieneke B 2013 *Meas Sci Technol* 24:024008
[43] Schröder A, Schanz D, Geisler R, Gesemann S 2016 18th International Symposium on Applications of Laser Techniques to Fluid Mechanics
[44] Mikheev NI, Dushin NS 2016 *Instruments and Experimental Techniques* **59(6)** 880–887.
[45] George WK, Hussein HJ 1991 *J. FluidMech.* **233** 1–23
[46] Schlatter P, Örlü R, Li Q, Brethouwer G, Fransson JHM, Johansson AV, Alfredsson PH, Henningson DS 2009 *Phys Fluids* **21(5)**:051702
[47] Schlatter P, Örlü R 2010 *J. Fluid Mech.* **659** 116-126

IOP Conf. Series: Journal of Physics: Conf. Series **899** (2017) 062005 doi:10.1088/1742-6596/899/6/062005

Pipe flow pattern downstream of local restrictions studied by an optical method

V M Molochnikov[1, 2], **D V Kratirov**[2], **A N Mikheev**[2]

[1] Kazan Scientific Center of RAS, Kazan, 420111, Russia
[2] Kazan National Research Technical University named after A.N.Tupolev, Kazan, 420111, Russia

e-mail: vmolochnikov@mail.ru

Abstract. An approach to the study of gas flow patterns in circular pipes using 2D methods (PIV, SIV) has been proposed. The approach is based on a special device which is similar to a closed test section of a wind tunnel. It allows measurements of flow parameters downstream of local flow restrictions in circular channels without any optical distortions caused by curved pipe walls. The test results are submitted together with some data on the flow pattern downstream of a gate valve at different channel blockage ratios.

1. Introduction

Flow ducts (channels) of power plants are complex systems with a large number of local flow restrictions (control valves, gates), bends, branch pipes, etc. These elements can significantly influence the channel flow pattern (velocity profile distortion, swirl, shedding of vortices of different scales). Such alteration often drastically changes the operation of individual elements of a power plant and the plant on the whole. For example, the velocity profile non-uniformity, flow swirling and vortex shedding from local restrictions can lead to further uncertainty or fluid flow metering system failures [1, 2]. To minimize the effect of local restrictions, special devices like straightening vanes and flow conditioners are often used. However, when their performance is estimated, the turbulent and vortex pattern of flow downstream of these devices is usually not studied.

In this context, when designing power engineering equipment and measurement systems, it is important to be able to estimate the flow pattern at the outlet of different devices, including the one downstream of local flow restrictions of different types. The application of numerical methods and software to computation of a large number of flow regimes in the ducts of complex geometry is very time-consuming and needs further detailed testing. Thus, a laboratory experiment seems to be the most reasonable approach to this problem.

The rapidly developing 2D optical methods of flow pattern investigation (PIV, PTV, SIV, etc.) offer great opportunities for application of this approach. These methods allow measurements of instantaneous vector fields of velocity and vorticity, from which the information on flow pattern and its evolution can be derived. These methods are based on frame-by-frame processing of photos or videos of tracer particles seeded into the flow. The videos are recorded in a light sheet generated by a pulse or continuous laser. However, experimental research of circular pipe flows by the optical method of PIV revealed the effect of the channel cross-section shape on the measurement uncertainty. The latter is the most pronounced near the wall, which is one of the most critical regions of flow. A laser beam reflected from a tracer particle travels through a transparent curved channel wall of different thickness depending on the distance from the pipe axis on its way to the lens. Moreover, the incidence angle of the laser beam striking the wall is different depending on the curvature of the latter.

In the case of liquid fluids, the light refraction problem can be solved almost entirely if the pipe section is placed into a tank filled with the same fluid or with the one of similar optical properties and is equipped with a plane transparent wall placed normal to the lens [3-8]. Authors [3] showed that this leads to significant reduction in spatial distortions. When studying the flow of fluid whose refractive index significantly differs from the corresponding index of the pipe material, this engineering solution is no longer valid. Making the walls thinner, one can obviously reduce the negative optical effects [7]. However, this only reduces the distorted area to some extent.

There are two main approaches to measurements of flow parameters in pipes using 2D methods. The first one is introduction of corrections for all statistical characteristics of flow in the region where optical distortions of measurements are observed, including the immediate vicinity of the wall. The corrections are derived from the comparison between the measured velocity profile and the well-known velocity law [9] for a developed pipe flow. Then they are used to correct the rest of the measured flow parameters. This approach can be applied to boundary layer measurements but is inapplicable to complex turbulent flows including the separated ones. Sometimes the velocity fields estimated by an optical method are reconstructed from the known local flow parameters obtained by non-optical methods [9–10]. The second approach includes estimation of optical distortion of flow field linear dimensions using calibration targets with further corrections when processing the measurement results. This approach is rather hard to implement. When 2D methods are used in axisymmetric channels, spurious reflection and flare in the near-wall region influence the measurement results [11]. There have been attempts at solving this problem using special masks applied to the flow images. However, this approach appeared to be ineffective, as the authors [11] failed to obtain reliable velocity fields near the wall of a conical diverging channel.

In the present paper, a special device similar to a wind tunnel test section enclosed in a chamber is proposed for reliable measurements of flow parameters in circular pipes. Such a device (called the Eiffel chamber in Russian literature and hereafter) has become a popular element of wind tunnels [12]. The experimental setup is described in Section 2. Section 3 elaborates on test experiments and measurements of flow pattern downstream of a gate valve.

2. Experimental facilities

2.1. Eiffel chamber

Figure 1 shows the diagram of the Eiffel chamber. It had a cylindrical casing *4* with the inner diameter of 450 mm equipped with a glass window *3*. The window allowed recording the flow patterns by a high-speed camera. A transparent slot *5* was cut in the casing sidewall to provide light sheet generation in the considered flow area. A light trap was installed opposite to the slot *5*, which prevented the laser beam from reflection to the flow region. The Eiffel chamber had a flange *1*, attached to an upstream straight run piping with the considered flow restriction, and a flange *2* attached to a downstream straight-run. The design of flanges *1* and *2* allowed installation of the Eiffel chamber in the measurement section with the diameter of $D = 50 - 300$ mm. Besides, the Eiffel chamber's casing could be rotated about its axis relative to the flanges. This provided the adjustment of measurement plane with respect to the considered local restriction. General view of the camera is given in figure 2.

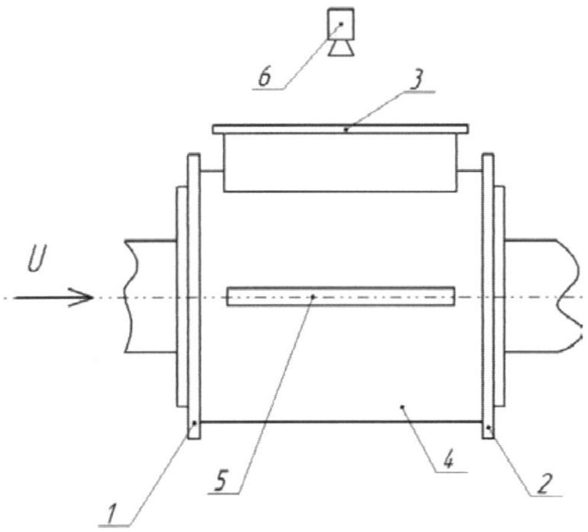

Figure 1. Diagram of the Eiffel chamber: *1*, *2* – flanges; *3* – window for flow pattern video recording; *4* – casing; *5* – transparent slot

Figure 2. General view of the Eiffel chamber. A laser beam trap is visible in the foreground

The pressure in the Eiffel chamber was approximately equal to the total pressure in the flow downstream of the considered local restriction. Thus, there was almost no jet expansion observed downstream of the chamber's inlet section. For this reason, the flow pattern downstream of the local restriction was similar to the jet pattern near the chamber inlet. This pattern could be measured, among other regions, in the immediate vicinity of the channel wall.

2.2. Experimental setup and procedure

The experimental setup schematic is shown in figure 3. Flow velocity fields in a smooth pipe and a pipe downstream of local restrictions were studied on this setup. The setup included a straight pipe *1*, local restriction in the form of a gate valve simulator *2*, the Eiffel chamber *3* installed immediately downstream of the gate valve, and a downstream straight-run pipe *4*. The air was sucked through the setup by a fan. The diameter of the pipe in which flow velocity fields were measured was $D = 150$ mm. The measurements were performed in two series. The first one was test experiments: velocity profiles were measured in the developed turbulent flow in the circular pipe. No gate valve simulator was installed in this series. Velocity profiles immediately downstream of the gate valve simulator were measured in the second series. The blockage ratio varied in these experiments (25, 50 or 75% of the pipe cross section was obstructed).

The instantaneous velocity vector fields and the profiles of statistical characteristics of flow derived from the latter were measured by Smoke Image Velocimetry (SIV) technique [13-14]. The technique is based on measurements of displacements of turbulent structures during a fixed time

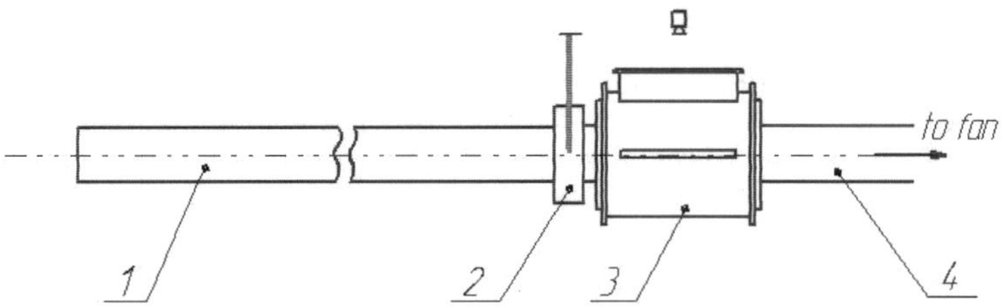

Figure 3. Schematic of experimental setup: *1* – straight-run piping; *2* – gate valve; *3* – Eiffel chamber; *4* – downstream straight-run

interval between the frames of a flow visualization video. The flow pattern was recorded using a high-speed monochrome Fastec HiSpec camera in a light sheet generated by a continuous laser. The flow was seeded with tracer particles, i.e. small glycerol droplets suspended in the flow. The particle diameter was 1-5 μm. The tracers were generated by an aerosol generator FOG 2010 Plus installed at the pipe *1* inlet (figure 3).

3. Results and discussion

Test experiments included the velocity profile measurements in the developed turbulent flow in the circular pipe at different Reynolds numbers. To that end, a 6.5-meter long (50 *D*) straight run circular pipe *1* (figure 3) was installed upstream of the Eiffel chamber without the gate valve *2*. An example of flow velocity profile $U_x(y/R)$ is shown in figure 4 for the Reynolds number Re = 2.06×10^4 based on the average velocity and pipe diameter.

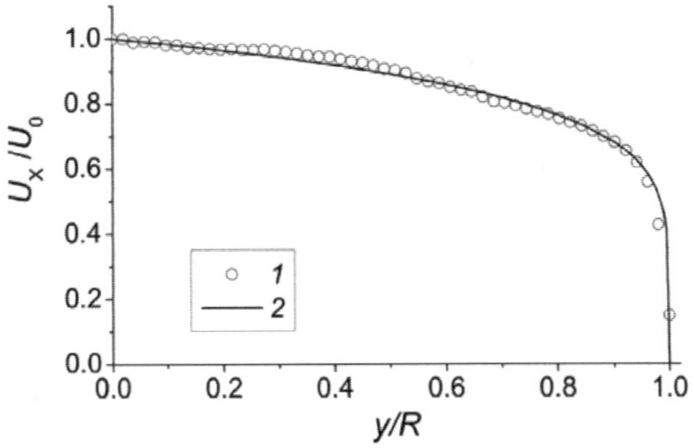

Figure 4. Velocity profile in the pipe: *1* – experiment; *2* – power law [15]

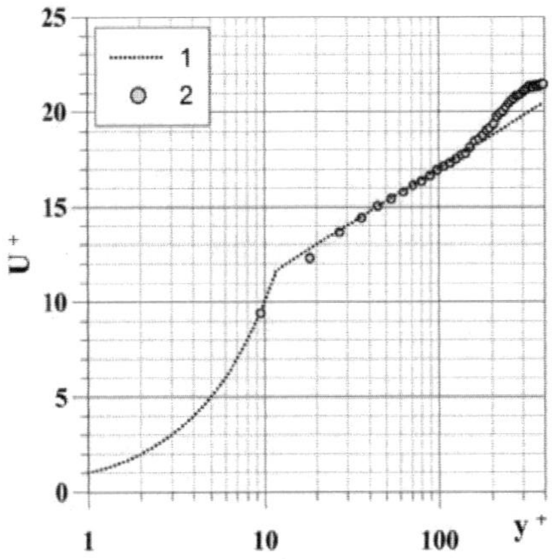

Figure 5. Velocity profile in wall coordinates: *1* – law of the wall [15]; *2* – experiment

Here, $R = D/2$. The figure shows good agreement between the velocity profile and the known power law $U_x/U_0 = (y/R)^{1/n}$ [16]. In the region where the law of the wall is applicable, the experimental results agree well with the classical relation $u^+ = 2.5\ln y^+ + 5.5$ (figure 5). Here, $u^+ = u/u_\tau$, $y^+ = yu_\tau/\nu$; $u_\tau = (\tau_w/\rho)^{0.5}$; ν is the kinematic viscosity. Wall shear stress τ_w was estimated according to Clauser [15]. Deviation of τ_w from its theoretical values for the developed turbulent pipe flow [16] was 2.1%.

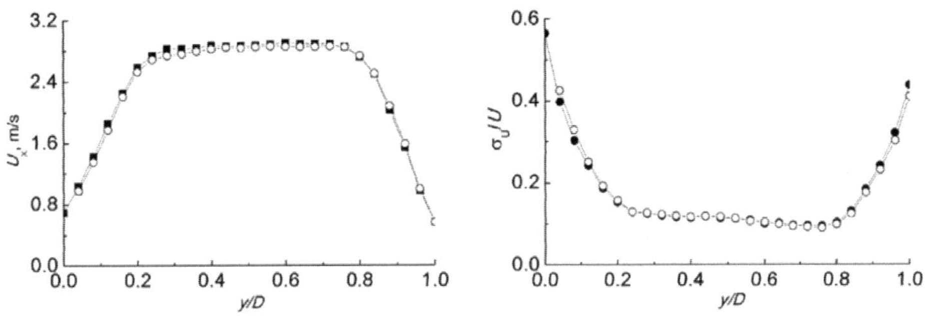

Figure 6. Profiles of velocity (*a*) and its turbulent fluctuations (*b*) downstream of the gate valve at 25% blockage ratio. Different marks denote repeated series of measurements

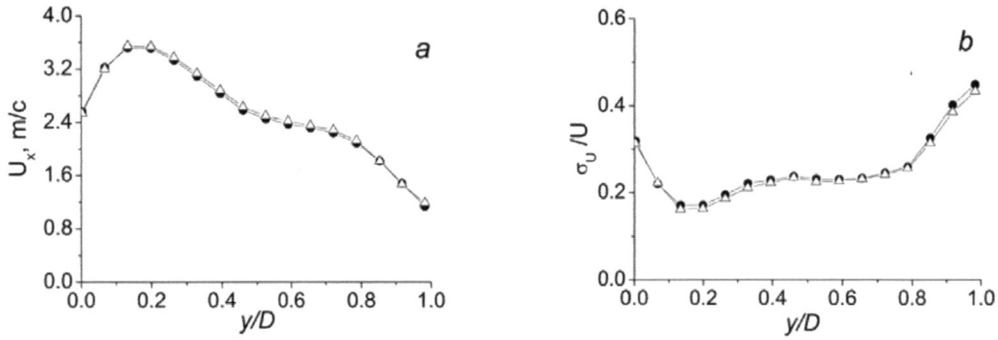

Figure 7. Profiles of velocity (*a*) and its turbulent fluctuations (*b*) downstream of the gate valve at 50% blockage ratio.

Flow pattern data downstream of the local restriction in the form of the gate valve simulator were obtained at the blockage ratios of 25, 50 and 75%. These data included profiles of the average streamwise velocity, U_x, and its turbulent fluctuations, σ_U, downstream of the gate valve. Figures 6–8 show some profiles of $U_x(y)$ and $\sigma_U(y)$ for the case when the gate moved normal to the measurement plane. As the figure shows, good reproducibility of results is observed.

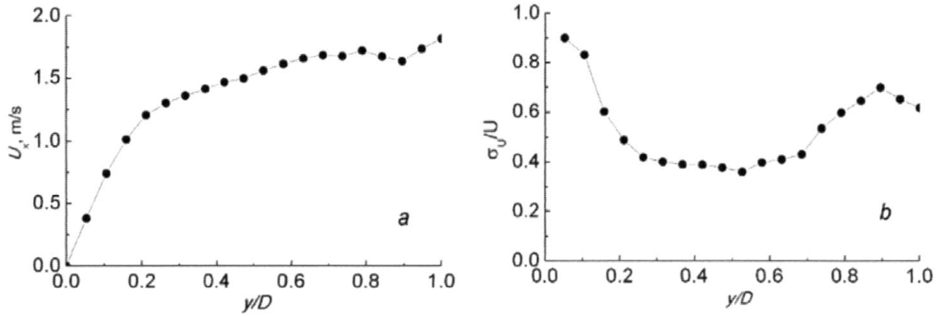

Figure 8. Profiles of velocity (*a*) and its turbulent fluctuations (*b*) downstream of the gate valve at 75% blockage ratio.

Conclusions

The results of experimental verification of the approach proposed by the authors to the study of gas flow patterns in circular pipes by the field methods based on the use of the Eiffel chamber analog are presented. On the canonical example of fully developed flow in a circular pipe, the satisfactory performance of the method is shown. The possibility of using the method for estimating the flow parameters in a circular pipe downstream of the gate valve simulator is shown.

References

[1] Faskhutdinov R E, Molochnikov V M, Mikheev N I and Kratirov D V 2009 *Trudy Academenergo* **1** 33–61 (in Russian).
[2] Kratirov D V, Mikheev N I, Molochnikov V M, Faskhutdinov R E and Fafurin V A 2012 *Vestnik Kazanskogo Tekhnol. Unversiteta* **21** 140-145 (in Russian).
[3] Sookdeo S, Siddiqui K 2010 *Sol. Energy* **84** 917–927.
[4] Garcia A, Solano J P, Vicente P G and Viedma A 2007 *Int. J. Heat Fluid Flow* **28** 516–525.
[5] Solano J P, Garcia A, Vicente P G and Viedma A 2011 *Appl. Therm. Eng.* **31** 2013–2024.
[6] Hellstrom L H O, Ganapathisubramani B and Smits A J 2015 *J. Fluid Mech.* **779** 701-715.

[7] Hellstrom L H O, Sinha A and Smits A J 2011 *Phys. Fluids* **23** 011703.
[8] Birvalskia M, Tummersa M J, Delfosa R and Henkesa R A W M 2014 *Int. J. Multiphase Flow* **62** 161–173.
[9] Tadeusz R F and Górecki G 2011 *Heat Transfer Engineering* **32:2** 109-126.
[10] Ramai L, Marko K A and Klick D 1982 *Proc. Combustion Institute* **19** 259–265.
[11] Kurkin E I and Shakhov V G *2011 Proc. of the 10th Int. Symp. on Experim. and Comput. Aero-thermodyn. Int. Flows. 4-7 July, Brussels, Belgium. ISAIF10-163*, 7.
[12] Pankhurst R C and Holder D W 1952 *Wind-Tunnel Technique* London Kent, United Kingdom.
[13] Mikheev N I and Dushin N S (2016) *Instruments and Experimental Techniques* **59** 882–89.
[14] Mikheev N I, Goltsman A E, Saushin I I, Dushina O A 2017 *Exp. Fluids* 58(8) 97
[15] Clauser F H (1954) *J. Aeronaut Sci.* **21(2)** 91–108.
[16] Schlichting H Boundary-Layer Theory 1955 *New York: McGraw-Hill* 535.

3D structure of flow in the near field of the quasi-two-dimensional turbulent jet

M V Shestakov[1,2], M P Tokarev[1,2] and D M Markovich[1,2,3]

[1]Kutateladze Institute of Thermophysics, 1, Lavrentev Ave., Novosibirsk, 630090, Russia
[2]Novosibirsk State University, Novosibirsk, 2, Pirogova str., Novosibirsk, 630090, Russia
[3]Institute of Power Engineering, Tomsk Polytechnic University, 30, Lenin Ave., Tomsk, 634050, Russia

E-mail: maximvsh@gmail.com

Abstract. The work explores the three-dimensional structure of the near field of a quasi-two-dimensional turbulent jet. The jet stream forms at an outflow from the rectangular channel with aspect ratio B/h equal to 1:2.5 into the channel, formed by two plane-parallel plates. The experiments were carried out for a wide range of Re number from 2 000 to 20 000. The measurements were performed using Tomographic PIV method with high time resolution. It is shown that for the whole investigated range of Re number in the near-field of the quasi-two-dimensional jet, there are formed secondary flows, caused by the presence of longitudinal vortex structures in the flow.

1. Introduction

Bounded turbulent jet flows are widely presented in various technical applications. Initially, the interest in the study of confined jet flows was due to the development of jet logical elements for military and space purposes [1, 2, 3, 4]. These works concerned the study of the effect of the bounding walls on the flow structure in the near field of the jet and aimed at the development of methods for the jet flow control. The studies have shown that the structure of the jet flow, top-bottom bounded by two plates, differs from that of the plane turbulent jets. The difference comes to the presence of secondary flows, formed near the bounding surfaces in the area of mixing layer development. The presence of the secondary flows was found due to the form of isotacs – lines of equal velocities, built in the jet cross sections. The distribution of isotacs in mixing layers had a saddle shape, i.e., a jet near the walls was wider than in the mid-plane. The same character of the confined jet flow was confirmed by visualizing with paint, supplied in different regions of the jet [5]. The isotac form widening is also realized near the wall in the wall jet flow [1]. The formation of the secondary flows is associated with reorientation of vortex filaments. Papers [3, 4] proposed a model of secondary flow forming on the basis of reorientation of the vortex filaments. The work [4] presents a spatial reconstruction of vortex structures, based on visualization of patterns and confirming the model of reorientation of vortex filaments. Authors of [6] experimentally studied the flow of a quasi-two-dimensional turbulent jet in a narrow channel with TomoPIV method [7]. In the near field of the quasi-two-dimensional jet they found longitudinal vortex structures, which according to the authors, make major contribution to the formation of secondary flows. In the study

IOP Conf. Series: Journal of Physics: Conf. Series **899** (2017) 062006 doi:10.1088/1742-6596/899/6/062006

by [8] the LES technique was used to numerically simulate the flow of the confined turbulent shallow jet. It was shown that in the near, middle and far fields of quasi-two-dimensional jets there are formed longitudinal vortex structures, whose linear size may reach about 10 h. Such longitudinal vortex structures are formed in confined shear flows at longitudinal streamlining of two closely located cylinders [9] and can significantly influence the processes of mass and heat transfer. The process of forming the longitudinal vortex structure is closely associated with the development of Kelvin – Helmholtz (K-H) vortical structures and their interaction with the bounding walls.

To understand the processes of formation of longitudinal vortex structures and to measure their influence on heat and mass transfer there is a need in 3D data on the dynamics of vortex structures in the near field of the jet where the process of their formation and development occur.

This paper presents experimental data on the three-dimensional structure of the near-field of the quasi-two-dimensional jet obtained, with high spatial resolution. The experiments were performed using Time Resolved Tomographic PIV method in a wide range of Re numbers $Re = 2*10^{-3} \div 2*10^{-4}$.

2. Experimental setup

Photograph of the experimental setup and diagram of the measuring system are shown in Fig. 2a. The experimental setup consists of a hydrodynamic circuit with a constant-level tank, pump, flow meter, and a work area. The work area is a narrow rectangular duct with the size of 307x270 mm^2, formed by two plane-parallel plates of Plexiglas, spaced 4 mm from each other. The nozzle is formed by two rectangular insets, set in the channel. Geometrical dimensions of the nozzle were as follows: width – 10 mm, length – 70 mm, and depth – 4 mm. The Reynolds number was calculated as $Re = h_d U_0/\nu$, where U_0 is the superficial velocity at the nozzle exit, and h_d is the hydraulic diameter of the nozzle.

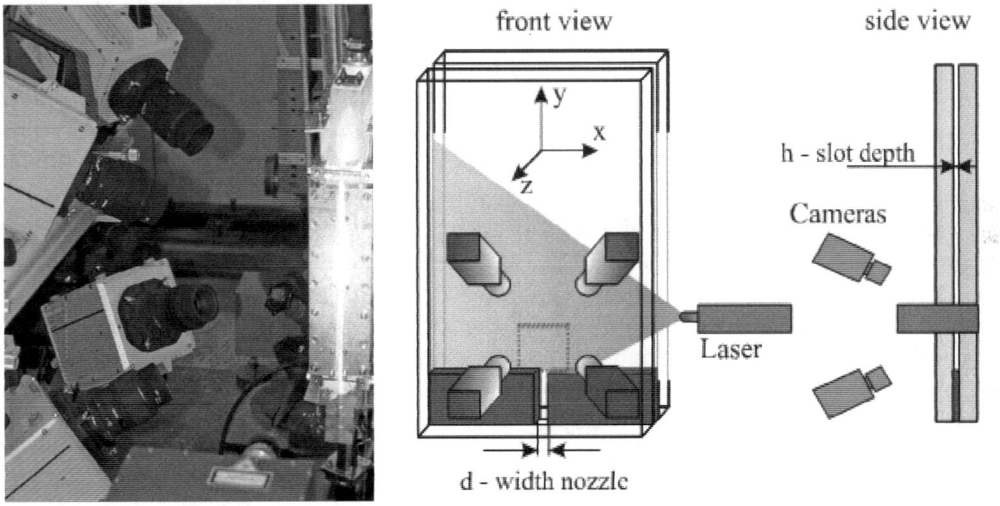

Figure 1. The photo (left) and the scheme (right) of experimental setup for TomoPIV measurements

The measuring system consisted of a high-speed dual Nd: YAG laser (Photonix DM-532-150 15 MJ, 15 kHz), four high speed CMOS cameras (Photron FASTCAM SA5 with resolution of 1024×1024 pixels, 12 bits with the shooting frequency of 7 kHz) and timing pulse generator Berkeley Nucleonics BNC 575. Laser sheet with 4 mm thickness, which covered the whole measuring volume between the plates, was formed with lens system. In the experiments, polyamide particles with a diameter of 50 microns were used. The optical systems of cameras used lenses SIGMA AF 105 mm f/2.8 EX DG MACRO. The volume of the measuring field was equal to 50×50×4 mm^3. The measurements were carried out in the area with dimensions of 8h×10h at the exit of the nozzle. The recording frequency in the experiments was equal to 10 kHz. Instantaneous three-component velocity distributions were calculated based on different successive images with a step depending on the number of Re. For Re equal to 20 000 the calculation was made on the adjacent images, for Re of 10 000 – in 1 image, for Re of 5 000 – in 2 images, and for Re 2 000 – in 5 images.

3. Calibration and data processing

3.1. Camera placement and calibration

To ensure equal parameters of the optical system of each camera, the cameras were located in the corners of a square in a plane, parallel to the central plane of the measuring volume (figure 1 right). The angles of camera arrangement in vertical and horizontal planes were equal to 30 degrees. In this configuration of the measuring system, cameras were located in four points on the sphere with the center, coinciding with the center of the measuring volume. As a result, identical conditions were provided for each camera, such as: rotation angles of the cameras, correction angles of Scheimpflug and the distance from cameras to the center of the measuring region. To align the focal plane and the matrix plane we used adapters, printed on a 3D printer and allowing one to set the lens at any angle in the range from 0 to 12 degrees in the desired plane. For calibration of the optical system we used a high-precision calibration target from Edmund Optics with dimensions of 50×50 mm2 and size of markers of 1 mm, located on a Cartesian grid with a step of 1 mm. As a result of the shallow depth of the channel of 4 mm, the shift of the calibration targets was difficult.

Therefore, to calibrate the measuring space it was necessary to displace the measuring system as a whole. The shift of the measuring system was carried out using a high-precision coordinate system and was controlled with an accuracy of 10 micrometers. The displacement of the measuring system was converted into a target shift, taking into account the difference between the refractive indexes of liquid and air. At that the channel wall was considered as a plane-parallel plate, not changing the direction of the beams. Measurement of markers of the calibration targets was performed in five parallel planes over a depth of the measuring volume.

3.2. Data processing

Data processing was carried out in the “ActualFlow” software. To improve the calibration parameters of the optical model there was self-calibration on pre-processed 500 images [10]. Preprocessing consisted of subtracting the median value, calculated for 21 000 images, applying the filter subtract minimum and a low-pass filter, and normalizing the intensity of particles on the average intensity in the image. As a result of self-calibration, the final average deviation did not exceed 0.05 pixels. The restored volume had dimensions of 506×659×62 voxels. The volumetric particle concentration was 8.5 particles/mm^3. Tomographic reconstruction was performed using the SMART algorithm [11]. Correlation analysis was realized using the iterative multigrid algorithm with continuous displacement of the measurement window. In calculating the velocity, four iterations took place: two iterations with a resolution of 64×64×32 voxels and two iterations with a resolution of 32×32×16 voxels. The last iteration was carried out with an overlap of 75%; as a result, the final size of the area for a single velocity vector amounted to 0.52×0.52×0.26 mm^3 with a distance of half the size between adjacent vectors.

4. Results and discussion

Figure 2 presents averaged three-dimensional velocity distributions measured in a near field of the quasi two-dimensional jet. Velocity averaging was performed by using 2 000 instantaneous three-dimensional velocity fields for Re = 2 000 and Re = 20 000 and 500 instantaneous three-dimensional velocity fields for Re = 10 000 and Re = 5 000. In the selected planes streamwise vorticity and streamwise velocity isolines are shown for averaged velocity distribution at different Re numbers. Red color corresponds to the clockwise rotation of the fluid while the blue color corresponds to counterclockwise rotation. One can see from the figure that the whole structure of the flow is similar at different Re numbers. There are some pairs of intensive secondary flows in the flow. The first and the second pairs of secondary flows are presented in the jet shear layer; the third pair is located in the core of the jet. Secondary flows are noticeable at the distance of 2h from the nozzle where Kelvin – Helmholtz vortical structures are larger than h. Secondary flow development comes with intensive fluid entrainment [8] which is also observable in our experiment. Fluid entrainment by K-H vortex structure near the channel walls leads to formation of longitudinal vortex structures which further interact with K-H vortex structures and are carried in the jet core. [12]. The central pair of secondary flows results from complex interaction of vortex structures developed in the nozzle channel and mixing layers. Secondary flows decay results from quasi-two-dimensional turbulent jet meandering and takes place at the distance of 8-9 h from the nozzle.

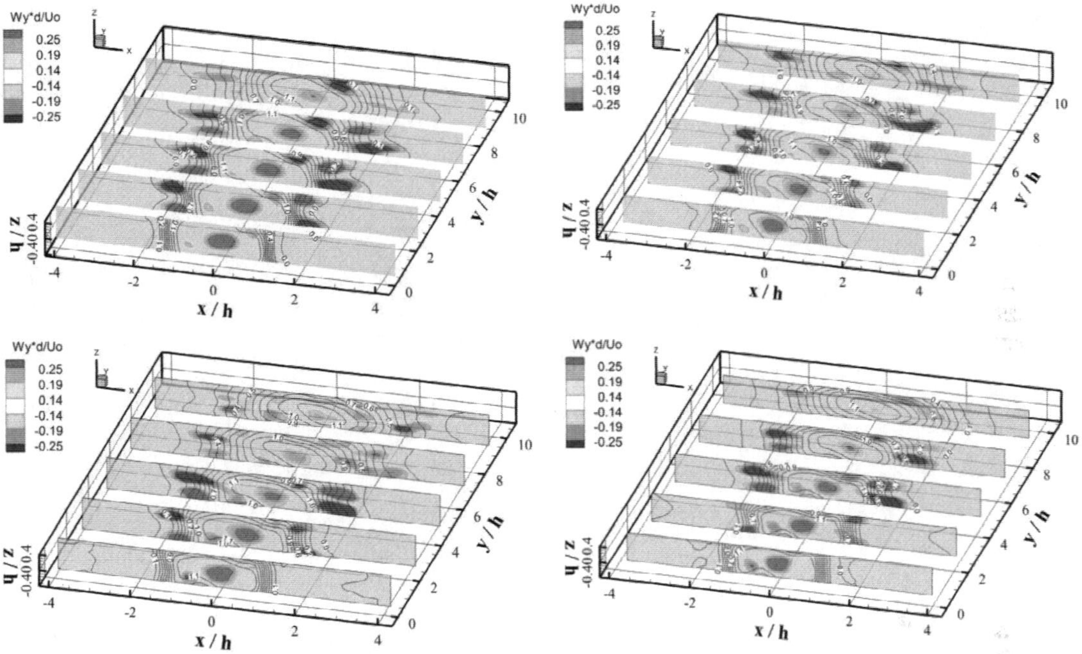

Figure 2. Streamwise vorticity for averaged velocity field and isolines of streamwise velocity a) Re 20 000, b) Re 10 000, c) Re 5 000, d) Re 2 000.

5. Conclusion

Tomographic PIV with high temporal resolution is applied for the study of the vortex structure formed in the near field of the quasi-two-dimensional turbulent jet. Experiments were carried out for a wide range of Re from 2 000 to 20 000. It was shown that for the whole range of Re numbers secondary flows form in the near field of the jet which is result of longitudinal vortex structure presence in the flow.

Acknowledgements

This work performed in the Kutateladze Institute of Thermophysics SB RAS is supported by a grant from the Russian Science Foundation (project No 14-19-01685).

References

[1] Foss J. F., Jones J. B, 1968 *J Fluids Eng.* **90** 241–248
[2] Holdeman J. and Foss J. 1975 *J Fluids Eng* **97** 342-352
[3] Nozaki T. et al. 1984 *Bull. Jpn. Soc. Mech. Eng.* **234** 2730
[4] Nozaki T. et al. 1986 *Proc of the 4th Inter. Symp. Flow Visual.* 495-499
[5] Dracos T, Giger M and Jirka G 1992 *JFM* **241** 587–614

IOP Conf. Series: Journal of Physics: Conf. Series **899** (2017) 062006 doi:10.1088/1742-6596/899/6/062006

[6] Bilsky A.V. et. al. 2012 *Proc of the 16th Inter. Symp. ALTFM*
[7] Scarano F. 2013 *Meas Sci Tech* **24** 012001
[8] Mullyadzhanov R., Ilyushin B., Hanjalic K. 2015 *Int. J. Heat Fluid Flow* **56** 284–289
[9] Meyer L. 2010 *Nucl. Eng. Des.* **240** 1575–1588
[10] Wieneke B. 2008 *Exp Fluids* **45** 549-456
[11] Bilsky A.V. et. al. 2013 *Meas Sci Tech* **24** 045301
[12] Shestakov M.V., Tokarev M.P., Markovich D. M. 2015 *Proc of 10th PSFVIP*

IOP Conf. Series: Journal of Physics: Conf. Series **899** (2017) 062007 doi:10.1088/1742-6596/899/6/062007

High-performance technique for initial velocity field calculation based on PIV images processing

D I Zaripov[1], **Renfu Li**[2]

[1] School of Mechanical Science and Engineering, Huazhong University of Science and Technology, Luoyu East Road 1037, Wuhan, China
[2] School of Aerospace Engineering, Huazhong University of Science and Technology, Luoyu East Road 1037, Wuhan, China

zaripov.d.i@mail.ru

Abstract. The high-performance technique is proposed in order to calculate the initial velocity field for planar Particle Image Velocimetry. The proposed technique is based on the use of Parallel Projection Correlation technique together with the region of interest that allows acceleration of calculation procedure and significantly increase the maximum value of measurable velocity using small-sized interrogation window at the first iteration. To show the performance of proposed technique, two synthetic test cases are considered in this investigation: one-dimensional uniform flow and linear shear flow. Accuracy analysis is carried out depending on different interrogation window sizes for two velocity components. The results of test cases show good performance of proposed technique in terms of accuracy. Finally, the proposed technique is applied together with the iterative multigrid window-deformation technique based on fractional interrogation window offset.

1. Introduction

Over the last three decades, the Particle Image Velocimetry (PIV) method has become a reliable tool for measurement of kinematic structure of fluid flow [1–3]. Wide variety of approaches is developed and implemented in order to achieve the high accuracy of obtained data such as velocity vectors and their derivatives. The development of PIV techniques that yield the high accuracy of data is the essential problem. At the same time, the problem of computational cost reduction has no less importance.

A Fast Fourier Transform Cross-Correlation (FFT-CC) approach is widely used when processing the PIV images. Its complexity in two-dimensional domain is $O(N^2\log_2 N^2)$ where N is the interrogation window (IW) size. This makes it much more preferable than the direct cross-correlation (2D-DCC) one which has the complexity of $O(N^4)$. The drawback of FFT-CC is that the IW in subsequent image cannot be slid over the region of interest (ROI) and its location must be fixed. Thus FFT-CC is commonly used for the predictor displacement field evaluation where particle images displacements are initially unknown and are supposed to be equal to zero [4]. It should be mentioned that 2D-DCC does not suffer from such a drawback and theoretically has no limitation for maximum measurable velocity [5] due to the possibility to use the ROI.

On the other hand FFT-CC imposes constraints on the IW size when considering gradient flow: typical IW size is 32×32 pix^2. According to [2], the further increase of IW size leads to high errors in determining of particle images displacements even though a low gradient shear flow is considered. According to well-known one-quarter rule [6], the maximum particle images displacement must be less than quarter of IW size, i.e. less than 8 and 16 pix for IW sizes 32×32 and 64×64 pix^2, respectively. This is a significant limitation of modern PIV techniques that use FFT-CC to calculate the initial velocity field.

Generally speaking, there are two approaches to calculate the initial velocity field: FFT-CC that allows acceleration of calculation but has limitation for maximum measurable velocity and 2D-DCC that has no mentioned limitation and gives more accurate results but is computationally intensive. The present work is dedicated to the novel approach that combines the advantages of both approaches. It is based on a Parallel Projection Correlation technique which has been first proposed in [7] when considering the Tomographic PIV in order to reduce the calculation time.

2. Description of technique for initial velocity field calculation

The technique of initial velocity field calculation consists of two basic operations and is schematically shown in figure 1. In the vicinity of grid nodes preliminarily applied to image k, small IW with sizes $N\times M$ pix^2 are selected. Then projections of

$$I_x(i)=\frac{1}{M}\sum_{j=1}^{M}I(i,j),\ i=\overline{1,N},\ j=\overline{1,M} \tag{1}$$

and

$$I_y(j)=\frac{1}{N}\sum_{i=1}^{N}I(i,j),\ i=\overline{1,N},\ j=\overline{1,M} \tag{2}$$

are considered instead of luminous intensity array of pixels $I(i, j)$: the elements of those are equal to arithmetic mean value of all elements in columns and rows of the array $I(i, j)$, respectively. As mentioned before, it is better to use ROI applied to image k+1 in order to have no limitation for maximum measurable velocity. Thus, the same procedure of calculating of projections $I_x(i, n, m)$ and $I_y(i, n, m)$ is conducted for IW within ROI with sizes $N_{\text{ROI}}\times M_{\text{ROI}}$ pix^2, where $n=[1,N_{\text{ROI}}-N+1]$ and $m=[1,M_{\text{ROI}}-M+1]$. Then the horizontal and vertical particle images displacements are determined comparing projections I_x and I_y of IW from two consecutive images k and k+1 as shown in figure 1. These displacements can be found by cross-correlating as

$$R_x(n,m)=\sum_{i=1}^{N}I_x(k,i)\cdot I_x(k+1,i,n,m),\ n=\overline{1,N_{\text{ROI}}-N+1},\ m=\overline{1,M_{\text{ROI}}-M+1} \tag{3}$$

and

$$R_y(n,m)=\sum_{j=1}^{M}I_y(k,j)\cdot I_y(k+1,j,n,m),\ n=\overline{1,N_{\text{ROI}}-N+1},\ m=\overline{1,M_{\text{ROI}}-M+1} \tag{4}$$

Then the search for the values n and m where the functions R_x or R_y have maximum values is conducted according to figure 1. These values correspond to the most probable integer displacements of particle images in pixels. Since the proposed technique deals with one-dimensional projections instead of two-dimensional luminous intensity array, it is proposed to use the abbreviation 1D-DCC for its definition.

Correlating the projections I_x and I_y instead of I yields the computational complexity of the order of $O(2Nn^2)$ for square IW and ROI. This makes 1D-DDC much more efficient in terms of calculation time than 2D-DCC with complexity of $O(N^2n^2)$. Thus, the theoretical speed-up of 1D-DCC compared to 2D-DCC is $N/2$. In practice, the speed-up is slightly less (see table 1). The computation was performed on Intel Core i5 with processor speed of 3.0 GHz. Speed-up estimation includes all overheads of cross-correlation map calculation, such as the creation or filling of appropriate arrays. As can be seen from table 1, the speed-up is almost independent of the ROI size that agrees with theoretical estimation. Moreover, the bigger IW size, the higher the speed-up.

Table 1. Speed-up of 1D-DIC compared to the 2D-DIC.

		N_{ROI} - N + 1, pix	
		16	32
N, pix	16	**6.4**	**6.5**
	32	**13.6**	**14.0**

However, it is not obvious what correlation map between R_x and R_y is better to determine n and m. In order to understand this, two test cases are further considered: one-dimensional uniform flow and linear shear flow.

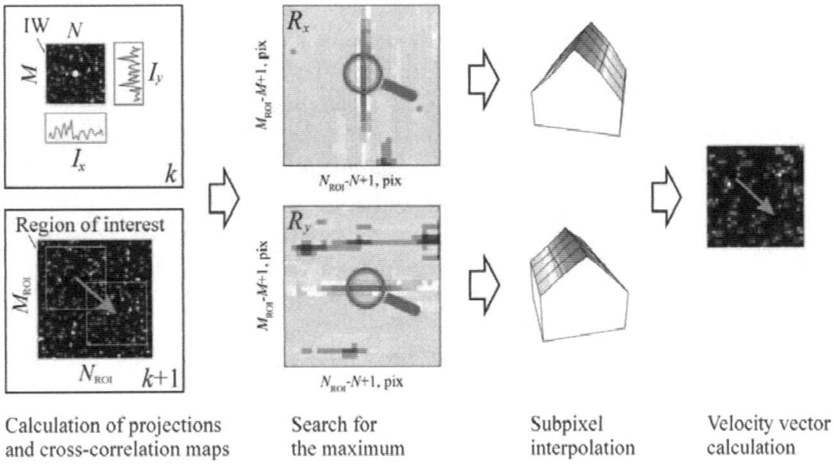

Figure 1. The algorithm of initial velocity field calculation.

3. Synthetic images generation

To quantitatively estimate the accuracy of 1D-DCC, synthetic images are generated using conventional approach to PIV image generation [2,8]. The luminous intensity of particle images is assumed to be continuous and distributed according to the Gaussian function, and the particle images borders are defined using e^{-2} level of maximal value of the Gaussian function. The luminous intensity of each pixel is defined by integration over the area occupied by the pixel [8,9]. Particle images with diameter d_p=2 pix are randomly and uniformly distributed over the whole image area with concentration C=0.076 particles/pix^2. To obtain the explicit dependence of the error on particle displacement values, noise is not added to the images.

Two synthetic test cases are considered in this research: one-dimensional uniform flow and linear shear flow. For both cases, a sequence of 35 images (for 34 displacement field) with the sizes 64×560 pix^2 is generated. Such image sizes allows the use of IW with sizes 16×16 and 32×32 pix^2 that are overlapped by 50%. The total number of IW placed in each image is 102 (3×34). Only the matrix of displacement vectors 1×32×32 placed in the middle windows are statistically analyzed even though 3×34×34 displacement vectors are calculated. The uniform vertical displacement of particles l_0 in the range from 0 to 2 pix with the increment of 0.1 pix is simulated. When shear flow is simulated, the particle images are displaced according to linear gradient shift from 0 to 0.3 pix/pix with the increment of 0.02 pix/pix resulting in zero mean displacement for statistically analyzed vectors. In order to make a comparative accuracy estimation between the proposed technique and other ones such as 2D-DCC and FFT-CC, the ROI sizes are chosen to be equal to 31×31 and 63×63 pix^2 for IW with sizes 16×16 and 32×32 pix^2, respectively (the ROI are placed at the center of IW).

The erroneous displacement vectors are detected using a 3×3×3 local median test proposed in [10] and replaced by the local average of the (accepted) neighbor vectors [11].

4. Results of 1D-DCC accuracy evaluation

Random and bias errors are analyzed for both particle images displacement directions: in flow direction (y-component of displacement) and perpendicular to flow direction (x-component of displacement). One should note that the particle images are not shifted in perpendicular to flow direction in simulation, i.e. the true values of their x-component of displacements are equal to zero. Only one iteration is applied when using 2D-DCC and 1D-DCC for both test cases.

The random and bias errors for uniform flow are plotted in figure 2 depending on IW size and particle images displacement. The comparative error estimation is conducted for data obtained by 2D-DCC and 1D-DCC. Moreover, particle images displacements are determined from both cross-correlation maps R_x and R_y when considering 1D-DCC. The results are compared with appropriate data obtained by the iterative multigrid technique with discrete IW offset [12] for synthetic images with the same particle images concentration and diameters.

Figure 2,*a* and figure 2,*b* illustrate the random error patterns which grow with IW size reduction. The difference of approximately one order of magnitude between random errors obtained from R_x and R_y is seen for both particle images displacement components. Besides, the random errors σ_v and σ_u have lower value if displacements are calculated from R_y and R_x, respectively. Such a behavior of random error distribution is explained by the shape of cross-correlation function (see figure 1). For instance, if cross-correlation function is obtained with the help of projection I_x, it has prolate shape in the vicinity of its maximum: broad cross-correlation peak in the y-direction and narrow peak in the x-direction (see top cross-correlation map R_x in figure 1). It is well-known [5] that broad cross-correlation peak reduces the accuracy of peak detection. Thus, it is better to calculate x-component of particle images displacement from R_x and, similarly, y-component of displacement from R_y.

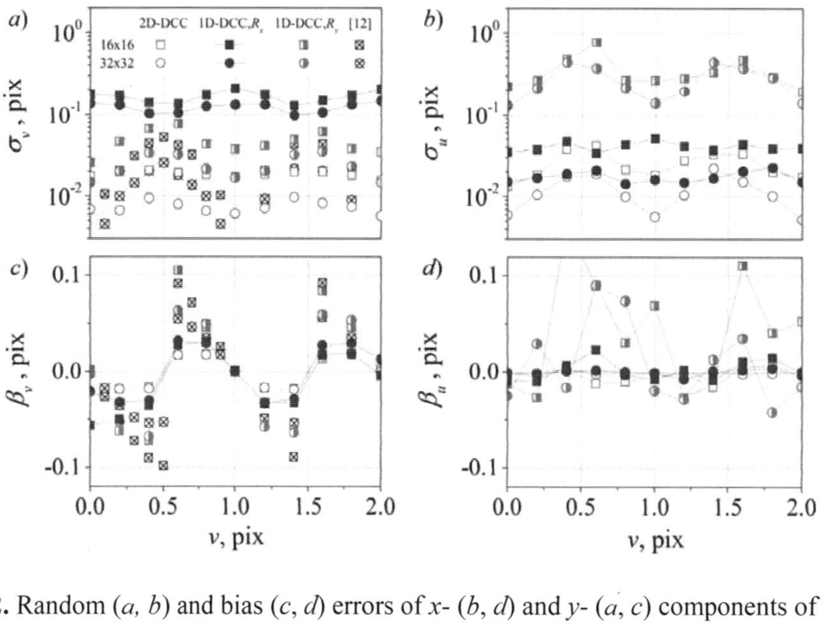

Figure 2. Random (*a, b*) and bias (*c, d*) errors of *x*- (*b, d*) and *y*- (*a, c*) components of particle images displacement for uniform flow.

Figure 2,*c* shows the periodic behavior of bias errors with a 1-pixel wave-length that is in accord with [12]. This periodicity appears due to the application of ROI that is similar to discrete IW offset used in [12]. The amplitude of the bias error patterns obtained by different techniques varies not much with the IW size and their values are in good agreement with [12]. Nevertheless, the bias errors β_v calculated from R_x are less than the same one obtained from R_y (see figure 2,*c*). No explanation is found yet for this result which will be further analyzed. Figure 2,*d* shows that *x*-component of particle images displacement is better calculated from R_x (note that in *x*-direction the particle images displacements are equal to zero).

In general, it is seen from figure 2 that 2D-DCC yields more accurate results compared to results obtained by 1D-DCC and the iterative multigrid technique with discrete IW offset [12]. However, if particle images displacements are properly obtained from appropriate cross-correlation maps, they are accurate within 0.1 pix that is quite sufficient for initial velocity field calculation.

Figure 3 illustrates the random and bias error patterns for shear flow. It is seen from figure 3,*a* that the *y*-component of particle images displacement is obtained more accurately from R_y when the displacement gradient is smaller than 0.2 and 0.1 pix/pix for IW sizes 16×16 and 32×32 pix^2, respectively. This behavior of random error distribution for low gradient shear flow is similar to the case of uniform flow and has been described above. However, when higher displacement gradient is considered, the cross-correlation map R_y yields worse results compared to R_x (see figure 3,*a*). This can be explained by the following. According to 1D-DCC, the result of particle images projecting on orthogonal sides of planar image is the one-dimensional field of illumination intensity where particle images are not single but may overlap yielding the homogeneous field. Thus, if particle images are displaced not uniformly in the direction perpendicular to the projecting side of image, the one-dimensional field can become unrecognizable. This effect becomes stronger, the higher the gradient is. Thus, it is better to use R_x when high gradient shear flow is considered (see figure 3,*a*). The influence of projecting procedure is more obvious when *x*-component of displacement is calculated (see figure 3,*b*): σ_u obtained from R_x is approximately two orders of magnitude less than the same one obtained from R_y and even less than the same one obtained by 2D-DCC. Not big difference in β_v is seen in figure 3,*c* in contrast to β_u shown in figure 3,*d* where R_x provides better results.

In general, the comparison of the random and bias errors for shear flow demonstrates that R_x yields better results in terms of accuracy than R_y especially for the high flow gradient.

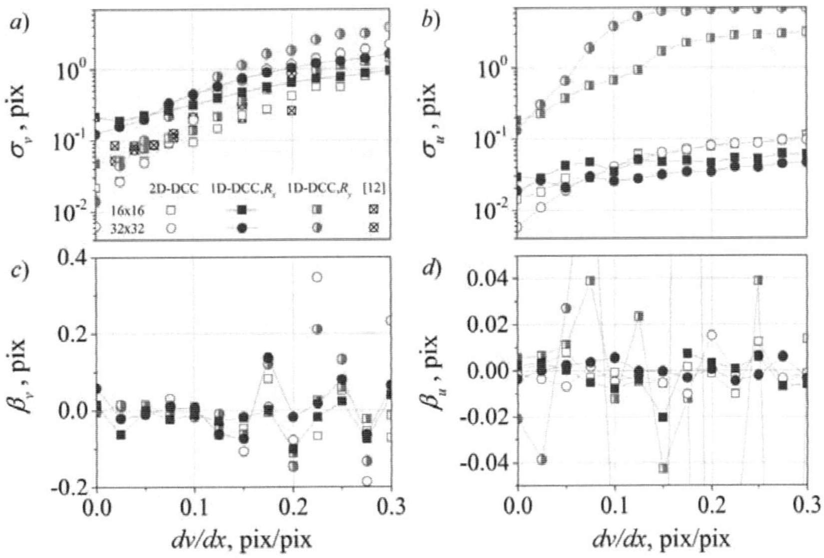

Figure 3. Random (*a, b*) and bias (*c, d*) errors of *x*- (*b, d*) and *y*- (*a, c*) components of particle images displacement for shear flow.

Finally, the application of 1D-DCC together with the iterative multigrid window-deformation (WIDIM) technique [12] is demonstrated. 1D-DCC based on R_x is used to obtain the initial particle images displacement field and then two additional iterations are executed by WIDIM. In order to interpolate the image deformed according to the predictor displacement field, the *sinc* function with kernel size 6×6 pix^2 is chosen. The results are compared with appropriate data from [12].

The random and bias errors for uniform flow are plotted in figure 4,*a* and figure 4,*c*, respectively, depending on IW size and particle images displacement. It is seen from figure 4,*a* that the random error is one order of magnitude higher than the same one obtained by WIDIM used FFT-CC [12]. Figure 4,*c* shows the nonmonotonic behavior of the bias errors as well as the same order of magnitude at their peak values. The periodic behavior of bias errors with a 1-pixel wave-length is in good agreement with [13] whereas no explanation has been found in [12] for periodic behavior with the 2-pixels wave-length. The amplitude of the bias error pattern is in accord with [12]. The results of shear flow test case shown in figure 4,*b* agree well with results from [12]. Nevertheless, further improvement in the proposed technique is needed to reach higher accuracy of data.

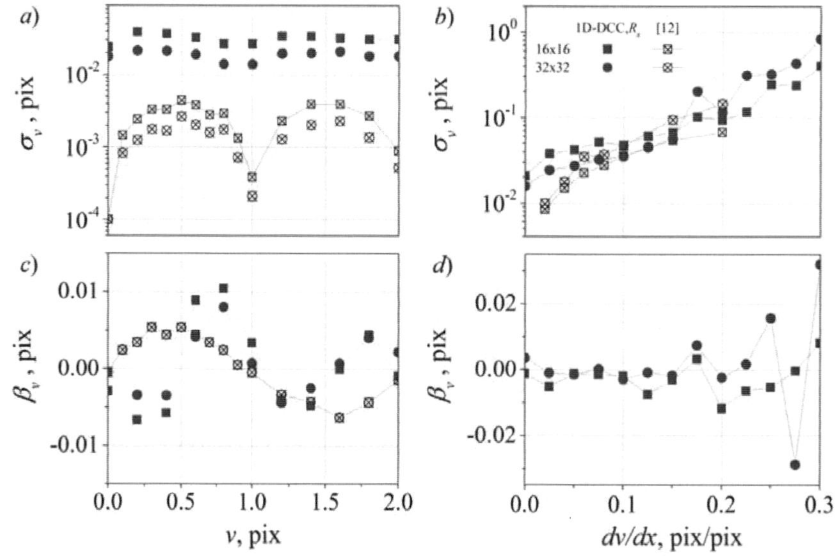

Figure 4. Random (*a, b*) and bias (*c, d*) errors of *y*-component of particle images displacement for uniform flow (*a, c*) and shear flow (*b, d*).

5. Conclusions

The high-performance 1D-DCC technique based on the PPC procedure is proposed in order to accelerate the cross-correlation function computation. It allows acceleration of cross-correlation function calculation up to *N*/2 times compared to 2D-DIC. It has been shown that, in practice, the speed-up is less. For example, for IW size of 32 pix, the cross-correlation procedure is accelerated about 14 times that is less compared to theoretical estimation of 16 times.

Two synthetic test cases have been considered to estimate the accuracy of the proposed technique: one-dimensional uniform flow and linear shear flow. It has been shown that the use of *x*- and *y*-projections of IW leads to prolate shapes of cross-correlation peaks in *y*- and *x*-directions, respectively, which adversely affects the accuracy of appropriate displacement components. In general, the comparison of the random and bias errors for uniform and shear flows demonstrates that the cross-correlation map R_x yields better results in terms of accuracy than R_y when particle images are displaced in *y*-direction. The results obtained by 1D-DCC together with WIDIM have shown that 1D-DCC can be recommended for the initial velocity field calculation.

References

[1] Adrian R J 2005 Twenty years of particle image velocimetry *Exp. Fluids* **39** 159–69
[2] Raffel M, Willert C and Kompenhans J 1998 *Particle Image Velocimetry*
[3] Kähler C J, Astarita T, Vlachos P P, Sakakibara J, Hain R, Discetti S, La Foy R and Cierpka C

2016 Main results of the 4th International PIV Challenge *Exp. Fluids* **57**
[4] Discetti S and Astarita T 2012 Fast 3D PIV with direct sparse cross-correlations *Exp. Fluids* **53** 1437–51
[5] Huang H T, Fiedler H E and Wang J J 1993 Limitation and improvement of PIV - Part II: Particle image distortion, a novel technique *Exp. Fluids* **15** 263–73
[6] Keane R D and Adrian R J 1992 Theory of cross-correlation analysis of PIV images *Appl. Sci. Res.* **49** 191–215
[7] Bilsky A V, Dulin V M, Lozhkin V A, Markovich D M and Tokarev M P 2011 Two-dimensional Correlation Algorithms for Tomographic PIV *9TH Int. Symp. Part. Image Velocim. - PIV11 Kobe, Japan, July 21-23, 2011* 1–4
[8] Scharnowski S and Kähler C J 2016 Estimation and optimization of loss-of-pair uncertainties based on PIV correlation functions *Exp. Fluids* **57** 1–11
[9] Lecordier B and Westerweel J 2004 The EUROPIV Synthetic Image Generator (S.I.G.) *Particle Image Velocimetry: Recent Improvements* ed M Stanislas, J Westerweel and J Kompenhans (Berlin, Heidelberg: Springer Berlin Heidelberg) 145–61
[10] Westerweel J and Scarano F 2005 Universal outlier detection for PIV data *Exp. Fluids* **39** 1096–100
[11] Westerweel J, Draad A A, van der Hoeven J G T and van Oord J 1996 Measurement of fully-developed turbulent pipe flow with digital particle image velocimetry *Exp. Fluids* **20** 165–77
[12] Scarano F and Riethmuller M L 2000 Advances in iterative multigrid PIV image processing *Exp. Fluids* **29** S051–60
[13] Kim B J and Sung H J 2006 A further assessment of interpolation schemes for window deformation in PIV *Exp. Fluids* **41** 499–511

IOP Conf. Series: Journal of Physics: Conf. Series **899** (2017) 072001 doi:10.1088/1742-6596/899/7/072001

Modeling of controlled synthesis of intermetallic coatings

A Knyazeva[1], O Kryukova[2]

[1] Professor, Tomsk Polytechnic University, Tomsk, Russia

[2] Senior researcher, ISPMS SB RAS, Tomsk, Russia

E-mail: anna-knyazeva@mail.ru

Composite materials and coatings creation in modern combined technologies is perspective direction in the science when exothermic synthesis and electron beam or laser treatment are used together. Similar multifactor technologies demand the employment of mathematical modeling. In this work, two-dimensional model of the formation on the ferrous substrate of intermetallic coating from the powders Ni and Al is suggested. The model takes into account the heating by moving energy source corresponding to electron beam and the staging of chemical conversion. Due to small thickness of the treated layer temperature and composition distributions along the coating depth are not analyzed in a first approximation. However, the possibility of intermixing of materials of the coating and substrate and corresponding chemical reactions are taken into consideration for the temperature close to the melting temperature. The model was realized numerically. The distributions of the temperature and concentrations of individual species and new phases are studied for different time moments. It was shown that multi-phase coating forms in the conditions under study. It is possible to control the phase composition of the coating varying the composition of initial powders and treatment parameters. Self-sustained regime does not realize because heat losses to the substrate are high. This corresponds to experimental data. Effective properties of the coating could be evaluated for quasistationary stage that establishes quickly for each pass of electron beam.

1. Introduction

Methods of powder technology [1] possess the special possibilities for obtaining the materials and compositions with given properties, when they are combined with concentrated energy sources [2]. One of them is the method of electron beam (EB) treatment of the material with the preliminary deposited layer where the coating synthesis is carried out [3]. This method allows integrating the advantage of EB – treatment and condensed phase conversions accompanied by chemical heat release. However the substrate deducted the heat from heated zone can affect the regimes of the reaction initiation and propagation in condensed phase similarly to inert inclusions [4]; inert rood [5] or connectable materials [6]. Similar conditions for multi phase coating synthesis are realized when laser irradiation is used [7,8]. But these methods are more energy consumed. EB-technologies are characterized by more high coefficient of input energy usage [2,9]. In any case, it is not possible practically to predict the phase composition of a treated layer based only on the reversible state diagrams and the results of the solution of pure thermal conductivity problems [10,11], when the irreversible conditions for the synthesis and complex heat exchange conditions with conjugate substances, chamber walls, environment, and etc. are not taken into consideration. Even such

contemporary method as LBE (Lattice Bolzmann Equation method) does not allow to analyze the total set of physical and chemical stages and stays an illustrative one [12-14]. It is necessary to develop special models including the technology conditions and physical – chemical phenomena accompanying the EB – treatment of the surfaces. The examples of such models are contained in [15-20]. So, in the papers [15, 16], in term of the model of EB surface treatment with modifying particles, it was shown that critical conditions exist dividing the homogeneous and composite coating formation. In the works [17, 18], the examples of the modeling of the process of intermetallic coating formation for the systems Ti-Al and Ti-Ni for different conditions of EB – heating. The model of laser sintering of two-component alloy was suggested in [19,20] taking into account the shrinkage of the powder layer and dependencies of thermal and optical properties on the porosity changing in time.
Here the mathematical model is presented for the synthesis process of multi-phase coating from the powders of Al and Ni on Fe-substrate at the conditions of EB-treatment taking into account key chemical stages.

2. Problem formulation

It is assumed that one-layered coating containing the mixture of Ni and Al powders was deposited on the surface of the detail in form of thin plate.
Effective energy source moves along the surface in the direction OX with some velocity *V*. Energy in the source is distributed corresponding to the law

$$q_e(x,y,t)=\begin{cases} q_0\, exp\left(-(x-Vt)^2/a_t^{\,2}\right), & y\le h_0;\\ 0, & y>h_0, \end{cases} \qquad (1)$$

where q_0 is the maximal powder density of the source (heat flux density), a_t is the effective radius of the source, h_0 is the scanning half-width. This effective source (1) corresponds to repetitive oscillations of electron beam [2,21], that is correct for the scanning frequency more then 50 hertz, and for the depth of the electron penetration more less then possible heated zone. For thermal thin plate or thin coating and thin surface layer of the substrate one can neglect temperature distribution along the depth similarly to [15-17]. Mass redistribution in this direction due to the diffusion is negligible small; however the melting of the substrate and coating lead to the intermixing and chemical interaction. Hydrodynamics of the melt is not analyzed explicitly. Hence the thermal part of the problem will include two-dimensional thermal conductivity equation

$$\begin{gathered}\left[h_1c_1\rho_1+h_2c_2\rho_2\right]\frac{\partial T}{\partial t}=\left[\frac{\partial}{\partial x}\bar{\lambda}\frac{\partial T}{\partial x}+\frac{\partial}{\partial y}\bar{\lambda}\frac{\partial T}{\partial y}\right]-\\ -\sigma\varepsilon_0\left(T^4-T_0^4\right)+q_e(x,y,t)+\left(h_2+h_1\right)\cdot W-\alpha_{eff}\left(T-T_0\right),\end{gathered} \qquad (2)$$

where $\bar{\lambda}=\lambda_{eff}h_2$; $\lambda_{eff}=\lambda_1+\lambda_2 h_2/h_1$ is effective thermal conductivity of material with the coating; λ_i, c_i, ρ_i, and h_i are the thermal conductivities, densities, heat capacities, and thicknesses of the surface layer of the substrate $i=1$ and coating $i=2$. Second term in (2) describes the heat losses from the surface according to Stephen-Boltzmann law; third term corresponds to the external heating, fourth term is the summary chemical heat release, and fifth term gives the effective heat losses to the depth of the substrate. Heat capacity changes sharply near the melting temperature that is taken into account by following way

$$\left(c_i\rho_i\right)=\left(c_i\rho_i\right)_{eff}+L_{m,i}\rho_{s,i}\delta\left(T-T_{m,i}\right), \qquad (3)$$

where

$$\left(c_i\rho_i\right)_{eff}=\begin{cases} c_{s,i}\rho_{s,i}, T<T_{m,i};\\ c_{L,i}\rho_{L,i}, T\ge T_{m,i}, \end{cases}$$

δ is Dirac delta-function; $L_{m,i}$ and $T_{m,i}$ are the heats an temperatures of the melting (crystallization) for substrate and coating; $i = L, s$; index «L» relates to the liquid phase; «s» relates to the solid phase.

Boundary and initial conditions

$$x = 0, H_x:\ \lambda \frac{\partial T}{\partial x} = 0;\quad y = 0, H_y:\ \lambda \frac{\partial T}{\partial y} = 0 \tag{4}$$

$$t = 0:\ T(x, y, 0) = T_0. \tag{5}$$

close to the thermal problem formulation.

To calculate summary heat release doe to chemical conversions it is necessary to solve kinetical sub-problem. Corresponding to state diagrams for the systems Fe-Al and Ni-Al [22,23] key reactions in the transient zone and in the coating are written as

$$3Ni + Al \rightarrow Ni_3Al \text{ (I)};\quad Ni + Al \rightarrow NiAl \text{ (II)}$$

$$3Fe + Al \rightarrow Fe_3Al \text{ (III)};\quad Fe + Al \rightarrow FeAl \text{ (IV)}$$

Reactions III and IV are possible in the case of intermixing above melting temperature of substrate. With designations of molar concentrations $y_1 = [\mathrm{Ni}]$; $y_2 = [\mathrm{Al}]$; $y_3 = [\mathrm{Fe}]$; $y_4 = [\mathrm{Ni_3Al}]$; $y_5 = [\mathrm{NiAl}]$; $y_6 = [\mathrm{Fe_3Al}]$; $y_7 = [\mathrm{FeAl}]$, mol/cm^3, kinetical equation's system will include

$$\frac{\partial y_k}{\partial t} = \omega_k$$

for the reagents and for the reaction's products. Here

$$\omega_1 = -3\varphi_1 - \varphi_2,\ \omega_2 = -\varphi_1 - \varphi_2 - \varphi_3 - \varphi_4,\ \omega_3 = -3\varphi_3 - \varphi_4,$$

$$\omega_4 = \varphi_1,\ \omega_5 = \varphi_2\ \ \omega_6 = \varphi_3\ \ \omega_7 = \varphi_4$$

Chemical reaction's rates are:

$$\varphi_1 = k_{01}\, exp(-E_{a1}/RT)\, y_1^3 y_2,\qquad \varphi_2 = k_{02}\, exp(-E_{a2}/RT)\, y_1 y_2,$$

$$\varphi_3 = k_{03}\, exp(-E_{a3}/RT)\, y_3^3 y_2,\qquad \varphi_4 = k_{04}\, exp(-E_{a4}/RT)\, y_3 y_2,$$

where k_{0i}, $E_{a,i}$ are pre-exponential factors and activation energies for reactions.

Initial concentrations for elements are $y_{01} = 0.035$ $y_{02} = 0.035$ $y_{03} = 0.141$ mol/cm^3.

The problem was solved numerically. Parameters were found in the literature or are calculated based on thermodynamics.

3. Results

The analysis of the results of numerical investigation shows that quasistationary regime establishes with time after some short non stationary stage together with source motion along the treated surface. This regime is characterized by the practically constant temperature (Fig.1) and shaped melting pool and heat affected zone (Fig.2). Melting pool is painted by dark gray color. Maximal size of the area of light gray color in the direction y corresponds to the size of heat affected zone. It corresponds to the temperature $T = 900\,K$. Plateau on the temperature curves corresponds the melting temperature Maximal temperature value , sizes of melting pool and heat affected zone depend on the powder density q_0, source velocity V, and sizes of layers. The cooling after the source passing goes slowly.

In some area of technological parameters (q_0, V, a_t, h_0), the external heating is accompanied by chemical conversions. Example of concentration distributions $C_i(x,0)$ for different time moments t and various values q_0 is shown from Fig. 3. The reaction zone propagates with constant velocity equal to source velocity. In quasi stationary stage, maximal values of new phase concentrations and maximal temperature are registered depending on q_0.

IOP Conf. Series: Journal of Physics: Conf. Series **899** (2017) 072001 doi:10.1088/1742-6596/899/7/072001

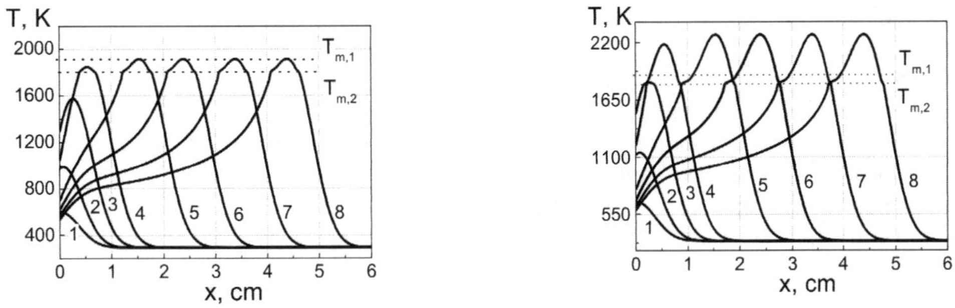

a b

Figure 1. Spatial distribution of the temperature $T(x,0)$ along the treated surface for sequential time moments: 1. – $t = 0.06$, 2. – $t = 0.24$, 3. – $t = 0.8$, 4. – $t = 1.6$, 5. – $t = 3.7$, 6. – $t = 5.4$, 7. – $t = 7.4$, 8. – $t = 9.4$ sec. a) – $q_0 = 9 \cdot 10^4$, b) – $q_0 = 1.1 \cdot 10^5$ W/cm^2

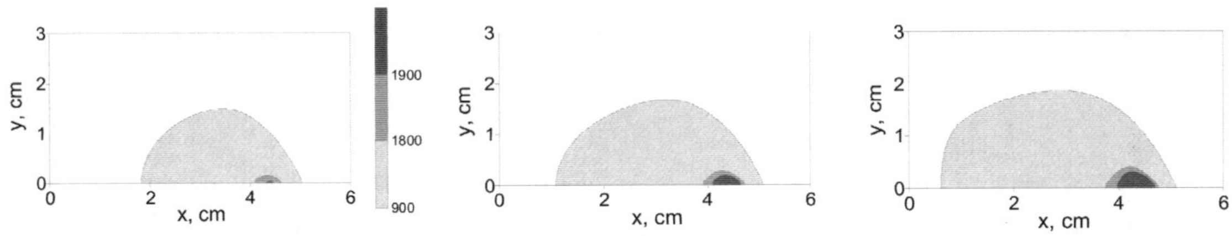

a b c

Figure 2. Influence of power density on the form and size of the melting pool (dark gray) and heat affected zone (light gray) to the quasi stationary regime establishment, $t = 9.4$ sec; a) – $q_0 = 9 \cdot 10^4$ W/cm^2, b) – $q_0 = 10^5$ W/cm^2, c) – $q_0 = 1.1 \cdot 10^5$ W/cm^2.

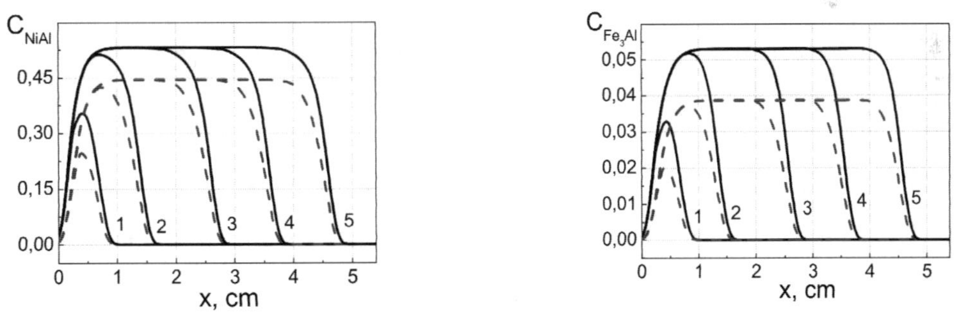

a b

Figure 3. Spatial distributions of some mass concentrations of phases for different time moments: 1. – $t = 1.6$ sec, 2. – $t = 3$ sec, 3. – $t = 5.4$, 4. – $t = 7.4$ sec, 5. – $t = 9.4$ sec; solid lines– $q_0 = 10^5$, dotted lines – $q_0 = 9 \cdot 10^4$ W/cm^2; V=0.5 cm/sec.

Relative molar concentrations in the boundary of heat affected zone depending on the time are demonstrated in Fig. 4 for various q_0. Using these data one can evaluate the interval of effective properties change for obtained multi phase coating [24,25]. It is possible to use the existing analytical models for effective physical properties of heterogeneous materials, including thermal conductivity, coefficient of thermal expansion, electric conductivity, and mechanical modulus. These different physical properties obey the same regularities and general expressions of governing laws.

A substantial number of theoretical (or analytical) models for effective physical properties have been proposed in the past of more than one hundred years, however the theoretical prediction of effective properties for multiphase material systems is very important nowadays to analysis and optimization of material creation. For example, using the simplest approach one can find the Voigt–Reuss bounds for the elastic modulus E and thermal conductivity coefficient

$$\lambda \ \langle b \rangle = \sum_{(i)} b_i x_i \ , \ \left\langle \frac{1}{b} \right\rangle = \sum_{(i)} \frac{x_i}{b_i}$$

where $b = E, \lambda, ...$ and $x_i = y_i \Big/ \sum_{(k)} y_k$ - volume fractions of phases. Index $"a"$ in the Fig.5 corresponds to above estimation, and index $"b"$ relates to below estimation.

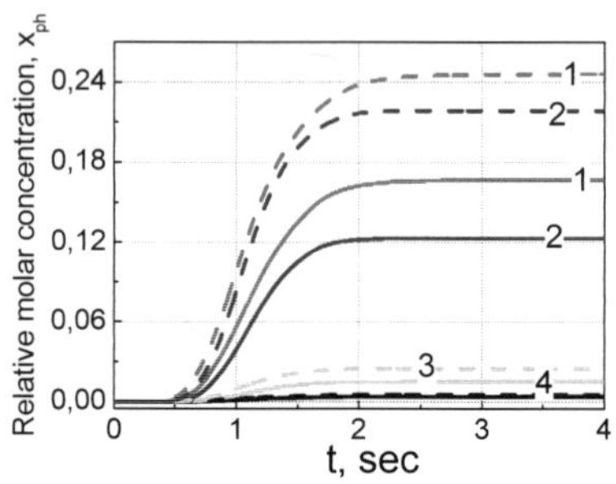

Figure 4. Phase concentration change in time in the point with coordinates x=1 cm, y=0 cm. Solid lines– $q_0 = 10^5$, dotted lines – $q_0 = 1.1 \cdot 10^5$ W/cm^2; V=0.5 cm/sec; 1. – NiAl, 2. – FeAl, 3. – Fe_3Al, 4. – Ni_3Al

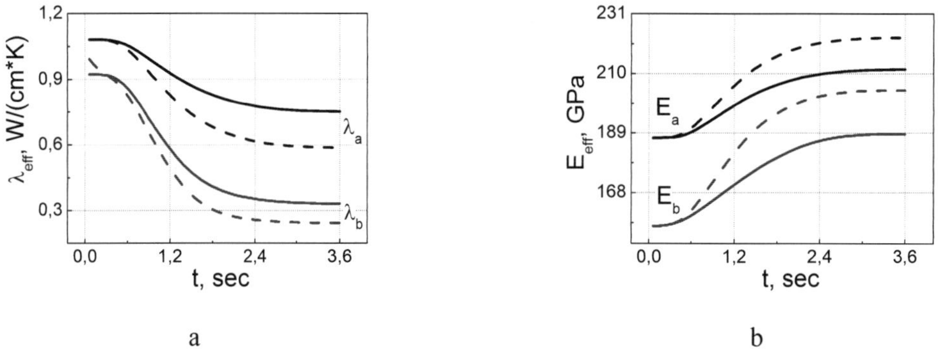

Figure 5. Thermal conductivity coefficient (a) and elasticity modulus (b) versus time. Solid lines– $q_0 = 10^5$, dotted lines – $q_0 = 1.1 \cdot 10^5$ W/cm^2

As a result for this case, for the flux $q_0 = 10^5$ W/cm^2 we have $189 \le \langle E \rangle \le 211$ GPa; $0{,}35 \le \langle \lambda \rangle \le 7{,}5$ W/(cm·K); and for $q_0 = 1.1 \cdot 10^5$ – $196 \le \langle E \rangle \le 224$ GPa; $0{,}22 \le \langle \lambda \rangle \le 0{,}59$ W/(cm·K)

4. Conclusion

So the model of surface layer formation on the ferric substrate from nickel and aluminium was suggested. It was shown that new phase's formation affect the temperature, melting pool and heat affected zone formation. The composition of surface layer depends on all technological parameters. It was suggested using known theoretical models to evaluate the effective properties of forming under electron beam action composites.

References

[1] Upadhyaya G S 2002 *Powder metallurgy technology* (England: Cambridge internationsl science publishing) p 160
[2] Schiller S, Panzer S. 1988 Thermal surface modification by electron beam high-speed scanning. *Ann. Rev. Mater. Sci.* **18** 121-140
[3] Shipko A A, Pobol I L and Urban I G 1995 *Strengthening of steels and alloys using electron-*

beam heating (in Russian), (Minsk: Nauka i technika) p 280
[4] Knyazeva A G, Chashchina A A. Numerical 2004 Study of the Problem of Thermal Ignition in a Thick - Walled Container *Comb., Expl., and Shock Waves* **40** (4) 432–437
[5] Prokof'ev V G, Pisklov A V, Smolyakov V K 2007 Effect of a heat-conducting element on the gasless combustion of cylindrical samples under nonadiabatic conditions *Comb., Expl., and Shock Waves* **43** (1) 56–61
[6] Aligozhina K A, Knyazeva A G 2014 Modeling of the Heat and Kinetic Phenomena Accompanying the Different Material Joining Using Solid-Phase Synthesis *Adv. Mater. Res.* **1040** 519-524
[7] Shishkovskii I V, Kuznetsov M V, Morozov Yu G, Parkin I P 2004 Laser-induced combustion synthesis of 3D functional materials: computer-aided design *J. of Mater. Chem.* **14** 3444-3448
[8] Quin L, Hu J, Cui C, Wang H, Guo Z 2008 Effect of Al content on reaction laser sintering of Ni-Al powders *Sci. of Sinter.* **40** 295-301
[9] Herderick E 2011 Additive Manufacturing of Metals: A Review, Materials Science and Technology (MS&T) *ASM International* **176252** 1413-1425
[10] Gusarov A V, Yadroitsev I, Bertrand Ph, Smurov I 2007 Heat transfer modeling and stability analysis of selective laser melting *App. Surf. Sci.* **254** 975-979
[11] Tiebing Chen Yuwen Zhang 2006 Three-Dimensional Modeling of Selective Laser Sintering of Two-Component Metal Powder Layers *J. of Manuf. Sci. and Eng.* **128** (2) 299-306
[12] Attar E 2011 *Simulation of Selective Electron Beam Melting Processes* (Erlangen: Doktor – ingenieur dissertation)
[13] Khairallah S A, Anderson A T, Rubenchik A, King W E 2016 Laser powder-bed fusion additive manufacturing: physics of complex melt flow and formation mechanisms of pores, spatter, and denudation zones *Acta Mat.* **108** 36-45
[14] Attar E, Körner C 2009 Lattice Boltzmann method for dynamic wetting problems *J. of Colloid and Interface Sci.* **335** (1) 84-93
[15] Kryukova O N, Knyazeva A G 2007 Critical phenomena in particle dissolution in the melt during electron-beam surfacing *J. of App. Mech. and Tech. Phy.* **48** (1) 109-118
[16] Kryukova O N, Knyazeva A G 2005 Comparative analysis of one and two domensional models of electron beam surfacing with modyfying partiles - in "*Mathematical modeling of system and processes*", Ed. Trusov V.P. (in Russian) / Perm: izd-vo PermGTU. 123-131
[17] Sorokova S N, Knyazeva A G 2008 Simulation of coating phase structure formation in solid phase synthesis assisted by electron-beam treatment *Theor. Found. of Chem. Eng.* **42** (4) 443-451
[18] Knyazeva A G, Sorokova S N 2009 Simulation of intermetallide synthesis on a cylindrical substrate *Fiz. mezomekh* **12** (5) 77-90 (in Russian)
[19] Knyazeva A G, Sharkeev Y P 2016 Thermal-kinetical model of laser shrinkage of Ti-Nb-alloy *AIP Conf. Proc.* **1783** 020094
[20] Knyazeva A G, Sharkeev Y P 2016 Temperature calculation for laser sintering of Titanium and Niobium taking into account properties change due to powder layer shrinkage *Key Eng. Mat.* **712** 220-225
[21] Rykalin N N, Zuev A A and Uglov A A 1978 *The foundation of electron - beam treatment of materials* (in Russian) (Moscow:Mashinostroenie) p 239
[22] Hansen M, Anderko K 1958 *Constitution of Binary Alloys* (Ney York: McGraw-Hill)
[23] Kubaschewski O 1982 *Iron-Binary Phase Diagrams* (Berlin: Springer-Verlag)
[24] Moran Wang, Ning Pan 2008 Predictions of effective physical properties of complex multiphase materials *Materials Science and Engineering R* **63** 1–30
[25] Torquato S 1991 Random Heterogeneous Media: Microstructure and Improved Bounds on Effective Properties *Appl. Mech. Rev.* **44** 37–76

Investigations of thermal barrier coatings of turbine parts using gas flame heating

A.R. Lepeshkin[1], N.G. Bichkov [1], O.I. Ilinskaja[2], V.V. Nazarov[3]

[1]Central Institute of Aviation Motors, Aviamotornaya 2, Moscow, 111116 Russia
[2]Moscow Aviation Institute - National Research University, volokolamskoe h. 4, Moscow, 125993, Russia
[3]Institute of Mechanics of Lomonosov Moscow State University, Michurinsk prospect 1, 119192, Russia

E-mail: lepeshkin.ar@gmail.com

Abstract. The development of methods for the calculated and experimental investigations thermal barrier coatings and thermal state of gas-turbine engine parts with a thermal barrier coatings is actual work. The gas flame heating was demonstrated to be effectively used during investigations of a thermal ceramic barrier coatings and thermal state of such gas-turbine engine parts with a TBC as the cooled turbine blades and vanes and combustion liner components. The gas-flame heating is considered to be preferable when investigating the gas-turbine engine parts with a TBC in the special cases when both the convective and radiant components of thermal flow are of great importance. The small-size rig with gas-flame flow made it possible to conduct the comparison investigations with the purpose of evaluating the efficiency of thermal protection of the ceramic deposited thermal barrier coatings on APS and EB techniques. The developed design-experiment method was introduced in bench tests of turbine blades and combustion liner components of gas turbine engines.

1. Introduction

Progress in gas-turbine engine (GTE) manufacturing is continuously linked with a rise of operating temperature of engine gas path elements, especially the turbine parts. More advanced cooling systems, structural materials and thermal barrier coatings (TBC) and other coatings provide the required life and strength reliability of these components. While engines are in use, the necessity arises to repair turbine blades and vanes with a TBC and combustion liners. At the same time, it is difficult to estimate the durability of the turbine blades and combustor components with a TBC and vanes because of the complexity of simulation of the damaging factors acting under service conditions and also because of problems in obtaining the input data required for making such estimations. Therefore, the development of methods for the calculated and experimental investigations thermal barrier coatings, thermophysical properties of TBC and thermal state engine parts with a TBC is of great importance. While conducting these investigations, the main tasks are the comparative estimation of the design and production (or repair) process solutions and verification of the methods of thermal calculation of the durability of engine parts with a TBC. To provide simulation of loading conditions for the hot engine parts with a TBC under service conditions, the test procedure shall ensure the possibility of cyclic surface heating of the object under testing (simulating its heating in hot gas flow) up to temperatures of 1150 °C and more at heating rates of 150-200 °C/s and subsequent cooling.

2. Ceramic thermal barrier coatings

For the purpose of providing the serviceability of high efficiency aircraft gas turbine engines and gas turbine plants of new generations, it is necessary to improve existing cooling systems, to design new refractory and ceramic high-temperature materials, and to enhance the protection of parts of the high-temperature section of gas turbine engines with the use of heat-resistant and refractory coatings [1-12]. Improvement of the internal heat removal system leads to the transformation of parts into heat exchangers, which is accompanied by an increase in the thermal stress and a decrease in the thermal cycle life. Currently widely used refractory materials based on nickel usually operate in gas turbine engines at maximum allowable temperatures. The gas temperature can be allowed to increase only in the case where care is taken to restrict the passage of heat flow through the wall of the part. The heat flow from the gas to the wall of the base material of the part can be considerably reduced by means of either using a well-organized protective cooling without ejection or depositing thermal barrier coatings on the surface of the most strongly heated regions of the part.

In recent years, works on introduction and practical use of thermal barrier ceramic coatings on parts of a high-temperature system of gas turbine engines have been carried out especially actively. The protection of the material of the part against the heat flux with heat-barrier coatings is most effective when the ceramic coatings used are based on ZrO_2 [4-6, 10]. The heat-protective effect of the thermal barrier ceramic coating reaches 100-120 °C under operating conditions. The heat-protective effect – decrease of metal temperature is function of thickness and heat conductivity of a thermal barrier ceramic coatings and thermal flows in a wall of a protected detail. The values of thermal flows on workers turbine GTE blades are in a range from 1.0 up to $(2\div2.5)\times10^6$ W/(m·K) in works [3-5]. In some cases the thermal flow makes 3×10^6 W/(m·K) and more [6]. In the given work with the use of system ANSYS the calculated investigations of influence of the specified factors on decrease of metal temperature of cooled blades were carried out and the estimations of the heat-protective effect of a thermal barrier ceramic coatings for cooled details have been obtained by Lepeshkin A.R. The results of calculated investigations are presented in figure 1 and figure 2.

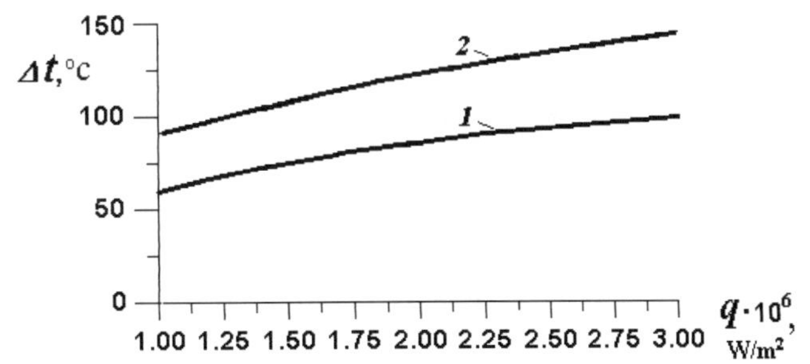

Figure 1. Values of decrease of metal temperature Δt on a surface of cooled GTE blades depending on thermal flows q at thickness h = 0,14 mm of ceramic coatings ZrO_2 and different heat conductivity: *1* – 1.5 W/(m·K); *2* – 0.8 W/(m·K)

The values of decrease of metal temperature on a surface of cooled GTE blades depending on The values of decrease of metal temperature on a surface of cooled GTE blades depending on thermal flows at thickness h = 0.14 mm of ceramic coatings and different heat conductivity coatings are shown in figure 1. The values of decrease of metal temperature on a surface of cooled GTE blades depending on thickness of ceramic coatings at gas thermal flow $q = 1.8\cdot10^6$ W/m^2 and different heat conductivity of coatings in figure 2. However, the questions regarding the thermal cyclic fatigue life are very problematic because the fracture strength of these coatings under tension is very low and thermal cycling usually leads to the appearance of alternating thermal stresses. Moreover, during operation of turbine blades, oxygen from an oxidizing medium (air, fuel combustion products) penetrates into the "ceramics-metal" interface.

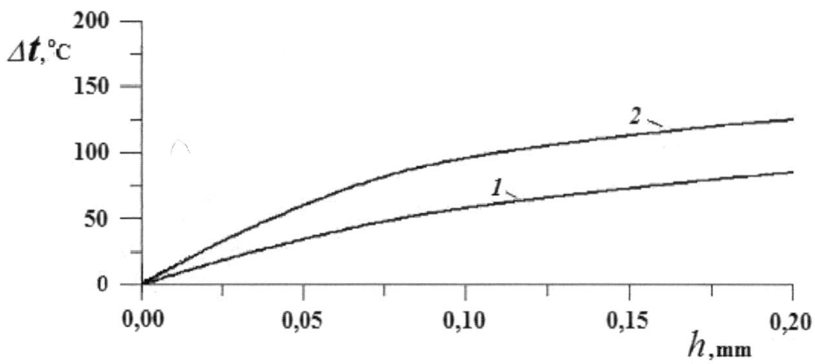

Figure. 2. Values of decrease of metal temperature *Δt* on a surface of cooled GTE blades depending on thickness *h* of ceramic coatings ZrO_2 at gas thermal flow $q = 1.8 \cdot 10^6$ W/m^2 and different heat conductivity of coatings: *1* – 1.5 W/(m·K); *2* – 0.8 W/(m·K)

The penetration of oxygen through the ceramic layer results in the oxidation of the sublayer. The formation of oxides gives rise to additional stresses and decreases the adhesion of the ceramic layer. Therefore, the above factors must be taken into account in the design of coatings. The efficiency of thermal protection of coatings depend not only on the thermophysical properties (figure1 and figure 2) but also on the technique used for depositing of the coating. Among numerous techniques currently employed for depositing of the coatings, the electron-beam technique provides the best thermal protection.

Development of thermal barrier coatings applied to cooled blades is one of the trends for improving gas turbines. Unlike aluminide protective coatings, the ceramic coatings not only protect blade surfaces from high-temperature oxidation and corrosion but also prevent base material softening at high temperatures. Thermal barrier coating application allows the reduction of the blade temperature and the significant increase in its service life. Under both steady-state and transient conditions, the application of ceramic TBC can diminish temperature gradients over the blade surfaces as well as reduce thermal stresses in them. A typical design of a TBC is presented in figure 3. The ceramic coating deposited directly on the superalloy surface does not show the required service life. Penetration of oxygen through the ceramic layer to the superalloy surface results in its quick oxidation and in spallation of the ceramic layer. That is why, as a rule, a TBC consists of at least two layers.

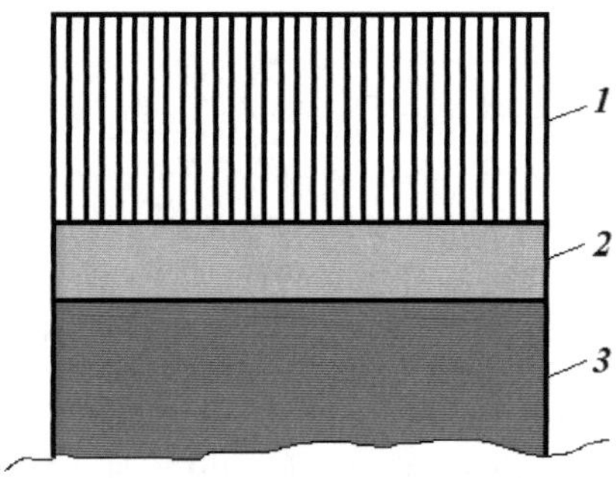

Figure 3. Thermal barrier coating system: *1* – ceramic coating ZrO_2-8%Y_2O_3, 2 – bond coating MCrAlY, *3* - superalloy

The outer zirconium oxide/yttrium oxide (ZrO2-Y_2O_3) system base ceramic layer can-be applied by two techniques (figure 4): air plasma spraying of powders (APS-technique) or vapor condensation at electron beam evaporation of ceramic pellets (EB-technique). For this system, ceramic coating service

IOP Conf. Series: Journal of Physics: Conf. Series **899** (2017) 072002 doi:10.1088/1742-6596/899/7/072002

life depends on Y_2O_3 content. The ZrO_2-(6 to 9%) Y_2O_3 compositions are usually applied, because they have demonstrated maximum service lives in the tests carried out [7, 9].
For providing the above-indicated heating conditions, there are various ways of heating such as gasdynamic heating and radiant heating, for example, in a reflective furnace electrical current (AC or DC) or induction heating with the use of high-frequency currents. The gasdynamic (flowing hot gas) heating has been used for more than 50 years. When using this method, a more accurate simulation of the heat exchange conditions from gas flow to the part is realized relevant to the gas-turbine engine. The rigs with gasdynamic heating enable a high heating rate to be provided to the part, to investigate the influence of oxidation in gas flow.

a) b)

Figure 4. Thermal barrier ceramic coating ZrO_2-Y_2O_3: a -APS-technique, b - EB-technique

3 Investigations of thermal barrier properties of ceramic coatings using gas-flame heating
For maintenance of competitiveness of aircraft engines it is necessary to raise a gas temperature over 1700 K in front of the turbine. Thus serviceability of details of a high-temperature gas can keep only at perfection of their heat-protection. Thus the serviceability of details of a high-temperature gas probably keep only at perfection of their heat-protection. It is known, that in world practice the ceramic heat-protective coatings on basis ZrO_2 are widely used. At the same time the data on heat conductivity and thermal conductivity and efficiency of a heat-protection of details with help thermal barrier ceramic coatings at their heating in a gas flow are rather limited. The characteristics of heat conductivity thermal barrier ceramic coatings have received at use of various known laboratory methods are inconsistent. Basically the preference is given the thermal barrier ceramic coatings have deposited on a plasma technology. At use of laser pulse heating it has been received that heat conductivity of plasma coatings approximately in 3 times is lower than at the coatings have deposited on the electron beam technology. The laser pulse method is inexpedient to use for determination of the temperature in part transparent ceramic coatings as the part of a beam flow warms up directly a metal on which it is deposited coating. The protective thin metal screen with thickness 10-15 μm deposited by researchers on surface of coating on the side of the laser at heating, itself starts to let out a beam flow. In real conditions the turbine blades and walls of combustion chambers are heated up by a gas flow. In the given work the developed technique by an objective estimation of efficiency of a heat-protective of metal with the help of coatings of plasma and electron beam technology is resulted at gas-flame heating of object on the developed rig. The essence of the given original technique protected by the patent RU will be that through the demountable specimen (collected from two halfs) the high-temperature gas flow (figure 5) is passed.

IOP Conf. Series: Journal of Physics: Conf. Series **899** (2017) 072002 doi:10.1088/1742-6596/899/7/072002

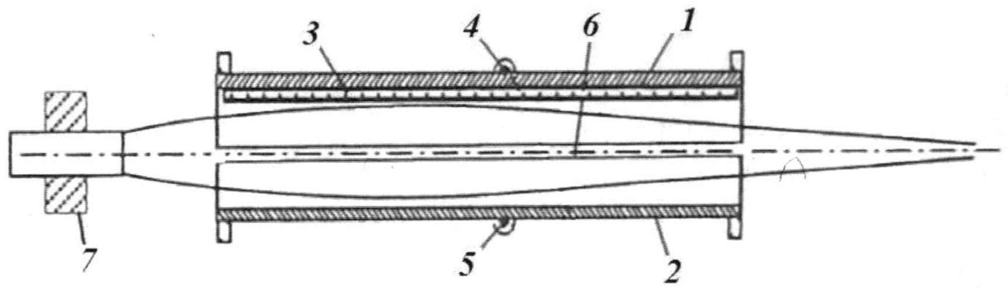

Figure 5. Sketch of specimen: *1* – specimen half with coating, *2* – specimen half without coating, *3* – coating, *4, 5* – thermocouples, *6* – axis of specimen (flame), *7* – burner

The investigations for evaluating the efficiency of thermal protection of materials of the turbine blades and parts with use TBC (received on electron beam technology and plasma technology) against the convective and radiant components of the high-temperature gas flow were conducted. In this case, it is recommended to conduct the tests at the small-size rig and use the small-size specimens whose surfaces during tests are accessible for inspecting the thermal state both by the contact and contactless methods. This rig in particular is usable effectively for conducting the comparison thermal barrier properties and thermocycles tests of various coatings. The rate of change of the temperature in a thermocycle reaches 100 °C/s. For performing these investigations, a test rig has been developed with gas-flame heating of model specimens. The gas generator is a water electrolysis device equipped with a control system; it provides the variable flammable gas flow. Hydrogen has a high combustion temperature and this fact ensures high-speed heating of the specimens. This test rig has a system for providing enrichment of the flammable gas with different fuels. This makes it possible to attain the required gas composition. While testing, the burner is installed fixed, however the attachment allows its position to be adjusted. The hollow specimen is of an axisymmetrical form. Before the test, the burner is installed in a way ensuring coincidence of the specimen axis with the flame torch axis in the process of heating. While investigating the efficiency of influence of ceramic coating on specimen temperature state, the unit with specimens was fixed. A special specimen construction was developed for these tests. The hollow specimen was cut longitudinally in two equal portions. The ceramic coating under investigation was applied on one half of the specimen, the other half remained uncoated. The thermocouples XA by diameter of 0.2 mm weld on an external surface of halfs of a compound specimen (figure 5) and are connected to recording computer system. The half of a specimen is protected from products of combustion by a coating it is warmed up less, than unprotected half. The difference of temperatures Δt of protected wall with coating and unprotected wall characterizes the efficiency and thermal conducting of the thermal barrier coatings. Heat insulating material was placed between them to exclude the influence of heat transfer through the contacting edges of the specimen halves. While heating, temperature was measured on the outer (opposite to flame torch) specimen side. Conditions for heating the inner surfaces of both specimen halves by flame were the same, but with a difference in heat protection efficiency the outer surface of the specimen with a TBC had a lower temperature than the surface without a TBC. The after of lighting of a combustible gas the heating of an internal walls of both halfs of model begins. The difference of temperatures on lateral side grows until the heat transfer from a hot surface of a wall to cold surface is less, than a heat-conducting from an external surface in an environment. At absence of the organized cooling lateral side of a wall the maximal difference of temperatures Δt_{max} outside of both of halfs corresponds to a gradient of temperatures on TBC under these conditions. In experiment Δt_{max} it is reached at temperature of a cold wall 600 °C. The results of investigations are presented on figure 6 and figure 7.

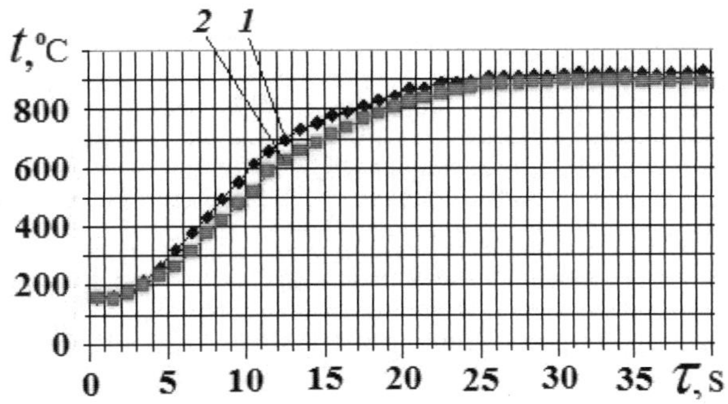

Figure 6. The temperature on an external (cold) surface specimen: *1* – without coating, *2* – with ceramic coating (APS technique)

The models with ceramic TBC ZrO_2 + 8%Y_2O_3 deposited on plasma technology were made of Ni-alloy. Other models with TBC of columnar structure deposited on electron beam technology were made of other Ni-alloy. Unprotected halfs of each model were made of the same material as halfs with TBC.

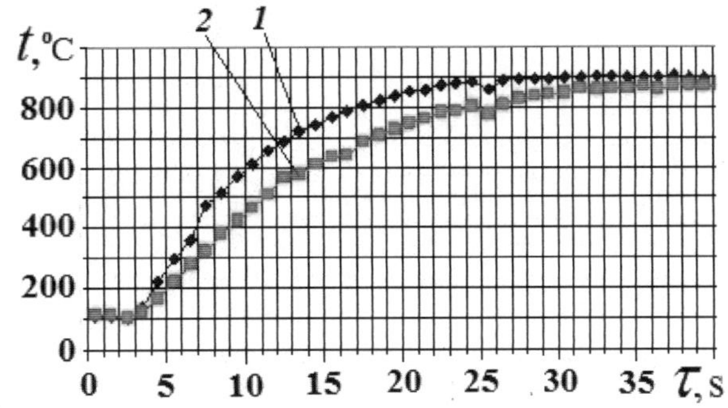

Figure 7. The temperature on an external (cold) surface specimen: *1* – without coating, *2* – with ceramic coating (EB technique)

The tests on each model were repeated some times for maintenance of reliability of determination of heat-protective efficiency. At retesting the model was unwrapped about the axis on 180 °. The difference of temperatures at repeated measurements did not exceed 10 °C. The maximal temperature on the "cold" side of a wall made 900 °C. The temperature of a gas flow made 1773 K. The experimental investigations have shown that the efficiency of decrease of metal temperature at gas-flame heating after deposited TBC by thickness δ = 120 μm on plasma and electron beam technologies make correspondingly $\Delta t_{мах}$ = 60-70 °C and $\Delta t_{мах}$ = 100-110 °C. By he received results it is possible to estimate the thermal conductivity EB ceramic coatings which on the average in 1.6 times is lower than at APS coatings. Thus the received results of experimental estimation of the thermal conductivity and decrease of wall temperature of heat-resistant materials after deposited TBC of ZrO_2 + 8%Y_2O_3 by thickness about 120 μm show that at gas-flame heating of models the investigated EB coating of columnar structure protects metal is better than the tested APS coating. The developed original method of the experimental determination of thermal conductivity and estimation of efficiency of the thermal protection of details with thermal barrier ceramic coatings at gas flame heating provides the reception of more exact data about thermophysic properties of ceramics under operating conditions of turbine details of aviation engines.

IOP Conf. Series: Journal of Physics: Conf. Series **899** (2017) 072002 doi:10.1088/1742-6596/899/7/072002

4. Conclusion

The gas flame heating was demonstrated to be effectively used during investigations of a thermal ceramic barrier coatings and thermal state of such gas-turbine engine parts with a TBC as the cooled turbine blades and vanes and combustion liner components. Gas-flame heating is considered to be preferable when investigating the gas-turbine engine parts with a TBC in the special cases when both the convective and radiant components of thermal flow are of great importance. The small-size rig with gas-flame flow made it possible to conduct the comparison investigations with the purpose of evaluating the efficiency of thermal protection of the ceramic deposited thermal barrier coatings on APS and EB techniques. The developed design-experiment method was introduced in bench tests of turbine blades and combustion liner components of gas turbine engines.

References

[1] Sotomonte SF, Pinzon CB and Vergara SG 2016 Growth of PEO ceramic coatings on AA 2024-T3 aluminium alloy *Journal of Physics: Conference Series* **687** 012037

[2] Sathyamoorthi S, Prabhakaran M and Muhammed SA 2016 Numerical investigation of ceramic coating on piston crown using Finite Element Analysis *International Journal of Scientific Engineering and Applied Science* (*IJSEAS*) **2**(4)

[3] Lima P, Bonarini A and Mataric M 2004 Application of Machine Learning *InTech* Vienna, Austria

[4] Li B, Xu Y and Choi J 2010 Applying Machine Learning Techniques *Proceedings of ASME 4th International Conference on Energy Sustainability* Phoenix, Arizona, USA 14-17

[5] Stringer J et al 1989 The Reactive Element Effect in High-Temperature Corrosion *High Temperature Corrosion* **2** 129-137

[6] Malashenko IS, Vashchilo NP, Belotserkovsky VA and Yakovchuk K.Y. 1997 Effect of Yttrium on Functional Characteristics of Vacuum Condensates and Protective Coatings MCrAlY and MCrAlY/ZrO2-8%Y2O3 at Thermocyclic Loading *Adv. Spec. Electrometali* **1** 24-33

[7] Miller RA, Garlick RG and Smialek JL 1983 Phase Distribution in Plasma-Sprayed Zirconia-Yttria *Am. Ceram. Soc. Bull.* **63**(12) 1355-1358

[8] Ingel RP, Lewis D, Bender BA and Rice RW 1984 Physical Microstructural and Thermomechanical Properties of ZrO2 Single Crystals *Advances in Ceramics* American Ceramics Society Inc. **12** 408-414

[9] Stecura S. 1986 Optimization of the Ni-Cr-Al-Y/ZrO2-Y2O3 Thermal Barrier System *Adv.Ceram. Mater.* **1** 68-76

[10] Ruckle DL and Duval DS 1984 Quench Cracked Ceramic Thermal Barrier Coatings U.S. *Patent* **4457948**

[11] Kuvaldin AB and Lepeshkin AR 2006 High speed Regimes of Heating and Thermal stresses in Articles *Novosibirsk*: *NGTU* 286

[12] Feng S, Lepeshkin AR et al 2016 Ceramic Coatings Applications in Engineering *Intech* 326

IOP Conf. Series: Journal of Physics: Conf. Series **899** (2017) 082001 doi:10.1088/1742-6596/899/8/082001

Simulation of partial discharges in cavities and streamers with high spatial resolution

D I Karpov[1], A L Kupershtokh[1], M B Meredova[1] and M V Zuev[1]

[1] Lavrentyev Institute of Hydrodynamics SB RAS, 15 Lavrentyev Prosp., Novosibirsk, 630090, Russia

E-mail: karpov@hydro.nsc.ru

Abstract. The electrical characteristics of partial discharges in cavities and streamer channels were simulated with the high spatial and temporal resolution. The parallel computations on graphic processing units were applied for the numerical solution of this system of equation. The current in an external circuit as well as the true charge during partial discharge were calculated for different positions of a single void and for two voids in electrode gap. The parallel algorithm for the stochastic growth of streamer in a dielectric was developed. The electric characteristics of the partial discharges due to the streamer growth were studied.

1. Introduction

Registration of the partial discharges (PD) activity in dielectrics subjected to the action of the high voltages is a one of the modern and effective methods of estimation of the reliability of high voltage equipments. The electrical strength of the air and other gases is by the order of magnitude lower than that of the condensed dielectric therefore PD occurs in cracks and in voids in solid dielectrics and in bubbles in liquids usually. There are two main types of the PD known [1,2]. The first type includes microdischarges in small cavities which always exist both on the electrode surface and in the bulk of the dielectrics. The second one corresponds to the PDs inside the channels of brunching structures (streamers) which grow in the bulk of the dielectrics or along the interface between two dielectrics.

To simulate the PDs of the first type, three models are known: the "equivalent circuit of capacitors", the "dipole model", and the model of the "complete calculation of the electric field inside the gap". The method of an equivalent circuit is still used in practical estimations of the isolation performance which is based on the consideration of an electrotechnical circuit consisting of discrete capacitors [3]. The drawbacks of this model are obvious, and the "dipole model" was proposed. The polarization of a cavity due to PD can be described with the model of the electric dipole roughly [4,5]. The electric field of this dipole changes the distribution of charges on the electrodes and the distribution of the electric field inside the gap. The properties of these two models are thoroughly discussed in the review [6]. The third model known at present is the model of the complete calculation of the electric field inside the gap according to the Poisson's equation which was realized qualitatively for the first time for a single cavity in solid dielectrics in [7]. The roles of the position of the cavity in an electrode gap and the size of the cavity in formation of the so called apparent Q_{app} and true charges Q_{true} of the PD were studied in [8]. The authors of this work used the static calculation of electric fields before and after the PD in order to estimate the value of $Q_{\rm app}$ and $Q_{\rm true}$. The conductive cavity was simulated with the high values of the dielectric permittivity ε .

The second type of the PD are the microdischarges inside the channels of branching structures (streamers), growing in the bulk of the dielectric or along the interface of two different dielectrics. The high electric field strength at the tip of the channels is provided by the transfer of electric charge along the channels from the electrode to the tips due to PDs that occur stochastically in different segments of the channels. The streamer development is accompanied by the flashes of light and the synchronous current pulses that are typical for the PD [9,10]. The model of the pulsed conductivity proposed in [11] allowed for the first time simulating this pulsed character of the currents. Most simulations with the models of stochastic growth of branching streamer structures [11-13] were performed on lattices with a rough spatial resolution because of the lack of the computational resources. In those simulations, the fine spatial structure of the streamer channels could not be represented with the real spatial resolution. Thus the electrical characteristics of the channels during the growth were described only qualitatively.

In the present work, we performed the simulation of the both types of the PDs in condensed dielectrics taking into account the dynamics of the electrical charges due to conductivity of the PD plasma. A computer programs were developed for parallel computations of these problems on very large spatial lattices. This allowed us simulating the PDs with very high resolutions in space and time.

2. The electrodynamic model

The electric field and the charge distribution in the gap between electrodes were calculated at each time step by solving the following system of equations. In the region occupied by dielectric

$$\mathrm{div}(\varepsilon \nabla \varphi) = 0 \ . \tag{1}$$

In the conducting regions of the electrode gap

$$\mathrm{div}(\varepsilon \nabla \varphi) = -4\pi\rho \ . \tag{2}$$

Here ε is the dielectric permittivity of a substance. For the conductive regions of the electrode gap (streamer channels or cavities) we used the equations

$$\frac{\partial \rho}{\partial t} = -\mathrm{div}\,\mathbf{j} \ , \qquad \mathbf{j} = \sigma \cdot \mathbf{E} \ , \qquad \mathbf{E} = -\nabla \varphi \ . \tag{3}$$

Here φ and $\mathbf{E}$ are the electric field potential and the electric strength, respectively, ρ is the electric charge density, σ is the average conductivity of plasma during PD, $\mathbf{j}$ is the current density.

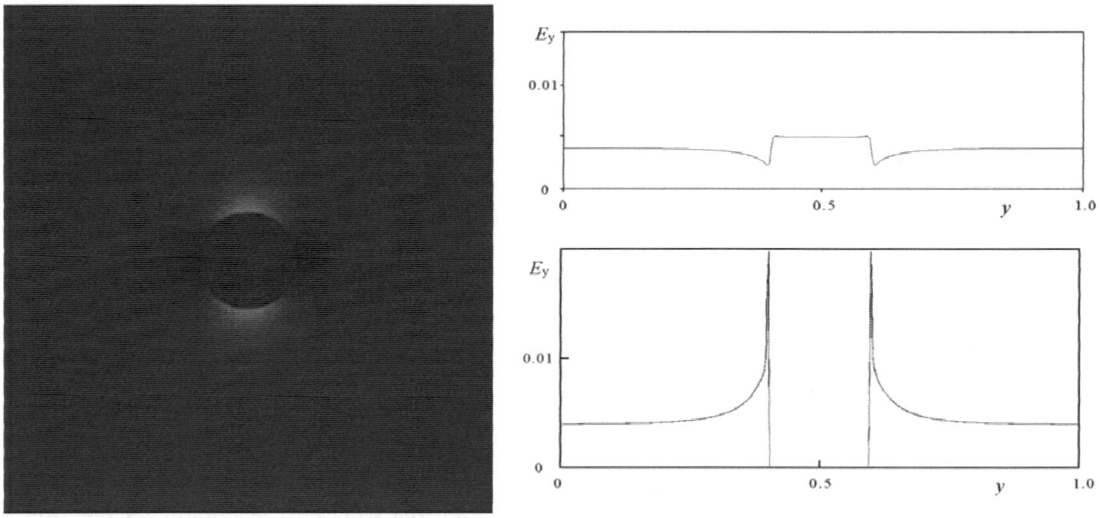

Figure 1. The field distribution $E_y(x,y)$ after the PD (left) and the plots $E_y(y)$ along symmetry axis before the PD (upper right) and after the PD (lower right) are shown. The lattice size is 256×256×256.

The duration of one PD is of the order of nanosecond. We neglected the injection of the charges to the dielectrics from the void-dielectric interface and the diffusion of the charges in the dielectric media on the time scales of several tens of nanoseconds. The boundary conditions were $\varphi = 1$ on the upper electrode and $\varphi = 0$ on the lower one. The values of φ changed linearly on the side surfaces of the simulation region. The calculation of the electric field distribution takes more than 90 per cent of the simulation time that imposes the restriction on the size of the lattice.

3. Field calculations with Graphic Processing Units

The conservative implicit in time finite-difference numerical method [12] was realized on NVIDIA Graphic Processing Units (GPU) for calculating the currents and the electric potential from the system (1) – (3). At each time step, the distribution of the potential was calculated by simple iterations. The CUDA programming technology with C language was used to implement the algorithm on GPU. The graphic card with 512 processor cores was used. Each lattice node was handled in its own thread. The blocks of 32 threads provided the maximum computing performance. The use of GPU accelerated the calculations by about 100 times. All the data were allocated in the fast global memory of GPU.

4. Simulations of partial discharges in voids with rigid walls

Two and three-dimensional calculations of the PD in one and two voids were performed. The spherical void with the rigid walls with the dielectric permittivity $\varepsilon = 1$ was placed at the different positions to the gap filled with the dielectric with $\varepsilon = 2$. In the first approximation, we did not take into account deformation of the bubble and the hydrodynamic processes related to the PD in it. The distribution of the vertical component of the electric field stress E_y in the central cross-section of the three-dimensional electrode gap is shown in figure 1 for the case of central positioning of one void as well as the plots of E_y along the symmetry axis before the PD and after the PD. The distances are reduced to the gap length d.

We used the Maxwellian relaxation time $\tau = \varepsilon / 4\pi\sigma$ as the scale of time. The magnitudes of current (as well as charges) can be calculated for the specific values of the applied voltage, the gap length d and the conductivity in clear way. Nevertheless, only the relative values of the charges are of importance in this study. Figure 2 shows the electric current in the external circuit and the true electric charge in the single bubble for the two cases. The bubble was placed on the electrode surface such as its center was in the central point of the electrode in the first case. In the second case, the position of the bubble was in the center of the electrode gap. The radiuses of the bubbles were the same and were equal to $R = 0.125d$. The PD current was of the order of the magnitude higher for the first case than for the second one although the difference in the values of the true charges was only about 15 per cent. We imply that true charge is the charge of the same sign that is generated in voids during the PD.

The magnitudes of PD charges were studied if the discharges occurred simultaneously in two voids. The voids were placed symmetrically at the same distances of $0.25d$ from the electrodes. The results of the simulations were compared with that for the single void of the same radius placed at the same distance of $0.25d$ from one of the electrode. The figure 3 shows the plots of the reduced current in the external circuit and the true charge reduced in one void and in two voids. The true charge Q_{true} and the charge registered in the external circuit Q_{app} were calculated at the different positions of the single void and for the two voids (table 1). The true charge per one void is represented in the last column of the table. The true charge in a void does not change noticeably with the position of the void. Nevertheless, the charge flowing in the external circuit increases significantly when the distance from the void to the nearest electrode becomes smaller. For two voids, the true charge per one void is practically the same as for the case of one void, but the value of Q_{app} increases considerably with comparison to the case of the single void. Nevertheless, the PD current increases only by about 25 per cent in this case (figure 3).

5. The model of streamer growth

We used the stochastic model of streamer growth developed earlier in order to simulate the PD in a streamer. The streamer is simulated with the conducting segments connecting neighbor nodes of the lattice. New elements of the streamer appeared with some probability that is a function of the local electric field. The probabilities of the appearance of the new segments were calculated using the following algorithm. Let us start from a node that belongs to the streamer structure. Find the value of the projection of the electric field E_i onto each of the possible direction to the neighbor nodes that belong to the dielectric state. If the Field Fluctuation Criterion (FFC) [11]

$$E_i > E_* - \delta_i , \tag{4}$$

was fulfilled then the segment between these nodes became the new conducting element.

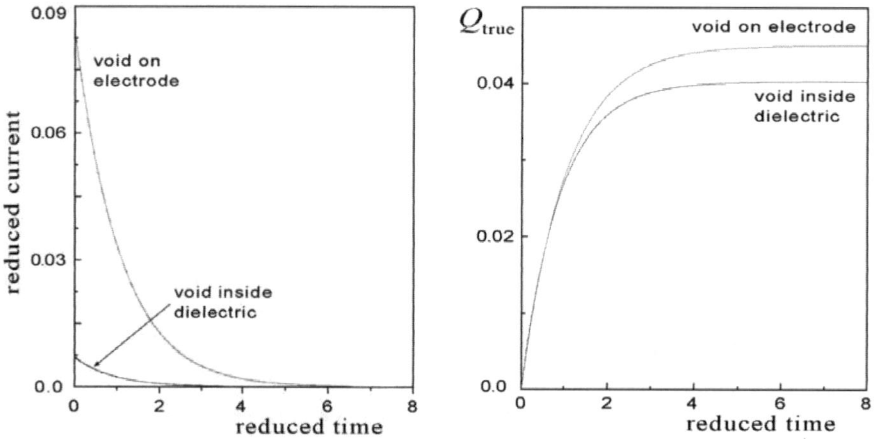

Figure 2. The electric current in a circuit (left) and the electric charge inside the cavity (right) for the positions of the single void on the electrode surface and in the center of the electrode gap.

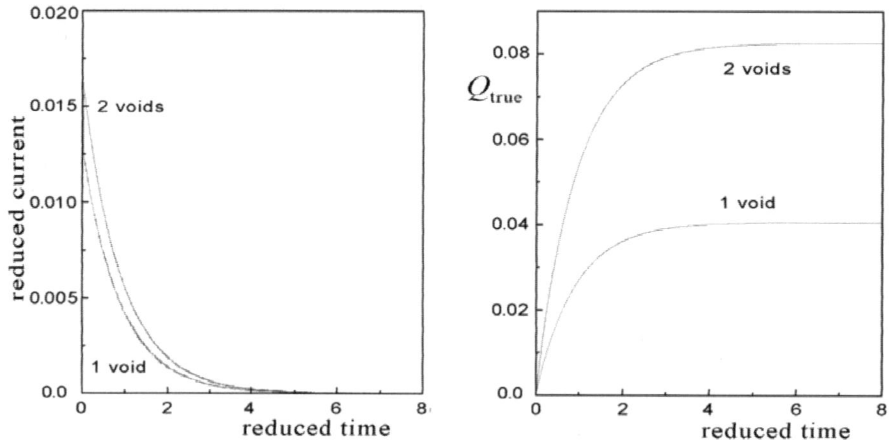

Figure 3. The electric current in a circuit (left) and the electric charge (right) in the two voids with comparison to the same characteristics for the single void.

Table 1. Calculated values of Q_{app} and Q_{true} at different position of the single void in the electrode gap and for the two symmetrically placed voids after single PD.

Positions of the centers of the voids	0.01 *d*	0.25 *d*	0.5 *d*	0.25 *d* and 0.75 *d*
Q_{app}	0.0224	0.00292	0.00157	0.0038
Q_{true}	0.045	0.0406	0.0403	0.0413

The distribution $\varphi(\delta) = \exp(-\delta / g) / g$ was used for the probabilities of the fluctuations δ that was equivalent to the random variable $\delta = -g\ln(\xi)$. Here, ξ is a random value uniformly distributed within the interval from 0 to 1. The parameter was $E_* = -g \cdot \ln(A \cdot \tau / h)$, where h is the lattice step, τ is the time step, A is the constant of the model related to the local velocity of streamer tip by

$$\upsilon(E) = A \cdot \exp(E / g). \quad (5)$$

Here g is the model parameter defining the rate of the velocity increase with the electric field stress.

6. The parallel algorithm for the growth of a branching streamer

The parallel algorithm was developed for the stochastic choice of the new segments of the streamer structure. The three-dimensional lattice was divided into the equal cubic blocks of sizes 4×4×4 lattice nodes. An indexing array $B[i]$ was additionally used with number of elements $N_{64} = N_x \times N_y \times N_z / 64$. Here N_x, N_y, and N_z are the sizes of the lattice. The zero value was assigned to the element $B[i]$ if there is no streamer nodes within the *i*-th block. The value of the *i*-th element was set to 1 if there is at least one of the ends of the segment belonging to the streamer structure inside the block. The treatment of the array $B[i]$ was carried out with the GPU. Each element was handled in one calculating thread. If $B[i] = 0$ the treatment of this block was cancelled. If $B[i] = 1$ the nodes of the lattice inside this block that belong to dielectric state were examined. If one of these nodes has a neighbor node that belongs to the streamer structure we checked if the FFC criterion (4) was fulfilled for these two nodes. If the condition (4) was true the dielectric node became the new site of streamer structure. The constant conductivity σ was assigned to this new segment.

7. PD current during initial stage of a streamer development

We simulated the streamer growth on the cubic lattice with the spatial spacing $h = 2.5$ μm. The calculations were performed on the lattice of the size of 386×386×386 nodes. We used the needle-plane electrode geometry with the needle tip radius of 2.5 μm and the distance from the tip to the plane electrode of 0.5 mm. The coefficient of the field non-uniformity was 33. The time step was $\tau = 10$ ps. The model parameters $g = 0.7$ MV/cm and $A = 1.4\ 10^{-7}$ m/s were obtained by the comparison of the simulations to the experimental data [14]. The dielectric permittivity was equal to 2 for the dielectric and 1 for plasma channel.

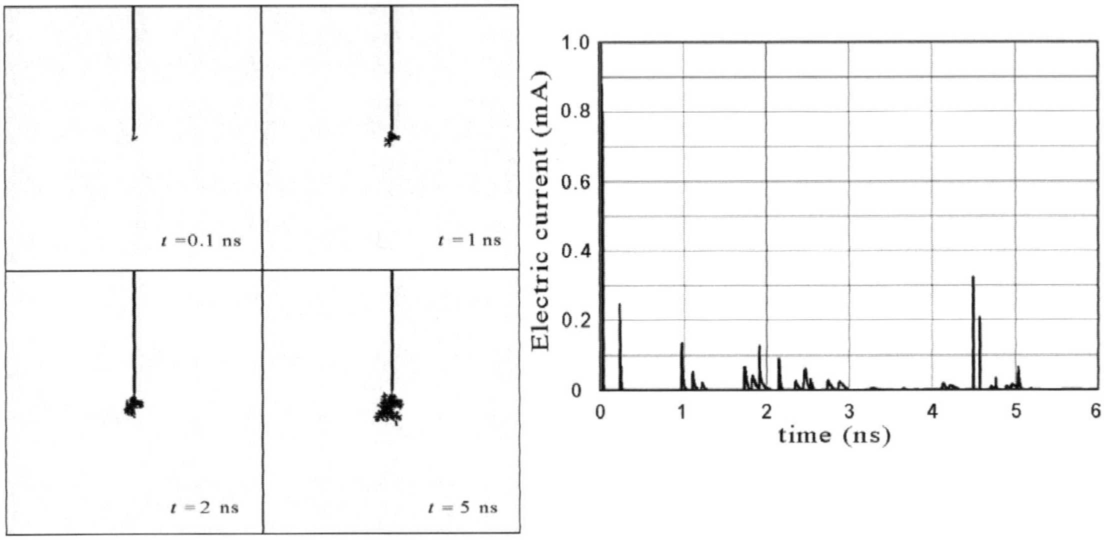

Figure 4. Initial stages of the growth of the streamer structure (left) and the pulses of the PD current accompanying the streamer growth in the external circuit. $V = 30$ kV.

The figure 4 shows the four shots of the streamer structure at the initial stage of the growth. The growth is stepwise. When the electric field is high enough is some region of the dielectric the new segments of the streamer channels grow in this region. Then the streamer stops until the electric field stress before the channel tip riches the value sufficient to the further growth due to charge relaxation in the channels. Each step of growth is accompanied by the current pulses of the magnitude that is significantly higher than the continuous component of the current during the stops. Figure 4 shows the PD current corresponding to the streamer growth. These magnitudes corresponds to the value of the conductivity of the channels averaged across the section $\sigma = 3.8$ S/m.

8. Conclusion

The simulations of the PDs in voids and in streamer channels in condensed dielectrics were performed with the use of parallel algorithms on GPU on large spatial lattices. Thus, the electrical characteristics of PDs were calculated more accurately than it was made before.

Acknowledgments
The study was supported by the Russian Scientific Foundation (grant No. 16-19-10229).

References

[1] Kreuger F H, Gulski E and Krivda A 1993 Classification of partial discharges *IEEE Transactions on Electrical Insulation* **28**(6) 917–31
[2] Niemeyer L 1995 A generalized approach to partial discharge modeling *IEEE Transactions on Dielectrics and Electrical Insulation* **2**(4) 510–28
[3] Gemant A, Von Philipoff W 1932 Die Funkenstrecke mit Vorkondensator *Zeitschrift für Technische Physik* **13**(9) 425–30
[4] Pedersen A 1986 Current pulses generated by discharges in voids in solid dielectrics. A field theoretical approach *IEEE Int. Symp. Electrical Insulation (IESI), 86CH2196-4-DEI*, pp 112.
[5] Pedersen A, Crichton G C and McAllister I W 1995 The functional relation between partial discharges and induced charge *IEEE Transactions on Dielectrics and Electrical Insulations* **2** 535–43
[6] Lemke E 2012 A critical review of partial-discharge models *IEEE Electrical Insulation Magazine* **28**(6) 11–6
[7] Wu K, Suzuoki Y and Dissado L A 2004 The contribution of discharge area variation to partial discharge patterns in disk-voids *J. Phys. D: Appl. Phys.* **37**(13) 1815–23
[8] Ovsyannikov A G and Korobeynikov S M 2016 "Apparent" and true charges of partial discharges *Proceedings of the 2016 IEEE International Conference on Dielectrics, ICD 2016, University of Montpellier: Montpellier, France.*
[9] Ushakov V Ya, Klimkin V F, Korobeynikov S M Impulse breakdown of liquids. – *Berlin: Springer-Verlag,* 2007
[10] Lesaint O, Saker A, Gournay P, Tobazéon R, Aubin J and Mailhot M 1998 Streamer propagation and breakdown under ac voltage in very large oil gaps *IEEE Transactions on Dielectrics and Electrical Insulations* **5**(3) 351–9.
[11] Kupershtokh A L 1992 Fluctuation model of the breakdown of liquid dielectrics Soviet *Technical Physics Letters* **18**(10) 647–49
[12] Karpov D I and Kupershtokh A L 1998 Models of streamers growth with "physical" time and fractal characteristics of streamer structure *Conference Record of the 1998 IEEE International Symposium on Electrical Insulation, IEEE No. 98CH36239 (Arlington, USA)* pp 607–10
[13] Kupershtokh A L and Karpov D I 2006 Simulation of the development of branching streamer structures in dielectric liquids with pulsed conductivity of channels *Technical Physics Letters* **32**(5) 406–9
[14] Lesaint O and Gournay P 1994 On the gaseous nature of positive filamentary streamers in hydrocarbon liquids. I and II *J. Phys D: Appl. Phys.* **27**(10) 2111–27

IOP Conf. Series: Journal of Physics: Conf. Series **899** (2017) 082002 doi:10.1088/1742-6596/899/8/082002

Investigation of charge dissipation in jet fuel in a dielectric fuel tank

E L Kitanin[1], P A Kravtsov[2], V A Trofimov[2], E E Kitanina[1] and D A Bondarenko[3]

[1] Peter the Great St. Petersburg Polytechnic University (SPbPU), Polytechnicheskaya st. 29, 195251, St. Petersburg, Russia
[2] St. Petersburg Nuclear Physics Institute (NRC "Kurchatov Institute" PNPI), Orlova roscha 1, 188300, Gatchina, Russia
[3] Moscow Aviation Institute (MAI), Volokolamskoe shosse 4, 125993, Moscow, Russia

E-mail: elkitanin@mail.ru

Abstract. The electrostatic charge dissipation process in jet fuel in a polypropylene tank was investigated experimentally. Groundable metallic terminals were installed in the tank walls to accelerate the dissipation process. Several sensors and an electrometer with a current measuring range from 10^{-11} to 10^{-3} A were specifically designed to study the dissipation rates. It was demonstrated that thanks to the sensors and the electrometer one can obtain reliable measurements of the dissipation rate and look at how it is influenced by the number and locations of the terminals. Conductivity of jet fuel and effective conductivity of the tank walls were investigated in addition. The experimental data agree well with the numerical simulation results obtained using COMSOL software package.

1. Introduction

The electrostatic fuel charging effect is extensively described both in theoretical [1] and practical [2, 3, 4, 5] terms. The relevance of dissipation studies is underpinned by broad use of polymers and composites in fuel tanks. This work focuses on experimental investigations of electrostatic charge dissipation in jet fuel in a tank made of a material which has much lower conductivity than fuel. In real environment, dissipation occurs as the charge escapes through the tank walls and through possible grounded points, such as the tank's external metallic elements exposed to environment. An experimental set-up was specifically designed to simulate the dissipation process in TS-1 jet fuel in a polypropylene tank by modeling the effects of possible grounded contacts on the dissipation rate. The principal dissipation rate evaluation method consisted in measuring the discharge current using probes immersed in fuel. For the method to be effectively used, we had to design and manufacture an electrometer capable of measuring currents as low as 10^{-11} A. An extensive series of experiments was conducted which provided a good insight into the impact of the grounded terminals on the dissipation rate and the effects of the resistance of the fuel surface layers bordering on the probes on the dissipation rate measurement accuracy. The charge sign was proved to have virtually no effect on the dissipation rate.

The dissipation process was simulated using COMSOL software package. Specific experiments were performed to measure the exact values of the fuel (TS-1) and tank wall conductivities required for the mathematical modeling.

The datasheet conductivity of TS-1 with no electrostatic agent added is $4 \cdot 10^{-12}$ S/m, which was confirmed by the measurement taken by the specifically designed instrument. The datasheet polypropylene tank conductivity value is from 10^{-15} to 10^{-14} S/m, whereas the effective conductivity measured in the experiment has a much lower value: $2.6 \cdot 10^{-16}$ S/m. These data were used in the dissipation process simulation with the aid of COMSOL software package. The simulated dissipation rates correlate well with the experimental values.

2. Experimental investigation

2.1. Experimental set-up

The experimental set-up diagram is given in figure 1. Fuel was either pumped from Tank 3 to Tank 1 or flowed from Tank 2 to Tank 1 due to the difference in levels.

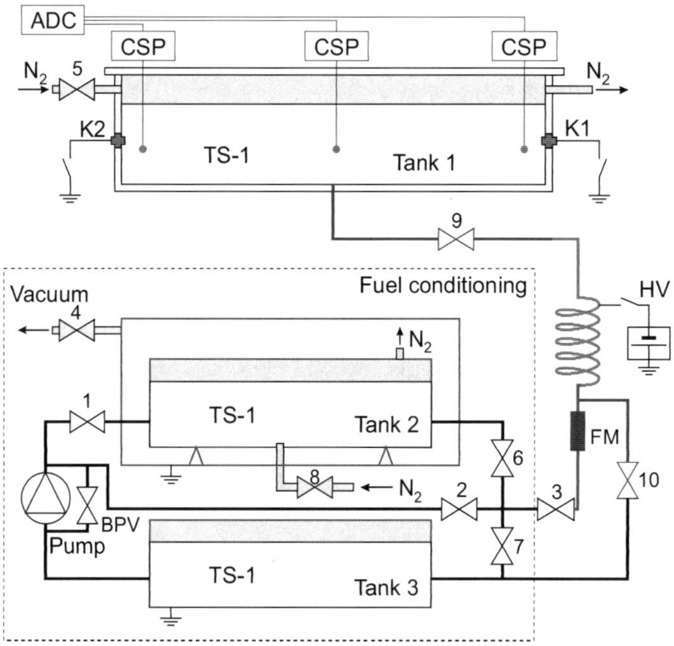

Figure 1. Experimental set-up block diagram: ADC – analog-digital converter block; CSP – charge-sensitive pre-amplifier; HV – high-voltage source; K1, K2 – ground terminals; FM – fuel flow meter; BPV– bypass pump valve.

The key element of the experimental set-up is a polypropylene tank with a volume V=0.48 m^3 (1200x800x500 mm) and a wall thickness of 10 mm (Tank 1). The tank has a polypropylene cover and the entire system is shielded by a grounded 3 mm thick steel sheet. The tank is fitted with eight groundable metallic contacts located on the walls and bottom and eight discharge current sensors. The cylinder-shaped current sensors are made of brass and have a diameter of 9 mm and a length of 10 mm. The sensors are connected to the discharge current meter (electrometer) by bronze conductors insulated from fuel and can be moved heightwise allowing the discharge current measurements to be taken at different distances from the tank bottom. The fuel column height was equal to 420 mm and the ullage was blown with nitrogen.

The fuel pipeline downstream of the valve 3 is made of polypropylene except for a copper section where high voltage is supplied to the fuel from an external voltage source. Groundable terminals (K1, K2, ..., K8) are installed in the tank walls to control the dissipation rate. Charged fuel is supplied through the valve 9 into the tank containing discharge current sensors and charge-sensitive pre-amplifiers (CSP). The CSP output signal is transmitted to the analog-digital converter and saved by the PC. The discharge current measurement system is described in more detail below. Once the experiment was over and the discharge current dropped by over 10 times, the fuel was discharged into Tank 3. The duration of the experiments varied from 1.5 to 25 hours depending on the number of grounded contacts. Besides, measurements were taken to register fuel temperature, nitrogen pressure in the ullage, fuel flow rate during tank filling, sensor depth and overall fuel level in the tank. A total of 11 series comprising 74 experiments were performed.

2.2. Discharge current measurement procedure

The charge-sensitive pre-amplifier converts the charge picked by the probe into a potential signal registered by the set-up's measurement system. The current running from the probe through the pre-amplifier should be much lower compared to typical currents in the system under study. In the pre-amplifier designed for the experiments and illustrated in figure 2, the measurement current was determined by the input bypass R_C (1 GOhm) and leakage current in the instrumentation amplifier INA116 ($3{\cdot}10^{-15}$ A). The amplifier's input circuits are provided with filters (R_FC_F) and gas dischargers R_A to ensure protection against high voltage.

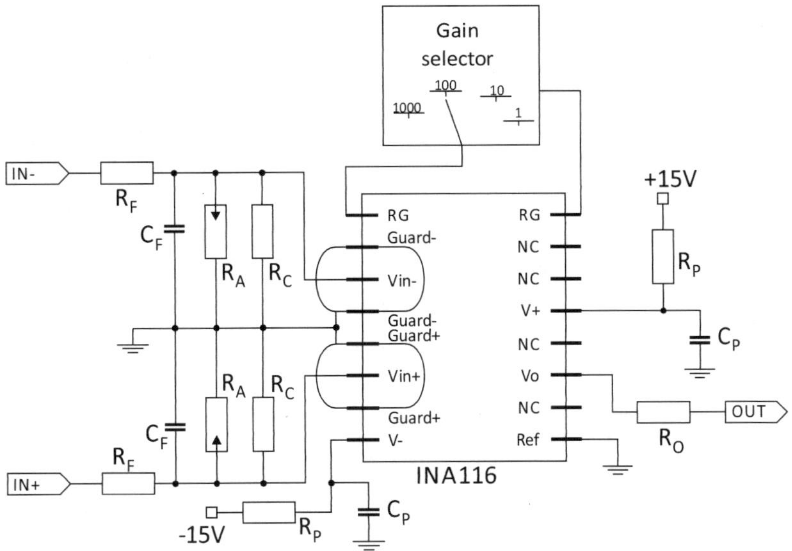

Figure 2. One channel of the charge-sensitive pre-amplifier.

Designed as a 9-channel instrument, the amplifier has sensitive input circuits attached to polycarbonate dielectric stands by point-to-point wiring. Each channel has a gain selector (1, 10, 100 and 1000). Each of the pre-amplifier's channels was calibrated using a specifically designed testbed. The calibration was performed in three input current ranges: 10^{-11} A, 10^{-10} A, and 10^{-9} A. Reciprocal influence of the different pre-amplifier channels was investigated in the course of calibration. The calibration curves demonstrated good linearity (0.1%) in all the input current and gain ranges.

2.3. Charge dissipation rate: experimental investigation results and their processing

Obtained as a result of the measurements was the dependency of the charge picked by the sensor from the time elapsed after the tank had been filled with fuel charged by a high-voltage source. The experimental data were processed by a dedicated software using ROOT software package [6]. A typical view of the dependency for one of the experiments is shown in figure 3.

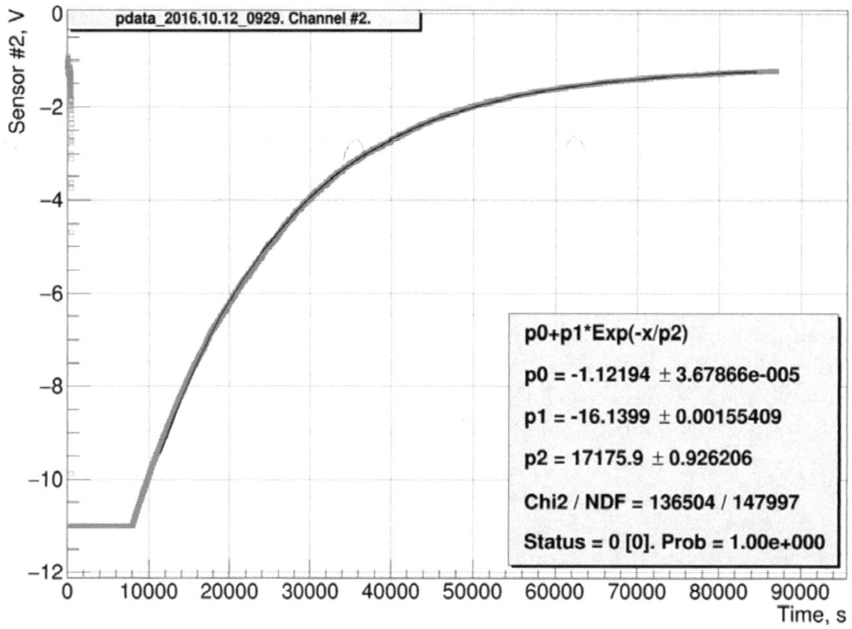

Figure 3. Charge dissipation in the experiment with a single sensor and no grounded contacts.

The problem of charge dissipation time profile in unlimited uniform medium was solved for a general case [1, 2]. Simultaneously solving the continuity equation and the field equations enabled describing the charge dissipation in the following form:

$$\rho(r,t) = \rho_0(r,0)\cdot e^{-t/\tau},\ \tau = \varepsilon\cdot\varepsilon_0/\sigma. \qquad (1)$$

Here $\rho(r,t)$ is charge density, $\rho_0(r,0)$ is the charge density at the initial time point, ε is the relative dielectric permittivity of the material, ε_0=8.8542·10^{-12} C^2/(N·m^2) is the absolute dielectric permittivity of vacuum, σ is the specific electric conductivity of the medium, *t* is time, and τ is the charge dissipation time.

In the experiment, the charge dissipates in a big although not infinite volume limited by the tank walls with electric conductivity σ_{wall} from 10^{-15} to 10^{-14} S/m. Therefore, the dissipation process was described using the same type of dependency (1) whereas the dissipation time was found by approximating the experimental data. Figure 3 depicts the results of the approximation which yielded a dissipation time of 17175.9±0.9 s in the absence of grounded terminals and with only one measurement probe used.

The probe has little effect on the dissipation rate for the walls with conductivity from 10^{-15} to 10^{-14} S/m due to high resistance of the fuel layer bordering on the small-sized probe. In the case of a ∅9 mm probe, the fuel resistance grows fast reaching 3.4 TOhm, which is comparable to the resistance of the walls with conductivity of 10^{-15} S/m. The experiments with one, two and eight probes enabled evaluating the effective electric conductivity of the tank's polypropylene walls: σ_{eff}=2.637·10^{-16} S/m is the level at which the probe has a noticeable effect on the dissipation rate. However, even such a strong effect can be neutralized by selecting a smaller probe. The fuel layer around the probe of a length of 5 mm and the diameter of 2.5 mm has resistance of 6.4 TOhm. Besides, if grounded terminals are connected, the dissipation rate is governed by their low resistance. Thus, the small probes are quite suitable for measuring dissipation rates in the experiment described here.

As suggested by figure 4a, increasing the number of terminals can significantly reduce the dissipation time. Moreover, measurements taken at different levels do not differ by more than 12%. Obviously, the charge is distributed over the fuel volume in a fairly uniform manner provided that σ_{fuel}>>σ_{wall}.

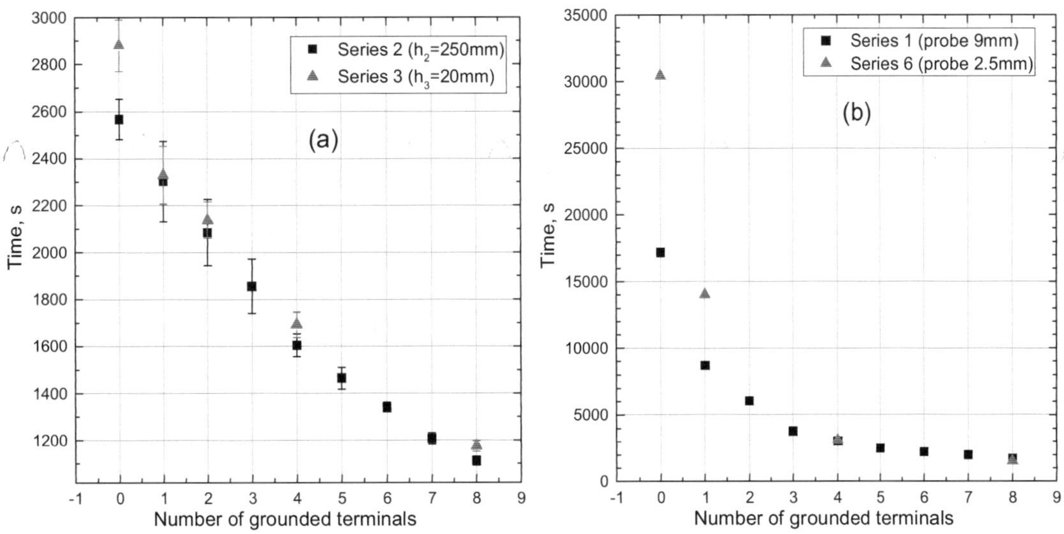

Figure 4. Effect of the number of grounded terminals on the dissipation rate. Two heights above the tank bottom: h_2=250 mm and h_3=20 mm (a). Two probes of different sizes (b).

In the absence of grounded terminals, the dissipation rates obtained from measurements with different probes differ significantly (figure 4b), because the charge dissipates both through the tank walls and the probes. With three or more grounded terminals, measurements taken with probes of different sizes yield virtually the same results.

3. Numerical simulation of the charge dissipation process in TS-1 in a polypropylene tank

The electrostatic charge dissipation process in TS-1 was simulated using COMSOL software package. The simulation model was designed so as to describe the experiment conditions as precisely as possible. The simulation model and the experiments alike used 8 cylindrical terminals placed on the tank walls and connectable to zero potential. Eight current probes were immersed in fuel. They could be connected to zero potential through the designated resistance. The tank's outer surface was grounded. The simulation model contained 48 charge volume density and potential measurement points corresponding to eight measurement probes placed at six different heights in the experiment.

The problem was solved in two steps. A simulation began with an electrostatic case with uniform charge volume density in fuel. The initial charge volume density was calculated based on the measurements of the current I taken during fuel charging, tank filling time τ and fuel volume V:

$$\rho_V = \frac{I \cdot \tau}{V} = \frac{2 \cdot 10^{-6} \cdot 1.2 \cdot 10^{3}}{0.5} = 4.8 \cdot 10^{-3}\ \text{C/m}^3.$$

The resulting potential distribution was used as the initial condition for solving the non-stationary dissipation case. The boundary conditions were the same as for the stationary case addressed at the first step: the tank's outer surfaces were grounded and the grounded terminals had the required configuration. Besides, the current probes' resistance was specified for the non-stationary case. Similarly to the experimental data, the numerical simulation results were saved to a text file and processed by a dedicated software using ROOT package [6].

Figure 5 compares the simulated dissipation rate to that obtained in the experimental series 2-5 with varied probe depths and external high voltage sign and an increase in the number of grounded terminals. The experimental points obtained in different experimental series are slightly shifted along the abscissa axis for the different series not to overlap. As indicated by figure 5, the differences in dissipation rates for all the series concerned lie within the same range (from 5 to 12%) as in the second series.

IOP Conf. Series: Journal of Physics: Conf. Series **899** (2017) 082002 doi:10.1088/1742-6596/899/8/082002

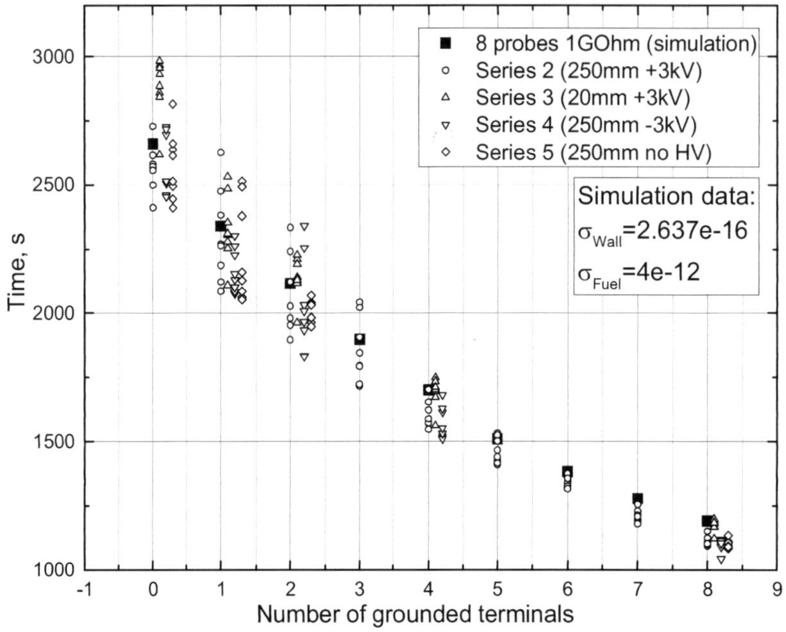

Figure 5. Comparison between series 2-5 and numerical simulation.

Thus the comparison shows that the experimental and simulated results differ by 12% maximum at room temperature. Smaller sensors display a bigger difference which is likely to be caused by the liquid moving in the experiment and unmoving in the simulation. It was noted in the experiments that the change in the fuel charge sign has little effect on the dissipation time. The location of a grounded terminal on the tank wall was demonstrated to have almost no effect on the dissipation rate.

4. Conclusion

Influence of the grounded metallic terminals on the charge dissipation process was investigated experimentally. Conductivity of jet fuel and effective conductivity of the tank walls were investigated in addition. It was shown that charge dissipation rate could be reliably measured using the charge-sensitive pre-amplifier. The experiments show that charge dissipation time exponentially decrease with increasing of the number of grounded terminals. The experimental data agree well with the numerical simulation results obtained using COMSOL software package.

References

[1] Meshkov I and Chirkov B 1987 *Electromagnetic Field. Part 1. Electricity and Magnetism* (Novosibirsk: Nauka)
[2] Dubovkin N et al. 1985 *Physical-Chemical and Operational Properties of Jet Fuels, Handbook* (Moscow: Khimiya) p 240
[3] Bard V and Kuzin A 1984 *Prevention of Accidents in Oil Processing and Petrochemical Production* (Moscow: Khimiya) p 248
[4] Grosu F, Bologa M *et al.* 2007 *Surface Engineering and Applied Electrochemistry* vol 5 N 5 (Allerton Press) p 16
[5] Mirotvorsky V and Stishkov I 1998 *Annual Report Conference on Electrical Insulation and Dielectric Phenomena (Atlanta)* vol 2 (IEEE Service Center) p 428
[6] ROOT *Data Analysis Framework* (https://root.cern.ch/)

IOP Conf. Series: Journal of Physics: Conf. Series **899** (2017) 082003 doi:10.1088/1742-6596/899/8/082003

Dynamics of bubbles in electric field

S M Korobeynikov[1,2], A G Ovsyannikov[2], A V Ridel[1,2] and D A Medvedev[1]

[1] Lavrentyev Institute of Hydrodynamics SB RAS, 15 Lavrentyev Prosp., Novosibirsk, 630090, Russia
[2] Novosibirsk State Technical University, 20 Karl Marx avenue, Novosibirsk, 630073, Russia

E-mail: ridel@ngs.ru

Abstract. We consider main topic related to the behavior of rising air bubbles in transformer oil under the action of alternating electric field. The data on the degree of deformation of bubbles are provided as well as the description of the experimental apparatus. The preliminary study of the development of partial discharges in rising helium bubbles in transformer oil under the action of alternating voltage is reported. Results of mesoscopic simulation of the dynamics of a bubble are presented.

1. Introduction

Partial discharges in transformer oil are known to be one of the causes of the gradual destruction of paper-oil insulation. Bubbles in the transformer which are formed due to different reasons including non-electric ones can be transported by the cooling system to the region of strong electric field. There, they can lead to the inception of partial discharges (PDs), and they also can initiate the breakdown of insulation. Mechanisms of inception and development of partial discharges in gaseous inclusions in solid insulation were investigated earlier [1], whereas the mechanisms of partial discharges in a liquid are still poorly investigated. Therefore, the goal of this work is the investigation of the behavior of gas bubbles in transformer oil under the action of strong alternating electric field.

2. Experimental setup

Experiments were carried out in a high-voltage box (figure 1) where the high-voltage transformer (1), the linking capacitor (2), the electric and optical registration systems, and the experimental apparatus (3) were placed.

Figure 1. High-voltage box.

The experimental apparatus (figure 2) was assembled on the optical bench (1) in order to ensure the collinearity of the whole registration system. All elements were mounted on optical carriages (2) which allowed one to provide the exact tuning of the optical registration system based on the high-speed camera Casio EX-F1 (6). Optical magnification of the object was provided by the lens system (5). The laser pointer (3) was used as the light source. The camera and the laser pointer were placed collinearly with the experimental cell (4) in order to avoid distortions of the image of the object under investigation.

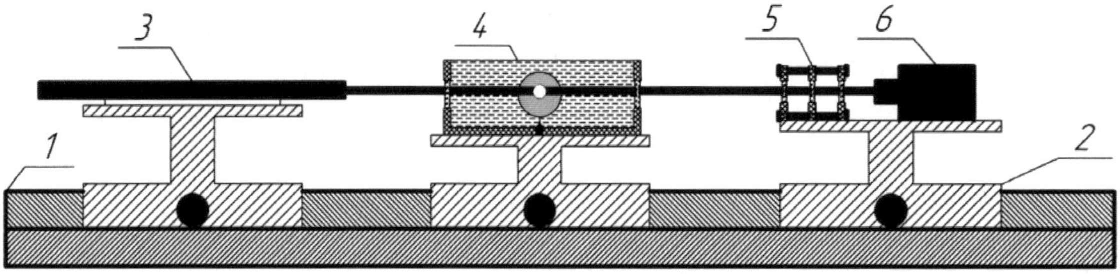

Figure 2. Experimental apparatus.

Experimental cell (figure 3) consisted of the case (1) made of PMMA, with windows (2) made of optical glass, two plane electrodes (3), and the system (4) for the injection of gas into the bulk of transformer oil. The interelectrode gap was equal to 6.8 mm.

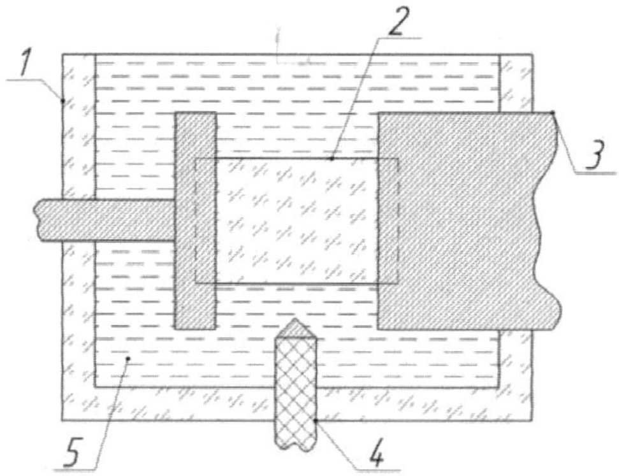

Figure 3. Sketch of the experimental cell.

3. Results

When attempting to inject gas to the region between electrodes, we observed the effect of the dielectrophoretic force which pushed the bubble out of the region of strong electric field. The rise of a bubble stopped at the boundary of electrodes, and later the bubble rose along the boundary of higher electric field magnitude. This problem was solved by the immediate injection of a bubble into the electrode region.

We observed in experiments the deformation of air bubbles under the action of alternating electric field. Figure 4 shows the deformation at the electric field strength (amplitude value) 4.4 kV/mm. The deformation proceeded with double frequency and without PD in accordance with the absolute value of the instant electric field magnitude. For example, a bubble of the diameter of 1.24 mm (figure 4*a*) elongated to 1.8 mm at the moment of maximum voltage (figure 4*d*). The experimentally obtained rise velocity of the bubble of this size was $V \sim 12$ cm/s.

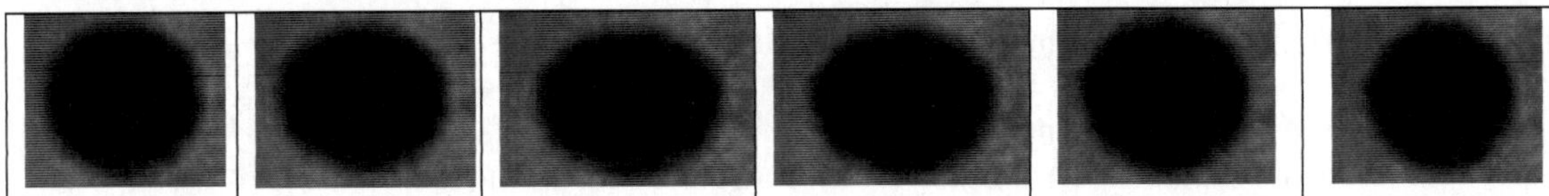

a	b	c	d	e	f

Figure 4. Deformation of the air bubble under the action of alternating voltage.

In the investigation of the development of partial discharges in rising bubbles, helium was used as a filling gas. There are several reasons for this choice. First, the electric strength of helium is lower, hence, the partial discharge in a bubble occurs at lower magnitude of electric field. Second, the numerical simulation of the development of partial discharge is simpler in the case of individual gas than for a mixture of gases such as air. Figure 5 shows the development of PD in the helium bubble of the diameter of 1.57 mm at the magnitude of electric field equal to 1.8 kV/mm (amplitude value).

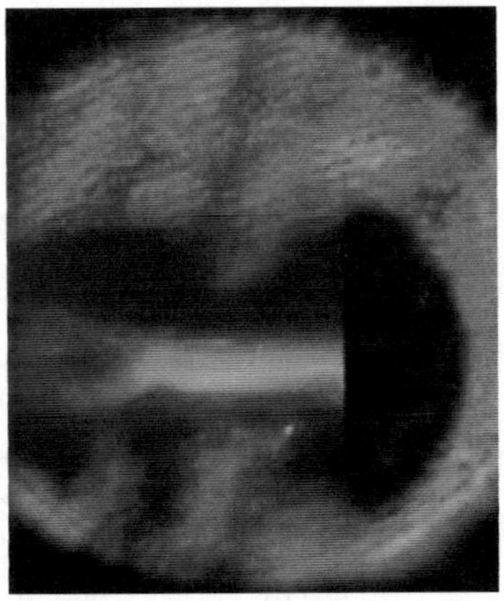

Figure 5. Partial discharge in the helium bubble.

4. Simulations of the dynamics of bubble

We investigated the dynamics of a gas-vapor bubble in a dielectric liquid placed in strong electric field. Lattice Boltzmann method (LBM) with possible phase transitions liquid-vapor and simulation of internal energy transfer, pressure work and latent heat of evaporation [2] was used to model the flow of dielectric fluid.

Electric potential φ is calculated from the Poisson's equation

$$\nabla \cdot (\varepsilon \nabla \varphi) = -q / \varepsilon_0.$$

Here, q is the electric charge density. This equation is solved numerically together with the equation of the electric charge transport

$$\partial q / \partial t + \nabla \cdot (q\mathbf{u}) = \nabla \cdot (\sigma \nabla \varphi)$$

using the time-implicit finite-difference scheme. Here, σ is the electric conductivity, and the electric current density is $\mathbf{j} = \sigma \mathbf{E} = -\sigma \nabla \varphi$. The advection and diffusion of electric charge is simulated using a second set of lattice Boltzmann distribution function (passive scalar) [3]. The local density of Joule heating $h = \mathbf{j} \cdot \mathbf{E} = \sigma E^2$ is added to the heat source term in the LBM.

The Helmholtz force acting on a charged fluid in an electric field is expressed by

$$\mathbf{F} = q\mathbf{E} - \frac{\varepsilon_0 E^2}{2} \nabla \varepsilon + \frac{\varepsilon_0}{2} \nabla \left[E^2 \rho \left(\frac{\partial \varepsilon}{\partial \rho} \right)_T \right].$$

The second term is the force acting on inhomogeneous dielectrics, the third term corresponds to the electrostriction force. This force is taken into account in LBM together with the interparticle forces.

All quantities are expressed in non-dimensional values (scaled by values in the critical point).

A vapor bubble was initially placed in the center of the calculation region. The electric field was applied in vertical direction, van der Waals equation of state was used for the fluid with the initial temperature $T = 0.8$. The Clausius–Mossotti formula for the electric permittivity was used with the permittivity of the initial liquid equal to 2. Boundary conditions were periodic in horizontal directions both for the flow and electric potential, and no-slip rigid walls with fixed electric potential at the top and bottom. The system was equilibrated during 5000 time steps before the electric field was applied.

We assumed that the matter can become conductive if the fluid density is sufficiently low $(\rho < 1.1\rho_g)$, and the magnitude of electric field is high enough $\left(E > 1.05\langle E \rangle\right)$. We simulated the evolution of a bubble in the electric field with the average magnitude $\langle E \rangle = 0.6$.

Distribution of fluid density and temperature is shown in figure 6 for two different time moments. Conductivity arises in the central part of the bubble leading to the heating of fluid and the enhanced evaporation. Later, the accumulated charge is advected by moving fluids which results in the injection of charge into liquid near the poles of the bubble. The motion of charge also produces the electric current in outer circuit.

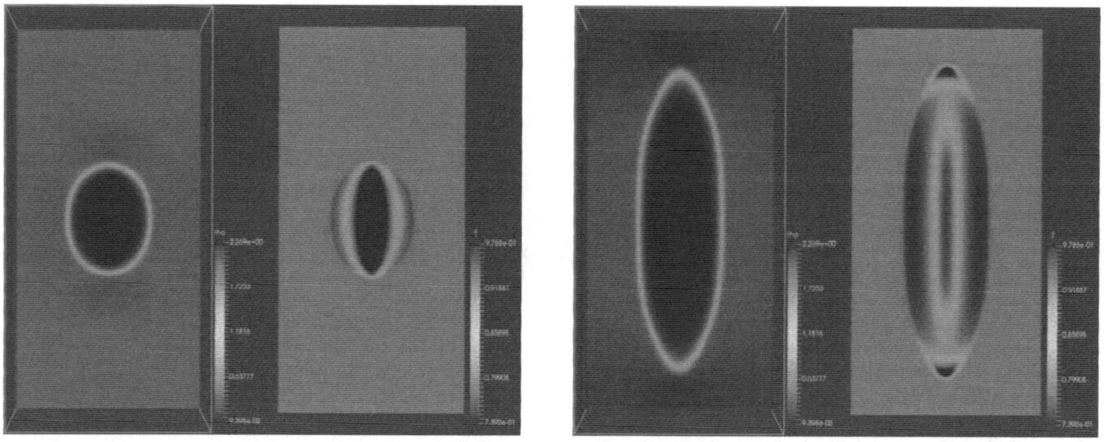

Figure 6. Distribution of density and temperature inside the bubble with conductivity. Time $t = 400$ (left) and $t = 1000$ (right).

5. Discussion

Using the optical registration, we found that the inception of a PD inside bubbles always results in the breakup of them. In our opinion, this can be connected with the fact that after the generation of the discharge plasma inside a bubble during the development of a PD, electric charges in plasma are separated, positive charges move to the negative electrode, negative charges towards the positive one. Then electric charges are deposited at the walls on the bubble. The Coulomb force acts at the walls leading to the sharp elongation of the bubble (by 2 times and more), and the instability of the elongated bubble leads to its breakup.

Note that PDs in rising bubbles do not correspond to the Paschen's curve due to the lack of initiating electrons inside the bubble. The value of $P{\cdot}d$ where P is the pressure inside the bubble, d is the bubble diameter is about 150 Pa·m. For helium, the voltage necessary for the breakdown of a bubble according to the Paschen's law is about 1.5 kV. The magnitude of electric field inside the bubble is about 1.2 times larger than the average field magnitude in the cell, then the average magnitude should be near 0.8 kV/mm, and the voltage necessary for the inception of PD should be 5–6 kV. This value is nearly twice lower than the voltage of PD inception observed experimentally.

Mesoscopic simulation showed that after the application of voltage, the bubble elongates, and the density inside decreases. The voltage drop along the bubble increases leading to higher probability of a breakdown. The heating of fluid, enhanced evaporation and injection of electric charge into the liquid near the poles are observed.

Acknowledgments
The study was supported by the Russian Scientific Foundation (grant No. 16-19-10229).

References

[1] Bartnicas R 2002 Partial discharges, their mechanism, detection and measurement *IEEE Trans. Dielectr. Electr. Insul.* **9**(5) 763–808
[2] Kupershtokh A L, Medvedev D A and Gribanov I I 2014 Modeling of thermal flows in a medium with phase transitions using the lattice Boltzmann method *Numerical Methods and Programming* **15** 317–28 (in Russian)
[3] Kupershtokh A L and Medvedev D A 2006 Lattice Boltzmann equation method in electrohydrodynamic problems *J. Electrostat.* **64**(7/9) 581–5

IOP Conf. Series: Journal of Physics: Conf. Series **899** (2017) 082004 doi:10.1088/1742-6596/899/8/082004

"Relay-race" mechanism of propagation of partial discharges in condensed dielectrics at linearly increasing voltage

A L Kupershtokh[1] and D I Karpov[1]

[1] Lavrentyev Institute of Hydrodynamics SB RAS, 15 Lavrentyev Prosp., Novosibirsk, 630090, Russia

E-mail: skn@hydro.nsc.ru

Abstract. A stochastic model of partial discharges (PDs) inside gas cavities in condensed dielectrics is developed. The equations for electric field potential and electric charge transfer are solved together for dielectric with gas voids. Computer simulations show the possibility of a "relay-race" mechanism of propagation of partial discharges in a linear chain of gas cavities. This mechanism can be realized if the spacing between cavities is relatively small and the dependence of probability function of partial discharges on electric field strength $r(E)$ is relatively sharp. In this case, the wave of partial discharges propagates along the chain of cavities. The PD waves can be initiated in the first cavity or in the last cavity in the chain. Occasionally, two PD waves can be observed together from the both ends of the chain. The sequence of partial discharges in the inclusions has a completely stochastic character at a relatively weak dependence $r(E)$ or if the mutual influence of cavities on each other is weak.

1. Introduction

When the high voltage is applied to an interelectrode gap filled with a dielectrics, the partial discharges (PDs) may occur that do not lead to the loss of the insulating properties of the dielectric as a whole. At this time, the pulses of the electric current are registered in the external circuit, the number and intensity of which can predict the probability of a breakdown of the gap. One type of partial discharges in condensed dielectrics is the electrical discharges inside the small gas inclusions (voids or bubbles) in solid or liquid dielectrics. One of the first attempts to simulate the stochastic nature of PDs using an equivalent electric circuit of a void was carried out in [1]. Later, this approach was developed in [2-4]. However, these studies did not take into account the spatio-temporal evolution of the electric field in the gap. The works [5-8] are devoted to the computer simulations of partial discharges in liquid and solid dielectrics with calculations of electric field distribution at every time step.

In present paper, the partial discharges in condensed dielectrics between two flat electrodes are studied. Gas inside the small cavities in solid dielectrics or inside the microbubbles in a liquid has an electric strength which is much lower than that of the condensed matter. Two-dimensional computer simulations of partial discharges in a chain of such gas inclusions located in condensed dielectric are carried out. The inclusions are located along the electric field line at the same close distance from each other.

The possibility of the wave of partial discharges in the chain of gas inclusions is demonstrated. In this case, the partial discharges occur sequentially one by one along the chain of insulation defects.

IOP Conf. Series: Journal of Physics: Conf. Series **899** (2017) 082004 doi:10.1088/1742-6596/899/8/082004

This interesting phenomenon can be named as a “relay-race” mechanism of propagation of partial discharges in the chain. This mechanism differs from “hopping spread streamers” in the medium with bubbles simulated in [9]. For "relay-race" mechanism, the streamers do not occur in the condensed phase but occur only inside the gas filled cavities. Moreover, it is most important that the “relay-race” mechanism of propagation can be realized even if the stochastic nature of the phenomenon is taken into account.

2. Partial discharges in a chain of gas inclusions spaced in condensed dielectric

At the same dimensions of these cavities in the bulk of the dielectric and at the same gas pressure inside them, the probability of micro-breakdowns inside the inclusions depends on the local electric field within them E_i.

The stochastic criterion MESTL (multi-element stochastic time lag) [10, 11] is used to describe the occurrence of micro-discharges in cavities. For all nonconducting cavities, the stochastic lag times t_i are calculated in accordance with the density distribution function for the probability of rare events

$$F(t_i) = r(E_i)\exp(-r(E_i)t_i). \quad (1)$$

Here, $r(E)$ is the sharply increasing function of the electric field. For small time step $\Delta t << 1/r(E)$, the probability of a micro-breakdown in a cavity is approximately equal to $f \approx r(E)\Delta t$. It is assumed that the micro-breakdown in a cavity occurs during the current time step if the condition $t_i < \Delta t$ is satisfied.

The function $r(E)$ on the local electric field inside the gas inclusion that describes the threshold character of partial discharges has the form [5]

$$r(E) = \begin{cases} 0 & \text{for} \quad E \le E_*, \\ \alpha(E - E_*) & \text{for} \quad E > E_*. \end{cases} \quad (2)$$

Here, E_* is the threshold electric field above which partial discharges in the gas inclusions are possible. The breakdown voltage of air in a void of size $d \sim 10$ microns is approximately equal to 350 V [12], which corresponds to the threshold field $E_* \approx 350$ kV/cm. The functions $r(E)$ for different values of α are shown in figure 1*b*.

A chain of several identical gas inclusions spaced along the electric field lines in condensed dielectric between two flat electrodes is studied (figure 1*a*). The spaces between all neighbor inclusions are equal to Δy.

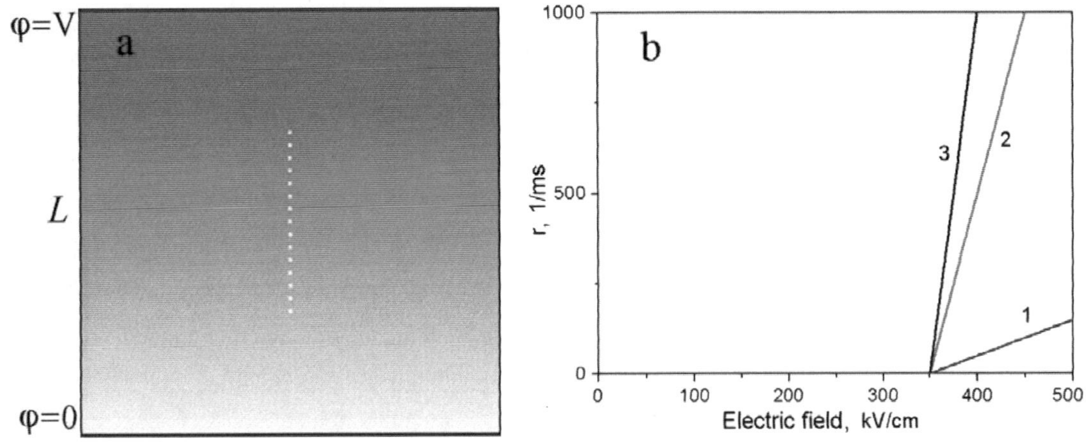

Figure 1. (a) Chain of gas cavities in the solid dielectric. $L =$ 1 mm. Lattice size 200×200. $\Delta y = 6h$. (b) Probability function $r(E)$ for PD in a cavity. Curves 1, 2 and 3 correspond to $\alpha =$ 1, 10 and 20 cm/(kV·ms).

The distribution of the electric field potential in the whole region between the flat electrodes (figure 1*a*) is calculated numerically at each time step by solving the Poisson's equation together for the electric field potential with the equations of the electric charge transfer inside inclusions during the partial discharges [6, 7]

$$\mathrm{div}(\varepsilon\varepsilon_0 \nabla\varphi) = -q, \tag{3}$$

$$\frac{\partial q}{\partial t} = -\mathrm{div}\,\mathbf{j}, \tag{4}$$

Here, ε is the permittivity of fluid, q is the electric charge density. The distribution of electric field strength is defined by the equation $\mathbf{E} = -\nabla\varphi$. It is assumed that the conductivity σ and the current density $\mathbf{j} = \sigma \cdot \mathbf{E}$ are nonzero only within gas inclusions during the partial discharges. The time-implicit finite-difference scheme is used to solve this system of equations [7, 10]. The computational lattice spacing is h. The periodic boundary conditions are used in x direction. The permittivity of the condensed dielectrics is assumed to be equal to $\varepsilon = 2$.

The values of electric field strength inside gas inclusions are greater than the electric field strength in condensed dielectric but are lower than the one inside a single isolated spherical cavity $E_0 = 3E\varepsilon/(2\varepsilon+1)$.

For linearly increasing voltage $V = \gamma t$ applied to the electrodes, the values of the electric field inside the first and the last gas inclusions in the chain become somewhat greater than the threshold value E_* at a certain moment of time (figure 2). Then, the probabilities of micro-discharge f during time step Δt inside these gas inclusions become noticeable. The rate of voltage rise is assumed to be equal to $\gamma = 10$ kV/ms.

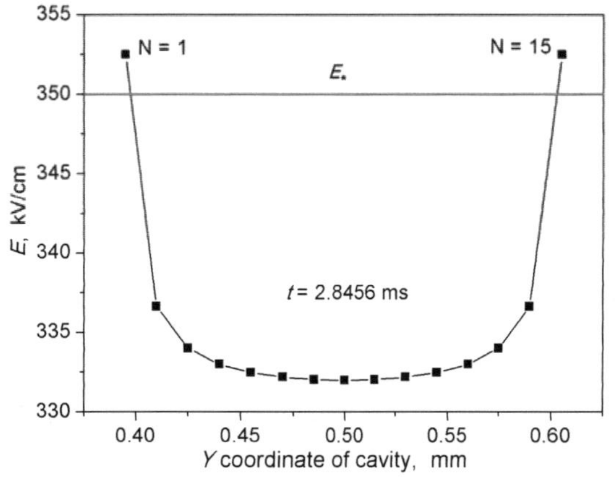

Figure 2. The values of electric field inside all gas inclusions in the chain just before the first partial discharge ($t = 2.8456$ ms). $L = 1$ mm. $E_* = 350$ kV/cm. $\Delta y = 3h$. $h = 5$ μm. $\alpha = 20$ cm/(kV·ms). Lattice size 200×200.

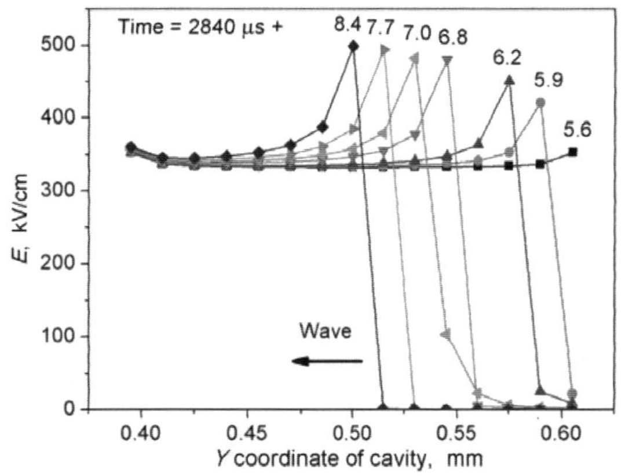

Figure 3. The electric field values inside the cavities as the wave of PDs propagates along the chain. $L = 1$ mm. $E_* = 350$ kV/cm. $\Delta y = 3h$. $h = 5$ μm. $\alpha = 20$ cm/(kV·ms). Lattice size 200×200.

IOP Conf. Series: Journal of Physics: Conf. Series **899** (2017) 082004 doi:10.1088/1742-6596/899/8/082004

One example of simulations is shown in figure 3. After the micro-discharge in the last gas inclusion ($N = 15$), this gas cavity becomes conductive, and the value of electric field in the neighbor gas cavity ($N = 14$) becomes considerably greater than the threshold value E_*. Hence, the probability of electrical breakdown in this gas inclusion increases significantly, and after short time, the micro-discharge occurs inside this gas cavity. Then, the process repeats for the next neighbor gas cavity ($N = 13$) and so on. As a result, the wave of partial discharges propagates along the chain of inclusions (figures 3 and 4). This interesting phenomenon can be named as a "relay-race" mechanism of propagation of partial discharges in the chain of insulation defects.

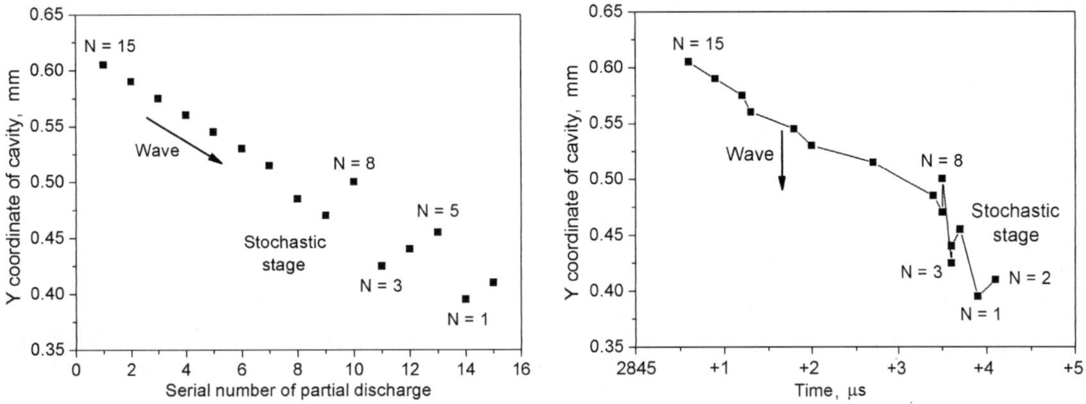

Figure 4. (a) The sequence of partial discharges in cavities. Wave of PDs and stochastic stage. (b) The sequence of partial discharges in time. $L = 1$ mm. $E_* = 350$ kV/cm. $\Delta y = 3h$. $h = 5$ μm. $\alpha = 20$ cm/(kV·ms). Lattice size 200×200.

The sequence of partial discharges becomes stochastic (figure 4) after the moment when the values of electric field strength inside the interior cavities become greater than the critical value (figure 5). As the wave of partial discharges propagates along the chain of cavities, the electric field increases inside the remaining cavities.

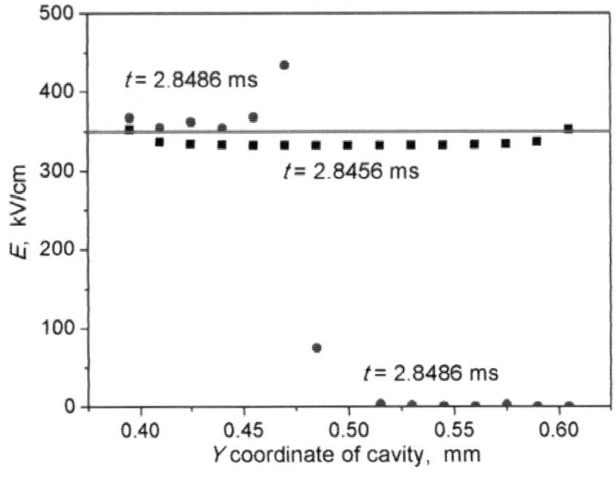

Figure 5. The values of electric field inside gas inclusions before the first PD (▪) and during the stochastic stage (●). $L = 1$ mm. $E_* = 350$ kV/cm. $\Delta y = 3h$. $h = 5$ μm. $\alpha = 20$ cm/(kV·ms). Lattice size 200×200.

The "relay-race" mechanism of propagation of partial discharges can be realized if the dependence $r(E)$ is sharp enough and the spaces between the cavities Δy are relatively small. For relatively small spacing Δy, the difference in values of electric field inside the cavities at the ends of the chain and in the central region of the chain is greater than for large distances (figure 6). After the micro-discharge inside a cavity, the electric field in the neighbor cavity increases significantly and becomes considerably greater than the threshold value E_*. The maximal values of electric field ahead of the

IOP Conf. Series: Journal of Physics: Conf. Series **899** (2017) 082004 doi:10.1088/1742-6596/899/8/082004

front of the waves of PDs at different spacing between cavities Δy are shown in figure 7. For relatively small spacing Δy, the enhancing of electric field in the neighbor cavity is more pronounced.

The PD waves can be initiated as at first cavity and at last cavity in the chain (figures 6 and 7). Occasionally, two PD waves can be observed together from the both ends of the chain (figure 7).

The sequence of partial discharges in the chain of cavities has a stochastic character at a relatively weak dependence $r(E)$ or if the spacing Δy is relatively large and the mutual influence of cavities on each other is weak. The stochastic regime of partial discharges in the chain of cavities at $\Delta y = 7h$ is shown in figure 8a. In this case, the mutual influence of cavities on each other is weak, and the values of electric field inside cavities have approximately the same values (figure 8b).

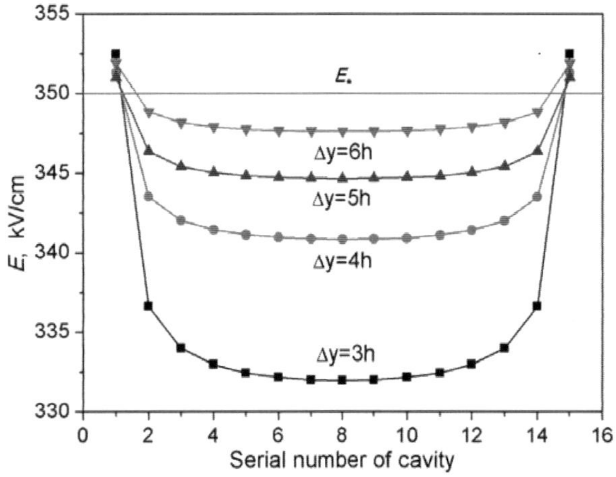

Figure 6. The values of electric field inside gas inclusions before the first PD in the chain at different spacing between cavities Δy. $L = 1$ mm. $E_* = 350$ kV/cm. $h = 5$ µm. $\alpha = 20$ cm/(kV·ms). Lattice size 200×200.

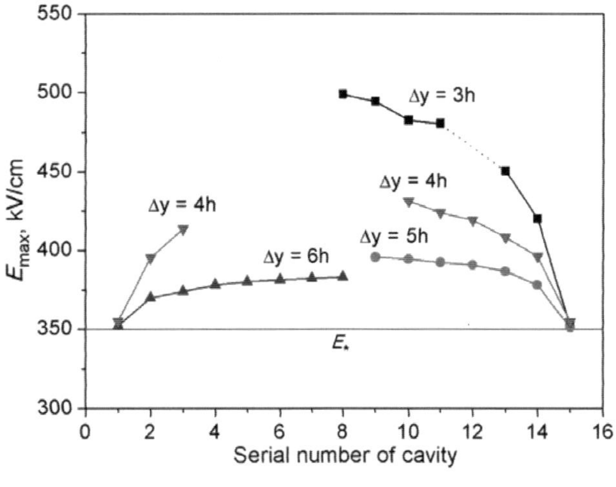

Figure 7. The maximal values of electric field ahead of the front of the waves of PD at different spacing between cavities Δy. $L = 1$ mm. $E_* = 350$ kV/cm. $h = 5$ µm. $\alpha = 20$ cm/(kV·ms). Lattice size 200×200.

3. Conclusion

The "relay-race" mechanism of propagation of partial discharges in a linear chain of gas inclusions at linearly increasing voltage is possible if the spacing between cavities Δy is relatively small. In this case, the wave of partial discharges propagates along the chain of cavities. The PD waves can be initiated as in the first cavity and in the last cavity in the chain. Occasionally, two PD waves can be observed together from the both ends of the chain. The sequence of partial discharges in the inclusions has a completely stochastic character at a relatively weak dependence $r(E)$ or if the mutual influence of cavities on each other is weak.

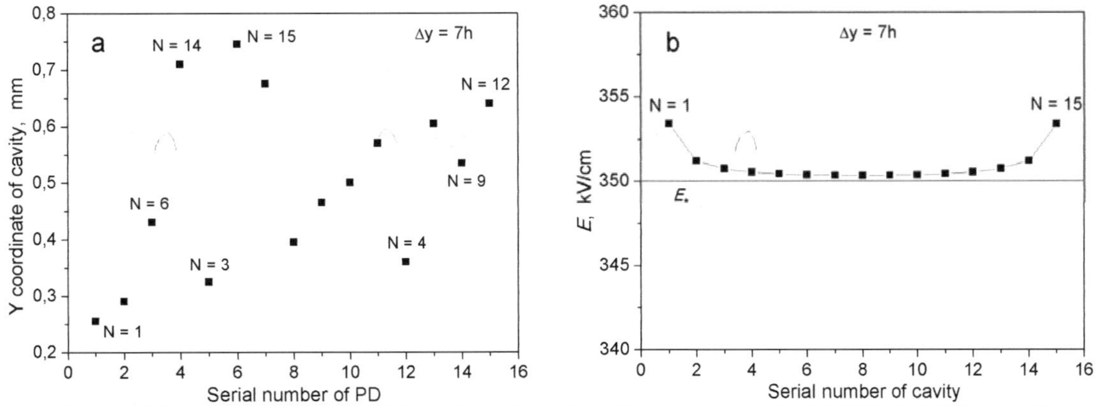

Figure 8. (a) The stochastic regime of partial discharges in the chain of cavities. (b) The values of electric field inside cavities before the first PD. $L = 1$ mm. $E_* = 350$ kV/cm. $\Delta y = 7h$. $h = 5$ μm. $\alpha = 20$ cm/(kV·ms). Lattice size 200×200.

Acknowledgments
The study was supported by the Russian Scientific Foundation (grant No. 16-19-10229).

References

[1] Hikita M, Yamada K, Nakamura A, Mizutani T, Oohasi A and Ieda M 1990 Measurements of partial discharges by computer and analysis of partial discharge distribution by the Monte Carlo Method *IEEE Trans. Electr. Insul.* **25**(3) 453–68

[2] Fruth B and Niemeyer L 1992 The importance of statistical characteristics of partial discharge data *IEEE Trans. Electr. Insul.* **27**(1) 60–9

[3] Cavallini A and Montanari G C 2006 Effect of supply voltage frequency on testing of insulation system *IEEE Trans. Dielectr. Electr. Insul.* **13**(1) 111–21

[4] Lemke E 2012 A critical review of partial-discharge models *IEEE Electrical Insulation Magazine* **28**(6) 11–6

[5] Wu K, Suzuoki Y and Dissado L A 2004 The contribution of discharge area variation to partial discharge patterns in disk-voids *J. Phys.* D*: Appl. Phys.* **37**(13) 1815–23

[6] Kupershtokh A L, Stamatelatos C P and Agoris D P 2006 Simulation of partial discharge activity in solid dielectrics under AC voltage *Technical Phys. Lett.* **32**(8) 680–3

[7] Kupershtokh A L, Karpov D I, Medvedev D A, Stamatelatos C P, Charalambakos V P, Pyrgioti E C and Agoris D P 2007 Stochastic models of partial discharge activity in solid and liquid dielectrics *IET Sci. Meas. Technol.* **1**(6) 303–11

[8] Kupershtokh A L and Karpov D I 2016 Simulation of waves of partial discharges in a chain of gas inclusions located in condensed dielectrics *J. Phys.: Conf. Ser.* **754**(10) 102006

[9] Babaeva N Yu, Tereshonok D V, Naidis G V and Smirnov B V 2016 Initiation of breakdown in strings of bubbles immersed in transformer oil and water: String orientation and proximity of bubbles *J. Phys.* D*: Appl. Phys.* **49**(2) 025202

[10] Karpov D I and Kupershtokh A L 1998 Models of streamers growth with "physical" time and fractal characteristics of streamer structures *Conf. Record 1998 IEEE Int. Symp. on Electrical Insulation, IEEE No. 98CH36239* vol 2 (Arlington, USA) pp 607–10

[11] Kupershtokh A L, Charalambakos V, Agoris D and Karpov D I 2001 Simulation of breakdown in air using cellular automata with streamer to leader transition *J. Phys.* D*: Appl. Phys.* **34**(6) 936–46

[12] Peschot A, Poulain C, Bonifaci N and Lesaint O 2015 Electrical breakdown voltage in micro- and submicrometer contact gaps (100 nm – 10 μm) in air and nitrogen *IEEE Holm Conference on Electrical Contacts* (San Diego, USA) pp 280–6

IOP Conf. Series: Journal of Physics: Conf. Series **899** (2017) 082005 doi:10.1088/1742-6596/899/8/082005

Inductive RF discharge in water vapor in toroidal geometry

R N Medvedev[1] and D S Churkin[2,3]

[1] Lavrentyev Institute of Hydrodynamics SB RAS, Lavrentyev Prosp. 15, 630090, Novosibirsk, Russia
[2] Institute of Laser Physics SB RAS, Lavrentyev Prosp. 13/3, 630090, Novosibirsk, Russia
[3] Novosibirsk State University, Pirogova St. 2, 630090, Novosibirsk, Russia

E-mail: ruslan@hydro.nsc.ru

Abstract. This paper presents the results of comparative experiments on the excitation of inductively coupled plasma in water vapor in cylindrical and toroidal discharge chambers having the same outer diameter. The plasma temperature in the cylindrical and toroidal chambers was determined using the Boltzmann plot method. With equal discharge power and pressure, the temperature in the toroidal chamber was found to be higher than that in the cylindrical chamber. This is attributed to the lower heat transfer and the absence of a near-axial layer in the toroidal chamber as compared with cylindrical geometry.

1. Introduction

Inductive RF discharge is used in many applications: coating, etching, and emission analysis of the composition of substances, purification and decontamination.

For effective use of inductive discharges in emission spectral analysis, it is necessary to provide the highest energy density with a minimum plasma volume in order to reach maximum temperature at minimum power consumption.

The temperature of inductively coupled plasma is determined by the electric power released in it and the heat loss, which is in turn determined by the heat transfer across the plasma surface and convection. The heat flux through the plasma surface is proportional to the surface area, so that to reduce the heat loss, it is necessary reduce the surface area.

Conventional inductive RF discharge devices are cylindrical tubes, on whose outer surface a spiral inductor is wound. Due to the skin effect, energy is released in the outer layer of the plasma near the tube wall, as indicated by the radial temperature distribution in the plasma volume [1]. The inner near-axial plasma layer is the least heated and of no interest to the excitation of spectra. In this regard, it has been proposed to optimize the inductive RF discharge by using discharge tubes of toroidal geometry.

In the case of a cylindrical chamber of diameter D, the plasma surface area is given by $S_C = \pi D(l + D/2)$, where l is the thickness of the plasma loop, which is slightly larger than the inductor width (about 10 mm). For a toroidal chamber, the surface area of the plasma loop is $S_R = \pi^2 Dd$, where D is the outer diameter (diameter of the axis) and d is the inner diameter (of the cross section) of the torus. For $d < (D+2l)/2\pi$, the toroidal geometry is more effective.

The toroidal geometry showed it's advantages in "tokamak"–type inductive closed-loop free-ferrite lamp [2]. The luminous efficacy of such type of lamps reaches 90 lumen per Watt. Some characteristics of inductive discharge with axial coil type were measured as plasma resistance, currents and electric field in plasma turn in the frequency range from 0.16 to 1 MHz. But there isn't any bibliographic data regarding using of coreless axial inductive discharge in spectroscopy under higher frequencies.

2. Experiment

The discharge was excited in a quartz tube 1 by a single-turn copper inductor 2 (figure 1). The inductor was supplied with a 50 MHz current from an RF source. The load power was 300 W. A vacuum pump was placed on one side of the quartz tube, and a vessel containing tap water and fitted with a metering valve on the other side. The vacuum system was similar to that shown in [3]. Before turning on the RF source, air was pumped out of the system to a pressure of ~ 0.1 mbar. After that, water vapor was fed to the system by means of the metering valve.

We used two types of quartz tubes: cylindrical (figure 1a) and toroidal (figure 1b). The dimensions of the cylindrical tube were $D = 40$ mm and $l \approx 20$ mm, and for the toroidal tube, $D = 40$ mm and $d = 8$ mm. Thus, $S_C = 5$ cm^2 and $S_R = 3.2$ cm^2. One end of the cylindrical tube was made flat to detect radiation from the thin plasma layer.

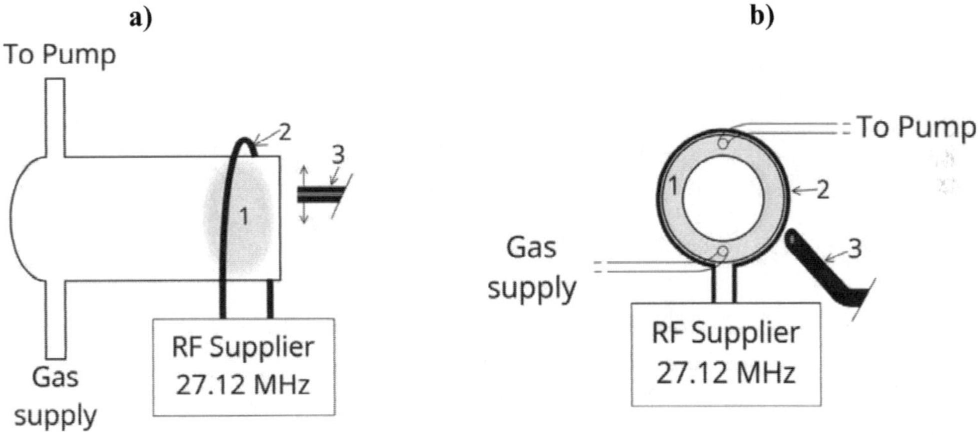

Figure 1. Experimental setup of cylindrical geometry (a) and toroidal geometry (b); 1 is the plasma discharge tube, 2 is the inductor, and 3 is the spectrometer lightguide.

The pressure in the discharge tube was monitored by a Testo 552 vacuum gauge and was from 0.1 to 1 mbar. The lower value was limited by the productivity of the vacuum pump and the upper one was limited by the maximal power of RF source. The stable inductive discharge under 300 W RF power exists up to 1 mbar of pressure.

The plasma radiation spectrum was recorded by a Kolibri-2 spectrometer (VMK Optoelektronika) with a spectral range from 180 to 1100 nm through a fiber-optic lightguide 3 having inner diameter 1 mm. In the case of the cylindrical discharge tube, the fiber was placed parallel to axis of the tube at different distances from the axis. Side light was limited using an additional light-proof tube into which optical fiber was inserted. In the case of toroidal geometry, the optical fiber was directed perpendicular to the plane of the axis of the quartz discharge tube.

3. Results and analysis

Figure 2 shows the emission spectra of the inductive discharge in cylindrical geometry with registration near the wall (figure 2a) and in toroidal geometry (figure 2b).

Comparison of the two spectra in figure 2 shows that the relative intensity of the H_β (486.1 nm) hydrogen line is higher in toroidal geometry than in cylindrical geometry. In addition, in toroidal geometry, the H_γ (434.1 nm) hydrogen line becomes visible, whereas, in cylindrical geometry, it is practically invisible. This indicates the greater importance of the plasma temperature for toroidal geometry.

The temperature was estimated by the Boltzmann plot method using the relation

$$\ln\left(\frac{I\lambda}{Ag}\right) = -\frac{E}{kT} + C, \quad (1)$$

where E is the energy of the upper (excited) level, I and λ are the intensity and wavelength of the corresponding spectral line, A and g are the transition probability (in 10^8/s) and the statistical weight of the excited level respectively, k is Boltzmann's constant, and C is a constant. The temperature can be determined knowing the slope of the graph plotted in the coordinates $\ln\left(\frac{I\lambda}{Ag}\right)$ and E and using the relation $T = -\frac{1}{k \cdot tg\varphi}$, where $tg\varphi$ is the slope of the graph. The method is described in detail in [4].

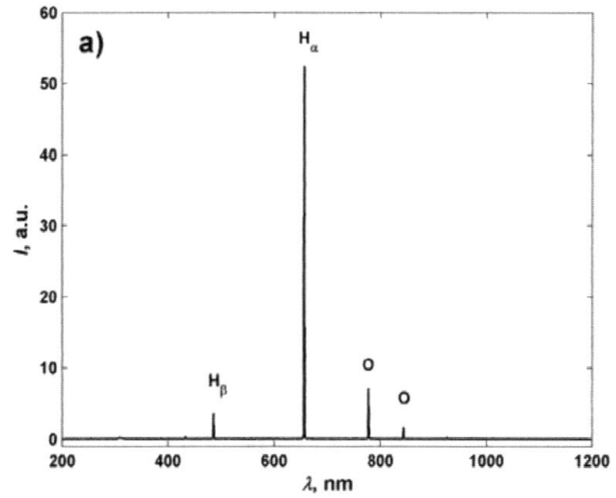

Figure 2. Inductive discharge spectra: near the wall in cylindrical geometry (a) and in toroidal geometry (b).

Figure 3 shows a graph of determining the temperature from the first three lines of the Balmer series (H_α, H_β, H_γ). The points in the graph were least-squares fitted to straight lines, after which the slope of each line was calculated and used to determine the plasma temperature. The graph shows three cases: a cylindrical tube with the optical fiber near the wall, a cylindrical tube with the optical fiber on the axis, and a toroidal tube.

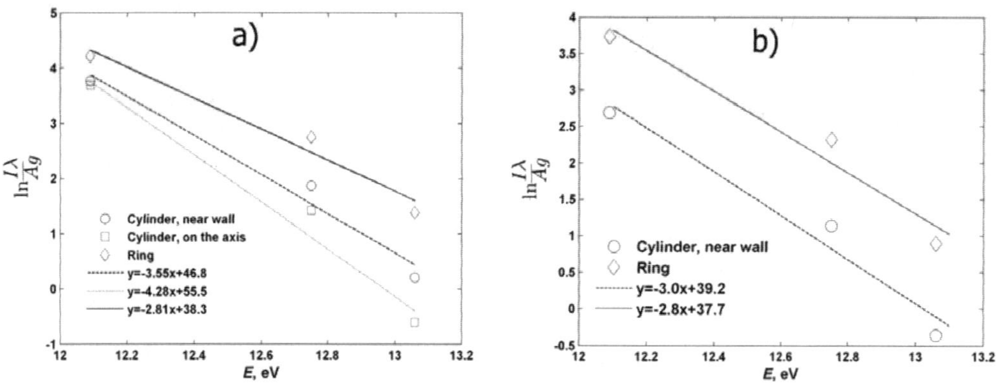

Figure 3. Boltzmann's plot computed from (1) for the first three lines of the hydrogen Balmer series under pressures: a) – 1 mbar, b) – 0.3 mbar. The dashed line corresponds to data ○, the dotted line – to data □, and the solid line – to data ◊.

For the cylindrical tube under pressure 1 mbar, the near-wall temperature was 3300 K, and the temperature on the axis was 2700 K. For the toroidal tube under the same pressure, the temperature was 4100 K. For 0.3 mbar of pressure the near-wall temperature in cylindrical tube was 3900 K while for the toroidal tube it was found to be equal 4200 K. It can be seen that the plasma temperature is higher in toroidal geometry than in cylindrical geometry. Also, it should be noted that for the lower pressure the mean temperature is higher.

From the figure 3, it can be concluded that the plasma is not in thermodynamic equilibrium (experimental points are not exactly at the approximation lines). In [5] it is experimentally shown that the pressure of the water plasma up to 3 mbar is not sufficient for elementary collisions to be in equilibrium. But the conclusion about the relativity of the temperatures still can be done.

Since, in toroidal geometry, the entire heat flux passes through the surface of the glass tube, this geometry is effective only for plasma-forming gases with thermal conductivity higher than the thermal conductivity of quartz glass, which at a temperature from 500 to 1000 K is about 2 $Wm^{-1}K^{-1}$. This condition is satisfied for water plasma and is not satisfied for, e.g., argon, whose thermal conductivity is an order of magnitude lower [6]. Therefore, to select the optimal geometry, one should take into account the type of the gas used.

4. Conclusions

The experimental investigation of the continuous inductive discharge in toroidal geometry with water vapour was carried out for the first time. The comparison with the classical cylindrical geometry showed that a toroidal discharge tube has advantages over a cylindrical one for the purpose of reaching the maximum plasma temperature for a given power.

The results can be used to develop and optimize inductive coupled plasma devices for specific applications.

Acknowledgment
This work was supported by the Russian Foundation for Basic Research (Grant No. 16-38-60039).

References

[1] Schafer J, Lyons W, Tong W G and Danehy P M 2009 Kinetic Temperature and Electron Density Measurement in an Inductively Coupled Plasma Torch using Degenerate Four-Wave Mixing *Applied Physics B* **95**(1) 161-8
[2] Popov O A and Chandler R 2002 Ferrite-free high power electrodeless fluorescent lamp operated at a frequency of 160–1000 kHz *Plasma Sources Sci. Technol.* **11** 218–27
[3] Medvedev R N, Churkin D S and Zarubin I A 2016 Inductive discharge as a method for direct detection of impurities in water *Journal of Physics: Conference Series* **754** 102007
[4] Hill S J 2007 *Inductively Coupled Plasma Spectrometry and its Applications* (Oxford: Blackwell Publishing) p 427
[5] Medvedev R N and Churkin D S 2017 An inductive RF discharge in water vapor for atomic-emission spectrometry *MATEC Web of Conferences* **115** 04001
[6] Tang Y Q and Trassy C 1986 Inductively coupled plasma: the role of water in axial excitation temperatures Spectrochimica Acta **41B**(1/2) 143-50

Influence of the external magnetic field on hydrodynamic structure of the electrovortex flow in hemispherical container

D A Vinogradov, I O Teplyakov[1], Yu P Ivochkin, I B Klementyeva

[1]Joint institute for high temperatures RAS, 125412 Moscow, Izhorskaya 13-2, Russia
E-mail: igor.teplyakov@mail.ru

Abstract. Electrovortex flow is formed as a result of interaction of the non-uniform electric current passing through the liquid metal with the own magnetic field of this current. Numerical calculation and measurements of the flow velocity in liquid metal (indium-gallium-tin alloy) in hemispherical container were carried out. Results on influence of the axial external magnetic field on the flow are presented.

1. Introduction

Electrovortex flow (EVF) is formed as a result of interaction of the non-uniform electric current passing through the liquid metal with the own magnetic field of this current [1]. Such flows significantly affect many processes in mechanical engineering (electro-welding) and electrometallurgy (electroslag remelting, various electric melting furnaces). In particular, electrovortex flow determines the hydrodynamic structure in the baths of DC-arc furnaces, which are increasingly used in industry [2].

Here is the brief review of works about electrovortex flows. The first articles associated with investigation of electrovortex flow were about flow calculation at the small currents using Stokes approximation [3,4]. Article [5] was devoted to possible reasons of rotation of the EVF. Problems connected with the influence of the external axial magnetic field were started to consider in [6,7,8]. Questions related to the surface deformation were present in [9].

In this paper we investigate the influence of the external axial magnetic field (MF) on the EVF of liquid metal in the hemispherical container. Let us consider the system where electric current propagates from the small hemispherical electrode through the conductive media (liquid metal) to the big hemispherical electrode. Electromagnetic force **F**=**J**×**B** (where **J** – current density and **B** – magnetic field) leads the liquid to a motion.

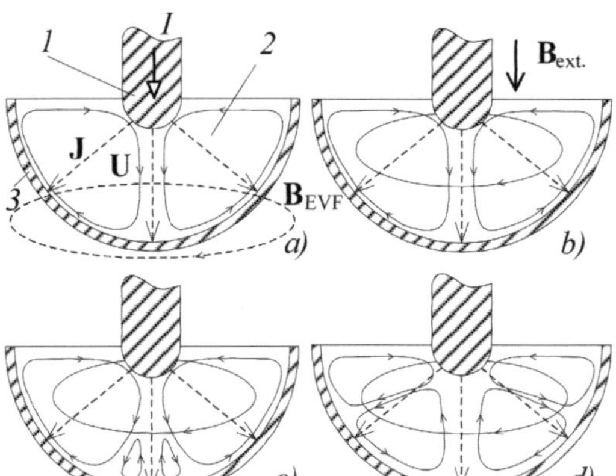

Fig.1. Elecrovortex structure. 1 - small electrode, 2 - liquid metal, 3 - big electrode. a) Toroidal EVF; b) occurrence of the azimuthal swirl; c) occurrence of the secondary vortex; d) double-vortices rotating EVF.

IOP Conf. Series: Journal of Physics: Conf. Series **899** (2017) 082006 doi:10.1088/1742-6596/899/8/082006

Usually it is assumed that in such system with the axially symmetric central electrode without the action of external magnetic fields EVF has the shape of a toroidal vortex (Fig.1a). External axial magnetic field in this geometry leads to the azimuthal swirl of the flow (Fig.1b) . And due to the azimuthal rotation, the secondary vortex rotating in a vertical plane opposite main EVF (Fig. 1d) and pushing it to the periphery of the bath occurs.

2. Numerical technique

The motion of the liquid is described by the Navier-Stokes equation:

$$\frac{\partial \mathbf{U}}{\partial t} + (\mathbf{U}\nabla)\mathbf{U} = -\frac{1}{\rho}\nabla p + \nu\Delta\mathbf{U} + \frac{1}{\rho}\mathbf{F},$$

where $\mathbf{U}$ is the velocity of the liquid, ρ is the density of the liquid, ν is the kinematic viscosity coefficient, p is the pressure and $\mathbf{F}$ is the electromagnetic force.

$$\mathbf{F} = \mathbf{J} \times (\mathbf{B} + \mathbf{B}_{ext}),$$

where $\mathbf{B}$ is the own magnetic field of the current and $\mathbf{B}_{ext}$ is external field. We consider case with axial external magnetic field so $\mathbf{B}_{ext}$ has the only z-component B_z.

At the small magnetic Reynolds number we can use the so-called electrodynamic approximation, according to which we neglect the influence of velocity on electric current in the liquid, so the density of the current $\mathbf{J}$ can be written in spherical coordinates as:

$$\mathbf{J} = \frac{I}{2\pi R^2}\mathbf{e}_R;$$

where I is the current. The magnetic field can be found from the Maxwell's equation:

$$\nabla \times \mathbf{B} = \mu_0 \mathbf{J}.$$

This equation can be rewritten in projection to R-coordinate :

$$\frac{1}{R\sin\theta}\frac{\partial\left(B_\phi \sin\theta\right)}{\partial\theta} = \frac{I}{2\pi R^2}.$$

So the magnetic field will be:

$$B_\phi = \frac{\mu_0 I(1-\cos\theta)}{2\pi R\sin\theta}$$

and the forces:

$$F_\theta = -\frac{\mu_0 I^2(1-\cos\theta)}{4\pi^2 R^3\sin\theta}; F_\phi = -\frac{IB\_z \sin\theta}{2\pi R^2}.$$

Using following scales for generalization Navier Stokes equation:

Table 1. Scales.

Length	R_2
Electric current density	$j_0 = I/R_2^2$
Magnetic field	$B_0 = \mu_0 I/R_2$
Electromagnetic force	$f_0 = \mu_0 I^2/R_2^3$
Velocity	$u_0 = \sqrt{f_0 R_2/\rho}$
Time	$t_0 = R_2/v_0$

IOP Conf. Series: Journal of Physics: Conf. Series **899** (2017) 082006 doi:10.1088/1742-6596/899/8/082006

Pressure	$p_0 = \rho v_0^2$
Modified Reynolds number	$\mathrm{Re} = \sqrt{f_0 R_2^3/\rho v_0^2} = I/v_0\sqrt{\mu_0/\rho}$
Parameter S	$S = Re^2 = \mu_0 I^2/\rho v^2$
Parameter N	$N = BIR_2/\rho v^2$
Parameter α	$\alpha = B_z/B_0 = N/S$

we can rewrite Navier Stokes equation in dimensionless form:

$$\frac{\partial \mathbf{U}}{\partial t} + (\mathbf{U}\nabla)\mathbf{U} = -\nabla p + \frac{1}{\sqrt{\mathrm{S}}}\Delta \mathbf{U} + \frac{\cos\theta - 1}{4\pi^2 R^3 \sin\theta}\mathbf{e}_\theta - \frac{\mathrm{N}}{\mathrm{S}}\frac{\sin\theta}{2\pi R^2}\mathbf{e}_\phi$$

To solve Navier–Stokes equation we use finite volume method on unstructured 2D-axisymmetric grid in cylindrical coordinates. In this coordinate system (r,φ,z) components of electromagnetic force have the following form:

$$F_r = -\frac{z(\sqrt{r^2+z^2} - z)}{4\pi^2(r^2+z^2)^2};$$

$$F_\varphi = -\frac{\mathrm{N}r}{2\pi(r^2+z^2)^{3/2}};$$

$$F_z = -\frac{(\sqrt{r^2+z^2} - z)}{4\pi^2(r^2+z^2)^2}.$$

Unsteady calculations of the velocity were carried out for different ratio of the small and big electrode: 1:10, 1:37.6, and 1:100. To obtain boundary line dividing one- and two- vortices zones in parameters I - B_{ext} (or α – S) the following algorithm was used: with some values of parameters S and α calculation was being carried out to get and to observe streamlines (Fig. 2).

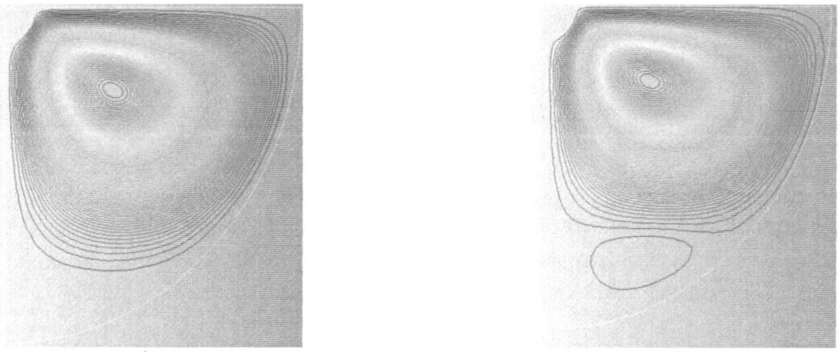

a) b)

Fig. 2. Calculated stream functions. a) single vortex system b) double vortex system

If we obtained the case **a** (one-vortex system) then parameter α was increased and calculation was repeated until reaching the case **b**. Then parameter α was decreased and calculations were carried out again until reaching of 5% accuracy on parameter α.

3. The experimental technique

Experimental studies were carried out at the setup shown in Fig.3. An eutectic indium-gallium-tin alloy (weight content: Ga-67%, In-20.55%, Sn-12.5%, physical properties: melting point +10.5°C , $\rho = 6482$ kg/m^3, ν=4.3e-7 m^2/s, σ=3.3e6 Sm, [10]) was used as the working liquid in the experiments. Alloy filled copper hemispherical container with the radius R_2=188 mm which also served as a large electrode. Small electrode - copper or steel cylinder with the radius R_1=2.5 mm with hemispherical tip was immersed into the alloy in the middle of the working bath. Power source developed on the basis of three-phase AC rectifier (I≤1500 A) was used to supply an experimental setup. To create an external longitudinal magnetic field the coil consisting of 15 turns of hollow copper tube (cooled with pumped water) was used. Coil power was supplied from a stabilized power source, providing smooth control of DC in the range from 0 to 100 A. This system allowed to get the magnetic field with induction B_z=5e-3 T in the middle of the working area at a current of ~100A.

The experiments were carried out with the fiber-optical transducers (Fig. 4) developed in our laboratory [11]. Transducer made of glass and is not sensitive to external electric and magnetic fields. The main characteristics of a typical transducer are in the table.

Table 2. Fiber-optical transducer main characteristics.

Average velocity measurement range	1÷50 cm/s
Frequency range	0÷200 Hz
Measuring volume	0.04×0.04×1 mm^3
Temperature range	<100°C

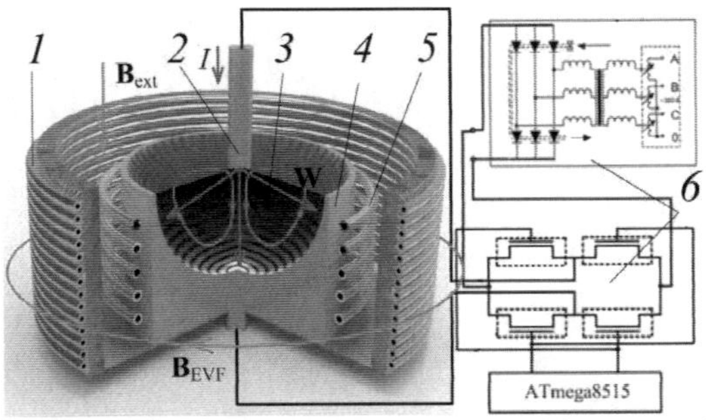

Fig. 3. Experimental setup. 1 - solenoid, 2 - small electrode, 3 - eutectic alloy In-Ga-Sn, 4 - hemispherical container, 5 - heat exchanger, 6- programmable power supply.

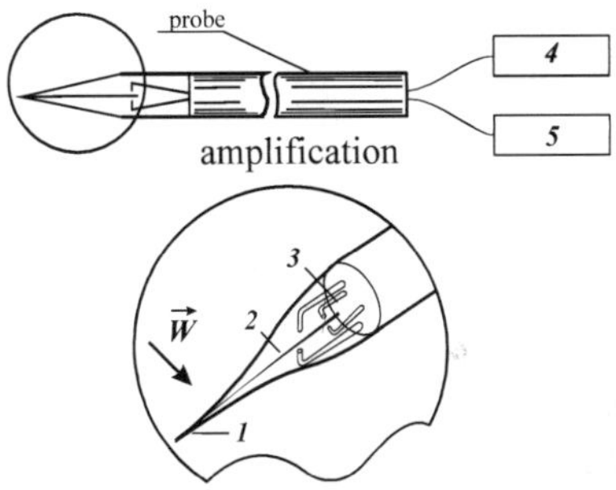

Fig. 4. Fiber-optical transducer. *1* -sensor, 2- pointer, *3* - light guides, *4* photodiode, *5* - photoreceiver.

The second vortex appears when the value of the velocity read by the fiber-optical transducer changes its sign.

4. Results

Typical results on measurements and calculations of the velocity are presented below. There are the oscillograms of axial velocity showing the evolution of the velocity on time (Fig. 5). First part of graph shows that in beginning velocity rises then (when azimathal rotation was developed) secondary upward flow suppresses main electrovortex flow and axial velocity degreases until zero and below.

IOP Conf. Series: Journal of Physics: Conf. Series **899** (2017) 082006 doi:10.1088/1742-6596/899/8/082006

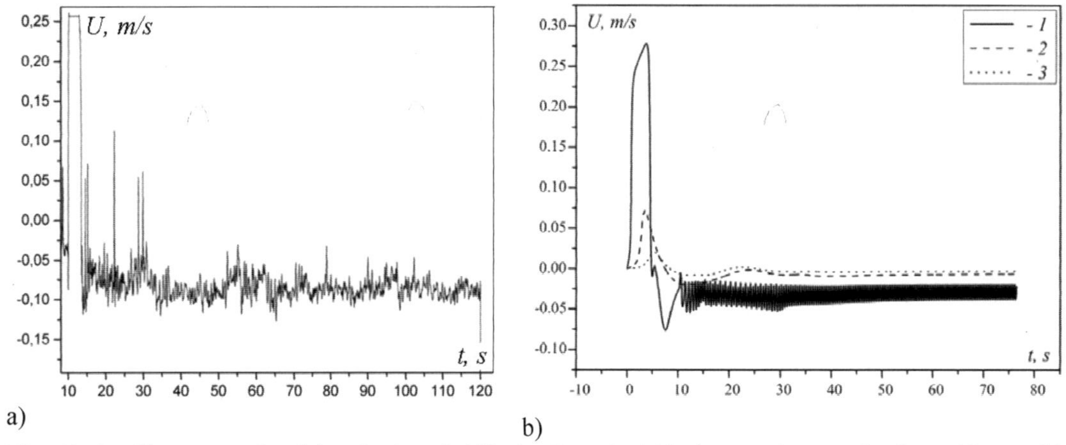

Fig. 5. Oscillograms of axial velocity. I=400 A, B_{ext}=5e-4 T a) experiment, depth z=*10* mm b) calculation, z - 1 - 10 mm, 2 – 30, 3 – 80.

Data on the time of the appearance of the second vortex are presented in Fig. 6. A bigger value of the external field leads to the fastest appearance of the second vortex.

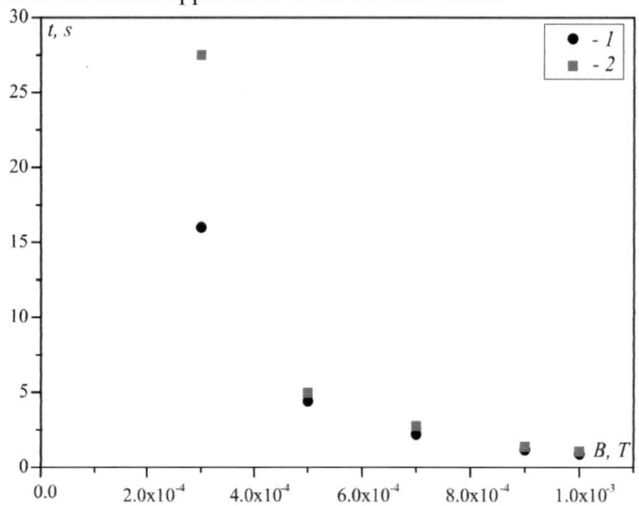

Fig. 6. Dependence of the time of appearance of the second vortex on the external magnetic field. I=400 A. 1 - Experiment, 2 - calculation.

Fig. 7 presents boundary line of zones with one and two vortices. With increasing of parameter S we need smaller α for formation of the second vortex. In dimensional form, with increasing of the electric current it requires stronger magnetic field.

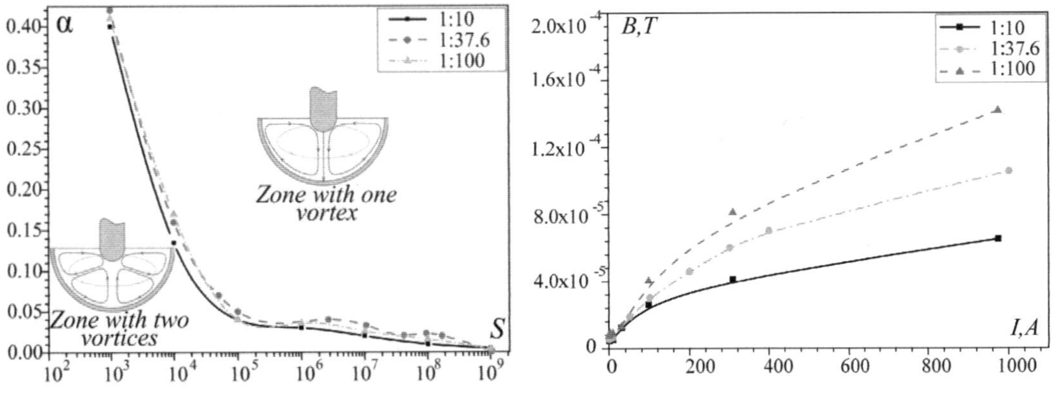

Fig. 7. Boundary line separating zones with one or two vortices in coordinates a) α-*S*, b) B_{ext}-*I*

5. Conclusions

Experiments and calculations of the axial velocity in electrovortex flow with presence of the external axial magnetic field were carried out. The boundary line separating one- and two-vortices zones and time of appearance of the second vortex were obtained. These results can be used for a priori estimates of the flow structure.

6. Acknowledgments
This work was supported by Russian Scientific Fund, grant №17-19-01745.

7. References

[1] V. Bojarevics, J. Frejbergs, E.I. Shilova, E.V. Shcherbinin. Electrically Induced Vortical Flows (Kluwer Academic Publishers, Dordrecht, Boston, London, 1989).

[2] I.M. Yachikov, O.I. Karandaeva, T.P. Larina. Simulation of electrovortex flows in the bath of the DC electric arc furnace. (Magnitogorsk, NMSTU, 2008).

[3] Sozou C., Pickering W. M.// J.Fluid Mech. 1976. 73. part 4. P. 641.

[4] Sozou C, Pickering W M 1978 Proc. R. Soc. Lond. A. 362 509.

[5] V. Bojarevics, E. Shcherbinin Azimuthal rotation in the axisymmetric meridional flow due to an electric-current source // J. Fluid Mech. (1983), vol. 126, pp. 413-430.

[6] Shatrov V., Gerbeth G. // Magnetohydrodynamics 2012. 48, N3. P. 469.

[7] V. G. Zhilin, Yu. P. Ivochkin, I. O. Teplyakov. The problem of swirling of axisymmetric electrovortex flows High Temperature // 2011, Vol. 49, №6, pp 927–929.

[8] Yu. P. Ivochkin, I. O. Teplyakov and D. A. Vinogradov. Investigation of self-oscillations in electrovortex flow of liquid metal, Vol. 52, No. 1/2, 277-286, 2016.

[9] A. Kharicha, I. Teplyakov, Yu. Ivochkin at al. // Experimental Thermal and Fluid Science, 2015. 62. P. 192.

[10] V. Ya. Prokhorenko, E. A. Ratushnyak, B. I. Stadnyk, V. I. Lakh, and A. M. Koval. Physical properties of thermometric alloy In-Ga-Sn. // High temperature vol.8. №2, 1970. p.374.

[11] Zhilin V.G. Optical–fiber Velocity and Pressure Transducers. (Hemisphere publishing corporation. New York, 1990).

Investigation of two-stage air-blown and air-steam-blown entrained-flow coal gasification

N A Abaimov[1]*, E B Butakov[2], A P Burdukov[2] and A F Ryzhkov[1]

[1] FSAEI HE "Ural Federal University named after the first President of Russia B. N. Yeltsin", Ekaterinburg, Russia
[2] Kutateladze Institute of Thermophysics SB RAS, Novosibirsk, Russia
*nick.sum41@mail.ru

Abstract. The aim of the paper is to compare effect of coal and steam supply to the second stage of the entrained-flow gasifiers on their main operating parameters. Two two-stage gasifiers of SB RAS Institute of Thermophysics were used as experimental units. The 5 MW unit has air-coal feeding to first and second stages (air-blown gasification). The 1 MW unit has air-coal feeding to first stage and steam feeding to second stage (air-steam-blown gasification). Experimental studies of air-blown and air-steam-blown gasification with various stoichiometric coefficients, flow rates, temperatures and supply points of secondary media (air-coal and steam) are carried out. Experimental and calculated (thermodynamic and CFD) methods have established that: in the air-blown gasification unit the increase of stoichiometric coefficients (secondary coal flow rate reduction) reduces the syngas heating value and increases the carbon conversion rate; in the air-steam-blown gasification unit the increase of the steam supply nozzle immersion to the gasifier leads to increase of the syngas heating value, the carbon conversion rate and decrease of the syngas H_2/CO ratio.

1. Introduction

Coal is the most wide-spread type of fossil fuel on the planet and has one of the first places in the world fuel and energy balances. Today coal is used either to generate heat and electricity, or to produce chemicals. One of the most promising energy technologies for the coal using is the integrated gasification combined cycle (IGCC) plant [1]. The main element of this plant is gasifier, in which coal gasification takes place – coal conversion into combustible gas (syngas). Oxygen or air is usually used as the reagent gas. Oxygen has high reactivity, so the vast majority of oxygen gasifiers have one reaction stage. The air contains 21% oxygen, so the gasification rate is low, which is usually compensated by using two-stage scheme gasifiers [2].

One of the most effective methods of the gasification process controlling is secondary media supply to the gasification chamber. Such media are usually coal, steam, carbon dioxide, combustion products, syngas, etc. The supply of secondary coal or steam is considered to be the most effective and technically simple measure. The use of coal as secondary media makes it possible to approximate the stoichiometric factor to the theoretically necessary value, and thus increase the gasifier cold gas efficiency. On the other hand, decrease of stoichiometric factor reduces the carbon conversion rate, so the method and location of coal supply to the gasifier have great importance. For example, many years development and operation experience of MHI two-stage air gasifier confirmed the effectiveness of radial secondary coal feed into high-heated products of primary coal incomplete combustion [3]. The steam supply is used to lower the temperature in the reaction space (to remove superheating) and to increase the fraction of hydrogen in the syngas due to the water gas shift reaction.

UrFU develop air-steam-blown two-stage coal gasifier for IGCC based on the MHI gasifier [4]. Basically from the MHI gasifier the UrFU gasifier differs by high-temperature steam supplying into the second stage together with the secondary coal, as well as additional heating of air up to 900 ° C.

The aim of the paper is to compare effect of coal and steam supply to the second stage of the entrained-flow gasifiers on their main operating parameters.

In view of the complexity and cost-effectiveness of gasifier experimental research, they are increasingly being replaced by numerical modeling [5]. The most effective is the approach combining experimental studies and the different dimensions (0-, 1-, 2-, 3-D) computational modeling [6].

To achieve this goal, the following tasks have been accomplished:

1) experimental studies of air-blown and air-steam-blown gasification have been carried out with the coal and steam supply into the gasifier second stage, respectively;

2) experiments results were processed and compared using zero-dimensional thermodynamic and three-dimensional CFD modeling;

3) effect of constructive and regime parameters on the gasifier operation is analyzed and compared.

2. Experimental study

Two two-stage gasifiers of SB RAS Institute of Thermophysics were used as experimental units. The 5 MW unit has air-coal feeding to first and second stages (air-blown gasification). The 1 MW unit has air-coal feeding to first stage and steam feeding to second stage (air-steam-blown gasification). Both gasifiers are cylindrical inside lined vessels, at one end of which is swirler of primary coal feeding. Axisymmetric with the reaction chamber into the end of the swirler is nozzle of secondary coal (air-blown gasification) and steam (air-steam-blown gasification) supply at the gasifier (figure 1).

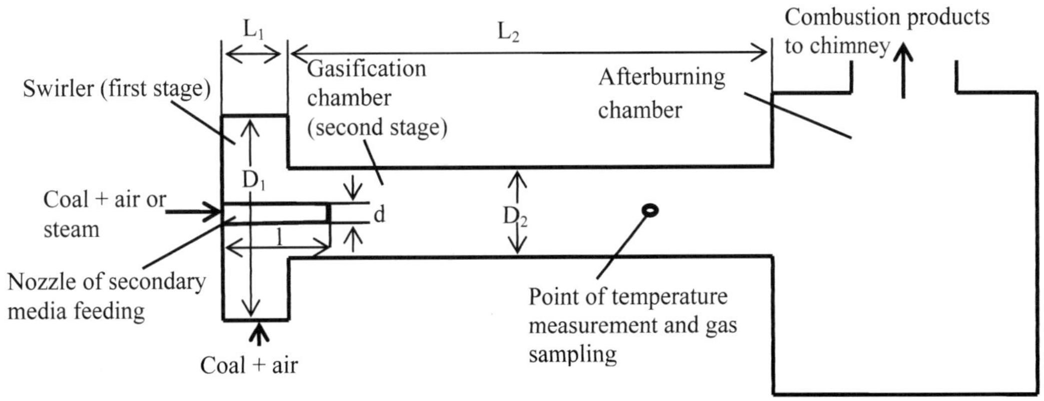

Figure 1. The scheme of experimental gasifiers.

The gasifiers geometry did not differ in principle, but the dimensions were different (table 1).

Table 1. Gasifiers dimensions.

Gasifier	L_1, m	L_2, m	l, m	D_1, m	D_2, m	d, m
Air-blown	0.14	1.5	0	0.5	0.32	0.05
Air-steam-blown	0.1	1.1	0.1-0.25	0.3	0.15	0.03

During air-blown gasification, Kuznetsky bituminous coal, and in the case of air-steam-blown gasification the Pereyaslovsky subbituminous coal was used (table 2).

Table 2. Proximate and ultimate analysis of used coals.

Coal	W^a, %	A^d, %	V^{daf}, %	C^{daf}, %	H^{daf}, %	N^{daf}, %	S^{daf}, %	O^{daf}, %	Q_s^{daf}, MJ/kg
Kuznetsky	5.4	22.3	44.7	75.57	5.66	1.78	0.55	16.44	27.7
Pereyaslovsky	9.1	7.2	46.6	74.6	5.1	1	0.3	19	28.9

To increase the reactivity of primary coal, coal mechanical activation was made by disintegrator mill to average coal particle diameter of 40 microns. Into the second stage of the unit (in the case of air-blown gasification), coal was fed after grinding in ball mill and had average diameter of 100 microns. Measurements of temperature and concentration were performed during quasi-stationary mode of gasifier operation. Further, the measured values were averaged over time of this mode. The gasifiers operation parameters are given in table 3.

Table 3. Gasifiers operation parameters.

Gasifier	Coal flow rate, kg/h	Air flow rate, Nm^3/h	Coal and air temperature, °C	Steam temperature, °C	Stoichiometric factor (α)	Temperature at measuring point, °C
Air	200-300	700-900	20-30	-	0.41-0.67	1400-1500°C
Air-steam	25-35	85-95	20-30	150-200	0.47	1000-1100°C

3. Calculation study

To process experimental data, thermodynamic modeling of syngas components equilibrium concentrations was carried out using the entropy maximization method [7]. The calculation implied idealized case, not taking into account unreacted carbon, incompleteness of reaction and mixing of all syngas components, and considered that the mixture reached complete chemical equilibrium state. Calculation without the above-described assumptions made it possible to estimate the carbon conversion rate by fitting the calculated concentrations to experimental ones.

During air-blown gasification experiment, the stoichiometric factor was varied α 0.41-0.67 (figure 2a), and during air-steam-blown - depth of nozzle immersion into gasifier l=0.1-0.25 m (figure 2b).

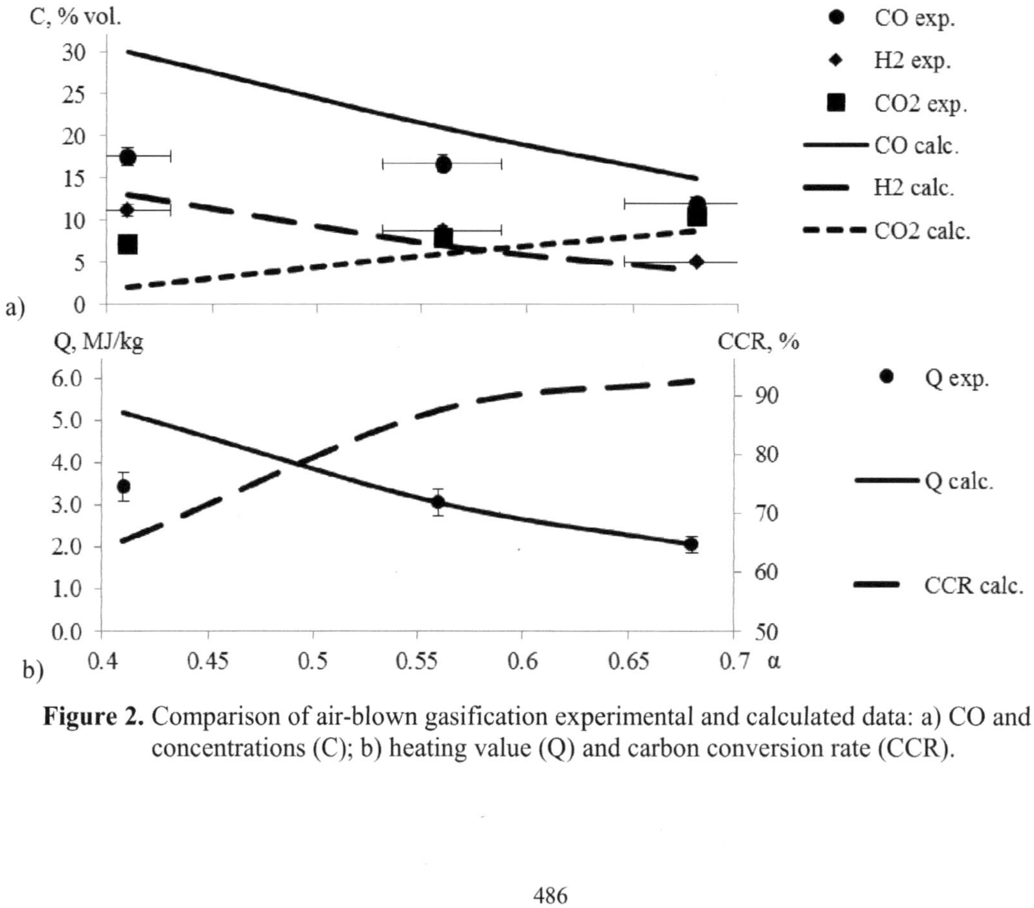

Figure 2. Comparison of air-blown gasification experimental and calculated data: a) CO and H_2 concentrations (C); b) heating value (Q) and carbon conversion rate (CCR).

Figure 3a shows that the experimental and calculated concentrations of combustible gases decrease with α increasing. The calculation overestimates the concentration of CO, which indicates noticeable amount of unreacted coal carbon. The syngas heating value decreases with α increasing and carbon conversion rate increases (figure 3b). This is due to the burn-out of CO and H_2 part and more complete combustion of carbon by excess oxygen.

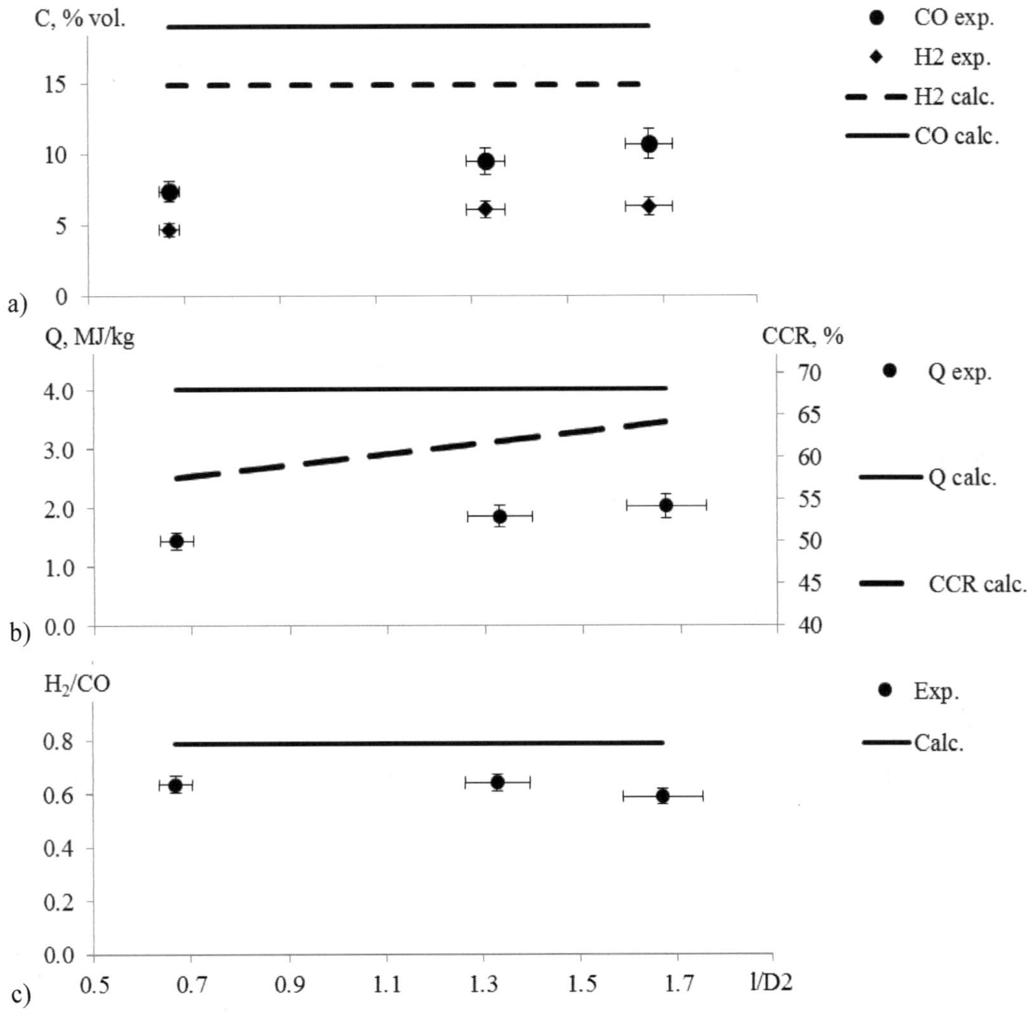

Figure 3. Comparison of air-steam-blown gasification experimental and calculated data: a) CO and H_2 concentrations; b) syngas heating value and the carbon conversion rate; c) syngas H_2/CO ratio.

The experimental and calculated data have differences in the figures 2 and 3 because, as mentioned above, calculation not taking into account unreacted carbon, incompleteness of reaction and mixing of all syngas components, and considered that the mixture reached complete chemical equilibrium state.

The calculation results do not depend on the depth of the steam nozzle immersion into the gasifier, since it does not take into account unreactivity of fuel carbon part. Experimental data show an increase of H_2 and CO concentrations, syngas heating value and the carbon conversion rate (figure 4b) as the depth of the steam nozzle immersion increases into the gasifier. This is due to the decrease of steam and reacting mixture contact time (steam takes less heat). The syngas H_2/CO ratio (figure 4c) decreases with increasing of steam nozzle immersion depth into the gasifier due to a decrease of the reaction time of the steam with fuel coke (steam gasification) and syngas (water gas shift reaction).

In addition to the thermodynamic calculations, CFD simulation of the operation of both gasifiers was performed using the previously verified CFD-model [8-10]. Special feature of the gasifiers is significant gradient of gas concentrations along the gasification chamber radius (figure 4). In the air-blown gasifier, radial CO concentration distribution in the cross section radius was experimentally determined (figure 4c).

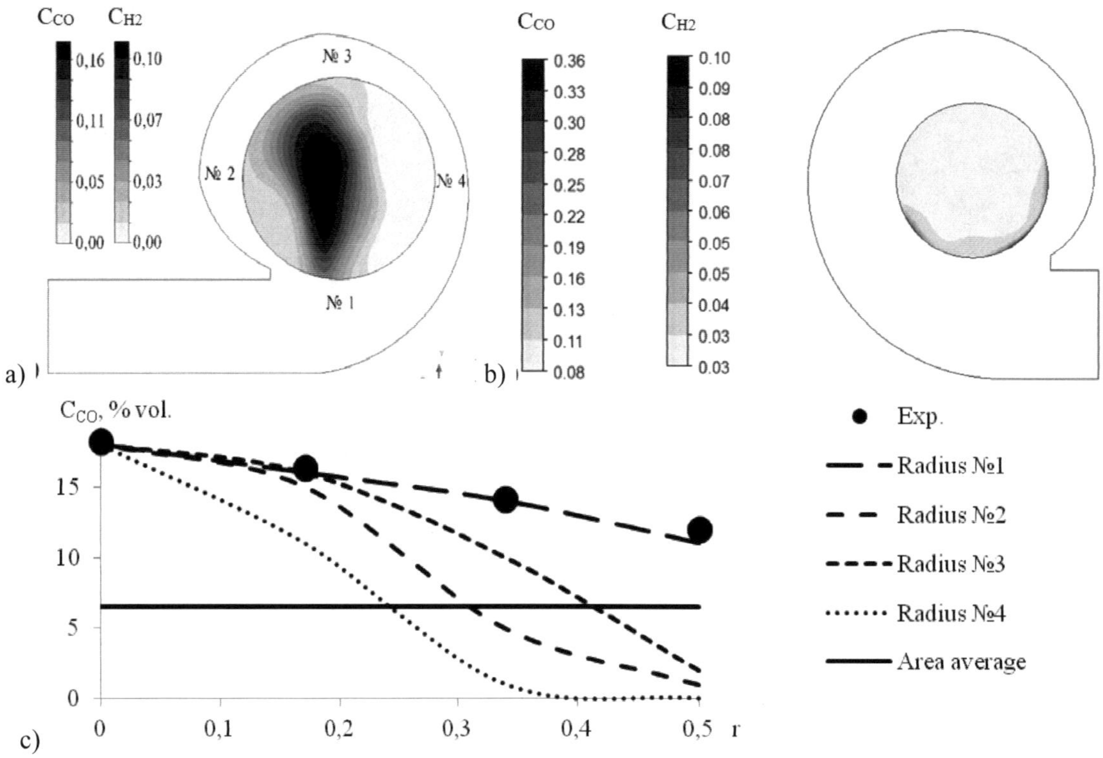

Figure 4. Distribution of CO and H_2 concentrations in the gas sampling cross section: a) air-blown gasifier; b) air-steam-blown gasifier; c) comparison of calculation and experiment CO concentration distribution of air-blown gasifier radius (r).

It can be seen from figure 4 that CO and H_2 concentrations distribution in the cross-section of air-blown and air-steam-blown gasifiers differ significantly. In the air-blown gasifier (figure 4a, c), the maximum values of combustible gases concentration are observed in the paraxial zone as the secondary coal is fed and gasified into the axial region of the gasifier. But in the air-steam-blown gasifier (figure 4b), maximum values of combustible gases concentration can be seen at the periphery of the gasification chamber because coal is fed to the gasifier through swirler that pushes coal to the gasifier periphery.

4. Conclusion

Experimental studies of air-blown and air-steam-blown gasification with various stoichiometric coefficients, flow rates, temperatures and supply points of secondary media (air+coal and steam) are carried out. Experimental and calculated (thermodynamic and CFD) methods have established that: in the air-blown gasification unit the increase of stoichiometric coefficients (secondary coal flow rate reduction) reduces the syngas heating value and increases the carbon conversion rate; in the air-steam-blown gasification unit the increase of the steam supply nozzle immersion to the gasifier leads to increase of the syngas heating value, the carbon conversion rate and decrease of the syngas H_2/CO ratio.

The resulted profiles of gas concentrations in the gasifier cross section indicate complex non-uniform character of the investigated processes dynamics and distribution. This is factor that complicates the experimental and calculated studies, but increases their significance. The investigation indicates the need for additional heating of the reacting media (air, steam) and flow swirling to improve the efficiency of the gasification process.

Acknowledgements
Experimental and computational studies of the air-blown gasifier (5 MW) was supported by the grant of Russian Science Foundation, project No.14-19-00524 (N.A. Abaimov and A.F. Ryzhkov).

Experimental and computational studies of the air-steam-blown gasifier (1 MW) was supported by the grant of Russian Foundation for Basic Research, project No. 16-38-50188 (N.A. Abaimov and A.P. Burdukov).

References

[1] Clean Coal Research Program, Technology Readiness Assessment - Analysis of Active Research Portfolio (United States Department of Energy, Office of Fossil Energy, December, 2012) / http://www.netl.doe.gov/File%20Library/Research/Coal/energy%20 systems /gasification/TRL-Comprehensive-Report_121112_FINAL_1.pdf

[2] Ryzhkov A F, Bogatova T F, Lingyan Z and Osipov P V 2016 Development of entrained-flow gasification technologies in the Asia-Pacific region (review) *Thermal Engineering* **63** pp 791-801

[3] Hashimoto T, Sakamoto K and Kitagawa Y 2009 Development of IGCC Commercial Plant with Air-blown Gasifier *Mitsubishi Heavy Industries Technical Review* **46-2** pp 1–5

[4] Abaimov N A, Osipov P V and Ryzhkov A F 2016 Experimental and computational study and development of the bituminous coal entrained-flow air-blown gasifier for IGCC *J. of Phys.: Conf. Series* **754** 112001

[5] Vascellari M, Roberts D G, Hla S S, Harris D J and Hasse C 2015 From laboratory-scale experiments to industrial-scale CFD simulations of entrained flow coal gasification *Fuel* **152** pp 58-73

[6] Lee H-H, Lee J-C, Joo Y-J, Oh M and Lee C-H 2014 Dynamic modeling of Shell entrained flow gasifier in an integrated gasification combined cycle process *Applied Energy* **131** pp 425-40

[7] Trusov B G 2002 Software TERRA for computing plasmachemical processes *Materials of 3rd Int. symp. on theoretical and applied plasma chemistry* (Ples: ISUCT) pp 217–8

[8] Abaimov N A and Ryzhkov A F 2015 Development of a model of entrained flow coal gasification and study of aerodynamic mechanisms of action on gasifier operation *Thermal Engineering* **62** pp 767–72

[9] Abaimov N A and Rizhkov A F 2015 Turbulence models verification for numerical Simulation of the cyclone chamber *Proc. of 5th Int. Conf. Heat and mass transfer and hydrodynamics in swirling flows* (Kazan: KAI) pp 429-37

[10] Chernetskiy M Y, Kuznetsov V A, Dekterev A A, Abaimov N A and Ryzhkov A F 2016 Comparative analysis of turbulence model effect on description of the processes of pulverized coal combustion at flow swirl *Thermophysics and aeromechanics* **23** pp 615-26

IOP Conf. Series: Journal of Physics: Conf. Series **899** (2017) 092002 doi:10.1088/1742-6596/899/9/092002

Gas dynamics of reactive gases in swirling-type furnace

A I Akhmetshina[1], G I Pavlov[1], A N Sabirzyanov[1], O A Tikhonov[1]

[1]Kazan National Research Technical University — KAI named after A. N. Tupolev

E-mail: galimova.alfiya@mail.ru

Abstract. It is known from the literature that for the complete reaction of two gases (fuel and oxidizer), it is necessary to fulfill three basic conditions: the stoichiometric ratio of reactive gases, qualitative mixing and ensuring the cooling of combustion products without "quenching". Of the above-stated conditions it is more difficult to organize a qualitative mixture formation. This physical process requires additional expenditure of energy flow. In this work we present the results of experimental and theoretical studies of the gas dynamics of a reactive gas mixture in a swirling-type furnace. The design scheme of the furnace includes two reaction zones for combustible components: the first zone is the zone of generation of combustible gases which composition is constant; the second zone of the furnace - zone of a homogeneous combustion reaction.

The problem of determining the necessary parameters is practically not solved because of the complexity of the physicochemical processes taking place in the first zone. The methods for calculating gas-dynamic parameters in homogeneous reacting mixtures have been sufficiently tested. The solution of gas dynamics problems in the second zone is classical and can be successfully solved using existing software packages.

The composition of the generator gases at the outlet from the first zone and the corresponding flow parameters were used as initial data. These parameters are determined by thermodynamic calculation under certain assumptions. The adequacy of the results of thermodynamic calculations and also the results of computational experiments in the second zone were verified by experiments.

A fire stand was designed for the experiments. The fire stand includes a model of a swirling-type furnace and a system for measuring the parameters of gases (velocity, temperature, composition of gases). A detailed description of the stand is described in work [1].

The complete geometric model is executed in the SIEMENS NX version 9.0 graphics package. The appearance of the furnace is shown in Figure 1a, the design of the internal parts of the boiler furnace without the elements of supplying secondary air into the reaction area is shown in Figure 1b.

IOP Conf. Series: Journal of Physics: Conf. Series **899** (2017) 092002 doi:10.1088/1742-6596/899/9/092002

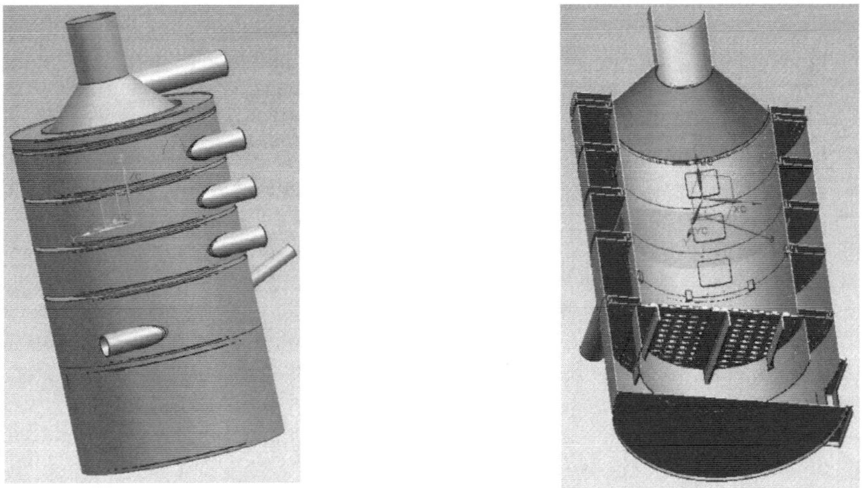

Figure1. The three-dimensional model of the furnace: (a) is a general view, (b) is a longitudinal section of the furnace without elements for supplying secondary air to the reaction zone

The swirling-type furnace has two reaction zones for the combustible components with air. In the first zone the combustion of a combustible substance takes place in a layer on the grate. Sleepers in the form of chips are used as fuel. This zone is characterized by such complex physicochemical processes as heating and thermal decomposition of wood chips, impregnating substances, partial combustion of the gas phase and so on. These phenomena occur due to the heat released during the combustion of a portion of the gas phase and coals. Air in the grate part of the furnace is supplied in an amount necessary for burning a small part of the combustible gases and afterburning of the solid residue (coals).

Further the gas mixture consisting of combustible gases of thermal decomposition of sleepers and combustion products of coals enters the second zone located above. Secondary air is supplied through different belts In the same zone. Air is supplied in the form of separate jets at different angles to the radius, thereby forming a vortex flow of different structures. Air in the furnace space is fed by a fan. The air supply system allows you to distribute the air through the combustion zones according to the set program. The zone from the grate to the gas phase where the primary heated air is supplied cannot be considered during the combustion process modeling. In the second zone air supply is carried out according to various design schemes: radially, tangentially and chordally.

To realize these schemes, air blocks in the form of cylindrical collectors with air nozzles (three blocks) are made which make up the lateral wall of the swirling-type furnace. The design of the furnace allows you to change the position of the blocks relative to each other and form 6 variants of air supply. In the body of the furnace there are technological hatches for inserting probes of measuring instruments into the investigated space of the furnace.

The following options of secondary air supply are considered:

1. option 1 - all units are tuned to the tangential air supply;
2. option 2: the first block - the tangential scheme; the second block is a radial scheme; the third block is the tangential scheme;
3. option 3 - the first block - the tangential scheme; the second block is a chordal scheme at an angle of 260 degrees to the tangent; the third block is a tangential diagram;
4. option 4 - the first block is the tangential scheme, the second block is a radial scheme; the third block is a chordal scheme at an angle of 260 to the tangent;
5. option5 - the first block is a tangential scheme; the second block is a radial scheme; the third block is a chordal at an angle of 450 degrees to the tangent;
6. option 6 - the first block is chordal scheme at an angle of 450 to the tangent; the second block is a radial scheme; the third block is the tangential scheme.

Numerical studies were based on the statistical approach (RANS) which is based on averaging the equations of motion. According to this model, all studied parameters are decomposed into the mean and turbulent components, and turbulence is modeled on the basis of semiempirical models. A laminar micro flame model for a non-blended mixture was used as a combustion model. The physical essence of the laminar micro flame model is the separate supply of components which leads to the fact that the combustion process is largely limited by mixing.However, in turbulent flames due to a considerable deformation of the flame front, the diffusion transfer rate increases substantially which leads to a noticeable deviation from the chemical equilibrium. To avoid large errors in the calculations, the transition from complete chemical equilibrium to partial is accomplished, when individual parts of the chemical process proceed with finite velocities and all the rest are infinitely fast. This approach is equivalent to the idea of a turbulent flame front, as an ensemble of laminar micro flames. The parameters of the laminar flame are preliminarily calculated and stored as a table, and then the laminar flame is introduced into the turbulent flow using statistical methods. A large number of computational experiments were carried out on the basis of the above-described model. The results of the experiments were reproduced in the form of fields and graphs.
Figure 2 shows the distribution of static temperature fields in the furnace with the supply of secondary air in two ways: Figure2a - option 1, Figure 2b - option 4.

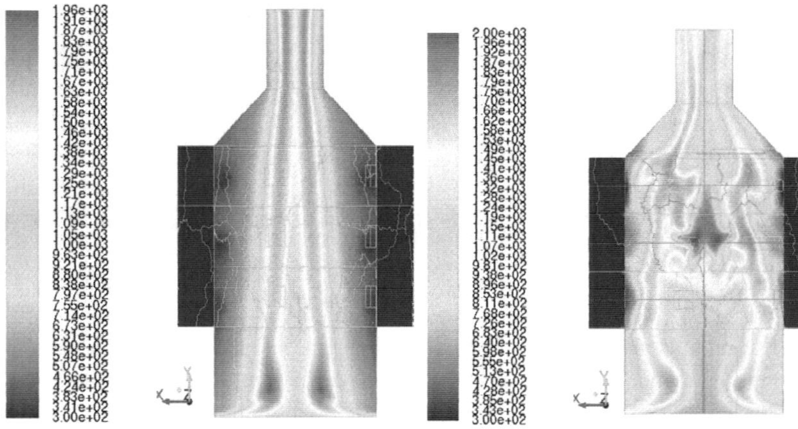

a) b)

Figure2. The distribution of static temperature fields in the z = 0 plane: (a) - according to the option 1, (b) - according to the option 4

It can be seen from Figure 2a that the temperatures in the combustion chamber are distributed as follows: the temperature near the chamber axis reaches 1960 ° K, near the walls is about 500 ° K, and in the collector 300 ° K. Other temperatures are observed in the picture shown in Figure 2b - the temperature in the greater part of the chamber volume is 1500-2000 ° K. When air is supplied by a combined method (variant No. 4), the temperature gradient in the furnace is greatly reduced along the radius and the axis. On the basis of the calculations obtained, it is possible to arrive at an unambiguous conclusion that a change in the scheme for supplying secondary air to the combustion chamber can be controlled by a temperature field. The change in the nature of the temperature field due to a change in the air supply to the furnace is explained by the influence of the intensity of the twist of the air flow, the effects of breaking up a single jet and the elimination of the conditions for cooling down the combustion products. This hypothesis is confirmed by the results of an analysis of the velocity field along the axis of the furnace which is shown in Figure 3.

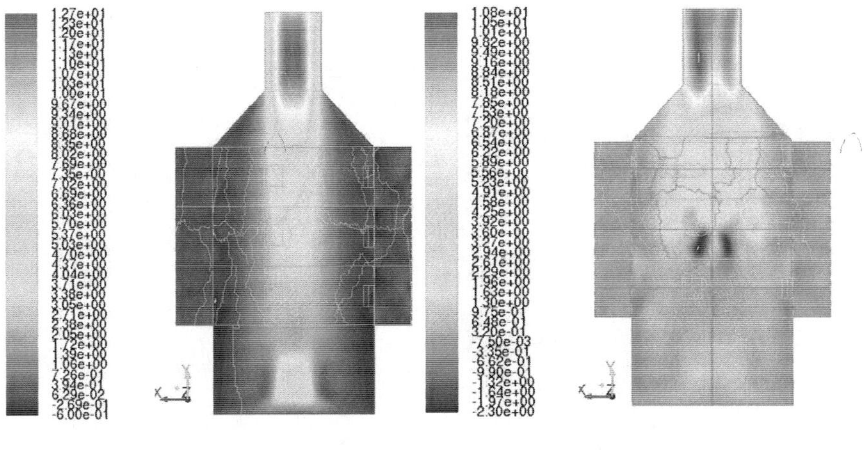

a) b)

Figure3. Distribution of the velocity component in the z = 0 plane: (a) -option 1, (b) - option 4

Figure3. shows the velocity distribution in the plane z = 0, which is determined by the intensity of the twist and the separation of the combustion products flow by secondary air. Option 1 is characterized by a distribution of the values of the velocity characteristic of a single swirling flow. It can be noted in variant 4 that there are practically no stagnant zones in the furnace - the reacting gas mixture moves from bottom to top with the same speed. Changes in gas-dynamic processes in the furnace affect the concentrations of combustion products. The results of calculations of the mole fractions of CO, CO2, SO2, H2S at the output section of the chamber are given in Table 1. According to the table values, it is possible to evaluate the effectiveness of one or another air supply to the furnace.

Experiments were carried out to verify the adequacy of the calculations' results. For this purpose a fire stand was created, in the article [1] is given a detailed description of it. During the experiments following parameters were obtained: the temperature of the reacting mixture in the combustion chamber, the content of harmful gases at the outlet from the furnace, and the excess air factor in the furnace.

Table 1. The results of experimental studies

Options	Total pressure, P , Pascal	Static temperature, T_s, K	Mole fractions				
			CO	CO_2	SO	SO_2	H_2S
1	12,070	1083,62	0,00961	0,06535	$1{,}9\cdot10^{-5}$	$2{,}14\cdot10^{-3}$	$1{,}3\cdot10^{-4}$
2	11,005	1234,98	0,00044	0,07707	$1{,}61\cdot10^{-6}$	$2{,}43\cdot10^{-3}$	$4{,}87\cdot10^{-7}$
3	11,229	1088,20	0,01049	0,07340	$2{,}65\cdot10^{-5}$	$2{,}19\cdot10^{-3}$	$1{,}39\cdot10^{-4}$
4	9,230	984,75	$2\cdot10^{-6}$	0,05308	$1{,}13\cdot10^{-8}$	$1{,}26\cdot10^{-3}$	$8{,}26\cdot10^{-9}$
5	10,819	1219,00	0,00184	0,07608	$5{,}88\cdot10^{-6}$	$2{,}45\cdot10^{-3}$	$5{,}53\cdot10^{-6}$
6	10,904	1234,47	0,00094	0,07720	$3{,}33\cdot10^{-6}$	$2{,}45\cdot10^{-3}$	$1{,}61\cdot10^{-6}$

The temperature was chosen as the physical parameter for comparing the calculated data with the experimental ones. This is justified by the fact that the temperature is an integral quantity that characterizes indirectly the influence of the studied factors (air jet flow schemes) on heat and mass transfer processes. The gas temperature was measured by an electronic device TK-1 with a maximum measurement range up to 1200 °C.

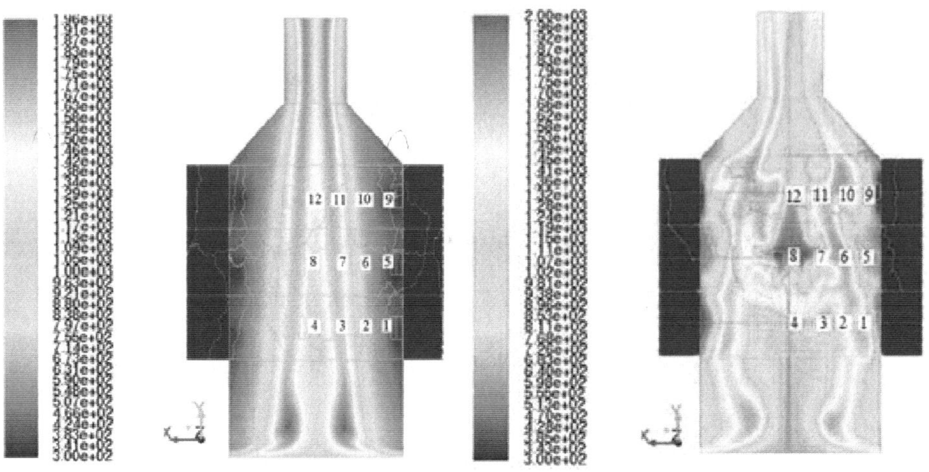

a) b)

Figure 4. The results of measurements of the gases temperature in the combustion chamber (a) - according to option 1: 1- 420 °K; 2 - 600 °K; 3- 1200 °K; 4-900 °K; 5-430 °K; 6 - 620 °K; 7 1200 ° K; 8-1010 °K; 9 -480°K; 10-570 °K; 11-1100°K; 12- 1250, (b) - according to the option 4: 1- 1280 °K; 2 - 1100 °K; 3- 1100 °K; 4-1100 °K; 5-730 °K; 6 - 1320 °K; 7- 1300 °K; 8-370 °K; 9-6800 K; 10-1370 °K; 11-1380°K; 12-1050.

The zones where the gas temperature was recorded are shown in numbers in Figure 4. From the analysis of temperature data obtained by the calculation method and by measurement, it can be concluded that the proposed theoretical model reproduces the qualitative picture of the temperature field in the combustion chamber with sufficient accuracy. At the studied points the difference between the measured temperature values and the calculated values reaches up to 35%. Large differences in the values of temperatures are typical for the central zone of the furnace. The gas dynamic processes proposed by the model are described more precisely near the walls of the furnace. Thus, the difference between the calculated temperature and measured at points 1,5, 9 is not more than 21%. The large temperature gradient in the combustion chamber along its radius shows that the tangential flow of air jets into the cavity of the combustion chamber adversely affects the intensification of heat and mass transfer and, as a result, the completeness of combustion of the gas phase.

It is possible to see from Figure 6b that the temperature gradient in the combustion chamber, both radially and along the axis of the chamber, differs greatly from the analogous parameter obtained for variant 1 in the case of air distribution according to variant 4. In this case the temperature field is more uniform. In the combustion chamber, there are no clearly defined regions with a low gas temperature. Analysis of the quantitative values of temperature indicates that optimal temperature levels are created in the combustion chamber for oxidation-reduction processes. The difference in the measured temperature values from the calculated values in the considered zones reaches up to 28%. Qualitatively, the distribution of the calculated temperature values in the combustion zone of the gas phase is in good agreement with the temperature field obtained experimentally. This allows us to judge the adequacy of the theoretical model proposed for the calculation of gas dynamic processes in the combustion chamber of the chosen design scheme.

Thereby, the numerical studies have established that the direction of the supply of air jets to the combustible gas stream exerts a strong influence on the combustion processes in the swirling-type furnace. Optimal directions of supply of air jets in relation to the direction of movement of products of thermal decomposition of wood waste are determined. The developed mathematical model makes it possible to simulate the formation of CO, NO, sulfur compounds, as well as the completeness of combustion of combustible gases depending on the flow rate, location and scheme of supply of air jets to the reaction zone. Optimal conditions are determined when the concentration of harmful substances at the outlet from the furnace and the temperature gradient in the combustion chamber are minimal. Experimentally measured values of temperatures confirm the reliability of the results of numerical studies with a certain error.

References

1. Pavlov, G. I., Kochergin, A. V., Akhmetshina A. I., R. R. Sirazieva Ecology and industry of Russia. T 20, No. 10, 2016. page 11-15

IOP Conf. Series: Journal of Physics: Conf. Series **899** (2017) 092003 doi:10.1088/1742-6596/899/9/092003

Simulation of heat and mass transfer in a shaft of plasma electric furnace, when utilizing technogeneous wastes

A S Anshakov[1,2], A I Aliferov[2], V A Sinitsyn[1,2], and P V Domarov[1,2]

[1] Kutateladze Institute of Thermophysics SB RAS, Novosibirsk, Russia
[2] Novosibirsk State Technical University, Novosibirsk, Russia

E-mail: Domaroff@yandex.ru, anshakov@itp.nsc.ru

Abstract. A mathematical model of heat and mass transfer between counter flows of gas and porous material of man-made waste is presented. The model takes into account the partial and total pressure of the vapor-gas mixture, the radiation heat exchange between the gas and solid phases, the chemical transformations in the individual components of the charge, and the moisture content of the waste is also taken into account. Heat and mass transfer in the chamber of a shaft plasma-ohmic electric furnace was studied for gasification of technogeneous wastes with the use of additional ohmic heating in the drying zone. The influence of ohmic heating on heat and mass transfer is shown not only in the drying zone, but also in the entire shaft of electric furnace.

1. Introduction

Among the environmental problems, the problem of disposing and neutralizing technogeneous waste, including organic ones, whose number is increasing year by year, is acute all over the world. Technogeneous wastes, dangerous for population and environment (solid domestic waste, bio-organic waste) increase annually by 2-3 billion tons per year (According to official data of the Ministry of Natural Resources of the Russian Federation http://rpn.gov.ru/node/854). One of the effective ways for solving this problem is waste gasification in the shaft plasma furnaces with pyrolysis and gasification of the organic part of waste using an electric-arc plasmatron for oxidizer heating and melting the inorganic part. The main disadvantage of these furnaces is high energy consumption. The development of technology for processing of technogeneous waste requires creation of the methods for two main tasks: calculation of heat and mass transfer between two interpenetrating environments, taking into account various phase and chemical transformations of the constituent components of waste, and improvement of energy efficiency. An increase in energy efficiency of such furnaces is currently possible through using combined plasma-ohmic heating for processing the technogeneous waste (TW).

Under the real conditions, waste humidity is up to 60%. During processing, the waste enters the loading zone in the wet state, and this moisture is kept only within the drying zone. While entering into the pyrolysis zone down the shaft, waste should be as dry as possible. Therefore, it is advisable to organize additional ohmic heating within the drying zone. The development of heat and mass transfer processes under the considered conditions is complicated by several processes: evaporation of moisture from waste, when the heated gas phase passes through it, and release of energy at ohmic heating in the TW material; exo- and endothermic chemical reactions on the material surface and in the gas phase that occur during waste gasification. In the high-temperature zone of electric furnace, a significant contribution to heat transfer can be made by thermal radiation.

To date, relatively large scientific experience on studying the process of heat and mass transfer in the porous media has been accumulated [1]. Among thorough studies of the processes of plasma gasification of wastes, we can distinguish the works [2, 3]. However, the electric current influence on heat and mass transfer processes has not been considered. The process of drying in the electric furnace chamber and its influence on the overall picture of temperature field distribution in the electric furnace chamber were not taken into account.

2. Mathematical model of heat and mass transfer

Simulation of heat and mass transfer in the charge is a challenge. The scheme of a shaft plasma-ohmic electric furnace for TW processing is shown in Fig. 1.

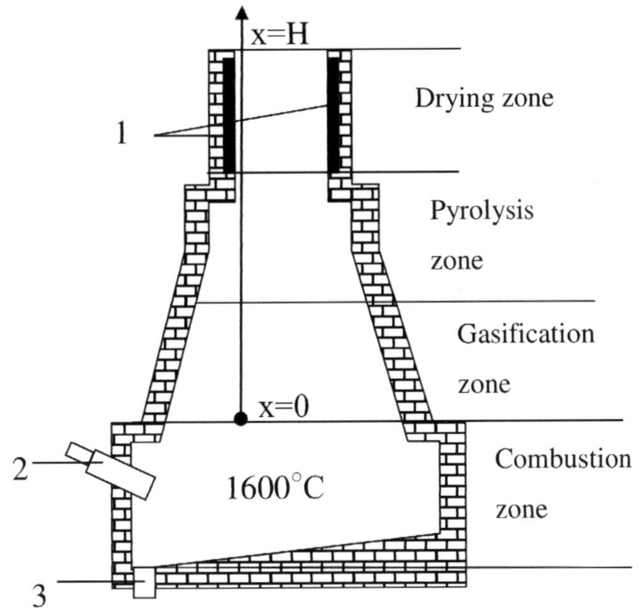

Figure 1. Scheme of plasma-ohmic furnace:
1 – electrodes; 2 – plasmatron; 3 – liquid slag removal

When modeling, it was taken into account that waste will contain moisture only in the loading zone and within the drying zone until all moisture evaporates under the influence of temperature. Thus, ohmic heating by the current of industrial frequency is possible only in the drying zone. Waste supplied to the pyrolysis zone is dehydrated and has low moisture content.

TW are fed to the upper part of the furnace from the top (x = H), and a gas flow heated in the lower part of the furnace by a plasmatron is raised from the bottom. Wastes, moving through the shaft, are subjected to drying by preheated gas and ohmic heating, pyrolysis, and gasification of the organic component. A part of waste, not subjected to gasification, turns into the melted slag in the lower part. The oxidizing gas, supplied from below, is saturated with an organic component and moisture, turning into the gaseous state.

When organizing a continuous process in the furnace chamber, the steady gas-dynamic and thermal conditions are formed. Therefore, the processes of heat and mass transfer in such a process in the system of counter-current flows of the solid and gas phases can be considered stationary. Due to heat and mass transfer, the temperature field is formed in the charge and gas phase.

Heat and mass transfer in the charge layer can be described by the following system of equations:

Energy equation in the charge

$$(1-m)\frac{d}{dx}\left[\left(\lambda_k+\frac{md}{\frac{(1-\varepsilon)d}{\lambda}+\frac{1}{\alpha_l}}\right)\right]\frac{dt}{dx}+q_{Vchem}(x)+q_{V_{em}}(x)-q_{tl}(x)-$$

$$-G_M c\frac{dt}{dx}+\left[r+c_{PD}(T-t)H\left(p_{DG}-\overset{*}{p_D}\right)\right]\frac{dG}{dx}+\alpha_V F(T-t)=0 \qquad (1)$$

where,

$$\frac{dG}{dx}=-F\frac{\alpha_V\left(\overset{*}{p_D}-p_{DG}\right)P}{\rho c_p\left(p-\overset{*}{p_D}\right)R_D T}=\frac{dG_M}{dx}=\frac{dG_G}{dx}$$

r is latent heat of vaporization, c_{PD} is heat capacity of vapor, p_{DG} is partial pressure of vapor, $H(x)=\begin{vmatrix}0, & x<0\\ 1, & x>0\end{vmatrix}$ is Heaviside function (single-phase function), F is area of shaft cross-section, ρ is density of vapor-gas mixture, c_P is heat capacity of vapor-gas mixture, p is total pressure of mixture, p_D^* is pressure of saturated vapor at charge temperature t, and c is heat capacity of wet charge.

- energy equation of the vapor-gas mixture:

$$G_G c_p\frac{dT}{dx}-\left[c_{PD}\frac{dG}{dx}H\left(\overset{*}{p_D}-p_{DG}\right)+\alpha_V F\right](T-t)=0 \qquad (2)$$

- equation of heat and mass transfer for the vapor-gas mixture

$$\rho w\frac{d\left(\frac{p_{DG}}{R_D T}\right)}{dx}+\frac{p_D}{R_D T}\frac{dw}{dx}=-F\frac{\alpha_V\left(\overset{*}{p_D}-p_{DG}\right)p}{\rho c_p\left(p-\overset{*}{p_D}\right)R_D T} \qquad (3)$$

- equation of motion for the entire mixture

$$\frac{p}{R_D T}w=-k\frac{dp}{dx} \qquad (4)$$

where R_D is gas constant of vapor, k is permeability coefficient of the medium.

Parameters, describing heat transfer in equations (1-2), are considered in detail in [4]. Mathematical description of the processes of electromagnetic field q_{Vem} in the furnace drying zone, continuous in space and time, are considered in [5].

The main calculation characteristics are as follows: t(x) is the temperature field in the charge; T(x) is the temperature field in the vapor-gas mixture; charge consumption $G_M(x)$ depending on drying (determined by partial pressure, total pressure); consumption of vapor-gas mixture $G_G(x)$ (determined by partial pressure, total pressure)

Auxiliary characteristics for calculation of drying are: field of partial vapor pressure $p_{DG;}$ field of total vapor pressure p; field of gas mixture velocity *w(x)*.

The system of equations (1-4) is non-linear due to consideration of parameters that are functions of temperature, coordinate or humidity. For example, such parameters are $q_{Vchem}(t)$, related to dependence of the chemical reactions rate on temperature [3], power release in the load due to ohmic heating as a function of humidity q_{Vem}; the field of gas mixture velocity $w(x)$, which is a function of temperature.

Therefore, an iterative method is used to solve this system. The numerical implementation of solution to the system of equations was carried out by the sweep method after conversion of the entire system to the finite-difference scheme. The accuracy of calculations was regulated by the error between the last and previous iteration.

3. Calculation results

Calculation results on distribution of the gas and charge temperature along the furnace shaft height, which demonstrate the effect of ohmic heating on the temperature regime inside the electric furnace,

IOP Conf. Series: Journal of Physics: Conf. Series **899** (2017) 092003 doi:10.1088/1742-6596/899/9/092003

are presented in Fig. 2. It can be seen that in the drying zone there is a noticeable release of energy inside the charge, dramatically increasing its temperature.

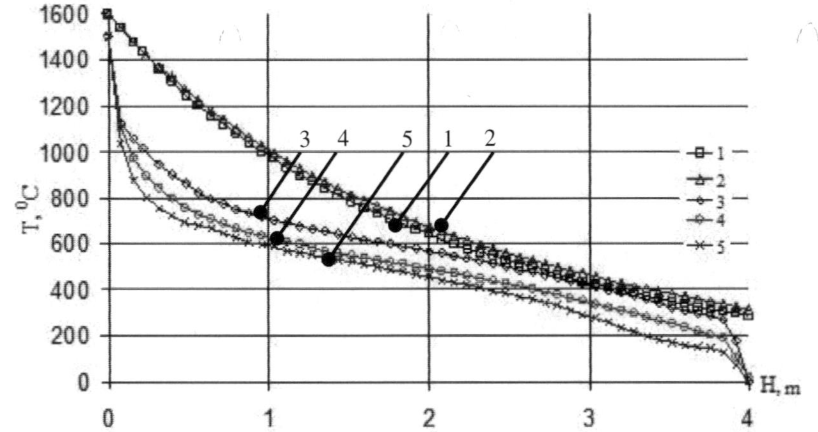

Figure 2. Distribution of charge and gas temperature in the space of electric furnace depending on waste humidity. Gas temperatures at humidity: 1 - 60%, 2 - 40%; charge temperatures at humidity: 3-60%, 4 - 40%, 5 - 20%

The release of energy at charge humidity of 60% leads to the fact that the charge temperature in the drying zone is practically equal to the gas temperature. This effect is especially obvious at the beginning of the drying zone because humidity of charge is supplemented with evaporating moisture from the waste in the form of vapor, released in the lower layers. Further, after the drying zone, a slight cessation in the temperature growth is observed in the gasification zone. This relates to ongoing endo- and exothermic reactions and an increase in the velocity of the gas mixture. An increase in the total pressure of the vapor-gas mixture in the electric furnace shaft indicates an increase in velocity w(x).

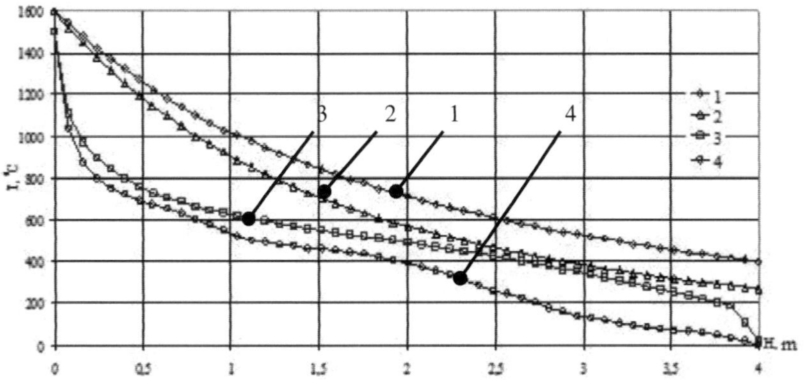

Figure 3 Temperature field distribution in the shaft of electric furnace with ohmic heating:
1 – gas temperature,
3 – charge temperature;
without ohmic heating
2 – gas temperature,
4 – charge temperature

Distribution of charge and gas temperature along the electric furnace shaft with and without ohmic heating at the same waste humidity (40%) is shown in Fig. 3. It is also necessary to note a small temperature difference between the gas and charge, especially in the drying zone, where heat and mass transfer between the charge and gas-vapor mixture is obvious unlike the other zones of the furnace shaft. This is due to the fact that in other zones of the shaft, there is no evaporation, and the enthalpy flow in gas is much higher than in the charge, so heat removal into the charge has a little effect on a change in the enthalpy flow and, hence, the gas temperature. The same effect is also observed in distribution of the temperature fields without ohmic heating.

4.Conclusions

Calculations of temperature distribution along the furnace height with consideration of ohmic heating and mass transfer effect have shown a significant effect on heat transfer inside the electric furnace shaft.

Inhomogeneous heat release due to ohmic heating with a pronounced effect at the beginning of the drying zone with a sharp increase in the charge temperature is shown.

Analysis of results shows that ohmic heating of wet waste in the upper part of the shaft electric furnace is an analog of the preliminary technological process of waste drying before loading into the furnace aimed at reduction of specific energy consumption for gasification of the organic component of mixed wastes. As it is shown in [6], it is possible to increase substantially the resource characteristics of plasmatron operation, reducing its loading [7].

ACKNOWLEDGMENTS

The work was financially supported by the Ministry of Education and Science of the Russian Federation according to the Subsidiary Agreement No.14.607.21.0118 (unique project identifier RFMEFI60715X0118)

References

[1] Georgiev, I.B., Mikhailov, B.I. 1991 High Energy Chem., **25** (4), 76-80.

[2] Lyubina Yu L and Suris A L 1999 Thermal Engineering **11** 64-67.

[3] Anshakov A S, Faleev V A, Danilenko A A et al. 2007 T and A **4** 607-616.

[4] Aliferov A I, Anshakov A S, Sinitsyn V A, Domarov P V, Vlasov D S 2011 ISJAEE. **10** 154–159.

[5] Aliferov A I, Anshakov A S, Sinicin V A, Domarov P V, Danilenko A A 2016 Journal of Phys.: Conf. Series **754** 112002

[6] Aliferov A I, Anshakov A S, Domarov P V, Faleev V A, Radko S I. 2016 Development of arc plasmatrons with long service life of IFOST-2016 **2** 37–40

[7] Zhukov, M.F., Timoshevskiy, A.N., Cherepanov, A.N. 1997 T and A **4** (2) 227-237.

Flare Systems Exploitation and Impact on Permafrost

M Yu Filimonov[1,2] **and N A Vaganova**[1,2]

[1] Ural Federal University, Yekaterinburg, Russia
[2] IMM UB RAS, Yekaterinburg, Russia

fmy@imm.uran.ru

Abstract. Mathematical models and numerical algorithms of horizontal and vertical flare systems exploitation in northern oil and gas fields located in permafrost zone are developed. Computations of long-term forecasts of permafrost degradation around such constructions have been carried out for various types of operation, including emergency situations, which cause a short-term increase in the heat flow near the ground surface, which leads to an additional soil temperature increasing.

1. Introduction

Permafrost occupies about 25% of the land area [1]. These territories occupy approximately 80%, 60% and 50% of the total area in Alaska, in Russia, and in Canada, respectively. In the permafrost zone of Russia, about 93% of Russian natural gas and 75% of oil are produced. An estimation of the reserves in the Arctic part of Russia consist of 7.6 billion tons of oil and 66.9 trillion cubic meters of gas.

When different technical systems are placed on a cluster sites used in the northern oil and gas fields, it is necessary to minimize the influence of various heat sources to the frozen ground. This is caused that thawing of permafrost can be accompanied by subsidence of the earth's surface and appearance of dangerous permafrost geological processes, called thermokarst, which leads to serious disasters. The problem of reducing the intensity of thermal interaction in the system “engineering facility (heat source) – soil” is of particular importance for solving energy conservation, environmental protection, safety, saving costs and improving the operational reliability of various technical systems. In this paper, the main attention is paid to investigation of the effect of vertical flare systems (VFS) and horizontal flare systems (HFS) on permafrost degradation and long-term forecast of the thawing zone dynamics under the flare systems.

2. The problem statement

Simulation of the processes of heat propagation in a permafrost soil reduces to solving a three-dimensional equation of contact (diffusion) heat conductivity with inhomogeneous coefficients in a domain, including the localized heat of the phase transition – an approach that allows to solve a Stefan-type problem without pre-determination of boundary of phase transition. Let consider a three-dimensional area large enough to avoid artificial influence of lateral and bottom boundaries. The equation for the temperature $T=T(t,x,y,z)$ in the soil has the form

$$\rho\left(c_v(T) + k\delta(T - T^*)\right)\frac{\partial T}{\partial t} = \operatorname{div}\left(\lambda(T)\ \operatorname{grad}\ T\right), \qquad (1)$$

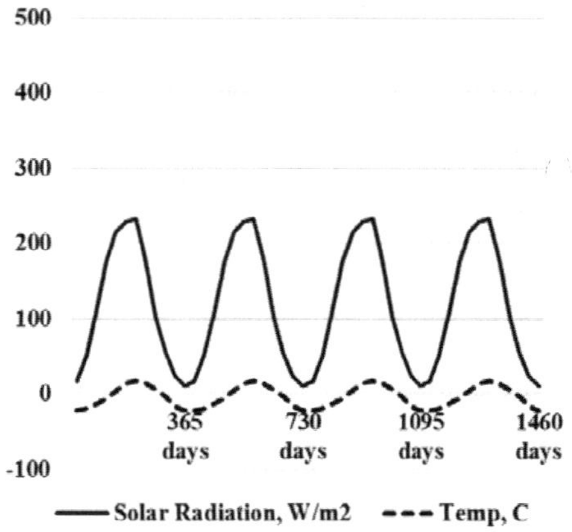

Fig. 1. The cyclic changes in solar radiation $q(t)$ (solid line) and air temperature T_{air} (dashed line) during 4 years.

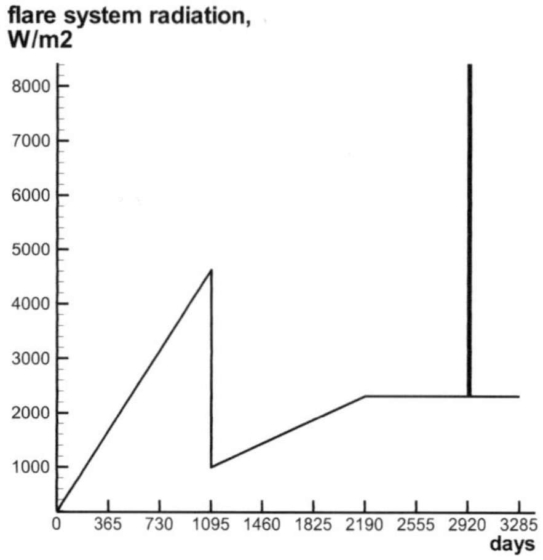

Fig. 2. The heat flux $f(x,y)$ from a vertical flare system.

where $\rho=\rho(x,y,z)$ is density [kg/m^3], $T^*=T^*(x,y,z)$ is temperature of phase transition, $c_v(T)=\begin{cases} c_1(x,y,z), \text{ for } T<T^*, \\ c_2(x,y,z), \text{ for } T>T^* \end{cases}$ is specific heat [J/kg K], $\lambda(T)=\begin{cases} \lambda_1(\text{x,y,z}), \text{ for } T<T^*, \\ \lambda_2(\text{x,y,z}), \text{ for } T>T^* \end{cases}$ is thermal conductivity [W/m K], $k=k(x,y,z)$ is specific heat of phase transition, δ is Dirac δ-function.

The coefficients included in equation (1) may vary at different points in the computational domain because of heterogeneity of the soil and possible presence of engineering structures. The ground surface is the main zone of formation of the natural thermal fields. On this surface the equation of balance of flows is used as a boundary condition, with taking into account the main climate factors: air

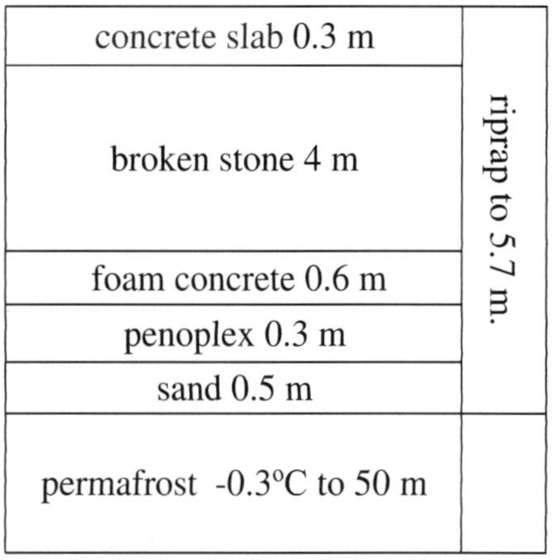

Fig. 3. A riprap for a vertical flare system

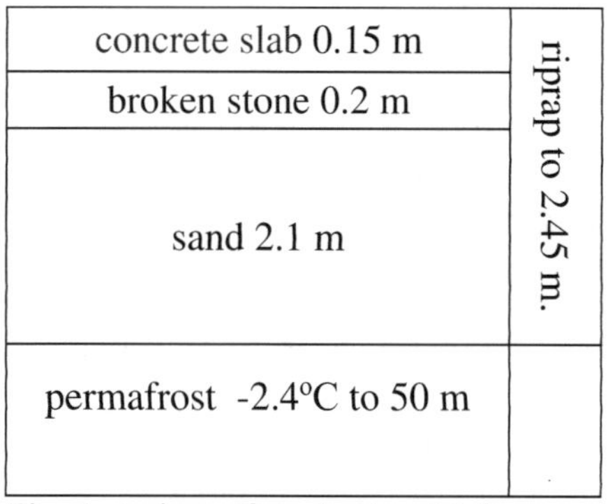

Fig. 4. A riprap for a horizontal flare system

temperature and solar radiation. In Fig. 1 the cyclic changes in solar radiation q and air temperature T_{air} during 4 years are shown.

We will consider two types of flare systems: horizontal flare system (HFS), when the flame is in contact with the ground, and vertical flare system (VFS), when the flame of burning gas does not

IOP Conf. Series: Journal of Physics: Conf. Series **899** (2017) 092004 doi:10.1088/1742-6596/899/9/092004

touch the ground, but effect on it by thermal radiation. For the position of HFS it is necessary to set a fixed temperature equal to the flame temperature. The zone of influence of the VFS flare may be approximated as a circle with radius R_f. As a boundary condition on the surface of the ground, the following equation of the balance of heat flows is used, which also takes the radiation from the flare:

$$f(x,y,t)+\alpha q+b(T_{air}-T\big|_{z=0})=\varepsilon\sigma(T^4-T_{air}^4)+\lambda\frac{\partial T}{\partial z}\bigg|_{z=0}, \qquad (2)$$

where $f(x,y,t)$ is a density of heat flow from the flame. It may be determined as

$$f(x,y)=\begin{cases}0, \text{ if } (x-x_f)^2+(y-y_f)^2>R_f,\\ \left(1-\left(\dfrac{\sqrt{(x-x_f)^2+(y-y_f)^2}}{R_f}\right)^2\right)P_f, \text{ if } (x-x_f)^2+(y-y_f)^2\le R_f.\end{cases}$$

The heat fluxes (for example, Fig. 2) from the flares has an extremely influence on the soil in compare with the natural heat sources: sun and air (Fig. 1). For example, in Fig. 2, the scenario of the operation of a vertical flare system is presented, taking into account an emergency discharge of gas in a short time interval.

Table 1. Thermal parameters of the riprap layers and the soil.

	Thermal conductivity, frozen W/(m K)	Thermal conductivity, meted, W/(m K)	Volumetric heat, frozen, kJ/(m³ K)	Volumetric heat, melted, kJ/(m³ K)	Heat of phase transition, kJ/(m³ K)
concrete slab	1.93	1.57	2150.0	3490.0	141504.0
broken stone	1.75	1.63	2160.0	2630.0	83616.0
sand	1.83	1.39	1536.0	1983.0	69505.8
permafrost soil	1.89-2.22	1.68-1.75	2200.0-2350.0	2750.0-2780.0	90983.7-10120.0

The numerical method is based on an algorithm that proved to be useful for finding thermal fields from underground pipelines [2-3], but taking into account the specifics associated with the possible phase transitions in the soil [4]. A finite-difference method is used that makes it possible to use the method of splitting with respect to spatial variables, for better organization of numerical calculations for various engineering objects operating in permafrost zones [5-13].

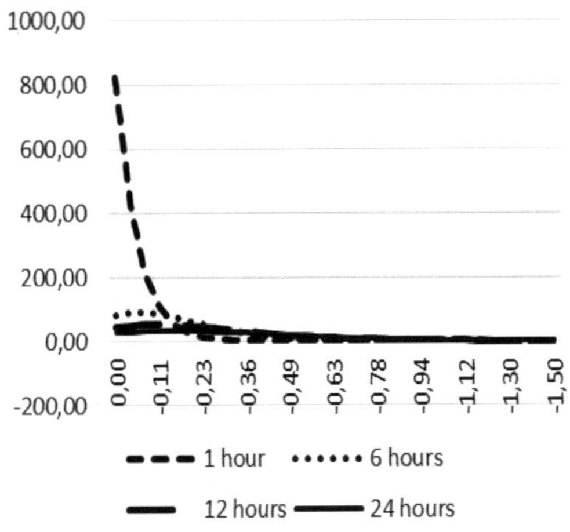

Fig. 5. Temperature in soil under HFS: heating during 1 hour and cooling in the following.

3. Numerical results

Numerical simulations were carried out in order to find optimal combination of materials for ripraps under the base of flare systems and to give a long-term forecast of the dynamics of the thawing zones during the start of the flare systems operation in March. A number of variants of such ripraps were considered, for example, in Fig. 3 and Fig. 4.

In the following, we will consider a permafrost soil with parameters presented in Table. 1.

To simulate the operation of the HFS, the temperature distribution from the flare on the surface of the ground is set in a rectangle area. The hottest area is 820°C in the center, then the temperature

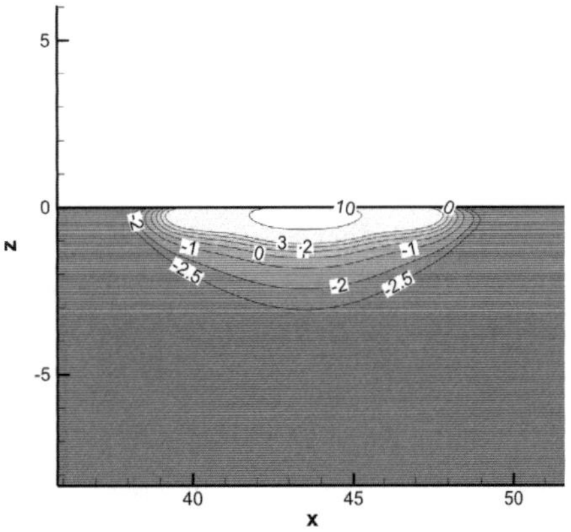

Fig. 6. Temperature field under HFS on March for 8 years of exploitation.

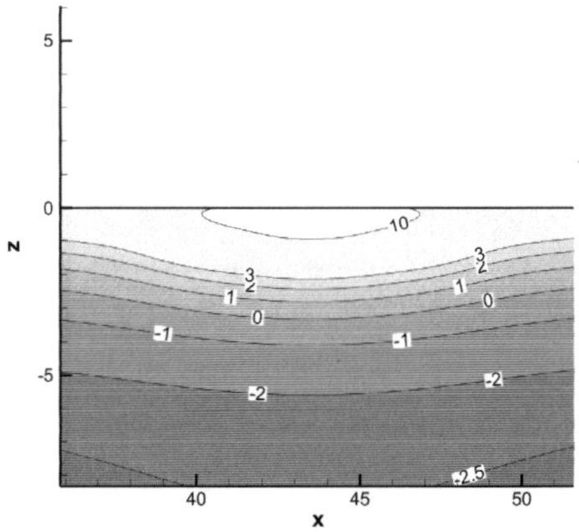

Fig. 7. Temperature field under HFS on September for 8 years of exploitation.

decreases to the boundary of the flare platform to the background temperature. We assume that the system operates in the following periodic mode: the gas flaring is carried out during 1 hour every 7 days. Respectively, the temperature increases to 820°C only in this short period and then the soil is under influence of natural thermal fields and is cooling. In Fig. 5 the temperature profiles under the center of HFS are shown for the cooling period.

This mode of operation leads to the computations have to be performed with variable time steps for the long-term forecasting of thawing zones. In Fig. 6 and Fig. 7 the thermal fields near the surface are shown on March and September, respectively, after simulation of 30 years of exploitation.

Now, let consider a vertical flare system operation by scenario, presented in Fig. 2. As a result of numerical simulations of thermal fields forming as a result of VFS operation, it was found, that on the "ground-air" surface in the zone of influence of VFS there still are under significant seasonal

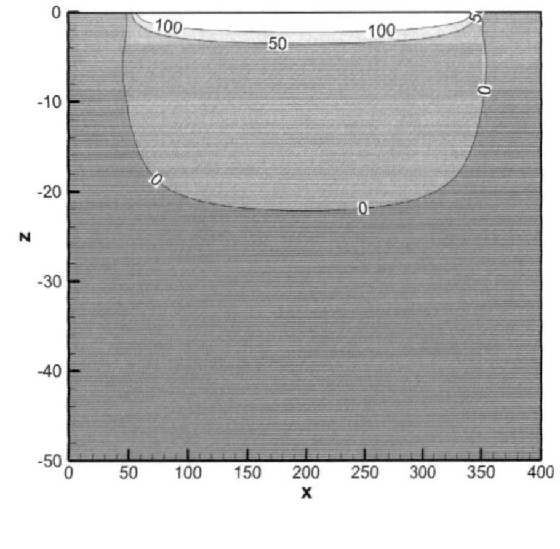

Fig. 8. Thawing lines in the soil without a riprap under HFS after 30 years of exploitation in March.

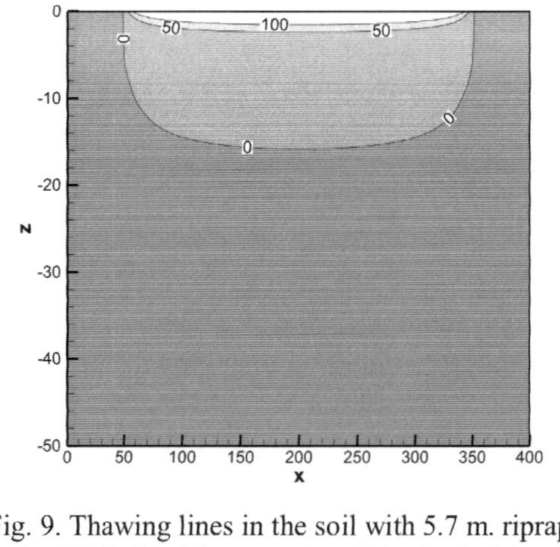

Fig. 9. Thawing lines in the soil with 5.7 m. riprap under HFS after 30 years of exploitation in March.

temperature fluctuations. The temperature on the surface of the zone of influence of the flame increases; the growth of the thawing zone stabilizes at 22-23 meters and 16-17 meters for the soil without a riprap (Fig. 8) and with a riprap (Fig. 9), respectively. Comparison of scenarios with an emergency flaring and an ordinary mode shows that the surface heating increases by 50°C and the effect of this is no more than 1°C after six months. The melted zone formation is a result more by the total power of the VFS supported for a long time. A short-term increase in the power of the VFS will only heat up near-surface layers of the soil. It is preferable to carry out emergency discharges for VFS during the cold season, in order to compensate for the additional heat.

4. Conclusion

The developed models and programs allow to carry out calculation of long-term effect on permafrost thawing from horizontal and vertical flare systems, taking into account various scenarios for emergency gas discharge. It is possible to make computations for a long time interval with variable time steps and, if necessary, to simulate any technological processes that take place in a limited period of time. The estimations of the temperature profiles in the ground can serve as a guide for constructing an optimal riprap under flare systems, and allow to determine the temperature in different layers of ground and to take into account heat resistance of insulations materials for stabilization of the thawing zones in permafrost.

Acknowledgments
This work was supported by RFBR projects (16-01-00401).

References

[1] Nelson F E, Anisimov O A and Shiklomanov N I 2002 *Natural Hazards* **26**(3) 203
[2] Vaganova N A 2008 *Proceedings of the Steklov Institute of Mathematics* **S1** S260
[3] Vaganova N 2014 (*AIP Conference Proceedings* vol 1631) (New York: AIP Publishing)
[4] Samarsky A A and Vabishchevich P N 1995 *Computational Heat Transfer, Volume 2, The Finite Difference Methodology* (Wiley)
[5] Filimonov M Y and Vaganova N A 2012 *9th International Pipeline Conference(IPC 2012) (ASME Conference Proceedings* vol 4) p 133
[6] Filimonov M Y and Vaganova N A 2013 *Applied Mathematical Sciences* **7** 7151
[7] Filimonov M and Vaganova N 2014 *Academic Journal of Science* **3** 121
[8] Vaganova N A and Filimonov M Y 2015 *Lect. Notes Comput. Sci.* **9045** 385
[9] Vaganova N A and Filimonov M Y 2015 (*AIP Conference Proceedings* vol 020016) (New York: AIP Publishing)
[10] Vaganova N and Filimonov M 2015 (*CEUR Workshop Proceedings* vol 1513) p 42
[11] Filimonov M Yu and Vaganova N A 2016 *Journal of Physics: Conference Series* **754** 112004
[12] Vaganova N and Filimonov M 2016 (*CEUR Workshop Proceedings* vol 1662) p 253
[13] Vaganova N A and Filimonov M Y 2016 (*AIP Conference Proceedings* vol 1789. 020019) (New York: AIP Publishing)

IOP Conf. Series: Journal of Physics: Conf. Series **899** (2017) 092005 doi:10.1088/1742-6596/899/9/092005

Validation of the thermal NOx emissions model from a gas fuel combustor under atmospheric pressure

P S Filippov, P Y Khudyakov and A F Ryzhkov

Ural Federal University, 19 Mira street, Ekaterinburg, Russia

E-mail: fps_proxi@mail.ru

Abstract: A description of various experimental and numerical investigation on NO_X emissions during the gas fuels combustion is presented. The selection of the validation object is justified. The NO_X formation and decomposition mechanisms are analyzed. The assembly of thermal NO_X formation and decomposition mechanisms was selected. Validation of the thermal NO_X emissions model by CFD-simulation was carried out. Validation is based on comparison of the numerical simulation results with well-known experimental data. The results of numerical simulation are analyzed. A method for improving the NO_X emissions computation from the gas fuel combustor is proposed.

1. Introduction

Currently, one of the most promising (for efficiency and environmental indicators) power generating system are integrated gasification combined cycles (IGCC). Approximately 2/3 of the IGCC output is production by the gas turbine, the remainder by steam-power cycle. The gas turbine fuel in IGCC is synthesis gas (syngas) produced by fossil fuel oxygen- or air-blown gasification (lower heating value (LHV) – 4 to 13 MJ/m^3 at 273 K, 1 atm) [1]. During the syngas combustion technologies development the much attention is paid to the environmental indicators of IGCC combustor, namely to nitrogen oxides (NOx) emissions.

The results of experimental investigation of NO_X emissions during the various gas fuels combustion are presented in [2, 3, 4]. Hasegawa et al. [2] experimentally evaluated the effect of nitrogen dilution of oxidant and gas fuel (a mixture of CO and H_2) on the rate of thermal and fuel NO_X emissions of two-stage combustor with diffusion burner. Experiments have shown that dilution significantly reduces the thermal NO_X emissions over a wide range of the equivalence ratio in the primary combustion zone (φ_p) and slightly affects the fuel NO_X emissions in the range $\varphi_p>1$. Feng et al. [3] shows the effect of the CO_2 content on the NO_X decomposition rate in methane combustion. Experiments have shown that the presence of CO_2 has little effect on the NO_X decomposition. Liu et al. [4] investigated the catalytic reduction of NO_X emissions due to injection of hydrazine hydrate into the hot exhaust gases. Authors identifies two temperature ranges of exhaust gases for effective denitrification: 820-960 K and 1260-1330 K.

At present, a CFD-methods are well established for investigation of IGCC combustors behavior, but before the commencement of work it is expedient to verify the NO_X formation and decomposition mechanisms.

Numerical studies of NO_X formation and decomposition mechanisms have been carried out in [5, 6, 7, 8]. Jurena et al. [5] numerically investigated the influence of the Magnussen constant of Eddy Dissipation combustion model on the gas combustion characteristics in diffusion burner. The investigation was carried out based on comparison of calculated results and experimental data. They showed that the optimal value of the Magnussen constant for predicting the gas combustion characteristics is 1.2. Goswami et al. [6], Volkov et al. [7] verified the thermal and fuel NO_X formation and decomposition mechanisms database. Bowman et al. [8] describes in detail the mechanisms of pollutant formation and decomposition during the gas combustion. Woo et al. [9] carried out the numerical and experimental studies of the fuel NO_X formation in the CH_4-NH_3 mixture combustion. The N_2/O_2 ratio in the oxidizer was varied in order to determine the regime of low-NO_X CH_4-NH_3 mixture combustion.

At present, in operating IGCC application the fossil fuel entrained-flow gasification technology and the cold gas cleanup (CGC) system for pollutant (fuel sulfur- (H_2S, COS) and nitrocompound (NH_3, HCN), mercury, dust, etc.) removed from syngas before combustion in gas turbine combustor. Fossil fuel entrained-flow gasification technologies provides the hydrocarbons (prompt NO_X sources) content in syngas does not exceed 2 vol.% [1]. CGC system provides fuel nitrocompounds (fuel NO_X sources) content in syngas <50 ppm [10]. Such negligible concentrations of hydrocarbons and fuel nitrocompounds in syngas provides to evaluate the IGCC combustor NO_X emissions by only thermal NO_X content in exhaust gases, without taking into account the prompt and fuel NOx availability.

In present work carried out the validation of the thermal NO_X emissions model (include the NO_X and other chemical compounds formation and decomposition mechanisms) in gas fuel combustor, which is similar in composition to the syngas of coal oxygen-blown gasification. Validation was carried out based on comparison of calculated results with well-known experimental data.

2. Method

Validation of the thermal NOx emissions model is based on a comparison of the calculated results (by Ansys CFX) and experimental data published by Hasegawa et al. [2]. According to Hasegawa et al. [2], in the experimental combustor (figure 1) organized two-stage supply of oxidizer (air). It organized to simulate the industrial gas turbine combustor operating conditions, namely to achieve the required flame length and temperature distribution in the longitudinal section of the combustor.

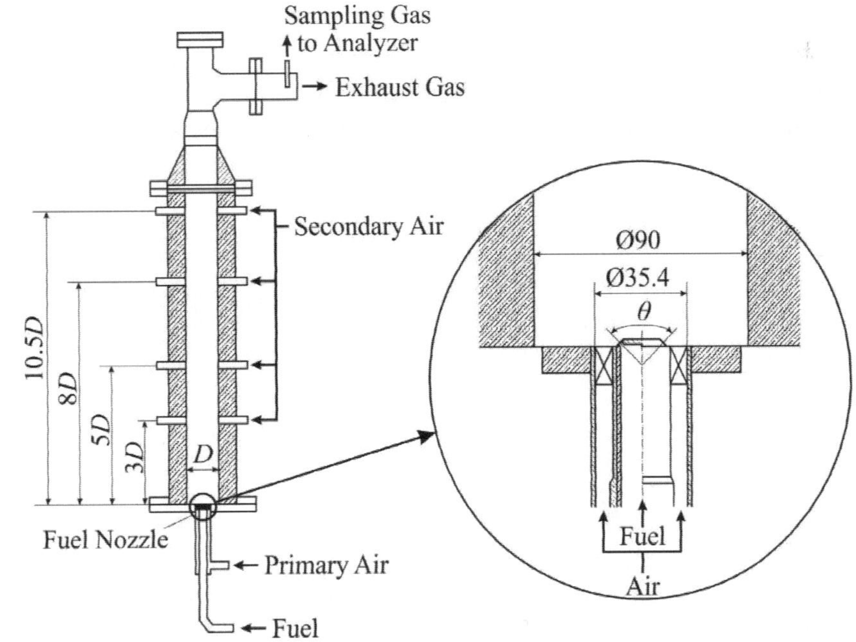

Figure 1. Scheme of experimental combustor [2].

Hasegawa et al [2] considered the influence of the equivalence ratio in the primary combustion zone (φ_p) of the experimental combustor on the thermal NOx emissions. In the experiments, the temperature of the exhaust gases was maintained at ~1500°C. The gas fuel flow rate and composition, the total air flow rate (the sum of the primary and secondary air) and the temperature of the reacting agent were unchanged. Gas fuel is a H_2-CO mixture with a ratio of H_2/CO=30/70 vol.%/vol.% («model syngas» of coal oxygen-blown gasification).

A CFD-model of the experimental combustor (figure 1) is created to the numerical simulation and the following parameters was set:

1) combustion model – Eddy Dissipation, adjusted by Jurena et al. [5];
2) turbulence model – Shear Stress Transport, based on work of Menter [11];
3) thermal radiation model – Monte Carlo;
4) combustion mechanisms – standard mechanisms of H_2 and CO combustion in terms of the radicals;
5) thermal NOx formation and decomposition mechanisms – adopted from Goswami et al. [6] (reactions №25-28 and 101). In numerical simulation, thermal NOx emissions were evaluated by calculated NO content in exhaust gases. Because NO content in exhaust gases during the gas fuel combustion is about 90% of all NOx [8].

The parameters of the reacting agents are setting according to the experiment [2]. The temperature of the exhaust gases was maintained at 1500±20°C. The equivalence ratio φ_p was varied in the range 0.5 to 2. The mass flow rate of gas fuel was set according to the experiment, the mass flow rates of primary and secondary air were determined by equations (1) and (2).

$$\varphi_p = \frac{\left(\dfrac{M_F}{M_{PA}}\right)^{act}}{\left(\dfrac{M_F}{M_A}\right)^{st}} \tag{1}$$

$$M_T = M_{PA} + M_{SA} \tag{2}$$

where M_F, M_{PA}, M_{SA}, M_T, M_A, – mass flow rates of gas fuel, primary air, secondary air, total air and stoichiometric air; $\left(\dfrac{M_F}{M_{PA}}\right)^{act}$ – actual fuel-air ratio in the primary combustion zone of experimental combustor; $\left(\dfrac{M_F}{M_A}\right)^{st}$ – stoichiometric fuel-air ratio in the primary combustion zone of experimental combustor.

3. Results

Because NO content in exhaust gases during the syngas combustion is about 90% of all NOx [8], the thermal NOx emissions were evaluated by calculated NO content in exhaust gases.

A comparison of calculated results and experimental data on NOx emissions depending on φ_p is presented in table 1.

Table 1. Comparison of calculated results and experimental data.

Equivalence ratio in the primary combustion zone (φ_p)		0.62	1	1.5	2
Excess air coefficient in the primary combustion zone (α_p)		1.6	1	0.7	0.5
$NOx_{th}(16\%O_2)$, ppm	Hasegawa et al. [2]	36	97.2	19	14
	Present work	41	118	35	16.8
Spread of computational results from experimental data, %		12	18	46	17

A visual comparison of the calculated results (predicted curve) and the experimental data [2] is shown in figure 2.

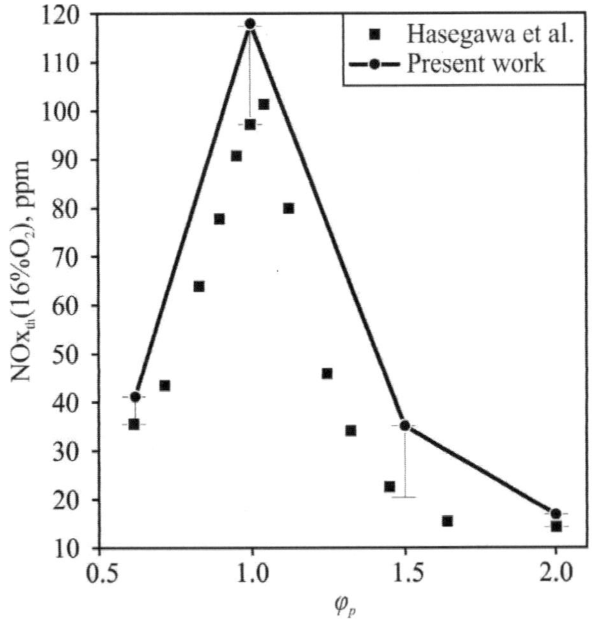

Figure 2. Comparison of the predicted curve and the experimental data [2].

The total RMS deviation of CFD-simulation was 0.15. The deviation of the predicted curve from the experimental data is observed in all points. Probably, the deviation is due to the CFD solver RMS deviation at prediction the flame maximum temperature coupled with the inaccuracy in determining and controlling the exhaust gases temperature or φ_p in the experiment.

4. Conclusions

Numerical simulation showed that the total RMS deviation in CFD prediction is 0.15. Generally, it is seen that the predicted curve of the combustor NO_X emissions is closely spaced to the experimental data. That indicates the possibility of further use of the developed CFD-model for the prediction of NO_X emissions from combustors of IGCC with CGC systems. For the prediction of NO_X emissions from combustors of IGCC with warm (WGC) or hot gas (HGC) cleanup systems, it is advisable to validation the fuel NO_X emissions model. Because in the IGCC with WGC or HGC systems the fuel nitrocompounds content in syngas before the gas turbine combustor can reach up to 1000-3000 ppm.

The selected assembly allows to predict NO_X emissions with satisfactory accuracy. In case of necessary to improve the accuracy of results, it is recommended to use a more detailed chemistry in terms of the effect of the NO_2, N_2O, etc. formation and decomposition on total NO_X emission from gas turbine combustor.

Acknowledgments

The reported study has been supported by RFBR, research project No. 16-38-00479

References

[1] Hasegawa T 2010 *Energies* **3** 335-449.
[2] Hasegawa T, Tamaru T 2007 *J. Eng. Gas Turbines Power* **129** 358-369
[3] Feng B, Ando T, Okazaki K 1998 *JSME Int. J., Ser. B* **41** 959-965
[4] Hong L, Chen D, Wang D 2012 *Chinese Journal of Environmental Science* **33** 2901-2908
[5] Jurena T, Belohradsky P, Skryja P, Hudak I 2016 *Chem. Eng. Trans.* **52** 1261-1266

[6] Goswami M, Volkov E N, Konnov A A, De Goey L, Bastiaans R 2008 *Low Emission Gas Turbine Technology for Hydrogen-rich Syngas* Netherlands
[7] Volkov E N, Konnov A, Gula M, Holtappels K, Burluka A A 2009 *Proceeding of the 4th European Combustion Meeting*
[8] Bowman C T 1975 *Prog. Energy Combust. Sci.* **1** 33-45
[9] Woo M, Choi B C, Ghoniem A F 2016 *Energy* **114** 961-972
[10] Woolcock P J, Brown R C 2013 *Biomass Bioenergy* **52** 54-84
[11] Menter F R 1994 *AIAA Journal* **32** 1598-1605

Research of the chemical activity of microgrinding coals of various metamorphism degree

A P Burdukov, E B Butakov and A V Kuznetsov

Kutateladze Institute of Thermophysics SB RAS, Novosibirsk, Russia

E-mail: temkansu@yandex.ru

Abstract. In this paper, we investigate the effect of mechanically activating grinding of coals of various degrees of metamorphism by two different methods - determination of the flash time in a vertical tubular furnace and thermogravimetric analysis. In the experiments, the coals that had been processed on a vibrating centrifugal mill and a disintegrator, aged for some time, were compared. The experiments showed a decrease in the ignition temperature of mechanically activated coals - deactivation of fuel, as well as the effect of mechanical activation on the further process of thermal-oxidative degradation.

Introduction

Coal is one of the most important energy sources in the world. According to forecasts, by 2020, the share of coal in the global energy sector will reach 50%, primarily due to the very likely reduction in oil and gas consumption, as well as the revision of the policy on the development of nuclear energy [1].

At present, most coal-fired power plants operate using petroleum products to start and color the boiler during operation. The cost of these petroleum products significantly exceeds the cost of the coal itself, and therefore the introduction of new technologies that allow to exclude petroleum products from the energy processes that take place at the CHPP (Central Heating and Power Plant) is topical [2, 3]. One of such technologies is the technology of mechanical activation of microgrinding coal obtained at mills-disintegrators. In this paper, coal studies are conducted using two different methods: determination of the self-ignition temperature of coals in a vertical tubular furnace and thermogravimetric analysis because the first method allows you to observe the processes of ignition of coal, and the results of the second one can conclude about the further process of thermal decomposition of coal.

Vertical furnace tube

An experimental setup was created for investigating the process of ignition of pulverized-coal fuel with its preliminary grinding in mills different in energy intensity and impact on coal matter. The layout of the stand is shown in Figure 1a, the general view in Figure 1b. The main elements of the structure are: furnace, coal feeder, starting mechanism, supply air heater [4].

The furnace is a vertical heat-insulated steel pipe 1 m long, with an internal diameter of 40 mm suspended vertically. Using the system of low-voltage transformers, electrical heating is carried out. Over the entire length of the combustion chamber, photodiodes and thermocouples with a pitch of 100

IOP Conf. Series: Journal of Physics: Conf. Series **899** (2017) 092006 doi:10.1088/1742-6596/899/9/092006

mm are located in special holes for recording the flash and temperature, respectively. To stabilize the temperature at the entrance to the furnace, as well as to remove stray convective currents and combustion products, a stream of air pre-heated to the temperature in the chamber with a velocity of 0-100 mm/sec is fed into the chamber.

The trigger mechanism consists of a magnetic valve and a chamber with a volume of 45 mm3. Above the valve there is a dust collector, where samples weighing 0.1-1g are poured. Then air is pumped into the chamber (1atm) and coal dust is injected into the furnace. The maximum temperature that can be obtained when using existing transformers is 1000 ° C. The arrival of dust into the combustion chamber is recorded by a microphone. The flash is recorded with thermocouples and photodiodes.

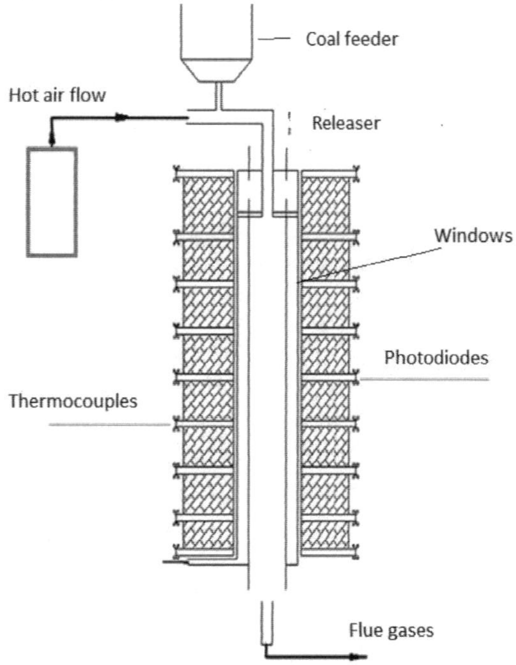

Figure 1a. Scheme of the experimental stand for studying the ignition of coal dust.

Figure 1b. Photo of the experimental stand.

In order to remove the external moisture, which can greatly affect the experimental results, the coal was pre-dried in a drying oven at 105 ° C for 4 hours before grinding.

Figure 2a show average times of ignition of coal dust of Kuznetsk coal for different temperatures and methods of grinding. It can be seen that the ignition time decreases with increasing temperature and a decrease in the size of the carbon particles. But it is important to note that the ignition time after grinding in the vibrating centrifugal mill is less than the ignition time after grinding in the disintegrator for all the cases under consideration. A similar result is observed for coal of a different degree of metamorphism - brown (Figure 2b). With increasing temperature, the differences in ignition timing become less. The results of the conducted experiments show that, after grinding in the disintegrator and some time holding, deactivation is possible a decrease in the reactivity of coal dust. It is known that the preliminary oxidation reduces the reactivity of coal [5]. The lower activity of aged coal is explained by the disappearance of active oxidation centers.

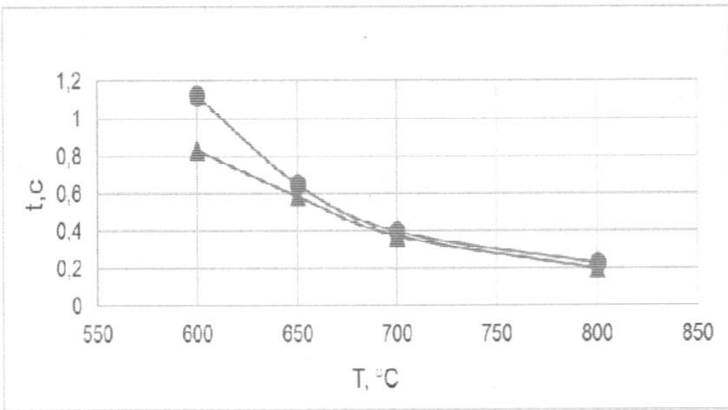

Figure 2a. Time of ignition of coal dust of Kuznetskiy coal in relation to temperature. o - after grinding the coal in the disintegrator, Δ - after grinding the coal in the vibrating centrifugal mill

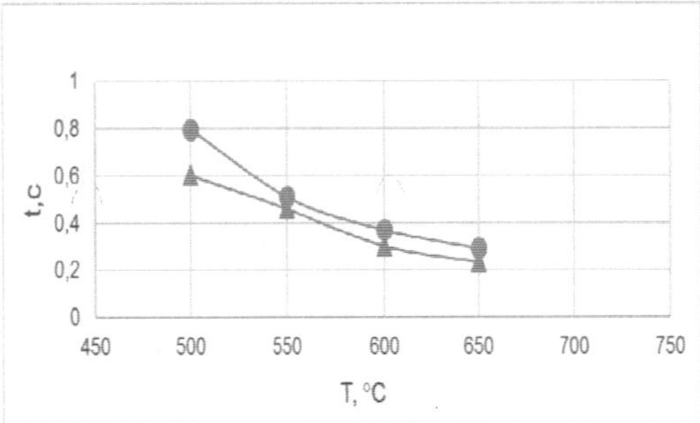

Figure 2b. Time of ignition of coal dust of brown coal in relation to temperature. o - after grinding the coal in the disintegrator, Δ - after grinding the coal in the vibrating centrifugal mill.

Crushing in the disintegrator allows to achieve a more complete disclosure of organomineral splices in comparison with the vibrational centrifugal mill and makes coal more active. But at the same time, the rate of oxidation in air increases at room temperature. Sifting of coal dust also intensifies the process of coal oxidation when interacting with air oxygen and causes the decrease of coal activity with time.

At the same time, the observed decrease in the reactivity of coal dust after grinding in the disintegrator - decontamination of coal, indirectly confirms the initial more complete disclosure of organomineral intergrowths compared to a vibrating centrifugal mill.

Thermogravimetric analysis

Synchronous thermal analysis (STA), which included simultaneous thermogravimetric determinations (TG), differential scanning calorimetry (DSC), and mass-spectrometric analysis of the recovered gas was performed on a STA 449F1 Jupiter® instrument combined with a quadrupole mass-spectrometer QMS 403D Aëolos® (Figure 3). The experiments were carried out in a temperature range of 30-700 ° C in an atmosphere of synthetic air (80% vol. Ar, 20% v / v O_2), gas flow rate 20 ml/min argon, 5 ml/min oxygen. Open crucibles of Al_2O_3 were used, the heating rate was 10 degree/min.

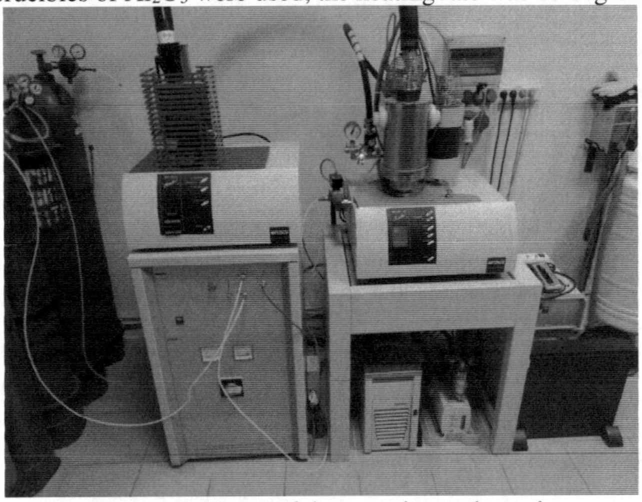

Figure 3. Photo of the experimental stand.

The processing of the experimental data was carried out using the Proteus analysis software package.

The experiments also used brown and hard coal, which were processed by a vibrating centrifugal mill and a disintegrator, aged for some time. Figures 4-5 show the results of thermogravimetric analysis.

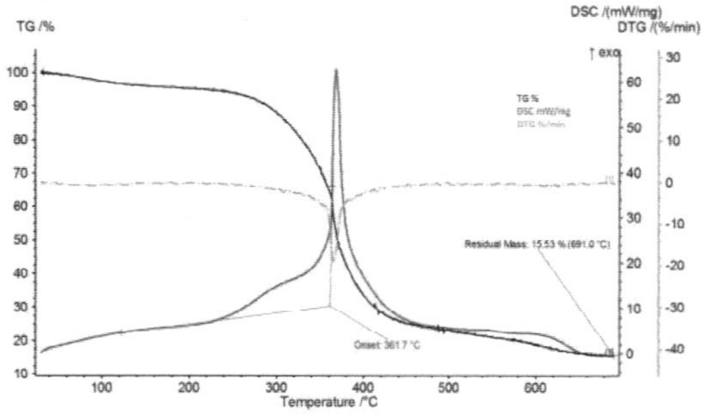

Figure 4a. Results of the derivatographic analysis of brown coal treated with a disintegrator. Black – TG, green – DTG, red – DSC.

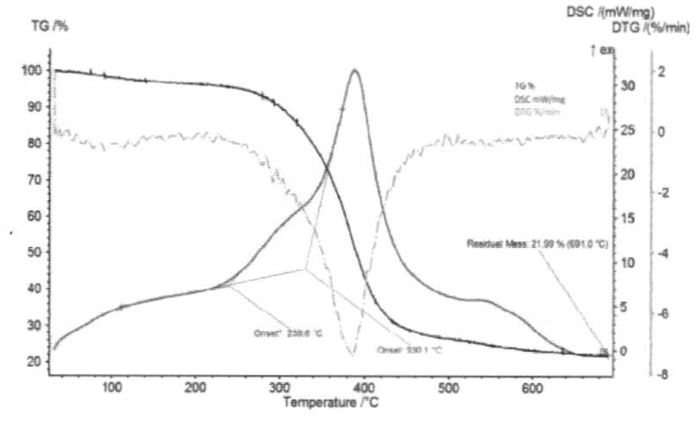

Figure 4b. Results of the derivatographic analysis of brown coal treated with a vibrating centrifugal mill. Black – TG, green – DTG, red – DSC.

The ignition temperature is determined by the method of tangents and is marked "Onset" [6], it is worth noting that for brown coals, the ignition temperature of coals treated with a disintegrator is also higher than the ignition temperature of coals treated with a vibrating centrifugal mill, which correlates with the results obtained in experiments on a tubular furnace. However, the main process of thermal decomposition takes place more efficiently in coals processed on a disintegrator, which is characterized by a more complete burning out of the samples (Residual Mass) and a more abrupt energy release (DSC curve).

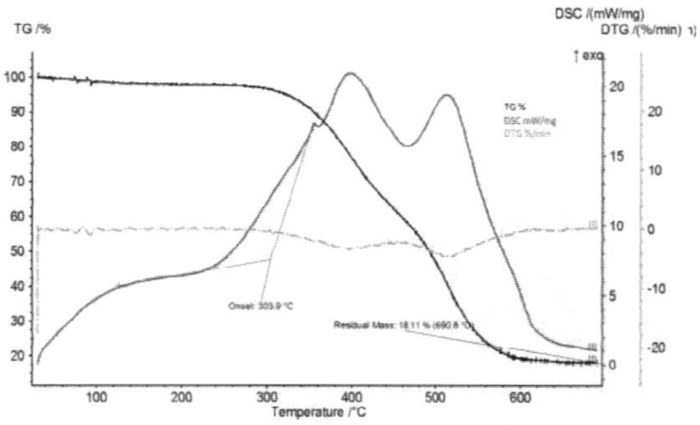

Figure 5a. Results of the derivatographic analysis of Kuznetskiy coal treated with a disintegrator. Black – TG, green – DTG, red – DSC.

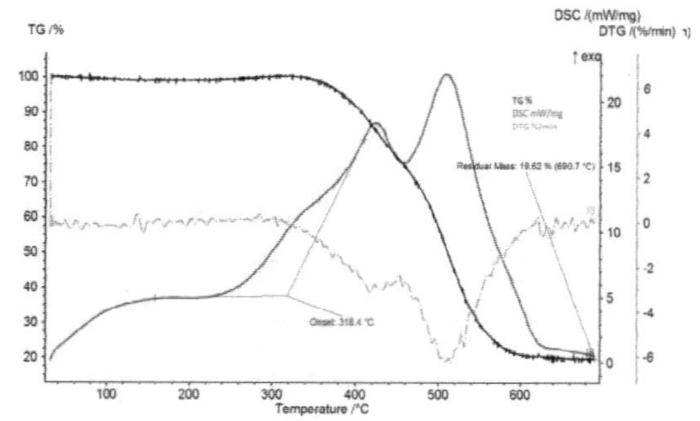

Figure 5b. Results of the derivatographic analysis of Kuznetskiy coal treated with a vibrating centrifugal mill. Black – TG, green – DTG, red – DSC.

For coal, the temperature determined by the method of tangents turned out to be slightly lower for coal treated on a disintegrator, but, as in the case of brown coal, positive changes in the dynamics of the thermal decomposition process are seen.

Conclusion

1. It should be noted the increase in the ignition time of mechanically activated coals in comparison with coals treated with a vibrating centrifugal mill.
2. The results of the experiments showed the effect of exposure after grinding on the ignition temperature of coals.
3. Derivatographic analysis data show positive changes in the behavior of the thermal decomposition process for mechanically activated microgrinding coals and a more complete burnout of coal dust samples.

Acknowledgments

The work was financially supported by the Russian Ministry of Education and Science, according to Subsidiary Agreement No. 14.607.21.00150 (unique project identifier RFMEFI60716X0150).

References

1. V.N. Churashev, G.V. Chernova Problems of substitution of fuel oil in heat power engineering // New technologies of fuel compression: their current state and future use: Proc. All-Russia. Scientific-techn. Seminar. - Moscow., 2001. - P. 262-269.
2. Burdukov A.P., Butakov E. B., Popov V.I., Chernetskiy M. Yu., Chernetskaya N. S. The use of mechanically activated micronized coal in thermal power engineering // Thermal Science. – 2016. - Vol. 20. – P. 23 – 33.
3. Burdukov A.P., Popov V.I., Lomovsky V.I, Yusupov T.S. Effect of mechanochemical activation on coal combustion in torch // Chemistry for Sustainable Development. – 2009. – Vol. 17. – P. 465-472.
4. Butakov E.B., Kuznetsov A.V. Research of influence of mechanical activaton on combustion of coal fuel // Bulletin of the Tomsk Polytechnic University. Geo Assets Engineering. – 2016. - Vol. 327, № 6
5. Beamish, B.B., Barakat, M.A., and St. George, J.D., 2001, Spontaneous combustion propensity of New Zealand coals under adiabatic conditions: International Journal of Coal Geology, v. 45, no. 2–3, p. 217–224, doi: 10.1016/S0166-5162(00)00034-3.
6. Yong Chen, Shigekatsu Mori, Wei-Ping Pan. Studying the mechanisms of ignition of coal particles by TG-DTA // Thermochimica Acta. – 1996. – V. 275. – P. 149–158.

IOP Conf. Series: Journal of Physics: Conf. Series **899** (2017) 092007 doi:10.1088/1742-6596/899/9/092007

Influence of $CuSO_4$ initiation additive on kinetic characteristics of coal oxidation and pyrolysis processes

K.B. Larioinov[1], I.V. Mishakov[1,2], A.A. Gromov[1,3], A.S. Matveev[1] and A.V. Zenkov[1]

[1]Tomsk Polytechnic University, Tomsk, Russian Federation
[2]Boreskov Institute of Catalysis SB RAS, Russian Federation
[3] Nurnberg Technical University Georg-Simon-Ohm, Nurnberg, Germany

e-mail: laryk070@gmail.com

Abstract. Coals oxidation and pyrolysis processes of different metamorphic grades containing 5%wt of initiation additive were studied. The experiment was performed by thermogravimetric analysis method at a heating rate of 2.5°C/min to a maximum temperature of 600°C in an atmosphere of air and nitrogen. Based on the results of TGA, an analytical evaluation of kinetic characteristics of pyrolysis and oxidation processes was conducted. It has been established that addition of $CuSO_4$ initiating agent leads to significant decrease in the initial temperature of oxidation and pyrolysis processes, which promotes the shift of the reaction to a low-temperature region. Change of the initial reaction temperature ΔT_i for oxidation process was 30°C, ΔT_d for pyrolysis was 45°C. Effect of initiation additive was also expressed in significant reduction of sample residence time within volatile substances sublimation zone: for oxidation Δt_i was 14 minutes, for pyrolysis Δt_d was 18 min. For modified sample characterized by lower stage of metamorphism, the decrease in oxidation time Δt_f was the greatest (22 min), which is explained by the high content of volatile substances. According to mass-spectrometric analysis, presence of SO_2 (intense peaks at 290°C and 440°C) was recorded in modified samples oxidation products, which is explained by partial hydrolysis of copper sulfate.

1. Introduction

The modern vector of coal-fired power industry development is aimed at technological improvement of equipment and processes taking place in it, with the purpose of improving ecological, technical and economic performance of fuel-burning plants [1]. One of the promising technologies is catalytic combustion of coal in boiler furnaces [2]. Catalytic additives application for oxidation processes intensification of solid fuels has a significant effect on the kinetics of the process. The shift of oxidation reaction to the low-temperature region helps to reduce emissions of nitrogen and sulfur oxides, while the average combustion rate increases [3]. Most of the research in the field of fuel catalytic conversion [1-4] is aimed at the use of rare earth and transition metal oxides, which promote a decrease in activation energy of volatile substances sublimation, which leads to a shift in the initial ignition temperature of coal. In particular, Gong [4] considered the effect of CeO_2 and Fe_2O_3 additives on oxidation kinetic characteristics of coals at different metamorphism stages. Thus, Liu et al. [5] mention that activation of metal oxides with carbon during layer decomposition appears in the high-temperature region beginning at 550°C. In studies by Zhao et al. [6], the effect of additives-precursors of metal oxide forms based on salts and bases was considered, which had a significant effect on the nature of the process in the low-temperature oxidation region. In previous paper [7] we carried out a study of the effect of additives with different nature ($NaNO_3$ and $CuSO_4$) on the kinetics of lignite and bituminous coal oxidation. It was shown that addition of nitrate additives promotes acceleration of volatile substances release and subsequent ignition in the region of lower temperatures comparing to the reference samples. In this paper, the effect of $CuSO_4$ initiation additive on the change in kinetic characteristics of bituminous coal oxidation and pyrolysis processes is considered.

2. Experimental section

2.1. Characteristic of researched samples

Samples of bituminous coal from "Osinnikovskaya" (sample 1) and "Listvyazhnaya" (sample 2) mines, located in Kemerovo region, were used within this work. Coarse samples of the initial coals (5-10 mm) were pulverized in a drum mill for 9 hours with equal mass ratio of grinding bodies and mill feed material. After grinding, the samples were fractionated on sieves with a mesh size of less than 80 μm. The particle size distribution was determined using Analysette 22 laser particle diffraction analyzer (Fritsch, Germany). The measurement results are shown in Table 1.

Table 1. Particle size distribution for samples of coal powders

Sample	X_{10}	X_{50}	X_{90}
1	4.6	20.4	57.6
2	3.6	21.1	53.2

The selected coal powders were dried in a drying oven at 105°C to a constant weight. Then, their physical characteristics were determined using standard methods [8]. The results are shown in Table 2.

Table 2. Physical characteristics of coals

Characteristic	Samples	
	#1	#2
Ash content, %wt	16	12
Moisture content, %wt	4	3
Volatile substances, %wt	15	29
Carbon, %wt	65	56
Humidity ratio, ml/g	2.7	2.9

As can be seen from Table 2, samples of coal #1 and #2 have comparable values of carbon amount and humidity ratio, while their content of volatile substances and ash differ significantly. Differences in characteristics are explained by different stage of metamorphism. It should be noted that morphology of the examined coal particles also differs significantly, which can be seen from the micrographs in Fig. 1, obtained with the SEM JSM-6460LV (JEOL, Japan) scanning electron microscope with an ion focused beam.

Particles in sample #1 have an elongated shape, close to cylindrical, with relatively uniform surface, while sample #2 consists of polyhedral spherical particles with amorphous structure and many surface cavities and cracks.

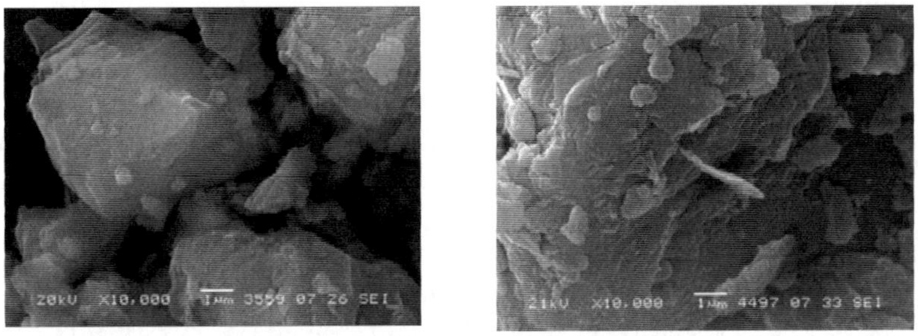

a b

Figure 1. REM images of bituminous coal samples:
a - sample #1, *b* - sample #2. REM data (x10000)

2.2. Application of initiation additive

$CuSO_4$ promoting additive was added into composition of coal samples by penetration method considering humidity ratio [9]. The content of copper sulfate in the prepared sample was 5%wt. Water-alcohol solution was used for penetration in the ratio H_2O/C_2H_5OH = 50:50. When the salt was completely dissolved, the solution was capillary applied to the prepared coal powders using mechanical dispenser. Further, the resulting slurries were placed in drying oven at 105°C for 20 hours. After drying, the agglomerated samples were pulverized until powdery state was achieved.

2.3. DTA and MS analyses

Research of coal samples thermal decomposition (oxidation) process was conducted using STA 449 C Jupiter synchronous thermal analyzer (Netzsch, Germany). The experiments were performed under the same conditions – at a heating rate of 2.5°C/min in corundum crucible with perforated cover (sample weight ~ 25 mg) in the temperature range 45-600°C. An inert gas, nitrogen, was used to organize the process in pyrolysis mode. A mixture of air (60 ml/min) and nitrogen (10 ml/min) was used as oxidizing medium. In the latter case, nitrogen was used as a shielding gas to ensure reliable operation of the analyzer and correct registration of the obtained data. For series of samples #1, calorimetric measurements (DSC) were performed to define thermal effects, as well as qualitative determination of oxidation products composition using STA attachment in the form of QMS 403 D Aeolos quadrupole mass spectrometer (Netzsch, Germany). All experiments were conducted at atmospheric pressure.

2.4. Kinetic analysis

Kinetic parameters of coal oxidation and pyrolysis were determined graphically (Fig. 2) using the obtained TG and DTG curves. In the oxidation mode (Fig. 2a), the points A and B on the TG curves characterize ignition temperature of exiting volatiles (T_i) and the end of combustible sample mass loss (T_f). Considering incomplete removal of volatile components during coal pyrolysis process (Fig. 2b) for given heating mode and selected maximum temperature, analysis of initiation additive effect was conducted taking into account temperature of sublimation beginning T_d (point A). In both cases, the characteristic point C on DTG curves reflects maximal value of reaction rate (w_{max}) and temperature at a given moment (T_1).

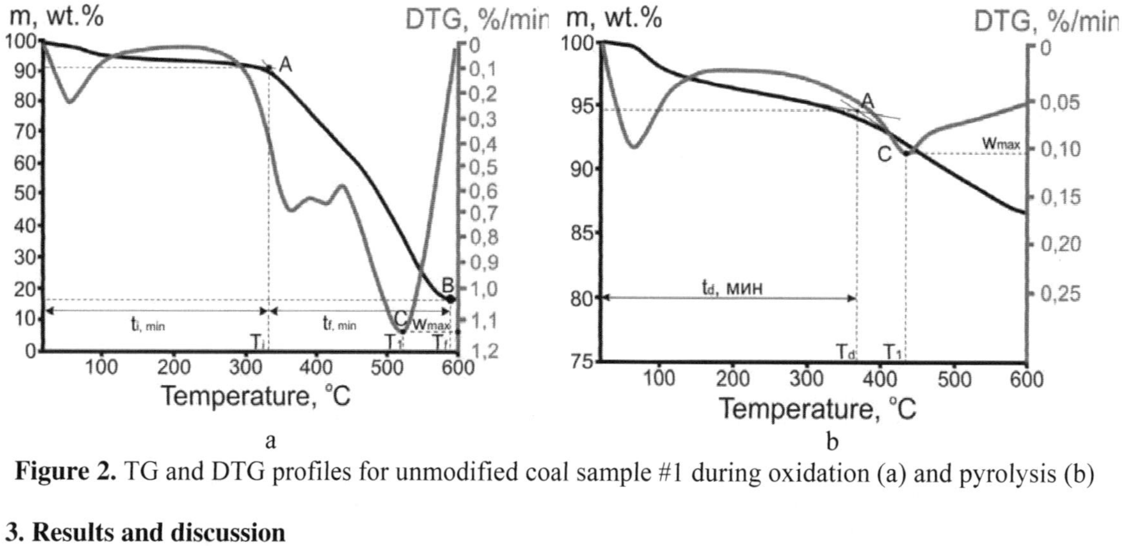

Figure 2. TG and DTG profiles for unmodified coal sample #1 during oxidation (a) and pyrolysis (b)

3. Results and discussion

3.1. Influence of initiation additive on coal oxidation and pyrolysis processes

Figure 3 shows experimental TG and DTG curves for oxidation and pyrolysis of coal samples, from which it follows that samples modified by $CuSO_4$ promoting additive have different character of the process in contrast to the reference samples.

IOP Conf. Series: Journal of Physics: Conf. Series **899** (2017) 092007 doi:10.1088/1742-6596/899/9/092007

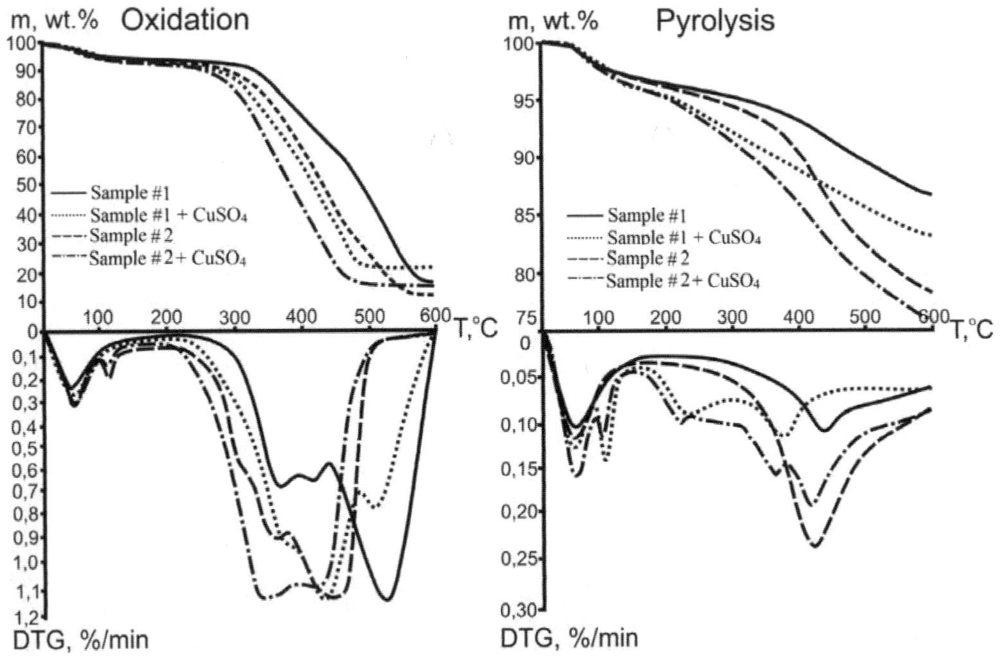

a b

Figure 3. TG and DTG curves for oxidation (a) and pyrolysis (b) of the researched coal samples

Obtained in oxidation mode of unmodified coal samples DTG curves (Figure 3a) have four distinct peaks. The first peak is due to evaporation of external moisture (up to 100°C), while the second and the third peaks (320-410°C) refer to combustion of volatile substances released from the surface and from the volume of particles as a result of their thermal destruction. The origin of the fourth peak (450-500°C) can be explained by intense burning of coke residue. In turn, two clearly pronounced maxima for test samples pyrolysis process associated with the removal of moisture (up to 100°C) and release of volatile substances (about 430°C) can be identified on DTG curves. It can be seen that series #2 samples pyrolysis occurs in a lower temperature range compared to samples of series #1, which is apparently explained by high content of volatile substances and presence of amorphous component in samples #2. Data analysis of samples containing 5%wt of copper sulfate shows that the presence of initiation additive leads to appearance of additional peak on DTG curve at T = 120°C, which is associated with dehydration of $CuSO_4{\cdot}5H_2O$ salt to form $CuSO_4{\cdot}H_2O$ monohydrate. The process is accompanied by marked decrease in the initial volatile substances release temperature and characteristic displacement of the curves corresponding to the process of coal oxidation and pyrolysis to the low-temperature region. Decrease in temperature values at reference points, where the rate of reactions has maximal value, also should be noted. It can be seen from Fig. 3a that DTG curves of modified samples oxidation process have monomodal form, which is explained by joint burning of volatile substances and coke residue formed on the surface of the particles. For the record, modified coal samples pyrolysis (Fig. 3b) is characterized by appearance of one more peak on DTG curves (220-240°C), which can be explained by partial destruction of copper sulfate as a result of hydrolysis.

3.2. Kinetic characteristics analysis

Kinetic characteristics of the process in oxidation and pyrolysis modes are presented in Tables 3 and 4, respectively.

Table 3. Kinetic characteristics of oxidation process

Sample	T_i,°C	T_f,°C	w_{max}, %/min	t_i, min	t_f, min	ΔT_i,°C	ΔT_f,°C	Δt_i, min	Δt_f, min
#1	330	590	1.17	114	104	35	85	14	20
#1+$CuSO_4$	295	505	1.13	100	84				
#2	300	570	1.15	102	108	25	80	10	22
#2+$CuSO_4$	275	490	1.11	92	86				

Table 4. Kinetic characteristics of pyrolysis process

sample	T_d,°C	w_{max}, %/min	t_d, min	ΔT_d,°C	Δt_d, min
#1	395	0.11	140	45	18
#1+$CuSO_4$	350	0.13	122		
#2	370	0.23	130	30	12
#2+$CuSO_4$	340	0.18	118		

Analyzing the obtained results (Tables 3 and 4), it can be concluded that addition of $CuSO_4$ initiation additive leads to a significant decrease of ignition temperature and sublimation of volatile substances. Maximal temperature decrease is observed for the samples, decomposition process of which took place in an inert medium (45°C for the sample {#1 + $CuSO_4$} and 30°C for the sample {#2 + $CuSO_4$}). In case of additionally modified samples #1 and #2 oxidation, the decrease in ignition temperature was 35°C and 25°C, respectively. The greatest temperature shift is observed for modified samples of series #1, characterized by a lower content of volatile substances (Table 1) and the nature of metamorphism. In [4] this effect is explained by relatively smaller number of longitudinal and transverse oxygen-containing bonds in molecular structure of high-carbon coals. It is important to note that the greatest intensity of change in the rate of mass loss coincides with the stages of oxidation and volatile substances release. A decrease in ignition temperature leads to a corresponding decrease in residence time of sample in the region of volatile substances sublimation t_i and t_d. It should also be noted that doping of the additive reduces total residence time of the sample, from the start of volatiles ignition to the mineral residue production t_f, which leads to an increase in average flow rate of the process. The observed temperature effects are most likely due to the weakening of oxygen-containing molecular bonds in the coal structure, which leads to a more active release of volatile substances. Figure 4 shows thermal effects and results of mass spectrometric analysis (m/z = 48, SO_2) recorded in oxidation mode of series #1 samples.

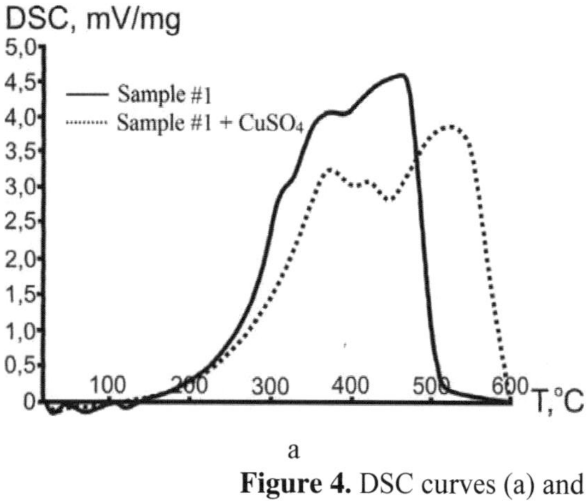

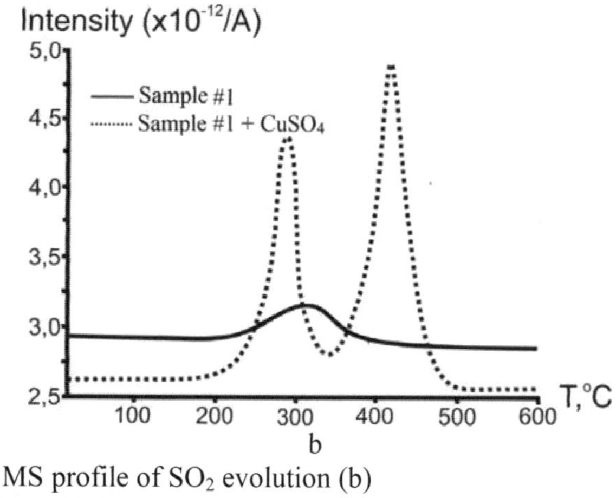

Figure 4. DSC curves (a) and MS profile of SO_2 evolution (b) for series #1 samples oxidation process

It can be seen that characteristic shape of DSC heat absorption and heat dissipation curves (Figure 4a) is largely similar to corresponding DTG profiles presented in Figure 3a. Endothermic and exothermic effects are associated with removal of coal external moisture, dehydration of salt and further oxidation of pyrolysis gas and coke residue. SO_x (m/z = 48) was detected in the gas-phase of modified sample oxidation products, and its isolation has bimodal nature (Fig. 4b). Two distinct peaks at temperatures of 290°C and 440°C are most likely due to the partial hydrolysis of $CuSO_4$ additive, thereby initiating more intensive process of coal oxidation.

4. Conclusion

The study shows that addition of $CuSO_4$ initiation additive (5% by weight) to composition of coal samples helps to reduce the temperature corresponding to the beginning of release and ignition of volatile substances, irrespective of the process mode (pyrolysis or oxidation). The greatest change in the initial temperature refers to the modified samples, thermal decomposition of which was conducted in inert medium. Most importantly, this effect appeared in series #1 modified samples, which have low content of volatile substances. In general, as a result of the modification, there is a decrease in samples residence time at the stage of volatile substances sublimation. As a result of oxidation time reduction of the samples combustible mass in the air, an increase in average reaction rate occurs, which facilitates shift of the process to lower temperatures. The result is associated with dehydration and partial hydrolysis of copper sulfate, as evidenced by appearance of SO_2 in reaction products. Destruction of initiation additive in carbon matrix pores contributes to intensification of mass and heat transfer processes, thereby accelerating the overall process of oxidation (pyrolysis).

5. References

[1] Cheng J., Zhou F., Xuan X., Liu J., Zhou J., Cen K. *Cascade Chain Catalysis of Coal Combustion by Na–Fe–Ca Composite Promoters from Industrial Wastes* // Fuel, 2016, Volume 181, Pages 820-826.

[2] Parmon V.N., Simonov A.D., Sadykov V.A., Tikhov S.F. *Catalytic Combustion: Achievements and Problems* // Physics of combustion and explosion [Kataliticheskoye szhiganiye: dostizheniya i problemy // Fizika goreniya i vzryva] 2015, vol. 51, №2, p. 5-13.

[3] Mochida I., Sakanishi K. *Catalysts for Coal Conversions of the Next Generation* // Fuel, 2000, Volume 79, Pages 221-228.

[4] Gong X., Guo Z., Wang Z. *Reactivity of Pulverized Coals during Combustion Catalyzed by CeO_2 and Fe_2O_3* // Combustion and Flame, 2010, Volume 157, Pages 351-356.

[5] Liu, Y., Che, D., Xu, T. *Catalytic Reduction of SO_2 during Combustion of Typical Chinese Coals* // Fuel Processing Technology Volume 79, Issue 2, 20 October 2002, Pages 157-169

[6] Gangwei Zhao, Weiqi Yu,Yunhan Xiao. *Study on Brown Coal Pyrolysis and Catalytic Pyrolysis* // Advanced Materials Research Vols. 236-238 (2011) pp 660-663.

[7] Larionov K.B., Mishakov I.V., Gromov A.A., Zenkov A.V. *Influence of $NaNO_3$ and $CuSO_4$ Catalytic Additives on Coal Oxidation Process Kinetic Dependencies* // MATEC Web of Conferences, 2017, 91, 01007.

[8] Slyusarskiy, K.V., Larionov, K.B., Osipov, V.I., Yankovsky, S.A., Gubin, V.E., Gromov, A.A. *Non-Isothermal Kinetic Study of Bituminous Coal and Lignite Conversion in Air and in Argon/Air Mixtures* // Fuel, 2017, 191, P. 383-392.

[9] Tokareva I.V., Mishakov I.V., Vedyagin A.A., Korneev D.V., Petuhkova E.S., Savvinova M.E. *Modification of Carbon Fibers for Reinforcing Pipe Polyethylene* // Composites and Nanostructures volume 6 №3 (2014) 158-167.

Acknowledgement

The research has been accomplished with financial support of the Russian Federation Ministry of Education and Science within implementation of the project №13.7644.2017/BCH

IOP Conf. Series: Journal of Physics: Conf. Series **899** (2017) 092008 doi:10.1088/1742-6596/899/9/092008

Modelling of porous biomass pyrolysis in screw reactor

Levin A.A., Kozlov A.N.

Department of thermal power system, Melentiev Energy Systems Institute (ESI) SB RAS, Irkutsk, Lermontova 130, Russian Federation

E-mail: levin@isem.irk.ru

Abstract. This paper is concerned with the development of a model of wood pyrolysis in a screw reactor as the first stage of the multistage gasification process. To prevent clinkering of particles and thermal inhomogeneities, screw-type transportation is used to transport fuel. In order to describe kinetics of pyrolysis and transport of volatiles within the wood particles and their transition to the gas phase we carried out the studies using a complex of synchronous thermal analysis. A detailed numerical modeling of pyrolyzer was performed with the Comsol Multiphysics software which makes it possible to optimize the design and operating parameters of the pyrolysis process in a screw reactor.

1. Introduction

One-stage processes of fuel gasification have been developed and thoroughly studied in Russia and other countries throughout the 20th century. They remain the basis for the creation and design of modern gasification equipment. Numerous studies showed that the one-stage gasification processes had already achieved the limit of perfection beyond which a considerable improvement in their operating characteristics turned out to be impossible or unprofitable.

The gasification technology meeting the contemporary level of technology development should correspond to the following criteria:

1) the gasifier should operate in automatic, mainly unmanned mode;
2) the gasifier operation should be stable and provide insignificant variations in gas composition and flow rate that randomly occur over time;
3) it should produce the minimum amount of liquid and solid waste;
4) the conversion process should be, if possible, little sensitive to changes in the properties of fuel biomass, i.e. moisture content and size of particles;
5) the efficiency of converting fuel chemical energy into gas chemical energy should reach 80-85%.

The staged gasification technology meets the current technical requirements most completely. In the first stage of the process the allothermal pyrolysis of biomass is organized using producer gas heat and exhaust gases of an engine unit. In this stage the volatiles (gas and tar) are produced from the fuel. In the second stage tar is decomposed under the action of heated air and steam outside the fuel bed. Further, the formed steam-gas mixture reacts with the charcoal in the third stage of the process.

In the last years there has been continuous evolution of the calculation models of pyrolysis plants. The research presented in [1] shows a mathematical description of the pyrolysis process. The authors devised a calculation model for the plant designed by them. The model consists of a set of stages of drying, thermal decomposition, interphase transitions and cooling. Thus, the total duration of the wood feedstock decomposition τ_m process was represented by a set of stages: heating τ_m, drying τ_d, thermal decomposition τ_{td}, and cooling of char coal τ_c. This method allowed the common model to be decomposed into sub models suitable for practical implementation that have different degrees of detail and respective form of equations (ordinary differential or in partial differential). A distinctive feature of this problem statement is the use of a one-dimension representation of distribution of parameters in the space of pyrolyzer chamber.

In the same way the authors of [1] describe all the above enumerated stages. The only assumption in their model was the absence of non-uniform distribution of parameters in the cross section. To verify their model the authors of [1] made an experimental bench since in their case, as in many other studies, verification of the constructed mathematical models is based on the unique experimental data. This complicates a comparative analysis of the proposed approaches to the selection of the best description of the processes that occur in similar plants.

A more thorough study on partial gasification, on the basis of a numerical modeling and application of thermogravimetrical analysis is presented in the recent publications [2, 3]. For example in the research [4] the authors applied a 3-D representation of the calculated region, which made it possible to more accurately determine variation in temperature profiles and gas motion velocities along the modeled furnace. Thus, a good correspondence was achieved between the calculated data and the experimentally measured process of fuel conversion. The analysis of the numerical modeling results that was carried out by the authors allowed them to make a conclusion on the effect of heat generated in the gasification process on 3d temperature fields and gas velocities inside the furnace. Thus, the detailed three-dimensional representation of an object in the numerical models is a necessary condition for the successful description of complex adjoining processes that occur in the course of gasification.

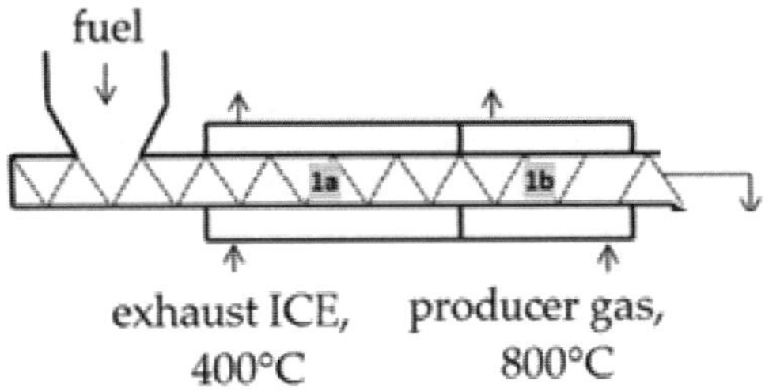

Figure 1.
A diagram of the screw reactor for multistage gasification..

Thus, the analysis of the existing studies shows the necessity to consider the detailed modeling of spatially distributed processes of heat-mass exchange, including the equations of kinetics of chemical reactions, considering natural constraints that follow from the laws of conservation. The most promising seems to be representation of heat-mass exchange equations in a 3d form. In this case the problem of a correct consideration of closing relationships in such a statement is not an exceptional characteristic of such detailed descriptions due to insufficient information on multi-parameter models that describe pyrolysis process. Thus, each new mathematical model requires verification involving results of mufti-factor experimental studies.

2. Mathematical model

The geometry used for modeling a screw pyrolyzer is presented in Fig.1. The interior space is filled with a solid mass by 40%, which should provide easy circulation of the produced gases around screw. The screw represents a spiral with a wall, 3 mm thick, which is wound on a hollow shaft. The interior space of the pyrolyzer with solid fuel and formed products is separated from heating gases by a wall, 5 mm thick. The screw represents a geometrically complex structure. There are 11 complete spiral turns with a pitch of 139 mm on the hollow steel shaft. An external diameter of the spiral made up 139 mm, the shaft diameter was 38.6 mm. In the calculation we took into account the body of the screw, shaft and external coating. All the objects are constructed from cylinders. The construction of all these objects caused the emergence of surfaces and regions of small size at the point of their intersection. The total number of components in the calculation mesh equals *N*. Optimal distribution and sizes of mesh cells were calculated with the Comsol Multiphysics Software on the basis of corresponding algorithms that ensure good convergence for the differential equations applied. The calculation mesh includes different number of cells:

For gas part it includes hot gas from internal combustion engine and gas inside the screw above the porous medium.

For a surface of solid elements it includes all solid surfaces.

For the others it includes solid volumetric parts such as screw, spiral, coating, etc.

The Comsol Multiphysics Software is used as a simulation environment. Heat exchange in the process of pyrolysis is simulated considering physical properties (porosity, permeability, etc.) of the medium [5]. The equation of a diffusion barrier for the plane separating the porous solid body from the gas space in the pyrolyzer is set as follows:

$$-nD_{s,i}\nabla c_{i,u} = \frac{D_{s,i}}{d_s}\left(c_{i,u} - c_{i,d}\right)$$
$$-nD_{s,i}\nabla c_{i,d} = \frac{D_{s,i}}{d_s}\left(c_{i,u} - c_{i,d}\right) \qquad (1)$$

where d_s – barrier thickness, mm; D_{si} – diffusion coefficient, m^2/s. In the calculations we assumed a bed thickness of 5 mm and a diffusion coefficient of 10^{-6} m^2/s.

The boundary conditions chosen to solve the problem are:

An inlet temperature of fuel is 20°C;

An inlet temperature of heating gases is 600÷800°C;

A velocity of fuel flow in pyrolyzer is 0.001 m/s;

An inlet velocity of heating gases is 0.1÷1 m/s.

3. Results and discussion

The calculated fields of the gas phase velocity u in the longitudinal cross section of the screw reactor are presented in Fig. 2. The highest values of u are observed over the surface of filled low-grade solid fuel.

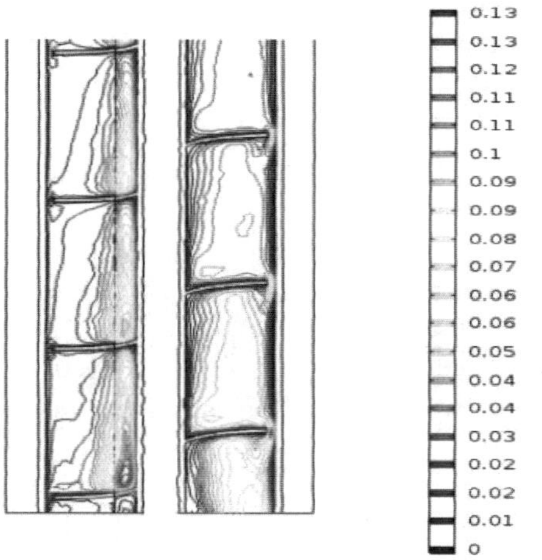

Figure 2.
The calculated isolines of velocity.

Markedly distinct velocity fields and as a consequence the diffusion coefficients characterizing transfer of the gas phase from the porous mass to the free space can be formed for different operating conditions. Figure 3 presents instantaneous values of gas flow velocities through the surface of the filled solid fuel mass. The effect of heating gas temperature on the diffusion processes of formed gas yield was analyzed. The calculations show a negligible effect of the boundary conditions on motion of the media in the region modeled by the diffusion thin bed. This fact suggests feasibility of using the constant coefficients to describe gas transition from the porous medium to the space free for gas flow in the screw reactor. The maximum change in the velocity was equal to ~1% at the change of the heating gas temperature by 140 K at the pyrolyzer inlet. Thereby, the general distribution type of values in the velocity field did not change.

The applied approach to determination of the fields of concentrations of substances formed in the course of substance reaction demonstrated satisfactory results of the calculation example. The highest

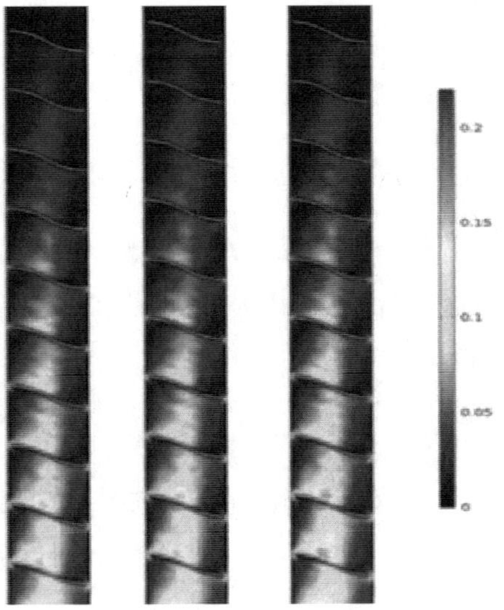

Figure 3. The fields of gas flow velocity on the surface.

concentration of steams formed during drying is observed in the zone of intensive heating – in the initial part of the screw reactor, and as they flow along the reactor the concentrations level off in the sections. As was mentioned in the work [21], the direct application of the Arrhenius modified models leads to the similar effects. The correct analysis of physical constraints requires more strict descriptions of the laws of concentration distribution which are based not only on the values of gas temperatures and velocities. One of the methods to solve this problem is to specify the initial concentration of associated water.

4. Conclusions

The model of heat transfer and aerodynamics of the pyrolyzer was developed for express optimization calculations. The temperature distribution in the pyrolyzer reactor was obtained in a wide range of operating conditions. Moreover, it is confirmed that in the considered thermal conditions the process of fuel conversion proceeds to completion. Inclusion of the kinetic block in the model of heat transfer made it possible to determine the concentration fields of the formed substances (steam, etc.).

Acknowledgments

The research was performed at Melentiev Energy Systems Institute SB RAS under the support of Russian Science Foundation (Grant № 16-19-10227) on the equipment of the multi-access scientific center "High temperature circuit".

References

[1] Grachev A.N., Safin R.G. Valeev I.A. (2006), Pyrolysis of waste wood processing enterprises // Chemistry and Chemical Engineering, 49, 10, 104–108.

[2] Wang G., Li, W., Li, B., & Chen, H. (2008). TG study on pyrolysis of biomass and its three components under syngas. Fuel, 87(4), 552-558. DOI 10.1016/j.fuel.2007.02.032.

[3] Han F., Meng A., Li Q., & Zhang Y. (2016). Thermal decomposition and evolved gas analysis (TG-MS) of lignite coals from Southwest China. Journal of the Energy Institute, 89(1), 94-100. DOI 10.1016/j.joei.2015.01.007.

[4] Meng X., de Jong W., Badri F. S., Benito P., Basile F., Verkooijen A. H. (2012). Combustion study of partially gasified willow and DDGS chars using TG analysis and COMSOL modeling. Biomass and bioenergy, 39, 356-369.

[5] Levin A.A., Shamansky V.A., Kozlov A.N. (2016) A model of pyrolysis in a staged scheme of low-grade solid fuel gasification *Journal of Physics Conference Series*, 754, 02206. DOI:10.1088/1742-6596/754/2/022006

IOP Conf. Series: Journal of Physics: Conf. Series **899** (2017) 092009 doi:10.1088/1742-6596/899/9/092009

Investigation of the additive effect of nanoparticles on the hydrodynamic characteristics of drilling fluids

E I Mikhienkova, A L Neverov, A V Matveev, V A Zhigarev and I V Kudryavcev

Siberian Federal University, Svobodniy 79/10, 660041, Krasnoyarsk, Russia

E-mail: mihienkova_evgeniya@mail.ru

Abstract. Investigations of the rheological properties of drilling fluids based on nanoparticles are carried out. The dependences of the effective viscosity and rheological parameters of fluids on the concentration, size and material of nanoparticles, and the physico-chemical properties of the model fluid are obtained. The study results of the filtration of drilling fluids with nanoparticles are obtained. Dependences of filtration characteristics of drilling fluids on concentration, size and material of nanoparticles are obtained.

1. Introduction

The drilling fluid circulating in the borehole can penetrate into cracks resulting from excessive drilling mud pressure into previously existing open cracks or into large cavities that have structural strength (such as large caverns or canals formed as a result of leaching). At present the main means of lost-circulation control is the flush of filler material in zone of loss, i.e. solid phase, which forms the colmatation zone at the boundary of the borehole - bed. The use of this method leads to a permeability decrease of the mud zone, followed by the formation of a barrier both at the flush and and the solid and colloidal phases contained in the drilling fluid. For the reservoir drilling-in the ability to penetrate the filtrate and solid particles into the reservoir pores is minimized, while optimizing the formulation of the drilling fluid, which is provided by the use of highly disperse stabilizing reagents and special colmatants [1].

At the same time, the filtration losses of the drilling fluid in the reservoir are one of the main causes of crack proliferation and the borehole instability. These processes significantly increase the costs and risks of drilling around the world. Filtration losses is the situation when the drilling fluid partially or completely flows into the reservoir through the areas called zone of loss [2].

Recently a studies have shown the effect of the use of nanoparticles additives in drilling fluids to improve their properties. Some papers have shown that composition optimizing and fillers properties in drilling fluid is a absorption prevention into permeable bed [3,4]. In others, on the contrary, nanoparticles reduce the injection rate by forming a material layer with an equal or higher ultimate strength on surface [5]. However, experiments are separated and not systemic. Results are often contradictory, so further research is needed that are carried out in this paper.

Special investigates to study filtration processes through a porous medium were carried out for a valuation of the colmatizing ability of materials. In our paper a complete research list of the properties of drilling fluids modified by nanoparticles was carried out.

2. Drilling fluids rheology

Viscosity and rheology of drilling fluids play the most important role for their application, since they depend on pressure drop during flushing-out of well, cutting transport efficiency, borehole stability and much more.

Thereby systematic studies of the adding nanoparticles effect on the viscosity and rheology of drilling fluids carried out in the paper.

Nanosuspensions are prepared on basis of model drilling fluids. Particles of silicon oxides are considered as nanoparticles. The particles concentration in solutions ranged from 0.25 to 2% by mass. The particle size is 10 and 100 nm. Powders of metal oxides were purchased from the company "New Powder Technologies", Tomsk. A standard two-step method was used to prepare the nanosuspension. The required quantity of powder is added to the fluid, after which the resulting suspension is mechanically mixed. To destroy the conglomerates of nanoparticles the suspensions are subjected to ultrasonic treatment of the Sapphire TC-10338 bath.

Clay drilling fluid on water base was used as the base fluid. The mass concentration of clay is 5%.

Rheology investigation of the created nanosuspensions was carried out by the OFITE HPHT rotational viscometer. This fully automated system is designed to determine the rheological properties of fracturing fluids, drilling fluids and oil-well cement at elevated temperatures (up to 260 ° C) and pressures (up to 17MPa). The range of shear stresses is 0-4000 D/cm^2. Speed range: 0.01 - 600 rpm. Speed control accuracy: 0.001 rpm. Shear rate range: 0.01 - 1022 s^{-1}.

As a result the dependences of the effective viscosity and rheological parameters of these solutions on the concentration, size and material of the nanoparticles, the physical and chemical properties of the model drilling fluid was obtained (Fig. 1.2).

It is shown that with the nanoparticles addition of 2% by mass, the power model exponent of the drilling fluid varies from n = 0.6 to 0.25 relative to the base solution. The structural viscosity increased from K = 0.07 Pa * c^n to K = 1.3 Pa * c^n.

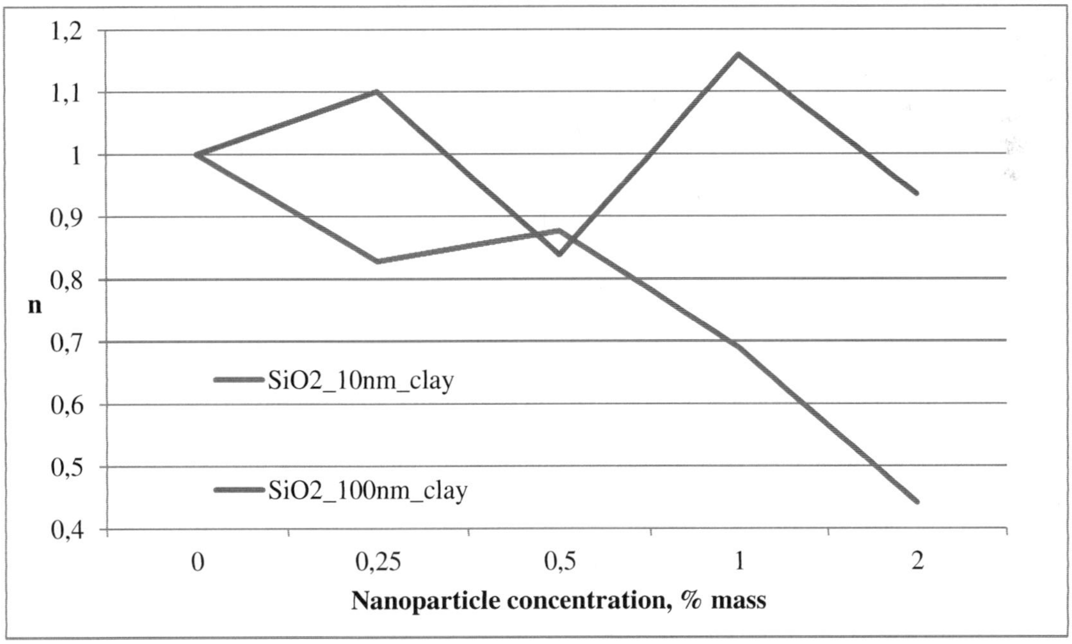

Fig.1 Dependence of power law index on the nanoparticles concentration

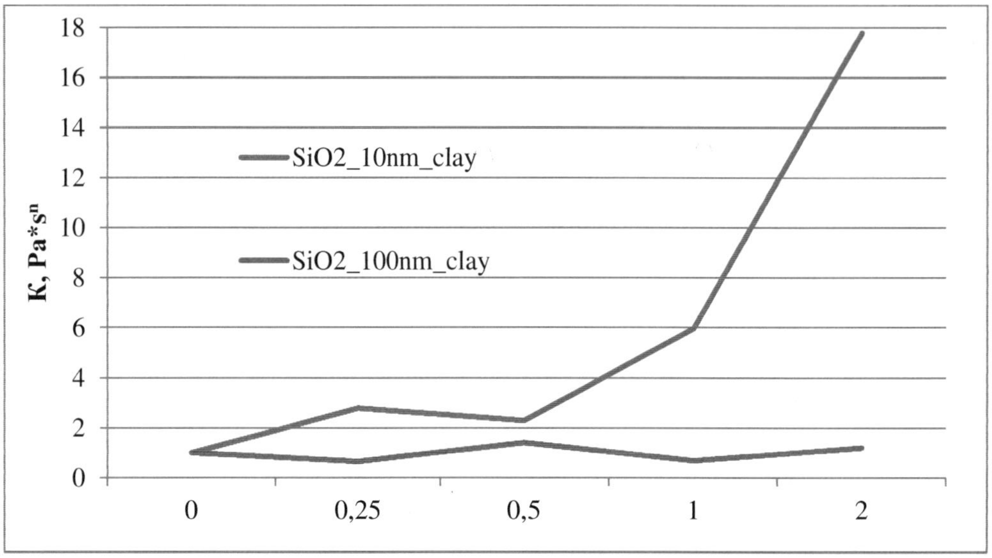

Fig.2 Dependence of consistency index on the nanoparticles concentration

3. Drilling fluids filtration

At present many studies performed on the filtration properties of drilling fluids with additives of nanoparticles [6-8]. In all these papers a significant effect of nanoparticles is noted. However, the experiments results are still contradictory: in some researchers the filtration losses decrease with the addition of nanoparticles, in others, on the contrary, ones increase. There is no clear explanation for this contradiction in the mechanisms of the influence of nanoparticles.

Thereby the filtration properties of drilling fluids with nanoparticles were systematically studied. These researches were conducted on standard paper filters (API standard). A 6-sectional press filter OFITE 140-50 was used, which has mud cells and a pressure building unit. The API standard (American Petroleum Institute) recommends filtering the solution through a filter for 30 minutes at a pressure of 7 atm. In this way the volume of percolating fluid through this filter is determined, which is called the filter loss.

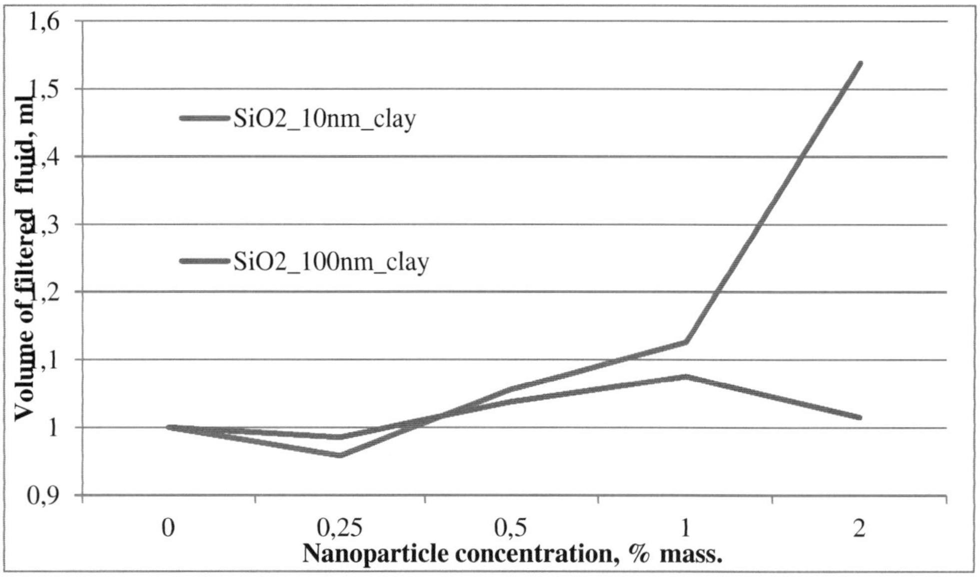

Fig.3 Filtration of water-based mud with the SiO_2 nanoparticles addition

Typical results of the experiment on the drilling fluid filtration with nanoparticles are shown in Fig. 3. It is shown that in this case the addition of nanoparticles with an average size of 10 nm leads to increase in the filtration losses by approximately 1.6 times. In addition it was shown that the filtration losses essentially depend on the size of the nanoparticles. Graph is shown the particles with an average size of 100 nm practically do not affect the filtration of drilling fluids. The dependence of the filtration properties on the particle size can be the reason for the observed divergences in different authors and requires further research.

4. Conclusion

The rheological and filtrational properties of drilling fluids with addition of various sizes nanoparticles are investigated. It is shown that as the concentration of nanoparticles increases, the power law index decreases, and the structural viscosity increases significantly. It was found that the nanoparticles effect on the properties of drilling fluid depends on the nanoparticles size. Influence of nanoparticles on the rheological and filtration properties of drilling fluids slackens with the increase ones. It is shown that filtering of the clay solution through the porous medium is greatly improved with the increase in the concentration of silicon oxide nanoparticles with an average particle size of 10 nm. In our opinion this is due to an improvement in the rheological properties of fluids with an increase the nanoparticles concentration. Thus, the relationship between the filtration losses and the rheology of drilling fluids with nanoparticles was established.

References

[1] Ivachev L.M. Combating the absorption of drilling fluid while drilling geological exploration wells. - Moscow: Nedra, 1982. - 293 p
[2] Cook D., Groukok F., Guo C., Hodder M., Van Ort E. Increasing the stability of the wellbore for preventing and eliminating mud absorption // Neftegazovoye Obozrenie, 2012. - No. 23 (4). - P. 36-49.
[3] Dodson T. Identifying NPT Risk. In: Proceedings of the Atlaneic Communication Drilling and Completing Trouble Zones Forum, Galveston, Texas, USA, 2010. Keynote presentation.
[4] Riveland F.A. Investigation of Nanoparticles for Enhanced Filtration Properties of Drillung Fluid. MsSci. Thesis. Trondheim, Norway: Norwegian University of Science and Technology, 2013, 79 p.
[5] Ulyasheva N.M. Influence of fillers on the permeability of the bottomhole formation zone // Construction of oil and gas wells on land and at sea. - M., 2012. - № 5. - P. 29-32.
[6] Riveland F.A. Investigation of Nanoparticles for Enhanced Filtration Properties of Drilling Fluid.MsSci.Thesis.Trondheim, Norway: Norwegian University of Science and Technology, 2013, 79p.
[7] Ismail A.R., Seong T.C., Buang N.A., Sulaiman W.R.W. Improve Perfomance of Water-based Drilling Fluids Using Nanoparticles. In: Proceedings of The 5thSriwijaya International Seminar on Energy and Envoronmental Science & Technology, Palembang, Indonesia, 2014b, p.43-47.
[8] Ragab A.M.S., Noah A. Reduction of Formation Damage and Fluid Loss using Nano-sized Silica Drilling Fluids. Petroleum Technology Development Journal, 2014, no. 2, p. 75-88.

The reported study was funded by Russian Science Foundation to the research project №17-79-20218.

Effect of mass transfer processes on accumulation and crystallization of boric acid in WWER core in emergency cases

A V Morozov, A V Pityk, S V Ragulin, A R Sahipgareev, A S Soshkina and A S Shlepkin

State Scientific Centre of the Russian Federation – Institute for Physics and Power Engineering named after A.I. Leypunsky, 1, Bondarenko Sq., Obninsk, 249033, Russian Federation

E-mail: sas@ippe.ru

Abstract. In this paper the processes of boric acid mass transfer in a WWER-TOI nuclear reactor in case of the accidents with main coolant circuit rupture and operation of passive safety systems (the hydro accumulators systems of the first, second and third stages, as well as the passive heat removal system) are considered. The results of the calculation of changes in the boric acid solution concentration in the core for the WWER emergency mode are presented. According to the results of the calculation a significant excess of the ultimate concentration of boric acid in accidents with main coolant circuit rupture after 43 hours of emergency mode is observed. The positive influence of the boric acid droplet entrainment on the processes of its crystallization and accumulation in the core is shown. The mass of boric acid deposits on the internals is determined. The received results allow concluding that the accumulation and crystallization of boric acid in the core may lead to blocking the flow cross section and to deterioration of heat removal from fuel rods. The necessity of an experimental studies of the processes of boric acid drop entrainment under conditions specific to the WWER emergency modes is shown.

1. Introduction

Currently, because of the modern international requirements to ensure safety of a new generation nuclear power plants, issues related to the possibility of a long-term (up to 72 hours) autonomous cooling of the nuclear reactor core in case of a beyond design basis accidents are of high priority. In the Russian Federation design of the WWER-TOI reactor plant installation, developed on the basis of the NPP-2006 project. In this project, nuclear reactor core cooling in case of a beyond design basis accidents with main circulation pipeline (MCP) rupture and loss of all AC sources within 72 hours is provided by the functioning of passive safety systems (PSS). The hydro accumulator systems of the first, second and third stage (HA-1, HA-2, HA-3) included in the PSS, due to the successive supply of boric acid solution with concentration of 16 g/kg to the reactor, jointly with passive heat removal system (PHRS), prevent draining of the nuclear reactor core and enable removal of residual heat from the core in case of loss of coolant accident (LOCA) with total loss of AC power [1-4].

Considering duration of the emergency process, boiling of boron-containing coolant and low content of boric acid in the steam, leaving the reactor, conditions of possible accumulation and subsequent crystallization of boric acid can occur in nuclear reactor core. These processes can lead to the appearance of flakes of boric acid in the coolant or deposit formation on the surfaces of fuel rods, which can lead to abnormality of core heat removal process. Consequently, studies related to the processes of mass transfer of boric acid in the WWER core in case of emergency take on important practical significance [5]. To estimate possibility of accumulation and crystallization of boric acid in the WWER reactor core, a calculation analysis of the change in boric acid concentration in the reactor in emergency mode has been done.

2. Determination of droplet entrainment value

Removal of boric acid from the reactor core in case of emergency can be carried out in two ways: by droplet entrainment in the steam-water mixture and due to steam solvent ability. There are a number of dependences that make it possible to determine the value of moisture of steam, leaving the reactor [6]. However, these dependencies cover a limited range of parameters and are not always applicable for calculating droplet entrainment of moisture from a WWER core in case of emergency mode. For example, in [6] there is the following formula for calculating water content of steam:

$$\omega = 2.75 \cdot 10^{8} \frac{N^{2.3}}{Ga^{1.1}\left[\rho''/(\rho'-\rho'')\right]^{0.25}}, \tag{1}$$

where $N = w_0^{''2}/(\varphi g H)$, $w_0^{''}$ – superficial steam velocity, φ – actual volumetric steam content, ρ' – fluid density, ρ'' – steam density, ω – water content of steam, Ga – Galilean number, H – height of steam rise, g – gravity factor.

However, this dependence is applicable for calculating water content of steam only when the steam rise is less than or equal to 1 m, which is significantly lower than in the WWER reactor.

In addition, there are dependencies that allow us to calculate the steam content in steam-water mixture flow x [6]:

$$x = 0.26(w_0^{''2}/g\sqrt{\sigma/g(\rho'-\rho'')})^{0.36}(\frac{\rho''}{\rho'-\rho''})^{0.12}, \tag{2}$$

where σ – liquid tension

The comparison shows that when calculating water content of steam and steam content at the WWER core outlet, operating in emergency mode, according to the above dependencies, divergence of results can reach 30%. Moreover, you must consider that calculation of the parameters of the steam-water mixture according to these formulas is possible only at a certain evaporation rate. In [6] boundary conditions are defined for which the dependences are valid, however, when beyond the boundary parameters, the calculating mechanisms are not proposed. The analysis showed that in the case of LOCA, evaporation rate is beyond the satisfiability limits of dependencies (1) and (2).

3. Calculation of the processes of mass transfer of boric acid in the nuclear reactor core

With regard to all stated above, use of dependencies (1) and (2) to calculate the droplet entrainment processes at the early stage of the accident is not possible. Consequently, during the calculation analysis, the entire emergency process was divided into two stages. During the early stage of an accident (Figure 1a), a very high steam content is observed in core, which causes the foamed liquid level to rise above the upper perforation of the reactor barrel, and the steam-water mixture splashes out from the nuclear reactor core into the reactor volume and then into the containment through the pipeline rupture. Duration of this stage is ~ 7 hours. Further, the intensity of the residual energy release decreases, the steam content in the liquid volume decreases and the level of the boric acid solution is set at the height of the broken pipe of MCP (the second stage of the accident - Figure 1b).

During the calculation, a number of assumptions were made, needed due either to the complexity of the processes occurring in the circuit or to inadequacy of data on the properties of boric acid solutions:

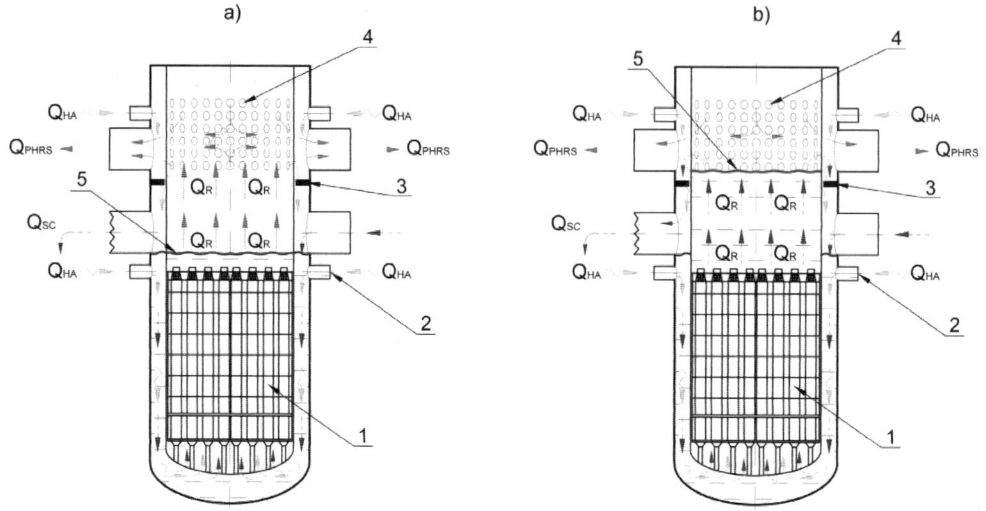

Figure 1. Mass transfer of boric acid in WWER reactor vessel in case of LOCA. a) – the initial stage of an accident, b) - the second stage of the accident: 1 – reactor core; 2 – emergency core cooling system nozzle; 3 – separation collar; 4 – reactor barrel perforation; 5 – liquid level; ⟶ – steam flow, - - -→ – condensate and coolant flow; ⁀⁀→ – inflow of boric acid solution from the hydro accumulator systems; Q_R – heat energy released in the reactor; Q_{SC} – energy of steam going through a leak into the containment; Q_{HA} – energy removed from the core by the liquid from hydro accumulators; Q_{PHRS} –heat energy removed from the core due to operation of the PHRS.

1. During the calculations, two volumes were allocated in the system: the volume of the reactor core and the volume of the reactor pressure chamber (RPC);
2. The thermal physical properties of boric acid solution, such as density and viscosity, were assumed to be equal to water parameters under appropriate conditions;
3. All water entering the core and RPC is at saturation temperature;
4. Pressure in the system was assumed to be constant and equal to 0.3 MPa;
5. All boric acid, initially located in the primary circuit, as well as entered into the reactor from the accumulators of the HA-1 system is carried out into a containment;

When calculating, it was assumed that the rated power of the reactor is $3.2 \cdot 10^9$ W. The change in residual heat removal (N_{RC}) in the reactor core is presented in [7]. The initial data for the calculations were the parameters of the accumulators of the passive flood of HA-2 and HA-3 [8].

At the initial stage of the emergency, when foamed liquid level is above the upper perforation of reactor barrel, intensity of the flow from the core into the RPC will be high, and mass of evaporated water during a period of time $\Delta\tau$ will be as follows:

$$\Delta m_S(\mathrm{H_2O}) = \frac{N_{\mathrm{RC}} \cdot \Delta\tau}{h'' - h_{\mathrm{RPC}}},$$

where h'' – enthalpy of steam at saturation temperature (kJ/kg), h_{RPC} – enthalpy of boric acid solution in RPC (kJ/kg).

Boric acid solution mass entered into the reactor from the hydro accumulators and mass of the condensate formed in the steam generator (SG) due to the operation of PHRS will be as follows:

$$\Delta m_{\mathrm{in}}(\text{solution}) = (G_{\mathrm{HA}} + G_{\mathrm{PHRS}})\Delta\tau,$$

where G_{HA} – flow rate of boric acid solution from HA (kg/s), G_{PHRS} – condensate flow rate from SG (kg/s).

Mass of boric acid entered the reactor from hydro accumulators will be as follows:

$$\Delta m_{\text{in}}(\text{H}_3\text{BO}_3) = G_{\text{HA}} \cdot C_{\text{HA}}(\text{H}_3\text{BO}_3) \cdot \Delta\tau,$$

where $C_{HA}(H_3BO_3)$ - concentration of boric acid in the hydro accumulators (kg/kg).

Since flow rate of the coolant entering the reactor exceeds the evaporation rate in the core, part of the solution will coming into the containment. Mass of boric acid solution entering the containment is following:

$$\Delta m_{\text{out}}(\text{solution}) = \Delta m_{\text{in}}(\text{solution}) - \Delta m_{\text{S}}(\text{H}_2\text{O}).$$

Accordingly, concentration of boric acid in the reactor core in the first stage of the emergency process is determined by the following formula:

$$C_i(\text{H}_3\text{BO}_3) = \frac{(m_{\text{RC}} + m_{\text{RPC}})C_{i-1} + \Delta \text{m}_{\text{in}}(\text{H}_3\text{BO}_3)}{m_{\text{RC}} + m_{\text{RPC}} + \Delta m_{\text{in}}(\text{solution}) - \Delta m_{\text{S}}(\text{H}_2\text{O})},$$

where m_{RC} и m_{RPC} – mass of boric acid solution in reactor core and RPC respectively (kg).

Over time, the residual heat removal will decrease, and at the same time foamed liquid level in the reactor will also fall. The distance from the top of the core to the upper perforation at core barrel is 2.77 m, and elevation difference between the lower generatrix of the "cold" branch of primary circuit and the upper row of perforation holes in the reactor barrel is 1.845 m. Then the critical steam content when foamed liquid level will stand above the upper row of holes in the barrel in case of emergency with a "cold" leg break is:

$$\varphi_{CR} = 1.845/2.77 = 0.666.$$

Steam content is determined by the formula from [6]:

$$\varphi = 0.26\left(\frac{w''^2}{g\sqrt{\sigma/g(\rho'-\rho'')}}\right)^{0.36}\left(\frac{\rho''}{\rho'-\rho''}\right)^{0.12},$$

where steam velocity $w'' = N_{RC}/r \cdot \rho'' \cdot F_{CS}$

In this model, the second stage of the emergency process begins after steam content drops below the critical one. At the second stage of the calculation, volumes of the reactor core and RPC are considered separately.

On the second stage of an accident energy of the residual heat will be spent on heating boric acid solution, entering the core from the RPC and evaporating it:

$$N_{\text{RC}} = G_N r + G_{12}\left(r + h' - h_{\text{RPC}}\right),$$

where G_N – steam flow rate (kg/s); G_{12} – flow rate of a solution of boric acid from RPC to the core (kg/s), r – specific heat of evaporation (kJ/kg), h' – enthalpy of water at saturation temperature (kJ/kg), h_{RPC} – enthalpy of boric acid solution in RPC (kJ/kg).

Mass flow rate of boric acid leaving the core and RPC due to droplet entrainment processes G_{Re} is:

$$G_{\text{Re}} = \omega G_m, \quad (3)$$

where G_m – steam-water mixture flow rate (kg/s).

Steam-water mixture leaving the reactor containing steam and drops of boric acid is replaced by a solution of boric acid coming from the RPC:

$$G_{12} = G_N + G_{\text{Re}}. \quad (4)$$

A change in the mass of boric acid in the volume of the core per time interval $\Delta\tau$ will be:

$$\Delta m_{\text{RC}}(\text{H}_3\text{BO}_3) = (G_{12}C_{\text{RPC}} - G_{\text{Re}}C_{\text{RC}})\Delta\tau$$

where C_{RPC} – concentration of boric acid solution in RPC (kg/kg), C_{RC} – concentration of boric acid solution in the reactor core (kg/kg).

As noted above, solution of boric acid from RPC will flow into the core with a flow rate G_{12}, and the excess of solution will be poured through the ruptured MCP into the containment. On the basis of this, it is possible to determine mass of the solution that has poured out into the the containment in time $\Delta\tau$ and a change in mass of boric acid in the volume of RPC for the same time interval:

$$G_{SC} = G_{HA} + G_{PHRS} + G_{Re} - G_N$$

$$\Delta m_{RPC}(H_3BO_3) = \left(G_{HA}C_{HA} + G_{Re}C_{RC} - G_{12}C_{RPC} - G_{SC}C_{RPC}\right)\Delta\tau$$

To determine evaporation rate, taking into account (3) and (4), we obtain formula for calculating the residual heat:

$$N_{RC} = G_N\left[\left(h'' - h_{RPC}\right) + \omega\left(h' - h_{RPC}\right)\right].$$

Hence the steam flow rate is the following:

$$G_N = \frac{N_{RC}}{\left(h'' - h_{RPC}\right) + \omega\left(h' - h_{RPC}\right)}.$$

Change in boric acid concentration in the nuclear reactor core over the $\Delta\tau$ will be the following:

$$\Delta C_{RC}(H_3BO_3) = \frac{\Delta m_{RC}(H_3BO_3)}{m_{RC}}.$$

4. Calculation results

The above stated dependencies are used to determine changes in boric acid concentration in RPC and the core. Variations in the parameters at the *i*-th step of the calculations over the time interval $\Delta\tau$ are calculated sequentially. The results are used for calculations at the next time interval. The results of calculation of change in boric acid solution concentration in the reactor core are shown in Figure 2.

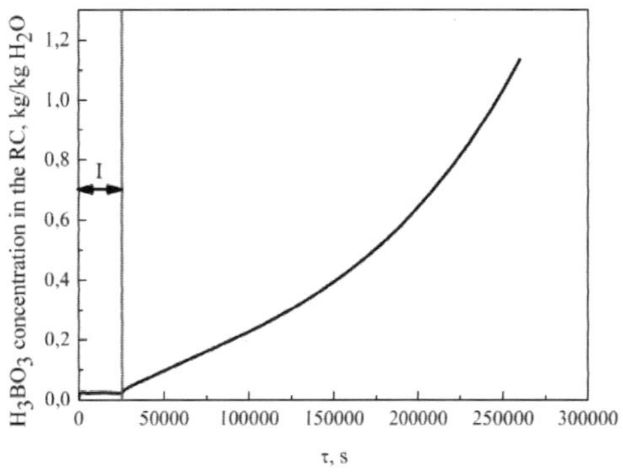

Figure 2. Change of boric acid solution concentration in the core (I – duration of the first stage of the emergency process).

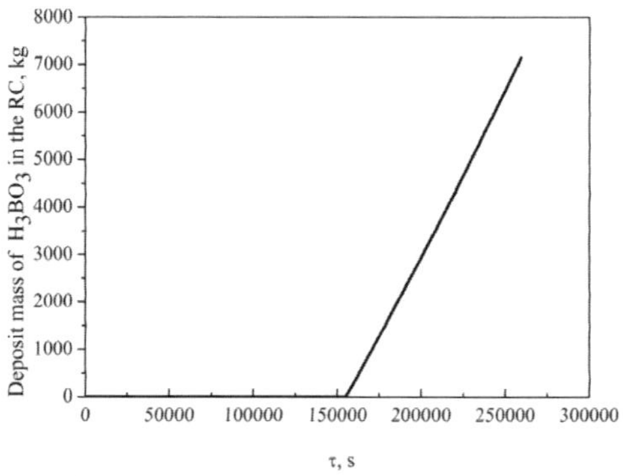

Figure. 3. Change of mass of boric acid deposits in the core during the whole emergency process.

As the graph shows, concentration of H_3BO_3 at the end of 72 hours will be ~ 1,13 kg/kg H_2O, which significantly exceeds the limiting concentration of boric acid (C_{lim}), which at water saturation temperature under consideration is about 0,415 kg/kg.

After reaching the solubility limit of boric acid in the reactor core, the process of H_3BO_3 crystallization starts (after about 43 hours of the initial of emergency process). As noted above, deposits of boric acid in the core has significant impact on heat removal during the emergency process, therefore, the calculation of the boric acid deposit mass was also performed during the calculation. Deposit mass ($m_{dep}(H_3BO_3)$) was calculated using the following formula:

$$m_{\mathrm{dep}}(\mathrm{H_3BO_3}) = m_{\mathrm{RC}} \cdot \left(C_{\mathrm{RC}}(\mathrm{H_3BO_3}) - C_{\mathrm{lim}}\right).$$

The calculation results are shown in Figure 3. As the graph shows, mass of boric acid deposits in the core to the end of 72 hours of emergency process is about 7150 kg.

5. Conclusion

Calculation of the processes of mass transfer of boric acid is of great practical importance in the analysis of emergency processes in WWER reactors of a new generation. The calculated estimation showed a significant excess of the limiting concentration of boric acid in the WWER-TOI reactor core in ~ 43 hours after start of emergency. This process can cause its crystallization on the fuel rods surface and deterioration of heat removal from the core. It should be noted that this calculation has been carried out using a number of conservative assumptions and did not take into account removal of boric acid from the nuclear reactor core due to dissolving ability of steam.

During the calculation analysis it was found that at the moment there is an uncertainty in the available dependencies for calculating moisture of steam and steam content of the medium at the outlet of the WWER core in case of emergency. These parameters are crucial for the analysis of the processes of mass transfer of boric acid. To eliminate existing uncertainties, experimental studies of the mass transfer processes of H_3BO_3 at the parameters specific to for emergency modes of NPPs with WWER are needed.

Acknowledgements
The study has been financed by grant of the Russian Science Foundation (project 16-19-10649).

References

[1] Kalyakin S G, Remizov O V, Morozov A V, Jur'ev J S and Klimanova J V 2003 *Izv. Vyssh. Uchebn. Zaved., Yad. Energ.* 2 94–101 (In Russian)
[2] Luk'yanov A A, Zaitsev A A, Morozov A V, Popova T V, Remizov O V, Tsyganok A A and Kalyakin D S 2010 *Izv. Vyssh. Uchebn. Zaved., Yad. Energ.* 4 172–82 (In Russian)
[3] Kalyakin S G, Sorokin A P, Pivovarov V A, Pometko R S, Selivanov J F, Morozov A V and Remizov O V 2014 *Atomnaya energiya* 116 4 241–46 (In Russian)
[4] Morozov A V and Remizov O V 2012 *Thermal Eng.* 59 5 359-64
[5] Shmal I I and Ivanov M A 2015 *Proc. 9 Scient. and Techn. Conf. "Safety assurance of NPPs with WWER" (Podolsk)* pp 25–29 (In Russian)
[6] Kutepov A M, Sterman L S and Styushin N G 1986 *Hydrodynamics and Heat Transfer during Steam Generation* (Moscow: Higher School) (In Russian)
[7] Kopytov I I, Kalyakin S G, Berkovich V M, Morozov A V and Remizov O V 2009 *Proc. 17 Int. Conf. on Nuclear Engineering (Brussels)* pp 735–743
[8] Morozov A V and Remizov O V 2012 *Thermal Eng.* 59 5 365-70

Formation of CO_2 hydrate in a porous reservoir at the liquid carbon dioxide injection taking into account its boiling

N G Musakaev[1,2] and M K Khasanov[3]

[1]Tyumen Branch of Khristianovich Institute of Theoretical and Applied Mechanics of SB RAS, Taymirskaya Str. 74, Tyumen 625026, Russia
[2]Industrial University of Tyumen, Volodarskogo Str. 38, Tyumen 625000, Russia
[3]Sterlitamak Branch of Bashkir State University, Lenina Str. 37, Sterlitamak 453103, Russia
E-mail: musakaev@ikz.ru

Abstract. The theoretical study results of the CO_2 hydrate formation process at the liquid carbon dioxide injection into the porous reservoir initially saturated with methane and water are presented. In the rectilinear-parallel approach the analytical solutions which describe the distributions of main parameters in the reservoir were built. On the basis of the obtained solutions, it is shown that, depending on the injection pressure of carbon dioxide, it is necessary to consider the frontal surface or the extended zone of the gas hydrate formation.

1. Introduction

A number of researchers identify the greenhouse effect among the main causes leading to climate change and global warming. Moreover, with the development of technology, the number of sources that provide the greenhouse effect in the atmosphere increases every year. The strategy to combat the intensification of the greenhouse effect should be to adopt a set of various measures. One of such measures is to reduce the concentration of carbon dioxide by the regular removal of CO_2 from the atmosphere and storing it in porous reservoirs, in particular, in depleted hydrocarbon deposits [1-5]. At long-term storage of CO_2 in gaseous or liquid state, there is a danger of its emission into the atmosphere. Therefore in a number of works the potential possibility of the CO_2 underground burial in a solid gas hydrate form is studied [6-9]. The main argument in favor of such disposal of carbon dioxide is the ability of hydrates to bind a significant amount of gas in a small volume and at a relatively low pressure [10].

In the presented work the mathematical model of the CO_2 hydrate formation at the liquid carbon dioxide injection into a reservoir taking into account the carbon dioxide boiling is studied.

2. Mathematical model

Assume that a reservoir, initially filled with methane and water with the initially saturation S_{w0}, occupy the half-space $x > 0$. The values of pressure p_0 and temperature T_0 in the reservoir at the initial time moment (before the liquid carbon dioxide injection) in the phase diagram are in the region of existence of carbon dioxide and its hydrate, and also in the region of existence of methane and water in the free (not gas hydrate) state:

$$t = 0, x > 0: \quad T = T_0, \quad p = p_0, \quad S_w = S_{w0}, \quad S_m = 1 - S_{w0}.$$

Here and below, the subscript $i = l, w, c, m, h$ corresponds to the parameters of the i-th phase (l is liquid carbon dioxide, w is water, c is gaseous carbon dioxide, m is methane, and h is the hydrate of carbon dioxide).

Assume that through the left boundary of the reservoir ($x = 0$) the carbon dioxide is being injected. The pressure p_e and the temperature T_e of the injected fluid correspond to the conditions of the CO_2 existence in the liquid state and they are constant at this boundary:

$$t > 0,\ x = 0:\quad T = T_e,\ p = p_e .$$

We will assume that the injection of liquid carbon dioxide is accompanied by its boiling and further formation of CO_2 hydrate.

In the considered problem, due to the continuous injection of carbon dioxide into the reservoir, the intensity of mass transfer caused by filtration in the permeable porous medium is significantly higher than the intensity of mass transfer caused by diffusion. Therefore, we shall neglect the diffusion mixing of carbon dioxide and methane. We also consider the front of the displacement of methane by carbon dioxide to be stable.

In the considered problem the fluid parameters at the left boundary of the reservoir and the pressure and temperature values in the porous medium at the initial time moment correspond to the conditions of the carbon dioxide hydrate formation. Therefore, it can be assumed that the formation of CO_2 hydrate occurs at the boundary of the displacement of methane by carbon dioxide. Consequently, three characteristic zones are formed in the reservoir. In the first (near) zone the pores are saturated with liquid carbon dioxide and CO_2 hydrate; in the second (intermediate) zone the gaseous carbon dioxide and CO_2 hydrate are present, and in the third (far) zone the pores are saturated with methane and water (fig.1).

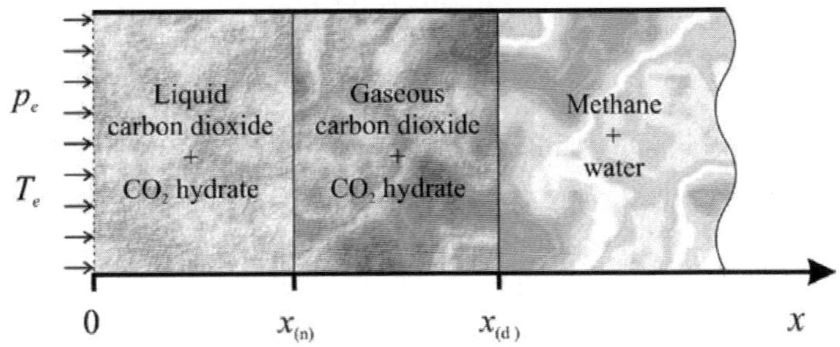

Figure 1. Scheme of the problem of the liquid carbon dioxide injection into a porous medium.

We adopt the following assumptions to describe the processes of heat and mass transfer when the carbon dioxide injection into a porous reservoir. The rock skeleton, water and hydrate are incompressible, the porosity ϕ is constant. The methane and carbon dioxide are the calorically perfect gases, and the liquid carbon dioxide is the elastic and slightly compressible liquid. Since in the considered cases the value of the initial water saturation does not exceed 0.2, the water will be assumed to be immobile. The porous medium skeleton and hydrate are also immobile.

The system of main equations describing, in the one-dimensional case, the processes of filtration and heat transfer in a porous medium is the laws of conservation of masses and energy, the Darcy law and the equation of gas state [11-15]:

$$\frac{\partial}{\partial t}(\phi\rho_i S_i)+\frac{\partial}{\partial x}(\phi\rho_i S_i v_i)=0, \quad (i = l, c, m) \tag{1}$$

$$\rho C\frac{\partial T_i}{\partial t}+\rho_i C_i \phi S_i v_i\frac{\partial T_i}{\partial x}=\lambda\frac{\partial}{\partial x}\left(\frac{\partial T_i}{\partial x}\right),$$

$$\phi S_i v_i = -\frac{k_i}{\mu_i}\frac{\partial p_i}{\partial x},$$

$$\rho_l = \rho_{l0}\left(1 + \beta(p_l - p_0)\right), \quad \rho_i = p_i / R_{gi} T_i\,, \ (i = c,\, m)$$

where t is the time; x is the coordinate; S_i, ρ_i, v_i, k_i, C_i and μ_i ($i = l$, c, m) are the pore saturation, density, velocity, relative permeability, mass heat capacity and the mobile fluid dynamic viscosity; R_{gc} and R_{gm} are the specific gas constants for carbon dioxide and methane; β is the coefficient of the volume compression of liquid carbon dioxide; ρ_{l0} is the true specific density of liquid carbon dioxide which is correspond to the pressure p_0; ρC and λ are the specific volumetric heat capacity and the coefficient of the thermal conductivity of the system. Since the main contribution to the values of ρC and λ is made by the corresponding parameters of the porous medium skeleton we shall consider them to be constant quantities.

Taking into account the Darcy law and the equation of state, the conditions for the balance of mass and heat at the boundary between the first and second zones (the boiling front) can be written in the following form:

$$-\frac{\rho_{l0} k_l}{\mu_l}\frac{\partial p_l}{\partial x} + \frac{k_c}{2\mu_c R_{gc} T_0}\frac{\partial p_c^2}{\partial x} = \phi(1 - S_h)\left(1 + \beta(p_{(n)} - p_0) - \frac{p_{(n)}}{\rho_{l0} R_{gc} T_0}\right)\dot{x}_{(n)}\,, \qquad (2)$$

$$\lambda\frac{\partial T_c}{\partial x} - \lambda\frac{\partial T_l}{\partial x} = \rho_{l0}\left(1 + \beta(p_{(n)} - p_0)\right)L_l\left(-\frac{k_l}{\mu_l}\frac{\partial p_l}{\partial x} - \phi(1 - S_h)\dot{x}_{(n)}\right).$$

Here L_l is the specific mass heat of evaporation of CO_2; $\dot{x}_{(n)}$ is the velocity of the moving of the carbon dioxide boiling boundary. The temperature and pressure at this boundary are assumed to be continuous.

The relationship between pressure and temperature at the boiling boundary in the case of small temperature differences can be represented in the form of the Clausius-Clapeyron curve approximation:

$$p_{(n)} = p_{c0}\left(\frac{\theta}{T_{c0}} - \frac{\theta}{T_{(n)}}\right), \qquad (3)$$

where p_{c0} is the equilibrium boiling pressure corresponding to the temperature T_{c0}; θ is the empirical parameter.

Taking into account the Darcy law and the equation of state, the balance and heat conditions at the boundary between the second and third zones can be written in the form:

$$-\frac{k_c}{\mu_c}\frac{\partial p_c}{\partial x} = \phi\left(\frac{\rho_h}{\rho_c} S_h G + S_c\right)\dot{x}_{(d)}\,, \qquad (4)$$

$$-\frac{k_m}{\mu_m}\frac{\partial p_m}{\partial x} = \phi S_m \dot{x}_{(d)}\,,$$

$$\phi S_h \rho_h (1 - G)\dot{x}_{(d)} = \phi S_w \rho_w \dot{x}_{(d)}\,,$$

$$\lambda\frac{\partial T_c}{\partial x} - \lambda\frac{\partial T_l}{\partial x} = \phi S_h \rho_h L_h \dot{x}_{(d)}\,,$$

where L_h is the specific mass heat of formation of carbon dioxide hydrate; ρ_w and ρ_h are the density of water and gas hydrate; S_w and S_h are water and hydrate pore saturation; G is the relative mass concentration of carbon dioxide in its hydrate; $\dot{x}_{(d)}$ is the speed of movement of the hydrate formation boundary. The temperature and pressure at this boundary are assumed to be continuous.

3. Calculation results

Fig. 2 shows the distributions of temperature and pressure in the reservoir three days after the start of the liquid carbon dioxide injection. For the parameters characterizing the system, the following values are assumed: $\phi = 0.1$, $S_{w0} = 0.2$, $p_0 = 3.0$ MPa, $T_e = 260$ K, $T_0 = 279$ K, $k = 5{\cdot}10^{-16}$ m^2, $G = 0.28$, $\lambda = 2.1$ W/(m·K), $\rho C = 2{\cdot}10^6$ J/(K·m^3), $\rho_h = 1100$ kg/m^3, $\rho_v = 1000$ kg/m^3, $\rho_{l0} = 890$ kg/m^3, $C_l = 2800$ J/(K·kg), $C_c = 800$ J/(K·kg), $C_m = 1560$ J/(K·kg), $R_{gc} = 189$ J/(K·kg), $R_{gm} = 520$ J/(K·kg), $\mu_l = 10^{-4}$ Pa·s, $\mu_c = 1.3{\cdot}10^{-5}$ Pa·s, $\mu_m = 10^{-5}$ Pa·s, $L_h = 4{\cdot}10^5$ J/kg, $L_l = 2.25{\cdot}10^5$ J/kg, $\theta = 1610$ K, $T_{c0} = 274$ K, $p_{c0} = 3.56$ MPa [7, 9, 14, 16].

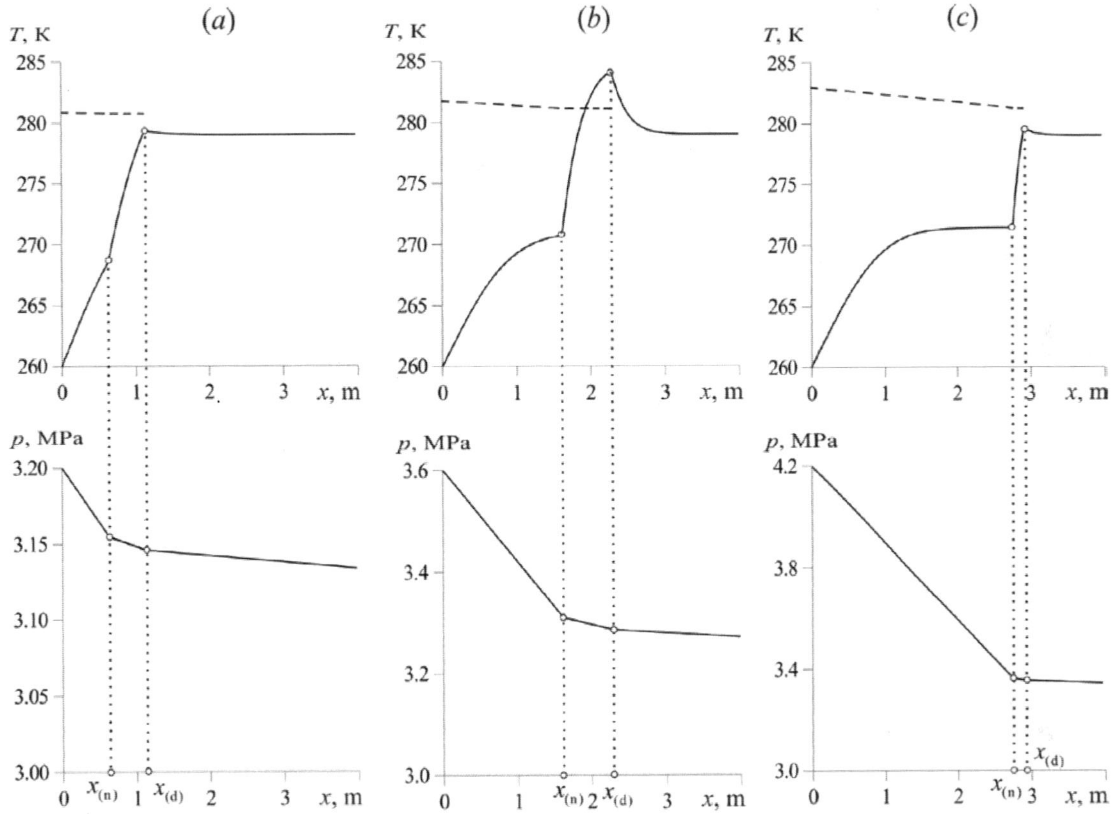

Figure 2. The temperature and pressure distributions at injection pressure p_e =3.2 (a), 3.6 (b) and 4.2 (c) MPa. The dashed line is the equilibrium temperature of the carbon dioxide hydrate decomposition.

The process of hydrate formation is accompanied by the release of latent heat of phase transitions. In this connection, the temperature in the regions saturated with carbon dioxide hydrate can rise above the equilibrium temperature of its dissociation. Fig. 2 shows that for the small value of the injection pressure of carbon dioxide $p_e = 3.2$ MPa (case a), the reservoir temperature in the regions saturated with hydrate is below the equilibrium dissociation temperature. This is due to the fact that in this case the gas hydrate zone is located close to the left boundary of the reservoir, the temperature of which is lower than the initial reservoir temperature T_0. Therefore, through this boundary, an efficient removal of the heat released during hydrate formation is ensured.

At the higher pressure $p_e = 3.6$ MPa (case b), the reservoir temperature in a portion of the region saturated with hydrate (immediately beyond the hydrate formation front) rises above the equilibrium temperature of its decomposition. This is due to the fact that in this case the hydrate formation front is already located far enough from the left boundary of the reservoir. Therefore, the efficiency of the heat removal through this boundary is reduced. This situation corresponds to the "overheating" of the gas hydrate. In this case it is necessary to consider the CO_2 hydrate formation not at the frontal surface, but in the extended (volumetric) region.

At even higher pressure of the carbon dioxide injection $p_e = 4.2$ MPa (case *c*), the temperature of the reservoir behind the hydrate formation front again drops below the equilibrium temperature of its dissociation. This is due to the fact that in this case the distance between the hydrate formation front and the boiling front is very small (fig. 2*c*). And since boiling is accompanied by absorption of latent heat of vaporization, this leads to cooling of the sections of the reservoir located near the boiling front. Consequently, in this case (as with the parameters corresponding to fig. 2*a*), the model with the frontal boundary of hydrate formation gives an adequate mathematical description of the process.

4. Conclusions

The mathematical model of the liquid carbon dioxide injection into the porous reservoir initially saturated with methane and water, taking into account the carbon dioxide boiling, was presented in this work. The features of this process were studied. It is established that, in a certain pressure range, for an adequate mathematical description of the process it is necessary to introduce an extended region of the formation of carbon dioxide hydrate in order to avoid the situation of the gas hydrate "overheating". It is shown that this situation is not realized in those cases when the hydrate formation front is located close enough either to the left boundary of the reservoir or to the carbon dioxide boiling front.

Acknowledgments

This work was supported by Grant of the President of the Russian Federation for State Support of Leading Scientific Schools of the Russian Federation (NSh-6987.2016.1) and Grant of the RFBR 17-51-540001.

References

[1] Oldenburg C M, Pruess R and Benson S M 2001 *Energy and Fuel* **15** 293–298
[2] Benson S M and Cole D R 2008 *Elements* **4** 325–331
[3] Chernov A A, Elistratov D S, Mezentsev I V, Meleshkin A V and Pil'nik A A 2017 *Int. J. Heat Mass Transfer* **108** 1320–1323
[4] Metz B, Davidson O, Coninck H C, Loos M and Meyer L A 2005 *IPCC special report on carbon dioxide capture and storage* (Cambridge: Cambridge University Press)
[5] Chernov A A, Pil'nik A A, Elistratov D S, Mezentsev I V, Meleshkin A V, Bartashevich M V and Vlasenko M G 2017 *Scientific Reports* **7** 40809
[6] Jadhawar P, Mohammadi A, Yang J and Tohidi B 2006 *Advances in the Geological Storage of Carbon Dioxide* vol 65 ed S Lombardi, L K Altunina and S E Beaubien (Dordrecht: Springer) 111–126
[7] Duchkov A D, Sokolova L S, Ayunov D E and Permyakov M E 2009 *Earth's Cryosphere* **13 (4)** 62–68
[8] Chuvilin E M and Guryeva O M 2009 *Earth's Cryosphere* **13 (3)** 70–79
[9] Shagapov V Sh, Khasanov M K and Musakaev N G 2016 *J. of Applied Mechanics and Technical Physics* **57 (6)** 1083–1092
[10] Byk S Sh, Makogon Y F and Fomina V I 1980 *Gas Hydrates* (Moscow: Khimiya)
[11] Khasanov M K 2015 *Thermophysics and Aeromechanics* **22 (2)** 245–255
[12] Tsypkin G G 2014 *Fluid Dynamics* **49 (6)** 789–795
[13] Nigmatulin R I 1991 *Dynamics of Multiphase Media* (New York: Hemisphere Publ. Corp.)
[14] Khasanov M K, Musakaev N G and Gimaltdinov I K 2015 *J. of Engineering Physics and Thermophysics* **88 (5)** 1052–61
[15] Khabibullin I L, Khamitov A T and Nazmutdinov F F 2014 High Temperature **52 (5)** 697–702
[16] Musakaev N G and Borodin S L 2016 *Heat and Mass Transfer* **52 (5)** 981–991

IOP Conf. Series: Journal of Physics: Conf. Series **899** (2017) 092012 doi:10.1088/1742-6596/899/9/092012

Electric conductivity of high explosives with carbon nanotubes

I A Rubtsov[1,2], **E R Pruuel**[1,2], **K A Ten**[1,2], **A O Kashkarov**[1,2] **and S I Kremenko**[1,2]

[1] Lavrentyev Institute of Hydrodynamics of the Siberian Branch of the Russian Academy of Sciences, 15 Lavrentyev Prosp., Novosibirsk, 630090, Russia
[2] Novosibirsk State University, 2 Pirogova Str., Novosibirsk, 630090, Russia

E-mail: rubtsov@hydro.nsc.ru

Abstract. The paper presents a technique for introducing carbon nanotubes into high explosives (HEs). For a number of explosives (trinitrotoluene, pentaerythritol tetranitrate, benzotrifuroxan), it was possible to achieve the appearance of conductivity by adding a small amount (up to 1% by mass) of single-walled carbon nanotubes TUBALL COAT_E H_2O (CNTs) produced by OCSiAl. Thus it is possible to reduce the sensitivity of explosives to static electricity by adding an insignificant part of conductive nanotubes. This will increase safety of HEs during production and application and will reduce the number of accidents.

1. Introduction

Presently, the problem with a static electricity in various spheres of human activity is becoming more and more actual, in particular, it is of great interest to obtain high explosives with conductive properties, because of the struggle with accumulating static charge is directly related to the safety of producing and the use of HEs.

Carbon nanotubes, which was discovered in 1991 [1] opened a wide range of new research lines. Theoretical predictions and experimental results have shown that carbon nanotubes demonstrate a range of unexpected properties: electrical, magnetic, optical and thermal [1–5]. Long thin conductive nanotubes easily can formed a contact network inside a non-conductive material. With a qualitative introduction of CNTs, one can expect a low percolation threshold. In [6–10] it was shown that addition of a small amount of CNTs to the dielectric leads to the appearance of conductivity in the composite. It is important to uniformly disperse of CNTs into the initial material, because, CNTs tend to stick together.

We used single-walled carbon nanotubes TUBALL COAT_E, produced by the OCSiAl company. Their diameter was 1.7 nm and length of about several microns.

As a basis for the composite material, initially non-conductive high explosives: trinitrotoluene (TNT), pentaerythritol tetranitrate (PETN), benzotrifuroxan (BTF) were used. For the uniform application of nanotubes, we used well-known technology of recrystallization (solvent/anti-solvent) of high explosives by dissolution in the acetone.

IOP Conf. Series: Journal of Physics: Conf. Series **899** (2017) 092012 doi:10.1088/1742-6596/899/9/092012

Figure 1. PETN with embedded nanotubes.

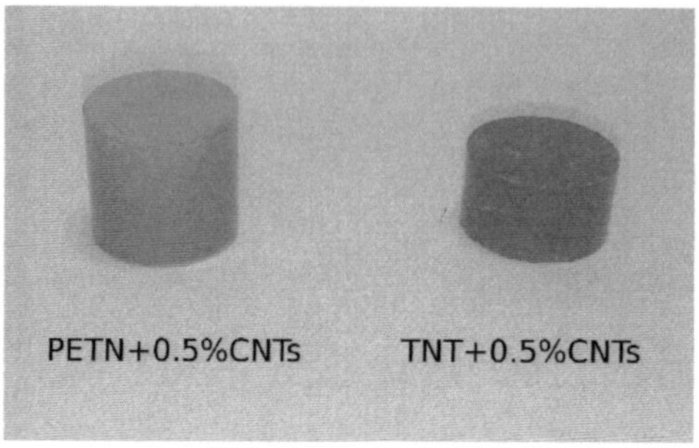

Figure 2. Pressed charges with embedded nanotubes.

2. The technique of nanotube insertion into the sample

To introduce nanotubes, the solution of explosives in acetone was poured into an aqueous suspension of nanotubes. In this case, the crystals of explosives were separated. The centers of crystallization were both CNTs and water. The efficiency of the first process is quite large. The dark suspension initially opaque became transparent.

The following step-by-step technology was used:

- HEs was dissolved in the acetone (1:2) at a temperature of about 50^oC.
- At intensive stirring into the aqueous suspension of CNTs (the concentration of nanotubes in the aqueous solution was 0.2%) the solution of explosives in acetone was poured in. As a result of the interaction, a gray precipitate of the explosive with captured CNTs disappeared.
- The precipitate was separated from the water and dried. The resulting gray powder is shown in figure 1.
- Then, the resulting material was pressed into a cylindrical charges of 19.6 mm diameter and of various lengths (figure 2).

To control the uniformity and dispersion of CNTs in the sample, photographs using an electron microscope (figures 3 and 4) were made. In the HEs particles the fibers and structures of CNTs are clearly visible.

This technique is possible to achieve a homogeneous introduction of conducting CNTs into the dielectric matrix of HEs. Good mixing allowed us preserving special properties of CNTs and achieve the conductivity of the composite with a low concentration of the additive.

3. Measurement of the conductivity of HEs and experimental results.

For conductivity measurement we used the pressed cylindrical charges which was clamped from the ends with copper electrodes, between which an electrical resistance R was measured (two-contact method). The conductivity was determined from the resistance using the following formula: $\sigma = h/(SR) = 4h/(\pi d^2 R)$, where h – length of the charge, d – diametr, S – area of its section. The experimental values of resistance and conductivity for a series of uniformly pressed samples are shown in table 1.

In Table 2 the average characteristics of the obtained materials from different HEs are shown. The obtained values of the conductivity of composites depend weakly on the matrix material and are mainly determined by the concentration of CNTs.

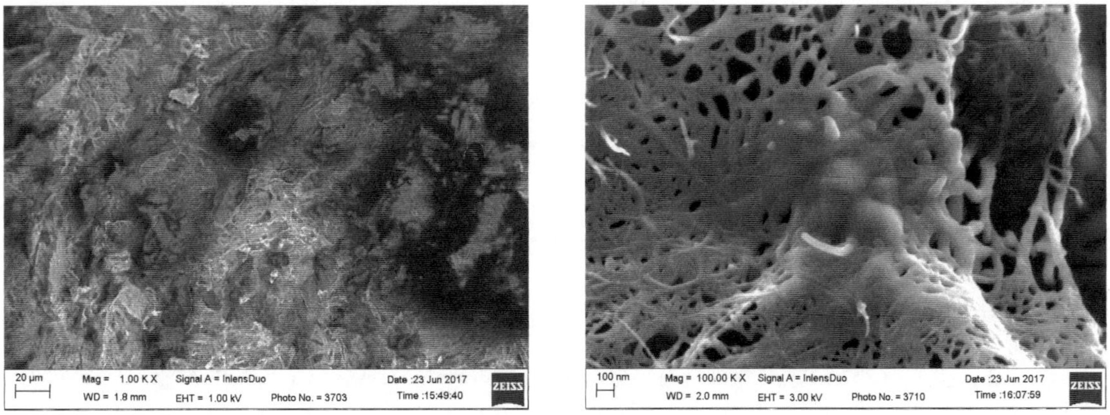

Figure 3. TNT with embedded nanotubes.

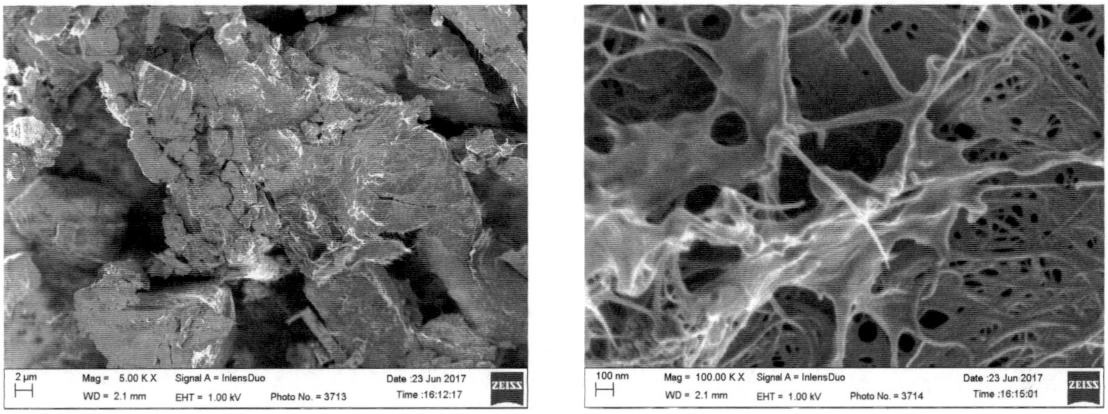

Figure 4. PETN with embedded nanotubes.

Table 1. Characteristics of pressed samples.

Composite	h, cm	d, cm	ρ, g/cc	R, Ω	$\sigma, 1/(\Omega cm)$
TNT+0.5% CNTs	2.53	1.96	1.60	10.1	0.083
TNT+0.5% CNTs	2.76	1.96	1.60	9.7	0.094
TNT+0.5% CNTs	1.86	1.96	1.52	13.2	0.047
TNT+0.5% CNTs	1.85	1.96	1.40	8.7	0.070
TNT+0.5% CNTs	1.84	1.96	1.31	5.4	0.112
PETN+0.5% CNTs	1.85	1.96	1.43	12.5	0.047
PETN+0.5% CNTs	2.13	1.96	1.66	12.3	0.057
PETN+0.5% CNTs	1.85	1.96	1.48	9.8	0.062
PETN+0.5% CNTs	1.85	1.96	1.48	8.2	0.074
BTF+0.5% CNTs	1.97	1.96	1.38	12.5	0.052

Table 2. Average characteristics of pressed materials.

Composite	ρ, g/cc	$\sigma, 1/(\Omega cm)$
TNT+0.5% CNTs	1.49	0.081
PETN+0.5% CNTs	1.51	0.060
BTF+0.5% CNTs	1.38	0.052

4. Conclusions

In this paper, we developed a simple technique for introducing CNTs into high explosive materials by recrystallization. It was possible to achieve the conductivity appearance of TNT, PETN and BTF at addition of 0.5% CNTs (the conduction increased more than by 10 orders). It is expected that the addition of such an amount of CNTs will not lead to a deterioration in the detonation properties of HEs. Conductivity of the composites is $\sigma = 0.08\ 1/(\Omega cm)$ for trinitrotoluene, $\sigma = 0.06\ 1/(\Omega cm)$ for pentaerythritol tetranitrate and $\sigma = 0.05\ 1/(\Omega cm)$ for benzotrifuroxan. It was possible to achieve relatively good homogeneity of the introduction of CNTs.

The technique, tested on HEs, will later be used to reduce the sensitivity of primary explosives to static electricity. This will increase the safety of using such explosives and products with them.

Acknowledgments

This work was supported by Russian Foundation for Basic Research (project No. 17-03-00251).

References

[1] Iijima S 1991 Helical microtubules of graphitic carbon *Nature* **354** 56-8 URL http://dx.doi.org/10.1038/354056a0

[2] Salvetat J P, Briggs G A D, Bonard J M, Bacsa R R, Kulik A J, Stöckli T, Burnham N A and Forró L 1999 Elastic and Shear Moduli of Single-Walled Carbon Nanotube Ropes *Physical Review Letters* **82** 944-7

[3] Srivastava D, Wei C and Cho K 2003 Nanomechanics of carbon nanotubes and composites *Applied Mechanics Reviews* **56** 215-29

[4] Buryachenko V, Roy A, Lafdi K, Anderson K and Chellapilla S 2005 Multi-scale mechanics of nanocomposites including interface: Experimental and numerical investigation *Composites Science and Technology* **65** 2435-65 ISSN 0266-3538 20th Anniversary Special Issue URL http://www.sciencedirect.com/science/article/pii/S026635380500299X

[5] Thostenson E T, Ren Z and Chou T W 2001 Advances in the science and technology of carbon nanotubes and their composites: a review *Composites Science and Technology* **61** 1899-912 ISSN 0266-3538 URL http://www.sciencedirect.com/science/article/pii/S026635380100094X

[6] Thostenson E T, Ziaee S and Chou T W 2009 Processing and electrical properties of carbon nanotube/vinyl ester nanocomposites *Composites Science and Technology* **69** 801-4 ISSN 0266-3538 oNR - Dynamic Failure and Durability URL http://www.sciencedirect.com/science/article/pii/S0266353808002522

[7] Gojny F H, Wichmann M H, Fiedler B, Kinloch I A, Bauhofer W, Windle A H and Schulte K 2006 Evaluation and identification of electrical and thermal conduction mechanisms in carbon nanotube/epoxy composites *Polymer* **47** 2036-45 ISSN 0032-3861 URL http://www.sciencedirect.com/science/article/pii/S0032386106000474

[8] Gryshchuk O, Karger-Kocsis J, Thomann R, Knya Z and Kiricsi I 2006 Multiwall carbon nanotube modified vinylester and vinylester based hybrid resins *Composites Part A: Applied Science and Manufacturing* **37** 1252-59 ISSN 1359-835X URL http://www.sciencedirect.com/science/article/pii/S1359835X05003465

[9] Bauhofer W and Kovacs J Z 2009 A review and analysis of electrical percolation in carbon nanotube polymer composites *Composites Science and Technology* **69** 1486-98 ISSN 0266-3538 cNT-NET 07 Special Issue with regular papers URL http://www.sciencedirect.com/science/article/pii/S026635380800239X

[10] Ayatollahi M, Shadlou S, Shokrieh M and Chitsazzadeh M 2011 Effect of multi-walled carbon nanotube aspect

ratio on mechanical and electrical properties of epoxy-based nanocomposites *Polymer Testing* **30** 548-56 ISSN 0142-9418 URL http://www.sciencedirect.com/science/article/pii/S0142941811000663

IOP Conf. Series: Journal of Physics: Conf. Series **899** (2017) 092013 doi:10.1088/1742-6596/899/9/092013

Analysis of the distribution of pollution in the Sea of Azov by the results of numerical simulation and data of satellite observations

T Ya Shul'ga[1]

[1]Marine Hydrophysical Institute, Russian Academy of Sciences, Kapinatskaya 2, Sevastopol, 299011, Russia

E-mail: shulgaty@mail.ru

Abstract. The paper discusses methods of sharing information obtained by remote sensing of the sea surface from space and model solutions. The estimation of the quality of the model prediction is made depending on the intervals between the assimilation of satellite data. Numerical hydrodynamic modelling of the Azov Sea water area is performed on the basis of the *POM* model for real atmospheric forcing *SKIRON* in the period from 2013 to 2014. The peculiarities of spatial and temporal dynamics of the pollution in the Sea of Azov have been studied on the basis of a joint analysis of numerical simulation data and space monitoring data from the *Terra*/*Aqua* satellites. Based on the developed new model algorithms, the results of numerical simulation and data of satellite observations on the state of water in the Sea of Azov water area are summarized.

1. Introduction

Ecological problems of the Sea of Azov are given special attention in connection with the continuing significant anthropogenic impact. The scale of pollution becomes threatening to the ecosystem and leads to extremely negative consequences [1]. The emergence of satellite systems that have a sufficiently high spatial resolution and ensure the daily arrival of data for any monitoring area makes it possible to monitor the state and pollution of the marine environment. However, in the situation of a catastrophic black oil spill in the Kerch Strait that occurred on November 11, 2007, the first radar satellite images were obtained only five days after the disaster. Data of the optical range, due to cloudy weather, were not informative. A joint analysis of the results of numerical simulation and satellite radar and optical images seems to be most effective, as it provides more complete information on the transport directions, sizes and concentration of contamination areas, at times of absence of images. The paper summarizes the results of numerical simulation and satellite data on the state of the waters of the Azov Sea for the period from 2013 to 2014. Based on the developed new model algorithms, a numerical study of the dynamics of pollution in the Sea of Azov is carried out.

2. Mathematical model. The model parameters

Numerical studies used the three-dimensional nonlinear hydrodynamic model *POM* (Princeton Ocean Model) [2], adapted to the conditions of the Azov basin, also applied to the study of marine pollution. The mathematical model is based on the equations of turbulent motion of a viscous fluid in the hydrostatic approximation [3]. The relations for the calculation of the vertical viscosity coefficients and the turbulent diffusion according to the semi-empirical model of Mellor-Yamada (level 2.5) are written as follows [4]. The coefficient of the horizontal viscosity is calculated using the Smagorinskii model of subgrid viscosity depending on the horizontal velocity gradients [5]. The projections of tangential wind stresses on a free surface are expressed in terms of its velocity values at a standard meteorological altitude, corrected for the coefficient of aero-dynamic resistance of the sea surface, which varies depending on the wind velocity [6]. The normal component of the velocity at the bottom is zero; the bottom tangential stresses are related to the speed according to the logarithmic law. The parameter of roughness is determined using the Grant-Madsen theory [7]. Zero fluid motion and zero fluctuations of the free surface before the atmospheric forcing are specified as the initial conditions. The choice of the integration steps over the temporal and spatial coordinates was performed according to the stability criterion for barotropic waves [8]. The resolutions of the model by latitude and longitude are $(1/59)° \times (1/84)°$, at which the linear sizes of the cell are 1.4 km; the number of sigma levels by the vertical is 11. The initial data for the simulation was the bathymetry map and the configuration of the shoreline of the Azov Sea, built on the basis of digitization of hydrographic maps and subsequent interpolation to the calculated grid. Calculation of the change in the impurity concentration is based on the solution of the transport and diffusion equation, for which conditions on the free surface and in the bottom layer are added to the dynamic boundary conditions for the absence of impurity fluxes [2].

2.1. *Information on the atmospheric forcing used in the numerical experiments*

As atmospheric forcing fields were used, the wind and the atmospheric pressure obtained according to the regional atmospheric model *SKIRON* for the period from 2013 to 2014, the Atmospheric model was created and developed at the University of Athens group of Atmospheric Modeling and Weather Forecasting Group [11]. This version of the model gives a detailed 72 h forecast meteorological parameters for the Azov, Black and Mediterranean Seas. At the first 48 h the output data is carried out after at the 2 h, further values are displayed after 6 h on a grid with a step of 0.1°. Data atmospheric model *SKIRON* was interpolated at the computational grid of the Azov Sea basin.

3. The reconstruction of the primary hydrooptical characteristics in the Sea of Azov according to the colour scanner

Satellite data for the 2013 and 2014 period of the *MODIS* device from the *Aqua* satellite [9] with data rejection according to the criteria [10] with a kilometer spatial resolution. To determine the features in the upper layer of the sea, two parameters are calculated. The first is the ratio of the normalized brightness of light emerging from under the water surface ($L_{WN}(\lambda)$), in two spectral channels: $index34 = L_{WN}(531) / L_{WN}(488)$, were $R_{RS}(531) = L_{WN}(531) / Fo(531)$ and $R_{RS}(488) = L_{WN}(488)/Fo(488)$ – the coefficients of the brightness of the sea with the central wavelengths of the spectral channels 531 and 488 nm. The physical essence of this parameter is that, in the first approximation, it characterizes the total absorption of all optically active substances contained in the upper layer of sea water. The solar constants *Fo* for the spectral channels under consideration can be found, for example in [10].

The second parameter is the backscattering coefficient of the particles of suspended matter at 555 nm ($b_{bp}(555)$), which allows one to observe the peculiarities of light scattering in the upper water layer; in general, this can be a suspension of biological origin (for example, coccolithophoride bloom); so inanimate suspension (for example, a mineral suspension, associated with river discharges or its ascent from the bottom as a result of a strong wind). Calculation of $b_{bp}(555)$ was carried out by the formula: $b_{bp}(555) = \{6.76\, L_{WN}(555) + 0.03\, [L_{WN}(555)]^3 + 3.4\, L_{WN}(555)\, [I_{510}]^{3.8} - 0.84\}\ 10^{-3}\ [\text{m}^{-1}]$, were $I_{510} = L_{WN}(555) / L_{WN}(510)$, $R_{RS}(555) = L_{WN}(555) / Fo(555)$, $R_{RS}(510) = L_{WN}(510)/Fo(510)$, R_{RS} – the luminance coefficient in the corresponding spectral channel, which is a satellite product of the second level, following [10]. The transition from *SeaWiFS* spectral channels to *MODIS* is based on the following expressions: $L_{WN}(555) = 0.9\, L_{WN}(547)$ and $I_{510} = 0.8\, index34 + 0.11$, were $L_{WN}(547) = R_{RS}(547)Fo(547)$.

IOP Conf. Series: Journal of Physics: Conf. Series **899** (2017) 092013 doi:10.1088/1742-6596/899/9/092013

3.1. *Analysis of time series of parameters index34 and $b_{bp}(555)$ according to the MODIS data*

In the course of the study, satellite images of the surface of the Azov Sea in the infrared range were analyzed for the period from 20013 to 2014, based on the processing of *MODIS* data of the second level (*Level* 2) of *R*2014.0. The resolution of remote data over time is due to the passage of the satellite over the Azov Sea area with the smallest time step (24 h). The most informative images, most free from the influence of cloudiness and the presence of omissions, are selected from the available satellite data. They are systematized into groups consisting of sequential images with the smallest time interval. Thus, six time groups were obtained, consisting of the most contrast satellite images with a discreteness of 1 to 2 days, which were used in the test calculations. All satellite data is pre-processed so that if there is a pair of images, they are combined into one snapshot. For example, a snapshot of the 4 group was obtained from two consecutive images on June 23, 2013 at the 9:35 and 11:50 (figure 1).

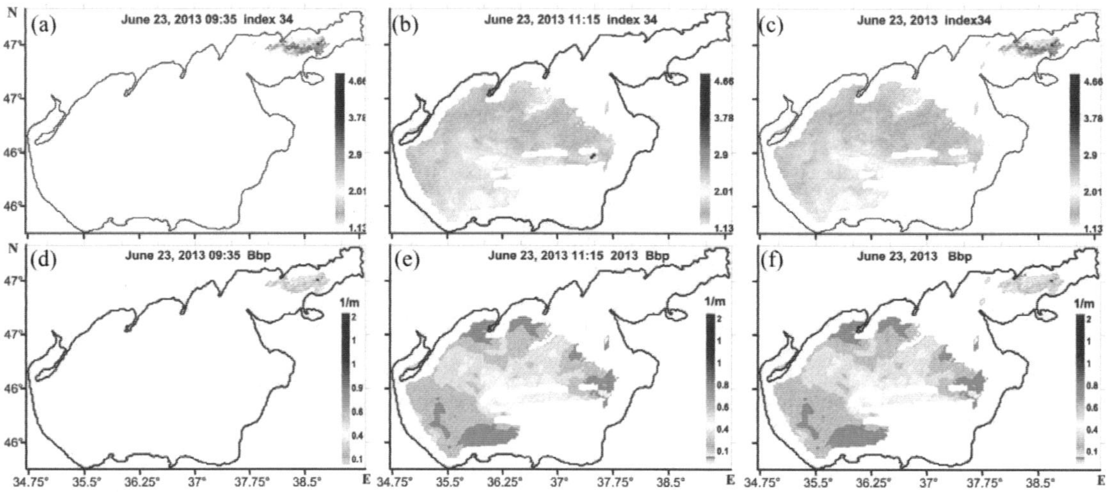

Figure 1. Combining satellite images in one.

4. Algorithm for the assimilation of observational data

The successive recursive algorithm of data assimilation in the problem of estimating the passive pollution concentration fields is based on Kalman's theory of optimal filtration [12]. At times when observations are not available, we believe that the assessment of the analysis coincides with the forecast for the model, and the covariance matrix of the analysis errors is equal to the forecast error covariance matrix. At the time t_k a vector of a priori estimate x_k^m is constructed based on the integration of the transport and diffusion equations to solve this problem. It is a short-term model forecast of the investigated parameter from the previous step of assimilation. Its dimension is equal to the number of points in the model space. The data of satellite observations constitute the observation vector y_k^0, whose dimension is not equal to the dimension of the vector x_k^m. The optimal estimate of the concentration of x_k^* from the observations and the model is based on the Kalman filter algorithm acting on the forecast-correction system (figure 2).

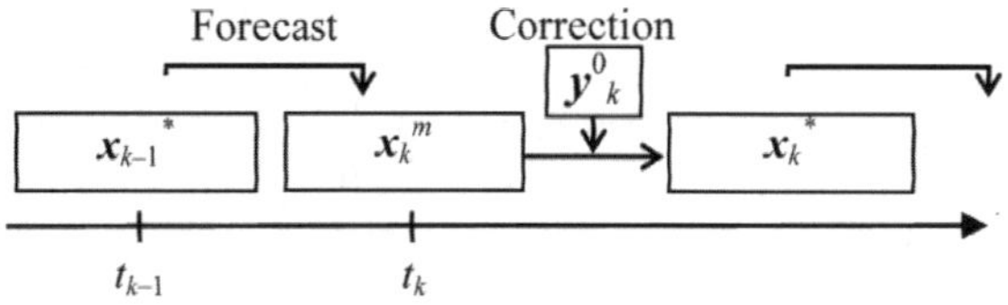

Figure 2. Data assimilation algorithm.

5. Results of experiments

The simulation was carried out at a real atmospheric exposure to *SKIRON* corresponding to the selected time period of a group of satellite images of the surface in the Sea of Azov. Numerical experiments were carried out in two scenarios: without further assimilation of satellite distributions of parameters *index*34 and $b_{bp}(555)$ and with the assimilation of distributions at times corresponding to the available satellite data.

5.1. The first experiment

In the course of the study, the propagation of these parameters was simulated with the assimilation of satellite data only at the initial time. Consider the satellite and model distributions of *index*34, referring to a group of images from April 26 to May 2, 2013 (figure 3).

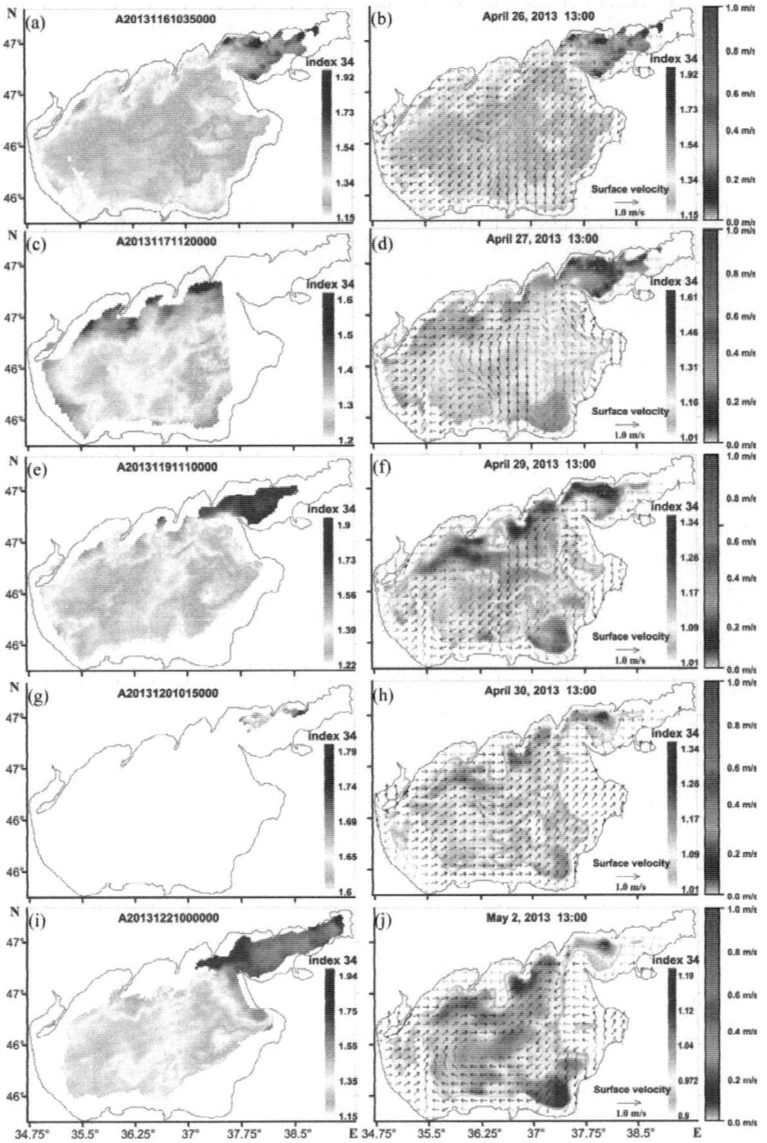

Figure 3. The distribution *index*34 in the surface layer in the Sea of Azov by satellite data (left column) and model data (right column).

IOP Conf. Series: Journal of Physics: Conf. Series **899** (2017) 092013 doi:10.1088/1742-6596/899/9/092013

Assimilation of observational data was carried out using the satellite image *A20131161035000*. The left column contains satellite images. In the right-hand side are the corresponding distributions of the *index*34 parameter, and also the surface velocities of the flows, constructed from the numerical simulation data for a near instant of time (the time difference does not exceed 2 h). Here, the white areas correspond to either the clouds or the gradient zones that were cut out during data processing. Date and local time are indicated for model distributions, the name of the *MODIS* file is given in the corresponding satellite data. The color scales closely correspond to each other in which satellite and model data are presented. In figure 3(a) shows the experimentally accepted satellite distribution of parameter *index*34 of April 26, 2013 at 10:35. Here regions with a high concentration index34 can be clearly traced. They spread from the coastal regions at the base of the Taganrog Bay towards the high seas. Analysis of wind speed showed a north-easterly wind speed of 10 – 12 m/s is developed 2 days before the point in time. Hydrodynamic scenario, confirm the results of the model. Surface velocity of currents directed towards the active wind. In a day, a fairly large area on the satellite image is occupied by cloudiness (figure 3(c)). The modeling data (figure 3(d)) allow us to estimate the character of the propagation of parameter *index*34 in this region. The direction of the wind changed to the west, a stream of currents appeared in the central part of the sea, a capturing impurity from the shore and transferring it to the center of the basin to the north and northeast. After 2 days, a large amount of contamination occurs in the Taganrog Bay area on a satellite image (figure 3(e)). The model distribution of this parameter, obtained without the subsequent assimilation of satellite data, has a significant discrepancy with the actual situation (figure 3(f)). The next satellite image, which corresponds on April 30, 2013 at 10:15, practically does not contain information (figure 3(g)). The corresponding model solution figure 3(h) makes it possible to estimate the spatial distribution of parameter *index*34. On the image relating to the date 02.05.2013 (figure 3(i)), in the region adjacent to Taganrog Bay, and near the coast of Berdyansk there is a region of the greatest concentration. The corresponding model distribution poorly reflects the real distribution as seen (figure 3(j)).

5.2. *The second experiment*

In the second numerical experiment, a consistent assimilation of observational data was carried out at times when information satellite information images were present. The correlation between the observational data and the simulation carried out for these two experiments was evaluated (figure 4).

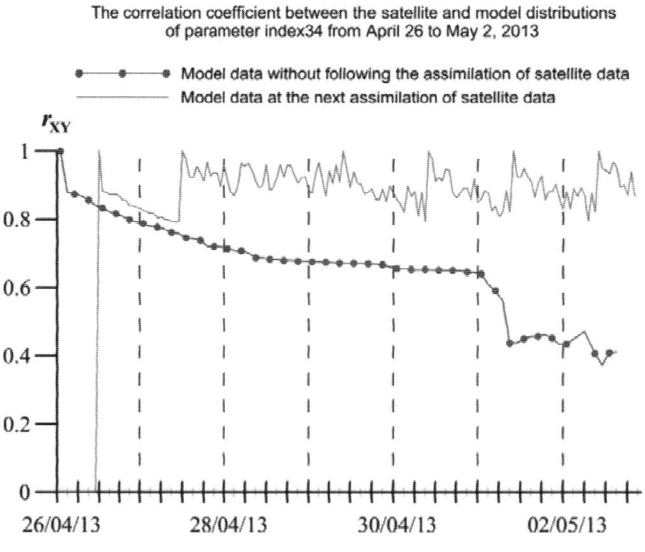

Figure 4. The correlation coefficient between the satellite and model distributions of the parameter *index*34 in the surface layer in the Sea of Azov.

The root-mean-square error in the concentration estimate was calculated during this experiment. This error was compared with the root-mean-square error of the concentration estimate without data assimilation. To solve the problems of assimilating the observations of the use of the parameters *index*34 and $b_{bp}(555)$, an algorithm based on the Kalman filter was applied. On the basis of the analysis of the conducted experiments it follows that the evaluation of the field of concentration of parameters *index*34 and $b_{bp}(555)$ with subsequent assimilation leads to a significant decrease in the mean square error and an increase in the correlation coefficient.

6. Conclusion

A system for assimilation of observational data is proposed, which is a set of applied programs designed to solve the problem of estimating the spread of the pollution in the Sea of Azov. The complex of programs implements the algorithm for assimilation of observational data and makes it possible to perform a prediction of the process of contamination spread in the absence of satellite imagery based on the model of transfer and diffusion of a passive impurity. Numerical experiments have shown the effectiveness of the algorithms proposed in this paper for estimating the evolution of the parameters *index*34 and $b_{bp}(555)$.

The work was carried out in the framework of the Marine Hydrophysical Institute (theme #0827-2014-0010).

References

[1] Matishov G G, Chikin A L, Berdnikov S V, Sheverdyaev I V, Kleschenkov A V, Kirillova E E 2013 Extreme flooding of the Don delta in spring 2013: chronology, conditions of formation and consequences *Bulletin of the SSC RAS* **10** (1) pp 17–24 (in Russian)
[2] Blumberg A F, Mellor G L 1987 A description of three dimensional coastal ocean circulation model *Three-Dimensional Coastal Ocean Models* ed N Heaps (American Geophysical Union vol 4) pp 1–16
[3] Cherkesov L V, Ivanov V A, Khartiev S M 1992 *Introduction into Hydrodynamics and Wave Theory* (St. Petersburg: Gidrometeoizdat) p 264 (in Russian)
[4] Mellor G L, Yamada T 1982 Development of a turbulence closure model for geophysical fluid problems *Rev. Geophys. Space Phys.* **20** pp 851–875
[5] Wannawong W, Humphries Usa W, Wongwises P, Vongvisessomjai S 2011 Mathematical Modeling of Storm Surge in Three Dimensional Primitive Equations *Inter. Comp. Math. Sci.* **5** pp 44–53
[6] Smagorinsky J 1963 General circulation experiments with primitive equations, I. The basic experiment *Mon. Weath. Rev.* **91** pp 99–164
[7] Grant W D, Madsen O S 1979 Combined wave and current interaction with a rough bottom *J. Geophys. Res.* **84** pp 1797–1808
[8] Courant R, Friedrichs K O, Lewy H 1967 On the partial difference equations of mathematical physics *J. IBM* March pp 215–234
[9] NASA Goddard Space Flight Center, Ocean Ecology Laboratory, Ocean Biology Processing Group. Moderate-resolution Imaging Spectroradiometer (MODIS) Aqua Ocean Color Data 2014 Reprocessing NASA OB.DAAC Greenbelt MD USA doi: 10.5067/AQUA/MODIS_OC.2014.0 accessed on 03/11/2017
[10] Suslin V, Churilova T 2016 A regional algorithm for separating light absorption by chlorophyll-a and coloured detrital matter in the Black Sea, using 480–560 nm bands from ocean colour scanners *International Journal of Remote Sensing* **37** (18) pp 4380–4400
[11] Kallos G, Nickovic S, Jovic D, Kakaliagou O, Papadopoulos A, Misirlis N, Boukas L, Mimikou N, Sakellaridis G, Papageorgiou J, Anadranistakis E, Manousakis M 1997 The regional weather forecasting system SKIRON and its capability for forecasting dust uptake and transport *Proceedings of WMO Conference on Dust Storms* pp 9
[12] Klimova E G 2003 Numerical experiments on the assimilation of meteorological data using a sub-optimal Kalman filter *Russian Meteorology and Hydrology* **10** pp 54–67 (in Russian)

IOP Conf. Series: Journal of Physics: Conf. Series **899** (2017) 092014 doi:10.1088/1742-6596/899/9/092014

Thermophysical properties of composite fuel based on T grade coal (Alardinskoe deposit) and timber industry wastes

S.A. Yankovsky[1], A.A. Tolokolnikov[1], V.E. Gubin[1], K.V. Slyusarskiy[1] and A.V. Zenkov[1]

[1]Tomsk Polytechnic University, Tomsk, Russian Federation

e-mail: Jankovsky@tpu.ru

Abstract. Results of experimental studies of composite fuel thermal decomposition processes based on T grade coal (Alardinskoe deposit) and timber industry wastes (fine wood) are presented. C, H, N, S weight percentage of each component of composite fuel was determined experimentally. It has been established that with an increase in wood concentration up to 50% in composite fuel, its energy characteristics decrease by less than 3.6%, while the yield of fly ash is 39.7%. An effective composite fuel composition has been defined as 50%/50%. Results of performed experimental studies suggest that it is possible to use composite fuels based on coal and wood at thermal power plants.

1. Introduction

The most common type of fuel for Thermal Power Plants is coal [1-2]. Requirements for energy characteristics of raw materials at Power Plants are high. The better its quality and the more remote transport component from coal mines, the more expensive final product cost is for consumers. Studies to reduce impact of coal plants emissions on the environment are becoming more urgent. In addition, there is a problem of greenhouse gas emissions (CO2, CH4, etc.). Solution of these issues is a global problem [2-4]. Many methods to reduce emissions during operation of coal-fired Thermal Power Plants have been investigated and proposed earlier, but most of them are expensive, which leads to an increase of production costs. A less expensive alternative may be co-combustion of coal and wood. It is known that wood is carbon-neutral fuel and has minimal impact on the environment during combustion. In papers [4-8], studies on the effect of component additions of waste sludge and sugar cane on slagging processes were conducted. Authors note that addition of these components increases deposition rate. Addition of wood, millet and straw components in the fuel flow reduces deposition rate, which confirms positive synergistic effect. Results obtained by the authors presented in [4-8] cannot be the basis for analyzing regularities of thermal decomposition processes of composite fuels with a change in ratio of wood and coal concentration in prepared mixture. Therefore, research in this area is relevant, and analysis of possibility of creating composite fuel based on coal and waste of timber industry (wood) with energy characteristics close to homogeneous coal and significantly better environmental parameters is of great interest.

2. Experimental section

2.1. Characteristic of researched samples

T grade coal from "Alardinskoe" deposit (particle size less than 80 microns) was used in experimental studies along with timber industry wastes from LLC Dzerzhinsky LPK, Tomsk (particle size less than 200 microns).

C, H, N, S weight percentage of each component in composite fuel was experimentally determined using Euro EA 3000 element analyzer. Results are shown in Table 1.

Table 1. Determination of C, H, N, S weight percentage

Sample	Measurement method	Mass fraction of the element, % wt.			
		C	H	N	S
1	CHNS analysis	53.3	6.9	not detected	not detected
2	CHNS analysis	67.6	3.5	2.9	not detected

It can be seen from table 1 that sulfur content in wood and coal was not detected, nitrogen is present only in coal, hydrogen content prevails in wood, and carbon in coal exceeds its concentration in wood in 1.3 times.

Preparation of composite fuel samples was conducted at different volumetric concentrations; compositions of composite fuels are presented in Table 2. Procedure of fuel samples preparation is described in [9].

Table 2. Concentration of researched fuel samples

Sample	Fuel sample
1	Wood
2	T grade bituminous coal
3	10 % – Wood, 90 % – T
4	25 % – Wood, 75 % – T
5	50 % – Wood, 50 % – T
6	75 % Wood, 25% – T
7	90 % – Wood, 10 % – T

Figure 1 represents photograph of composite fuel sample surface with 50%/50% concentration with x200 zoom.

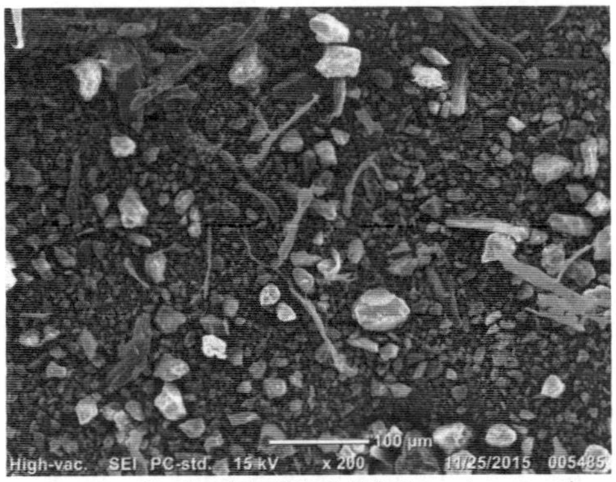

Figure 1. REM images of composite fuel with 50 %/50% concentration of coal and wood, sample 5 REM data (x200)

Figure 1 shows equal distribution of coal and wood in the sample. Components of coal have more rounded shape, and components of wood are oblong.

2.2. Experimental section

Experimental research was conducted to determine calorific value, ash content, moisture content and yield of volatile substances of the studied composite fuels (samples 3, 4, 5, 7).
Figure 2 presents experimentally determined dependence of composite fuel calorific value from increase of wood concentration.

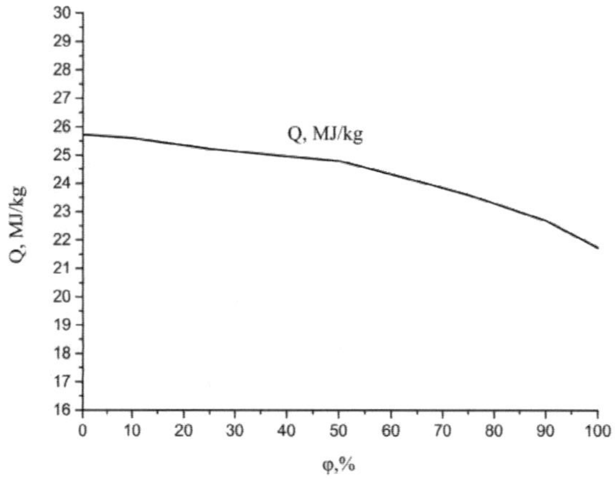

Figure 2. Wood concentration effect on calorific value of composite fuel in relation to homogeneous coal (φ – wood content)

It can be seen from Fig. 2 that when wood concentration is increased to 50%, calorific value of the fuel decreases by less than 3.6% with respect to homogeneous coal. With increase of wood components up to 90%, calorific value reduces by 11.8%. Analysis of conducted experiment allows determining qualitative concentration of composite fuel – 50%/50%.
Figure 3 provides dependence of the change in composite fuels ash content from increase of wood component.

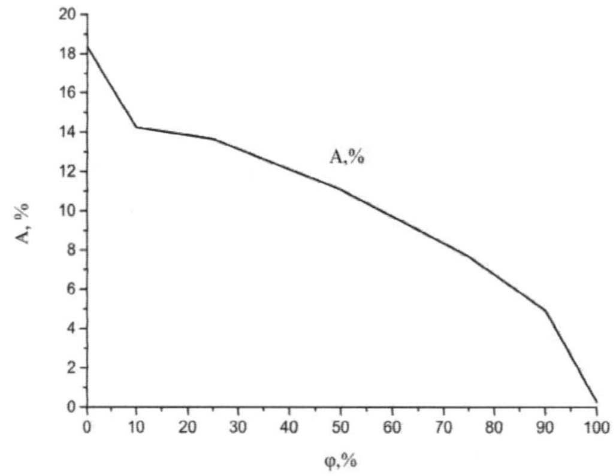

Figure 3. Wood concentration effect on composite fuel ash content in relation to homogeneous coal (φ – wood content)

Results of the research, presented in Figure 3, allow estimating decrease of composite fuels ash content as a function of wood concentration increase in homogeneous coal. Increase in wood component to 50% leads to decrease of ash content by 39.7%; with increase of wood up to 90% ash content decreases by at least 73%. Analysis of presented experiments allows confirming synergetic effect of coal and wood cooperation. The most effective composition of composite fuel for combustion in power boilers furnaces corresponds to concentration of 50%/50%.

An important part of fuel's energy component assessment is determination of volatile substances yield. Figure 4 shows experimentally established dependence of volatile substances yield from increase of wood component in composite fuel (relatively to homogeneous coal).

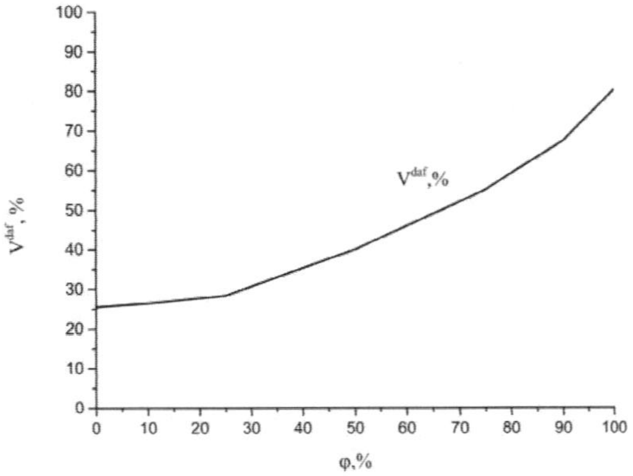

Figure 4. Wood concentration effect on composite fuel volatiles yield in relation to homogeneous coal (φ – wood content)

Analysis of Figure 4 allows determining that increase of wood concentration in composite fuel leads to significant increase in yield of volatiles. The most effective concentration of the components is 50%/50%. Yield of volatile substances increases no more than by 57% at this concentration. Further increase of wood component will result in decrease of obtained synergistic effect.

3. Conclusion

Concentration composition of composite fuel has been experimentally determined as 50%/50%, at which kinetic and energy characteristics are similar to the initial homogeneous coal. Obtained synergistic effect suggests that this component composition of wood and T grade coal is promising for combustion in boiler furnaces.

4. References

[1] Abreu, P., Casaca, C., Costa, M. Ash deposition during the co-firing of bituminous coal with pine sawdust and olive stones in a laboratory furnace. // Fuel, 2010, Volume 89, Issue 12, December 2010, p. 4040 – 4048 .

[2] Heinzel T, Siegle V, Spliethoff H, Hein KRG. Investigation of slagging in pulverized fuel co-combustion of biomass and coal at a pilot-scale test facility // Fuel Process Technol 1998, vol. 54, p. 109 – 125.

[3] Jenkins BM, Thy P, Turn SQ, Blevins LG, Baxter LL, Jakeway LA, et al. Composition and microstructure of ash deposits from co-firing biomass and coal. In: Proceedings of the 10th biennial bioenergy conference, Boise, Idaho; 2002.

[4] Van der Stelt MJC, Gerhauser H, Kiel JHA, Ptasinski KJ. Biomass upgrading by torrefaction for the production of biofuels: a review. Biomass Bioenergy. 2011, vol. 35, p. 3748 – 3762.

[5] Nunes LJR, Matias JCO, Catalo JPS. A review on torrefied biomass pellets as a sustainable alternative to coal in power generation. Renew Sustain Energy Rev. 2014, vol. 40, p. 153 – 160.

[6] Robinson AL, Junker H, Baxter LL. Pilot-scale investigation of the influence of coal-biomass cofiring on ash deposition. Energy Fuels 2002, vol. 16, p. 343 – 355.

[7] Annamalai K, Sweeten J, Freeman M, Mathur M, O'Dowd W, Walbert G, et al. Co-firing of coal and cattle feedlot biomass (FB) Fuels, Part III: fouling results from a 500,000 BTU/h pilot plant scale boiler burner. Fuel 2003, vol. 82, p. 1195 – 1200.

[8] Turn SQ, Jenkins BM, Jakeway LA, Blevins LG, Williams RB, Rubenstein G, et al. Test results from sugar cane bagasse and high fiber cane co-fired with fossil fuels. Biomass Bioenergy 2006, vol. 30, p. 565 – 574.

[9] Yankovskiy, S., Matveev, A., Tolokolnikov, A., Zenkov, A. Research of wood waste concentration influence on composite fuel energy characteristics based on low-grade 2b coal from "borodinskoe" deposit // MATEC Web of Conference, Volume 92, 21 December 2016, Article number 01036.

Acknowledgement
The research has been accomplished with financial support of the Russian Federation Ministry of Education and Science within implementation of the project №13.7644.2017/BCH

Investigation of aerodynamic structure of isothermal swirl flow in a two-stage burner

R R Yusupov[1,2], D V Krasinsky[1] and S I Shtork[1,2]

[1] Kutateladze Institute of Thermophysics SB RAS, Novosibirsk, Russia
[2] Novosibirsk State University, Novosibirsk, Russia

Abstract. The work is devoted to experimental and numerical study of aerodynamic structure of a swirl flow in isothermal model of a vortex burner device that is characterized by the fluid flow supply via two sequentially-mounted tangential swirlers. Depending on the way of the flow supply into the second-stage swirler, either co-swirl or counter-swirl of two flows can be realized. The effect of the second-stage supply direction on the resulting aerodynamic structure has been investigated. Using LDA measurement system the profiles of averaged axial and tangential velocity components were obtained. Experiments have shown that in the co-swirl case the flow inside the vortex burner model is characterized by strong non-uniformity, while in the counter-swirl regime a rapid mixing of the flows from the first and second stages occurs, resulting in a uniform distribution of the flow across the chamber section. Numerical simulation of 3D isothermal turbulent flow has been performed for the counter-swirl regime using the differential Reynolds stress model in the time-dependent formulation. Using the *Q*-criterion for the identification of vortices in numerical data arrays, the evolution of large-scale vortex structures of the swirl flow inside the vortex chamber has been visualized, indicating the presence of two spiral-shape vortex filaments in the vortex chamber. The periodic character of dynamics of these vortex structures has been revealed.

1. Introduction

Swirl flows are widely used in many technical devices, including those for mixing liquids, spraying liquid fuels and stabilizing the flames. The actual task for nowadays is the study of the behavior of co-axial swirl flows [1]. For combustion chambers a characteristic feature is the swirl imparted to the flow. The flow swirling contributes to an enhancing the fuel and oxidizer mixing and improving the flame stabilization that occur due to the formation of a central recirculation zone. The degree of swirl of the flow generally has a strong effect on the formation of turbulence in the flow, which in turn characterizes the region of mixing of co-axial swirl flows [2].

The present work is devoted to an isothermal modeling the aerodynamic flow structure in a two-stage vortex burner with a tangential swirl of the flow. In view of the geometry simplicity, the burners with a tangential or scroll swirlers are often used in practice for example for the combustion of fine pulverized coal [3]. In turn, the two-stage design of the burners can be used in the schemes for burning pulverized-coal fuel, when the first ignition stage is fed with more reactive fuel subjected to mechanochemical activation and the second stage is used for combustion of the lower-reactive coal fuel prepared by the regular method [4].

Isothermal modeling, which allows to determine with relatively low costs the direction of the burner design optimization with respect to operating modes and geometry of the working area, was widely used earlier for the development of power plants [5]. This approach remains to be relevant and is currently supplemented by the capabilities of modern non-intrusive optical systems (laser Doppler anemometry – LDA, Particle image velocimetry - PIV) for a detailed analysis of internal structure of the flows [6, 7]. Moreover, the physical experiment provides also an empirical information for verifying the results of mathematical modeling [8, 9].

2. Experimental setup

A hydrodynamic model of a two-stage vortex burner is used as a working area (figure 1). To provide optical access the model chamber is made of transparent material - Plexiglas. The inlet part of the working chamber consists of two sequentially connected swirlers with a diameter of 185 mm, in which a swirling of the flow is realized by means of a tangential inlet through pipes with an internal diameter of 40 mm. The fluid flow initially gains a swirl on the swirler (1) and then passes through the nozzle into the area of sudden expansion. Later in this area it interacts with a flow that acquires a swirl on the second swirler (3). It is possible to change the swirl of the flow in the second stage of the model by closing or opening the corresponding supply pipes with the help of stopcocks. Depending on the way of fluid supply to the second swirler, co-swirl (white arrows) and counter-swirl (dark arrows) of two flows are realized between the swirlers of the first and second stages of the working section. The resultant flow from the cylindrical zone of 104-mm diameter (4) is fed to the outlet section (5), and it leaves it through four symmetrical pipes with the inner diameter of 20 mm.

The total flow rate in the working area for this series of experiments was of the order of 10 m^3/h, which correspond to a turbulent flow regime with a Reynolds number of the order of 10^5. These conditions characteristic for the operation of practical devices correspond to the self-similarity region when the flow structure does not depend on the Reynolds number [5, 8].

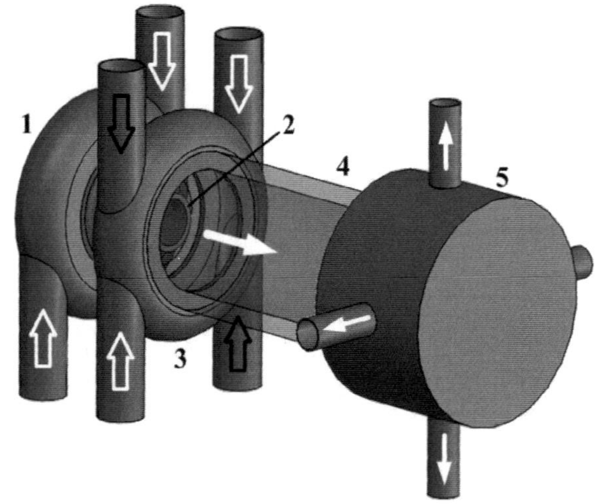

Figure 1. Scheme of hydrodynamic model of two-stage burner. 1 and 3 – swirlers of the first and second stages of the burner, respectively, 2 – shaped nozzle, 4 – zone of sudden expansion, 5 – outlet section with exit pipes and deswirler blades.

3. Results and discussion

3.1. LDA-measurements

To acquire quantitative information about the flow field, the two-component laser Doppler anemometer (LDA) LAD-06i was used. Time-averaged axial and tangential velocity components were measured along one line in the cross-section of the cylindrical zone of sudden expansion (figure 3). The characteristic distance from the exit of the second-stage flow swirler to the measured cross section shown in figure 2. The presented results were obtained for the fluid flow rate of 4 m^3/h at the first swirler and 6 m^3/h at the second swirler. The distance from the central axis x is normalized to radius R_0 of the cylindrical area of sudden expansion (figure 1 (4)).

Figure 3 shows the profiles of the axial and tangential velocity components. Axial velocity distributions show that in cases with co-swirl flows the formation of a non-uniform axial flow distribution in the transversal cross-section with an intense flow along the walls and a broad region of stagnant flow along the axis of the chamber takes place. The jet flow from the nozzle of the first stage of the swirler partially fills the central gap so that the axial velocity profiles near the axis have a backward bend. In the regime with counter-swirl flows, the resulting flow structure is characterized by a fairly uniform distribution of the axial velocity along the transversal cross section.

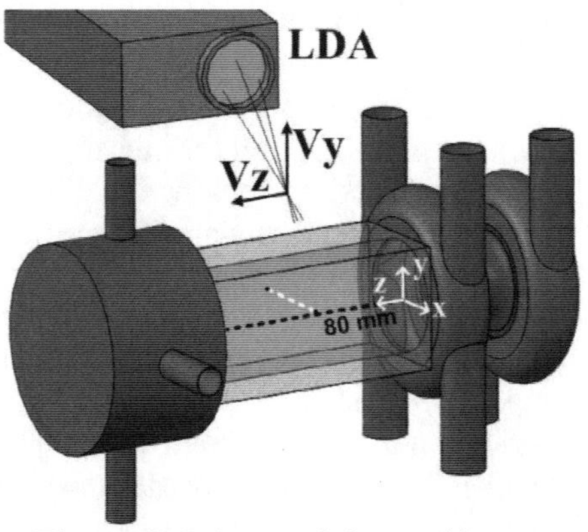

Figure 2. Scheme of the working area layout for LDA-measurements.

In counter-swirl mode the flow of the second stage rotates the resulting flow in the opposite direction. Due to the effect of mutual suppression of the swirl, the level of tangential velocities becomes lower. And the swirl is more localized near the channel wall, and the central region has a very weak level of swirl, with a tendency to change the sign of rotation.

Thus, the measurement results show that the counter-swirl of the flows in the first and second stages contributes to the formation of a more optimal aerodynamic flow structure characterized by a uniform distribution of the axial velocity along the cross section in combination with a sufficiently pronounced general rotational motion of the flow in the working section.

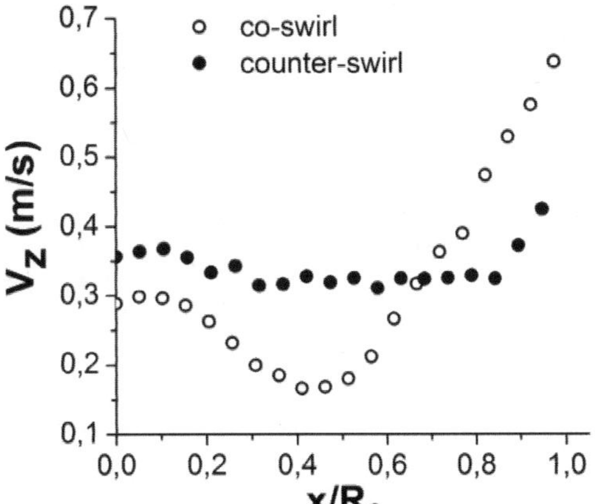

(a)

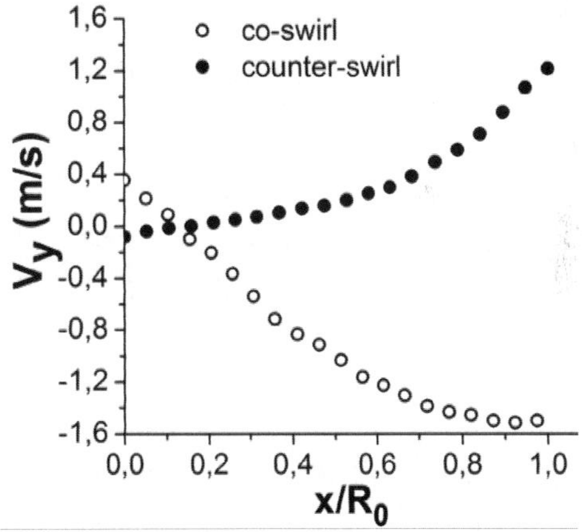

(b)

Figure 3. Profiles of velocities in a vortex chamber for regimes of co-swirl and counter-swirl of flows. (a) - axial velocity component, (b) - tangential velocity component.

3.2. Numerical predictions

Numerical modelling of isothermal turbulent flow in the studied laboratory model of the swirl burner has been performed in the case of counter-swirl burner regime. In the framework of unsteady RANS (URANS) second-moment closure approach to turbulence modelling, the differential Reynolds stress model (DRSM) based on the linear pressure-strain correlation (LRR) and the transport equation for ω (r.m.s. vorticity of turbulent eddies) has been applied. This DRSM LRR-ω model [10] allows to account for non-isotropy of turbulent field and therefore to improve the accuracy of turbulence modelling [10] in the complex 3D swirl flow. Inlet conditions prescribed in numerical simulation correspond to those used in experiments on the counter-swirl flow regime. The numerical 3D solution was obtained on unstructured computational grid consisting of $1.45 \cdot 10^6$ polyhedral cells. At each time step $\Delta t = 2$ ms up to 50 sub-iterations were done using the SIMPLEC algorithm [11] for pressure-velocity decoupling until the convergence of discretized system of governing equations. Computations have been carried out with the use of CFD package ANSYS FLUENT.

From numerical simulation the ensemble of instantaneous flow fields has been computed for the period of ~24 s of the flow evolution, and the fields of mean and turbulent quantities of the flow have been obtained by the time-averaging over the period of 6.3 s. In figure 4 the comparison between the numerical and experimental results is shown for the profiles of mean axial and tangential velocity components plotted at cross-section z=80 mm of the vortex chamber. With the use of known Q-criterion of a vortex identification [12], visualization of large-scale vortex structures and their evolution inside the swirl burner model has been performed – a typical 3D picture of isosurfaces $Q = 750$ s^{-2} educed from computational data (at time moment $t = 22.8$ s) is demonstrated in figure 5. The visualization analysis allowed to reveal the presence and evolution of two spiral-form vortex core filaments in the vortex chamber, which have also been observed in experiments. The spectra obtained from URANS simulation (see figure 6) indicate a periodic character of dynamics of these vortex structures, and the typical frequency $f_p = 3.945$ Hz manifesting the maximal pulsations of total pressure in the vortex chamber has been found from the spectra analysis.

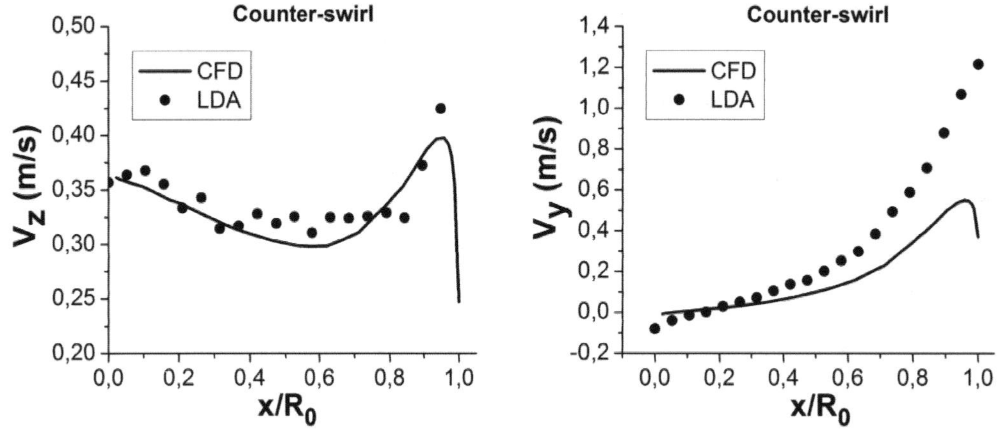

(a) (b)

Figure 4. Comparison of the velocity profiles at cross-section z = 80 mm for the counter-swirl regime of flows in LDA measurements and numerical simulation. (a) - axial velocity component, (b) - tangential velocity component.

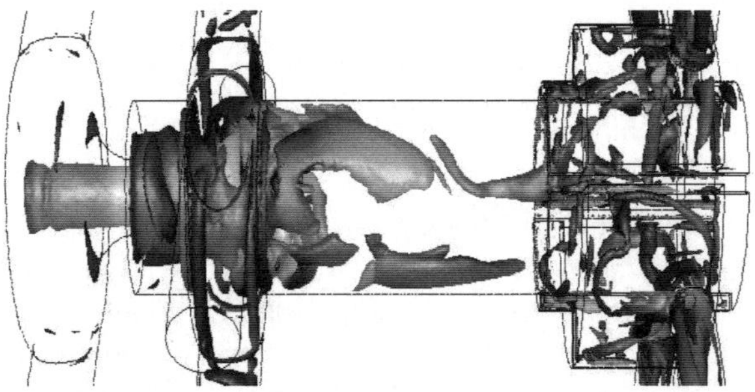

Figure 5. Isosurfaces of Q-criterion (Q = 750 s^{-2}) in the swirl burner model at time t = 22.8 s (URANS simulation).

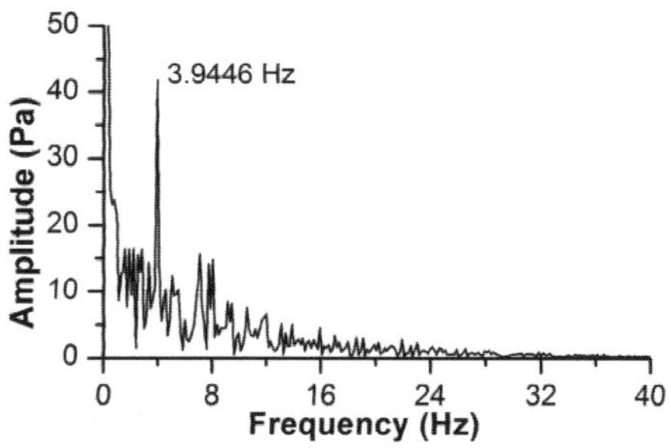

Figure 6. Spectrum of total pressure pulsations in the vortex chamber (URANS simulation).

4. Conclusions

The effect of the second-stage supply direction on the resulting aerodynamic structure of a swirl flow in the laboratory isothermal model of a swirl burner has been investigated. Using LDA measurement system the profiles of axial and tangential mean velocity components have been obtained. Experiments have shown that in the co-swirl case the flow inside the model vortex chamber is characterized by strong non-uniformity, namely, the primary fluid flow moves along the side wall, but an extended stagnant region is formed along the chamber axis. On the contrary, in the counter-swirl regime a rapid mixing of the flows from the first and second stages occurs, resulting in a more uniform distribution of the flow across the chamber section. Therefore the counter-swirl mode appears to be more preferable for the use in a two-stage burner device.

Numerical simulation of 3D isothermal turbulent flow in the studied laboratory model of two-stage burner has been performed for the counter-swirl regime, using URANS approach with DRSM LRR-ω model. The comparison between the numerical and experimental results has shown a consistency of numerical predictions with LDA measurements on axial mean velocity component, however some quantitative discrepancy in the profile of tangential mean velocity component is observed. With the use of known Q-criterion of a vortex identification, the evolution of large-scale vortex structures of the swirl flow inside the vortex chamber has been visualized from numerical data, indicating the presence of two spiral-shape vortex filaments in the vortex chamber. The periodic character of dynamics of these vortex structures has been revealed and the frequency of maximal pulsations of total pressure inside the vortex chamber has been found from the spectra analysis.

5. References

[1] Volshanik V V and Orekhov G V 2015 *Vestnik MGSU* **7** pp 87–104
[2] Mehta J, Shin H, Wisler D 1989 *Mean Velocity and Turbulent Flow-Field Characteristics Inside an Advanced Combustor Swirl Cup* (Nevada: 27th Aerospace Sciences Meeting)
[3] Burdukov A P, Chernetskii M Yu , Dekterev A A, Anufriev I S, Strizhak P A and Grebenkov P Yu 2016 *Thermal Engineering* **63** pp 61–67
[4] Alekseenko S V, Burdukov A P, Popov V I, Popov Yu S and Shtork S I 2012 Method of burning of micronized coal and regularly milled coal in pulverized coal burner and design for the method realization Patent RF no 2460941
[5] Kutateladze S S, Lyakhovskii D N and Permyakov V A 1966 Modeling of thermal power equipment (Moscow: Energiya) p 350 (in Russian)
[6] Anufriev I S, Kuznetsov G V, Shadrin E Yu, Sharypov O V, Anikin Yu A, Dvoinishnikov S V, Kulikov D V and Rakhmanov V V 2013 *Bulletin of the Tomsk Polytechnic University Geo Assets Engineering* **322**(4) pp 31-35
[7] Anufriev I S, Strizhak P A, Chernetskii M Yu, Shadrin E Yu and Sharypov O V 2015 *Tech. Phys. Lett.* **41**(8) pp 727–730
[8] Gesheva E S, Litvinov I V, Shtork S I and Alekseenko C V 2014 *Thermal Engineering* **61**(9) pp 649–657
[9] Alekseenko S V, Anufriev I S, Glavniy V G, Krasinsky D V, Rakhmanov V V, Salomatov V V and Shadrin E Yu 2016 *Heat Transfer Research* **47**(7) pp 653–667
[10] Hanjalic K and Launder B 2011 *Modelling Turbulence in Engineering and the Environment: Second-Moment Routes to Closure* (Cambridge University Press)
[11] Van Doormaal J P and Raithby G D 1984 *Heat Transfer* **7**(2) pp 147–163
[12] Dubief Y and Delcayre F 2000 *Journal of Turbulence* **1**(11)

Acknowledgments

This work was financially supported by the Russian Ministry of Education and Science, according to Subsidiary Agreement No. 14.607.21.0150.

IOP Conf. Series: Journal of Physics: Conf. Series **899** (2017) 092016 doi:10.1088/1742-6596/899/9/092016

Studying laminar flows of power-law fluids in the annular channel with eccentricity

V A Zhigarev[1], A L Neverov[1], D V Guzei[1,2], M I Pryazhnikov[1,2]

[1] Siberian Federal University, 79 Svobodny pr., Krasnoyarsk, 660041, Russia
[2] Institute of Thermophysics SB RAS, 1 Acad. Lavrentiev pr., Novosibirsk, 630090, Russia

E-mail: zhigarev.vladimir@yandex.ru

Abstract. The paper deals with numerical and experimental investigation of non-Newtonian flow of modeling drilling fluids in the annular channel. The Reynolds number was ranged from 100 to 1500. The parameters of the power-law model of drilling fluids were varied within the following ranges: n=0.43-0.49, K=0.22-0.89. The eccentricity was changed from 0 to 1. We have measured pressure drop in the annular channel and compared calculations with experimental data, achieving good agreement between calculations and experiment.

1. Introduction

Despite the huge number of theoretical, numerical, and experimental studies devoted to the drilling fluid flows in the boreholes, today it is impossible to fully provide the necessary information on all flow characteristics in the borehole within required wide range of drilling parameters. In practice, this flow is usually complicated by the presence of eccentricity e between axes of the outer and inner pipes of the annual channel as well as rotation of the inner pipe. In addition, drilling mud is usually non-Newtonian fluid. In the general case, no exact analytical solutions describing such a complicated flow are known, though there are many approximate and asymptotic solutions. Since no quite a satisfactory solution was received over the recent time, this problem is still being actively studied.

The first studies dealing with laminar flow of non-Newtonian fluids in annular channels were carried out by Volarovich and Gutkin [1], who suggested the approximation of analytical solution of the enforced flow of Bingham fluid in a concentric annular channel.

Fredrickson, together with Bird [2] and Laird [3], obtained the first exact solutions of this problem for power-law and Bingham fluids. Asymptotic solutions, taking into account the eccentricity were also proposed for these rheological models. This was carried out by Hanks [4] for power-law fluids, and Guckes [5] for Bingham fluid.

Many works devoted to the numerical modeling of such flows were appeared in recent years. First of all, it is worth noting the following studies: Hussain and Sharif [6] dealt with simulation of power-law and Herschel–Bulkley fluids flow in eccentric channel with a partial blockage and consideration of the rotation of the inner tube; another work was carried out by Escudier M. P., Gouldson I. W., Oliveira P. J., Pinho F. T. [7], where the authors studied the effect of the inner tube rotation.

At that, verification of these mathematical models requires experimental data, which at the moment are insufficient. This is especially true for non-Newtonian flows in channels with eccentricity. The laminar flow of power-law fluid in an annular channel without and with eccentricity was study in this work.

2. Experimental technique

An experimental test bench to verification the numerical algorithm of the motion of a solution of a non-Newtonian fluid in an annular channel (see figure 1) was created. The stand is a closed loop. The fluid is supplied from the tank (1) to the working section by a centrifugal pump (2). The flow rate is regulated by valves (3-4) and controlled by a flowmeter (5). Then the liquid hit the work area (6). The working area is an annular channel made of stainless steel, which is a pipe-in-pipe system. The outer diameter of the inner tube (D_{in}) is 12.5 mm, the inner diameter of the outer tube (D_{out}) is 21 mm, so the hydraulic diameter of the channel is 8.5 mm.

Measurements of pressure drop in the measuring section of the annular channel were carried out using a differential manometer (7). To measure the pressure drop in the outer tube, chokes were installed, to which a differential pressure gauge was connected. In the experiments, the length of the measuring section was 0.5 m. Before the measuring section, the inlet section of the flow stabilization was organized.

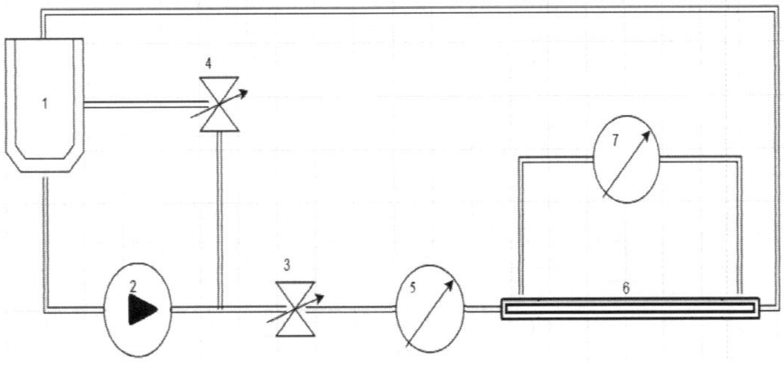

Figure 1. Schematic of the experimental test bench.

Experiments were carried out using preliminary prepared model drilling fluids. At that when preparing drilling fluids we used the recipe and the ingredients which are usually used for real drilling fluids. The following ingredients were used for the preparation of solutions: "GAMMAKSAN" – xanthan biopolymer (produced by Mirrico Company); low viscosity polyanionic cellulose (PAC–LV "OSNOPAK–N" TU 2231-001-70896713-2004, produced by Mirrico Company); liquid glass with density of 1.33-1.45 g/cm^3; sodium chloride (GOST 4233-77), potassium chloride (GOST 4234-77), ammonium chloride (GOST 3773-72); POLY–SAL – a high quality starch (MI SWACO, Schlumberger Company); PLATINUM PAC – polyanionic cellulose (MI SWACO, Schlumberger Company); FLOTROL – starch (MI SWACO, Schlumberger Company); THRUTROL – specially modified derivatives of starch (MI SWACO, Schlumberger Company). Distilled water served the basis for drilling fluid prepared in laboratory conditions. The compositions of solutions are given in Table 1. The rheological properties of drilling fluids were measured by means of a rotational viscometer produced by OFITE MODEL 900 VISCOMETER company, fluid shear rate was varied from 50 to 1022 s^{-1}. Viscosity measurement showed that the rheology of the selected solutions was well described by a power-law model $\tau = K \cdot \gamma^n$. The rheological properties of the solutions are given in table 1. The density of the solutions was equal to 1100 kg/m^3.

Table 1. The rheological properties of model solutions

	Composition of the solution, % wt.			Rheological parameters of power-law fluid ($\tau = K \cdot \gamma^n$)	
No.	THRUTROL	FLOTPOL	NaCl	*n*	*K, Pa·s^n*
1	0.25	–	9	0.4871	0.2240
2	0.5	–	–	0.4317	0.5289
3	–	0.5	–	0.4300	0.8900

3. Mathematical model

In the general case, a viscous fluid flow is described by a system of Navier–Stokes equations consisting of the mass conservation equation:

$$\frac{\partial \rho}{\partial t} + \nabla(\rho \mathbf{v}) = 0$$

and the equations of motion or momentum conservation law:

$$\frac{\partial(\rho \mathbf{v})}{\partial t} + \nabla(\rho \mathbf{v} \cdot \mathbf{v}) = -\nabla p + \nabla(\boldsymbol{\tau}) + \mathbf{F}$$

where $\mathbf{v}$ – is the fluid velocity vector, $\boldsymbol{\tau}$ – is the tensor of viscous stresses, F – is the body forces vector, p – is the static pressure, ρ – is the fluid density. Since in most cases the drilling mud is non-Newtonian fluid, for simulating non-Newtonian flows we used well-known approach [8-9], in which the medium is considered as a nonlinear viscous fluid with the introduction of effective fluid viscosity $\mu(\dot{\gamma})$ which in general is dependent on shear rate. At that, the tensor of viscous stresses $\boldsymbol{\tau}$ is determined as follows:

$$\boldsymbol{\tau} = \mu D$$

The components of the strain velocity tensor $\mathbf{D}$ are of the form:

$$\mathbf{D}_{ij} = \frac{\partial u_i}{\partial x_j} + \frac{\partial u_j}{\partial x_i},$$

shear rate $\dot{\gamma}$ is the second invariant of the strain velocity tensor:

$$\dot{\gamma} = \sqrt{\frac{1}{2}\mathbf{D} \cdot \mathbf{D}}.$$

Depending on the drilling fluid rheology, the effective viscosity is: $\mu(\dot{\gamma}) = K$, for Newtonian medium (K – is the molecular viscosity of the fluid), $\mu(\dot{\gamma}) = K\dot{\gamma}^{n-1}$ for a power-law model, $\mu(\dot{\gamma}) = \frac{K\dot{\gamma} + \boldsymbol{\tau}_0}{\dot{\gamma}}$ for Bingham Plastic model, $\mu(\dot{\gamma}) = \frac{K\dot{\gamma}^n + \boldsymbol{\tau}_0}{\dot{\gamma}}$ for Herschel– Bulkley model, where n and K – are the coefficients of the rheological models, τ_0 – is the yield limit of viscoplastic fluid.

In our calculation we used structured grid consisting of 30 nodes along the radius, 120 nodes along the circumference, and 30 nodes along the channel length. The flow rate values obtained in the experiment were set at the inlet to the channel.

Figure 2 shows the calculated axial velocity profiles in the annular channel for these solutions. It is seen that with decrease in power exponent n velocity profile becomes flatter. In the general case this should contribute to the better removal of cuttings during drilling.

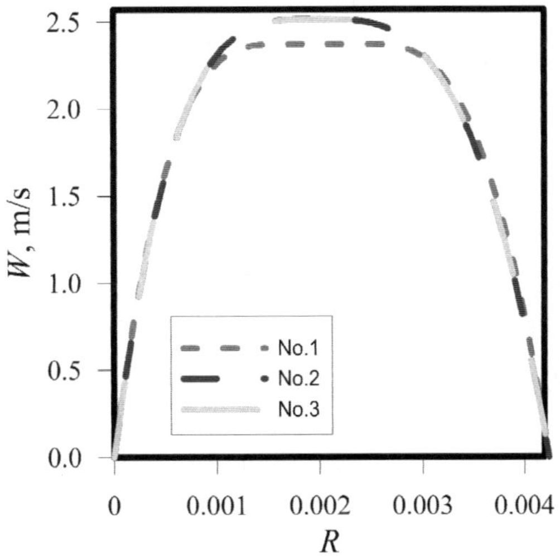

Figure 2. The velocity profiles in the annular channel (D_{out}=21mm, D_{in}=12.5mm) for various drilling fluids for G=0.473 kg/s and e=0.

Comparison of calculated and experimental values of the pressure drop in concentric annular channel for the considered solutions is given in figure 3. As can be seen, all cases show very good agreement between calculations and experiments. The maximum discrepancy does not exceed 10%.

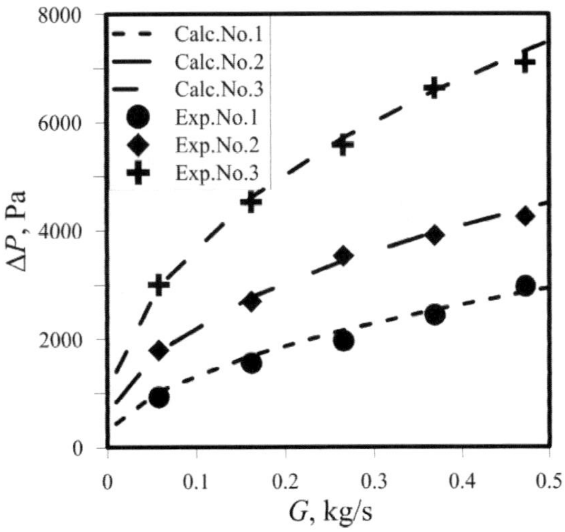

Figure 3. Pressure drop in the concentric annular channel (D_{out}=21mm, D_{in}=12.5mm) versus fluid flow rate. Solid lines correspond to calculation, symbols are experiment.

Further we investigated the effect of eccentricity on flow parameters and the pressure drop. Figure 4 shows the behavior of axial velocity of drilling fluid No.3 at the flow rate G=0.473 kg/s depending on increasing eccentricity. As seen when the inner tube is displaced to the wall of the outer pipe, the flow of the drilling fluid is redistributed. With increasing eccentricity, the width of the annular gap in the wide part increases. In this zone, the maximum flow rate is observed. In the narrow part of the annular channel, zones with a reduced velocity are formed. The redistribution of the velocity in the channel affects the pressure drop that is shown in figure5. As is obvious, the presence of the eccentricity leads to an almost twofold drop in the differential pressure in the channel. The experiment shows the same behavior. The calculated pressure drop for eccentricity *e* equal to 1 (where the inner pipe rests on the wall of the outer tube) is in a good agreement with the corresponding experimental value.

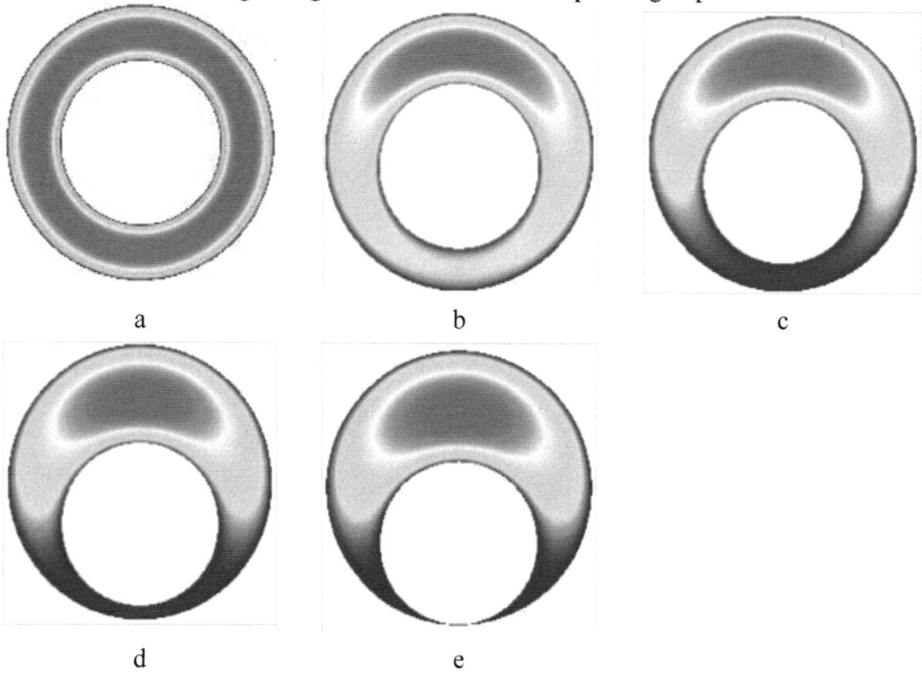

Figure 4. The isolines of the velocity modulus in the annular channel (D_{out}=21mm, D_{in}=12.5mm) for *G*=0.473 kg/s and different eccentricities, a) *e*=0 b) *e*=0.25 c) *e*=0.5 d) *e*=0.75 e) *e*=1.

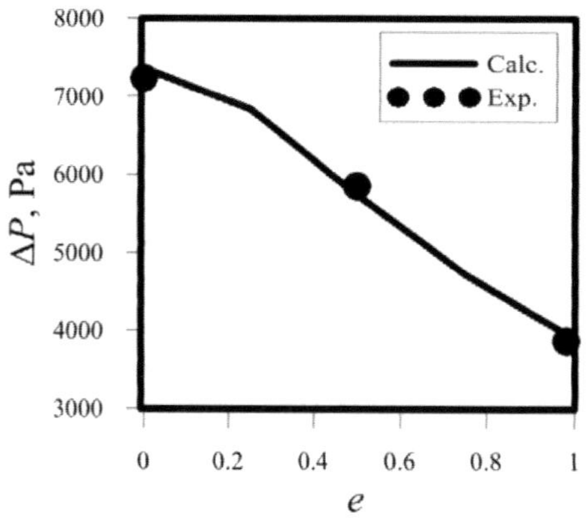

Figure 5. The pressure drop in the drilling fluid No.3 versus the eccentricity for flow rate *G*=0.473 kg/s. Solid lines correspond to calculation, symbols are experiment.

4. Conclusion

The numerical and experimental investigation of laminar flow of power-law fluid has been conducted in an annular channel with the eccentricity. Three model solutions with properties close to real drilling fluids were examined. Experimental values of pressure drop are obtained depending on the rheological properties of drilling fluids, flow rates, and eccentricity. The obtained experimental data were used for verification of the proposed numerical methodology. The calculated data coincided with good accuracy with the experimental data. In the future we plan to extend these studies to the case of turbulent flows.

Acknowledgments
The reported study was funded by Russian Foundation for Basic Research, Government of Krasnoyarsk Territory, Krasnoyarsk Region Science and Technology Support Fund to the research project № 16-41-243034.

References

[1] Volarovich M P 1949 *Tech. Phys.* **16** (3) 321-328.
[2] Fredrickson A G and Bird R B 1958 *Ind. Eng. Chem.* **50** 347-352.
[3] Laird, W M 1957 Slurry *Ind. Eng. Chem.* **49** (1) 138–141.
[4] Hanks R W 1979 *Ind. Eng. Chem. Process Des. Dev.* **18** (3) 488–493.
[5] Guckes T L 1975 *J. Eng. Ind.* **97** (2) 498–506.
[6] Hussein Q E and Sharif M A R 1997 *J. Energy Res. Tech.* **120** 201–207.
[7] Escudier M P, Gouldson I W, Oliveira P J and Pinho F T 2000 *Int. J. Heat and Fluid Flow.* **21** 92–103.
[8] Launder B E. 1972 *Lectures in mathematical models of turbulence* (London: Academic press).
[9] Gavrilov A A, Dekterev A A, Minakov A V and Rudyak V Y 2011 *J. of Appl. and Ind. Math.* **5** (4) 559 – 568.